N. R. SRIVATSA

OPTIMIZATION: THEORY AND PRACTICE

McGraw-Hill Chemical Engineering Series

THE SERIES

Anderson and Wenzel—*Introduction to Chemical Engineering*
Aries and Newton—*Chemical Engineering Cost Estimation*
Badger and Banchero—*Introduction to Chemical Engineering*
Bennett and Myers—*Momentum, Heat, and Mass Transfer*
Beveridge and Schechter—*Optimization: Theory and Practice*
Clarke and Davidson—*Manual for Process Engineering Calculations*
Coughanowr and Koppel—*Process Systems Analysis and Control*
Dodge—*Chemical Engineering Thermodynamics*
Griswold—*Fuels, Combustion, and Furnaces*
Groggins—*Unit Processes in Organic Synthesis*
Harriott—*Process Control*
Henley and Bieber—*Chemical Engineering Calculations*
Johnson—*Automatic Process Control*
Johnstone and Thring—*Pilot Plants, Models, and Scale-up Methods in Chemical Engineering*
Katz, Cornell, Kobayashi, Poettmann, Vary, Elenbaas, and Weinaug—*Handbook of Natural Gas Engineering*
Knudsen and Katz—*Fluid Dynamics and Heat Transfer*
Lapidus—*Digital Computation for Chemical Engineers*
Leva—*Fluidization*
Lewis, Radasch, and Lewis—*Industrial Stoichiometry*
Mantell—*Adsorption*
Mantell—*Electrochemical Engineering*
McAdams—*Heat Transmission*
McCabe and Smith, J. C.—*Unit Operations of Chemical Engineering*
Mickley, Sherwood, and Reed—*Applied Mathematics in Chemical Engineering*
Nelson—*Petroleum Refinery Engineering*
Perry (Editor)—*Chemical Engineers' Handbook*
Peters—*Elementary Chemical Engineering*
Peters and Timmerhaus—*Plant Design and Economics for Chemical Engineers*
Reid and Sherwood—*The Properties of Gases and Liquids*
Robinson and Gilliland—*Elements of Fractional Distillation*
Schechter—*The Variational Method in Engineering*
Schmidt and Marlies—*Principles of High-polymer Theory and Practice*
Schweyer—*Process Engineering Economics*
Sherwood and Pigford—*Absorption and Extraction*
Shreve—*Chemical Process Industries*
Smith, B. D.—*Design of Equilibrium Stage Processes*
Smith, J. M.—*Chemical Engineering Kinetics*
Smith, J. M., and Van Ness—*Introduction to Chemical Engineering Thermodynamics*
Treybal—*Liquid Extraction*
Treybal—*Mass-transfer Operations*
Van Winkle—*Distillation*
Vilbrandt and Dryden—*Chemical Engineering Plant Design*
Volk—*Applied Statistics for Engineers*
Walas—*Reaction Kinetics for Chemical Engineers*
Walker, Lewis, McAdams, and Gilliland—*Principles of Chemical Engineering*
Williams and Johnson—*Stoichiometry for Chemical Engineers*
Wilson and Ries—*Principles of Chemical Engineering Thermodynamics*

OPTIMIZATION: THEORY AND PRACTICE

GORDON S. G. BEVERIDGE and ROBERT S. SCHECHTER

Senior Lecturer
University of Strathclyde
Glasgow, Scotland
and Visiting Professor
The University of Texas

Professor
The University of Texas

McGraw-Hill Book Company

New York St. Louis San Francisco
London Sydney Toronto
Mexico Panama

OPTIMIZATION: THEORY AND PRACTICE

Library of Congress Catalog Card Number 69-13596

07-005128-3

11 12 13 14 15 16 KPKP 8321

This book was set in Baskerville by The Universities Press and printed on permanent paper and bound by Kingsport Press.
The designer was Richard Paul Kluga;
the drawings were done by J. & R. Technical Services, Inc.
The editors were B. J. Clark and Peter R. Karsten.
Adam Jacobs supervised the production.

To Trudy and Mary Ethel

"They also serve who only stand and wait"
John Milton ("On His Blindness")

PREFACE

In recent years applied science has achieved a high level of competence in industrial, scientific, and military design. This progress has been aided by a corresponding growth in computer science. In the past, designs incorporating large margins of safety were acceptable, but now, under increasing competitive pressures, with the growing reservoir of reliable models and the widespread availability of large and rapid computing facilities, a real demand has been created for methods which systematically improve performance. The past decade has seen, as a result, the explosive growth of a class of mathematical methods known, collectively, as optimization techniques.

To date many publications have appeared, attaining a high degree of competence with respect to specific applications. The names of Bellman, Dantzig, and Pontryagin stand among the protagonists, with the developments of dynamic programming, linear programming, and the maximum principle to their credit. These advances are supported by the works of many other eminent contributors.

Much of the development has been at a research level, but now that attention is being focused on applications, there is a real need for a more general understanding. Optimization is an aspect of engineering which is often neglected in academic training, and may in fact never be considered from a fundamental viewpoint at any stage in one's professional education. As formal instruction develops to remedy this deficiency, the need arises for suitable introductory texts covering the various methods of optimization and their applicability. Apart from some edited compilations, however, there has been little attempt to develop a unified presentation of the underlying principles of optimization. This need has prompted the preparation of this volume, which presents a broad but unified coverage of the fundamentals at an introductory level.

In general, this presentation demonstrates the interrelationships between the various facets of optimization. These aspects range from the differential calculus through direct search and mathematical programming techniques to the more specialized game theory and decision theory required when competition is present. The integrated approach is seen, for instance, in the discussion of multidimensional numerical search techniques. Each search may be characterized by the two essential features of a distance and direction of movement. These, together with a further classification based on whether or not the gradient is required, have provided the framework within which search methods are presented. In this context the similarities and differences, the advantages and disadvantages, and the range of applicabilities and failures of all search techniques can be clearly understood. Thus such well-known search methods as Rosen's gradient projection and Zoutendijk's feasible directions are seen to stem from the same basic concept, namely, local linearization. A second example of the interrelationship of methods is the evolution from the Lagrangian formulation of such diverse techniques as the so-called discrete maximum principle, the maximum principle of Pontryagin, duals in linear problems, the Kuhn-Tucker conditions, steepest ascent, the gradient projection, and other important techniques.

It is interesting that the concepts underlying mathematical optimization require no more than an understanding of the calculus, so that this level of mathematical achievement is a sufficient prerequisite for full comprehension of the material presented here. To discuss the topics at this level, rigorous mathematical proofs have been replaced by plausibility arguments which will satisfy most applied scientists and permit an extension of the reasoning to other cases. The emphasis throughout has been on understanding and the applications while maintaining precision, so that all the results are correctly stated and properly qualified.

The presentation is intended to serve not only as the background for a formal course, but is also designed to facilitate self-instruction. The contents of this book have been the basis of a series of courses offered over the past six years at both senior and postgraduate levels at the University of Edinburgh, the University of Texas at Austin, and the Heriot-Watt University at Edinburgh. In addition, special courses for industrialists have been given in Texas and Scotland, using this material.

The book is divided into three parts. Part One discusses the organization for optimization. The factors influencing the definition of suitable objectives and their subsequent formulation are discussed in detail. The techniques of optimization are presented in Part Two, the two basic approaches of analytical and numerical optimization being investigated for functions of a single variable, many variables, and an infinite number of variables. Some practical aspects of optimization are examined in Part Three, with particular attention paid to the decomposition of the more complicated systems.

A large number of examples are provided throughout the text as an aid to attaining a working proficiency in applying the concepts introduced in each section. Many problems are also included, together with extensive representative references. The latter provide a reservoir of additional information, although this does not represent an exhaustive review.

The contents of the various chapters are now considered.

In *Chapter I*, optimization and the need for optimization are defined.

In *Chapter II*, the problem of organizing for optimization is discussed in general terms, the coverage including the choice of suitable criteria of selection, the structure of systems and their relation to the problem of optimization, the types of models, the simulation of the objective function in terms of the system variables, and the definition of restrictions.

In *Chapter III*, the various types of models are considered, with particular emphasis on the interchangeability of the black-box with the analytical type of model. Curve fitting, which plays a prominent role in the treatment of models, is also an important aspect of numerical search, and this facet of model studies is emphasized.

In *Chapter IV*, the very important relationship between a function, its admissible region, and optimization is presented. Although much of the material may be familiar, new emphasis is given to various aspects, especially those which influence optimization, as in the definition of potential locations of local and global optima.

In *Chapter V*, the means of finding the optimum of a function of a single variable are developed. Here the necessary conditions defining stationary points and the associated sufficiency conditions are given. The arguments are important, since the same approach is subsequently used in connection with the optimization of functions of many variables and functions of continuous variables. The problems of boundaries are also considered.

In *Chapter VI*, various one-dimensional numerical optimum-seeking methods are discussed. These methods, which search for the optimum iteratively, are required in the case of complicated analytical functions and black-box models. In particular, the Powell gradient and the direct search method of Swann are described in connection with unrestricted functions; while for constrained search, it is shown that the Fibonacci method guarantees the most efficient use of experiments.

In *Chapter VII*, the analytical techniques required to handle multi-dimensional functions are studied, the methods being categorized with respect to unrestricted, equality-restricted, or inequality-restricted problems. In unrestricted search the main emphasis is placed on the sufficiency conditions, whereas in constrained optimization, the finding of the possible locations of optima, using constrained variations, penalty functions, direct substitution, or Lagrangian multipliers, receives the bulk of the attention. Sufficiency conditions are discussed, based on constrained variations, and the important additional necessary conditions of Kuhn and Tucker are also developed. Special cases of practical importance such as linear programming, geometric programming, and duals are included in the presentation.

In *Chapter VIII*, multidimensional methods of numerical search are presented within a framework, their position depending on whether the search is restricted or whether the gradient of the function is available. For unrestricted direct search when gradients are not available, the Sequential Simplex and the methods of Rosenbreck and Powell have a definite utility. Powell's conjugate direction technique has a special interest, since it possesses quadratic convergence. The best methods using gradients, such as that of Fletcher and Reeves, also use conjugate directions. For restricted functions the Complex direct search method of Box and the gradient methods of Rosen and Zoutendijk are also described.

In *Chapter IX*, the calculus of variations, namely, the problem of optimizing functions of continuous variables, is considered. For unrestricted problems a differential equation, the Euler-Lagrange equation, is found as the necessary condition for the existence of an optimum. In restricted cases the method of Lagrangian multipliers is shown to be effective in developing appropriate necessary conditions. For some problems additional necessary conditions are found, these being known as the maximum principle of Pontryagin.

In *Chapter X*, direct calculations based on the integral formulations found when optimizing with respect to continuous variables are presented. This approach, known as the Ritz method, is very useful in those cases for which the Euler-Lagrange equation determining the stationary points is too complex to be readily solved. This is analogous with the need for numerical search methods in multidimensions. Restricted problems can also be handled using the method of Lagrangian multipliers.

In *Chapter XI*, some of the difficulties encountered in the optimization of practical systems are noted. A general critique of the available methods is presented. In particular, problems of scaling, convergence, multiple optima, and sensitivity are examined.

In *Chapter XII*, the special approach to optimization required when some of the variables are controlled by a hostile or neutral agency are summarized with the basic ideas of game theory and decision theory.

In *Chapter XIII*, the important concepts of decomposition, by which a complex problem is reduced to a series of subproblems, are introduced. The most highly developed techniques, those of splitting and dynamic programming, are carefully studied, together with selected counterexamples designed to show that the problem is not a trivial one. Methods of nonfeasible decomposition are also presented.

For an introductory senior-level course it is possible to omit Chapters III, IX, X, XII, and parts of XIII without destroying the continuity of the presentation.

Arguments have sometimes been put forth attempting to minimize the importance of optimization in present-day industrial activities. Thus one hears that the lack of realistic system models makes the study of optimization techniques academic, that the simplifications assumed in the theoretical analysis cannot be realized in practice, or that data are too inaccurate for meaningful calculation.

In reply it need only be noted that precisely the same system models are required in optimization as in design, so that it is certainly true that unrealistic results would be obtained with inadequate design techniques. In many cases, however, good design procedures are available, and optimization under these circumstances would be quite advantageous. Simplifications unattainable in practice can be justified by noting that an optimization solution can provide an ultimate limit against which practical achievement can be judged. A similar situation is experienced when equilibrium thermodynamics imposes a bound on the ultimate performance of irreversible processes. The validity of any "optimal solution" depends, of course, on the accuracy of the numbers used to generate the answer. Optimization can again, however, play the role of providing limits, the latter being associated with a particular set of data, and the data being varied as desired to produce a range of solutions to test the sensitivity of the optimum.

To coauthor a book it is necessary to be able to exchange ideas on a continuing basis for more than brief time periods. On several occasions one of the authors has served as a visiting professor at the University of Texas, while the University of Edinburgh and the Heriot-Watt University have similarly tendered the same opportunity to the other author. Without such support this book could not have been prepared, and the authors wish to express their thanks to these universities and, in particular, to Dr .J. J. McKetta, Dean of Engineering at the University of Texas in Austin, who was responsible for a number of the arrangements. Many fruitful conversations with colleagues in academic and industrial life are acknowledged, in particular with Mr. M. D. Wynne of I.C.I. Ltd. Acknowledgement is also made to Miss Margaret Nichol, who typed most of this manuscript.

Gordon S. G. Beveridge *Robert S. Schechter*

CONTENTS

PART ONE

THE ORGANIZATION FOR OPTIMIZATION

The optimization of any system, process, or project cannot be accomplished without first assembling the appropriate information in a suitable form. The step of organizing for optimization is therefore a critical one and is now considered with particular reference to the problems of models and their interaction.

CHAPTER I

AN INTRODUCTION TO OPTIMIZATION

1 OPTIMIZATION AND ITS BACKGROUND

Almost any problem in the design, operation, and analysis of manufacturing plants and industrial processes, and any associated problem such as production scheduling, can be reduced in the final analysis to the problem of determining the largest or smallest value of a function of several variables. Since optimization is the collective process of finding the set of conditions required to achieve the best result from a given situation, it follows that the techniques of optimization must be brought to bear on every task of practical importance. The subject, seemingly simple, is in fact a complex one, and can best be mastered by viewing the components of an optimization study before carrying out the complete task. Before defining these components it is perhaps instructive to note the scope of the problem. It will be seen on careful analysis that the organization of the problem will itself fall into the realm of optimization.

1.1 Applications of Optimization

Optimization in its broadest sense can cover a wide range of examples and applications. Some typical applications might be found in the determination of:

1. The shortest route taken by a salesman visiting different cities during one tour.
2. The design of a single unit or entire plant to produce a fixed quantity of material of given specifications at minimum cost or maximum revenue.
3. The operation of a unit or plant whose performance is to be adjusted to give a maximum yield or minimum cost, as, for example, in the estimation of the best temperature profile through a continuous tubular reactor in order to obtain a maximum yield of a particular product.
4. The color schemes and type of music which enable female workers to produce work more efficiently under the apparently monotonous conditions of mass production.

5. The statistical evaluation of plant data to build empirical models as a basis for improvement, as in the fitting of mathematical curves to experimental data to obtain the most accurate representation.
6. The planned maintenance and replacement of equipment to reduce operating costs.
7. The minimization of inventory charges, whether in a plant with stores of products, raw materials, and spares or in the supply of air hostesses required at any airport terminal and the number undergoing training.
8. The allocation of resources or services among several processes or activities to increase overall efficiency.
9. The distribution of production throughout the nationwide facilities of a company in relation to its markets and raw material sources in order to reduce its transportation costs.
10. The controlling of waiting or idle time and queuing in production lines in order to reduce costs.
11. The planning of strategies to gain a greater share of a competitor's market or to reduce his effectiveness.
12. The planning and scheduling of efficient construction.

1.2 Why Optimize?

Why do we want to optimize at all? In most aspects of industrial life continual improvement is an important feature. Thus we may want to get the largest production from given raw materials, the greatest profit from a fixed investment, and so on. Optimization is a formal presentation of these ideas.

Improvement can be regarded from two viewpoints. Economic improvement provides an overall framework in which a given situation must be examined, since all problems must be considered within a financial structure. Many problems, however, may not be directly related to the company finances, and for this reason many improvements will be carried out on a technical basis. Even here, however, we should not lose sight of the economic criteria.

In many cases it might be argued that the use of formal optimization methods is not worthwhile. Thus, for example, data may be inaccurate or the assumptions used in constructing a model for equipment design may be oversimplified. These situations are certainly found in practice, and therefore optimization results cannot be blindly accepted. In addition, it may be found that it is almost impossible to achieve the stated optimal conditions. An excellent illustration of this situation is the design of a continuous flow tubular reactor. Here it can be shown, as in the work of Horn [1961a] and Denbigh [1958], that the best temperature profile along the reactor is a continuous function which results in, say, a 70 percent yield. This is an idealistic solution, since a temperature changing continuously along the length of the system according to a predefined function is in practice most difficult, if not impossible, to achieve. A few simple calculations may then show that two smaller reactors, each operating isothermally at different temperature levels, will produce, say, 69 percent yield. We now have a result which lies close to the ultimate possible, and one which is practicably feasible. Optimization studies, while thus sometimes seemingly rather idealistic and detached, can therefore also provide knowledge on the limits of possible improvement of the system under study. For this reason, optimization is valuable in indicating the maximum system potential and in producing confidence for the adoption of the modified, more easily attained designs.

2 ESSENTIAL FEATURES IN OPTIMIZATION

Irrespective of the application, several salient features can be distinguished in most optimization studies. Let us now examine some of these.

2.1 Objectives

Any problem investigated in an optimization analysis will have as its objective the improvement of the system or systems. It should be fairly obvious that, in order to improve any system, it is essential that at least one solution be obtainable for that system. In other words, by defining the inputs to a system, we can find the resulting output. If this is not possible, we cannot design or operate or control the system, far less optimize it.

If a system is completely defined by a set of specified inputs, the output will be fixed. For example, the volume of a cylindrical vessel would be determined completely if its diameter and height were given. No improvement can be carried out on this system unless one of the specifications is relaxed. When a system is not completely defined by specified inputs, it is called *underdetermined*, and results in the possibility, at least in principle, of an infinite number of solutions. This is a necessary condition before optimization can be used in a system analysis.

No single answer is normally found to any problem, and it is therefore necessary to choose the "best" solution for a given problem from the multitude of possible solutions. How can this be achieved? First, it is necessary to define the objective of the study. This may vary from one problem to another, but for industrial applications it might be either economic or technical. Such economic aims as maximum profit or minimum costs are common, while possible technical objectives might include the largest yield of a particular product from a reactor or the smallest surface area for a given volume of vessel. Normally, however, most industrial optimization has to be carried out within an economic framework. Until some such objective is decided upon, no improvement can be attained from any study since there will be, otherwise, no basis for comparison.

It should also be noted that not all problems are capable of quantitative analysis, as in example 4 in Sec. 1.1, although all have the general objective of creating a better set of conditions to achieve the best possible results. In general, however, it is necessary to have a quantitative measure in order to provide a comparative basis, and thus all problems must be formulated so as to present all their information in a quantitative form, even though relative numbers have to be allocated in an empirical manner to evaluate a set of conditions. This quantitative form is the objective function by which the criterion of selection is related to the system variables.

2.2 Competing Influences

A characteristic feature of many such problems—especially in plant design and operation—is the presence of conflicting or opposing influences. Thus, for example, a process may be operated with a large number of low-paid personnel (as with labor in some of the underdeveloped nations) who might produce a lower grade of product due to the variable quality caused by uneven control. The use of automation, on the other hand, may upgrade the product by keeping a close check on its quality, with its highest level in the computer-controlled system (e.g., Williams [1959–1965]), but this will introduce higher maintenance labor charges. This latter cost will then, in conjunction with the cost of manual and production labor, produce a total labor cost which goes through a minimum as the degree of automation increases, as shown in Fig. I-1. This minimum has been produced by the presence of the two contributing, but competing costs. (It is realized, of course, that other factors will apply here, such as the increased value of the product, so that it is unlikely that the point of minimum labor and maintenance cost would be chosen.)

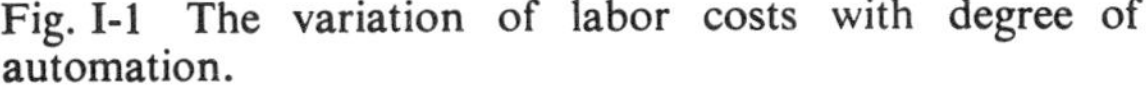
Fig. I-1 The variation of labor costs with degree of automation.

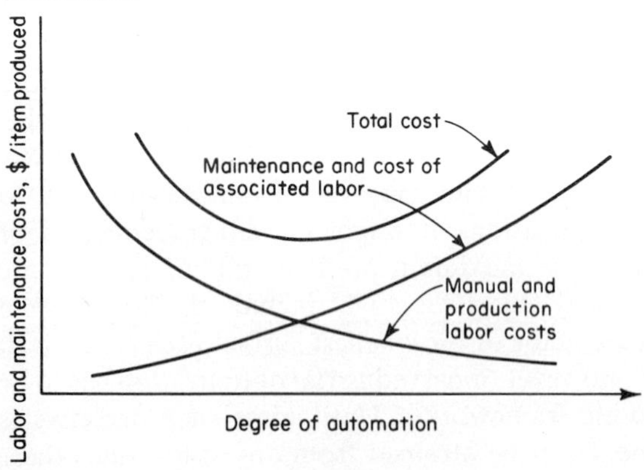

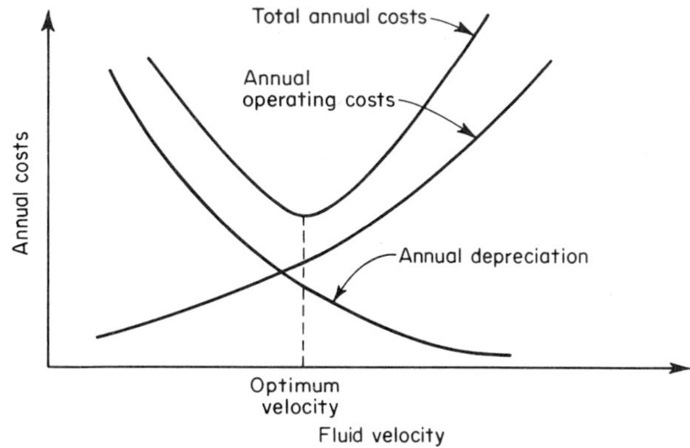

Fig. I-2 Annual operating and capital costs for a heat exchanger of fixed duty.

Normally, we are not interested in such particular costs, but will consider the total costs in any project. In this we must include not only the direct operating cost, but also the capital cost, both on the same basis (thus the capital expenditure may be included in the annual operating cost by the use of some depreciation method which signifies the recovery of capital—the initial outlay required for the original purchase and construction and operation of the plant—on a yearly base, with or without the inclusion of the effect of interest). A simple example of this is the choice of a heat exchanger to carry out a specific cooling duty at a fixed temperature level. Its position in a complex process is such that no value may be reasonably assigned to its function, so that we cannot consider an associated income, but must confine our attention to the relative costs of operation between the various alternative designs. In particular applications the significant independent variable of optimization might be taken to be the fluid flow velocity. The higher the velocity, the larger will be the heat transfer coefficients and the smaller will be the exchange surface required, thus reducing the capital cost. However, such increased flows will result in higher pumping and running costs, thus increasing the operating costs. The opposing cost components will therefore have the same conflicting effect as seen above, shown here in Fig. I-2. The sum of the two contributions, the total annual cost, will thus show the characteristic minimum, giving an optimum operating condition for the fluid velocity.

Many other examples will be obvious, some being described later. In all such situations there must be some form of compromise. A lower cost of operation might be desirable, but since this may be countered by an increased capital cost, some balancing of the two opposing influences is essential. It will be seen that the point of balance will be largely determined by the form of objective chosen.

2.3 Restrictions

Although optimization studies may be required to achieve the best result from a given situation, meaning the *absolute best*, this may not always be attainable due to the imposition of restrictions.

If the problem is unrestricted, we may take as an example the case of a batch extraction unit in which the largest amount of extraction will be attained with an infinite amount of solvent. We then have an absolute best in the sense of the total amount of solute which can possibly be removed. When restrictions are imposed, as on the availability of solvent, a smaller amount will be extracted. We can still, however, regard this latter condition as an optimum; but it is not an absolute optimum, only one relative to the restriction. A simple example will clarify this point.

Example I-1: Storage vessel of fixed surface area

Determine the dimensions of a totally enclosed storage vessel to contain a maximum volume of liquid, V, if the surface area of the containing walls is to be a given value, A.

If no further restriction is stated (the first constraint being that of fixed area), a spherical shape could be chosen, and be shown (Example IX-8) to have the greatest volume V^* of all regular shapes. Its volume would then be

$$V^* = \frac{A^{3/2}}{6\sqrt{\pi}}$$

On the other hand, if the choice of vessel is restricted to a cylindrical shape, the maximum volume will be

$$V = \frac{A^{3/2}}{3\sqrt{6\pi}}$$

which is less than that for the unrestricted situation. It will be seen that the optimum volume achieved will depend upon the presence of restrictions. The restricted value, here for a cylindrical vessel, can be further improved only if the restriction on the vessel shape is removed.

While Example I-1 is a particular type of illustration, it will be obvious that we cannot always look for an absolute minimum or maximum, but must obtain a restricted optimum—the best possible in the circumstances, which might improve if the constraining restriction were removed. The general question of how restrictions arise is examined in Chap. II, Sec. 6.

3 LEVELS OF OPTIMIZATION

Having seen the types of problem encountered in optimization studies and considered some salient features, let us now return to the general problem of plant design and operation, a problem which need not be specifically defined at this stage, but which will be illustrated by general reference to chemical plants.

In the manufacturing industries, as Singer [1962] has noted, the project may be of varying complexity, as follows: (1) a combination or complex of factories, works, or companies, for example, individual oil refineries within an oil company, which itself may form a component of a group of companies; (2) a combination of unit processes, as in a chemical works or oil refinery in which a network of individual plants using interconnected feeds and products yields a range of products from a given range of raw materials; (3) a combination of components, such as distillation towers, filters, heat exchangers, reactors, and the like, arranged in a unit process, such as an acetic acid plant; (4) a specific component, such as a heat exchanger.

Whatever the level of complexity, the project may, in addition, be analyzed from the point of view of design, operation, or control. For example, at the design stage, at the third level above, the available capital might be used to build a plant for the production of a salable chemical. Here the sale may be within the company or on the open market, but the same arguments will apply to both cases, since, if a chemical cannot be produced cheaply enough within a company, it will be bought outside, unless there are other reasons preventing this, such as a desire for control of sources, quality, etc. In this plant design some or all of the points presented in Table I-1 must be examined.

TABLE I-1 TYPES OF DESIGN PROBLEM

Design at the management level
Evaluation of a given project; comparison between projects; capital allocation among projects

New projects
Choice of product; choice of plant capacity; choice of chemical route; choice of raw materials; choice of site location

Process and flowsheeting
Choice of process; choice of process flowsheet; arrangement of flowsheet

Equipment specification
Selection of type of equipment for a given flowsheet duty; choice of materials of construction; sizing of equipment units; layout

Operating conditions
Time-dependent and time-independent process units; control systems; maintenance and equipment replacement policies; scheduling of production; allocation of raw materials; storage inventories for raw materials, intermediates, and products; shipping, transportation, and distribution

Optimization problems can be found at all these levels, as can be seen from the examples in this text. It will be noted that the scope of the optimization problem can be the entire business enterprise, a process, a single operation, a single component in that operation, or any intermediate stage between these. The depth of the study may involve generalizations indicating only trends, or it may examine the minute detail, depending upon the use to which the results will be put, the availability of accurate data, and the time available in which to carry out the study. While the technology may be chemical, business, or any other type, optimization methods can be applied in any individual stage. For the best results, any optimization study ought to be included from the earliest period in the design analysis. It should thus be taken into account from the initiation of a project and be regarded as an integral part of the design process, and not treated as a series of small isolated maximization or minimization problems. Actually, improvements can be carried out at any isolated stage in the design, as in the simple illustrations quoted previously, but if this is done, it does not follow that the optimal result for the integrated process will be found.

We have mentioned the complete design and optimization as being the highest level. Correspondingly, the problem of design and operation of the chemical plant can be entered as successively lower levels, but after entering at a specific level, for example, at the discussion of equipment sizes, we must consider and optimize all successive stages in order to achieve the optimum conditions.

The ultimate in design technique, and hence in optimization, is the synthesis of the complete reaction system, with the optimal control, operation, and choice of equipment having been made and the items arranged in the most suitable choice of flowsheet and chemical route to the desired product, all this being carried out as an integrated scheme. The degree of complexity of this synthesis is associated with the level of the design and optimization attempted, being most complex at the higher levels. In general, it may therefore be stated that we must solve the problem of choosing a set of design and operating variables in such a way as to optimize some suitable criterion of selection while still satisfying all the constraints of the types indicated. In its complete generality, this problem is often so complex that it becomes impossible to tackle, either because sufficient information is not available to allow its formulation in precise terms or simply because of the enormous number of variables and constraints to be considered, often as nonlinear functions. Indeed, it might be said that the aim of engineering design studies is to put the designer in possession of all the necessary physical, chemical, mechanical, and economic data required to formulate such a problem with the full precision and confidence needed, whether the situation is at the design or the operation stage in a plant or process. On a much less ambitious scale than that just outlined, it has proved possible to solve general problems of this type by restricting their scope and by making a number of simplifying assumptions. The present state of optimization theory and application is such that the most complex process, that of the complete project, has not yet been handled in detail. Optimization of a chosen flowsheet and comparison of this with other flowsheets is the most complex type of situation considered as yet. The design of individual items of process equipment, such as converters, distillation columns, and heat exchangers, and the numerical optimization of complex mathematical functions are the two main areas in which optimization is at present most effective.

In order even partially to achieve these aims, we must familiarize ourselves with the various available methods of optimization as described in Part Two. Part Three will consider the application of these techniques to more complex examples, where the interest is on the system, and not on the mode of optimization. Before considering these in detail, however, let us consider the general problem of how to tackle an optimization study. This is the subject of the next two chapters.

CHAPTER II

ORGANIZATION OF OPTIMIZATION PROBLEMS

1 INTRODUCTION

The structure of most problems in optimization is amazingly similar. Indeed, it is this similarity that has given rise to the interdisciplinary exchange which has characterized the present progress in this field. Thus chemical engineers, traffic engineers, and biologists may all have a deep interest in precisely the same mathematical problem, each with a different application in mind. Moreover, this similarity in structure permits the development of a logical framework within which any problem can be studied. This chapter is intended to elucidate those principles by which systems, complex or simple, can be organized for optimization.

In order to design or operate any system it is necessary to obtain a suitable model representing the system, to obtain solutions, to choose a suitable objective criterion which will guide any decisions, and to find a method of optimization.

The model may be one of two types. First, a *mathematical model* may be postulated to represent the process system, in which analytical relationships, together with appropriate restrictions, define the response of the process. Alternatively, the process may already exist, and in this case the response to various input signals may be obtained by experimentation. A model developed from this type of experimental information is called a *black-box model*, since it is not necessary to know the governing equations, but only the range of inputs to be selected. Sometimes, especially for more complex mathematical models, a system adequately represented by a mathematical model may be treated in the second manner, so that outputs such as profits or yields are determined numerically for a given set of input values. These outputs are treated in the same way as black-box outputs.

Once the model has been defined and a solution is obtainable, optimization techniques may be selected, assuming that suitable selection criteria are available and simulation is achieved. The criteria essentially determine the information required in the optimization. Thus an investigation of the optimum conditions yielding the most profit at the design stage necessitates finding the capital, the operating costs, and the income for the process. On the other hand, if a minimum cost were the only object, it would not be necessary to include the project income. Similarly, the capital and the design variables of an existing system would not be required. The optimization—once a given problem has been set in the form of an objective function, determined either by numerical data from a black-box model or from mathematical functions, subject to various equality or inequality restrictions—may be of many different forms, ranging from a straightforward comparison between two results to the most elegant of mathematical techniques. The important feature is, however, the obtaining of an improved solution. It should be noted that optimization may not only modify the system variables, but in addition can alter the system itself. Thus one type of reactor might be replaced by another (a batch by a stirred tank flow reactor, and so on), or the arrangement of the flowsheet may be changed.

A realistic model is the most important factor in any optimization study. It is of little use to carry out improvement studies based on a model which does not adequately represent the existing or proposed system. Although the model will not be discussed, except as required to provide a basis for optimization problems, the importance of the choice of the model cannot be emphasized too strongly. In the same vein, it should be noted that too complex a representation creates grave problems either in optimization or in model building. Such detail would, however, be required in the final stages of an engineering design. Simplified models often can be used to represent the problem adequately over small ranges of the independent variables.

While an infinite variety of problems may have to be considered, most of these can be handled by the same general procedure.

1.1 General Procedure for Optimization

Since the optimization problem may usually be written in some mathematical form, it follows that the actual physical situation need be taken into account only when constructing the model. The optimization problem is then a purely mathematical one. Several features may be distinguished which allow a common approach to be formulated, irrespective of the particular application investigated. These common features include the use of distinguishing criteria; the need for a mathematical relationship linking these criteria to the set of system variables, i.e., a model; the requirement that certain of the system variables be controllable; the existence of constraints limiting the range of variation of some or all of the variables; and the importance of being able to calculate a first solution.

The general objective in optimization is to choose a set of values of the independent variables, subject to various restrictions which will produce the desired optimum response for the particular problem under examination.

A general approach or procedure can be listed before being discussed in detail.

General procedure

1. Define a suitable objective for the problem under examination (Sec. 2).
2. Examine the restrictions imposed upon the problem by external agencies (Sec. 6).
3. Choose a system or systems for study (Sec. 3).
4. Examine the structure of each system and the interrelationship of the system elements and streams (Sec. 4).
5. Construct a model for the system. This is the technical design stage which allows the objective to be defined in terms of the system variables (Sec. 5).
6. Examine and define the internal restrictions placed upon the system variables (Sec. 6).
7. Carry out the simulation by expressing the objective in terms of the system variables, using the system model (Sec. 7). This is the objective function.
8. Analyze the problem and reduce it to its essential features. This reduction is necessary in many cases to allow optimization to be attempted (Sec. 8).
9. Verify that the proposed model in fact represents the system being studied (Sec. 9).
10. Determine the optimum solution for the system and discuss the nature of the optimum conditions (Sec. 10 and Part Two).
11. Using the information thus obtained, repeat this procedure until a satisfactory result is found (Sec. 11).

While the following sections discuss each of these steps in more detail, with some reference to problems found in the field of chemical engineering, these notions have a much wider field of application.

2 THE OBJECTIVE

It has been seen that many different problems may be encountered. All these, however, have the feature that, while at least one solution exists, there are usually an infinite number of solutions possible. The aim of optimization is the selection, out of the multiplicity of potential solutions, of that solution which is the best with respect to some well-defined criterion. The choice of this criterion, the objective, is therefore an essential step in any study.

Many criteria have been applied to guide the determination of the best conditions under which to design or operate a system. The nature of these criteria depends upon the project in hand, differing perhaps at the research, development, and design stages. The effect of imposing different objectives is, however, to produce different optimal solutions. In general, economic criteria should be used, although technical forms are common.

In many realistic situations, more than one criterion may need to be satisfied. Thus it may be necessary to obtain a certain profit level in a given process with a minimum investment, or to achieve a maximum product yield with minimum concentration of some undesirable product. With such plural objectives there arises the possibility of conflict, so that in the final analysis there may need to be some sacrifice of one or other standard.

An important part of the final analysis is, therefore, an examination of the interdependence of the criteria, so that the full consequences of selecting a particular objective may be seen. Normally, one criterion is chosen as the primary objective, with the others as secondary targets to be satisfied if possible. In general, the main intention is the maximization of profit to be achieved at the smallest feasible capital investment and at the most satisfactory level of the other responses.

2.1 Criteria of Selection

The two most important criteria are the level of profit (Sec. 2.1.1) and level of investment (Sec. 2.1.2) or their equivalent forms. Normally, technical criteria (Sec. 2.1.3) should be some reduced form of the economic criteria.

2.1.1 Profit criteria

When a system exists, the investment is fixed and the maximum profit is the usual objective. When certain other variables are held constant, the profit function may in turn reduce to the sales income or costs in a system, or at a still lower level to an equivalent technical form such as a reaction yield or catalyst weight. In such systems the investment value can be omitted if the profit or cost function is in terms of relative values, but would be needed if absolute values were required. Aris [1961d] and Van Krevelen and Hoftyzer [1962] describe some applications.

In its simplest form the profit, or its equivalent, represents an average annual value, and no account is taken of the time periods involved. In fact, the value of money depends in part on the period within which it is received. Thus \$1 received today should be worth more than \$1 received after n further periods of time, because today's \$1 can be invested and will be worth $\$(1 + R)^n$ in n periods time. Thus we may say that a return of \$1 in n years will not have the same importance as \$1 in the first few years of operation in assessing the relative value of profits received over a period of years. The time value of money can be taken into account by the equivalence expressions

$$S = A(1 + R)^n = B\frac{(1 + R)^n - 1}{R}$$

where S is the future worth of the present worth A after n periods, at an interest rate R, and B is the annual payment which is equivalent to A or S (see, for example, Schweyer [1955]). These expressions can be used to reduce cash flows occurring at different periods to equivalent values at a stated time, thus allowing direct comparison. Here A, the present worth, and B, regular annual cost, might be used as suitable criteria. The rate of interest or return, R, which equates to zero the sum of cash in- and outflows reduced to their equivalent values at some stated time, is another criterion now in common use, being the basis of the discounted cash-flow technique (Merrett and Sykes [1966]).

2.1.2 *Profit—investment criteria*

In the general case we must consider both the system profit P and the investment I, the latter being the sum of the capital cost of installing equipment and the working capital to allow start-up. It is unsatisfactory to consider only a profit objective function when investment is a variable.[1] Since any investment will require an annual return, the profit, we are thus interested in the relationship of P to I.

The simplest profit-investment criterion requires the maximum of the absolute rate of return, R, on investment. Here $R = P/I$. Similar forms include the payout time ($n = I/P$), the venture profit function ($P - \alpha I$) of Happel [1958], the net realization (operating costs $+ \alpha I$)†, and others which involve the time value of money, for example, the discounted cash flow technique (Merrett and Sykes [1966]).

2.1.3 *Technical criteria*

In certain cases, the economic criteria can reduce to equivalent technical forms. Such criteria might include all the design parameters—overall size, weight, holdup, fluid flow, chemical product yield, impurity levels—but regardless of which is chosen, it should be used within an economic framework. While there is a large volume of literature devoted to technical

[1] The inclusion of depreciation as a cost in the profit function does not account for the capital investment. Depreciation is a method of recovering the original capital over the life of the project without accounting for the interest expected on this investment, and it is therefore an operating cost (McEwan and Beveridge [1965]).

† Here α is the minimum acceptable alternative rate of return on investment, which for most companies has a value from 15 to 25 percent (Williams [1964]) before tax.

optimization (Van Krevelen and Hoftyzer [1962], Horn [1961b], and Denbigh [1958] being typical references), ideally these forms should be used only if they are derived from economic criteria. When the objective is solely technical, as in quality control or in maintaining safety standards, the profit or equivalent economic measure can be retained as a restriction on the solution to avoid impractical results.

A few purely technical criteria do exist, however, mainly in mathematical applications, such as the fitting of surfaces to data with minimum error.

2.2 Optimization and the Objective Function

The choice of an appropriate objective is not straightforward, although it is usually dictated by company policy.

Most objectives will be normally minimized or maximized as appropriate. If investment is included, however, misleading results might be obtained, unless it is realized that both absolute and marginal rates of return must be examined. Thus, even though the absolute rate of return is acceptable (operating in the region $I_1 \leq I \leq I_5$ of Fig. II-1), the need for an adequate incremental or marginal rate of return greater than some specified minimum value α, obtainable from an alternative investment, makes it more profitable to operate the process below I_3. Any additional capital should then be invested in the alternative investment, yielding the total profit-investment

Fig. II-1 Selection criteria: I_1, minimum investment to yield acceptable returns; I_2, maximum profitability, P/I; I_3, maximum investment, yielding acceptable marginal return; I_4, investment to yield maximum profit; I_5, maximum investment to yield acceptable return.

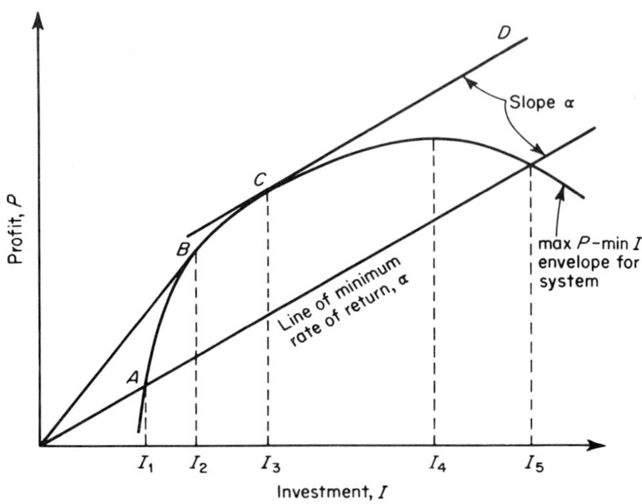

relation, as given by curve *ABCD*. This procedure shows that operation under maximum profit is less advantageous. Normal operation would preferably be in the region *ABC*, with *B* yielding the maximum profitability.

Such a *P-I* envelope is the upper limit of a set of *P-I* results determined for various alternative investments in different systems or in different arrangements within a given system. Such alternatives exist, for example, in the case when they represent different chemical processes or various flowsheet arrangements. However, even if a single project is being evaluated, the results can be compared with the alternative offered by making no investment and leaving the capital in a bank or similar source of income.

Since the sales income is essentially a variable, dependent upon external agencies, such as competition, markets, and advertising, it is often preferable not to include this, but to use the net realization criterion (Sec. 2.1.2) in project evaluation. This allows the maximum competitive position to be found, and the sales income to be determined separately.

While most objectives are to be maximized or minimized, there are situations in which expected values must be optimized when the system is stochastic in nature. Uncertainty due to competition can also require such approaches as the minimization of the maximum loss. These and other practical difficulties in criterion definition and use are discussed in Chap. XI.

3 THE SYSTEM

A system or systems must now be defined for the problem under examination. Thus, if we wish to store a liquid in a closed vessel, we might specify vertical or horizontal cylindrical tanks, spherical vessels, cone-roofed cylindrical tanks, or rectangular tanks. Each of these might be considered as a possible system, or alternatively, we might consider an arbitrary vessel shape and suitably modify this during optimization. Such an approach is desirable, but is often not feasible. If the problem merely specified the need for producing a given product, such as phenol, the required systems would be found in the different chemical routes and processes available, and in the various flowsheets and flowsheet arrangements possible for any given process (Chap. XI, Sec. 3). The choice of a system is of course subject to any restrictions imposed by an external agency (Sec. 6.1.1). Recent literature on methods for synthesizing system structure includes that of Rudd [1968] and Masso and Rudd [1968].

4 THE STRUCTURE OF A SYSTEM

The objective function must, ultimately, be expressed in terms of those system variables which can be manipulated or which are controlled by an external agency. This task can often be facilitated by preparing an adequate representation of the system (Sec. 4.1) and by understanding the nature of the component parts of the system (Secs. 4.2 and 4.3).

4.1 Representation of the Structure

There are a number of ways in which the important topological features of a system can be represented. Generally, a block diagram is used (Sec. 4.1.1), but with computer calculations other forms are sometimes preferable (Sec. 4.1.2).

4.1.1 Block diagrams

The structure of the system is taken to mean the arrangement and interrelationship of the various units or components constituting the total system. This can be represented by a block diagram such as is presented in Fig. II-2. Each block might represent such items as an oil refinery, a distillation column, or a tray within a distillation column, depending upon the nature of the particular problem and the level of optimization. The whole diagram represents the system. The interconnecting lines show the system streams,

Fig. II-2 System structure.

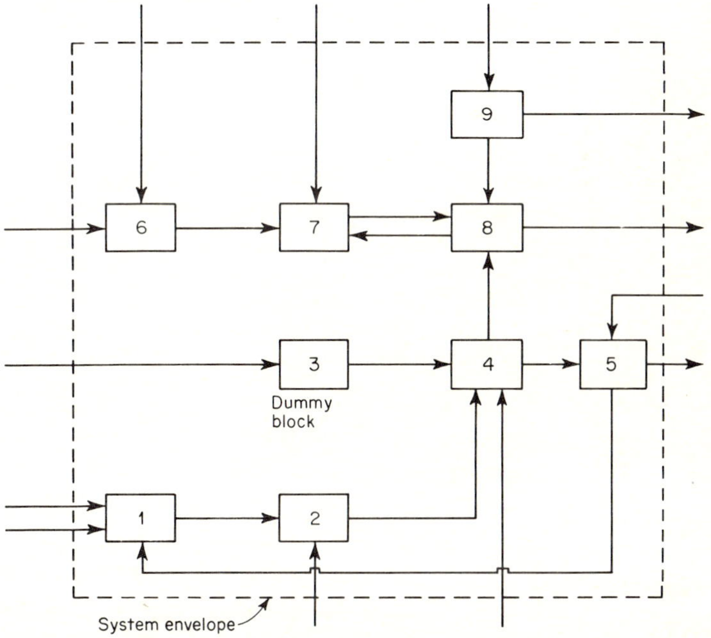

which could, for example, be fluid streams, lines of information communication, heat flow, or even the flow of money into and out of a business operation.

While such diagrams can represent the spatial arrangement of equipment linked by physical streams, they can also be used to show time variations, with each block representing a physical unit at a given stage in time. Examples include a batch reactor in which the heat input is changed after finite time intervals and a plant in which money is spent annually over a period of years for such items as planned maintenance.

These conventional flow diagrams or block diagrams provide information about the lines of communication between various parts of the system. This is an essential step in the construction of an objective function.

4.1.2 Mathematical representation

The diagrammatic representation is convenient even for complicated systems as long as there is some human interpretation during calculation. Different forms, however, must be considered for computer calculations.

Application of matrices to convey the same information as a block diagram have been reported by Rudd and Stewart [1965] and Himmelblau [1966]. The latter, for example, constructs a matrix which consists of as many rows and columns as there are blocks in the structure diagram. The element $a_{i,j}$ of this matrix is set equal to unity if there is a flow from block i to block j. Otherwise the element is set equal to zero. By defining certain arithmetic operations, one can, for example, determine at which stage recycles enter the system. Obviously, such a representation can readily be used with computers. We shall, however, also use a similar representation in the optimization of complex systems (Chap. XIII, Sec. 5.2.1).

An alternative representation is the listing of units and the interconnecting links. This may be done, for example, by listing the inputs and outputs of a given unit and by numbering units when there is more than one of the same kind. Thus a link would be completely represented by a typical statement:

Input 2 of heat exchanger 14 is output 4 of plate column 3

Since every unit has at least one input or output, a list of such statements will also specify all the units in the system, and thus provide a complete description of the structure. A problem of economic storage of such information in the computer does arise, but this is outside the scope of our discussion (Sargent and Westerberg [1964]). This approach has also been discussed, for example, by Lee, Christensen, and Rudd [1966] and Christensen and Rudd [1968].

The list processing technique seems more efficient than the matrix approach and is now finding wider application in design computation.

4.2 System Streams

Three types of system streams can be distinguished. Those streams which enter the system from the outside are called *inputs*. Similarly, *output* streams are those leaving a unit, conveying product from the system. The *interlinking* streams both originate and terminate in the system, connecting the various units. Examples of each type are shown in Fig. II-3.

To evaluate the objective function, all these streams must be defined. This definition can be achieved by specifying several variables associated with each stream. Thus, in the analysis of a chemical reactor, the feed stream might be specified. This can be done only if the composition, temperature, and flow rate of the stream are all given. These quantities, when given, specify the state of that stream. In other cases a single variable might be sufficient to represent the conditions within a stream, as in the case of the flow of investment capital or the rate of fresh catalyst addition to a reaction system.

The state of the jth interlinking stream may be denoted by the vector $\mathbf{p}_j$. Here $\mathbf{p}_j$ consists of a list $\{p_{1,j}, p_{2,j}, \ldots, p_{P_j,j}\}$ of the P_j system variables required to specify the state of the jth process stream. A component of $\mathbf{p}_j$, say, $p_{i,j}$, might, for example, be a flow rate or a temperature. Here the vector notation is merely a convenience indicating the several variables which may be required to determine the state of a single process stream. The state of a given stream is completely specified when all the system variables contributing to the objective function are given. Thus, in the design of a heat exchanger, a stream of material may be specified by its temperature and flow rate, but its composition is not particularly important. However, in the design of a chemical reactor, the stream composition would be one of the most significant quantities. Similarly, the state of the kth input stream can be represented by $\mathbf{q}_k$, and the lth output stream by $\mathbf{r}_l$. Such a notation may be used directly or replaced in appropriate cases by mnemonic symbols (T for temperature, etc.) or a general variable, x_i.

4.3 The Block Element

Optimization cannot proceed until the behavior of each of the system units, represented by a block, is successfully modeled, as described in Sec. 5. The existence of such a technical design implies that if we are given the states of all the entering interlinking streams $\mathbf{p}_j$ and all the inputs $\mathbf{q}_k$, the outputs $\mathbf{r}_l$ are determined either exactly or in terms of stochastic functions.

4.3.1 The transformation equation

The model has to be found in the form of a transformation equation which expresses the outputs in terms of the inputs. Thus, for a

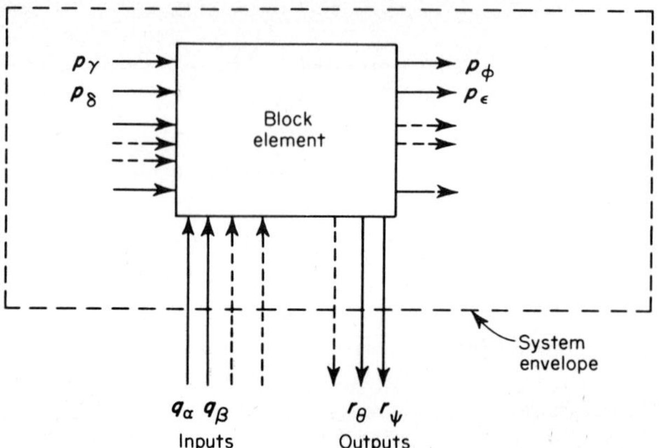

Fig. II-3 Block element with system inputs and outputs.

deterministic system, we can write that

$$\begin{aligned}
\mathbf{p}_\phi &= \mathbf{T}_\phi(\mathbf{q}_\alpha, \mathbf{q}_\beta, \ldots, \mathbf{p}_\gamma, \mathbf{p}_\delta, \ldots) \\
\mathbf{p}_\epsilon &= \mathbf{T}_\epsilon(\mathbf{q}_\alpha, \mathbf{q}_\beta, \ldots, \mathbf{p}_\gamma, \mathbf{p}_\delta, \ldots) \\
&\cdots\cdots\cdots\cdots\cdots\cdots \\
\mathbf{r}_\theta &= \mathbf{T}_\theta(\mathbf{q}_\alpha, \mathbf{q}_\beta, \ldots, \mathbf{p}_\gamma, \mathbf{p}_\delta, \ldots) \\
\mathbf{r}_\psi &= \mathbf{T}_\psi(\mathbf{q}_\alpha, \mathbf{q}_\beta, \ldots, \mathbf{p}_\gamma, \mathbf{p}_\delta, \ldots) \\
&\cdots\cdots\cdots\cdots\cdots\cdots
\end{aligned} \tag{4.3-1}$$

for the system block depicted in Fig. II-3.

4.3.2 *Inversion of the transformation equation*

It may sometimes be necessary or desirable to fix the output of a particular block at a given value. In this case it might be convenient if one of the inputs could be expressed in terms of the other inputs and the specified output. Unfortunately, such an inversion of the transformation equation is often impossible, and any analysis relying upon the existence of such an inverse is impractical.

Suppose that the transformation equation is

$$\mathbf{r}_1 = \mathbf{T}_1(\mathbf{q}_1, \mathbf{q}_2) \tag{4.3-2}$$

Such a relationship is seen in Fig. II-4, each vector consisting of a single component. Thus, for each feasible value of q_1 and q_2, the inputs, there is a corresponding value of the product, r_1. The converse is not necessarily true. For if r_1 is required to lie at the level $r_1 = a$, as represented by the horizontal

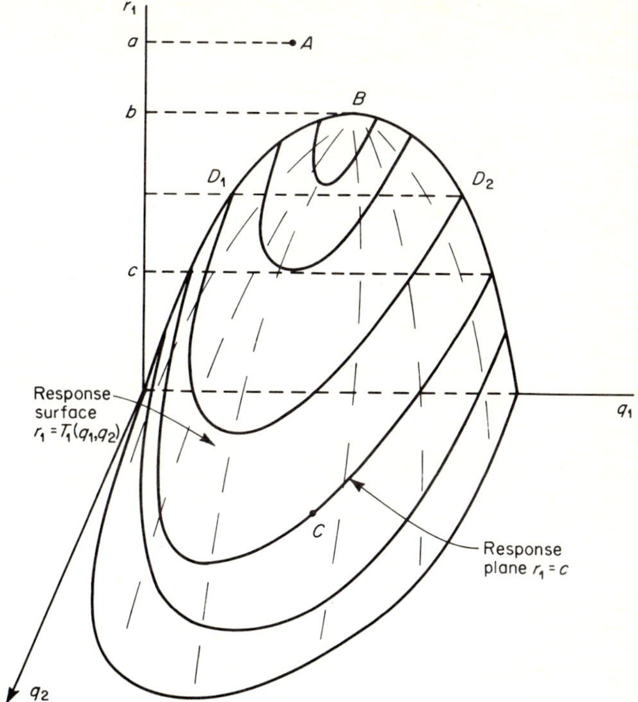

Fig. II-4 Two-dimensional response surface.

plane passing through point A, then no combination of q_1 and q_2 will give this response. The system is not, therefore, workable, since the specifications cannot be satisfied. This is not a very interesting case, because if operation is not possible, there can be no optimization.

If the demand for r_1 is placed at level $r_1 = b$, the specification can be met only by fixing the values of both q_1 and q_3 as represented by point B. This would be an example of a single specification determining the values of two input variables.

The more usual case is that found when the demand is set at a level $r_1 = c$ and there are a number of combinations of q_1 and q_2 which will yield this result. These values will lie on the curve of intersection between the plane $r_1 = c$ and the response surface $r_1 = T_1(q_1,q_2)$, point C being one position on this admissible curve. The important feature to be noted here is the fact that the specification of the product r_1 at a level c would reduce the number of degrees of freedom by one since q_1 and q_2 cannot both be varied freely.

A final point is that Eq. (4.3-2) gives a unique r_1 for each q_1 and q_2. However, the converse is not true; i.e., an r_1 and q_2 do not uniquely determine a q_1. Indeed, in viewing Fig. II-4 for the case $q_2 = 0$, we see that two values of q_1

yield the same r_1, as represented by points D_1 and D_2. This lack of uniqueness makes it difficult to carry out general inversions of the transformation equations.

4.4 System Variables and the Objective Function

The complete system as depicted in Fig. II-2 consists of a set of block elements, or units, interconnected by system streams. This complete system receives certain inputs and produces certain outputs. Since each element of the system has a technical design basis, or model, given by a set of transformation equations, the outputs are determined once the inputs are specified. It is this determination which permits us to discuss the freedom of the system.

4.4.1 *Degrees of freedom*

Since the input specification determines the output for the system, the number of degrees of freedom, defined as the number of input variables which can be altered independently, leading to changes in the output, can be represented by

$$D = \sum_{k=1}^{Q} Q_k - S - L \tag{4.4-1}$$

where D = number of degrees of freedom
Q_k = dimensionality of kth input, $\mathbf{q}_k$
S = number of input components not under control of optimizer
L = number of degrees of freedom lost by specifications imposed on products
Q = number of inputs

This equation, although of no real computational importance, expresses a most important concept of optimization.

Clearly, the first term on the right-hand side represents the sum of all possible input variables which determine the system's behavior. On reflection, however, we shall see that certain of these variables, S in number, are beyond the control of the optimizer and cannot be freely manipulated. These variables, controlled by an external agency, are discussed in Sec. 6.1.1. Of the total number of input components, only $\sum_{k=1}^{Q} Q_k - S$ will therefore be available for variation.

The external agency can also specify some of the output properties. To meet these specifications we shall have to use one or more of our manipulative inputs. Since the transformation equations cannot be inverted, the number of manipulated variables, L, required to bring the product to specification cannot be determined without refering to the particular problem. Indeed, L may change as the operating level of the system changes. This anomaly was discussed in Sec. 4.3.2. Its importance here is in the recognition that there is a loss in degrees of freedom which is at least as large as the number of product specifications.

4.4.2 *Underdetermined systems*

Underdetermined systems, with $D > 0$, are essential in optimization; otherwise a range of solutions is not possible. Such problems have in principle an infinite number of solutions, and the objective of optimization is the selection of the one which is best with respect to some given criterion. If $D = 0$, there will be only a single solution to the system. This is called a *defined system*. In an overdefined system ($D < 0$), there are more specified inputs than exist, and no solution can be obtained.

Thus, in a continuous stirred tank reactor, the specification of the feed vector, the reactor volume, and the rate of heat input would completely define the system outputs. There would then be only a single solution, and if further variables, such as the system exit composition, were also fixed, a solution would be generally impossible. Another example is found in the design of a distillation column. Here, if the feed specification, pressure, desired split of key components, and the reflux ratio are given, the system is completely specified. except for design details. The number of plates, temperatures, and compositions of the non-key components are all physically dependent variables, and are therefore fixed when the above variables are given. The additional specification of any from the second group would result in an unsolvable model (see, for example, Kwauk [1956]).

4.4.3 *The form of the objective function*

Once the objective of the system has been chosen, it can be related to the state of the system streams by a process called *simulation* (Sec. 7). The objective then has some functional form as follows:

$$y = y(\mathbf{q}_1, \mathbf{q}_2, \ldots, \mathbf{q}_Q, \mathbf{p}_1, \mathbf{p}_2, \ldots, \mathbf{p}_P, \mathbf{r}_1, \mathbf{r}_2, \ldots, \mathbf{r}_R) \qquad (4.4\text{-}2)$$

Of course, only D of the components determining the objective function can be manipulated. A particular choice of D variables then provides a unique value of y. It is misleading, and often complicating, to always select the variables to be manipulated from the $\sum_{k=1}^{Q} Q_k$ components of the input variables. Thus, from a calculational or experimental viewpoint, on occasion the composition of some interlinking stream may be more conveniently varied rather than an input variable. During the process of optimization, the stream composition may in this case be considered as a system variable, and the input variable then constrained to take on certain values, depending on the composition. Of course, after the optimization study was completed, the desirable operating conditions would actually be achieved by altering the input value. The means of setting the optimum conditions in the process plant is of no importance here. Our concern is to find the optimum conditions, and not to worry about how these are attained, for by imposing adequate constraints, we are assured that the solutions are practical. In setting up a general optimization problem, we may therefore not attempt to distinguish between the types of variables encountered, but call all the system variables x_i, for $i = 1, 2, \ldots, n$, n being the degrees of freedom if no restrictions are imposed. In general, we should then write Eq. (4.4-2) as

$$y = y(x_1, x_2, \ldots, x_n) = y(\mathbf{x})$$

In Part Three, however, where specific case studies are presented, it may be convenient to invoke the notation of Eq. (4.4-2).

5 THE UNIT MODEL, OR TECHNICAL DESIGN BASIS

It has already been noted that a system must be represented by a relationship between the system variables. Such a relationship, given by the transformation equations and forming the model, is required for each unit of the structure diagram. The derivation of these models is an essential step, since the objective function cannot otherwise be expressed in terms of the system variables. In other words, the response of the objective function to changes in variables must be computable for optimization to begin.

The model for a given unit must define the outputs of that unit for a given set of inputs. If, as may happen in certain reactor systems, several results are possible, depending on the mode of start-up, this must be specified. Optimization depends on the outputs being unambiguously defined for a given set of inputs.

The necessary model can be either analytical or black-box in nature. An analytical model is one which is expressed in mathematical terms, whereas in a black-box model, the response of the unit to a particular input is determined by a suitable experiment or numerical computation. A black-box model could thus be an existing physical system such as a process unit, a collection of mathematical formulas, or data, from which definite results can be derived for a specified input.

A full development of the derivation, interaction, and characteristics of these two types of models is presented in Chap. III.

6 SYSTEM RESTRICTIONS

In most practical operations, it is unsafe to assume that the system variables can be altered without constraint and that the system itself can operate independently of all external influences. It is uncommon to find an unrestricted problem. Of course, as we shall see, it is sometimes possible to use restrictions to eliminate certain variables, and thus end up with an equivalent unrestricted system.

6.1 The Origin of Restrictions

It is convenient, although not necessary, in discussing the origin of a restriction, to distinguish between those restrictions controlled internally by the optimizer and those controlled externally by an agency outside his jurisdiction.

6.1.1 External restrictions

The external restrictions imposed upon a system may be defined as the rigid, or uncontrollable, specifications set by an agency outside our jurisdiction. This external agency might include management, competing companies, the government, or nature. Some typical external restrictions are:

1. The total quantity of product to be produced is limited by management.
2. The sales and market analysis group may impose an upper limit upon the production rate due to a limited sales potential, or a total market is fixed and production up to that limit is controlled by the action of competitors in the same field.
3. The quality of the product is specified, either by legal specification (as with drugs) or by customer or company specification.
4. The quantities of raw material available might be restricted and of a specified quality.
5. The laws and regulations imposed by governmental agencies often give rise to restrictions, such as the size of packages handled by the post office, thus affecting distribution costs.
6. The system heat losses and the temperature of cooling water are partly governed by the daily and seasonal weather fluctuations.
7. The plant steam pressures are determined by existing facilities.

Constraints such as these must be recognized and accommodated in any optimization model. Thus, if the production rate R of a specific product is limited by a maximum capacity estimated to be D^L, then $R \leq D^L$, in which, generally, the output R is a variable and D^L is fixed. R would then usually be expressed in terms of the input system variables governing the production rate, so that the inequality restriction would become

$$R(x_1, x_2, \ldots, x_n) \leq D^L \tag{6.1-1}$$

a relation governing the otherwise free choice of the system variables. It ought to be noted that restrictions are not necessarily invariant. Thus D^L is an estimated quantity, and it might vary with time or change with the use of more accurate estimating methods. It is then advisable to treat the restriction D^L as a variable, and not to bury the restriction in the mathematical formulation as a numerical constant. Only if its identity is preserved can further revision be incorporated.

6.1.2 *Internal restrictions*

In addition to the external constraints which may be imposed, important restrictions arise within the system. These internal constraints are rigid specifications imposed by the designer, an important group being those relationships arising from the conservation of mass, momentum, or energy, or the rate expressions, which link the response to the inputs. Others are found in certain design relations. Thus, in the design of a cylindrical tank for a given volume, we may write down the restrictive expression linking the diameter x_1, the length x_2, and the volume x_3 as

$$\frac{\pi}{4} x_1^2 x_2 = x_3 \tag{6.1-2}$$

In addition, a restriction may be placed upon the ratio of the dimensions x_1/x_2, so that it lies in some reasonable range. Either of the situations shown in Fig. II-5 would be ridiculous for such a storage vessel, so that we might write

$$R_L \leq \frac{x_1}{x_2} \leq R^L \tag{6.1-3}$$

This type of restriction would normally be incorporated in the design by the common sense of the designer, but it would be necessary to state it if the complete design were to be produced from some computer program.

Another common restriction is that found when an independent variable may be altered only over a given range. Thus a temperature may be restricted by upper and lower bounds:

$$T_L \leq T \leq T^L \tag{6.1-4}$$

the upper limit T^L being fixed by the maximum temperature which can be withstood by the materials of construction, and the lower limit T_L by the cooling capacity of the plant services. In addition, the economics of mass

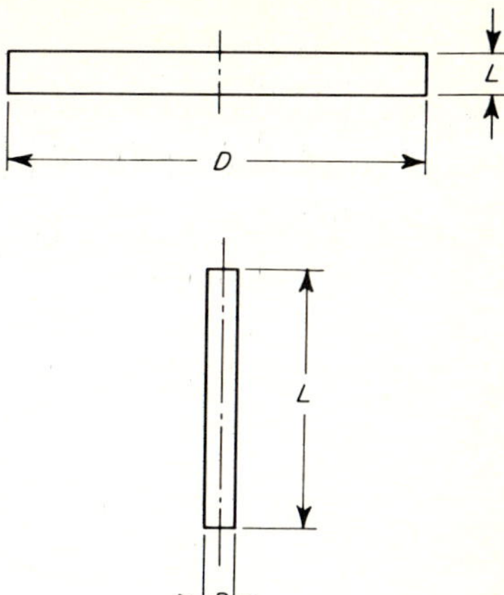

Fig. II-5 Unusual dimension ratios for cylindrical storage vessels.

production restrict many items to standard sizes, additional expense normally ruling out the use of nonstandard sizes. Thus many variables, such as pipe diameters or steam lagging thicknesses, will take on only discrete values.

These restrictions are rigidly imposed once the system has been defined. They are, however, capable of change. Thus a new material of construction with greater temperature stability might be chosen, so that the previous upper limit T^L can be extended.

6.2 General Form of Restrictions

While it is convenient to introduce the restrictions in terms of their internal or external source, there is in fact no hard-and-fast division between the two types. Thus an internal restriction for one system might become an external restriction for another. The type of restriction thus depends upon the level of optimization, as discussed in Chap. I, Sec. 3, but this is an unimportant distinction since it does not affect the mathematical formulation. Internal restrictions may, however, be altered as a further stage in optimization by changing the system.

Mathematically, all restrictions can be written in the equality form

$$g_j = g_j(x_1, x_2, \ldots, x_n) = 0 \tag{6.2-1}$$

or as an inequality,

$$g_k = g_k(x_1, x_2, \ldots, x_n) \leq 0 \tag{6.2-2}$$

Thus Eq. (6.1-2) can be represented by the equality restriction in the form of

$$g_1 = \frac{\pi}{4} x_1^2 x_2 - x_3 = 0 \tag{6.2-3}$$

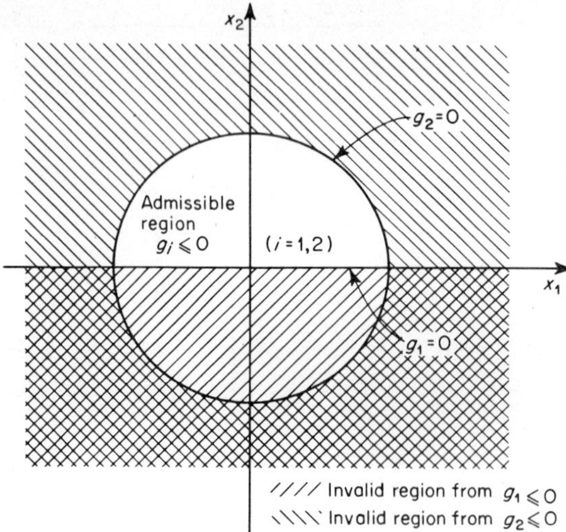

Fig. II-6 An admissible search region.

Similarly, the inequalities of Eqs. (6.1-1), (6.1-3), and (6.1-4) can be represented by Eq. (6.2-2). Thus the relations

$$g_1 = T - T^L \leq 0$$
$$g_2 = T_L - T \leq 0$$

are together equivalent to Eq. (6.1-4).

Restrictions of the form $g_k(\mathbf{x}) \leq 0$ define a region in which acceptable values of $x_1, x_2, \ldots, x_n$ are found. This region has boundaries which are defined by the locus of points satisfying the equality given by the upper limit of these restrictions, that is, $g_k = 0$. Thus, in a problem having two degrees of freedom, x_1 and x_2, the restrictions

$$g_1 = -x_2 \leq 0$$

and

$$g_2 = -1 + x_1^2 + x_2^2 \leq 0$$

might be imposed upon the solution. Then, as shown in Fig. II-6, the admissible values of x_1 and x_2 must be located in the semicircular region defined by letting $g_1 = 0$ and $g_2 = 0$, that is,

$$x_1^2 + x_2^2 = 1$$

and

$$x_2 = 0$$

Points satisfying these two equalities are points on the boundary, and points satisfying both the inequalities, $g_1 < 0$ and $g_2 < 0$, are interior points lying within the admissible region. All other solutions are excluded by the restrictions from consideration in the optimization. Thus the absolute optimum result might never be achieved, as in the simple illustration of Example

I-1, due to the presence of restrictions. In such situations we must settle for the best possible optimum for these circumstances.

7 SIMULATION—THE EXPRESSION OF THE OBJECTIVE IN TERMS OF THE SYSTEM VARIABLES

At this stage in our preparation for optimization we have our basic aims well in mind and have developed the required models representing each block depicted in the structure diagram. The next step is to transform the objective into a function of the independent variables, so that for each set of these variables there corresponds a particular numerical value of this objective function. Only by the use of such a function can the optimization be attempted by a well-defined series of operations.

How do we go about translating a knowledge of the system response into an objective function? Much of the work has already been completed. If the objective is technical in nature, the simulation is essentially completed. The technical variables, such as weight, fuel consumption, and yields, all appear directly in at least one of the transformation equations. In fact, the technical objective is normally one of the outputs and as such is given directly.

On the other hand, for an economic objective, simulation is more complex, although not usually difficult. The first step is the reduction of the criterion to its economic components. Thus a profit is found from sales income less various costs, while investment is dependent upon the purchased costs of individual components. These component costs, in a second step, can be related to appropriate technical variables through suitable cost data. This means that, for given system inputs, the technical variables, the costs, and the desired objective function values can be evaluated.

Simulation is thus not difficult if a system model is available. The basic approach may be illustrated in the following examples, the final form of the objective function being expressed as $y = y(\mathbf{x})$, subject to a set of restrictions of the forms $g_j(\mathbf{x}) = 0$ and $g_k(\mathbf{x}) \leq 0$, where $\mathbf{x}$ represents the n variables $\{x_1, x_2, \ldots, x_n\}$. y may be presented in many forms, depending upon the model types used in simulation. Tables of data, graphical relationships, and nomograms, as well as mathematical formulas, may all contribute to the objective function evaluation.

Example II-1: Simulation of an economic criterion

Let us find the profit from a given chemical process in terms of the process variables, assuming that models exist for each unit in the process.

First, the capital investment for the process: this is the sum of the fixed and working capital (Sec. 2.1.2). There are various methods of cost estimation (Vilbrandt and Dryden [1959]), but a convenient approach is the use of the expression

Total capital investment
$= (\text{total cost of items of equipment delivered and installed by supplier})\phi_I$
$\quad + (\text{total cost of items delivered to site but not installed by supplier})\phi_D$

Here the factors ϕ account for all the other plant construction costs, including piping, buildings, instrumentation, foundations, services, and contractors' costs, typical values being given in Table II-1. Typical

TABLE II-1 CAPITAL COST ESTIMATION FACTORS[1]

Type of plant	*Delivered-equipment cost factor* ϕ_D	*Installed-equipment cost factor* ϕ_I
Solids-handling plant	3.10	2.16
Fluid and solid processing	3.63	2.56
Fluid processing	4.74	3.30

[1] Lang [1948].

costs of equipment items can be found in various journals and texts (Bauman [1964], Aries and Newton [1955], and Vilbrandt and Dryden [1959]) as functions of some characteristic dimension. These may be updated in cost by multiplying by an appropriate cost index. Since each item of equipment can be sized for each set of process variables, the total capital investment can be estimated. It should be noted that even when considering the investment or cost for a single item, this would be greater than the purchased cost of the item by the factor ϕ, to allow for the associated overheads and cost of accessories.

Second, we may estimate the process profit. In general, the profit equals the value of the salable products less the total costs of operation, overheads, and depreciation. The operating costs, such as cost of utilities, can often be found tabulated for typical processes, and the raw-material costs can be added to these. Overheads, too, can be accounted for by factors (Vilbrandt and Dryden [1959]). The depreciation cost is on average about 10 percent of the fixed capital, which is approximately 80 percent of the total capital investment for the process. Once the process variables have been specified, the various costs can be derived and the profit function evaluated.

Example II-2: Simulation of a tubular reactor

Let us indicate the approach to specify the profit and capital investment associated with carrying out a single chemical reaction $\sum_i A_i = 0$ in a continuous tubular reactor.

The first step in any simulation study is the formulation of the technical design basis, or model, of the system. Here we need consider only the steady-state conservation balances for mass and energy and momentum. From Eqs. (2.1-2) and (2.2-4), Chap. III, the yield of product from a plug flow reactor, $0 \leq l \leq \theta$, is

$$v(c_{i,\theta} - c_{i,0}) = \int_0^\theta a_i r(c_1, c_2, \ldots, c_n, T)\, dl$$

where c_i are the concentrations of components A_i, and T is the reaction temperature, which may vary with position l along the reactor. The associated energy balance is found from Eq. (2.2-2), Chap. III, to be

$$\int_0^\theta [r(-\Delta H) + h(T - T_a)]\, dl = vc_p(T_\theta - T_0)\rho$$

h being the heat-transfer coefficient at the reactor wall. Further expressions might be produced relating the pressure changes through the system to the process stream velocity v.

For any given size of reactor, any specific temperature variation along its length, and a specified fluid stream velocity, the system will now be completely specified, and its operating and installation costs can be estimated.

The fixed capital required to finance the installation of the reactor and its associated equipment is directly related to the equipment dimensions. Thus the estimated cost I_R of purchasing the reactor itself can be found from published information. If a recent cost is not available for I_R, updating might be necessary, obtained through the use of cost indices. This can be done for each item of equipment required, and the total delivered or installed equipment cost can be found. The fixed capital cost can then be found upon multiplying by a suitable factor, ϕ, from Table II-1. Addition of an allowance of about 25 percent of the fixed capital cost to provide working capital then yields the desired total capital investment expression. For a fixed throughput, this will here be a function of the reactor length θ and the stream velocity v.

The system profit will equal the income from sale of products less the operating costs and depreciation allowance for the fixed capital employed. The income is given by the value of the product and is thus determined by the amount sold and the price obtained. The operating costs include the cost of supplying or removing heat at any point along the reactor, the raw material costs, and the cost of providing energy for transferring material through the circuit (a function of the stream velocity and the reactor dimensions), together with any fixed charges such as management overheads. The capital used has to be paid back over a fixed period, often taken as 10 years, so that the depreciation allowance is usually about one-tenth of the fixed capital investment.

Such a procedure, of which fuller details can be found in such texts as that of Vilbrandt and Dryden [1959], can be adopted to obtain the profit and the investment for the system, using pertinent economic data in conjunction with a suitable system model. For purely technical problems, the model can often directly provide the necessary simulation. The simulation in the present example thus defines economic objectives in terms of the system variables—the reactor temperature, the reactor dimensions, and the stream velocity—for use in subsequent optimization studies.

7.1 Note on Stochastic Objective Functions

So far it has been inherently assumed, in using the technical design basis to obtain an objective function in terms of the various independent variables, that once the latter are defined, a single value of the objective function will emerge. This is not always so. The objective function may be either exact or stochastic in nature, the probabilistic result being due to a stochastic model for the design basis or to the nature of the data input.

The distinction between the two cases may be illustrated by considering an example (Johnston [1963]) in which the volume of sales, y, of a particular chemical is determined largely by its price x, all other system variables being considered unimportant in their effect on the sales. For an exact or determinate relationship to hold between the sales and price, the expression

$$y = y(x)$$

implies that a unique or exact value of sales, y, will be obtained for any given or specific price x. On the other hand, a stochastic relation implies that the average or expected sales (over a given period) will depend upon the price level x, but for any given price the actual sales will fluctuate around the expected value, with probabilities u for fluctuations of various sizes. Thus we may write the sales objective function y as

$$y = y(x,u)$$

This stochastic element in the design model and its simulation arises here from variations in market demand.

Stochastic systems have been discussed, for example, by Bellman [1957, chap. 2], Aris [1961d], Rudd [1962a–c], Bellman and Dreyfus [1962], Graves and Wolfe [1963], and Roberts [1964b].

The actual values of y will fluctuate with known probability about an "expected," or mean, value. Most stochastic problems can be handled in a similar manner to "exact," or deterministically defined, problems if the probabilities are included in the system specifications. An alternative procedure is the use of expected values in the objective function. When problems involve repeated actions at frequent intervals, such a criterion can be expected to give plausible results. On the other hand, when a decision occurs only once, difficulties may arise. Such problems are common in practice, an illustration being the single investment of large capital sums in new industrial plants, these, usually, not being repeatable in similar circumstances. Suitable criteria for these "one-shot" risk situations have been considered by Schlaifer [1961] and others, based on the work of Von Neumann and Morgenstern [1953]. In many cases, however, such difficulties can be avoided by considering objectives in terms of company operations, rather than individual projects, so that the larger complex can provide a cushion for chance failures.

8 SIMPLIFICATION

At this stage in the setting up of the problem, the objective can be found as a function which has a well-defined value for each set of system variables. It may be found that the number of variables to be handled is very large, this usually being the case at the higher levels of optimization, as described in Chap. I, Sec. 3. This may prevent an analysis being attempted unless some simplification takes place. Simplification may take three forms.

First, all those variables which, when altered in value, produce negligible changes in the objective function can be dropped from further consideration. These variables would, of course, have to be retained in the design specifications and would represent a constant value affecting the order of magnitude of the objective function values. An interesting example is found in the economic analysis of a system using heat exchangers. Although the heat-exchanger design variables may include the flow rate and temperature of the fluids, the material, number, length, diameter, and wall thickness of the tube passes, shell diameter, type of shell, number of baffles, etc., it will, in fact, be found that, for a given type of heat exchanger, the cost of the unit is essentially dependent upon only the heat-transfer surface area. This, then, is the only variable which needs to be considered in the analysis of this portion of the system; all the other variables which are certainly required for a design specification can be neglected as far as a cost-optimization study is concerned. The slight increase in accuracy gained by including all the design variables is outbalanced by the greatly increased difficulty of analysis in a multivariable system. The aim in this reduction is to leave the objective function expressed only in terms of those variables whose variation significantly affects its value.

Second, the problem may be divided into stages, so that optimization can be carried out with respect to smaller numbers of variables at any stage in the calculation. Thus the division of a problem with 20 independent variables into four problems, each with five variables, will reduce the necessary computation of the objective function by about a third, or even more. Thus, using the rule-of-thumb relation

$$p = \tfrac{1}{2}(N + 1)(N + 2)$$

of Cochran [1960] for finding the number of points, p, required to estimate a second-degree response surface which is a function of N variables, we get

N, no. of variables	3	5	10	20	50
p, no. of necessary calculations	10	21	66	231	1,326

On this basis, with digital computer time at, say, \$400/hr and a single point calculation requiring 5 min, a problem involving 20 variables would cost about \$8,000 to fit a second-degree response surface for further use in optimization. Since large problems may need much more time to calculate even one solution or point, and may involve several hundred variables, much analysis and simplification is justified before the optimization calculation is attempted. It should also be noted that the setting up and "debugging" of the computer program, and the interpretation of the results, is also a power function of the number of variables, and this is an even greater reason for simplifying the original model of the process.

It should also be noted—and this is an important point—that if we are concerned with an overall project, we cannot in many cases design or operate an individual unit or process which is part of the total project at an optimum level found by examining only that isolated individual unit. The objective function for the whole project must be considered by optimizing the whole project, or by considering specific units only in some carefully defined circumstances. Problems will be given later showing how such individual optimization may lead to erroneous values of the optimum objective function. The correct approach to such decomposition is discussed in Chap. XIII.

Third, most practical problems entail a very large number of independent variables, constrained by a large number of restrictions. Generally, some degree of simplification is required, since optimization becomes difficult for more than a few variables (Part Two), unless the objective function has a very special form, such as being linear. Many of the variables may therefore have to be held constant for optimization to proceed, leading to a quasi optimization in which improvement is achieved in extremely difficult situations. A simplification may also be achieved by linearization, yielding results valid over a limited range. The essential point to be made here is that, even if the original problem, stated in its most general terms, is too costly to solve, one can still profitably use optimization techniques by considering a

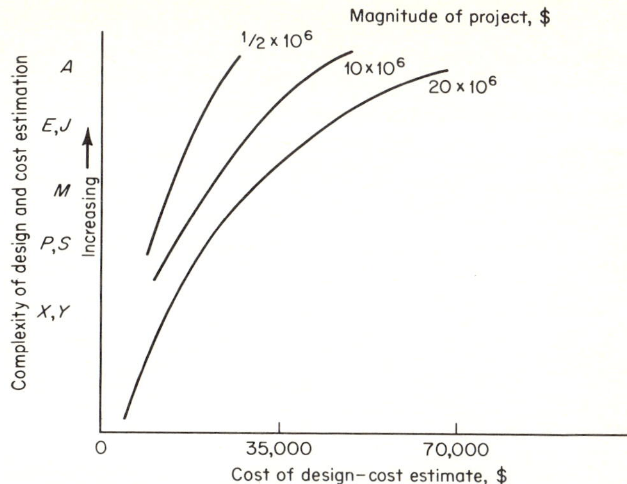

Fig. II-7 The cost of carrying out a design-cost estimate.

reduced form of the problem, ensuring, of course, that the reduction has not destroyed the realism of the model.

In general, the simplification and analysis step requires at least one detailed calculation of the response or objective. This is usually done in greater depth than will be attempted in the final simulation, to ensure that a valid solution is being obtained.

Quite apart from the organizational and mathematical problems of optimization, the cost of obtaining any solution is closely allied to the complexity of a given problem. Both of these are functions of the number of simultaneous independent variables and external parameters involved. Nichols [1951] has given some figures for the cost of economic analyses for chemical plant (while these are for design and economic evaluation, they give some indication of the type of costs involved in an optimization study), the costs for different magnitudes of projects being given for varying degrees of complexity in cost estimating, as shown[1] in Fig. II-7. Obviously, the appropriate detail to be included in a study of a given problem is an optimization problem in itself. There is little point in spending more money optimizing than can be received from the improvements gained by applying the results.

[1] Here *A* is the detailed cost estimate of all items, based upon quotations; *E* and *J* are factored estimates (based on the total equipment cost) to give the installed cost, and the additional items such as buildings, utilities are estimated separately and added in; *M* is the detailed factor method in which only the delivered or installed equipment cost is found and all the other costs are found as factors of the total equipment cost; *P* and *S* require only the total equipment cost, this being multiplied by the single factor to account for all other costs; and *X* and *Y* are ratio methods based only on factors and a knowledge of sales, etc. Therefore, these methods range from the complex to the trivial.

The cost and the complexity of a problem therefore normally make it desirable to reduce the dimensionality in any calculation by reducing the number of variables involved, by breaking down the problem, and by simplifying the mathematical representation.

9 VERIFICATION

Verification is listed separately to emphasize its importance. In practice, it is part of the simulation procedure.

A great deal of effort must be expended in determining whether or not the simulation expressions adequately represent the response of the real system. Otherwise, no amount of optimization can produce correct answers. If the system is still in the design stage, this verification can only be inferred. Van Heerden [1961], perhaps with tongue in cheek, has made some pertinent remarks regarding elegant optimization techniques and unreliable or unknown data, this aspect being especially important when the methods are complex and expensive in time and money.

10 OPTIMIZATION

Having expressed the objective function in terms of the independent variables and the associated set of restrictions, it is now necessary to use a suitable optimization technique to find the desired optimum conditions. These methods are the subject of Part Two.

Some of the practical problems involved in optimization are discussed in Part Three.

11 ITERATION

Once an optimum set of conditions has been determined, it may be necessary to evaluate the underlying assumptions required in its solution. In purely mathematical analyses such as carried out in Part Two, this may not be necessary, but in most practical applications any optimization will require several iterative attempts before the problem can be suitably defined, the appropriate model formulated, and the system optimized.

The objective of the analysis may also define the type of model or calculation procedure to be used in the optimization study. Thus the study might be an initial appraisal of many vastly different projects to decide which is worthy of further investigation. Then, for example, a quick check of the raw material costs, based upon an optimistic reaction yield, even 100 percent, may be sufficient, when compared (1) with the value of the product (neglecting purification and operating costs) and (2) with raw material costs for alternative chemical routes, to dismiss a process from further consideration. Other calculations may be aimed at determining the magnitude of an optimum, the elimination of variables having little significance in the problem, the selection of items for suboptimization, the definition of data requirements, the selection of methods of simulation, the full-scale analysis and optimization of a system, or the removal of a critical restriction. The effect of the lack of data and the sensitivity of the value and location of the optimum to changes in data may also have to be considered. Obviously, less accurate and simpler representations can be used at the earlier stages of the investigation, and once a problem has been carefully defined and delineated, it may be possible to use linearized models valid for closely specified regions. The validity of these assumptions must, however, be established.

Iteration may therefore affect all stages in the general procedure for optimization.

Problems

II-1 Show that \$100 invested today at 5 percent per annum is worth \$162.60 in 10 years.

II-2 What is the present worth of a system which originally, 3 years ago, cost \$1,000 and which will require \$1,050 and \$1,100 replacements after 5 and 10 years, respectively?

II-3 Indicate the main units in a block diagram representing the production of sulfuric acid, or a similar product, showing the main process streams and heat flows.

II-4 The XYZ Process and Chemical Company has a new process for catalytically hydrogenating benzene to cyclohexane. In this process the feed benzene is mixed with makeup hydrogen and an amount of recycled hydrogen, and the resulting mixture heated to reaction temperatures. The reaction is exothermic, and at the conditions of operation, complete hydrogenation of benzene is achieved in each pass.

The effluent from the reactor is cooled and subsequently flashed. The excess hydrogen is recycled to the reactor so as to minimize the requirements for fresh makeup hydrogen. The liquid cyclohexane product is further purified by stripping of light gases in a stripper and sent to storage.

a. Draw a block diagram representing this process and discuss possible arrangements to utilize the heat of reaction so as to reduce steam costs.

b. Assuming that the system described above is an existing one, what studies would you recommend to improve its economic performance? A plant producing 15 tons of finished cyclohexane per hour will require about 6,500 lb of steam (400 psia, saturated) to strip the light gases from the hexane product. This is the only steam requirement, assuming that the reactants are heated by a product recycle. Also, 387 kw is required for compression, and 450 gpm cooling water is required to further cool products before flashing.

c. Assuming this unit is a part of a larger manufacturing plant which is to be optimized, write the mathematical model for this unit, explaining why the reaction kinetics are not important and clearly denoting the external feeds.

II-5 Discuss the degrees of freedom available in a two-sphere system when movement is constrained by the surfaces of both spheres and the spheres (*a*) touch, (*b*) intersect, and (*c*) do not touch or intersect.

II-6 It is desired to cool 3,000 lb/hr of a gas (heat capacity = 0.3 $\text{Btu/lb}_m \cdot {}^\circ\text{F}$) from 195 to 90°F, using cooling water at 80°F. Water costs \$0.20/1,000 ft^3, and the annual fixed charges on the cooling surface are \$0.50/ft^2 of inside surface. The overall heat-transfer coefficient can be assumed to be 8 Btu/ft^2 · hr · °F.

Make a plot showing both the yearly cost of the cooling water and the heat-exchange surface against the hourly use of cooling water. If minimum cost, considering only these costs, is the objective function, what rate of cooling-water flow would you select? What competing factors give rise to a minimum cost?

II-7 It is desired to cool a gas (heat capacity = 0.3 Btu/lb$_m$ · °F) from 195 to 90°F, using cooling water at 80°F. Water costs \$0.20/1,000 ft^3, and the annual fixed charges on the cooling surface are \$0.50/ft^2 of inside surface, with an inside diameter of 0.0875 ft. The power costs are such that a mean velocity 9,000 lb$_m$/ft^2 · hr for water inside the tubes is optimum, and a corresponding overall heat-transfer coefficient of 8 Btu/ft^2 · hr · °F is realized when the gas rate is 3,000 lb/hr.

Make a plot showing both the yearly cost of the cooling water and the exchanger as a function of the outlet water temperature. What is the outlet temperature giving the minimum cost? Why is the answer without practical interest? What additional information must be given to have a result of practical interest? Given the additional information, is the problem, then, one properly suited for optimization? Discuss this latter question within the framework of underdetermined and overdetermined systems.

II-8 Derive a model for the cone-roofed storage-tank system (a vertical cylinder capped by a vertical cone), and hence find an objective function and restrictions for minimizing the surface area subject to a fixed storage volume.

II-9 Using suitable published information (for example, Vilbrandt and Dryden [1959] and Bauman [1964]), estimate the installation cost of a steel-tube and steel-shell heat exchanger of 3,000 ft^2 transfer area. What would its present cost be? What is the total cost of the project if this unit represents the main equipment installed?

II-10 Set up the model for the catalytic-reaction system described by Hills [1963].

II-11 Set up the model for an adiabatic catalytic reactor for SO_2 oxidation, using the kinetic data of Calderbank [1953].

CHAPTER III

SYSTEM MODELS

1 THE INTERRELATION OF MODELS

As noted in the preceding chapter, each unit in the system must be modeled in such a way that the output of that unit is determined by well-established rules once all the inputs are specified. Such relationships must be available before optimization can proceed. As suggested in Chap. II, Sec. 5, the model can be one of two forms, analytical or black-box, although it is not possible, or even desirable, to make a clear-cut distinction between the two types. Indeed, these labels are suggestive of a philosophy of approach, rather than representative of a precise definition. The two models are quite interchangeable, Fig. III-1 depicting the free communication between them. The route by which analytical models can be treated as the black-box type, as well as the means of converting the black-box outputs into analytical relationships, is shown.

In the analytical approach certain mechanisms, which are postulated as the controlling features, are used in conjunction with the accepted laws of conservation (matter, energy, momentum, and electric charge) to derive a self-consistent set of equations, where self-consistency means that the equations define uniquely the outputs for a specified set of inputs. We can use the resultant equations in two ways.

First, it may be possible, by a series of both physical and mathematical simplifications, to reduce these equations so that they are suitable for use in conjunction with the analytical methods of optimization. Here we require equations which can be differentiated, since analytical methods rely on the search for points at which the first derivative vanishes. The possibility of simplification and solution by analytical means is indicated by the vertical path on the left-hand side of Fig. III-1 and is discussed in Sec. 2.

The second way in which the analytical model can be used is as a black-box system. This route, shown in Fig. III-1, is usually selected when the analytical model is complex and cannot be reduced, either on physical grounds or by mathematical arguments, to a simpler valid form. In this situation, common in most practical applications, the model is used to generate black-box type

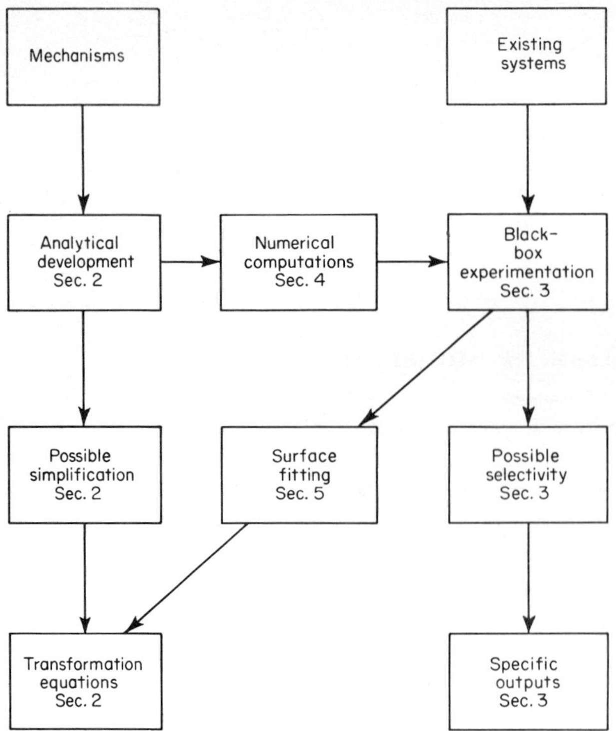

Fig. III-1 Interrelationship between models.

outputs by applying appropriate numerical techniques and using fast computing machines (Sec. 4). These outputs are completely determined by the particular inputs to the system.

On the right-hand side of Fig. III-1, there is a second vertical sequence paralleling the analytical path. Here the requirement is for an existing system, so that outputs can be measured for a given set of inputs. Various methods of obtaining the outputs can be used (Sec. 3). The relevant outputs (relevant in the sense of contributing to the objective function) must be screened from the data. These outputs provide the basis for numerical optimization.

The data can, however, be used in a second way. If a sufficient number of them are available, surface fitting techniques can be used to produce an analytical model which is representative of the system behavior over a certain restricted range of inputs. This analytical model can then be manipulated in the same way as one developed from the mechanistic approach and is, for optimization purposes, quite indistinguishable, provided the independent variables are not allowed to range beyond their region of validity. Surface fitting is a very important tool in the process of optimization, and this subject has been developed into rather specialized techniques, primarily by the statisticians. Because of its importance, a discussion of surface fitting is presented in this chapter (Sec. 5).

2 THE ANALYTICAL APPROACH

If the system to be optimized does not yet exist, or if experimentation on an existing system is not feasible, due to high costs or for other practical reasons, the only approach is through an analytical model. For simple systems there is perhaps a systematic approach to modeling (see, for example, Himmelblau and Bischoff [1968]); complex systems, however, must be investigated on an individual basis, and hopefully, information such as pilot-plant data and scaling laws will be available. It is in fact reasonable to expect that such information will be available, since the same model required in the optimization is also needed in the design of the system. Therefore model building is not just a requirement of optimization, but is also demanded by the designer. The main point being made in this discussion is that adequate models for optimization are presently available in all cases for which adequate designs are possible, and we can therefore dismiss the pessimists who suggest that one should first solve the problem of developing adequate models, to the exclusion of optimization.

In this section we are interested in investigating the structure of models, especially emphasizing those aspects which relate particularly to the problem of optimization. It is necessary to distinguish between two types of analytical models, namely, the distributed and lumped systems (Sec. 2.1), and to consider the evolution of a model using the mechanistic approach (Sec. 2.2). These steps give models which are not usually in the form of the desired transformation equations (Sec. 2.3), and in most cases of interest it is impossible to develop that particular structure. Lastly, we shall consider the problem of simplifying the analytical model, using both physical and mathematical arguments (Sec. 2.4).

2.1 Distributed and Lumped Systems

The models for simple systems are usually derived by incorporating the important mechanisms into a balance, or conservation, equation. This procedure usually leads to a relationship between the inputs and the outputs, but not necessarily in the form of a transformation equation, where the outputs are explicitly given as functions of the inputs.

The analytical model will fall within one of two categories, depending on the nature of the system being considered and the detail needed to achieve the desired understanding. If the relevant system variables are independent

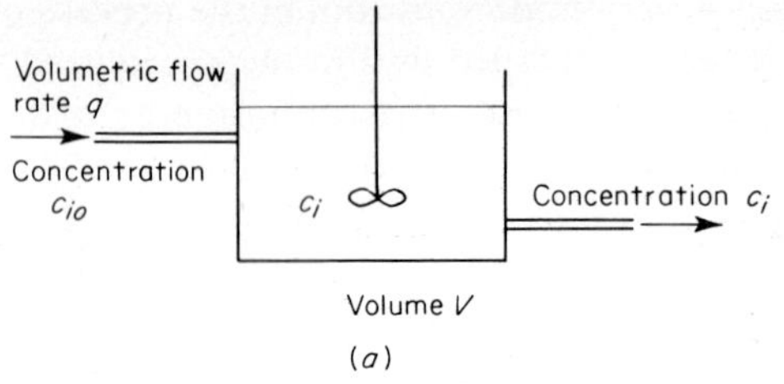

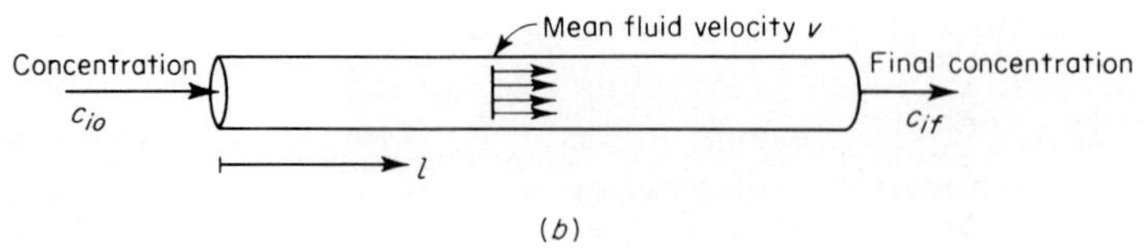

Fig. III-2 Models for continuous flow reactors. (*a*) Stirred tank; (*b*) tubular.

of spatial position within the unit, the unit is called a *lumped-parameter system.*[1] *Distributed systems*, on the other hand, are those in which the system variables vary from point to point within the unit. As illustrated in Fig. III-2*b*, the concentration varies along the axis of a tubular reactor, and hence such a unit is a distributed one. It will be noted that the continuous unit is the limit of a series of discrete lumped stages as the number of these becomes infinitely large.

Example III-1: The difference between lumped and distributed chemical reactors

Let us model both the tubular and the stirred tank reactors shown in Fig. III-2. The equation defining the transient behavior of the stirred tank reactor is

$$V\frac{dc_i}{dt} = -r_iV + qc_{i0} - qc_i \tag{2.1-1}$$

where V = volume of reactor

c_i = concentration of substance A_i

c_{i0} = inlet concentration of A_i

q = volumetric feed rate

t = time

r_i = rate of reaction of substance A_i per unit of volume

[1] For a rather comprehensive discussion distinguishing between lumped- and distributed-parameter systems, the reader is referred to Himmelblau and Bischoff [1968, chap. 3.].

The essential feature which we wish to emphasize is that the concentration c_i does not vary from point to point, but is only a function of time. In fact, if we carefully investigated the concentrations within the reactor, we should find that they do indeed vary to some extent from point to point at a given instant of time. Therefore Eq. (2.1-1) is an approximation to the true solution.

The equation describing the time-dependent behavior of the tubular reactor is

$$\frac{\partial c_i}{\partial t} + v \frac{\partial c_i}{\partial l} = -r_i \tag{2.1-2}$$

In this equation v is the velocity of the fluid through the reactor. Comparing Eq. (2.1-1) and (2.1-2), we see that in the case of the tubular reactor, additional derivatives appear, owing to the spatial dependence of the concentration in this distributed system.

We see that the basic difference between the lumped- and distributed-parameter models resides in the fact that the lumped model reduces to an algebraic equation in the steady-state (time-independent case), whereas the distributed system is represented by a differential equation even when the system is operating in the steady state.

2.2 The Role of Mechanisms

The unit to be modeled may be any one of a variety of operations, processes, or businesses. It is therefore impossible for us to consider more than the most general facets of model building if our remarks are to pertain to all these possible systems.

In general, we can assert that the simple systems are given mathematical representation by imposing balances on certain quantities, together with a number of simplifying assumptions. Such equations take the form

$$\begin{aligned}\frac{\partial}{\partial t}\text{(property)} &= \text{flow rate of property in} - \text{flow rate of property out} \\ &\quad + \text{rate of generation of property}\end{aligned} \tag{2.2-1}$$

for lumped systems, and

$$\begin{aligned}\frac{\partial}{\partial t}\text{(property)} + \frac{\partial}{\partial l}\text{(convection of property)} &= \text{divergence (flux of property)} \\ &\quad + \text{rate of generation of property}\end{aligned} \tag{2.2-2}$$

for a distributed system.

Such balance equations usually have a number of physical approximations tacitly implied in their structure. The stirred tank reactor is only approximately described by Eq. (2.1-1) since volume changes on reaction, imperfect mixing, and diffusion in both the inflow and outflow lines are all neglected in writing this equation. It is interesting to note that both Eqs. (2.1-1) and (2.1-2) are balance equations having precisely the form of the general equations (2.2-1) and (2.2-2). The rate of chemical reaction is seen, then, as a negative rate of generation, or equivalently, a rate of disappearance. Generation terms appear in the equations, since A_i is not conserved.

Let us now define the role of mechanisms in the determination of the system model. Mechanisms serve to define both the rates of generation and the fluxes of that property being balanced in terms of all the system properties. The generation term could, for example, relate to the generation of money if a business system were the unit of interest. The mechanistic approach requires that the rate of generation of money, then, be expressed in terms of the amount of money present in the system, and thus, in this case, generation would simply be the rate of interest times the amount of money. The rate of generation of a substance by chemical reaction can often be modeled by the equations detailed in the following example.

Example III-2: The generation term in chemical reaction

In Example III-1, we saw that the balances for both the stirred and the tubular reactors include a term $-r_i$, appearing in Eqs. (2.1-1) and (2.1-2). Let us see how this generation term might possibly be modeled. Suppose that the stoichiometric equation can be written

$$a_1A_1 + a_2A_2 + \cdots + a_nA_n = 0$$

where the a_i are the stoichiometric coefficients, and the A_i symbolize the various substances formed by chemical reaction.[1] The rate of this reaction can be quite generally represented as

$$r = k_1(T)\prod_{i=1}^{n} c_i^{\gamma_i} - k_2(T)\prod_{i=1}^{n} c_i^{\beta_i} \tag{2.2-3}$$

where r = rate of reaction per unit volume

c_i = concentration of substances A_i

γ_i, β_i = constants

k_1, k_2 = reaction velocity constants, dependent upon temperature T.

The rate of reaction is defined so that the rate of disappearance of

[1] Thus, for the reaction

$$3N_2O_4 + 2H_2O \rightarrow 4HNO_3 + 2NO$$

$A_1 = N_2O_4$, $A_2 = H_2O$, $A_3 = HNO_3$, and $A_4 = NO$. Moreover, $a_1 = -3$, $a_2 = -2$, $a_3 = +4$, $a_4 = +2$.

substance A_i per unit volume r_i is given by

$$r_i = -a_i r \tag{2.2-4}$$

The γ_i and β_1 are usually (but not necessarily) simple multiples of the stoichiometric coefficients. For a detailed discussion of the methods of constructing kinetic rate expressions and the experimental techniques involved in evaluating the parameters, the reader is referred to specialized texts on kinetics (see, for example, Laidler [1965], Frost and Pearson [1961]).

An experimental determination of the relevant constants, k_1, k_2, γ_i, and β_i, in any one system would then be valid for the same chemical reaction in other systems. In the previous example we saw that the mechanistic approach derives its real value from the fact that experimental data obtained in a particular way can be generally applied. Experimental data are always required at some stage in the construction of a model, and the best we can hope for is that the required experiments have already been performed and the results are available to us. Thus diffusion coefficients, thermal conductivities, and electrical conductivities would all serve to define the flux of some property in terms of the system properties (the flux of mass is given by the diffusion coefficient multiplied by the gradient of the concentration), and these are reported in the literature for a variety of cases. A fuller account of transport properties and their use in balance equations is given by Bird, Stewart, and Lightfoot [1960], and the discussion of the role of the mechanism in modeling is described by Himmelblau and Bischoff [1968].

We should note that the tactics described above are likely to be fruitful only when dealing with rather simple systems, and for the more complex cases black-box experimentation is likely to be the sole means of obtaining a valid representation for some time to come.

2.3 Transformation Equations

In Chap. II, Sec. 4.3.1, we saw that the desirable structure for optimization is an expression which explicitly gives the output as a function of the inputs. To achieve this structure, some mathematical rearrangement, and perhaps simplification of the analytical model, is required. If the unit is lumped, then, in the steady state, an algebraic equation must be manipulated to obtain the transformation equations. This rearrangement of the model is always possible in principle, but in most practical situations it is too complex to be attempted.

Even in the steady state we must integrate a differential equation to obtain the transformation equation modeling a distributed system. Again we note that in practice this is not usually possible.

Example III-3: The transformation equations for chemical reaction

In Examples III-1 and III-2 we considered the simple stirred tank and tubular reactors, having developed both the material balances and a mechanism defining the reaction rate. Let us use these results to find the transformation equations describing the steady-state behavior of these reactors for the simple reaction

$$-A_1 + A_2 = 0$$

Assuming $k_2 = 0$, $\gamma_1 = 1$, and $\gamma_2 = 0$, Eq. (2.2-3) becomes

$$r_1 = -a_1 r = k_1 c_1$$

Substituting this expression into Eq. (2.1-1), we find, on neglecting the time derivatives, that

$$-Vk_1c_1 + qc_{10} - qc_1 = 0$$

This algebraic equation can easily be manipulated to obtain the transformation equation. Thus the output c_1 is given by

$$c_1 = \frac{qc_{10}}{k_1 V + q}$$

For the same reaction and in the steady state, the equation defining the tubular reactor Eq. (2.1-2), reduces to

$$v \frac{dc_1}{dl} = -k_1 c_1$$

Integrating this differential equation gives the transformation equation

$$c_1 = c_{10} e^{-lk_1/v}$$

This example demonstrates the procedure required to develop the transformation equation. It is not difficult to visualize a model which renders this procedure all but impossible. Any optimization procedures, such as those discussed in Chap. XIII, Sec. 5, requiring the transformation equations, are therefore likely to have limited application.

2.4 Simplification of the Analytical Model

Analytical models are not usually found having the structure of a transformation equation, and further simplification of the model is often desirable. This simplification can result from either a further valid physical (Sec. 2.4.1) or a mathematical (Sec. 2.4.2) approximation.

2.4.1 *Physical simplification*

The steps necessary to simplify a model on physical grounds are obviously specific to each particular problem, and we cannot hope to give any general rules applicable to all situations. Replacing a distributed system with a sequential series of lumped units and representing the effect of turbulent fluctuations with an eddy viscosity are both physical approximations which, when used, may simplify considerably the task of obtaining input-output relationships. Clearly, this is a route to simplification which we should consider every time we are faced with a complex model. The essential point to remember is that the approximation must be valid: i.e., the resultant equations must still represent the physical system with sufficient accuracy.

2.4.2 *Mathematical simplification*

The second approach to simplification is a purely analytical one. The task facing us can be seen by noting that the model is usually expressed as an equation of the type

$$F(p,\mathbf{q}) = F(p, q_1, q_2, \ldots, q_n) = 0 \tag{2.4-1}$$

where p is the output, and the q_i are inputs. Here the output is an implicit function of the inputs, and we seek an explicit expression of the form

$$p = f(\mathbf{q}) \tag{2.4-2}$$

In general, we cannot rearrange Eq. (2.4-1) so as to find this explicit form (Sec. 2.3). Some progress can be made if we are willing to settle for a result which is applicable in a narrow range about some operating or base point, say, $\{p_0,\mathbf{q}_0\}$. The most direct approach is to use a Taylor's series expansion (Chap. VII, Sec. 2.2), namely,

$$\begin{aligned} F(p,\mathbf{q}) = F_0 &+ \left(\frac{\partial F}{\partial p}\right)_0 \Delta p + \sum_{i=1}^{n} \left(\frac{\partial F}{\partial q_i}\right)_0 \Delta q_i + \frac{1}{2}\left(\frac{\partial^2 F}{\partial p^2}\right)_0 \Delta p^2 \\ &+ \sum_{i=1}^{n} \left(\frac{\partial^2 F}{\partial p\, \partial q_i}\right)_0 \Delta p\, \Delta q_i + \frac{1}{2}\sum_{i=1}^{n}\sum_{j=1}^{n} \left(\frac{\partial^2 F}{\partial q_i\, \partial q_j}\right)_0 \Delta q_i\, \Delta q_j \\ &+ \frac{1}{6}\left(\frac{\partial^3 F}{\partial p^3}\right)_0 \Delta p^3 + \cdots \end{aligned} \tag{2.4-3}$$

where $\Delta p = p - p_0$, and $\Delta q_i = q_i - q_{i,0}$.

The base point certainly satisfies Eq. (2.4-1), and so do all the points of interest to us, and therefore the Taylor's series expansion reduces to

$$\begin{aligned} \left(\frac{\partial F}{\partial p}\right)_0 \Delta p &+ \sum_{i=1}^{n} \left(\frac{\partial F}{\partial q_i}\right)_0 \Delta q_i + \frac{1}{2}\left(\frac{\partial^2 F}{\partial p^2}\right)_0 \Delta p^2 + \sum_{i=1}^{n} \left(\frac{\partial^2 F}{\partial p\, \partial q_i}\right)_0 \Delta p\, \Delta q_i \\ &+ \frac{1}{2}\sum_{i=1}^{n}\sum_{j=1}^{n} \left(\frac{\partial^2 F}{\partial q_i\, \partial q_j}\right)_0 \Delta q_i\, \Delta q_j = 0 \end{aligned} \tag{2.4-4}$$

where we have retained terms of second order; i.e., terms for which the deviations from the base point are taken to powers higher than the second are neglected. Thus we should expect this quadratic form, obtained by truncating the series, to represent a good approximation only if the deviations from the base point are kept small. If it is required to investigate large deviations, we had best use Eq. (2.4-1).

The quadratic approximation can be rearranged so as to give an implicit relation like Eq. (2.4-2).

Example III-4: A quadratic approximation

If the model for a unit is given as

$$F = -p + q_1 e^{-q_1 p/10} = 0$$

let us find an approximation for the transformation equation by using the Taylor's series expansion, assuming that the operating point is $p_0 = 1.486$ and $q_{1,0} = 2$. Note that the form of the model given is not a transformation equation since p appears in both terms of the equation. We start by forming the derivatives

$$\frac{\partial F}{\partial q_1} = \left(1 - \frac{q_1 p}{10}\right) e^{-pq_1/10}$$

$$\frac{\partial F}{\partial p} = -1 - \frac{q_1^2}{10} e^{-pq_1/10}$$

$$\frac{\partial^2 F}{\partial p\, \partial q_1} = \left(-\frac{2q_1}{10} + \frac{q_1^2 p}{100}\right) e^{-pq_1/10}$$

$$\frac{\partial^2 F}{\partial p^2} = \frac{q_1^3}{100} e^{-pq_1/10}$$

$$\frac{\partial^2 F}{\partial q_1^2} = \left(-\frac{2p}{10} + \frac{q_1 p^2}{100}\right) e^{-pq_1/10}$$

Evaluating these derivatives at the operating point, we find that

$$(\partial F/\partial q_1)_0 = 0.522; \qquad (\partial F/\partial p)_0 = -1.297; \qquad (\partial^2 F/\partial p\, \partial q_1)_0 = -0.253;$$
$$(\partial^2 F/\partial p^2)_0 = 0.059; \qquad (\partial^2 F/\partial q_1^2)_0 = -0.188.$$

Equation (2.4-4) becomes for this special case,

$$0 = -1.297\Delta p + 0.522\Delta q_1 + 0.0295\Delta p^2 - 0.253\Delta p\, \Delta q_1 - 0.094\Delta q_1^2$$

where

$$\Delta p = p - p_0 = p - 1.486 \qquad \text{and} \qquad \Delta q_1 = q_1 - q_{1,0} = q_1 - 2$$

Solving this quadratic equation for Δp gives

$$\Delta p = \frac{\left[\begin{array}{l}43.96 + 8.576\Delta q_1 \\ \qquad - \sqrt{(43.96 + 8.576\Delta q_1)^2 + 12.75\Delta q_1{}^2 - 70.78\Delta q_1}\end{array}\right]}{2}$$

The negative sign was selected before the square-root radical because Δp must vanish when Δq_1 does. This expression is an approximation to the transformation equation, valid only in the region near the base point, as shown in Fig. III-3. Here we see the value of p as determined by the approximation compared with the solution of the original equation, this being computed by the Newton-Raphson method, as shown in Example III-7. The approximation begins to fail badly for values of q_1 greater than 4, since $\Delta q_1 \geq 2$. For these values of Δq_1 the third-power terms are simply not negligible compared with these of the second power.

Equation (2.4-4) can be further truncated by neglecting terms of second order so as to obtain an equation which is linear in Δp and Δq_1. This is called *local linearization* and is very useful in developing methods of numerical search (Chap. VIII) and in formulating problems which can then be solved by the methods of linear programming (Chap. VII, Sec. 4.4), but of course the range of validity is even more restrictive than that of the quadratic approximation.

Example III-5: Local linearization

If the same expression given in Example III-4 is approximated by a linear approximation, we find from a Taylor's expansion (Eq. 2.4-3) that

$$F = F_0 + \left(\frac{\partial F}{\partial p}\right)_0 \Delta p + \left(\frac{\partial F}{\partial q_1}\right)_0 \Delta q_1$$

Using the numerical values of the first derivatives evaluated in Example III-4, we get

$$F = 0 = 0 - 1.297\Delta p + 0.522\Delta q_1$$

provided the base point is the same as in Example III-4. Solving for Δp, the linear approximation becomes

$$\Delta p = -0.402\Delta q_1$$

This expression is tangent to the true model at the base point, as shown in Fig. III-3. The fit of the approximation is therefore dependent on the curvature of the true expression about the base point.

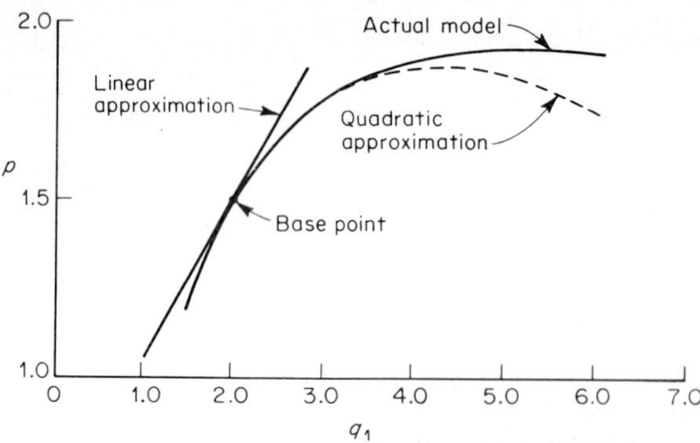

Fig. III-3 The validity of approximations near the base point.

One last word of caution appears to be in order. While the system equations should be examined with a view to neglecting unimportant terms, or to making physical simplifications, or to making mathematical approximations, care must be taken to avoid those simplifications which might change the character of the model. Checks on the model (Chap. II, Sec. 9) must be carried out if at all possible, using exact solutions of the original equations, and preferably against actual experimental data.

Another approach to simplification of the analytical model, and the one usually taken, is to use the model as a black-box, obtaining a number of outputs for a range of inputs, and then to produce a simplified analytical model by curve fitting (Sec. 5). Other methods of approximation have been discussed by Berezin and Zhidkev [1965].

3 BLACK-BOX APPROACH

The methods used to extract the appropriate responses from a black-box model depend largely on the nature of the system and the desired information. Thus the experimentation on an operating plant must be quite different from that on a small pilot plant or on a computer simulation. If, moreover, the response is time-dependent or is subject to experiment errors, then quite different lines of attack may be required.

In general, black-box experimentation provides a response corresponding to each set of inputs. This output may be used directly in numerical search, as described in Chaps. VI and VIII, or it can be converted into an analytical expression by surface fitting techniques (Sec. 5).

In this usual procedure of curve fitting, the experiments are normally required to be carried out at specified locations. Changing the inputs will, of course, alter the response, and in many cases, as in plant operation, this might not be allowable, due, for example, to strict quality control. Box

[1957] has proposed a method which he called "evolutionary operation," whereby a process is optimized during its operation. In this method the plant is not run at a fixed set of operating conditions, as is the usual operating procedure. Instead, small systematic modifications are imposed, following a preplanned and continuously repeated cycle. Hopefully, the modifications in the operating conditions are small enough so that their effect will not significantly affect the quality of the product. (See also Box and Chanmugan [1962], and Chanmugan and Jenkins [1963].)

If the time response of the system is important, the process of experimentation is far more complex, for no longer is it sufficient simply to specify a single output response for a given input. Indeed, a continuum of outputs will result from a certain continuum of inputs. If the system is linear,[1] then the output can be represented as (Chang [1961])

$$p_0(t) = \int_0^t W(t - \tau) p_i(\tau)\, d\tau \tag{3-1}$$

where p_0 = response

p_i = input

$W(t)$ = a characteristic of the system, known as a *weighting function* (see, for example, Truxal [1955])

Once $W(t)$ is determined, any output can be computed for a given input. Thus the objective of a testing program might be to determine W. This can be accomplished by forcing the system with a series of inputs, observing the outputs, and attempting to invert Eq. (3-1), solving for W. One very convenient input is the pulse, for then the resulting output is $W(t)$. A second very widely used input is the sinusoidal forcing function. If the response is measured over a range of frequencies, $W(t)$ can easily be determined by taking the inverse Fourier transform of the output. Henley et al. [1962] give a review of these techniques and a survey of reported applications.

If the system is not linear, the form expressed in Eq. (3-1) will not be found. The problem is then quite complex. If the disturbances about some steady state of interest are small, the nonlinear system may behave linearly, and Eq. (3-1) is again applied. However, general techniques are not available for characterizing nonlinear systems. Each new task represents a new challenge.

4 THE ANALYTICAL MODEL AS A BLACK-BOX

We have seen that the analytical model may not be in a form which is convenient for the application of analytical methods such as those discussed in Chaps. V, VII, and IX. If possible, simplifications are made, but in

[1] A linear system is one which has the following property: If the output response to input b is B, and c is C, then the output response to input $b + c$ is $B + C$.

practice, this is often a most complex step. In these complex cases, hope of achieving an analytical solution is abandoned, and numerical search is used to optimize. It is this application of the analytical model that we wish to discuss in this section.

The analytical model can be either distributed or lumped. If it is of the latter type, the model becomes an algebraic equation, and in general the problem of determining the outputs becomes one of root finding (Sec. 4.1); however, if the system is distributed, a differential equation must be solved (Sec. 4.2) even in the time-independent case, and in the time-dependent case one must solve a partial differential equation (Sec. 4.2).

4.1 The Solution of Algebraic Equations

Since lumped systems are described by algebraic equations, we may need to solve for an output p, given an implicit equation

$$F(p,\mathbf{q}) = 0 \tag{4.1-1}$$

and a particular set of values for $\mathbf{q} = \{q_1, q_2, \ldots, q_n\}$. The problem then reduces to the finding of a root (or roots), say, p^*, of the equation $f(p) = 0$. Many methods are found in the literature, two representative ones being discussed below.

It should be noted, before beginning our discussion of the methods selected from the classical numerical analysis, that a second approach to the problem would be to find the p which minimizes the function $[f(p)]^2$, thus recasting the problem as one of optimization. Any of the techniques described in Chap. VI for use in unrestricted search would therefore be applicable to its solution.

4.1.1 The method of interpolation

The simplest approach to determining a root, but practicable only in the case of a single variable, is the graphical method of solution. Thus we might plot $f(p)$ against p to determine the points at which it crosses the p axis, as shown in Fig. III-4.

The method of interpolation, or false position (see, for example, Hildebrand [1956] or Scarborough [1955]), is a logical extension of the graphical method. Thus we may choose locations p_1 and p_2 in Fig. III-4, so as to straddle the estimated position of the desired root. If an estimate of the desired root is not available, the root must be straddled by making a series of calculations, starting from a point known to be less than the root, and computing $f(p)$ for a sequence of positions moving steadily in the direction of the root. This sequential series of calculations is continued until a root is straddled, as evidenced by a change in the sign of $f(p)$.

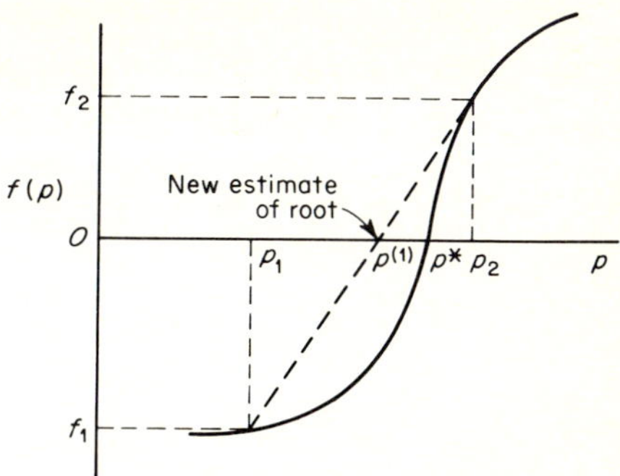

Fig. III-4 Interpolation to determine new estimate of root.

Once two points are located, as p_1 and p_2 of Fig. III-4, an estimate of the root is obtained by interpolation. Using the proportionality of the sides of similar triangles, we can write for the geometry shown in Fig. III-4

$$\frac{p^{(1)} - p_1}{-f_1} = \frac{p_2 - p_1}{f_2 - f_1}$$

Solving for $p^{(1)}$, our new estimate of the root, we find

$$p^{(1)} = p_1 - \frac{p_2 - p_1}{f_2 - f_1} f_1 \tag{4.1-2}$$

This new estimate is then used in place of either p_1 or p_2, depending on the sign of $f(p^{(1)})$, replacing the one having the same sign. Since we still straddle the root after p_1 or p_2 is replaced by $p^{(1)}$, in accordance with the rule described above, a new estimate of the root can be obtained by again applying Eq. (4.1-2). This iterative procedure will converge to the straddled root.

Example III-6: The method of interpolation

In Example III-4 we approximated the equation

$$F = -p + q_1 e^{-q_1 p/10} = 0$$

by a quadratic function and determined p as a function of q_1. The expansion was about the operating point $q_{1,0} = 2.0$ and $p_0 = 1.486$, but we did not attempt to describe how these conditions satisfying $F = 0$ might be found. Let us find the appropriate p corresponding to $q_1 = 2$. The equation reduces to

$$f(p) = -p + 2e^{-0.2p} = 0$$

We wish to find the value p^* satisfying this equation, using the method of interpolation. We start our calculations at $p = 0$, and, moving in unit steps, we find that $f(0) = 2$, $f(1) = 0.6374$, and $f(2) = -0.6594$. There is a root between $p = 1$ and $p = 2$ because $f(p)$ has changed sign. Designate $p_1 = 1$ and $p_2 = 2$, and, using Eq. (4.1-2), we obtain the new estimate of the root as

$$p^{(1)} = 1 - 0.6374 \frac{2 - 1}{-0.6594 - 0.6374} = 1.492$$

The value of $f(p^{(1)})$ is -0.0074, very nearly a solution, but we need a better approximation. Since $f(p^{(1)})$ is negative, we shall replace the p_2 of the previous iteration by $p^{(1)}$, both of these estimates of the root giving negative values of $f(p)$. Therefore, in the new iteration, $p_1 = 1$, with $f_1 = 0.6374$ and $p_2 = 1.492$ and $f_2 = -0.0074$. From Eq. (4.1-2) we find $p^{(2)} = 1.4859$, with $f(p^{(2)}) = -0.0001$. This is sufficiently accurate for our needs. Obviously, a better approximation can be obtained by continuing the calculations.

4.1.2 Newton-Raphson method

Another systematic method of approaching this problem is to use the Newton-Raphson method, which starts with an assumed solution, say, p_0, and attempts to determine an improved approximation based on a knowledge of the gradient of $f(p)$. The essence of the method is best understood by referring to Fig. III-5. The most recent guess or estimate of a root is p_i, at which point the function and its derivative have values f_i and f_i', respectively. Our problem is to determine a better estimate of the root based on a knowledge of f_i and f_i'. Since f_i' is the tangent of the angle that the dashed line constructed tangential to the curve at the point p_i makes with the axis, we see that, if the new estimate is to be the intersection of the tangent line with

Fig. III-5 Geometric basis of the Newton-Raphson method.

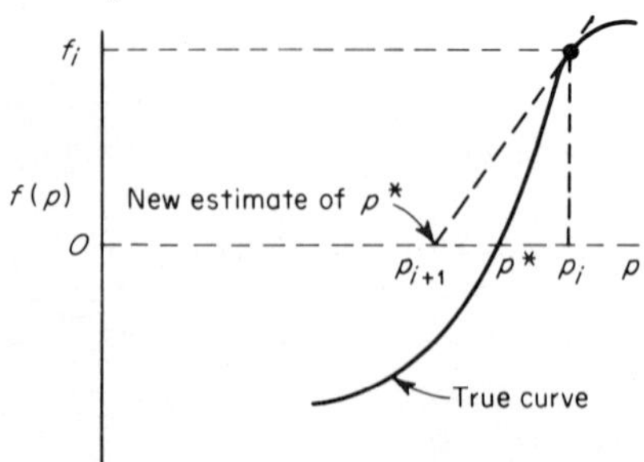

the axis, then, by construction,

$$f_i' = \frac{f_i}{p_i - p_{i+1}}$$

or solving for the new approximation,

$$p_{i+1} = p_i - \frac{f_i}{f_i'} \tag{4.1-3}$$

This equation is to be used repeatedly until a sufficiently accurate estimate of the root is obtained.

Example III-7: Application of the Newton-Raphson method

To calculate the true curve shown in Fig. III-3 relating to Example III-4, we found it necessary to have roots of the equation

$$F = q_1 e^{-q_1 p/10} - p = 0$$

for various values of q_1. Let us find the root for $q_1 = 2$. For this case

$$f(p) = 2e^{-0.2p} - p$$

and

$$f'(p) = -0.4e^{-0.2p} - 1$$

For this case Eq. (4.1-3) becomes

$$p_{i+1} = p_i + \frac{2e^{-0.2p_i} - p_i}{1 + 0.4e^{-0.2p_i}}$$

Let us start with an assumed guess $p_0 = 1$, and since $e^{-0.2} = 0.8187$, then

$$p_1 = 1 + \frac{2(0.8187) - 1}{1 + 0.4(0.8187)} = 1.4802$$

This new guess is very nearly the root, but we shall improve this value by a new iteration. With $p_1 = 1.4802$ and $e^{-0.296} = 0.7438$, we have

$$p_2 = 1.4802 + \frac{2(0.7438) - 1.4802}{1 + 0.4(0.7438)} = 1.4859$$

which is sufficiently accurate for our purposes.

The Newton-Raphson method does not always converge. Indeed, one can easily sketch a curve for which the process oscillates between two points. Such a curve is shown in Fig. III-6. If p_0 is the initial guess, then, following the tangent at that point to the axis, the second approximation is p_1. However, following the tangent from the curve back to the axis, we find p_0 as the third approximation. Clearly, this will never converge, for the computed values will cycle between p_0 and p_1.

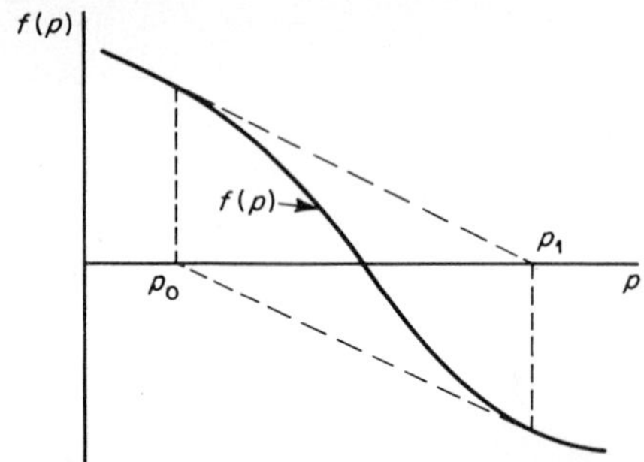

Fig. III-6 Failure of the Newton-Raphson method.

If the $f'(p)$ and $f''(p)$ do not change sign in the interval (p_0,p^*), and if f_0' and f_0'' have the same sign, then the Newton-Raphson method will converge. If these conditions are not satisfied, it may not converge. In all cases the method depends upon having a sufficiently good approximation to the root in the first case. Moreover, if there are multiple roots, the method will probably converge to the nearest root. If the conditions described above are met with respect to the nearest root, convergence to that root is assured.

Example III-8: Multiple roots and the Newton-Raphson method

Let us find the roots of the equation

$$f(p) = 3{,}000 - 100p^2 - 4p^5 - 6p^6 = 0 \tag{4.1-4}$$

by the Newton-Raphson technique.

Before investigating the roots, let us look at the form of the function $f(p)$. This function, together with both its first and second derivatives, is shown in Fig. III-7, where we see that there are two roots to be found. According to the sufficiency conditions for convergence of the Newton-Raphson method, any initial guess for p_0 greater than zero will surely yield the particular root located between 2 and 3, since in this region $(p > 0)$ neither $f'(p)$ nor $f''(p)$ changes sign, and they both have the same sign, as can be seen by inspection of Fig. III-7. On the other hand, a starting value in the region $p < 0$ may not converge to the negative root of the given polynomial, because here the sufficiency conditions are not satisfied, $f'(p)$ being of different sign than $f''(p)$. We therefore seek two roots, with some question about the convergence to the negative one.

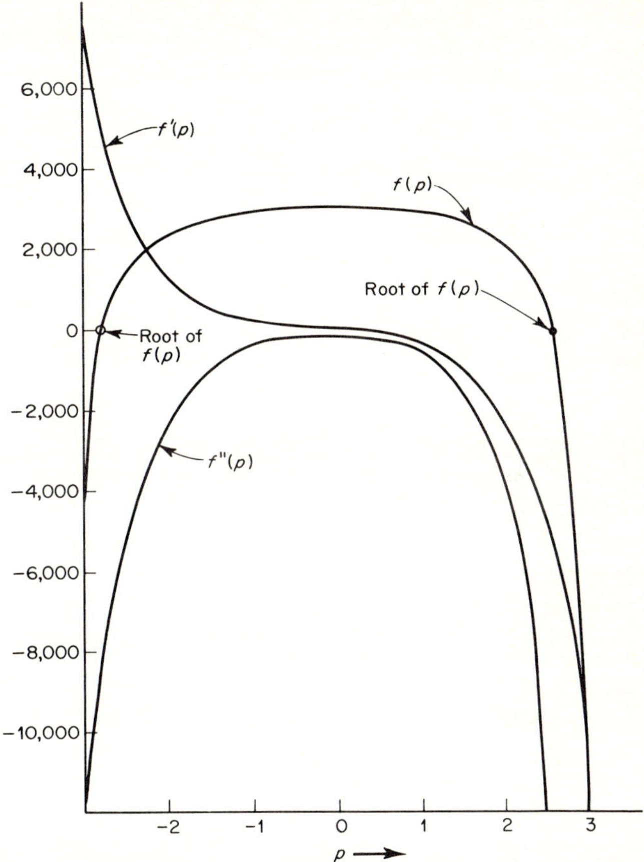

Fig. III-7 A profile of the function $f(p) = 3{,}000 - 100p^2 - 4p^5 - 6p^6$.

Having seen that two solutions exist, let us determine these using the Newton-Raphson method. The recursive relationship is

$$p_{i+1} = p_i - \frac{3{,}000 - 100p_i^2 - 4p_i^5 - 6p_i^6}{4p_i(50 + 5p_i^3 + 9p_i^4)}$$

Starting with $p_0 = 2$, the results presented in Table III-1 are computed. After four iterations we find the first root as 2.5992, which is correct to five significant figures. Let us now try to find the root along the negative axis, starting with $p_0 = -2$. The series of calculations given in Table III-2 shows the convergence to the second root. Both roots are found by using the Newton-Raphson method, the convergence being to the

TABLE III-1 THE CONVERGENCE OF THE NEWTON-RAPHSON METHOD TO A ROOT OF EQ. (4.1-4)

i	p_i	$f(p_i)$	$f'(p_i)$	p_{i+1}
0	2.0	2,088	−2,192	2.9526
1	2.9526	−2,744.3	−10,188	2.6832
2	2.6832	−515.3	−6,580.1	2.6049
3	2.6049	−32.7	−5,759.4	2.5992
4	2.5992	−0.2	−5,703.4	2.5992

TABLE III-2 THE CONVERGENCE OF THE NEWTON-RAPHSON METHOD TO A SECOND ROOT OF EQ. (4.1-4)

i	p_i	$f(p_i)$	$f'(p_i)$	p_{i+1}
0	−2.0	−2,344.0	−1,232.0	−3.9026
1	−3.9026	−16,099	28,730	−3.3423
2	−3.3423	−4,812.3	13,187	−2.9773
3	−2.9773	−1,029.3	7,446.1	−2.8391
4	−2.8391	−210.4	5,909.0	−2.8035
5	−2.8035	−6.3	5,559.7	−2.8024
6	−2.8024	−0.0	5,548.9	−2.8024

nearest one to the initial estimate. Therefore, to find all the roots of an algebraic expression, we must have some estimate of the locations. It is easy to see how one or more multiple roots can be overlooked.

4.2 Solution of Ordinary Differential Equations

The determination of the output for given inputs to a distributed system requires the integration of a differential equation, as does the same task with respect to the time-dependent lumped systems. Most systems of practical interest are described by equations too complex to be integrated by analytical means, and we therefore turn to numerical methods, treating the model as a black-box. Two representative methods of solving ordinary differential equations will be described here.

4.2.1 *The Euler method*

The Euler method (see, for example, Scarborough [1955]) is a relatively simple method of integrating ordinary differential equations. Consider the differential equation

$$\frac{dp}{dl} = f(p,l) \tag{4.2-1}$$

The Euler method is a single step integration which depends on the Taylor's expansion of $p(l)$ about some starting point, say, l_0, in the following way:

$$p(l) = p_0 + \frac{dp}{dl}(l - l_0) + \frac{1}{2}\frac{d^2p}{dl^2}(l - l_0)^2 + \cdots \tag{4.2-2}$$

If $l - l_0$ is defined to be h, and if h^2 is small compared with h, then, from Eqs. (4.2-1) and (4.2-2), we find

$$p(l_0 + h) \equiv p_h = p_0 + h\frac{dp}{dl} = p_0 + hf(p_0,l_0) \tag{4.2-3}$$

Equation (4.2-3) can be used to compute the value of p at $l = l_0 + h$, knowing the value $p = p_0$ at $l = l_0$. The entire process can be repeated, finding values of p at discrete locations separated by a distance h along the path of integration.

The geometric meaning of Eq. (4.2-3) is obvious. Given a point p_0 on the graph of the true solution, as shown in Fig. III-8, a straight-line approximation is used over the interval h to replace the curving trajectory. The accumulated errors can be substantial, as shown in Example III-9. The interval h must be kept very small to ensure reasonable accuracy. Because of its inherent inaccuracy, the Euler method is seldom used in digital computation, except as a starting routine or in the form of the modified Euler method (Kunz [1957]). However, because of its simplicity, this method is frequently used for crude calculations performed, for example, on a desk calculator.

Fig. III-8 Estimation of p_h by the Euler method.

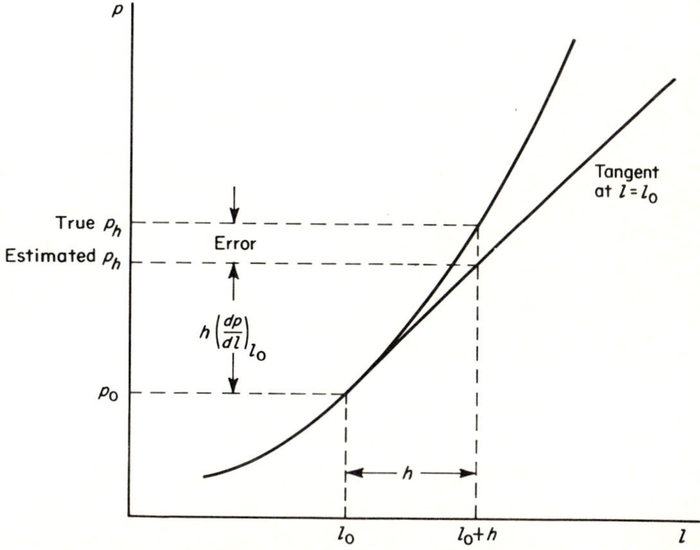

Example III-9: Euler method

Solve the differential equation (in the range $0 \leq l \leq 1$)

$$\frac{dp}{dl} = p^2 \tag{4.2-4}$$

for the initial condition $p = 0.5$ at $l = 0$, using the Euler method. From Eq. (4.2-3), letting $h = 0.1$,

$$p_{0.1} = p_0 + p_0^2 h = 0.5 + (0.5)^2(0.1) = 0.525$$

Continuing,

$$p_{0.2} = p_{0.1} + p_{0.1}^2\, h = 0.525 + (0.525)^2(0.1) = 0.5526$$

the complete numerical solution being tabulated in Table III-3, together with the exact solution. It is seen that the errors accumulate because each point depends on the succeeding value. The final error at $l = 1$ is about 6 percent, rather large for the final stages of a numerical calculation, but acceptable as an initial estimate.

4.2.2 *The Runge-Kutta method*

A better and more accurate method, which in fact is the most widely used in computations on high-speed computers, is the Runge-Kutta (Kunz [1957]) method or some modification of it. The method can be summarized as follows, the new value of p being estimated from

$$p_h = p_0 + \tfrac{1}{6}(k_1 + 2k_2 + 2k_3 + k_4) \tag{4.2-5}$$

where $k_1 = hf(p_0, l_0)$

$$k_2 = hf\left(p_0 + \frac{k_1}{2}, l_0 + \frac{h}{2}\right)$$

$$k_3 = hf\left(p_0 + \frac{k_2}{2}, l_0 + \frac{h}{2}\right)$$

$$k_4 = hf(p_0 + k_3, l_0 + h)$$

This is a better, but correspondingly more complex, approximation than either the Euler or the modified Euler method. It is somewhat laborious for hand computation.

Example III-10: Runge-Kutta method

Let us repeat Example III-9 using the Runge-Kutta method. For the first step, from Eq. (4.2-5), with $h = 0.1$,

$$k_1 = 0.1(0.5)^2 = 0.025$$

$$k_2 = 0.1(0.5 + 0.0125)^2 = 0.02627$$

$$k_3 = 0.1(0.5 + 0.01313)^2 = 0.02633$$

$$k_4 = 0.1(0.5 + 0.02633)^2 = 0.02770$$

$p_{0.1} = 0.5 + \frac{1}{6}[0.025 + 2(0.02627) + 2(0.02633) + 0.02770] = 0.5263$. Repeating this calculation in a stepwise fashion, the results shown in Table III-3 can be obtained. Clearly, the answers are much superior

TABLE III-3 SOLUTION OF EQ. (4.2-4)

	Exact $p = \dfrac{l}{2-l}$	*Euler method* (*Example* III-9) $h = 0.1$	*Runge-Kutta method* (*Example* III-10) $h = 0.1$
0	0.5000	0.5000	0.5000
0.1	0.5263	0.5250	0.5263
0.2	0.5556	0.5526	0.5555
0.3	0.5882	0.5831	0.5882
0.4	0.6250	0.6171	0.6250
0.5	0.6667	0.6552	0.6666
0.6	0.7143	0.6981	0.7143
0.7	0.7692	0.7468	0.7692
0.8	0.8333	0.8026	0.8333
0.9	0.9091	0.8670	0.9090
1.0	1.0000	0.9422	1.0001

to those found by the Euler method, and it is in fact difficult in the present illustration to show any difference between the numerical and exact solutions. The computational work is, however, correspondingly greater than in the previous approach.

4.2.3 *Other methods*

There are many other methods of solving differential equations numerically, such as the Milne method, the Adams-Moulton method (Lapidus [1962]), and variational methods (Schechter [1967], Kantorovich and Krylov [1958], and also Chap. X). However, those presented here are sufficient to indicate the nature of the process and are powerful enough for the calculations that we present later.

A second method of solving differential equations is by analog methods (Smith and Wood [1959], Korn and Korn [1964]).

4.3 Solution of Partial Differential Equations

The solution of partial differential equations by numerical methods is indeed a very important and complex subject. The essential ideas entail replacing the operations of partial differentiation by finite difference operators. The finite difference equations representing a particular partial differential equation form a set (a large set) of algebraic equations which can in principle be solved using large digital computers. The techniques of iteration necessary to solve a large set of algebraic equations are often subject to certain inherent instabilities, and the entire process is complex. A number of texts are available, such as those of Booth [1955], Lapidus [1962], Southwell [1956], and Kantorovich and Krylov [1958].

4.4 Monte Carlo Methods

In some cases the inputs to the system are defined only in some statistical sense. In these situations it is customary both to determine the response of the system to the average expected values of the inputs and, in addition, to investigate a possible range of values of the inputs, thus singling out the most critical of these. This sensitivity analysis is an important part of every design and optimization study, but it has one weakness; namely, the probability of a given response is not determined. The sensitivity analysis therefore defines the range of expected outputs, but does not provide a measure of the likelihood of a particular output or range of outputs. Monte Carlo methods are designed to provide an estimate of this probability (Meyer [1956]), the essential idea being to make repeated computations, using inputs selected at random from populations (tables of random numbers, for example) which have the same statistical properties expected of each particular input. After performing a large number of such computations, a statistical distribution of the output can be estimated. Note that, to use the Monte Carlo methods in a meaningful way, one must have a good model, information about the statistics of the inputs, and finally, the means of making repeated computations. Hess and Quigley [1963] have used these techniques in examining the problem of management decisions in the face of uncertainty.

5 THE ANALYTICAL MODEL FROM BLACK-BOX EXPERIMENTATION–SURFACE FITTING

Surface fitting is defined here as the procedure of representing a set of black-box outputs in terms of the inputs by a mathematical relationship. In most cases this mathematical expression will yield approximations to the observed responses, but the representation is valid if the approximation is of sufficient quality so as to preserve the essential features of the unit response for use in optimization.

In developing an approximation designed to "fit" a certain set of experimental data, two questions of prime importance must be considered: What mathematical form is best suited to this particular set of data? Are there experimental errors in the data? If the exact mathematical form is known and the experimental errors negligible, the process of surface fitting considered in Sec. 5.2 is simple in principle, although the necessary calculations may well be complex.

On the other hand, it is more likely that the form of the true response surface will not be known, or if it is known, complexity may preclude its use, except numerically along the upper route shown in Fig. III-1. Therefore the true response surface is usually to be approximated by a simpler relationship. The nature of such approximations is considered in Sec. 5.1, and the mathematical process involved in obtaining a particular result is discussed in Secs. 5.2 to 5.4.

Before embarking on our program, it should be noted that the methods of surface fitting have wider application in optimization than as a means of simulating manageable objective functions, although this is itself an exceedingly important aspect. In the intermediate and final stages of any numerical search for an optimum, it will be seen from Chap. VIII that it is often desirable to carry out a local exploration. In general, this may be achieved by fitting a surface to the outputs obtained for a suitable set of preselected inputs and then examining the fitted surface to find locations of improved response.

5.1 Representation of the Transformation Equation

The transformation function relates the outputs of a system to the inputs, the latter being either interconnecting or external (Chap. II, Sec. 4.2). So far we have used the symbols $\mathbf{p}$ and $\mathbf{q}$ to denote interconnecting streams and external inputs, respectively. Continuing this practice here will complicate the notation considerably, so that it is convenient to represent the transformation function[1] as

$$p = p(\mathbf{x}) \equiv p(x_1, x_2, \ldots, x_n) \tag{5.1-1}$$

where an x_i stands for a particular input, be it external or interconnecting.

[1] In the present discussion, the measured response will be represented by the symbol $\hat{p}$; the true response determined as the mean value of a large number of experiments, if experimental errors are present, will be denoted by p; and the response determined by the mathematical approximation will be called $\underline{p}$. The difference between $\hat{p}$ and p is of course due entirely to experimental error, since $\hat{p}$ is the measured response, whereas the difference between p and $\underline{p}$ may be due to both experimental and approximation errors. Such errors are here entirely due to measurement, and not to inherent random disturbances in the process itself.

If the true response function is too complex, or if it is unknown in the case of an existing system, our knowledge of the system is usually contained in a table of data showing outputs for a range of inputs. We should like to represent these data, and hence the system, by an analytical approximation having sufficient accuracy so that it can be used in the optimization, rather than experiment with the system itself.

What form should the approximating function take? A number of possible forms have been discussed by Berezin and Zhidkev [1965], but from our viewpoint generalized polynomials will provide an adequate representation, and in addition have some real advantages in computational simplicity and susceptibility to statistical analysis. Other functions may well possess other advantages, but these depend on the criterion used to judge the quality of fit, as well as the nature of the true response. A generalized polynomial approximation would have the form

$$\underline{p}(\mathbf{x}) = \beta_0 + \beta_1 x_1 + \beta_2 x_2 + \cdots + \beta_n x_n + \beta_{1,1} x_1^2 + \beta_{1,2} x_1 x_2 + \cdots + \beta_{n,n} x_n^2 + \beta_{1,1,1} x_1^3 + \beta_{1,1,2} x_1^2 x_2 + \cdots + \beta_{n,n,n} x_n^3 + \cdots \quad (5.1\text{-}2)$$

and the accuracy of the representation depends to some extent upon the number of terms included in the polynomial and on the choice of the coefficients, β. The degree of any term is defined as the number of variables multiplied together, which in Eq. (5.1-2) is also represented by the number of subscripts on a particular coefficient. Since $\beta_{1,1,2}$ has three subscripts, its term $\beta_{1,1,2} x_1^2 x_2$ is thus of the third degree. The degree of the polynomial is the highest degree of any term appearing on the right-hand side. If Eq. (5.1-2) were therefore truncated as follows:

$$\underline{p} = \beta_0 + \beta_1 x_1 + \beta_{2,2} x_2^2$$

the polynomial would be of the second degree.

Once the form of the approximation has been selected, the essential problem is to choose the coefficients β so as to best represent the true response. As noted above, selecting a polynomial of certain degree provides real computational advantages in the evaluation of the coefficients. Before pursuing this important aspect, it should, however, be noted that the polynomial representation has in fact a certain theoretical justification, provided the range of the inputs is restricted, because the polynomial can be viewed as being a Taylor's series expansion (Sec. 2.4.2) to the true response function. Thus, for a response depending on a single variable, this expansion about the base point x_0 is

$$p(x) = p_0 + \left(\frac{\partial p}{\partial x}\right)_0 (x - x_0) + \frac{1}{2}\left(\frac{\partial^2 p}{\partial x^2}\right)_0 (x - x_0)^2 + \cdots \quad (5.1\text{-}3)$$

By retaining a sufficient number of terms, any smooth function (Chap. V, Sec. 2.1) can be represented by such a polynomial in a restricted region. If the range of interest is limited so that the quantity $x - x_0$ is small, then Eq. (5.1-3) reduces to

$$p(x) \approx \underline{p}(x) = \left[p_0 - \left(\frac{\partial p}{\partial x}\right)_0 x_0\right] + \left(\frac{\partial p}{\partial x}\right)_0 x = \beta_0 + \beta_1 x$$

Thus the approximation $\underline{p}(x)$ to the true response function $p(x)$ is valid over a limited range of x. We can therefore conclude that a first-degree polynomial will "fit" over a limited range provided β_0 and β_1 are selected properly.

Clearly, the same conclusions are possible if we retain terms of second order in the Taylor series, except that the range of validity will have been extended (Sec. 2.4.2). Certainly, the larger the range of variation, the greater will be the number of terms necessary to represent the model correctly. These same remarks also apply to the models of higher dimensionality.

The polynomial representation becomes applicable over a wider range of inputs as the degree is increased. The amount of data required to determine the coefficients will also increase markedly. The least number of experiments required to determine the coefficients is given in Table III-4. Owing

TABLE III-4 THE LEAST NUMBER OF EXPERIMENTS REQUIRED TO DETERMINE THE POLYNOMIAL COEFFICIENTS

Number of variables	*Degree of fitted equation* 1	2	3	4
2	3	6	10	15
3	4	10	20	35
4	5	15	35	70
5	6	21	56	126

to the enormous increase in the minimum number of experiments required to evaluate the coefficients as the degree of the polynomial increases, one normally agrees to restrict the inputs to a limited region, so that first-degree, or at the most second-degree, polynomials provide an adequate representation. In these circumstances surface fitting can be used profitably.

5.2 The Least-squares Concept in Surface Fitting

Now that we have agreed to express the transformation equation in the form of a polynomial such as Eq. (5.1-2), our task is to assign values to the parameters $\beta_i, \beta_{i,j}, \ldots$. How should we choose these constants? If the true transformation equation is a polynomial of precisely the same degree as that used as the approximation, and if there are no experimental errors, then, by making as many measurements as there are parameters appearing in the approximation, the parameters could be determined. This procedure is demonstrated in the following example.

Example III-11: Fitting a two-dimensional plane surface

Suppose it is required to approximate a response by

$$\underline{p} = \beta_0 + \beta_1 x_1 + \beta_2 x_2 \tag{5.2-1}$$

Since there are three coefficients, β_i, a minimum of three experiments is required to evaluate these coefficients (Table III-4). Let us suppose that

TABLE III-5 MEASURED RESPONSES FOR EXAMPLE III-11

Experiment	*Inputs*		*Measured response*
j	$x_{1,j}$	$x_{2,j}$	$\hat{p}$
1	1.0	1.5	4.571
2	0.8	1.1	4.383
3	1.6	0.5	4.607

the measured responses are as listed in Table III-5.

If $p = \hat{p}$ at these three points, Eq. (5.2-1) can be written for each experiment as

$$4.571 = \beta_0 + 1.0\beta_1 + 1.5\beta_2$$

$$4.383 = \beta_0 + 0.8\beta_1 + 1.1\beta_2$$

$$4.607 = \beta_0 + 1.6\beta_1 + 0.5\beta_2$$

Obviously, these three equations can be solved for the three unknowns, giving

$$\beta_0 = 3.751 \qquad \beta_1 = 0.46 \qquad \beta_2 = 0.24$$

to determine the approximate form of the response function as the plane surface

$$\underline{p} = 3.751 + 0.46x_1 + 0.24x_2$$

This method is not, however, normally recommended, because in the presence of experimental error a very poor approximation may result, even if the true response surface has the same form as the approximating function. Moreover, such a procedure can also be quite misleading if the true response function differs from that of the approximating polynomial. Another difficulty arises if there are more experiments than coefficients. Which of the experiments should be selected to determine these coefficients? Certainly, the fit would vary greatly with the particular set of experiments chosen as representative. This situation can, however, be turned to our advantage.

We have seen that an approximating function may fail to represent the observed responses for two basic reasons: first, that the data used to determine the approximation may include experimental error, and second, that the mathematical form of the approximating function may not correspond to the true, but usually unknown, function. In spite of these difficulties, it seems reasonable for us to expect to be in a position to find an adequate representation of the true response surface, given sufficient experimental data and assuming that the range of inputs is limited.

Let us consider an approximating function $\underline{p}$, where

$$\underline{p} = \underline{p}(\mathbf{x},\boldsymbol{\beta}) = \underline{p}(x_1, x_2, \ldots, x_n; \beta_0, \beta_1, \ldots, \beta_s)$$

with $s + 1$ coefficients to be determined. Suppose that r experiments, where $r > s$, have been performed, each experiment yielding a response $\hat{p}$, so that we have r responses associated with the r sets of inputs $\mathbf{x}_1, \mathbf{x}_2, \ldots, \mathbf{x}_r$. Here, of course, if the experiment were to be repeated for any one of these inputs, a different $\hat{p}$ might result, owing to experimental error.

Let us now define a residual R_j for the jth experiment as the difference

$$R_j = \hat{p}_j - \underline{p}(\mathbf{x}_j,\boldsymbol{\beta}) \tag{5.2-2}$$

between the observed response $\hat{p}$ and that predicted by the approximating function. The parameters β_i obviously influence the magnitude of the residual, since the approximation is altered by changing the numerical values of these coefficients. The residual can, of course, be positive or negative. Let us form these residuals for each of the r experiments and add them in such a way that each contributes a positive quantity to the sum. Any number of possibilities exist. For example,

$$M = \sum_{j=1}^{r} |R_j| \tag{5.2-3}$$

where M is the sum of the absolute values of the residuals. An alternative summation is given by the sum E of the squares

$$E = \sum_{j=1}^{r} R_j^2 \tag{5.2-4}$$

while yet other summations may be envisaged. Clearly, the sum will depend on the numerical values of $\beta_0, \beta_1, \ldots, \beta_s$, due to this dependence of each individual residual. How should we choose the numerical values of β_i? The "fit" of any approximation is improved if the total of all residuals is reduced. It is reasonable to select them so that M, E, or some other suitable sum is minimized.

The sum of the squares, E, is usually chosen as the criterion because of ease of computation and because of certain statistical features (see, for example, Davies [1956]). The coefficients $\boldsymbol{\beta}$ might be selected so as to minimize the sum of the squares of the residuals. If E is to be a minimum, the necessary conditions (Chap. VII, Secs. 2 and 3) are then that

$$\frac{\partial E}{\partial \beta_k} = 0 \qquad \text{for } k = 0, 1, 2, \ldots, s \tag{5.2-5}$$

providing the necessary relations defining the desired coefficients $\boldsymbol{\beta}$. Let us now consider the two special cases of fitting plane (Sec. 5.3) and quadratic (Sec. 5.4) surfaces.

5.3 Fitting of Plane Surfaces

In this section we shall use the criterion of least squares in the fitting of plane surfaces. Due to the seeming complexity of the results, it is convenient to introduce this section by examining, initially, those responses which depend on only a single input.

5.3.1 *One-dimensional responses*

Suppose that the true function $p(x_1)$ is to be approximated by the first-degree polynomial

$$\underline{p} = \beta_0 + \beta_1 x_1 \tag{5.3-1}$$

To obtain symmetry with respect to the coefficients, let us rewrite this as

$$\underline{p} = \beta_0 x_0 + \beta_1 x_1 \tag{5.3-2}$$

where x_0 is unity for all experiments. Let us now form the residual for a given experiment as

$$R_j = \hat{p}(x_{1,j}) - \underline{p}(x_{1,j}) = \hat{p}(x_{1,j}) - \beta_0 x_{0,j} - \beta_1 x_{1,j}$$

If there are r experiments, Eq. (5.2-4) becomes

$$E = \sum_{j=1}^{r} (\hat{p}_j - \beta_0 x_{0,j} - \beta_1 x_{1,j})^2 \tag{5.3-3}$$

If we wish to select the β_0 and β_1 so as to minimize E, then, applying Eq. (5.2-5), we find the following equation for $\partial E/\partial \beta_0 = 0$:

$$\sum_{j=1}^{r} \hat{p}_j x_{0,j} - \beta_0 \sum_{j=1}^{r} x_{0,j}^2 - \beta_1 \sum_{j=1}^{r} x_{1,j} x_{0,j} = 0 \tag{5.3-4}$$

and for $\partial E/\partial \beta_1 = 0$, we get

$$\sum_{j=1}^{r} \hat{p}_j x_{1,j} - \beta_0 \sum_{j=1}^{r} x_{1,j} x_{0,j} - \beta_1 \sum_{j=1}^{r} x_{1,j}^2 = 0 \tag{5.3-5}$$

These equations can be written more concisely as

$$C_{0,0}\beta_0 + C_{0,1}\beta_1 = D_0 \tag{5.3-6a}$$

$$C_{1,0}\beta_0 + C_{1,1}\beta_1 = D_1 \tag{5.3-6b}$$

where

$$C_{i,k} = \sum_{j=1}^{r} x_{i,j}x_{k,j} = C_{k,i} \tag{5.3-7a}$$

$$D_i = \sum_{j=1}^{r} x_{i,j}\hat{p}_j \tag{5.3-7b}$$

These two equations can be solved simultaneously to determine the values of β_0 and β_1, which minimize the sum of the squares of the residuals.

Example III-12: Least-square fit in one dimension

Values of the responses measured in a series of five experiments are given in Table III-6. Let us represent this system by a first-degree polynomial.

TABLE III-6 RESPONSE DATA FITTED IN EXAMPLE III-12

Experiment j	*Input* $x_{1,j}$	*Measured response* $\hat{p}_j$	*Approximate response* $\underline{p}_j$	*Residual* $R_j = \hat{p}_j - \underline{p}_j$
1	5	5.90	6.01	−0.11
2	1	1.90	2.01	−0.11
3	4	5.10	5.01	0.09
4	2	3.10	3.01	0.09
5	3	4.05	4.01	0.04

First, let us form[1] from Eq. (5.3-7)

$$C_{1,0} = \sum_{j=1}^{5} x_{1,j}x_{0,j} = 15 \qquad C_{0,0} = \sum_{j=1}^{5} x_{0,j}^2 = 5$$

$$C_{1,1} = \sum_{j=1}^{5} x_{1,j}^2 = 55 \qquad D_1 = \sum_{j=1}^{5} x_{1,j}\hat{p}_j = 70.15$$

$$D_0 = \sum_{j=1}^{5} x_{0,j}\hat{p}_j = 20.05$$

Thus Eqs. (5.3-6) become, in this particular example,

$$5\beta_0 + 15\beta_1 = 20.05$$

$$15\beta_0 + 55\beta_1 = 70.15$$

[1] Remembering that $x_{0,j} = 1$ for all j.

so that $\beta_0 = 1.010$ and $\beta_1 = 1.000$. The first-degree approximating polynomial then has the form

$$\underline{p} = 1.010 + 1.000x_1$$

The quality of the approximation can also be examined by calculating the residuals as listed in Table III-6. Therefore

$$E = \sum_{j=1}^{5} R_j^2 = 0.042$$

The values of β_0 and β_1 found here make E smaller than any other pair of numbers.

As we have said, the residual is attributable to experimental error and to the curvature of the true response surface. If the true surface were actually a plane or nearly plane surface, the approximation $\underline{p}$ could be more nearly the true answer than the measured response $\hat{p}$, the more so the larger the number of experiments performed.

5.3.2 *The general plane surface*

If the response depends upon inputs $\mathbf{x}$, a series of r experiments can be performed for the particular points $\mathbf{x}_1, \mathbf{x}_2, \ldots, \mathbf{x}_r$, and the outputs $\hat{p}_1, \hat{p}_2, \ldots, \hat{p}_r$ observed. The coefficients defining the plane surface can then be obtained by a least-square criterion. In this case the approximating function is represented by an equation of the form

$$\underline{p}(\mathbf{x}) = \sum_{i=0}^{n} \beta_i x_i \tag{5.3-8}$$

where again $x_0 = 1$ for all experiments.

The value of E as defined by Eq. (5.2-4) is, clearly,

$$E = \sum_{j=1}^{r} (\hat{p}_j - \underline{p}_j)^2 = \sum_{j=1}^{r} (\hat{p}_j - \sum_{i=0}^{n} \beta_i x_{i,j})^2$$

Applying Eqs. (5.2-5), we may find the necessary conditions to minimize E and to define the β_i by the set of $n + 1$ equations,

$$\begin{array}{c} C_{0,0}\beta_0 + C_{0,1}\beta_1 + C_{0,2}\beta_2 + \cdots + C_{0,n}\beta_n = D_0 \\ C_{1,0}\beta_0 + C_{1,1}\beta_1 + C_{1,2}\beta_2 + \cdots + C_{1,n}\beta_n = D_1 \\ \cdots\cdots\cdots\cdots\cdots\cdots\cdots\cdots \\ C_{n,0}\beta_0 + C_{n,1}\beta_1 + C_{n,2}\beta_2 + \cdots + C_{n,n}\beta_n = D_n \end{array}$$

or $$\sum_{k=1}^{n} C_{i,k}\beta_k = D_i \qquad \text{for } i = 0, 1, \ldots, n \tag{5.3-9}$$

where $$C_{i,k} = \sum_{j=1}^{r} x_{i,j}x_{k,j} = C_{k,i}$$

$$D_i = \sum_{j=1}^{r} x_{i,j}\hat{p}_j$$

There are clearly $n + 1$ equations and $n + 1$ coefficients to be found. These equations must therefore be solved simultaneously, as shown in the next example.

Example III-13: Least-squares fit in two dimensions

Let us find a first-degree polynomial which will best represent, in the sense of the best least-squares fit, the data shown in the first four columns of Table III-7. Here

$$C_{0,0} = \sum_{j=1}^{8} x_{0,j}x_{0,j} = 8 \qquad C_{0,1} = \sum_{j=1}^{8} x_{0,j}x_{1,j} = 5$$

$$C_{0,2} = \sum_{j=1}^{8} x_{0,j}x_{2,j} = 14 \qquad C_{1,1} = \sum_{j=1}^{8} x_{1,j}x_{1,j} = 7$$

$$C_{1,2} = \sum_{j=1}^{8} x_{1,j}x_{2,j} = 10 \qquad C_{2,2} = \sum_{j=1}^{8} x_{2,j}x_{2,j} = 32$$

$$D_0 = \sum_{j=1}^{8} x_{0,j}\hat{p}_j = 1 \qquad D_1 = \sum_{j=1}^{8} x_{1,j}\hat{p}_j = 82$$

$$D_2 = \sum_{j=1}^{8} x_{2,j}\hat{p}_j = 76$$

Therefore, substituting in the general equations (5.3-9), we get

$$8\beta_0 + 5\beta_1 + 14\beta_2 = 1$$

$$5\beta_0 + 7\beta_1 + 10\beta_2 = 82$$

$$14\beta_0 + 10\beta_1 + 32\beta_2 = 76$$

which can be solved to give $\beta_0 = -23.47$, $\beta_1 = 18.82$, $\beta_2 = 6.76$. The form of the desired approximation is then

$$\underline{p} = -23.47 + 18.82x_1 + 6.76x_2$$

This is a reasonable, but not really good, approximation to the data, as will be seen by comparing the approximate and observed responses

TABLE III-7 RESPONSE DATA FITTED IN EXAMPLE III-13

Experiment j	*Input* $x_{1,j}$	$x_{2,j}$	*Measured response* $\hat{p}_j$	*Approximate response* $\underline{p}_j$
1	0	0	−25	−23.47
2	0	1	−16	−16.71
3	0	2	−9	−9.95
4	0	3	−4	−3.18
5	1	1	3	2.11
6	1	2	10	8.87
7	1	3	15	15.64
8	2	2	27	27.69

presented in the last two columns of Table III-7. The accuracy of presentation will, however, be sufficient for many purposes, especially within the region of the original experiments.

5.3.3 *Fitting an n-dimensional plane surface using a fixed base point*

In treating a particular unit, physical arguments may lead to a known response for a specific set of inputs. Very often, for example, if all the inputs vanish, it is then known that the output must also vanish—a knowledge arising from the physical aspects of the system under examination. In addition, in certain numerical search applications, it is desirable to force the surface to have a certain response at a particular base point.

In both of these cases, the approximating function has to yield a stated value, $\hat{p}_0$, at a given base point, $\mathbf{x}_0$. For a plane surface the approximating function will be

$$\underline{p} = \sum_{i=0}^{n} \beta_i x_i = \beta_0 + \sum_{i=1}^{n} \beta_i x_i \tag{5.3-10}$$

remembering that $x_0 = 1$. If we substitute for the values at the base point, Eq. (5.3-10) gives

$$\underline{p}_0 = \hat{p}_0 = \beta_0 + \sum_{i=1}^{n} \beta_i x_{i,0} \tag{5.3-11}$$

where $\hat{p}_0$ is the known response at $\mathbf{x}_0$. Clearly, one of the coefficients can be eliminated using this result. In other words, if the surface is forced to exhibit a certain response, a degree of freedom will then be lost. It is most convenient to eliminate β_0, so that, using Eq. (5.3-11), we get

$$\beta_0 = \hat{p}_0 - \sum_{i=1}^{n} \beta_i x_{i,0}$$

which, upon substitution into Eq. (5.3-10), yields

$$\underline{p} = \hat{p}_0 + \sum_{i=1}^{n} \beta_i(x_i - x_{i,0}) \qquad (5.3\text{-}12)$$

We see that Eq. (5.3-12) has only n, rather than $n + 1$, coefficients.

Defining $\Delta\hat{p}$ as $\Delta\hat{p} = \hat{p} - \hat{p}_0$, the residual would then be

$$R_j = \Delta\hat{p}_j - \Delta\underline{p}_j$$

where

$$\Delta\underline{p}_j = \underline{p}_j - \hat{p}_0$$

The least-squares argument can now be repeated, and the β_j determined from the resulting n equations,

$$\sum_{k=1}^{n} \tilde{C}_{i,k}\beta_k = \tilde{D}_i \qquad \text{for } i = 1, 2, \ldots, n \qquad (5.3\text{-}13)$$

where the $\tilde{C}_{i,k}$ are expressed in terms of the deviations of the coordinates from the base point. Thus

$$\tilde{C}_{i,k} = \sum_{j=1}^{r} \zeta_{i,j}\zeta_{k,j} \qquad (5.3\text{-}14)$$

in which $\zeta_{i,k}$ are the deviations from the base point, given by

$$\zeta_{i,k} = x_{i,k} - x_{i,0}$$

The $\tilde{D}_i$ can also be defined in terms of the deviations as

$$\tilde{D}_i = \sum_{j=1}^{r} \zeta_{i,j}\,\Delta\hat{p}_j \qquad (5.3\text{-}15)$$

Example III-14: Least-squares fit to include a prescribed point

Consider the responses found in Example III-12. Let us again seek an approximation in terms of a first-degree polynomial, but the polynomial will now be forced to yield the value $\hat{p}_0 = 1.9$ at the base point $x_{1,0} = 1.0$.

Thus we can compile the information given in the first five columns of Table III-8. To obtain the approximating relationship we may

TABLE III-8 RESPONSE DATA FOR EXAMPLE III-14

					Derived Results	
Experiment j	*Input* $x_{1,j}$	*Deviation* $\zeta_{1,j} = x_{1,j} - x_{1,0}$	*Response* $\hat{p}_j$	*Relative response* $\Delta\hat{p}_j = \hat{p}_j - \hat{p}_0$	*Approximate response* $\Delta\underline{p}_j$	*Residual* $R_j = \Delta\hat{p}_j - \Delta\underline{p}_j = \hat{p}_j - \underline{p}_j$
Base point 1	1	0	1.90	0		
2	5	4	5.90	4.00	4.148	−0.148
3	4	3	5.10	3.20	3.111	0.089
4	2	1	3.10	1.20	1.037	0.163
5	3	2	4.05	2.15	2.074	0.076

proceed as follows: On omitting the first experiment listed, since it has already been used in defining the base point, we may form, using Eqs. (5.3-14) and (5.3-15),

$$\tilde{C}_{1,1} = \sum_{j=2}^{4} \zeta_{1,j}\zeta_{1,j} = 30$$

$$\tilde{D}_1 = \sum_{j=2}^{4} \zeta_{1,j}\,\Delta\hat{p}_j = 31.1$$

Hence, from Eq. (5.3-13),

$$\beta_1 = \frac{31.1}{30} = 1.037$$

and the approximating function is

$$\Delta \underline{p} = 1.037\zeta_1$$

or

$$\underline{p} - 1.9 = 1.037(x_{1,j} - 1.0)$$

We may also calculate the residuals R_j, as listed in Table III-8. Here we can again easily compute E to be 0.062. Note that this is larger than the value of 0.042 found in Example III-12, because of the additional restriction imposed here.

An important principle arises from Example III-14, one which will continue to manifest itself, namely, that a restricted minimum cannot be less than an unrestricted one. In this example we have restricted the curve to pass through a certain point. Therefore we should expect the resulting minimum to be the same or larger than for the unrestricted case.

Example III-15: Fitting a relative surface

Let us consider Example III-13, this time taking the sixth point ($x_{1,0} = 1$, $x_{2,0} = 2$, $\hat{p}_0 = 10$) as a base point. We may transform the data of Table III-7 into the more suitable forms listed in Table III-9 by

TABLE III-9 RESPONSE DATA FOR EXAMPLE III-15

Experiment j	*Input* $x_{1,j}$	$x_{2,j}$	*Response* $\hat{p}_j$	*Coordinate variations* $\zeta_{1,j}$	$\zeta_{2,j}$	*Relative response* $\Delta\hat{p}_j$
1	0	0	−25	−1	−2	−35
2	0	1	−16	−1	−1	−26
3	0	2	−9	−1	0	−19
4	0	3	−4	−1	1	−14
5	1	1	3	0	−1	−7
Base point	1	2	10	0	0	0
6	1	3	15	0	1	5
7	2	2	27	1	0	17

the use of relative responses and variables. It will be seen that the row corresponding to the base point is omitted from further consideration. This is simply a matter of convenience, since it has already been included and, if used in the following sums, would not change the result. From Eqs. (5.3-14) and (5.3-15), we get

$$\tilde{C}_{1,1} = 5 \qquad \tilde{C}_{1,2} = 2 \qquad \tilde{C}_{2,2} = 8 \qquad \tilde{D}_1 = 111 \qquad \tilde{D}_2 = 94$$

so that

$$5\beta_1 + 2\beta_2 = 111$$

$$2\beta_1 + 8\beta_2 = 94$$

Solving, we get

$$\beta_1 = 19.44 \qquad \text{and} \qquad \beta_2 = 6.89$$

The first-degree polynomial is then, in relative terms,

$$\Delta \underline{p} = 19.44\zeta_1 + 6.89\zeta_2$$

This result can be compared with that found in Example III-13 when no constraint was placed upon the system.

5.3.4 *Orthogonal design and plane surfaces*

While the linear equations (5.3-13) can be solved to determine the n coefficients β_i, it is sometimes convenient and of merit statistically to designate the locations of the experiments in advance. In particular, if the experimental locations are selected in such a way that all but one of the $\tilde{C}_{r,s}$ vanish in each of the equations (5.3-13), the solution for the coefficients β_i can then be immediately written. A number of combinations of experimental locations will give this desired configuration. It is obvious that such a design can be realized only if planned experiments are carried out. If they are not so preplanned, the set of experimental results will in general have to be analyzed using Eqs. (5.3-13).

Let us now consider a suitable design. An experimental design specifies the locations, relative to a base point, at which the new experiments must be evaluated. If all the coefficients $\tilde{C}_{r,s}$ in which $r \neq s$ vanish and only coefficients $\tilde{C}_{r,s}$, with $r = s$, remain, then there will be only one such coefficient in each equation. The n equations (5.3-13) under these circumstances reduce to

$$\begin{aligned}
\tilde{C}_{1,1}\beta_1 \qquad\qquad\qquad &= \tilde{D}_1 \\
\tilde{C}_{2,2}\beta_2 \qquad\qquad &= \tilde{D}_2 \\
\cdot \qquad\qquad &\quad \cdot \\
\cdot \qquad &\quad \cdot \\
\cdot \quad &\quad \cdot \\
\tilde{C}_{n,n}\beta_n &= \tilde{D}_n
\end{aligned}$$

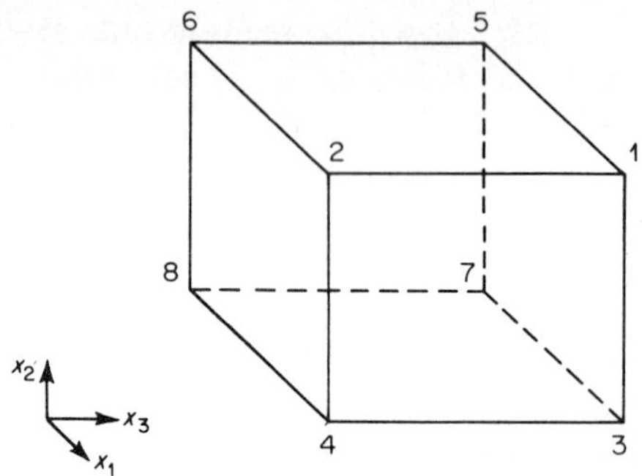

Fig. III-9 An orthogonal experimental design in three dimensions.

yielding immediately the desired values of the coefficients, β_i, in the form

$$\beta_i = \frac{\tilde{D}_i}{\tilde{C}_{i,i}} \qquad \text{for } i = 1, 2, \ldots, n \tag{5.3-16}$$

Any experiment design in which all the $\tilde{C}_{r,s}$ are zero for $r \neq s$ is acceptable. A design of points, where $\Delta\hat{p}$ is to be calculated or measured, is said to be an *orthogonal* design if it has the property that the nondiagonal terms vanish. As a very simple example, consider a set of four points lying at the corners of a square, distributed symmetrically about a base point in a two-dimensional system. The coordinates of these points relative to the base point are those listed in Table III-10. Thus

$$\begin{aligned}\tilde{C}_{1,2} &= \sum_{j=1}^{r} \zeta_{1,j}\zeta_{2,j} = \alpha^2 - \alpha^2 + \alpha^2 - \alpha^2 = 0 \\ &= \tilde{C}_{2,1}\end{aligned}$$

This is therefore an orthogonal design in two dimensions, in the sense just defined. It can be similarly shown that the eight points at the corners of a cube symmetrically placed about the starting point from an orthogonal design in three dimensions (Fig. III-9). For a discussion of orthogonal design with larger numbers of variables, see, for example, Davies [1956, chap. 11]. It is obviously easy to test whether any proposed assembly or design is orthogonal.

Example III-16: Orthogonal designs

To fit a plane surface to a set of experimental data obtained from an orthogonal design, a suitable set of experimental locations would be the 2^2 locations, shown in Table III-10. Taking an arbitrary distance $\alpha = 0.5$ about a base point {2,2}, the experiments should then be

TABLE III-10 AN ORTHOGONAL EXPERIMENTAL DESIGN

Experiment j	Location $\zeta_{1,j}$	$\zeta_{2,j}$
1	α	α
2	α	$-\alpha$
3	$-\alpha$	α
4	$-\alpha$	$-\alpha$

carried out at the four locations $\{\mathbf{x}_j\}$ for $j = 1, 2, 3, 4$, as shown in Table III-11. If the experimental results at those locations are as indicated in the table, a surface having a response $\hat{p}_0 = 8.0$ at the base point $\{2.0,2.0\}$

TABLE III-11 THE APPLICATION OF AN ORTHOGONAL DESIGN SHOWN IN EXAMPLE III-16

Experiment j	Location $\zeta_{1,j}$	$\zeta_{2,j}$	Location of experiments $x_{1,j}$	$x_{2,j}$	Experimental output $\hat{p}_j$	Relative output $\Delta\hat{p}_j = \hat{p}_j - 8.0$
0	0	0	2.0	2.0	8.0	0
1	0.5	0.5	2.5	2.5	9.1	1.1
2	0.5	−0.5	2.5	1.5	7.7	−0.3
3	−0.5	0.5	1.5	2.5	8.1	0.1
4	−0.5	−0.5	1.5	1.5	6.9	−1.1

can then be fitted by forming the relative variables ζ_i and the corresponding response $\Delta\hat{p}$, as shown in the table. Then, from Eqs. (5.3-14), we get

$$\tilde{C}_{1,1} = \sum_{j=1}^{4} (\zeta_{1,j})^2 = (\tfrac{1}{2})^2 + (\tfrac{1}{2})^2 + (\tfrac{1}{2})^2 + (\tfrac{1}{2})^2 = 1$$

$$\tilde{C}_{2,2} = \sum_{j=1}^{4} \zeta_{2,j}^2 = 1$$

$$\tilde{C}_{1,2} = \tilde{C}_{2,1} = \sum_{j=1}^{4} \zeta_{1,j}\zeta_{2,j} = 0$$

and from Eq. (5.3-15),

$$\tilde{D}_1 = \sum_{j=1}^{4} \zeta_{1,j}\,\Delta\hat{p}_j = 0.55 - 0.15 - 0.05 + 0.55 = 0.9$$

$$\tilde{D}_2 = \sum_{j=1}^{4} \zeta_{2,j}\,\Delta\hat{p}_j = 0.55 + 0.15 + 0.05 + 0.55 = 1.3$$

so that, from Eqs. (5.3-16),

$$\beta_1 = \frac{\tilde{D}_1}{\tilde{C}_{1,1}} = 0.9$$

$$\beta_2 = \frac{\tilde{D}_2}{\tilde{C}_{2,2}} = 1.3$$

This therefore produces the desired plane surface passing through the base point {2.0,2.0} as

$$\Delta \underline{p} = 0.9\zeta_1 + 1.3\zeta_2$$

or in absolute terms,

$$\underline{p} - 8.0 = 0.9(x_1 - 2) + 1.3(x_2 - 2)$$

or

$$\underline{p} = 3.6 + 0.9x_1 + 1.3x_2$$

5.3.5 *Scaling*

To simplify the presentation of various experiment designs, it is convenient to introduce the concept of a level. The usual situation in black-box experimentation is that the region of interest is centered about a particular set of inputs, such as those at some base point, $\mathbf{x}_0$. The relevant response is the difference between the response at some point within the region of interest and that of the base point. To use curve-fitting techniques properly, this region must be well defined, and it is therefore possible to associate with each independent variable a scale of "distance." This concept of scale is particularly important when, as in most physical problems, there is no intrinsic metric in the n-dimensional space of independent variables. Thus, if one of the independent variables represents temperature and another volume, the basic measure of temperature can be changed without altering that of the volume. This lack of a single yardstick applicable to all the n dimensions gives rise to some difficulties in the process of optimization, these being discussed in Chap. XI, Sec. 5.3.1.

For our purposes the presentation of experiment designs is simplified if we modify our definition of a relative coordinate to the form

$$\zeta_i = \frac{x_i - x_{i,0}}{S_i} \tag{5.3-17}$$

Here S_i is the scale of x_i, being a number having the same dimensions as x_i and a magnitude which is of the same order of magnitude as the range of interest. In our previous work, S_i was unity.

Equation (5.3-17) permits us to express the experimental locations in terms of dimensionless experimental locations, giving a generality to the tables which follow.

Example III-17: Scaling

If an experimental system produces data valid within the temperature range 100 to 300°C, indicate a suitable scale for the temperature variable T.

The mean temperature is 200°C, so that the temperature variation about this value is ± 100°C. We might therefore use a scaled temperature ζ, such that

$$\zeta = \frac{T - 200}{100}$$

where $S_i = 100$°C. This then allows values of ζ to vary only within the range $-1 \leq \zeta \leq 1$. In particular, in a given study, it may be convenient to choose values of ζ at $+1, 0$, and -1, so that this scaling would then require experiments to be placed at 100, 200, and 300°C. Obviously, this choice of scale is not unique.

Two orthogonal designs which are suitable for determining the constants of a first-degree polynomial (plane surface) are the two-level factorial designs and the fractional factorial designs (see, for example, Davies [1956, chaps. 7 and 10]). One such two-level factorial design is found if the response function is measured along each x_i direction in n-dimensional space at positions $+1$ and -1 from the base point, so that the origin for the variables ζ_i is the midpoint of the design and the coordinates of the experimental points consist only of -1's and $+1$'s. Thus, in a three-dimensional problem, a 2^3 design would be chosen, with the experimental points located at the vertices of a cube, as shown in Fig. III-9, the variables being $\zeta_i = \pm 1$, as listed in Table III-12. If the variables x_i do not interact with one another, then, as discussed by Davies [1956, chap. 10], it is necessary only to carry out

TABLE III-12 THREE-DIMENSIONAL ORTHOGONAL DESIGN

Experiment j	*Location* $\zeta_{1,j}$	$\zeta_{2,j}$	$\zeta_{3,j}$
1	1	1	1
2	1	1	−1
3	1	−1	1
4	1	−1	−1
5	−1	1	1
6	−1	1	−1
7	−1	−1	1
8	−1	−1	−1

half of these experiments, i.e., four for a three-dimensional design. This is called a *half-replicate design.* For example, we could choose points 2, 3, 5, 8 (so that $\zeta_1\zeta_2\zeta_3 = -1$) or 1, 4, 6, 7 (so that $\zeta_1\zeta_2\zeta_3 = +1$), either set forming a regular tetrahedron.

5.4 Quadratic Surfaces

The techniques of fitting higher-order polynomials are also based upon the concept of least squares, as set forth in Sec. 5.3 for plane surfaces. We shall therefore summarize only the salient results at this point.

5.4.1 *General problem of fitting quadratic surfaces*

The approximating quadratic polynomial

$$\Delta \underline{p} = \underline{p} - \hat{p}_0 = \beta_1\zeta_1 + \beta_2\zeta_2 + \cdots + \beta_n\zeta_n + \beta_{1,1}\zeta_1^2 + \beta_{1,2}\zeta_1\zeta_2 + \cdots + \beta_{n,n}\zeta_n^2$$

written about a base point has a larger range of validity than the plane surface, and also plays a rather special role in the final stages of a numerical search. There are $n + [n(n+1)]/2$ coefficients to be determined, and the number of experiments, r, must therefore be in excess of this number.

Proceeding in a manner similar to that used for a plane surface in Sec. 5.3.1, we may form the sum of the squares of the residuals,

$$E = \sum_{j=1}^{r} (\Delta\hat{p} - \Delta\underline{p})^2$$

This will be minimized if the coefficients β_i and $\beta_{r,s}$ are chosen so as to satisfy the following two sets of algebraic equations. First, by differentiating E with respect to the β_i and equating the result to zero, we get the n equations

$$\beta_1\tilde{C}_{1,i} + \beta_2\tilde{C}_{2,i} + \cdots + \beta_n\tilde{C}_{n,i} + \beta_{1,1}\tilde{C}_{1,1,i} + \cdots + \beta_{n,n}\tilde{C}_{n,n,i} = \tilde{D}_i \tag{5.4-1}$$

for $i = 1, 2, \ldots, n$. Second, by equating the $[n(n+1)]/2$ derivatives of E with respect to $\beta_{r,s}$ to zero, we get

$$\beta_1\tilde{C}_{1,i,k} + \beta_2\tilde{C}_{2,i,k} + \cdots + \beta_n\tilde{C}_{n,i,k} + \beta_{1,1}\tilde{C}_{1,1,i,k} + \beta_{1,2}\tilde{C}_{1,2,i,k} + \cdots + \beta_{n,n}\tilde{C}_{n,n,i,k} = \tilde{D}_{i,k} \tag{5.4-2}$$

for all combinations of i and k ($i, k = 1, 2, \ldots, n$) which lead to distinct equations. Here, as before,

$$\tilde{C}_{i,k} = \sum_{j=1}^{r} \zeta_{i,j}\zeta_{k,j} \tag{5.4-3a}$$

$$\tilde{C}_{p,i,k} = \sum_{j=1}^{r} \zeta_{p,j}\zeta_{i,j}\zeta_{k,j} \tag{5.4-3b}$$

$$\tilde{C}_{p,q,i,k} = \sum_{j=1}^{r} \zeta_{p,j}\zeta_{q,j}\zeta_{i,j}\zeta_{k,j} \tag{5.4-3c}$$

$$\tilde{D}_{i} = \sum_{j=1}^{r} \zeta_{i,j}\,\Delta\hat{p}_j \tag{5.4-3d}$$

$$\tilde{D}_{i,k} = \sum_{j=1}^{r} \zeta_{i,j}\zeta_{k,j}\,\Delta\hat{p}_j \tag{5.4-3e}$$

Equations (5.4-2) are linear in the coefficients β_i and $\beta_{i,k}$, and hence the only increase in difficulty of fitting a quadratic surface compared with a plane surface simply stems from the enlarged number of simultaneous equations to be solved. We are now in a position to determine the coefficients β_i and $\beta_{i,k}$ (for $i, k = 1, 2, \ldots, n$) from any set of data over r experimental locations.

Example III-18: Definition of the quadratic coefficients

For the case of three dimensions, let us write out the equations defining the coefficients β_i and $\beta_{i,k}$ required in fitting a quadratic approximation. The necessary equations to define the β_i and $\beta_{i,k}$ are found from Eq. (5.4-1) and (5.4-2) for $i, k = 1, 2$, and 3. Thus we have

$$\beta_1\tilde{C}_{1,1} + \beta_2\tilde{C}_{2,1} + \beta_3\tilde{C}_{3,1} + \beta_{1,1}\tilde{C}_{1,1,1} + \beta_{1,2}\tilde{C}_{1,2,1} + \beta_{1,3}\tilde{C}_{1,3,1} + \beta_{2,2}\tilde{C}_{2,2,1} + \beta_{2,3}\tilde{C}_{2,3,1} + \beta_{3,3}\tilde{C}_{3,3,1} = \tilde{D}_1$$

$$\beta_1\tilde{C}_{1,2} + \beta_2\tilde{C}_{2,2} + \cdots + \beta_{1,1}\tilde{C}_{1,1,2} + \beta_{1,2}\tilde{C}_{1,2,2} + \cdots + \beta_{3,3}\tilde{C}_{3,3,2} = \tilde{D}_2$$

$$\beta_1\tilde{C}_{1,3} + \beta_2\tilde{C}_{2,3} + \cdots + \beta_{1,1}\tilde{C}_{1,1,3} + \beta_{1,2}\tilde{C}_{1,2,3} + \cdots + \beta_{3,3}\tilde{C}_{3,3,3} = \tilde{D}_3$$

and

$$\beta_1\tilde{C}_{1,1,1} + \beta_2\tilde{C}_{2,1,1} + \beta_3\tilde{C}_{3,1,1} + \beta_{1,1}\tilde{C}_{1,1,1,1} + \beta_{2,2}\tilde{C}_{2,2,1,1} + \beta_{3,3}\tilde{C}_{3,3,1,1} + \beta_{1,2}\tilde{C}_{1,2,1,1} + \beta_{1,3}\tilde{C}_{1,3,1,1} + \beta_{2,3}\tilde{C}_{2,3,1,1} = \tilde{D}_{1,1}$$

$$\beta_1\tilde{C}_{1,2,2} + \beta_2\tilde{C}_{2,2,2} + \cdots = \tilde{D}_{2,2}$$

$$\beta_1\tilde{C}_{1,3,3} + \beta_2\tilde{C}_{2,3,3} + \cdots = \tilde{D}_{3,3}$$

$$\beta_1\tilde{C}_{1,1,2} + \beta_2\tilde{C}_{2,1,2} + \cdots = \tilde{D}_{1,2}$$

$$\beta_1\tilde{C}_{1,1,3} + \beta_2\tilde{C}_{2,1,3} + \cdots = \tilde{D}_{1,3}$$

$$\beta_1\tilde{C}_{1,2,3} + \beta_2\tilde{C}_{2,2,3} + \cdots = \tilde{D}_{2,3}$$

where $\tilde{C}_{i,k}$, $\tilde{C}_{p,i,k}$, $\tilde{C}_{p,q,i,k}$, $\tilde{D}_i$, and $\tilde{D}_{i,k}$ have the meanings defined by Eqs. (5.4-3).

For any set of data at r locations, we could now calculate the coefficients β_i and $\beta_{r,s}$ for the required second-degree approximation to the actual surface.

This would be the only possible approach if the experimenter had no control over his experimental conditions. However, just as in the case of plane surfaces, we may normally choose any reasonable set of points at which the r experiments or calculations are to be obtained. This of course means that the experimental design may have to be chosen before the experiments are carried out.

5.4.2 *Orthogonal designs for quadratic surfaces*

Let us now consider a suitable choice of experimental locations so that the estimates of the values of the coefficients β_i and $\beta_{i,k}$ can be more easily attained.

Two-level Factor Experiments The orthogonal design using only two levels $\zeta_i = \pm\alpha$ is called a 2^n *factorial design*, since 2^n experiments are required. In two dimensions we should then choose the ζ_1 and ζ_2 as given in Table III-10, while in three dimensions we should use the $\zeta_{i,j}$ as shown in Table III-12 with $\alpha = 1$. Upon substitution for the $\zeta_{i,k}$ in the definitions of the coefficients $\tilde{C}$, as given in Eqs. (5.4-3), we get

$$\tilde{C}_{i,k} = \sum_{j=1}^{r} \zeta_{i,j}\zeta_{k,j} = \begin{cases} 0 & \text{if } i \neq k \\ 2^n\alpha^2 & \text{if } i = k \end{cases}$$

where i and k are any pair from $i, k = 1, 2, \ldots, n$. Thus

$$\tilde{C}_{1,1} = \sum_{j=1}^{r} \zeta_{1,j}^2 = r\alpha^2 = 2^n\alpha^2$$

$$\tilde{C}_{1,2} = \sum_{j=1}^{r} \zeta_{1,j}\zeta_{2,j} = \frac{r}{2}\alpha^2 - \frac{r}{2}\alpha^2 = 0$$

$$\tilde{C}_{2,2} = \sum_{j=1}^{r} \zeta_{2,j}^2 = r\alpha^2 = 2^n\alpha^2$$

Similarly,

$$\tilde{C}_{p,i,k} = \sum_{j=1}^{r} \zeta_{p,j}\zeta_{i,j}\zeta_{k,j} = 0$$

for all values of p, i, and k which are different from each other. Finally,

$$\tilde{C}_{p,q,i,k} = \sum_{j=1}^{r} \zeta_{p,j}\zeta_{q,j}\zeta_{i,j}\zeta_{k,j} = 0$$

for $p = q = i \neq k$, any one being different from the other three, and

$$\tilde{C}_{p,q,i,k} = 2^n\alpha^4$$

for any pair equal, that is, $p = q$ and $i = k$, including $p = q = i = k$. In other words, the nonzero terms are

$$\tilde{C}_{p,p} = 2^n\alpha^2 \tag{5.4-4a}$$

$$\tilde{C}_{p,p,q,q} = 2^n\alpha^4 \qquad \text{(including } p = q\text{)} \tag{5.4-4b}$$

Example III-19: An orthogonal design

Choose an orthogonal design and evaluate the $\tilde{C}$ coefficients in two dimensions. If $\alpha = 1$, then $\zeta_i = \pm 1$. There will be $r = 2^n = 4$ experiments, placed at

ζ_1	ζ_2
1	1
1	−1
−1	1
−1	−1

with resulting $\tilde{C}$ values of

$$\tilde{C}_{1,1} = 4$$
$$\tilde{C}_{2,2} = 4$$
$$\tilde{C}_{1,2} = \tilde{C}_{2,1} = 0$$
$$\tilde{C}_{1,1,1} = 0$$
$$\tilde{C}_{1,1,2} = \tilde{C}_{1,2,1} = \tilde{C}_{2,1,1} = 0$$
$$\tilde{C}_{2,2,2} = 0$$
$$\tilde{C}_{1,2,2} = \tilde{C}_{2,1,2} = \tilde{C}_{2,2,1} = 0$$
$$\tilde{C}_{1,1,1,1} = 4$$
$$\tilde{C}_{2,2,2,2} = 4$$
$$\tilde{C}_{1,1,2,2} = \tilde{C}_{2,2,1,1} = \tilde{C}_{1,2,1,2} = \tilde{C}_{2,1,2,1} = 4$$
$$\tilde{C}_{1,1,1,2} \cdots = 0$$
$$\tilde{C}_{1,2,2,2} \cdots = 0$$

Since we are now in a position to choose the experiments and calculate the $\tilde{C}$ coefficient values, those may now be substituted from Eq. (5.4-4) with

Eqs. (5.4-1) and (5.4-2), getting the three sets of equations defining β_i and $\beta_{i,k}$:

$$\left.\begin{aligned} \beta_1\tilde{C}_{1,1} &= \tilde{D}_1 \\ \beta_2\tilde{C}_{2,2} &= \tilde{D}_2 \\ &\vdots \\ \beta_n\tilde{C}_{n,n} &= \tilde{D}_n \end{aligned}\right\} n \text{ equations} \tag{5.4-5a}$$

$$\left.\begin{aligned} \beta_{1,1}\tilde{C}_{1,1,1,1} + \beta_{2,2}\tilde{C}_{2,2,1,1} + \cdots + \beta_{n,n}\tilde{C}_{n,n,1,1} &= \tilde{D}_{1,1} \\ \beta_{1,1}\tilde{C}_{1,1,2,2} + \beta_{2,2}\tilde{C}_{2,2,2,2} + \cdots + \beta_{n,n}\tilde{C}_{n,n,2,2} &= \tilde{D}_{2,2} \\ \cdots\cdots\cdots\cdots\cdots\cdots\cdots\cdots\cdots \\ \beta_{1,1}\tilde{C}_{1,1,n,n} + \beta_{2,2}\tilde{C}_{2,2,n,n} + \cdots + \beta_{n,n}\tilde{C}_{n,n,n,n} &= \tilde{D}_{n,n} \end{aligned}\right\} n \text{ equations} \tag{5.4-5b}$$

$$\left.\begin{aligned} \beta_{1,2}\tilde{C}_{1,2,1,2} &= \tilde{D}_{1,2} \\ \beta_{1,3}\tilde{C}_{1,3,1,3} &= \tilde{D}_{1,3} \\ &\vdots \\ \beta_{p,q}\tilde{C}_{p,q,p,q} &= \tilde{D}_{p,q} \end{aligned}\right\} \frac{n(n-1)}{2} \text{ equations for } p \neq q \tag{5.4-5c}$$

Hence, from Eqs. (5.4-5*a*),

$$\beta_i = \frac{\tilde{D}_i}{\tilde{C}_{i,i}} = \frac{\sum_{j=1}^{r} \zeta_{i,j}\Delta\hat{p}_j}{\sum_{j=1}^{r} \zeta_{i,j}^2} = \frac{\sum_{j=1}^{r} \zeta_{i,j}\Delta\hat{p}_j}{2^n\alpha^2} \tag{5.4-6a}$$

and from Eqs. (5.4-5*c*), for $p \neq q$,

$$\beta_{p,q} = \frac{\tilde{D}_{p,q}}{\tilde{C}_{p,q,p,q}} = \frac{\sum_{j=1}^{r} \zeta_{p,j}\zeta_{q,j}\Delta\hat{p}_j}{\sum_{j=1}^{r} \zeta_{p,j}^2\,\zeta_{q,j}^2} = \frac{\sum_{j=1}^{r} \zeta_{p,j}\zeta_{q,j}\Delta\hat{p}_j}{2^n\alpha^4} \tag{5.4-6b}$$

However, the remaining equations (5.4-5*b*) containing $\beta_{p,p}$ reduce to

$$\begin{aligned} \beta_{1,1} + \beta_{2,2} + \cdots + \beta_{n,n} &= \frac{\tilde{D}_{1,1}}{2^n\alpha^4} \\ \beta_{1,1} + \beta_{2,2} + \cdots + \beta_{n,n} &= \frac{\tilde{D}_{2,2}}{2^n\alpha^4} \\ \cdots\cdots\cdots\cdots\cdots\cdots \\ \beta_{1,1} + \beta_{2,2} + \cdots + \beta_{n,n} &= \frac{\tilde{D}_{n,n}}{2^n\alpha^4} \end{aligned} \tag{5.4-7}$$

By definition,

$$\tilde{D}_{i,i} = \sum_{j=1}^{r} \zeta_{i,j}^2 \, \Delta\hat{p}_j = \alpha^2 \sum_{j=1}^{r} \Delta\hat{p}_j \qquad \text{for all } i = 1, 2, \ldots, n$$

so that the n equations (5.4-7) reduce in fact to a single equation

$$\beta_{1,1} + \beta_{2,2} + \cdots + \beta_{n,n} = \frac{\sum_{j=1}^{r} \Delta\hat{p}_j}{2^n\alpha^2}$$

which yields no information about the individual values of $\beta_{p,p}$. Further experimentation is required to determine these quantities. This parallelism of least-square equations exhibited by Eqs. (5.4-7) is a result of the levels at which experimentation was carried out. In a sense, it is the "price" paid to get into a position in which the computation of the other coefficients is very simple.

Example III-20: A two-level factorial design

Let us fit a quadratic surface, as far as we can, using a 2^2 type design on the data given in Table III-11. Here

$$\beta_1 = \frac{\tilde{D}_1}{4(0.5)^2} = \tilde{D}_1 = \sum_{1}^{4} \zeta_{1,j} \, \Delta\hat{p}_j = 0.90$$

$$\beta_2 = \frac{\tilde{D}_2}{4(0.5)^2} = \tilde{D}_2 = \sum_{1}^{4} \zeta_{2,j} \, \Delta\hat{p}_j = 1.30$$

$$\beta_{1,2} = \frac{\tilde{D}_{1,2}}{4(0.5)^4} = 4\tilde{D}_{1,2} = 4\sum_{1}^{4} \zeta_{1,j}\zeta_{2,j} \, \Delta\hat{p}_j = 0.2$$

$$\beta_{1,1} + \beta_{2,2} = \frac{\sum_{1}^{4} \Delta\hat{p}_j}{4(0.5)^2} = -0.20$$

While the complete surface

$$\Delta\underline{p} = (\beta_1\zeta_1 + \beta_2\zeta_2) + (\beta_{1,2}\zeta_1\zeta_2 + \beta_{1,1}\zeta_1{}^2 + \beta_{2,2}\zeta_2{}^2)$$

cannot be fully specified, an estimate of the magnitude of $\beta_{1,1}$ and $\beta_{2,2}$ together is possible.

Once the first two groups of coefficients, β_i and $\beta_{i,k}$, have been evaluated, let us consider an additional set of experiments to be chosen and used to evaluate the remaining $\beta_{p,p}$ coefficients. This can be done most easily by using either a three-level factorial or a modified two-level design.

Three-level Factorial Design As we saw earlier, a two-level factorial design allows $[n(n + 1)]/2$ of the required $[n(n + 3)]/2$ coefficients to be determined.

A further set of experiments is needed in the orthogonal design to complete the fitting of the second-degree surface. A complete three-level factorial design of 3^n experiments could be used to estimate those second-order effects.

Example III-21: Three-level orthogonal design

Let us find the resultant equations relating the coefficients of a second-degree polynomial in two dimensions, given a complete three-level factorial design such as that of Table III-13. It will be seen that the

TABLE III-13 3^n FACTORIAL DESIGN FOR TWO DIMENSIONS

Experiment j	*Location* $\zeta_{1,j}$	$\zeta_{2,j}$
0	0	0
1	1	1
2	1	−1
3	−1	1
4	−1	−1
5	0	1
6	0	−1
7	1	0
8	−1	0

points 1 to 4 are the full two-level factorial, the additional experiments 5 to 8 being required to investigate the mid-level, when some $\zeta_i = 0$.

Referring to Eqs. (5.4-1) to (5.4-3), we see that

$$\tilde{C}_{1,1} = 6 \qquad \tilde{C}_{1,2} = 0 \qquad \tilde{C}_{2,2} = 6$$
$$\tilde{C}_{1,1,1} = 0 \qquad \tilde{C}_{1,1,2} = 0 \qquad \tilde{C}_{1,2,2} = 0 \qquad \tilde{C}_{2,2,2} = 0$$
$$\tilde{C}_{1,1,1,1} = 6 \qquad \tilde{C}_{1,1,1,2} = 0 \qquad \tilde{C}_{1,1,2,2} = 4 \qquad \tilde{C}_{1,2,2,2} = 0$$
$$\tilde{C}_{2,2,2,2} = 6$$

Therefore, once the $\tilde{D}_i$ and $\tilde{D}_{i,k}$ are found from Eqs. (5.4-3*d*) and (5.4-3*e*), we can get the β_i from Eqs. (5.4-6*a*) and (5.4-6*b*) as

$$6\beta_1 = \tilde{D}_1 \qquad 6\beta_2 = \tilde{D}_2 \qquad 4\beta_{1,2} = \tilde{D}_{1,2}$$

We can also have, from the additional experiments and Eqs. (5.4-5*b*) that

$$6\beta_{1,1} + 4\beta_{2,2} = \tilde{D}_{1,1}$$
$$4\beta_{1,1} + 6\beta_{2,2} = \tilde{D}_{2,2}$$

which can be used to determine $\beta_{1,1}$ and $\beta_{2,2}$, once the $\tilde{D}_{i,k}$ are found from Eq. (5.4-3*e*).

Thus all the coefficients of the second-order polynomial can be determined from a three-level factorial experiment.

We see that a three-level factorial design could thus be used to obtain estimates of the second-order coefficients. If there are n dimensions, 3^n experiments will be required. If n is greater than 2, however, the number of experiments needed often greatly exceeds the number of coefficients to be estimated. For example, with four variables, the design required $3^4 = 81$ experiments, but the number of coefficients of second and lower order is only 15 (Table III-4), made up of one at the design center (β_0), four for linear effects ($\beta_1, \ldots, \beta_4$), four for the quadratic effects ($\beta_{1,1}, \ldots, \beta_{4,4}$), and six for the interaction effects ($\beta_{1,2}, \ldots, \beta_{3,4}$). Thus, if the experimental error is small, three-level designs are not usually used, and alternative designs have to be sought which will give estimates of the coefficients through second order without requiring a large number of experiments in excess of the number of constants to be determined. Such designs, called *composite designs*, are very useful in sequential experimentation.

Modified 2^n, or Composite, Design Box and Wilson [1951] give two useful composite designs. These consist of a basic factorial 2^n assembly, with additional points situated either symmetrically around the center of the factorial (*central-composite design*) or near to one corner of it (*noncentral-composite design*).

The procedure is, first, to choose a two-level design (2^n) so that all effects of first order and all interactions of second order can be estimated, as was done on p. 86, getting β_i and $\beta_{i,k}$ for $i \neq k$. This design is then supplemented by further points to allow an estimation of the quadratic terms $\beta_{i,i}$.

Examples of those designs for three dimensions are shown in Tables III-14 and III-15 and in Fig. III-10 (Box and Wilson [1951] and Davies [1956, chap. 11]). These may be generalized to n dimensions. Thus, in the general case for central-composite designs, $2n + 1$ additional points have to be added to the basic two-level design, one added at the center and the remaining $2n$ in pairs along the coordinate axes at $\pm\alpha_1, \pm\alpha_2, \ldots, \pm\alpha_n$, respectively. Similarly, n extra experiments are required for the general noncentral-composite design, as shown in Fig. III-10. The advantages of the latter approach are due to the fewer extra points required, but its main disadvantage is that the quadratic effects are not very precisely determined.

By a suitable choice[1] of the α_i, orthogonal designs can be again chosen.

[1] For example, in the central-composite design, we may put $\alpha_1 = \alpha_2 = \cdots = \alpha$, where α is given by

n	2	3	4	5
α	1.000	1.215	1.414	1.547

TABLE III-14 A THREE-FACTOR CENTRAL-COMPOSITE DESIGN

Experiment j	$\zeta_{1,j}$	$\zeta_{2,j}$	$\zeta_{3,j}$	
1	1	1	1	
2	1	1	−1	
3	1	−1	1	
4	1	−1	−1	Basic two-level
5	−1	1	1	design
6	−1	1	−1	
7	−1	−1	1	
8	−1	−1	−1	
9	α_1	0	0	
10	$-\alpha_1$	0	0	
11	0	α_2	0	Additional points to
12	0	$-\alpha_2$	0	complete design
13	0	0	α_3	
14	0	0	$-\alpha_3$	
15	0	0	0	

TABLE III-15 A THREE-FACTOR NONCENTRAL-COMPOSITE DESIGN

Experiment	$\zeta_{1,j}$	$\zeta_{2,j}$	$\zeta_{3,j}$	
1	1	1	1	
2	1	1	−1	
3	1	−1	1	
4	1	−1	−1	Basic two-level
5	−1	1	1	design
6	−1	1	−1	
7	−1	−1	1	
8	−1	−1	−1	
9	α_1	1	1	Additional points to
10	1	α_2	1	complete design
11	1	1	α_3	
9′	α_1	1	−1	Alternative
10′	1	α_2	−1	additional set, to
11′	1	1	$-\alpha_3$	replace points 9, 10, 11

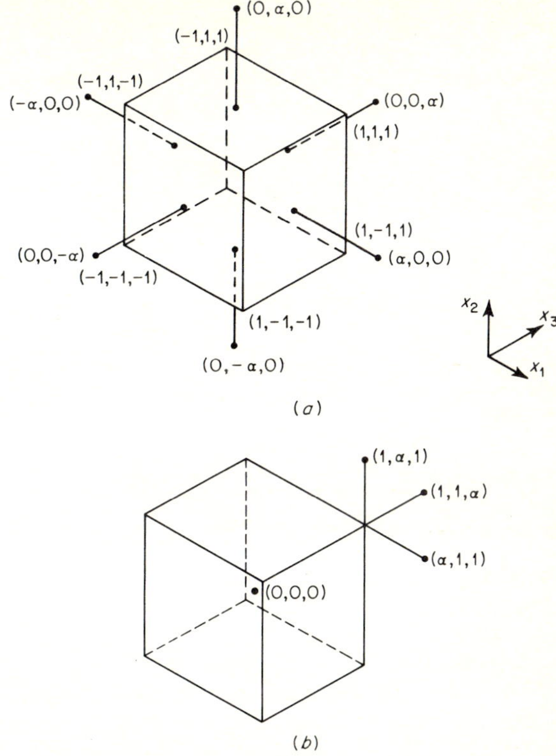

Fig. III-10 Composite designs. (*a*) Central-composite design; (*b*) noncentral-composite design.

It is not essential, however, that the distance α_2 be chosen to make the design completely orthogonal, or indeed to have all the α_i equal. The effect of this alteration has been discussed, for example, by Box [1956].

Example III-22: A composite design

Data have been produced from an experimental investigation, with the experimental locations so chosen in the region

$$\text{Temperature } T = 150 \pm 10°\text{C}$$
$$\text{Concentration } c = 25 \pm 3 \text{ lb/ft}^3$$
$$\text{Time } \theta = 10 \pm 2 \text{ hr}$$

that a central-composite orthogonal design in three dimensions (Fig. III-10*a*) may be used to find a quadratic representation of the data.

If

$$\zeta_1 = \frac{T - 150}{10} \qquad \zeta_2 = \frac{c - 25}{3} \qquad \zeta_3 = \frac{\theta - 10}{2}$$

TABLE III-16 EXPERIMENTAL DATA FOR EXAMPLE III-22

Experiment	$\zeta_{1,j}$	$\zeta_{2,j}$	$\zeta_{3,j}$	$\hat{p}_j$
1	1	1	1	80.3
2	1	1	−1	87.0
3	1	−1	1	90.8
4	1	−1	−1	82.3
5	−1	1	1	88.6
6	−1	1	−1	89.6
7	−1	−1	1	84.3
8	−1	−1	−1	74.0
9	α	0	0	88.1
10	$-\alpha$	0	0	79.4
11	0	α	0	83.3
12	0	$-\alpha$	0	76.5
13	0	0	α	87.2
14	0	0	$-\alpha$	77.1
15	0	0	0	85.8

$\alpha = 1.215$

let us find the quadratic approximation to the responses, using the central-composite design and data shown in Table III-16. The appropriate equations (5.4-5) defining the β_i and $\beta_{i,k}$ (written out in full for $n = 3$ in Example III-18) reduce to

$$\beta_1\tilde{C}_{1,1} = \tilde{D}_1$$
$$\beta_2\tilde{C}_{2,2} = \tilde{D}_2$$
$$\beta_2\tilde{C}_{3,3} = \tilde{D}_3$$

and

$$\beta_{1,1}\tilde{C}_{1,1,1,1} + \beta_{2,2}\tilde{C}_{2,2,1,1} + \beta_{3,3}\tilde{C}_{3,3,1,1} = \tilde{D}_{1,1}$$
$$\beta_{1,1}\tilde{C}_{1,1,2,2} + \beta_{2,2}\tilde{C}_{2,2,2,2} + \beta_{3,3}\tilde{C}_{3,3,2,2} = \tilde{D}_{2,2}$$
$$\beta_{1,1}\tilde{C}_{1,1,3,3} + \beta_{2,2}\tilde{C}_{2,2,3,3} + \beta_{3,3}\tilde{C}_{3,3,3,3} = \tilde{D}_{3,3}$$
$$\beta_{1,2}\tilde{C}_{1,2,1,2} = \tilde{D}_{1,2}$$
$$\beta_{1,3}\tilde{C}_{1,3,1,3} = \tilde{D}_{1,3}$$
$$\beta_{2,3}\tilde{C}_{2,3,2,3} = \tilde{D}_{2,3}$$

where $\tilde{C}_{i,i} = 8 + 2\alpha^2$, $\tilde{C}_{i,i,i,i} = 8 + 2\alpha^4$, and $\tilde{C}_{i,i,k,k} = 8$, and all other $\tilde{C}$'s are zero. For the present data, with $\alpha = 1.215$, we therefore get, from Eqs. (5.4-3*d*) and (5.4-3*e*), defining the $\tilde{D}_i$ and $\tilde{D}_{i,k}$,

$$\tilde{C}_{i,i} = 10.95 \qquad \tilde{D}_1 = 14.47 \qquad \tilde{D}_{1,1} = 924.17 \qquad \tilde{D}_{1,2} = -25.70$$
$$\tilde{C}_{i,i,i,i} = 17.64 \qquad \tilde{D}_2 = 22.36 \qquad \tilde{D}_{2,2} = 912.80 \qquad \tilde{D}_{1,3} = -7.50$$
$$\tilde{C}_{i,i,k,k} = 8.0 \qquad \tilde{D}_3 = 23.37 \qquad \tilde{D}_{3,3} = 919.44 \qquad \tilde{D}_{2,3} = -26.5$$

Substitution in the appropriate equations above then yields the desired coefficients:

$$\begin{array}{lll} \beta_1 = 1.321 & \beta_{1,1} = 27.87 & \beta_{1,2} = -3.213 \\ \beta_2 = 2.042 & \beta_{2,2} = 26.69 & \beta_{1,3} = -0.938 \\ \beta_3 = 2.134 & \beta_{3,3} = 27.38 & \beta_{2,3} = -3.313 \end{array}$$

so that the quadratic surface representing the data is

$$\Delta \underline{p} = \beta_1\zeta_1 + \beta_2\zeta_2 + \beta_3\zeta_3 + \beta_{1,1}\zeta_1^2 + \beta_{2,2}\zeta_2^2 + \beta_{3,3}\zeta_3^2 + \beta_{1,2}\zeta_1\zeta_2 + \beta_{1,3}\zeta_1\zeta_3 + \beta_{2,3}\zeta_2\zeta_3$$

the ζ_i being defined relative to the base point $\{150,25,10\}$.

5.5 Summary of Surface Fitting

Surface fitting represents a possible route from black-box data to an analytical expression.[1] The criterion of least squares is an important guide in estimating the coefficients introduced into an approximating mathematical form. It was seen that the polynomial represents an extremely convenient form, since the resulting least-square equations are linear in the coefficients, even though the polynomial may be quadratic in form, and since by an appropriate design of experiments, computations can be greatly simplified.

If it is intended to represent the response surface, linking inputs and outputs, over a wide range of experimental conditions, a polynomial of high degree will be required to obtain an adequate representation. However, it is often wasteful in time and calculations to fit such a polynomial, since this requires a disproportionate number of experiments. Most investigations use a fairly simple form of first- or second-degree surface approximation, which is intended strictly as a local representation. As we shall see in Chap. VIII, where the attendant simplicity in calculation is a major advantage, a plane is often used in initial studies, while a quadratic surface may be used in the final stages of some of the numerical search optimization methods.

6 CONCLUSIONS

Throughout this chapter we have emphasized the interchangeability between the black-box and analytical models. Although used in quite different ways, the fact that black-box data are frequently used to generate an analytical model, and analytical models are often used in black-box calculations, means that we need not make a clear distinction between the two in

[1] Some recent work includes that of Box and Lucas [1959], Carr [1960], Hunter and Mezaki [1964], Lapidus and Peterson [1965], Ball and Groenweghe [1966], Hunter and Atkinson [1966], Hunter [1967], Draper and Smith [1966], Hunter and Hoff [1967], Box and Hill [1967], Draper and Hunter [1967], Atkinson and Hunter [1968], Hill, Hunter, and Wichern [1968], Mezaki and Butt [1968], Kittrel and Erjavec [1968], Hunter, Hill, and Henson [1969].

optimization. We have also considered some of the details arising from such conversions.

If the system is lumped, then in the steady state an algebraic equation must normally be solved to obtain the outputs from a given set of inputs. In general, the method of interpolation is suitable for finding the roots of a complex expression which cannot be easily differentiated, and moreover, this method always converges, once the root has been straddled. These two features make the method particularly attractive, although the Newton-Raphson method, used when the first derivative is available, may converge more speedily. The Newton-Raphson method can also be extended to more than one dimension, whereas this is not possible with the former method.

If the system is distributed, or time-dependent-lumped, the output is determined by a differential equation, which must be integrated. Of the methods given here, the Runge-Kutta appears to be the most satisfactory, but other methods are available in the literature on numerical analysis.

The task of converting from black-box information to an analytical model is solved by fitting the available data to a polynomial using the least-square criterion. This method tends to give a fit of the true response surface even in the face of experimental error, and if the number of experiments is larger than the number of coefficients to be determined in the polynomial, the polynomial should be more reliable than a single experiment within the region of interest. The problem of surface fitting is greatly simplified if the experimental locations can be selected in advance so that the experimental design is orthogonal.

Problems

III-1 The volumetric flow through a valve is determined experimentally to be

$$Q = R(p_1 - p_2)$$

in which $p_1 - p_2$ is the pressure difference acting across the valve. Based on this empirical expression, obtain an analytical model representing the time-dependent behavior of the system shown in Fig. III-11. The cross-sectional areas of both tanks 1 and 2 are equal to A. If the level in tank 2 is the quantity of interest, develop a transformation equation for steady-state operation. Is this system a lumped- or distributed-parameter system?

III-2 The bank pays a rate of interest α dollars per unit of time. Assuming, for simplicity, that this interest is paid continuously, develop an analytical model representing the bank balance $S(t)$, in relationship to the rate of monetary input $q_i(t)$ and output rate $q_0(t)$. If at a certain point in time $S = 10^6$ dollars and q_i and q_0 are constant and equal to 5×10^4 and 10^5 dollars/year, respectively, find S after 20 years if $\alpha = 0.06$ dollar/year. How much is this value reduced if $\alpha = 0.05$ instead? (*Hint:* To solve this problem use a balance equation in the form proposed in Sec. 2.2.)

III-3 A moist solid is dried in contact with a dry warm stream of air. The solid moves through the dryer on a conveyor belt moving at U ft/min. If the drying rate is proportional to the moisture content of the solid, obtain a model expressing the moisture content

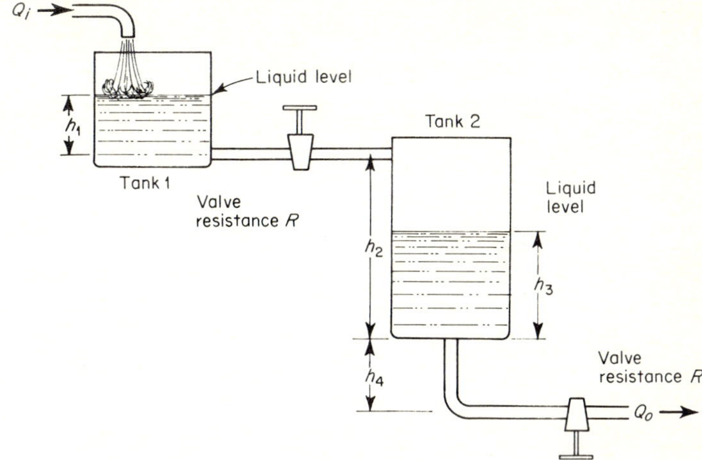

Fig. III-11 A series of storage tanks.

of the solid as a function of distance measured from the inlet of the dryer. Is this a lumped- or distributed-parameter system?

III-4 Suppose that, instead of the simple proportionality given in Prob. III-3, the rate of drying expressed as rate of water loss per pound of solid is

$$r = \alpha V(w - w^*)$$

where α, w^* = constants

V = component of velocity of air stream normal to direction of motion of the solids

w = mass of water per unit mass of solid

Using this rate expression, model the transient start-up of the dryer described in Prob. III-3. The partial differential equation can easily be solved by Laplace transforms.

III-5 The transformation equations relate the outputs to the inputs explicitly. Such a form may not be easily obtained. Consider, for example, the task of relating the heat loss per unit of length to the thickness of insulation surrounding a circular tube. Assume that the temperature of the inside surface of the insulation (this neglects temperature drop in the tube) is fixed, as is the surrounding air temperature. If the heat-transfer coefficient from the outer surface of the insulation to the air is given by

$$h = C(T_0 - T)^{1/4}$$

where C is a constant, T_0 is the outside surface temperature, and T is the air temperature and is constant, develop an expression relating the heat loss to the insulation thickness. Can this be put into the form of a transformation equation?

III-6 Expand $\sqrt{1 + 2x}$ about $x = 0$ by means of a Taylor series. What is the error for $x = 0.05$ if terms of second order and higher are neglected? Calculate the error if terms of third order and higher are neglected.

III-7 Expand $\ln x$ into a series about $x = 1$ by means of a Taylor series. What is the error at $x = 1.03$ if $\ln x$ is approximated, omitting the contribution of terms third-order and higher?

III-8 Suppose $y(x)$ is the solution of the differential equation

$$\frac{dy}{dx} = 2xy \qquad \text{with } y = 1 \text{ at } x = 1$$

Express y as a power series in $x - 1$, finding the first five a_i values in the expression

$$y(x) = \sum_{i=0}^{\infty} a_i(x-1)^i$$

III-9 Suppose that the function $y(x)$ is the solution of the differential equation

$$\frac{dy}{dx} = 3x^2 + y + 1 \qquad \text{with } y = 0 \text{ when } x = 0$$

Find the first seven values of a_i in the series

$$y(x) = \sum_{i=0}^{\infty} a_i x^i$$

III-10 Find a real root of the equation

$$xe^x - 2 = 0$$

III-11 Using the Newton-Raphson method, find to within four significant figures the root of

$$\sin x - \frac{x+1}{x-1} = 0$$

III-12 Using the method of interpolation, estimate the roots of

$$t^3 - 5t^2 - 6t + 4 = 0$$

III-13 Repeat Prob. III-12 using the Newton-Raphson method.

III-14 Find the roots of

$$x^3 - 3x^2 + x - 5 = 0$$

III-15 Compute approximate values of p at $l = 1.1$, 1.2, 1.3, and 1.4 for the differential equation

$$\frac{dp}{dl} = p + l \qquad \text{with } p = 0 \text{ at } l = 1$$

using the Euler method and a value of $h = 0.1$. Compare your results with the exact solution.

III-16 Repeat Prob. III-15 using the Runge-Kutta method.

III-17 Compute approximate values of p at $l = 0.1$ and 0.2 for the differential equation

$$\frac{dp}{dl} = p + l^2 \qquad \text{with } p = 1 \text{ at } l = 0$$

using the modified Euler method. Compare the approximate results with these obtained by integrating the differential equation.

III-18 Repeat Prob. III-17 using the Runge-Kutta method.

III-19 Obtain an approximation for the p defined by the differential equation

$$\frac{d^2p}{dl^2} + l\frac{dp}{dl} + p = l \qquad \text{with } p = \frac{dp}{dl} = 0 \text{ at } l = 1$$

Find values of p at $l = 1.1$, 1.2, 1.3, and 1.4. (*Hint:* Reduce the second-order equation to a pair of first-order differential equations by letting $dp/dl = Z$, say.)

III-20 Obtain an approximation, to slide-rule accuracy, for the p defined by the nonlinear differential equation

$$\frac{d^2p}{dl^2} + p\frac{dp}{dl} + p = l \qquad \text{with } p = \frac{dp}{dl} = 1 \text{ at } l = 0$$

Find values of p at $l = 0.1$, 0.2, 0.3, and 0.4.

III-21 Evaluate the integral

$$I = \int_0^1 y(x)\,dx \qquad \text{with } y(x) \equiv x^2$$

by a Monte Carlo method as follows: Plot the $y(x)$ function. Divide both the x and y axes (in the range $0 \leqslant x,\ y \leqslant 1$) into nine equal divisions. Looking in a phone book, obtain the last two digits of a number of consecutive listings (50 should do). These two digits locate an $x - y$ coordinate. If the coordinate falls beneath the $y - x$ curve, mark a yes; if not, no. The integral should be given by

$$I = \frac{\text{number of yes trials}}{\text{total number of trials}}$$

III-22 Suppose that an output p is given in terms of the inputs q_1 and q_2 as follows:

$$p = 4 \sin\left(\frac{\pi}{2} q_1 q_2\right) + q_1^2 + q_2^2$$

It is also known that q_1 can be either 0 or 1 with equal probability. If, however, q_1 is 0, then q_2 can be either 3 or 4 with equal probability, and if q_1 is 1, then q_2 is either 1 or 2 with equal probability. To compute precisely the expected value of p with such inputs is usually a difficult task. Use the Monte Carlo method to estimate the expected values of p. Two flips of a coin will serve to generate a pair q_1 and q_2. (For example, heads-heads could be equated to $q_1 = 1$, $q_2 = 4$.) After determining p about 25 times, find an estimate of the expected value. (Compare your value with $E\{p\} = 9$.)

III-23 Consider the plate shown. The surfaces of the plate are maintained at the indicated temperatures. Estimate the temperature of point A by the following Monte Carlo

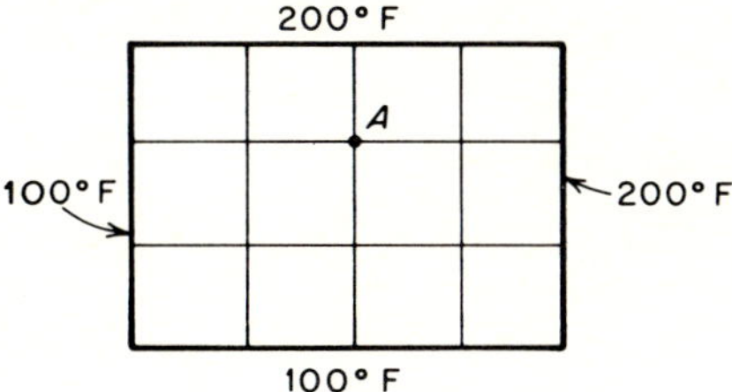

method: Flip a coin twice, the first flip dictating whether the move will be positive or negative and the second the direction x or y. Move to the neighboring node point indicated by the two flips and repeat the process. Each time a boundary is encountered, record the temperature of that boundary and return to A. The temperature at point A is determined by averaging the recorded temperatures.

III-24 Repeat Prob. III-24, but replace the surface nearest point A, which is now at 200°F, by an insulated boundary. Each time this boundary is encountered, rebound off it perpendicularly.

III-25 The response $\hat{p}$ of a black-box system to changes in the input x_1 is as follows:

i	$x_{1,i}$	$\hat{p}$
1	1	5
2	2	17
3	3	24
4	4	33
5	5	41

Represent this system by an expression of the form $\underline{p} = \beta_0 + \beta_1 x_1$ and compute the residual at each point.

III-26 Repeat Prob. III-25, with the curve fitting done in such a way that the line passes through the point $\hat{p} = 33$, $x_1 = 4$. Compute the residuals for each point. Compare the sum of the square of the residuals found here with that found in Prob. III-25.

III-27 Represent $\hat{p}$ as a cubic function, given the following data:

i	$x_{1,i}$	$\hat{p}_i$
1	−4	−35.1
2	−2	15.2
3	−1	16.0
4	0	8.8
5	1	0.1
6	3	0.2
7	4	21.4
8	6	135

Compute the residual at each point.

III-28 Fit the following data by the quadratic expression $\underline{p} = \beta_0 + \beta_1 x_1 + \beta_{11} x_1^2$, ensuring that the curve passes through the point $x_1 = 1, \hat{p} = 2$.

i	$x_{1,i}$	$\hat{p}$
1	−3	17
2	−2	8
3	−1	3
4	0	1
5	1	1
6	2	7
7	3	15

Compute the residuals at each point.

III-29 Hunter [1960] cites a numerical example exploring the response of a chemical reactor to changes in the residence time and the temperature. The values given by Hunter are as follows:

i	T, °C	θ, *min*	$\hat{p}_i$
1	160	60	60.3
2	150	50	54.3
3	150	70	60.3
4	170	50	62.2
5	170	70	68.0
6	160	60	62.3
7	150	50	56.5
8	150	70	61.7
9	170	50	64.6
10	170	70	66.8

Note that repeated experiments do not give the same response, indicating that experimental errors are included in the result.

The temperature is studied at 10°C intervals, and the residence time varies in increments of 10 min. This suggests a scaling of the experiments in the following way:

$$x_{1,i} = \frac{T_i - 160°\text{C}}{10°\text{C}} \qquad \text{and} \qquad x_{2,i} = \frac{\theta_i - 60\text{ min}}{10\text{ min}}$$

Tabulate the experimental results with respect to the system response to changes in the scaled variables, and determine by least squares the parameters in the equation

$$\underline{p} = \beta_0 + \beta_1 x_1 + \beta_2 x_2$$

III-30 Repeat Prob. III-29, but force the reactor response to be 61.3 when the temperature is 160°C and the residence time is 60 min.

III-31 Is the experiment described in Prob. III-29 an orthogonal one?

III-32 Is the following experimental design orthogonal for plane surface fitting?

i	$x_{1,i}$	$x_{2,i}$	$x_{3,i}$
1	1	1	1
2	−1	1	1
3	−1	−1	1
4	−1	−1	−1
5	1	1	−1
6	1	−1	−1

III-33 Suppose that the response of an experiment is given by $p = 2 + 3x_1 - 2x_2 + x_3$. Using a two-level experiment design, tabulate the response of this system if each experiment is assumed to be in error by ±10 percent, the sign being selected by a flip of the coin.

Thus, if $x_1 = x_3 = x_2 = 1$, the response would be 4 ± 0.4, the sign being determined by flipping a coin.

After compiling the set of "experiments," determine, using the tabulated results, the constants in the equation

$$\underline{p} = \beta_0 + \beta_1 x_1 + \beta_2 x_2 + \beta_3 x_3$$

The difference between the original true equation and that found by fitting the response data is due entirely to experimental error. This difference could, of course, be reduced by further experimentation.

III-34 Using a three-factor central-composite design (Table III-14), tabulate the response of a system whose transformation equation is

$$p = 2 + 3x_1 + 2x_2 - 3x_3 + x_1 x_2^2$$

Fit the results with a second-degree polynomial.

III-35 Repeat Prob. III-34, adding ± 5 percent error, the sign of the error being determined by the flip of a coin.

PART TWO

OPTIMIZATION TECHNIQUES

In general, there are two approaches to the problem of optimizing a function subject to various restrictions. After a preliminary discussion of functions and search regions, we consider, successively, the analytical and numerical approaches to the optimization of one-dimensional, multidimensional, and infinitely dimensional (continuous variables) functions.

CHAPTER IV

FUNCTIONS, REGIONS, AND OPTIMIZATION

1 INTRODUCTION

Let us now turn to the general problem of finding the optimum. The first part of this text was devoted to the development of an objective function representing vastly different applications, each involving different degrees of complexity. However, once the objective function y has been defined in terms of the independent variables $\mathbf{x}$, where $\mathbf{x}$ represents all the variables $x_1, x_2, \ldots, x_n$, as

$$y = y(x_1, x_2, \ldots, x_n) = y(\mathbf{x})$$

subject to m restrictions of the form

$$g_j = g_j(\mathbf{x}) = 0$$

or

$$g_j = g_j(\mathbf{x}) \leq 0$$

the problem is independent of the actual application. Here $g_j(\mathbf{x})$ can be a variable, a function, or a functional as described in Chap. II, Sec. 6. The sole considerations are now mathematical, and the optimization techniques to be used are determined by the mathematical structure of the objective function and the associated restrictions. Let us now examine some general features of the objective function which are important in achieving an optimal state.

2 FUNCTIONS

It might seem superfluous to give consideration to functional continuity and to discontinuities in functions or their derivatives. However, these concepts are basic to the problem of optimization, and it is therefore essential to come to an agreement on terminology. Before discussing the pertinent features of functions, let us consider the problem of representation.

2.1 Representation of a Function

A function can be represented in several ways. In general, the dependent variable is expressed in terms of the independent variables, $y(\mathbf{x})$ being said to

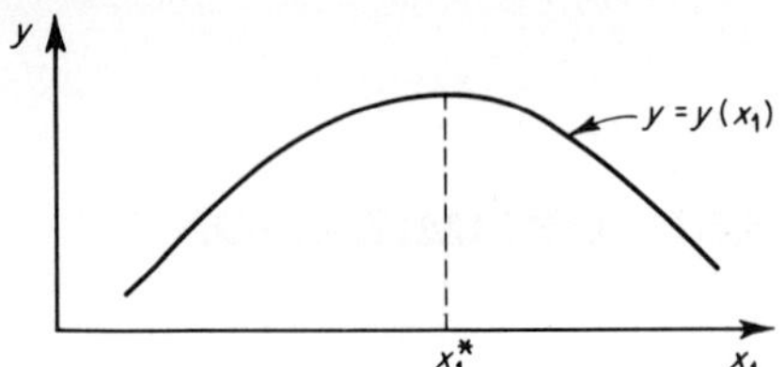

Fig. IV-1 One-dimensional representation of a unimodal function which is also concave.

be a function of $\mathbf{x}$, defined in some region, if we have any rule which allows us to determine y for any set of $\mathbf{x}$ in this region. Any analysis of the function is then carried out by examining this expression.

One approach is by the use of graphical methods. If this is possible, it allows an insight into the nature of the function, which is most valuable in any study, since little can replace a feeling for the type of result expected. In the graphical approach, the independent variables $x_1, x_2, \ldots, x_n$ can be viewed as coordinates in an n-dimensional space, and the objective function is some mapping of these coordinates. This is easily shown for one and two dimensions, as indicated in Figs. IV-1 and IV-2, respectively. In the former, the function $y(x_1)$ is a curve, while in the latter $y(x_1, x_2)$ can be readily visualized as a surface. In both cases the "n-dimensional surface" is generated by calculating the value of y for each set of values of the independent variables. The shape of the resulting response surface is quite important, not only in determining the optimum, but in examining the sensitivity of the optimum to changes in the independent variables. An alternative representation is the contour map shown in Fig. IV-3 for a surface in two variables. Here the equation $y(\mathbf{x}) = K$, a constant, defines the relationship between x_1 and x_2,

Fig. IV-2 Two-dimensional surface representation.

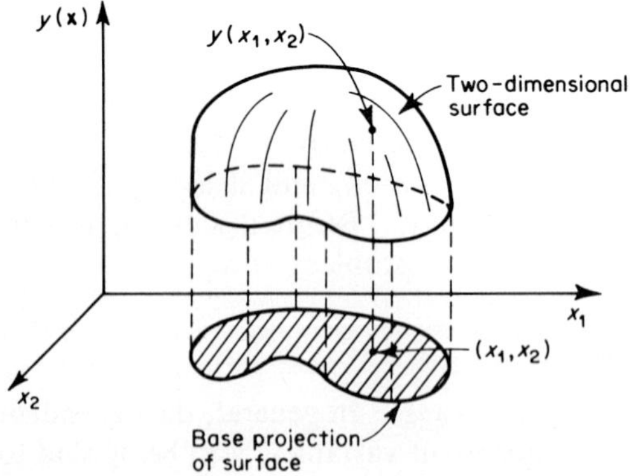

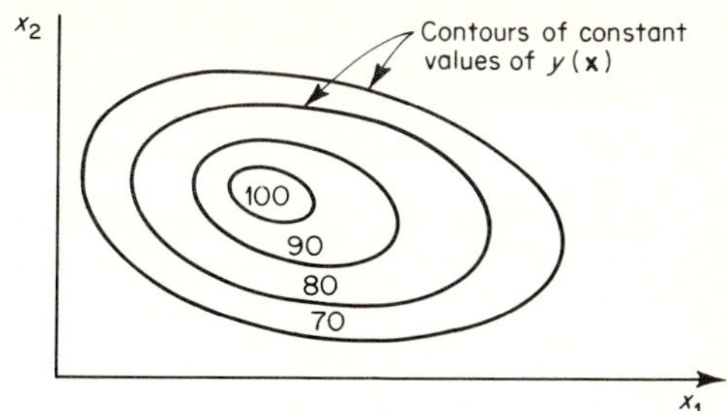

Fig. IV-3 Two-dimensional contour representation.

shown by a curve or contour on the map. A representative set of these curves gives a very useful picture of the function, it being noted that Fig. IV-1 is equivalent to a possible cross section through Fig. IV-3 for a given x_2. In the general case, however, the desirable geometric visualization becomes impossible, and we must fall back on algebraic or other methods. These, however, do not allow such an easy interpretation of a function.

Although higher-dimensional systems cannot be represented geometrically, it is convenient to discuss and visualize the objective function as a geometric surface—a practice which seems to be derived from the work of statisticians. This lends considerable simplification to the concepts, even if the mathematical labor in optimization is not diminished, and allows one to discuss the scaling of mountains or the descent into valleys in order to attain the optimum at the mountain peak or valley bottom, even if these are in an n-dimensional terrain. In general, within this text, most procedures will be illustrated by a simplified geometric representation to indicate the significant features.

2.2 Function Continuity

The deterministic objective function y, where $y = y(\mathbf{x})$, implies a definite value of the function for any given set of variables $\mathbf{x}$. This function may be continuous, discontinuous, or discrete in form, each of these forms giving rise to peculiarities in the search for an optimum. The procedure in using a stochastic objective function has been suggested in Chap. II, Sec. 7.1, and will not be considered separately.

2.2.1 Continuous functions

A function may be termed continuous at some point $\mathbf{x}$ if

$$y(\mathbf{x}) = \lim_{\mathbf{h}\to 0} y(\mathbf{x} + \mathbf{h})$$

for all possible ways in which all the **h** can approach zero. This is not a rigorous definition, for which texts such as Brand [1958, p. 89] should be consulted, but it will suffice for our purposes. In the case of a single variable, this definition simply implies that the function may be represented by an unbroken curve in the region of interest, as shown in Fig. IV-1.

2.2.2 *Discontinuous functions*

The most common form of discontinuity in a function is the jump discontinuity, which is represented in one dimension in Fig. IV-4. One discontinuity is shown at $x = x_0$, and for the limit of $y(x + h)$ as $h \to 0$ with $h \geq 0$, the y approaches one value, $y(x_0^+)$. For the limit of $y(x + h)$ as $h \to 0$ with $h \leq 0$, the objective function approaches a different value, $y(x_0^-)$. The function y has therefore a different value at the discontinuity, x_0, depending upon the direction of approach.

In most physical processes, change always take place at a finite rate, so that, physically, a discontinuous function is unlikely. In such processes, however, as the slicing of a filter cake from the cloth of a continuous drum filter, the change from final thickness to zero thickness is very rapid in comparison with the slow buildup of cake, so that we might well think of it as a jump discontinuity, as shown in Fig. IV-4. A similar situation would be found in a storage warehouse in which the inventory of finished stock would be continuously renewed by production but the removal of loads by rail or road might be considered as an instantaneous reduction in stock inventory. In both illustrations, if the changes were not so rapid, discontinuities in the derivative of the function would be experienced at the beginning and end of the change, but the function itself might not be discontinuous. Such a situation is also shown in Fig. IV-4, the discontinuity in the derivative resulting in a series of high and low peaks in the function. It will be seen that, as

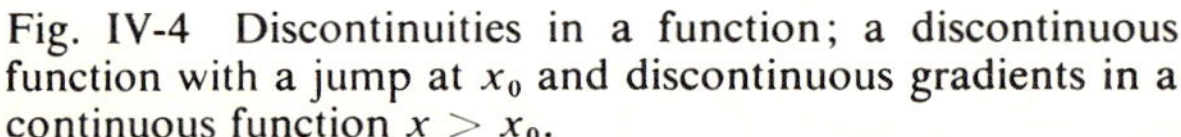

Fig. IV-4 Discontinuities in a function; a discontinuous function with a jump at x_0 and discontinuous gradients in a continuous function $x > x_0$.

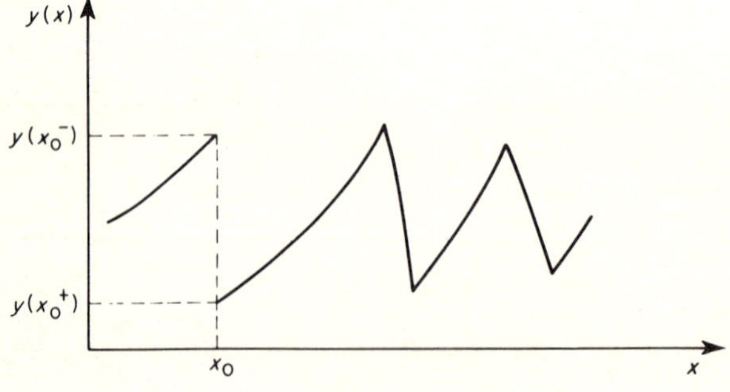

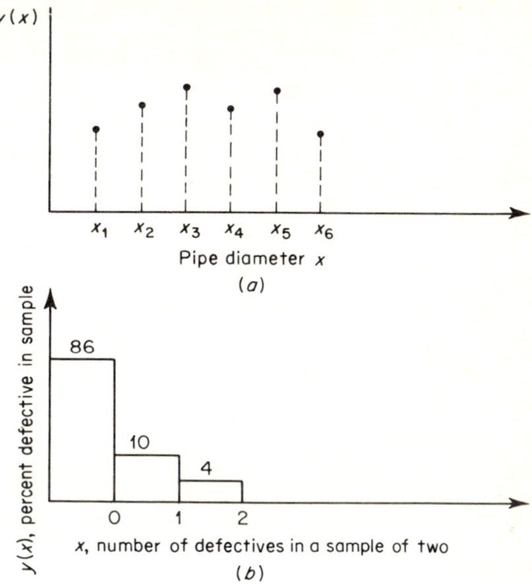

Fig. IV-5 Discrete-valued functions.

the period of change decreases, adjacent pairs of discontinuities in the derivative will come together, yielding the jump discontinuity of the function at, for example, the point of cakeremoval or stockdispatch dates.

2.2.3 *Discrete functions*

Another form of discontinuous function is found when the function is valid only at discrete values or sets of values of the variables. Thus, for example, if some function y, such as the pressure drop through a fixed pipe length when fluid is passed at a fixed flow rate, is plotted against the pipe diameter, a continuous function would not result, since pipes are normally available only in standard diameters. A discrete-valued function such as shown in Fig. IV-5*a* would thus be found. Another illustration would be found in plotting a histogram of the results of a test in which the number of defective items in a sample of a certain size was plotted as shown in Fig. IV-5*b*.

2.3 **Function Behavior**

We may define certain terms which are useful in describing the behavior of functions.

A function is said to be *monotonic-increasing* along a given path when $y(x_2) > y(x_1)$, and *monotonic-decreasing* when $y(x_2) < y(x_1)$, both for all

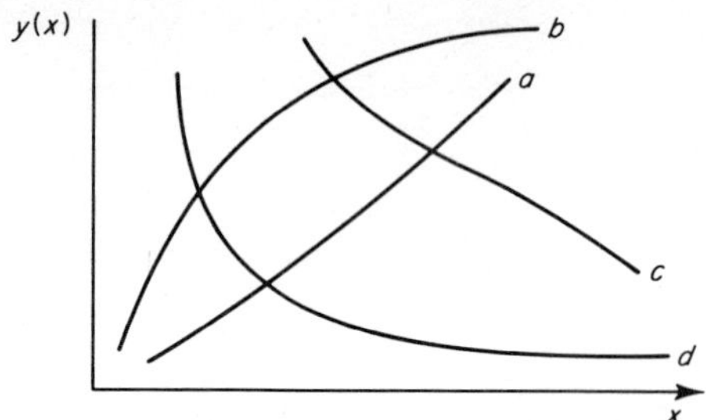

Fig. IV-6 Monotonic functions. (*a*) Monotonic-increasing; (*b*) monotonic-nondecreasing; (*c*) monotonic-decreasing; (*d*) monotonic-non-increasing.

$x_2 > x_1$. For the situation in which $y(x_2) \geq y(x_1)$ and $y(x_2) \leq y(x_1)$ for $x_2 > x_1$, these are, respectively, called *monotonic-nondecreasing* and *monotonic-nonincreasing*. Here a path is defined as being any continuous curve between any two points, both the points and the curve lying wholly within the surface region. The function $y = x^2$ is thus monotonic-increasing for $x > 0$, but is monotonic-decreasing for the region $x < 0$. A practical illustration may be found in filtration, where the thickness of the cake may be represented by y and time by x, so that the temporal change of cake thickness may then be represented by the monotonic y-x relation, as shown in Fig. IV-6, curve b. An example of a monotonic-nonincreasing function would be the variation in height of liquid in a tank with time, after flow commences through a discharge orifice in the base of the tank (curve d of Fig. IV-6).

If the independent variables are altered in a given direction,[1] and if the resulting values of the function are found initially to increase up to a maximum and then to fall away again, the function is termed a *unimodal function.* Similarly, it may fall to a minimum and then rise upon further change of the system variables. In such functions, as the name suggests, there is only a single peak. A practical example is found in the determination of the product yield from a first-order reversible reaction, $A \rightleftharpoons B$. As shown in Fig. IV-1, the yield of B, y, is a unimodal function of temperature, x. A very common form of unimodal function is found in the economic evaluation of a system when the limiting envelope of the profit function is plotted against the system investment, as shown, for example, in Fig. II-1. Figure IV-3 illustrates a unimodal function in two dimensions.

Although functions with more than one local maximum (or minimum) are found, a system with two such points or peaks being called *bimodal* (Fig. IV-7), such functions must be examined region by region, the regions being selected so that the objective function is likely to be unimodal in each. Some

[1] Direction is defined in Chap. VIII, Sec. 1.4.

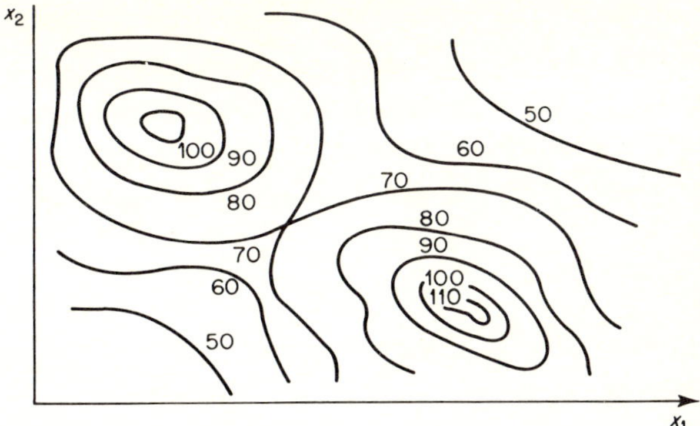

Fig. IV-7 Bimodal function in two dimensions.

examples of unimodal functions are indicated in Fig. IV-8. While all only have one optimum, a global value, in the interval under examination, the unimodal function will be seen to include not only the smooth continuous concave case of Fig. IV-8*a*, but also the discontinuities of function or gradient, including the discrete-valued functions of Fig. IV-8*d*. The assumption of unimodality is not very restrictive and holds in many cases, being most useful in the application of numerical search techniques. Since, in a unimodal function, the value of the dependent variable strictly decreases (increases) as the value of the independent variables changes in any direction from the

Fig. IV-8 Unimodal functions; note that (*a*) is also concave and (*b*) is convex for $x < x_1$ and concave for $x > x_1$.

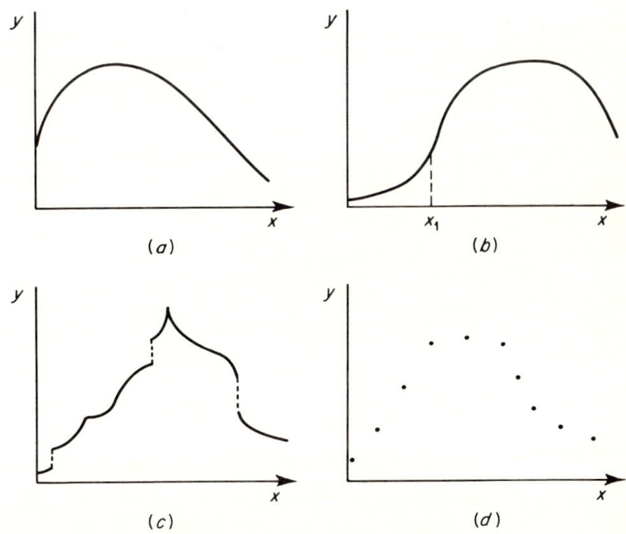

maximum (minimum), unimodality can be defined in the one-dimensional case as

$$\begin{aligned} y(x_{1,1}) < y(x_{1,2}) < y(x_1^*) \qquad & \text{if } x_{1,1} < x_{1,2} < x_1^* \\ y(x_1^*) > y(x_{1,3}) > y(x_{1,4}) \qquad & \text{if } x_1^* < x_{1,3} < x_{1,4} \end{aligned} \tag{2.3-1}$$

for a maximum at x_1^*, holding for all admissible values of x_1, a similar definition holding for a minimum (Bellman and Dreyfus [1962, p. 152]). In the general case, we can say that a function $y(\mathbf{x})$ is unimodal if, for every pair of points $\mathbf{x}_1$ and $\mathbf{x}_2$ in the valid region, there exists a path from $\mathbf{x}_1$ through $\mathbf{x}^*$ to $\mathbf{x}_2$ along which the function is unimodal according to the definition of Eq. (2.3-1), where the $x_{1,i}$ must be replaced by a parameter measuring the distance along the path. When the function is unimodal along all straight lines passing through the optimum, it is termed *strongly unimodal* (Wilde [1964]). In the n-dimensional problem, we shall normally consider only the concave and convex forms of unimodal function.

Concave and *convex functions* are special cases of unimodal functions. Thus, while Fig. IV-8*a* illustrates a concave unimodal function, the functions of Fig. IV-8*b* to *d* are not concave. This can be seen by considering the definition of a concave function (Margenau and Murphy [1964, p. 209])

$$y[(1-\theta)\mathbf{x}_1 + \theta\mathbf{x}_2] \geq (1-\theta)y(\mathbf{x}_1) + \theta y(\mathbf{x}_2) \tag{2.3-2}$$

for all $0 \leq \theta \leq 1$ and any two points $\mathbf{x}_1$ and $\mathbf{x}_2$ in n-dimensional space. A convex function will have the same form of definition, but with the inequality sign reversed. Intuitively, the surface $y(\mathbf{x})$ is concave if a line segment joining any two points on the surface lies on or below the surface. (Similarly, for a convex function, the line would lie on or above the surface, as indicated in Fig. IV-9.) This can be shown to be true by developing a somewhat more restrictive concavity definition. Expanding the left-hand side of

Fig. IV-9 A convex function showing the secant gradient to be greater than the function gradient.

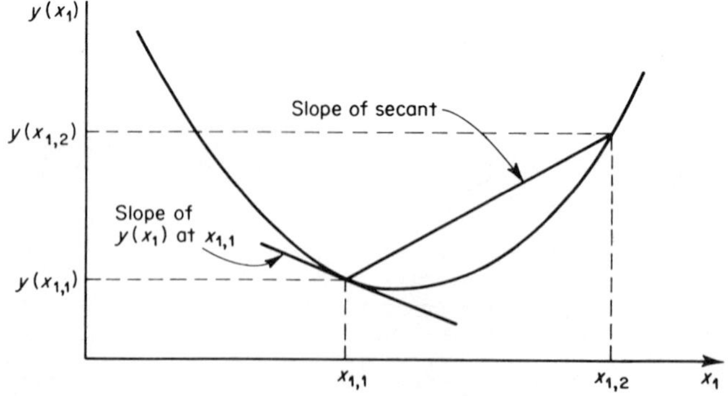

inequality (2.3-2), using a Taylor expansion (Chap. VII, Sec. 2.2), we get, for small values of θ, that

$$y[(1-\theta)\mathbf{x}_1 + \theta\mathbf{x}_2] = y[\mathbf{x}_1 + \theta(\mathbf{x}_2 - \mathbf{x}_1)]$$
$$= y(\mathbf{x}_1) + \theta \sum_{i=1}^{n} \left(\frac{\partial y}{\partial x_i}\right)_{\mathbf{x}_1} (x_{i,2} - x_{i,1})$$

provided that the second-order terms in the expansion can be neglected. Substituting this approximation into inequality (2.3-2) then gives the more restrictive condition of concavity as

$$y(\mathbf{x}_2) \leq y(\mathbf{x}_1) + \sum_{i=1}^{n} \left(\frac{\partial y}{\partial x_i}\right)_{\mathbf{x}_1} (x_{i,2} - x_{i,1}) \qquad (2.3\text{-}3)$$

A similar form holds for convex functions, but with the inequality sign reversed. Thus, in one dimension, we have for the convex case

$$y(x_{1,2}) \geq y(x_{1,1}) + \left(\frac{\partial y}{\partial x_1}\right)_{x_{1,1}} (x_{1,2} - x_{1,1})$$

or

$$\left(\frac{\partial y}{\partial x_1}\right)_{x_{1,1}} \leq \frac{y(x_{1,2}) - y(x_{1,1})}{x_{1,2} - x_{1,1}} \qquad \text{for } x_{1,2} > x_{1,1}$$

Thus the slope of y at the point $x_{1,1}$ is less than the slope of the secant through $x_{1,1}$ and $x_{1,2}$, as indicated in Fig. IV-9. This confirms our earlier intuitive statement on convexity and concavity of functions.

While the function shown in Fig. IV-8*b* is a unimodal function, it is in fact neither a concave nor convex function. However, in regions it is locally convex (region $0 \leq x_1 \leq x_{1,1}$) or locally concave (region $x_1 \geq x_{1,1}$). A linear function is both concave and convex, although it is neither strictly concave nor convex, since, for the latter conditions, the defining inequality would not include the equality condition.

Example IV-1: A concave quadratic function

Let us examine the form of the quadratic function

$$y(\mathbf{x}) = 20 + 25x_1 + 5x_2 - 6x_1{}^2 - 3x_2{}^2 \qquad (2.3\text{-}4)$$

Now, for any two points $\mathbf{x}_1$ and $\mathbf{x}_2$,

$$\theta y(\mathbf{x}_2) = \theta(20 + 25x_{1,2} + 5x_{2,2} - 6x_{1,2}^2 - 3x_{2,2}^2)$$
$$(1-\theta)y(\mathbf{x}_1) = (1-\theta)(20 + 25x_{1,1} + 5x_{2,1} - 6x_{1,1}^2 - 3x_{2,1}^2)$$
$$\begin{aligned} y[(1-\theta)\mathbf{x}_1 + \theta\mathbf{x}_2] &= y[\mathbf{x}_1 + \theta(\mathbf{x}_2 - \mathbf{x}_1)] \\ &= 20 + 25[x_{1,1} + \theta(x_{1,2} - x_{1,1})] \\ &\quad + 5[x_{2,1} + \theta(x_{2,2} - x_{2,1})] \\ &\quad - 6[x_{1,1} + \theta(x_{1,2} - x_{1,1})]^2 \\ &\quad - 3[x_{2,1} + \theta(x_{2,2} - x_{2,1})]^2 \end{aligned}$$

so that, after simplification, we can write

$$A \equiv y[(1-\theta)\mathbf{x}_1 + \theta\mathbf{x}_2] - \theta y(\mathbf{x}_2) - (1-\theta)y(\mathbf{x}_1)$$
$$= -\theta(1-\theta)[-6(x_{1,2} - x_{1,1})^2 - 3(x_{2,2} - x_{2,1})^2]$$

Since $0 \le \theta \le 1$, then $\theta(1-\theta) \ge 0$, and hence $A \ge 0$. However, from inequality (2.3-2), this is the condition for concavity. The quadratic function given by Eq. (2.3-4) is thus concave.

It can be shown (for example, see Hadley [1964, p. 85]) for functions which can be expressed in the separable form

$$y(\mathbf{x}) = \sum_{j=1}^{k} y_i(\mathbf{x})$$

that if the $y_i(\mathbf{x})$ are concave (convex) functions, the function $y(\mathbf{x})$ is also concave (convex).

Example IV-2: The form of separable functions

Let us show that

$$y = 3x_1^2 + 8x_2^2 + 9x_3^2 + x_4^2 \tag{2.3-5}$$

is a convex function.

We can write this in the form

$$y(\mathbf{x}) = \sum_{i=1}^{4} y_i(\mathbf{x})$$

where $y_i(\mathbf{x}) = A_i x_i^2$, A_i being a positive constant. Now y_i can be shown to be a convex function by use of inequality (2.3-2), since

$$y_i[(1-\theta)\mathbf{x}_1 + \theta\mathbf{x}_2] - \theta y_i(\mathbf{x}_2) - (1-\theta)y_i(\mathbf{x}_1)$$
$$= A_i[(1-\theta)x_{i,1} + \theta x_{i,2}]^2 - A_i\theta x_{i,2}^2 - A_i(1-\theta)x_{i,1}^2$$

which becomes upon reduction

$$= -A_i(1-\theta)\theta(x_{i,1} + x_{i,2})^2 \le 0$$

Since each term on the right-hand side of Eq. (2.3-5) is convex, the sum of these terms must thus also be convex, providing the desired solution.

3 REGION OF SEARCH

There are very few problems in optimization in which the search is unrestricted, the restrictions being imposed upon the choice of values of the independent variables or upon the objective function itself. In general, the restrictions can be expressed in the form

$$g_k(\mathbf{x}) = 0 \qquad \text{for } k = 1, 2, \ldots, l \tag{3-1}$$

and

$$g_k(\mathbf{x}) \le 0 \qquad \text{for } k = l+1, l+2, \ldots, m \tag{3-2}$$

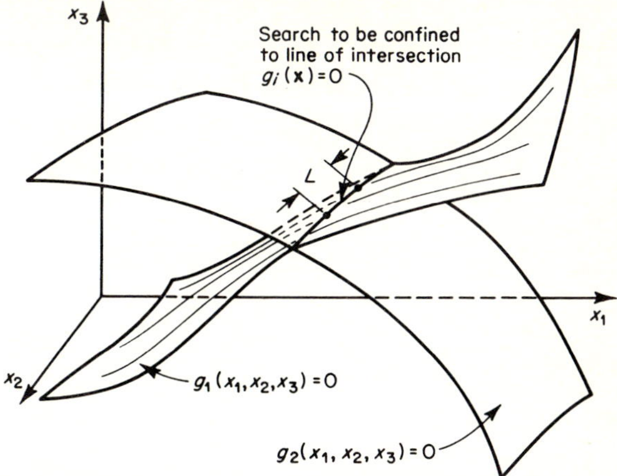

Fig. IV-10 Restricted region of search imposed by equalities.

i.e., the restrictions are clearly either equalities or inequalities. The equalities force us to select values of $\mathbf{x}$ which lie along the intersections of the l surfaces in an n-dimensional space. This situation is illustrated in Fig. IV-10, where two such surfaces are shown in a three-dimensional space. Of course, if the desired solution must lie on both surfaces simultaneously, then clearly the points along the line of intersection are the only admissible points. The imposition of these two equality restrictions therefore limits the search to a line. If, in addition to these equalities, two inequality constraints are imposed, then the search may well be restricted to a portion of the line, say, L in Fig. IV-10. Thus, with two equality restrictions and two inequalities, the region of the search is a line segment. The line is segmented by the boundaries imposed by the inequality restrictions. If the problem is well posed, the region of search is not empty; i.e., the surfaces must intersect at least once, or else the two equalities cannot be simultaneously satisfied, and the system is overdefined.

A second situation is depicted in Fig. IV-11. Here a single equality restriction is imposed, forcing the search to be confined to a surface. An additional inequality restriction gives rise to boundaries, leaving the final region of search as an area contained upon the surface. If the inequality restriction is identically satisfied, that is, $g_2 = 0$, then the required location is said to be on the boundary of the region.

The type of region defined by any set of restrictions is important. A *convex region* is one such as that shown in Fig. IV-12*a*. In this type of region, the straight line joining any two points lying on the region boundaries must lie entirely within the region. A region such as that depicted in Fig. IV-12*b* or *c* obviously does not meet this condition, and might be termed a

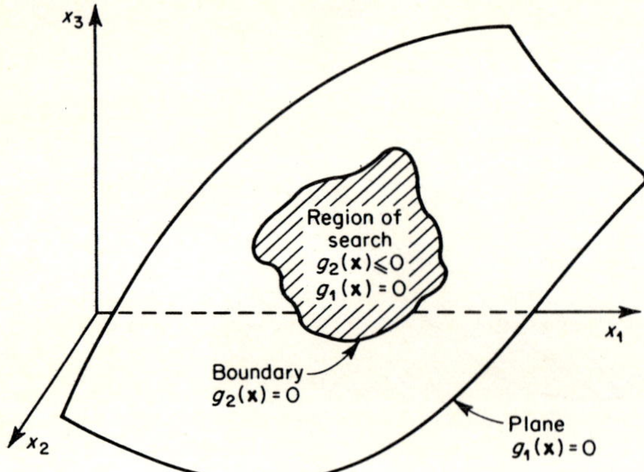

Fig. IV-11 Restricted region of search imposed by equality and inequality constraints.

Fig. IV-12 Regions. (*a*) A convex region formed from a set of constraints $\sum_{i=1}^{n} a_{j,i}x_i \leq b_j$; (*b*), (*c*) nonconvex regions.

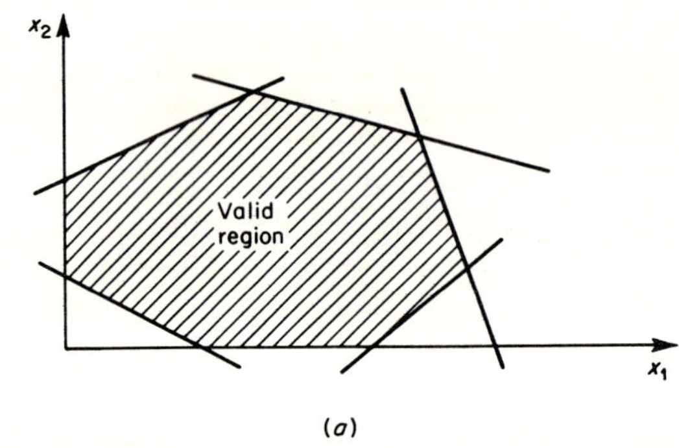

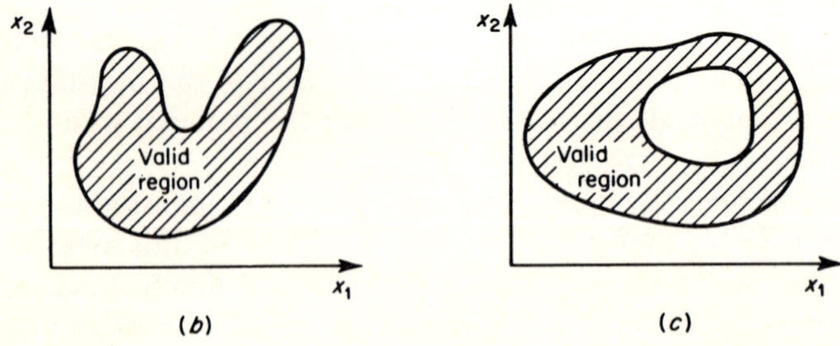

nonconvex region. The distinction between such regions in optimization will be shown in Sec. 4.4.

4 CONDITIONS FOR OPTIMUM VALUES

Optimum conditions can be found only at particular points within a region of search. In searching for the optimum of a certain function, it is thus only necessary to search among these specific locations. It is therefore important to examine all feasible locations for an optimum, and in the final analysis compare the values of the objective function at each feasible point with each other.

4.1 Continuous Functions

As long as a function is continuous, there will be a maximum or minimum value of that function, either in the interior of the permitted region or at the boundary of that region. The existence of a maximum and a minimum is guaranteed by a theorem of Weierstrass:[1] Every function which is continuous in a closed domain possesses a largest and smallest value either in the interior or on the boundary of that domain. The validity of this statement is intuitively evident on looking, for example, at the one-dimensional examples shown in Fig. IV-13. Clearly, not all these internal situations are suitable for treatment by the differential calculus, but in general the location of the optima in the interior of a region may be determined by the theorem: A continuous function $y(\mathbf{x})$ of n independent variables, $x_1, x_2, \ldots, x_n$, will attain a maximum or minimum in the interior of a region only at these values of the variables, x_i, for which either the n partial derivatives, y_{x_i}, vanish simultaneously (a stationary point) or at which one or more of these

[1] See, for example, Courant and Hilbert [1953, p. 164].

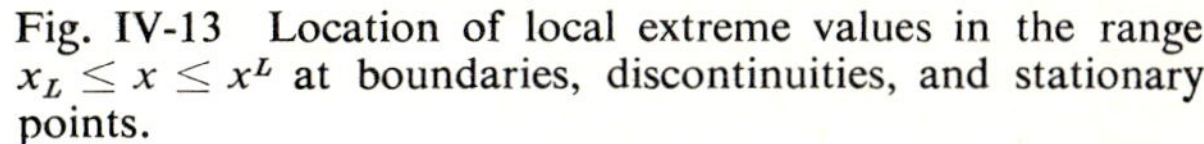
Fig. IV-13 Location of local extreme values in the range $x_L \leq x \leq x^L$ at boundaries, discontinuities, and stationary points.

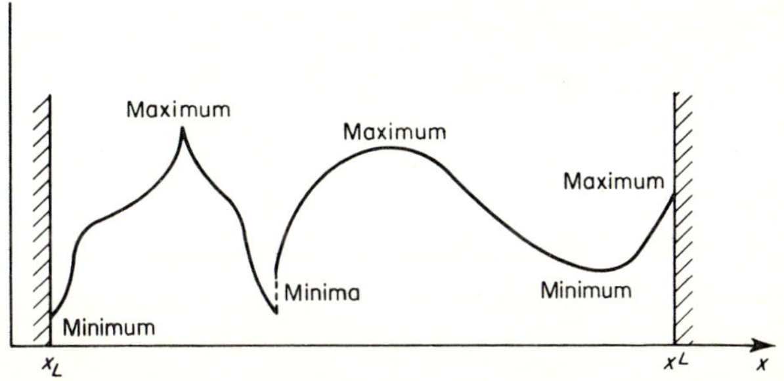

derivatives cease to exist at a discontinuity (Edelbaum [1962] and Sokolnikoff and Redheffer [1966]).

We therefore require to look for an optimum value only under three possible conditions:

1. In the interior of a region at a stationary point where all the continuous first derivatives are simultaneously zero.
2. In the interior of the region at discontinuities of one or more first derivatives.
3. At the boundary to the region.

Let us now look at each of these conditions.

4.1.1 The stationary point

The search for an optimum of a continuous function having continuous derivatives in the range of interest may be by analytical or numerical search methods. In the first, the stationary points are found by the simultaneous solution of the n equations obtained by setting the n partial derivatives equal to zero. Having attained this stage, it is then necessary to examine each set of points to find out if, in fact, the stationary points are extreme values, some examples of stationary points being shown in Fig. IV-14, not all of which are extrema. This problem will be examined further in both Chaps. V and VII. Methods of searching numerically for the optimum region are discussed in Chaps. VI and VIII.

4.1.2 Interior discontinuities

Extreme values may also be found at points in the interior of the region at which the partial derivatives are discontinuous. Here again, discontinuities need not necessarily provide extreme values, as will be seen from Fig. IV-8c. Such discontinuities are often handled by placing artificial boundaries at them and then treating the system in parts.

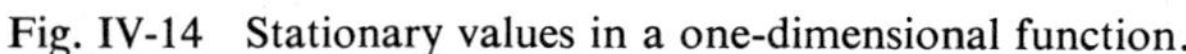

Fig. IV-14 Stationary values in a one-dimensional function.

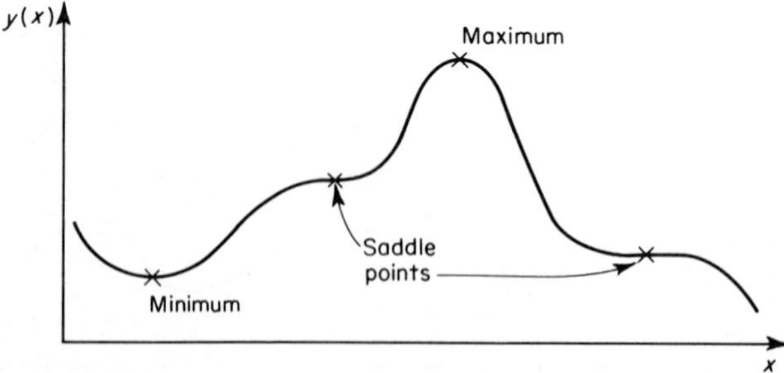

4.1.3 Boundaries

When the search for an optimum position is being conducted along the boundary of a region, as would be the situation when handling monotonic functions, we may use some search method or, by modifying the problem, incorporate the boundary restrictions in some modified objective function and use analytical methods. Such boundaries arise, of course, from the restrictions of the form $g(\mathbf{x}) \leq 0$ or $g(\mathbf{x}) = 0$ (Chap. II, Sec. 6). It has been noted by Edelbaum [1962] that if we are dealing with an n-dimensional problem, the search for an optimum will generally lead to several problems in which the boundary restrictions produce searches in $n - 1, n - 2, \ldots, 2, 1$ dimensions. Thus, for example, if we have a region in three dimensions defined by the surface area of a cube, we must search the sides in two dimensions, the edges in one dimension, and the interior in three dimensions, while if the restricted region were defined by a sphere, we should have to search only one surface and in the interior.

4.2 Discontinuous Functions

In this type of problem, as illustrated for one dimension by Fig. IV-4, it is best to divide up the search region by placing additional boundaries at the discontinuities, and then using the above methods for handling the continuous sections of the function. These sectional results can then be compared with each other and with the function values at the imposed boundaries.

4.3 Discrete Functions

Here we must normally compare each of the values of the discrete-valued function at each point for which the function exists. If the general form of the function is, however, known in advance, some economy in the search can be effected, for instance, by the use of unimodality in logical searches in one dimension.

4.4 Multiple Extreme Values of a Function

It is obvious that, with certain functions (for example, that shown in Fig. IV-7 or in Fig. IV-14), more than one extreme value will be found. These will be local extreme values, since small variations in the defining variables, $\mathbf{x}$, will cause the objective function value to be less desirable than that at the optimum point. Larger variations might, however, cause improvements in the objective function, since some local extreme values will prove better than others. For the general case, therefore, we must not only

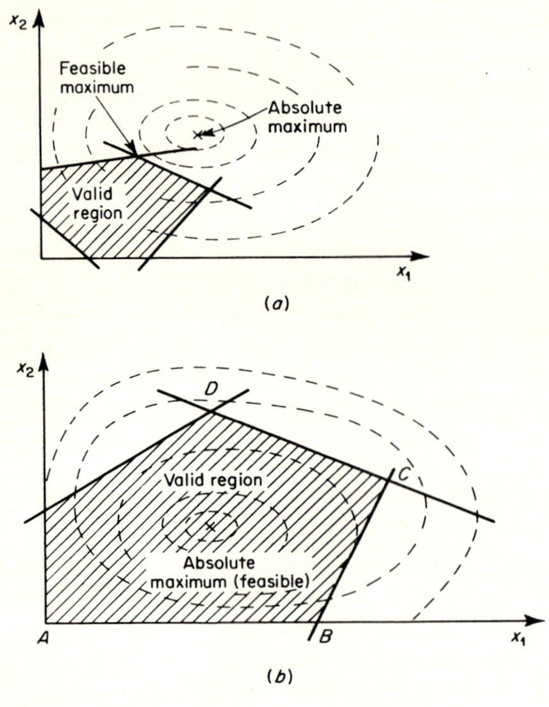

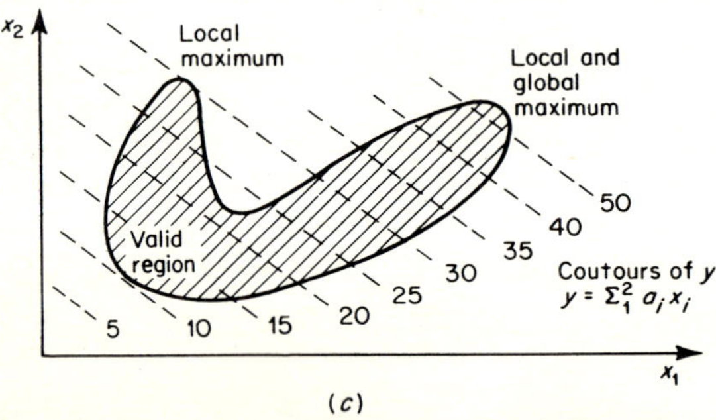

Fig. IV-15 Relationship between functions and regions in optimization. (*a*) Maximization of a concave function, with the absolute maximum outside the valid convex region; (*b*) maximization of a concave function, with the absolute maximum inside the valid convex region and with local minima at *A*, *B*, *C*, and *D*; (*c*) maximization of a linear function, with a nonconvex valid region.

determine the local extreme values, but compare them, so that a globally valid result is found.

This comparative approach is not always necessary, since, with certain combinations of function and restricted-region behavior, there will be only a single optimum. Let us first consider the problem of searching a concave function restricted by a convex region. It will be seen from the two-dimensional illustration in Fig. IV-15*a* or *b* that only one maximum will be found in the system. This must thus be the global maximum. Similarly, for a convex function in a convex region, there will only be a single minimum. On the other hand, more than one minimum can exist in a concave function, as seen in Fig. IV-15*b*, and similarly, more than one maximum can be found for a convex function. We can thus conclude that a global optimum can be found directly in the minimization of a convex function, or the maximization of a concave function, when the region of search is convex. The linear function is a special case, since it is both a convex and concave function, its global optimum being always directly found at the boundary when the region is convex. Such knowledge of well-behaved functions can prove advantageous in devising search routines.

We can see that a function which is suitable for maximization need not be well behaved for minimization, or vice versa. Thus the minimization of concave functions and the maximization of convex functions are very likely to involve more than one optimum. Objective functions of the wrong configuration can lead to the same problems, involving local optima, as are found with nonconvex regions. The use of nonconvex regions can produce local optima even if the functions are well behaved. Thus, as shown in Fig. IV-15*c*, more than one local maximum might be found when a nonconvex region restricts the search of a linear function.

5 APPROACH TO OPTIMIZATION

It will be seen, after perusal of Part Two, that there is no general method of calculation for handling a search for an optimum value. In fact, many methods have been developed solely for the efficient solution of a given class of problem. We should therefore not only concern ourselves with the model and simulation of our problem, but should, in addition, investigate the class of problem encountered. This may well, in many cases, dictate the types of optimization technique available or the most efficient type, if more than one is feasible.

5.1 Optimization Techniques

The necessity of determining a maximum or minimum in optimization might lead one to think of the elementary differential calculus approach to the problem of finding a stationary value of the function by considering the variation of the independent variables under the control of the designer or operator. However, in most realistic problems, the situation is normally complicated, due to the shape of a valid search region created by the system restrictions; the complexity and behavior of the functions under examination, and the possibility of certain functions being noncontinuous or nondifferentiable in the range of interest.

The general problem of finding the largest or smallest value of a function of one or many variables, when the function may be linear or nonlinear, discrete-valued or continuous, exact or stochastic, restricted or unconstrained, is a very difficult one, and is still under development. For this reason many types of optimization methods have been suggested, usually with a particular group of applications in mind. In general, however, the optimum can exist only at one or more of the following conditions, since it cannot exist elsewhere: a stationary point, a discontinuity in the gradient or the function, or a boundary. We should note, however, that these points do not all need to be extreme values, and once we have found their locations, we must find out if the corresponding objective function is in fact an extreme value. However, the search has now been reduced to a search for the location of these types of situation and a subsequent comparison of function values at these locations. Since we shall not know in advance at which type of position the global optimum will lie, except in a few well-behaved cases, it is necessary to examine all three possibilities, and any search must examine both the interior of the permitted region and its boundary.

The functions under examination can be of different forms. For example, in a multivariable system with many n-dimensional mountain peaks, the search for a maximum entails the search for the stationary values (as well as examination of other locations), a search which can sometimes be done analytically. Indeed, the analytical approach is the only one which will yield information on all the stationary-point locations within the region. These methods will, however, require assumptions as to the continuity of the function and its derivatives. Such analytical techniques are discussed in Chaps. V and VII for the respective cases of single and multiple variables. The presence of discontinuities can be handled by the use of additional restrictions at each discontinuity. The effect of restrictions in the general case is to include the possibility of an optimal solution lying on the boundary of the valid region. A special case of this arises (Chap. VII) when the function is a linear combination of variables, since the required location must then lie on a boundary. If analytical methods are not convenient, as when a black-box model is producing the technical design data or when the expressions are too complex for convenient mathematical analysis, numerical search methods are required. This is by far the commonest approach, the search being again for a global optimum chosen from the local optima at stationary points, discontinuities, or boundaries. This numerical approach can be likened to starting off in a mountain valley, but blindfolded. The search will involve the climb to the highest peak, but this must be achieved by repeated test moves. Once the local peak has been attained, there is no way of ascertaining by that search whether or not it is the highest peak of the whole mountain range except by starting from a new base location and again looking for another peak. Such search methods are discussed in Chaps. VI and VIII for single and multiple variables, respectively.

The only rigorous method of determining the global optimum, and the only way of handling discontinuities, is by the direct comparison of the function values at the various local optima, once these have been achieved.

While most problems encountered will have several variables, the analysis of single-variable functions is fairly detailed in this presentation, in Chaps. V and VI, for two main reasons. First, the basic notions of both analytical and numerical searches are, respectively, the same, whether for single or multi-variable systems, so that the one-dimensional problems, being more easily understood, can serve as an introduction. Second, even in the general case, the problem can often be reduced to a series of one-dimensional searches.

When problems involve not a finite number of variables, as above, but an infinite number as represented by a continuous variable, a different type of approach is necessary. This might be considered as the limiting case, in which the number of independent variables becomes infinite, and instead of having to determine each of these, the variables can be replaced by some function, $x(t)$. We now have the problem of choosing the appropriate form of the function $x(t)$ in some interval t_1 to t_2 for the variable t so as to make the objective function y, where $y = y[x(t)]$, as large or as small as possible. This type of problem is discussed in Chaps. IX and X.

The methods discussed in Part Two are sufficient to cover most applications. Usually, a range of techniques is potentially available in a given situation, since there is no universal approach to optimization. The special problems associated with the more complex systems often found in practice have been deferred to Part Three.

Problems

IV-1 Find the values $y(0^+)$ and $y(0^-)$ of the function

$$y(x) = \frac{e^{1/x}}{1 + e^{1/x}}$$

Is $y(x)$ continuous?

IV-2 Write the equations for any boundaries which must be imposed on the function $y = |x_1 + x_2|$ to divide the two-dimensional space into subregions in each of which y has continuous derivatives. Determine the location of any discontinuities in the derivative of the function.

IV-3 Is the function

$$y(x) = \frac{2x - 3}{1 + e^{1/2x-3}}$$

continuous? Where are the discontinuities in the first derivative?

IV-4 Define typical regions in which the following functions are found to be (*a*) monotonic-increasing, (*b*) monotonic-decreasing, (*c*) unimodal, (*d*) bimodal, (*e*) concave, (*f*) convex:

$$y = \sin x$$
$$y = x^2$$
$$y = e^x$$
$$y = x$$

IV-5 Is the function

$$y = \begin{cases} -x^2 & \text{for } -\infty \le x \le 0 \\ x^2 & \text{for } 0 \le x \le 1 \\ e^{x-1} & \text{for } 1 \le x \le \infty \end{cases}$$

continuous? Does this function have continuous derivatives? Is it monotonic? Is the function unimodal?

IV-6 Prove that the function

$$y = \begin{cases} x_1^2 + x_2^2 & \text{for } x_1^2 + x_2^2 \le 1 \\ 2e^{x_1^2+x_2^2-1} & \text{for } x_1^2 + x_2^2 \ge 1 \end{cases}$$

is unimodal.

IV-7 Is the function

$$y = 3x_1^2 - 2x_1x_2 + 2x_2^2$$

convex?

IV-8 Is the function

$$y = x_1^2 - 6x_1x_2 + x_2^2$$

convex?

IV-9 Is the function

$$y = \sum_i c_i e^{b_i x_i}$$

a convex function when c_i, b_i, and x_i are all nonnegative?

IV-10 The search in a three-dimensional region is constrained by the equalities

$$g_1 = 2x_1 + 3x_2 - 6x_3 - 6 = 0$$
$$g_2 = x_1^2 + 2x_2^2 - x_3^2 - 2 = 0$$

and

$$g_3 = -x_3 \le 0$$

Specify the coordinates of a number of feasible points, if these exist.

IV-11 If the restrictions in Prob. IV-10 are altered to the inequalities $g_i \le 0$, specify some feasible points. Sketch the feasible search region.

IV-12 Is the feasible region defined by

$$g_1 = x_2^2 - x_1 + 3 < 0$$

and

$$g_2 = x_1 + x_2 - 5 < 0$$

a convex region?

IV-13 Is the feasible region defined by

$$g_1 = x_2^2 - x_1 + 3 < 0$$
$$g_2 = x_1^2 + x_2^2 - 16 < 0$$

a convex region?

IV-14 Is the function

$$y = \begin{cases} a(x - \alpha) & x \geq \alpha \\ b(x - \alpha) & x \leq \alpha \end{cases}$$

convex when $a > 0$, $b < 0$ (Hadley [1964])?

IV-15 Repeat Prob. IV-13 for $g_1 < 0$ and $g_2 > 0$.

IV-16 Repeat Prob. IV-2 for $y = |x_1^2 - x_2^2|$.

CHAPTER V

FUNCTIONS OF A SINGLE VARIABLE—ANALYTICAL METHODS

1 INTRODUCTION

If the dependence of the objective function on the independent variables can be expressed by a mathematical relationship, certain equations can be developed which aid the search for an optimum. These subsidiary equations arise from a consideration of an infinitesimal region, and thus require the use of the differential calculus. This approach allows those optima lying at stationary points in the interior of a region to be located by analytical methods. Such locations must then, in the general problem, be compared with the results obtained from the other feasible optima locations, namely, discontinuities and boundaries. The analytical approach thus allows part of the search for potential locations of optima to be defined by subsidiary equations, in contrast to the numerical searches discussed in the next chapter, which require a series of trial and improvement steps.

As the optimum can lie in the interior of a region at a stationary point or at a discontinuity, or at the region boundary, only certain cases can be treated using analytical methods. In particular, this analysis is usually restricted to interior points for which the objective function and its derivatives are continuous (Sec. 2). An optimum at other conditions must be determined by searching among the possible locations (Sec. 3).

2 UNRESTRICTED FUNCTIONS OF A SINGLE VARIABLE

Suppose that the objective function, which may have been obtained from either a mathematical model or a fitting of response data, depends only on a single independent variable, that it is continuous, and that certain of its derivatives are also continuous. For this situation we are able to develop expressions which can be used to locate the possible locations of an optimum, and also allow the character of the function to be checked at these positions. Let us now develop these conditions, which are both necessary and sufficient for the existence of a local optimum in these circumstances.

2.1 Necessary Conditions for a Stationary Value

Using Taylor's theorem, which states that if $y(x)$ and its first $j - 1$ derivatives[1] are continuous in the range $\alpha \leq x \leq \beta$ and its jth derivative exists in the range $\alpha < x < \beta$, then we may write for any point in this range that

$$y(x) = y(a) + (x - a)y'(a) + \cdots + \frac{(x-a)^{j-1}}{(j-1)!} y^{(j-1)}(a) + \frac{(x-a)^j}{j!} y^{(j)}(\zeta) \tag{2.1-1}$$

where ζ lies between x and a.

Letting $x = a + h$, we then have that

$$y(a+h) - y(a) = hy'(a) + \frac{h^2}{2!} y''(a) + \cdots + \frac{h^j}{j!} y^{(j)}(\zeta) \tag{2.1-2}$$

[1] A function is differentiable at the point x_0 if the limit

$$\lim_{h \to 0} \frac{y(x_0 + h) - y(x_0)}{h}$$

exists as h tends to zero. Graphically, this expression is the chord slope AB, as shown in Fig. V-1, and the statement $h \to 0$ implies that B approaches A. Clearly, at points of discontinuity in the function or its gradient, as illustrated in Fig. IV-8c, the function is not differentiable.

Notation: When a function is differentiable, the above limit is written

$$y'(x_0) \quad \text{or} \quad \left(\frac{dy}{dx}\right)_{x=x_0} \quad \text{or} \quad y_x\big|_{x_0} \quad \text{or} \quad y_x(x_0)$$

and is called the *first derivative of* $y(x)$.

The higher derivatives may be formed by continuing this limiting process. Thus

$$y''(x_0) = \lim_{h \to 0} \frac{y'(x_0 + h) - y'(x_0)}{h}$$

and in general, for the nth derivative, we get

$$y^{(n)}(x_0) = \lim_{h \to 0} \frac{y^{(n-1)}(x_0 + h) - y^{(n-1)}(x_0)}{h}$$

Fig. V-1 Function differentiation.

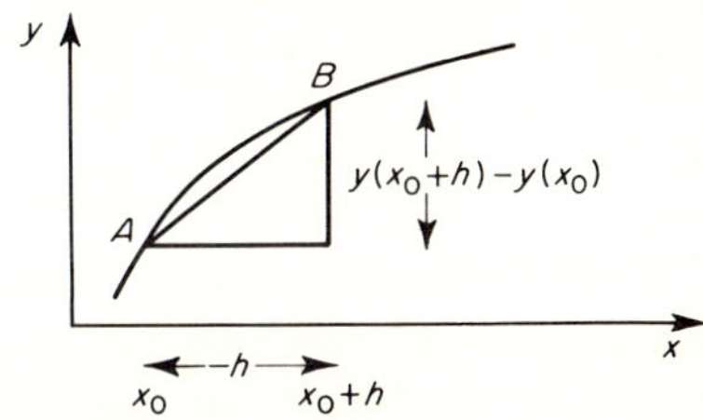

Let us suppose that $y(x)$ has a local minimum at the point $x = a$ and at the same point $y'(a) \neq 0$. We propose to show that these two conditions are contradictory and that y' must equal zero at a point where y is a minimum. The argument is simple. Consider a value of h which is so small that the term $hy'(a)$ dominates the right-hand side of Eq. (2.1-2), and, this being the case, we find the contradiction, for if $y'(a) \neq 0$, then it is certainly possible to select the sign of h so that $hy'(a) < 0$. This condition implies that $y(a) > y(a + h)$, and $y(a)$ cannot be a minimum. Thus, if $y(a)$ is to be a minimum, it is necessary that $y'(a)$ vanish. The point at which $y'(a) = 0$ is called a *stationary point*.

It will be obvious from similar arguments that a maximum is also possible only at a stationary point, excluding of course discontinuities and boundaries.

2.2 Sufficient Conditions for a Local Optimum to Exist at a Stationary Value

Let us now consider the further conditions required to determinè the character of a stationary point. Let us first of all look for a minimum value.

When the Taylor expansion, Eq. (2.1-2), is applied at a stationary point, it reduces, if $y''(a) \neq 0$, to

$$y(a + h) - y(a) = \tfrac{1}{2}h^2y''(a) + 0(h^3) \tag{2.2-1}$$

since the second-order term will be the dominant term on the right-hand side for sufficiently small values of h, the higher-order terms being grouped as the remainder term, $0(h^3)$. If $y''(a)$ is greater than zero, we see that the right-hand side of Eq. (2.2-1) is positive for all sufficiently small values of h, so that $y(a + h) > y(a)$. This is thus the required sufficiency condition for a minimum to exist.

A similar development can be carried through for a maximum, finding that $y''(a) < 0$ is a sufficient condition.

If $y''(a)$ is zero, then $y'''(a)$ must also be zero, for otherwise the h^3 term would change sign with h, and the first significant term $[(h^3/6)y'''(a)]$ would be

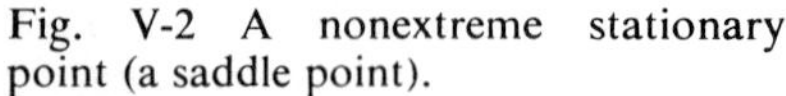

Fig. V-2 A nonextreme stationary point (a saddle point).

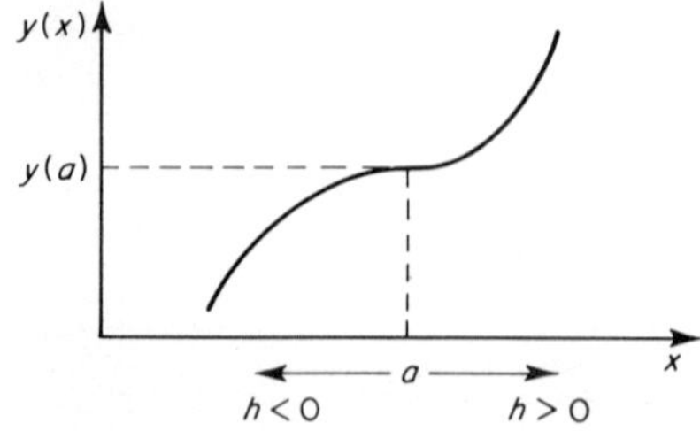

incompatible with the assumption that $y(a)$ was a minimum or a maximum. Such a nonzero condition would indicate that an inflection, or *saddle point*, exists in the function at the point $x = a$, as shown in Fig. V-2.

For $y'(a) = 0$, $y''(a) = 0$, $y'''(a) = 0$, and $y^{iv}(a) \neq 0$, an argument similar to that for the first even derivative shows that we must have $y^{iv}(a) > 0$ for a minimum and $y^{iv}(a) < 0$ for a maximum to exist at $x = a$. If $y^{iv}(a) = 0$, we must investigate still higher derivatives, but the same type of argument will apply.

2.3 Summary of the Necessary and Sufficient Conditions Required for a Stationary Point to Exist and to Be an Extreme Value

In the search for the maximum or minimum of an unrestricted, well-behaved function, we must first determine the location of the stationary points, if they exist, using the equation $y' = 0$, and then check each one individually to determine whether it is in fact an optimal value.

It is evident that the first nonvanishing derivative must be an even one, and that it must be positive for a minimum and negative for a maximum to occur. These conditions are necessary and sufficient as long as the turning value at $x = a$ occurs within a range where $y(x)$ is defined by a continuous function. We may thus write that $y(a)$ is a local extreme value of $y(x)$ if, for some $k \geq 1$,

$$y'(a) = y''(a) = \cdots = y^{(2k-1)}(a) = 0 \tag{2.3-1a}$$

$$y^{(2k)}(a) \neq 0 \tag{2.3-1b}$$

and $y^{(2k+1)}(a)$ exists. It will be a minimum or a maximum according as to whether $y^{(2k)}(a)$ is greater or smaller than zero. If $y^{(2k)}(a) = 0$ and $y^{(2k+1)}(a) \neq 0$, then the point a is a stationary point, but is not an optimum point. Table V-1 summarizes the results of some tests at a stationary point.

Example V-1: Insulation costs

The cost c_1 of purchase and installation of a steam pipe lagging may be represented by

$$c_1 = ax + b \tag{2.3-2}$$

and the cost c_2 of the heat loss from the piping may be given by

$$c_2 = \frac{c}{x} + d \tag{2.3-3}$$

both c_1 and c_2 being expressed on the same basis. Here a, b, c, and d are known positive constants, and x is the lagging thickness. Then the total

TABLE V-1 SUMMARY OF TESTS AT STATIONARY VALUES OF A CURVE $y(x)$

$y'(a)$	$y''(a)$	$y'''(a)$	$y^{iv}(a)$	*Nature of point*, $x = a$
0	$-$	Exists	Anything	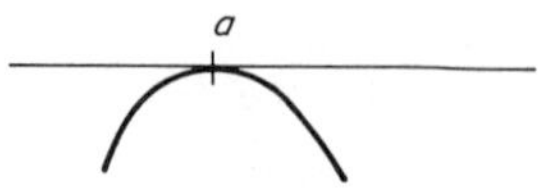 a
0	$+$	Exists	Anything	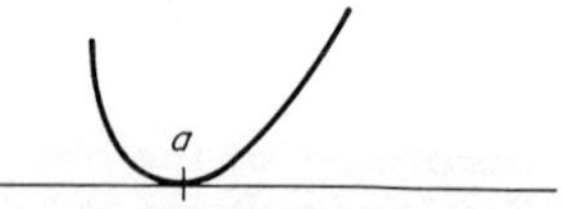 a
0	0	$+$	Exists	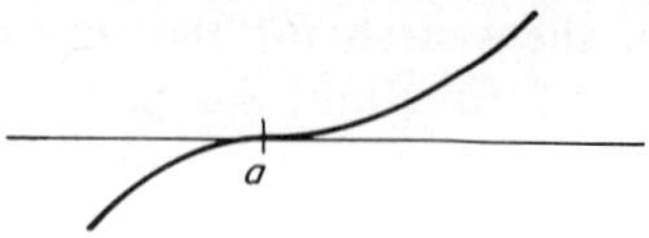 a
0	0	$-$	Exists	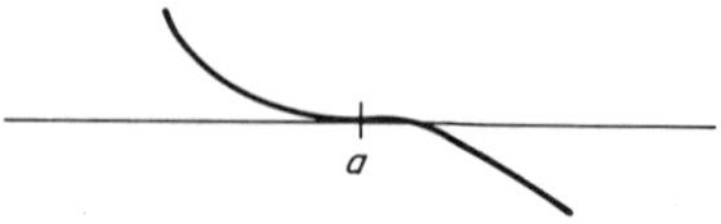 a
0	0	0	$+$	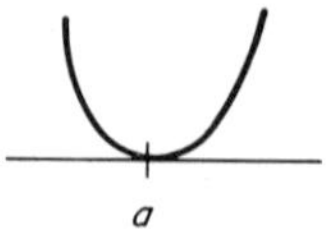 a $y^{v}(a)$ exists
0	0	0	$-$	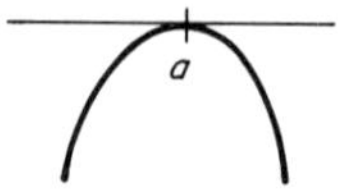 a $y^{v}(a)$ exists
0	0	0	0	Examine $y^{v}(a)$

cost y of the lagging system is the sum of these two costs,

$$y = ax + b + \frac{c}{x} + d \tag{2.3-4}$$

This is a continuous convex function of a single variable x, as shown in Fig. V-3. The minimum cost of operation can then be found.

For such a continuous function, with no boundaries or discontinuities, the optimum must lie at a stationary point, defined by

$$y' = \frac{dy}{dx} = 0$$

Here
$$y' = \frac{dy}{dx} = a - \frac{c}{x^2}$$

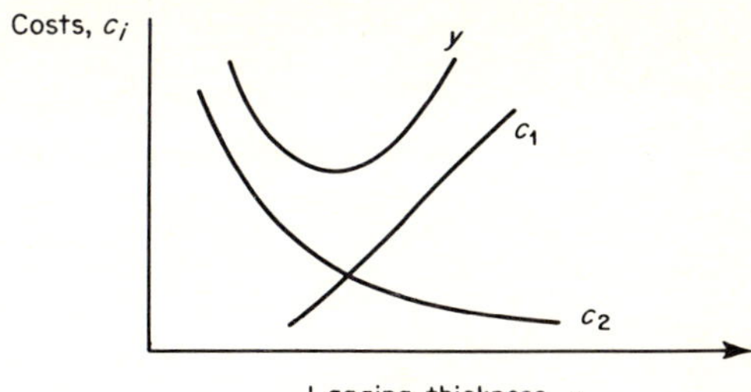

Fig. V-3 Steam-pipe insulation costs. $y = c_1 + c_2$, where c_1 represents lagging cost in dollars per year and c_2 represents heat-loss costs in dollars per year.

which is zero when

$$x = \left(\frac{c}{a}\right)^{1/2}$$

the positive root being taken, since only positive thicknesses make physical sense. Thus the only physically feasible stationary value of y exists at $x = (c/a)^{1/2}$, yielding a y value of

$$y = 2(ca)^{1/2} + b + d$$

If the feasible stationary point is to be a minimum, the first non-vanishing derivative must be both an even derivative and positive, as required by Eq. (2.3-1). Here

$$y'' = \frac{d^2y}{dx^2} = \frac{2c}{x^3}$$

has a value of $2c(a/c)^{3/2}$ at the stationary point and is clearly positive. The stationary point thus represents a minimum—the desired result.

Example V-2: The necessity of considering higher derivatives

Find the maximum of the objective function

$$y = 1 + 8x + 2x^2 - {}^{10}\!/_3x^3 - {}^1\!/_4x^4 + {}^4\!/_5x^5 - {}^1\!/_6x^6$$

The first derivative

$$y' = 8 + 4x - 10x^2 - x^3 + 4x^4 - x^5 = (1 + x)^2(2 - x)^3$$

vanishes[1] at $x = -1$ and $x = 2$, defining two stationary points. To

[1] In most cases the determination of the location of any stationary points will probably require numerical root-finding methods for the solution of the equation $y'(x) = 0$ (Chap. III).

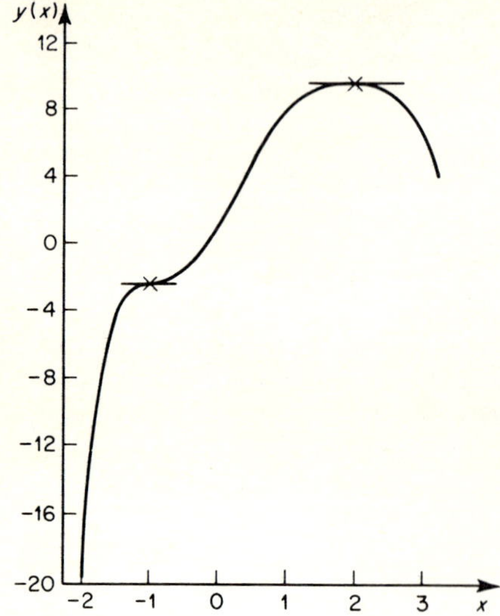

Fig. V-4 The function of Example V-2.

determine if either or both of these points is a maximum of the objective function y, consider the second derivative,

$$y'' = (1 - 5x)(1 + x)(2 - x)^2$$

This vanishes for both $x = -1$ and $x = 2$. The test (Sec. 2.3) is therefore inconclusive, and higher derivatives must be computed.

$$y''' = -2(2 - x)(5 + 4x - 10x^2)$$

Here y''', an odd derivative, does not vanish for $x = -1$, and therefore this stationary point in neither a maximum nor a minimum [Eq. (2.3-1)]. Indeed, as shown in Fig. V-4, the point $x = -1$ is a saddle point, or point of inflection. Since y''' vanishes at $x = 2$, the fourth derivative must be examined:

$$y^{iv} = -6(1 - 16x + 10x^2)$$

Since this is negative ($= -54$) at $x = 2$, the stationary point at $x = 2$ is thus a maximum, the fourth derivative, an even derivative, being negative, even though all lower-order derivatives vanish. This point at $x = 2$ is also shown in Fig. V-4. No others exist, and consequently the maximum value of y occurs at $x = 2$.

3 RESTRICTED FUNCTIONS

In the unrestricted search, the optimum was always found at a stationary point, provided the function and its first derivative were continuous everywhere. If there are restrictions imposed on the permissible values of the independent variable, the optimum may not then be at a stationary point, but lie at the restriction boundary. Moreover, if the first derivative is discontinuous, the optimum may also be found at a point other than a stationary point (Chap. IV, Sec. 4). We should thus look not only at stationary points, but at boundaries and discontinuities.

3.1 Boundaries

If an objective function is to be optimized subject to an equality restriction

$$g(x) = 0 \tag{3.1-1}$$

this is equivalent to specifying one or more values of x, the roots of Eq. (3.1-1). The optimization analysis then reduces to a comparison of the objective function values at each of these possible values of x.

On the other hand, an inequality restriction

$$g(x) \leq 0 \tag{3.1-2}$$

limits the range of x only. Thus, if

$$g = x - x^L \leq 0$$

values of x smaller or equal to x^L will be the only feasible choice. One way to handle this type of problem is to consider the function unrestricted and choose only these stationary values which occur for feasible values of x. Finally, the optimal values at stationary points must be compared with the objective function evaluated on the boundary. More elegant methods are available for multidimensional systems (Chap. VII) and might be used here.

Example V-3: Boundaries and the stationary point

Let us determine the maximum value of the function

$$y = \frac{x^5}{5} - \frac{5x^4}{2} + \frac{35x^3}{3} - 25x^2 + 24x - 4$$

(*a*) subject to the restriction that

$$g = x^2 - (4.1)^2 = 0$$

and (*b*) subject to the restriction that

$$g = x^2 - (4.1)^2 \leq 0$$

Case a If $x^2 = 4.1^2$, then $x = \pm 4.1$, so that only two values of the objective function are feasible,

$$y(-4.1) = -2{,}264.88$$
$$y(+4.1) = 3.50$$

the latter being the desired maximum.

Case b When an inequality restriction is imposed, the range of possible values of x is limited. Here only x values inside the range

$$-4.1 \leq x \leq 4.1$$

may be chosen. A check on the stationary values of the unrestricted objective function shows that four such points exist at $x = 1, 2, 3$, and 4, since

$$y' = (x - 1)(x - 2)(x - 3)(x - 4)$$

Since

$$y'' = (x - 1)(x - 2)(2x - 7) + (x - 3)(x - 4)(2x - 3)$$

we get the results shown in Table V-2. The objective function is plotted as a function of x in Fig. V-5.

TABLE V-2 CLASSIFICATION OF THE STATIONARY POINTS

x	y	y'	y''	*Type*
1	4.37	0	−6	max
2	3.73	0	2	min
3	4.10	0	−2	max
4	3.47	0	6	min

Two locations are maxima, and of these, that at $x = 1.0$ is the larger. Since the function increases monotonically for $x > 4$, the upper boundary at $x = 4.1$ must be tested. We find that

$$y(4.1) = 3.50$$

The global maximum thus lies at $x = 1.0$, a stationary point.

Fig. V-5 The function of Example V-3.

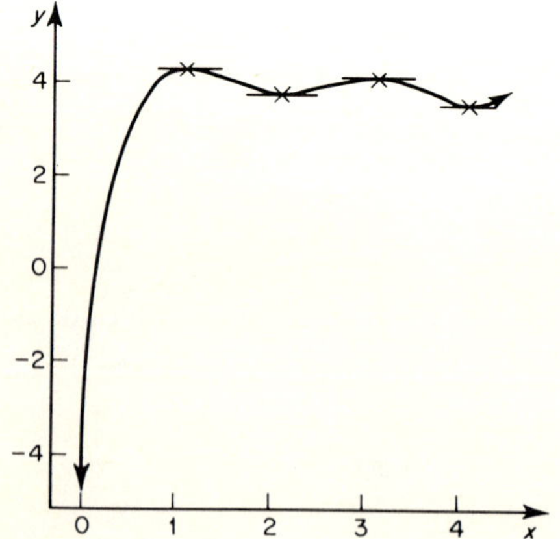

A boundary may on occasion seem a complicating feature, but in certain cases it eases the solution. The simplest illustration of this is with a linear function

$$y = ax + b$$

If a is a positive constant, the y will be a maximum at the upper limit of x, that is, x^L, and the minimum will be at its lower limit, x_L. This feature in the n-dimensional case will provide the basis for the analysis known as *linear programming*, in which only the boundaries need to be examined (Chap. VII, Sec. 4.4). Any monotonic function will show this feature, an example being found in the endothermic reaction of Prob. V-9.

3.2 Discontinuous Functions and Derivatives

While analytical methods are applicable in presence of boundaries (constraints), these methods fail when the objective function ceases both to be continuous and to have continuous derivatives in the region of interest. The easiest approach in presence of discontinuities is the imposition of an artificial boundary at each discontinuity, thus dividing the region of interest into subregions in which the functions and their first derivative will be continuous. The procedure of Sec. 3.1 is then applicable. Hancock [1960] has developed necessary conditions for points with discontinuous first derivatives to be optimal. However, we shall not present his work, since the procedure suggested here, while not as elegant, is manageable without introducing any new results.

Example V-4: Gradient discontinuity

The function

$$y = |x^3|$$

clearly has a minimum value at $x = 0$. Note that both y' and y'' vanish at $x = 0$ and that y''' does not, in fact, vanish. If the criterion given by Eq. (2.3-1) is applied, the point $x = 0$ would therefore be rejected as a possible minimum. This seeming paradox arises because the objective function written here does not satisfy all the basic premises necessary to develop Eq. (2.3-1). In particular, $y'''(x)$ is discontinuous at $x = 0$, and the criterion given by Eq. (2.3-1) is valid for functions which are continuous and have $2k$ continuous derivatives. That $y'''(x)$ is discontinuous can be seen as follows: The objective function can be represented as

$$y = \begin{cases} x^3 & \text{for } x \geq 0 \\ -x^3 & \text{for } x \leq 0 \end{cases}$$

The values of y''' evaluated at $x = 0$ are then 6 or -6, depending upon whether it is approached from a higher or lower value of x.

This is an illustration of an optimum lying at a discontinuity. It can be best handled by inserting a boundary plane at the discontinuity and examining each zone and the boundary separately.

Example V-5: Cylindrical storage tank

Let us determine the optimum dimensions of a closed cylindrical tank which is to be used to store a given volume of liquid. The cost of the side and end walls is denoted as SK and EK dollars per unit area, respectively, where E and S depend on the radius R and the height L of the tank, as shown in Table V-3, and where K is a constant (Amundson et al. [1960]).

TABLE V-3 COSTS AT STATIONARY VALUES

Region	*Restrictions on valid range of:*		*Relative costs of steel plate*		*Possible stationary values*			
					$V = \frac{\pi}{2}$		$V = \frac{\pi}{8}$	
	R	L	E	S	R	$L\left(=\frac{1}{2R^2}\right)$	R	$L\left(=\frac{1}{8R^2}\right)$
1	≤ 1	≤ 1	1	1	0.63	1.26†	0.40	0.80
2	>1	<1	2	1.5	0.57‡		0.36‡	
3	<1	>1	1	1.5	0.72	0.96†	0.45	0.61†
4	>1	>1	2	2	0.63‡		0.40‡	

† Not valid value of L.
‡ Not valid value of R.

The technical design basis, or model, for the tank then gives the volume V as

$$V = \pi R^2 L \tag{3.2-1}$$

and the areas of the end wall A_E and side wall A_S are

$$A_E = 2\pi R^2$$

and

$$A_S = 2\pi RL$$

The total cost can then be written in terms of the independent variables as

$$C_T = A_E EK + A_S SK = 2\pi K(R^2E + RLS)$$

Since the volume is fixed, the total cost can be written in terms of a single variable R as follows:

$$C_T = 2\pi K\left(R^2E + \frac{VS}{\pi R}\right)$$

in which L has been eliminated, using Eq. (3.2-1).

Letting $y = C_T/2\pi K$, we can write the desired objective in terms of a single variable R as

$$y = R^2E + \frac{V}{\pi R}S$$

If the R were unrestricted and y continuous, the optimum would have to occur at a stationary value, determined by

$$\frac{dy}{dR} = 2ER - \frac{VS}{\pi R^2} = 0 \qquad (3.2\text{-}2)$$

Solving gives the location of the stationary points at

$$R = \left(\frac{VS}{2\pi E}\right)^{1/3} \qquad (3.2\text{-}3)$$

However, y is not continuous, owing to the discontinuous nature of E and S (Table V-3). We shall therefore adopt the following approach: First S and E will be assumed to be constant within a specified region, and the value of R and L at the stationary point corresponding to a pair of values for S and E can be computed. The R and L at the stationary point might not, however, be values of R and L corresponding to the assumed costs. If this is the case, we shall say that the stationary point falls outside of the permitted region. In other words, this feasible region does not possess a stationary point. On the other hand, if the values of R and L at the stationary point correspond to the assumed values of E and S, the stationary point is termed *valid.* This can best be understood by referring to Table V-3, in which the calculations for two volumes, $V = \pi/2$ and $V = \pi/8$, are given. Shown in columns 6 to 9 are the values of R and L, which yield a stationary point corresponding to the values of E and S appearing in the same line of the table. It will be seen that the values of R marked by ‡ fall outside the permitted range, and those values of L denoted by † are likewise invalid. For $V = \pi/2$, there is no stationary point in any permitted region, but one exists for $V = \pi/8$ at $R = 0.40$ and $L = 0.80$, with a value of $y = 0.42$.

Since the objective function is discontinuous, the artificial boundaries at $R = 1$ and $L = 1$ must also be searched, and indeed, since there are no valid stationary points for $V = \pi/2$ the optimum must then be at a boundary which lies at either $R = 1$ or $L = 1$ or both. The objective function must be evaluated at these boundaries. This has been done, and the results are tabulated in Table V-4. Since the function goes to infinity as R goes to very large values or toward zero, no other positions need be considered.

TABLE V-4 COSTS AT BOUNDARIES

L	R	*Region*	y	L	R	*Region*	y
For $V = \frac{\pi}{2}$				For $V = \frac{\pi}{8}$			
1.00^{+}†	0.71^{-}	3	1.56	1.00^{+}	0.35^{-}	3	0.66
1.00^{-}	0.71^{+}	1	1.21	1.00^{-}	0.35^{+}	1	0.48
0.50^{+}	1.00^{-}	1	1.50	0.13^{+}	1.00^{-}	1	1.13
0.50^{-}	1.00^{+}	2	2.75	0.13	1.00^{+}	2	2.19

† The signs following numbers are required to indicate direction of approach to discontinuity. For example, 1^{+} denotes a number slightly larger than 1.

Examining Table V-4, we see that the global minimum for $V = \pi/2$ is $y = 1.21$, and so the optimal dimensions are $R = 0.71$ and $L = 1.0$.

For $V = \pi/8$, the minimum cost at a boundary is 0.48. The value of y at the stationary point is, however, 0.42, and thus the optimum lies at the stationary point $R = 0.40$ and $L = 0.80$.

This example is intended to show the need to exercise great care in treating optimization problems in which the function, or its first derivative, is discontinuous, since, as noted earlier, an optimum can lie at a stationary point, boundary, or discontinuity.

4 GENERAL PROCEDURE FOR HANDLING ONE-DIMENSIONAL FUNCTIONS

The preceding problems should indicate that care must be taken in searching for a maximum or a minimum when using the elementary calculus method, even when it is quite obvious that the function is continuous.

A general procedure can, however, be formulated to handle the search for optimum conditions for continuous functions, discontinuous functions, and for functions which have discontinuous gradients.

1. Divide the complete region of search into intervals in each of which a single formula will apply and in which the function $y(x)$ and its gradient $y'(x)$ are continuous. This means that boundary planes must be inserted at all internal points at which the function possesses a discontinuity in the function or its gradient.

2. Within each region, in which we are now dealing with a continuous, continuously differentiable function, search for the positions of all the stationary values of this function by determining the real roots of $y'(x) = 0$. Determine the nature of each stationary point. Calculate the value of the function at each stationary point which is a maximum or minimum as required. These points are *local* optima.
3. Find the values of the function at each boundary plane, whether they determine the original region or the subdivided regions.
4. Since the extreme values of the function must exist at either stationary values or at discontinuities or boundary planes (both the latter being found at boundaries, using the present procedure), a comparison of all the function values found in steps 2 and 3 will allow the greatest or smallest value to be selected. This final value will be the *global*, or *absolute*, optimum.

Such extensive precautions may not always be necessary, but their importance should be recognized, especially when the functions have discontinuities.

5 CONCLUSIONS

If the objective function is expressed as a mathematical function of a single variable, either as a result of mathematical modeling or curve fitting of numerical data, the methods developed in this chapter are applicable although they may not always be practical. The optimum is known to lie at either a stationary point, boundary, or discontinuity. Moreover, we have developed a criterion for testing the nature of the stationary point (Sec. 2.3) so that the maxima, minima, and points of inflection can be distinguished. These results must then, however, be compared directly with the results from boundary or discontinuity searches.

The analytical methods are subject to some criticism for the following reasons:

1. If there are a number of discontinuities, a number of subregions must be defined, and this may lead to a number of separate searches (see Example V-5).
2. The function must be readily differentiable in order to apply the analytical methods. This may not always be the case.
3. Even if the derivatives are easily computed, it may be difficult to find the roots of the equation determining the stationary points. This is perhaps the greatest difficulty found in using the analytical methods.

The analytical methods, therefore, are truly applicable to the simplest problems. In the next chapter we shall develop methods of numerical search which circumvent many of the above difficulties and which are easily programmed for computation of a very wide class of problem. However, the numerical search methods have the disadvantage of requiring exploration of the entire region, whereas the analytical methods can specify some of the points (namely, those at stationary points) which need to be compared with the results at the boundaries and discontinuities to find the global optimum. Indeed, there are advantages and disadvantages to each optimization technique proposed here. One must in each case select the method best suited to the particular problem in hand.

Problems

V-1 Find the locations and determine the nature of the stationary values of the unrestricted functions

a.
$$y = -3x^4 + 10x^3 - 20$$

b.
$$y = x^4 - 8x^3 + 24x^2 - 32x + 16$$

V-2 Find the minimum of

$$y = x^x \qquad \text{for } x > 0$$

Be careful to ensure that the stationary point found is in fact a minimum. Check the boundary point to be certain that it is not a better point.

V-3 Find the location of the global minimum of

$$y = x^2 - e^{1/x^2}$$

Is the minimum at a stationary point, discontinuity, or boundary?

V-4 Find the minimum and maximum values of the function

$$y = \frac{e^{1/x}}{1 + e^{1/x}}$$

Do these occur at a stationary point, discontinuity, or boundary?

V-5 Find the area of the greatest rectangle that can be described inside an equilateral triangle and having one of its sides upon its base (Hancock [1960]).

V-6 A length of wire is to be cut into two parts. One portion is to be bent into the form of a circle, and the other into the form of a square. In what ratio must the wire be cut if the sum of the areas enclosed by the circle and the square is the least possible?

V-7 It is required to replace a length of piping within a process plant. Since the site is cramped, the pipe run must be brought into the site in lengths and welded *in situ*. Obviously, to reduce the cost of welding, the pipe lengths should be brought in, in sections which are as long as possible. If the head room is low and the pipe has to be carried horizontally, what lengths should be introduced if the restriction is the passage of the pipe around a corner from a 10-ft wide corridor into a 6-ft-wide corridor?

V-8 A two-stage reversible adiabatic gas compressor compresses a gas from an initial pressure p_1 to a final pressure p_3. If the fixed capital charges on the compressors are assumed to be essentially independent of the interstage pressure p_2 employed, and if the total work done is given by

$$W = NRT_1 \frac{K}{K-1}\left[\left(\frac{p_2}{p_1}\right)^{(K-1)/K} - 2 + \left(\frac{p_3}{p_2}\right)^{(K-1)/K}\right]$$

determine the optimum operating interstage pressure for which the total power requirement is a minimum for a given gas flow rate. Here the gas enters at T_1 and is cooled to T_1 by an interstage cooler, N is the pound moles of gas compressed, R the molal gas constant, K the ratio of specific heat at constant pressure to that at constant volume for the compressed gas, and W is the total work done.

V-9 Find the reaction temperature T which maximizes the rate r of a reversible reaction for

a. The exothermic case, with $E_2 > E_1$
b. The endothermic case, with $E_1 > E_2$, where the compositions c_i are fixed and

$$r = k_1(T)\prod_{i=1}^{n} c_i^{\gamma_i} - k_2(T)\prod_{i=1}^{n} c_i^{\beta_i}$$

$$k_i(T) = B_i\, e^{-E_i/RT}$$

with R, B_i, and E_i as defined parameters.

V-10 Determine the maximum value of the function

$$y = x^2 - x + 1$$

if the range of x is restricted by the relation

$$g = x^2 - 4 \leq 0$$

V-11 Find the minimum of the expression

$$y = 10x^3 - 4x^2 + 3x - 1$$

if $-3 \leq x \leq 11$.

V-12 Find the minimum and maximum values of the function

$$y = 2x^3 - 5x^2 - 8x + 3$$

if $-3 \leq x \leq 3$.

V-13 The safe distance, in feet, between automobile (center to center) on a certain highway has been found experimentally to be

$$L = 30 + 0.07\,V^{3/2}$$

with $40 \leq V \leq 70$. Here V is the speed of the car in miles per hour. Find the safe maximum capacity of the highway per lane.

V-14 Two cables of equal length may be referred to as AB and CD. The strain which any section of AB can stand varies as the cube of its distance from B, and the strain which any section of CD can tolerate varies as the cube of its distance from D. The stress carried at A is four times that at C. If the cables are woven together so that A and D coincide, find the weakest point in the composite cable.

V-15 A uniform sheet of metal, weight 180 lb, is in the form of an isosceles triangle ABC, with its base BC suspended horizontally and a weight of 20 lb attached to A which lies 12 ft below BC. PQ is a line parallel to BC cutting AB, AC at P, Q and at a distance x ft from A.

Find in terms of x the weight of the part below PQ and the strain per foot length of PQ, given $BC = 6$ ft. For what position of PQ is this least?

V-16 A cross-channel ferry is constructed so as to transport a fixed number of tons across each way per day. If the cost of construction of the ferry without the engines varies as the load, and the cost of the engines varies as the product of the load and the cube of the speed, prove that the total cost of construction is least when twice as much money is spent on the ferry as on the engines. (Neglect the time of loading and unloading, and assume the ferry runs continuously.)

CHAPTER VI

FUNCTIONS OF A SINGLE VARIABLE—NUMERICAL METHODS

1 INTRODUCTION

The location of an optimum can sometimes be found analytically (Chap. V, Sec. 5), but this requires the continuity of the function and its derivatives, together with a solution for the points at which the first derivatives vanish. Because of these restrictive conditions, alternative methods have been suggested, commonly called *numerical search methods.* The general principle behind all these is suggested by the expression "homing in," used by Boas [1963a, b]. These searches essentially require the determination of the value of the objective function, and sometimes its gradient, at successive locations. These are evaluated either directly from an analytical expression or by experimentation on an existing system.

Numerous search techniques exist, but all are based upon the same approach: a base point is chosen (represented by any possible solution to the problem, not necessarily the best), and each search method then selects a new set of independent variables and tests the objective function to see if the new set gives a better value of the objective. Based upon this comparison, another set of variables is chosen, the choice being such that wasteful calculations or experiments are eliminated as far as possible, and the objective function recalculated and retested. This is discussed in more detail in Chap. VIII, Sec. 1.3, the various search methods differing essentially only in their choice of a new search location.

1.1 Case Study Methods

When the choice of a new location has to be made at random, or at best intuitively, it is likely that the resultant search will not be particularly efficient in the number of calculations required to achieve a given accuracy. Such an approach might, however, have to be adopted if common variables are not found between the two locations or systems under comparison. Thus, even in analytical optimization, we had to compare various local optima existing at stationary points, discontinuities, or boundaries, as in Example V-5, where a discontinuous objective function had to be examined at several locations before a global result was found. In addition, case studies are required in the comparisons of two or more distinctly different systems designed to achieve the same objective. Thus, for example, in determining which type of reactor system to be best, the continuous stirred tank, the batch, the tubular, or various combinations of reactors would all be possible alternatives. The best results obtainable for each possible reactor system must then be compared to find the most desirable one. This is a case study, and no other method is feasible.

1.2 Logical Methods

Normally, we should prefer a more logical approach to optimization, the choice of new search locations in a common variable being made in such a manner as to reduce the effort involved. Such methods start at some base point and, by iterative processes, determine a new set of values for the variables. If the new set enhances the objective function, the new set is adopted as the new base point, and the search is continued from this point, as before. On the other hand, if the new set does not improve the objective function, various rules are followed to turn the search in the proper direction or to stop it. A common feature in most techniques is the use of an accelerating factor which speeds the search during periods of success. These logical methods will allow the position of the optimum to be found within a given accuracy, not as an exact position. Most, however, also require an assumption of unimodality (Sec. 2.1) to ensure that there is only one optimum in each region of search, so that any searches must be confined to local regions. This implies that these numerical search methods will find only local optima, which need not be global values (Sec. 2.3). On these occasions, when all the experiments have to be planned in advance, such a cyclic approach may not be possible (Sec. 2.2).

Many search methods exist, some of those for unrestricted regions being described in Sec. 3, those for restricted regions in Sec. 4. Due to the multiplicity of possible search techniques, almost one to every investigator, it would be helpful to be able to say that some were better than others in the sense of quickness of achieving the desired result or of economy in the number of calculations or experiments required. This is not generally possible except when working in a restricted one-dimensional search.

The convergence of the search to the desired optimum and the criteria for stopping a search are important considerations. These are discussed in Chap. XI, Sec. 5.3.2.

To improve the search still further, the position of the optimum may be estimated within the final range of uncertainty, the region in which it is known that the optimum will lie, by fitting a mathematical relation to the experimental points, and then carrying out an analytical optimization (Sec. 6). By this approach the position and the magnitude of the optimum can be found more accurately in the final stages without carrying out additional searches once sufficient information has been obtained to allow the empirical relation to be fitted.

The one-dimensional search methods, while of value in themselves, also find application in the analysis of problems of greater dimensionality. Here it is often found that one-dimensional searches must be carried out once a direction of search has been selected, and we shall find in Chap. VIII that an efficient one-dimensional search routine is most important.

1.3 Graphical Methods

The procedure of calculating a number of values of a function and then plotting them is perhaps the most elementary method of finding an optimum. This approach can be used for case study or the more logical techniques. While we shall use graphical methods only to illustrate specific points during numerical optimizations, we might note two important points required in carrying out optimum searches, using these methods. First, small enough intervals between points must be used if the graph is to reflect the salient features of the objective function. Second, the region near the minimum or maximum must be enlarged, and another graph should be drawn to find the optimum with greater accuracy.

Example VI-1: Cylindrical storage tank

In Example V-5 we considered the optimum dimensions of a closed cylinder for a given volume, given the cost of the side and end walls per unit area. Let us repeat this problem using a graphical method in order to illustrate the points raised above. We previously found the objective function to be

$$y = \frac{C_T}{2\pi K} = ER^2 + \frac{SV}{\pi R}$$

a function of R only. This was to be minimized subject to the restriction that $V = \pi R^2 L$, which allows the required dimensions to be determined.

For the case of $V = \pi/2$, it was found that the objective function could be written

$$y = ER^2 + \frac{S}{2R}$$

Since $R^2L = 0.5$, we may determine the valid regions of R and the corresponding forms of the objective function, as shown in Table VI-1. It is found that of the four regions originally specified in the $R - L$ space,

TABLE VI-1 FEASIBLE SEARCH REGIONS IN EXAMPLE VI-1

Region	*Possible combinations of R and L*	*Permitted region of R, since* $L = 1/2R^2$	*Relative costs*		y
			E	S	
1	$R \leq 1$ $L \leq 1$	$0.707 \leq R \leq 1$	1	1	$R^2 + 1/2R$
2	$R > 1$ $L \leq 1$	$R > 1$	2	1.5	$2R^2 + 3/4R$
3	$R < 1$ $L > 1$	$R < 0.707$	1	1.5	$R + 3/4R$
4	$R > 1$ $L > 1$	Not feasible	2	2	

the fourth region is not feasible, since for $R > 1$ and $L > 1$, the latter inequality gives $L = 1/2R^2 > 1$ or $R < 0.707$, which contradicts the first limit. Our attention must therefore be concentrated on the three regions $R < 0.707$ (region 3); $0.707 \leq R \leq 1$ (region 1) and $R > 1$ (region 2). Within each region it is evident that the objective function y is a sufficiently smooth curve so that a few points will produce the graph shown in Fig. VI-1*a*, the curve *ABCDEF* having been produced by considering the value of y at the boundary points and at a few intermediate positions. It is easy enough to see, by comparing the local optima (values of y at B, C, and E) that the value at point C gives the lowest cost. This is the point $\{R = 0.707, L = 1/2R^2 = 1\}$ at which $S = 1$, $E = 1$, giving a minimum cost of $y = 1.207$.

Fig. VI-1 The dimensions of a storage vessel. (*a*) $V = \pi/2$, (*b*) $V = \pi/8$ (line XYZ being an approximation).

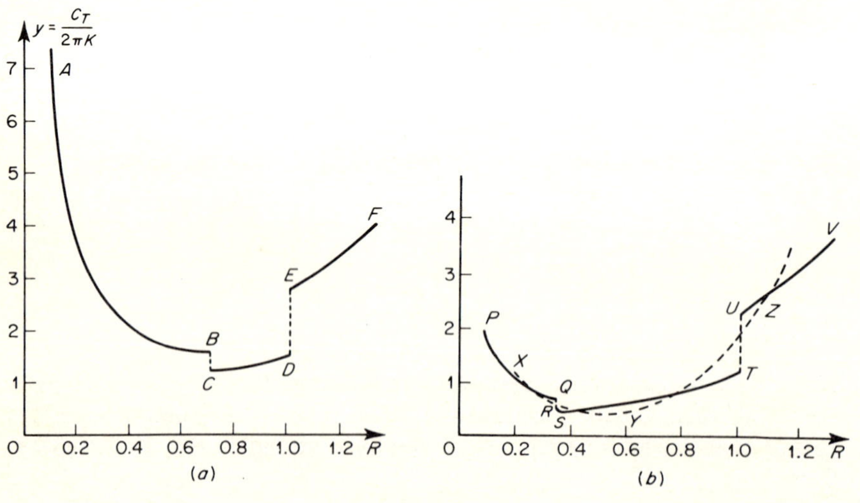

The above y-R relation should be compared with the results found by the analytical method for constant V values in Example V-5.

Other values of V may be examined. Thus, for $V = \pi/8$, we get

$$y = \begin{cases} R^2 + \dfrac{0.187}{R} & 0 \le R \le 0.354 \\ R^2 + \dfrac{0.125}{R} & 0.354 \le R \le 1 \\ 2R^2 + \dfrac{0.187}{R} & 1 < R \end{cases}$$

giving the curve $PQRSTUV$, with the minimum no longer at a point of discontinuity, but at the point S. The accuracy in finding the position of S depends on the number of values calculated in this region and on the scale of the graph. It is obvious that the interval between adjacent points must be small enough to catch the significant features of the graph. If we had calculated only the points at $R = 0.25$, 0.5, 0.75, 1 for the second case, and not noticed the discontinuity at $R = 0.354$, we should have obtained the graph XYZ, which is very different, with a minimum near $R = 0.5$.

Usually, the minimum has to be found by conducting another search, close to the approximately determined optimum point. Since the function has a fairly flat curve, the minimum value of the function may be estimated with greater accuracy than the value of the independent variable at which it occurs. However, it is not very important that the latter be found so accurately, since the flatness of the curve indicates that the function will not vary much on either side of the correct location. On the other hand, the extreme value that is easy to locate might lie at the sharp corner, where the derivative is discontinuous at a discontinuity or a boundary, as in our first illustration.

2 SOME COMMON FEATURES IN LOGICAL SEARCH

Let us now consider some of the common features required in searching one-dimensional functions.

2.1 Unimodality in Numerical Search

It should be strongly emphasized that all logical search techniques can be used only on unimodal functions. Any multimodal function must be treated locally in unimodal regions. In Chap. IV, Sec. 2.3, a unimodal

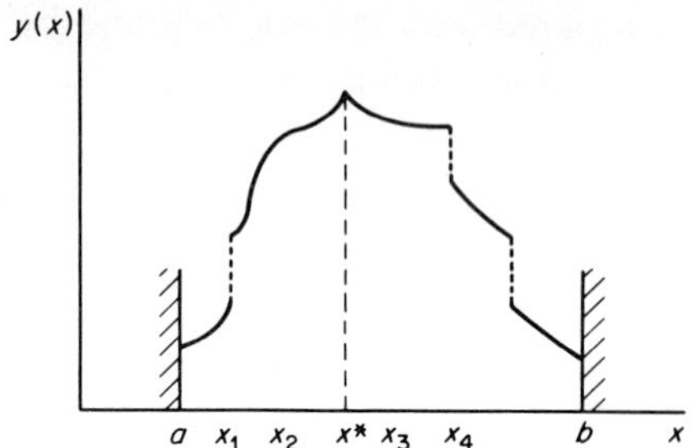

Fig. VI-2 A unimodal function within the search region $a \leq x \leq b$.

function exhibiting a maximum was defined by the following set of inequalities:

$$y(x_1) < y(x_2) < y(x^*) \qquad \text{if } x_1 < x_2 < x^* \tag{2.1-1}$$

and[1]

$$y(x^*) > y(x_3) > y(x_4) \qquad \text{if } x^* < x_3 < x_4 \tag{2.1-2}$$

the point x^* being the location of the optimum. Similar arguments apply for a minimum. Such a function, depicted in Fig. VI-2, need not, of course, be smooth or even continuous.

Since optimization problems in one dimension are generally a result of two competing factors, as mentioned in Chap. I, Sec. 2.2, it is not uncommon for the objective function to be known to be unimodal without having detailed knowledge of the exact relationship between two competing factors, especially if both are monotonic, one increasing and the other decreasing. Indeed, the property of unimodality seems to be the rule rather than the exception.

The validity of any search technique depends upon a knowledge of the general behavior of the particular function being examined. The particular expression for the function, however, does not need to be known, only its general shape. Since we do not know the function $y(x)$ in advance, the only way to obtain information about $y(x)$ will be to measure or calculate its value at different values of the independent variables. Thus we need to have prior knowledge of the general shape of the function, although not necessarily of its values.

Let us confine our attention to *unimodal functions*. For these, since the value of the dependent variable worsens as the value of the independent variable changes in either direction from the optimum, we know that there will be only one optimum within the interval of search $a \leq x \leq b$. Thus we might have the relationship shown in Fig. VI-2. Here the maximum value of y lies at $x = x^*$.

[1] While in general we shall refer to the jth location in the kth stage of a search by the vector $\mathbf{x}_j^{(k)}$, with components $x_{1,j}^{(k)}, x_{2,j}^{(k)}, \ldots, x_{n,j}^{(k)}$, for one-dimensional problems the vector will contain only a single component, and we can use $x_j^{(k)}$, the superscript being added only if necessary to distinguish between stages in a search.

2.1.1 *Direct search*

Our general procedure in any direct search will be to make a number of measurements of $y(x)$ (assuming, meanwhile, a negligible experimental error in the measurement) at values of x anywhere in the region $a \leq x \leq b$. At the end of the set of measurements we wish to be in a position to say that $y(x^*)$ will lie in a certain subinterval of (a, b).

The principle underlying the method will be seen if we consider two measurements placed at any two positions x_1 and x_2 within the range (a,b) so that

$$a \leq x_1 < x_2 \leq b$$

There are now three possible results from this set of two experiments, as indicated in Fig. VI-3. In case 1, with $y(x_2) > y(x_1)$, we have discovered that there is a value of $y(x)$ at an $x > x_1$, giving a value greater than $y(x_1)$, and it *must* therefore follow from the property of unimodality that $y(x^*)$, the maximum, cannot lie at $x \leq x_1$. Our desired maximum value must thus lie in the double interval

$$x_1 < x^* \leq b$$

Note that we are unable to specify the position of x^* within this remaining interval, but can only delineate regions in which it cannot lie. By a similar reasoning for case 3, with $y(x_1) > y(x_2)$, the desired maximum will be found to lie in the double interval

$$a \leq x^* < x_2$$

For case 2, in which the two y values are equal, the optimum will lie in the center interval. Normally, this situation would be handled along with one of the other two cases, since the center interval is included in both of the previous results.

Thus, at any time, if we make more than one measurement, or an additional measurement if some already exist, we may apply the above reasoning, based upon unimodality, to reduce the interval in which we are certain to find the optimum value $y(x^*)$.

Fig. VI-3 Unimodality in numerical search. (*a*) $y(x_1) < y(x_2)$; (*b*) $y(x_1) = y(x_2)$; (*c*) $y(x_1) > y(x_2)$.

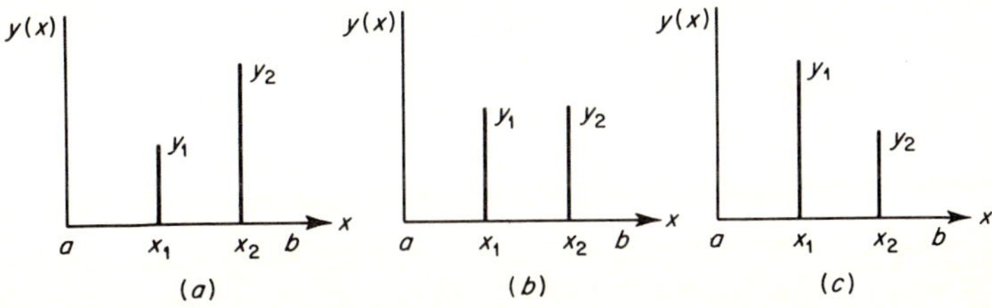

Example VI-2: Region elimination

If all the experimental work is planned in advance and it is found that $y(x_1) < y(x_2)$ and $y(x_2) > y(x_3)$, what can be said about the position of the maximum value of $y(x)$?

From the first inequality we can say that the optimum position will not lie in the range $x \leq x_1$. From the second inequality, we see that it will not lie in the region $x \geq x_3$. We may then conclude that the maximum value must lie in the double interval given by $x_1 < x^* < x_3$.

2.1.2 *Gradient search*

If the function derivative is available, this additional information allows the optimum to be isolated with fewer experiments. Thus, with one experiment within a given search interval, a knowledge of the gradient, together with the assumption of unimodality, allows us to isolate the desired optimum to within one of the resulting subintervals.

2.2 Approach to Search

All search methods can be classified as sequential or preplanned.

A *preplanned search* is one in which all the experiments are completely scheduled in advance, and the results do not and cannot influence the course of these experiments. This situation is sometimes unavoidable, for example, in the placing of thermocouples during the design of a bed reactor to facilitate later search for a "hot spot." In general, however, it is preferable to examine the results of one set of experiments before positioning and carrying out another set. This is the method of sequential search.

The *sequential search methods* decide, after each set of experiments, where the most promising areas of search are located, and the search is continued in the most favorable region. The basic notion is that each trial will eliminate a portion of the region being searched. In this way the exploration is confined to successively smaller sections, until the final interval of uncertainty is quite small and within the desired limits. This interval of uncertainty is the region as yet unexplored but which is known to contain the optimum.

2.3 Multiple Optima

While many objective functions possess the property of unimodality, precautions must be taken to ensure the validity of this assumption. If more than one optimum is suspected, the base point should be changed and the search for an optimum repeated. If the repeated calculation yields the

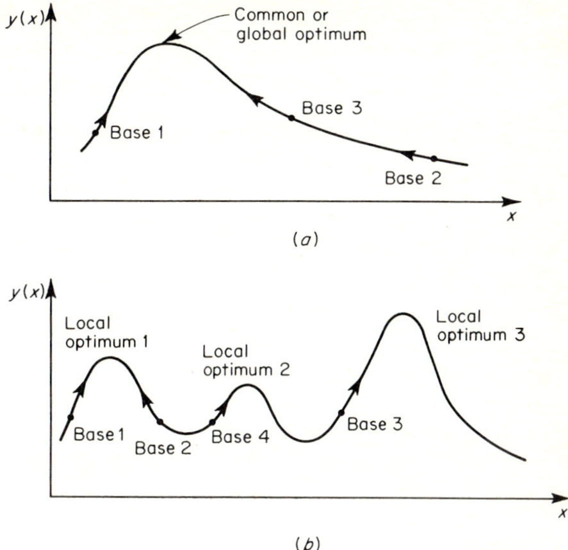

Fig. VI-4 Search of unimodal and multimodal functions. (*a*) One mode; (*b*) more than one mode.

same result, then the optimum can, with a degree of confidence which depends on the number of repetitions, be called a global optimum. If several local optima are found, the various values of the objective function must be compared to select the best one. This may be illustrated in Fig. VI-4, in which the arrows indicate the direction of search from a particular base point. Such a bimodal example has been discussed, for example, by Aris, Rudd, and Amundson [1960].

There is, of course, no substitute for an understanding of the form of a function, for it is always possible to think of a particular function which would confuse and defeat a given search routine. The main purpose of any preliminary calculation is to divide the original region so as to locate the subregions within which the function is unimodal. Once we have obtained such a region containing only a unimodal function, any of the search methods of this chapter may be used with impunity.

3 NUMERICAL SEARCH WITH UNRESTRICTED FUNCTIONS

The practical solutions to most problems encountered lie within restricted ranges of variables. In some cases, however, especially when conducting a one-dimensional search along a chosen direction in some of the n-dimensional direct searches, as described in Chap. VIII, the range is unknown. Starting from a given base point, movement is known to lie only along the specified direction; this is our open-ended, or unrestricted, problem.

What, then, is our best approach to this search for an optimum in one dimension?

3.1 A basic Gradient-free Approach to Unrestricted Search

Let us initially carry out a search using a fixed step size (Sec. 3.1.1). While this is easy to use, it is not very efficient, and it would be preferable to use an improved form by incorporating an acceleration method (Sec. 3.2).

3.1.1 Fixed step size

One approach is to use a fixed step size and to advance from the initial base point in a favorable direction. This is not particularly efficient, but it is a method which might have to be used, for example, in problems involving integration. Thus, in optimization with respect to an upper limit of integration, we might wish to maximize y with respect to x, where

$$y = \int_0^x f(\lambda)\, d\lambda$$

The method is based upon the assumption of unimodality (Sec. 2.1) and is a sequential one. Starting from a base point (obtained as the end point in a previous search, or arbitrarily defined) such as point $x_0^{(1)}$ of Fig. VI-5, and assuming a step size s in x, point $x_1^{(1)} = x_0^{(1)} + s$ can then be tested. If

Fig. VI-5 Open-ended search, showing the use of stages.

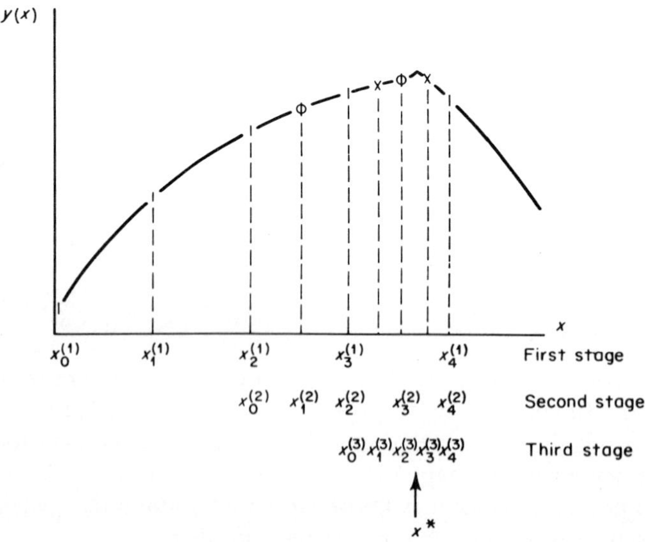

$y(x_1^{(1)})$ is greater than $y(x_0^{(1)})$, and we are looking for a maximum, the assumption of unimodality allows us to say that the desired maximum will not lie at $x < x_0^{(1)}$, and so the search can be continued using $x_1^{(1)}$ as a new base point. The search thus continues along points $x_2^{(1)}$ $x_3^{(1)}, \ldots,$ using the unimodality assumption when testing each pair of experiments. This is continued until the final point, such as $x_4^{(1)}$, shows a decrease. The search is then terminated at point $x_4^{(1)}$, since it will be known from the assumption of unimodality that the desired maximum will lie in the final double interval of search. If $y(x_1^{(1)})$ is less favorable than $y(x_0^{(1)})$, the search should be carried out in the reverse direction ($x_{-1}^{(1)}$, etc.). If it happens that both $y(x_{-1}^{(1)})$ and $y(x_1^{(1)})$ are less favorable, this implies that the desired result will lie in the double interval $x_{-1}^{(1)} < x^{(1)} < x_1^{(1)}$.

Having located the local optimum to an accuracy of within two step sizes in the first stage, the search can be stopped or continued, depending upon the final accuracy required. For greater precision, the search can be repeated using the same rules, commencing at a new base point situated at the beginning of the final interval and using a reduced step size (such as half the previous size). Several cycles of this approach are indicated in Fig. VI-5, it being obvious that not all succeeding experiments need to be carried out, since some of the experimental locations will lie in already tested positions. The final location of the optimum will here lie in the double interval $x_2^{(3)}$ to $x_4^{(3)}$.

While this approach involves many calculations, it is easily programmed for a computer. This is an example of a sequential approach with a decision being made after every experiment so that past results are utilized in the placing of the next experiment. If, on the other hand, all experiments had to be preplanned, this approach would not be possible, and the experiments would have to be laid out at intervals corresponding to the final accuracy desired.

Example VI-3: Search of an open-ended function

Determine the minimum of

$$y = (x - 100)^2$$

imagining that the function is so complex that differentiation is not possible. A direct search is therefore called for, and there is no initial indication as to the location of the optimum, since the range of x is unrestricted. The search could successfully proceed as follows:

1. Choose an initial arbitrary base point, say, $x_0^{(1)} = 0$.
2. Evaluate the objective function at this point, finding for this particular case $y(x_0^{(1)}) = 10{,}000$.

3. Change the base point by an increment, say, $s = -3$. Thus we could first test $x_1^{(1)} = -3$.

Since $y(-3) > y(0)$ and the function is known to be unimodal, the minimum cannot, therefore, lie in the region $x < -3$. We must then investigate the other direction, starting with $x_2^{(1)} = 3$. Continuing this reasoning, a sequence of tests will yield the results shown in Table VI-2*a*.

TABLE VI-2 STAGES IN AN OPEN-ENDED SEARCH WITH FIXED STEP SIZE

(*a*) *First-stage search* ($s = 3$)

k	0	1	2	3	4	...	32	33	34	35
$x_k^{(1)}$	0	−3	3	6	9	...	93	96	99	102
$y(x_k^{(1)})$	10,000	10,609	9,409	8,836	8,281	...	49	16	1	4

(*b*) *Second-stage search* ($s = 0.3$)

k	0	1	2	3	4
$x_k^{(2)}$	99.0	99.3	99.6	99.9	100.2
$y(x_k^{(2)})$	1.00	0.49	0.16	0.01	0.04

Since the minimum is obviously in the range $96 < x < 102$, the step size could be reduced to $s = 0.3$, with results as shown in Table VI-2*b*. Here the starting point $x_0^{(2)}$ is placed at $x_{34}^{(1)}$. The optimum now lies in the range $99.6 < x < 100.2$, and a further reduction of the step size could be used to determine the optimum within a closer accuracy (given by double the final step size).

Let us now show that the use of such fixed step sizes is not the best approach to this type of optimization.

3.1.2 *Acceleration*

The basic approach described above is not particularly efficient, its fundamental weakness common to all fixed-step-size techniques for unrestricted search residing in the fact that the position of the desired optimum could lie at some unknown point in the range $-\infty < x < \infty$.

Suppose, for example, $x^* = 10^6$ and a unit step size were selected. Starting from the origin, 1,000,001 experiments would be required to bracket the position of the optimum to within a length of twice the step size. This, of course, is an extreme example, but a thoughtful consideration of more realistic possibilities forces one to conclude that the use of a fixed step size is not practicable in sequential search. A much better technique is to increase the step size with each successive move as long as this results in an improvement of the objective function.

There are two distinct classes of search techniques to be studied. The classification depends on whether or not the first derivative of the objective function is readily available to aid the search. There are certainly many cases in which the model is of the black-box type, thus requiring several experiments to establish the value of the derivative at some point. In such cases a *direct search* technique (Sec. 3.2), one not requiring derivatives, is highly recommended. If, on the other hand, the objective function is differentiable, then very effective *gradient* methods (Sec. 3.3) for numerical search can be brought to bear on the problem.

It should be noted that the search, whether direct or gradient, must use some type of acceleration of the step size to be really effective. This is an essential feature of these sequential searches.

3.2 Direct Search with Acceleration

Our problem is to find the maximum of an objective function $y(x)$, where x is unbounded by constraints. The basic approach may be divided for convenience into three sections: an accelerated search, which will determine the optimum within a certain region (Sec. 3.2.1), a choice of the initial step size (Sec. 3.2.2), and a further improvement in the estimated position of the optimum (Sec. 3.2.3).

3.2.1 The determination of the location of the optimum

In any direct search, any two objective function evaluations can be used to eliminate a region from further exploration. Thus, if we seek a maximum, with $x_1 > x_0$ and $y(x_1)$ greater than $y(x_0)$, we can say from an assumption of unimodality that the maximum will not lie in the region $x < x_0$. We shall always conclude that the optimum lies in a double interval, here, x_0 to x_1 and x_1 to ∞. Proceeding in this manner, as before, we may add new experiments, one at a time, and with each one further reduce the interval still to be searched. Finally, we shall conclude with the situation as depicted in Fig. VI-6, in which the maximum is contained within the double interval x_α to x_γ, since $y(x_\alpha) < y(x_\beta)$ and $y(x_\beta) > y(x_\gamma)$. Let us now consider how to achieve this final state more efficiently than by using a fixed step size in the search.

Let us assume that we start from a base point x_0, with an initial step length s. The discussion of the rather critical choice of this length will be deferred to Sec. 3.2.2. The choice of x_0 is rather arbitrary, but it will be our best

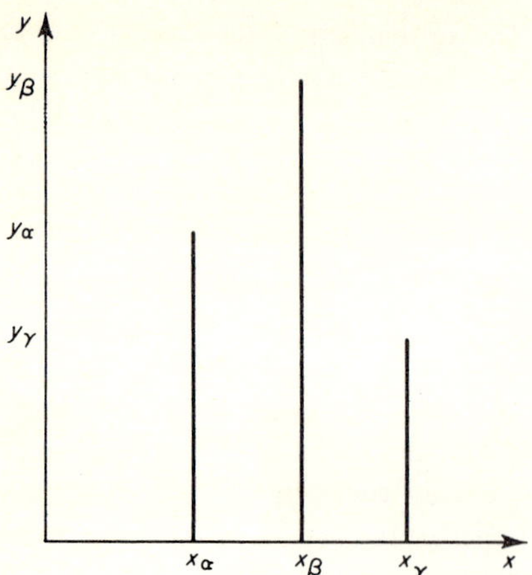

Fig. VI-6 Final stage of a direct search.

guess at the location of the optimum, and is thus an initial approximation to the desired result. Let us evaluate the objective function at x_0 and $x_1 = x_0 + s$. Three possible situations result:

If $y(x_0) = y(x_1)$, then the maximum lies in the interval $x_0 < x < x_1$.

If $y(x_1) > y(x_0)$, then the maximum will lie at $x > x_0$, and we must search along the x axis in the direction of increasing s. The method of Swann [1964] requires that the objective function be evaluated at the sequence of points $x_0;\ x_1 = x_0 + s;\ x_2 = x_1 + 2s;\ x_3 = x_2 + 4s;\ \cdots;\ x_k = x_{k-1} + 2^{k-1}s$, where x_k is the first position tested which fails to yield an increase in the value of the objective function; i.e.,

$$y_0 < y_1 < y_2 \cdots < y_{k-1}$$

but

$$y_{k-1} \geq y_k$$

This situation is depicted in Fig. VI-7a, where, for a unimodal function, the maximum is shown to lie between x_{k-2} and x_k. Note that we have advanced in the direction of increasing x, using a step which is doubled in length after each experiment. This procedure represents a form of acceleration. The distance $d/2$ between points x_{k-2} and x_{k-1} is given by

$$\frac{d}{2} = x_{k-1} - x_{k-2} = 2^{k-2}s \tag{3.2-1}$$

and the following interval is, of course, twice this distance, as depicted in Fig. VI-7a. The maximum is thus located within the range $x_{k-2} < x < x_k$, a double interval of total length $3d/2 = 3(2^{k-2}s)$.

For the third case, $y(x_0) > y(x_1)$, we shall eliminate the region $x > x_1$ and search at decreasing values of x. No different arguments need be presented

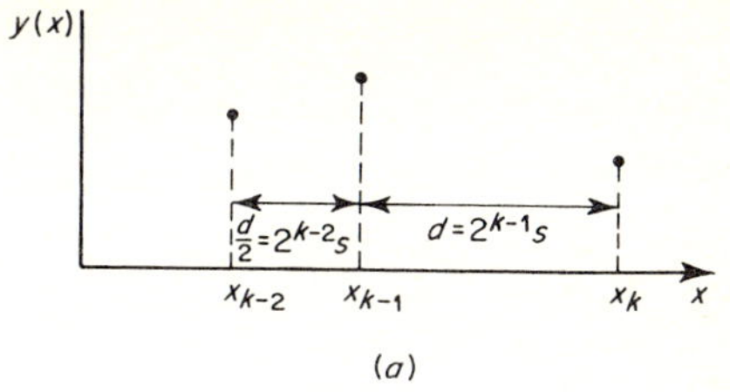

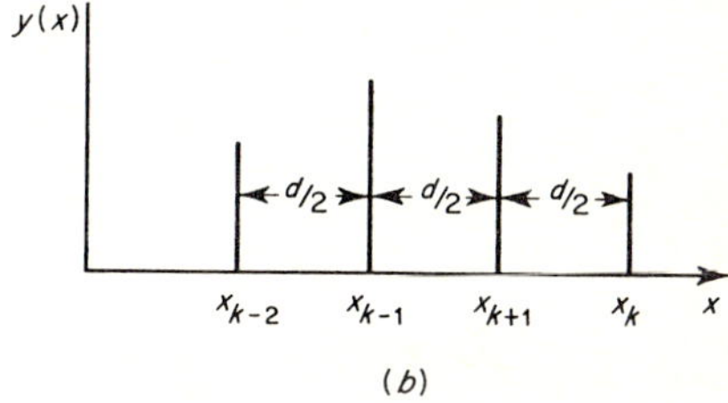

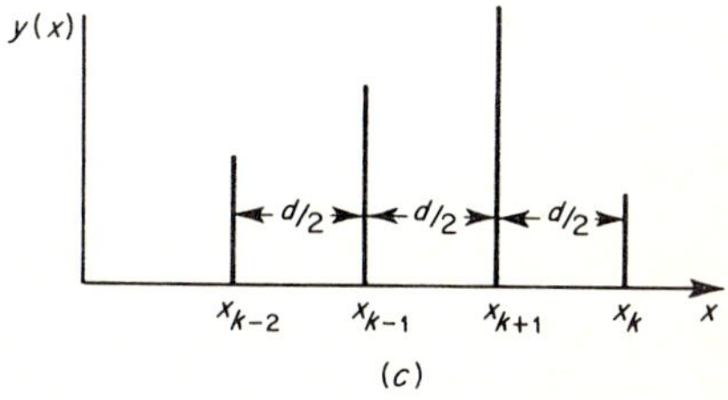

Fig. VI-7 The final stage modification of Swann: basic method (*a*) and the use of an extra experiment yielding (*b*) $y(x_{k+1}) < y(x_{k-1})$ or (*c*) $y(x_{k+1}) > y(x_{k-1})$.

since we can now redefine $x_0 = x_1$ and $x_1 = x_0$ and search at locations $x_k = x_{k-1} - 2^{k-1}s$ as before, stopping when $y(x_k)$ is the first value tested which fails to improve the objective function.

We now have determined the location of the optimum to within a restricted region, using a method of acceleration.

Example VI-4: Acceleration in an open-ended search

Let us reconsider Example VI-3. Since no information is, supposedly, available on the location of the optimum, let us arbitrarily start at the origin with a step size s of 3.0. Doubling the step size after each successful experiment yields the results shown in Table VI-3. The search is stopped at $k = 6$, since the objective function is found to increase in value after passing a low value. The desired minimum is thus restricted to the double interval $45 < x < 189$, with the best approximation as $y(93) = 49$.

TABLE VI-3 ACCELERATION IN AN OPEN-SIDED SEARCH

k	0	1	2	3	4	5	6
x_k	0	3	9	21	45	93	189
$y(x_k)$	10,000	9,409	8,281	6,241	3,025	49	7,921

The reduction in the number of experiments required compared with the first stage of the fixed-step procedure of Example VI-3 will be obvious. The actual saving will of course depend upon the choice of step size made. Various methods of further improving the accuracy of the location of the minimum are discussed in Sec. 3.2.3.

3.2.2 *The choice of an initial step size*

The first problem in any search is the choice of a step size. This is important, since it will largely govern the number of experiments required. From our previous discussion, the total distance moved in seeking a maximum in k moves is the sum of the terms $\sum_{i=0}^{k-1} 2^i s$, when on the kth move the objective function will decrease. Then, clearly,[1]

$$x^* - x_0 < x_k - x_0 = \left(\sum_{i=0}^{k-1} 2^i\right) s = (2^k - 1)s \tag{3.2.3}$$

This inequality can be rearranged to give the number of moves as

$$k > \frac{\ln\,[1 + (x^* - x_0)/s]}{\ln 2} \tag{3.2-4}$$

[1] The series $\sum_{i=0}^{k-1} 2^i$ is a geometric series which can be summed as follows: Let

$$s(k) = \sum_{i=0}^{k-1} 2^i \tag{3.2-2}$$

Then

$$s(k-1) = \sum_{i=0}^{k-2} 2^i$$

The difference between the two series is given by

$$s(k) - s(k-1) = 2^{k-1}$$

We also have, from Eq. (3.2-2),

$$2s(k-1) = s(k) - 1$$

These two results can be combined to eliminate $s(k-1)$, so that

$$s(k) = 2^k - 1$$

which is the desired result.

Since k represents the number of moves, $k + 1$ experiments must have been performed, because the objective function had also to be found at the base point x_0. Thus $N = k + 1$ experiments have to be used to define a final region of length $\frac{3}{2}d$ containing the desired optimum. This length is

$$\tfrac{3}{2}d(k) = 3(2^{k-2}s)$$

from Eq. (3.2-1), k being defined by inequality (3.2-4).

Assuming that $x^* - x_0$ is 10^6, an extreme case, the results of using different initial step sizes, s, are presented in Table VI-4. The seemingly erratic

TABLE IV-4 RELATIONSHIP BETWEEN NUMBER OF EXPERIMENTS AND STEP SIZE

$x^* - x_0$	s	$N = k + 1$	$\frac{3}{2}d(k)$
10^6	1	21	786,432
10^6	10	18	983,040
10^6	10^2	15	1,228,800
10^6	10^3	11	768,000
10^6	10^4	8	960,000
10^6	10^5	5	1,200,000
10^6	0.9×10^6	3	2,700,000

N here is the minimum value found when x^* lies in the second portion of the final double interval. If x^*lay in the other portion, N would be increased by 1, with a corresponding increase in the interval of uncertainty by a factor of 2.

variation in the interval of uncertainty is due to the fact that, for certain values of s, x^* is almost, but not quite, reached. The next step therefore advances far beyond the optimal point. This is seen to be one of the weaknesses of the method of acceleration. In fact, the region of uncertainty is not particularly sensitive to the value of s, as is seen in Table VI-4, and it need not, therefore, be considered in making our decision as to a suitable choice of s. We should rather consider the number, $k + 1$ of experiments required. It will be seen from the table and inequality (3.2-4) that the smallest number of experiments is needed when s and $x^* - x_0$ are of the same order of magnitude. Let us refer to this distance as the *scale* of the problem; i.e., the scale S is defined as

$$S = x^* - x_0$$

Obviously, the order of magnitude of the scale is not always known, although in a large number of physical problems some estimate of its value can be made. A second method of evaluating the scale is to have an idea of the magnitude of the maximum. Then, as will be discussed in Sec. 3.3, one can estimate the scale using the first derivative. In the total absence of any information or intuitive basis, the scale can be estimated by fitting Eq.

(6.1-1) to the first three experiments, whether the maximum had been bracketed or not. Once the constants in Eq. (6.1-1) have been evaluated, the scale can be determined from Eq. (6.1-3) as

$$x^* - x_0 = \tfrac{1}{2}\left(x_1 - x_0 - \frac{\beta_1}{\beta_{11}}\right) \approx S \tag{3.2-5}$$

where β_1 and β_{11} are determined as in Sec. 6. Indeed, Powell [1964] uses an extrapolation of the quadratic form to bracket the optimum rather than use the stepping procedure. The accuracy of both the present estimate and the Powell search method depends upon the objective function being almost quadratic.

The choice of the step size should thus be, roughly, of the same order of magnitude as the scale, as might have been deduced intuitively.

3.2.3 *A closer estimation of the optimum*

Having carried out a $k + 1$ experiment search and defined the optimum within a known region, we may further improve our estimation of its location by one of three methods: fitting a response curve, starting a new unrestricted search, or using the methods of Sec. 4 for restricted search.

Fitting a Response Curve Since we defined the final region of uncertainty by three experiments at x_{k-2}, x_{k-1}, and x_k, as shown in Fig. VI-7*a*, we might fit an approximating second-order polynomial through these points, as discussed in Sec. 6.1. The resulting equation is

$$y = \beta_0 + \beta_1(x - x_{k-2}) + \beta_{11}(x - x_{k-2})(x - x_{k-1})$$

with the β coefficients defined by Eq. (6.1-2). The estimated location of the optimum is then, from Eq. (6.1-3),

$$x^* = \tfrac{1}{2}\left(x_{k-2} + x_{k-1} - \frac{\beta_1}{\beta_{11}}\right)$$

If the objective function were a quadratic, this would yield an accurate location for the optimum. In any case we have made a closer estimation without incurring further experiments.

The approach of Swann [1964], a modification of the method of Smith [1962], is essentially that described in Secs. 3.2.1 and 3.2.2, with an additional stage. Having performed the $k + 1$ experiments to define the optimum region, a further experiment is added midway between x_{k-1} and x_k, as shown in Fig. VI-7*b* and *c*, i.e., at

$$x_{k+1} = x_{k-1} + \frac{d}{2} \tag{3.2-6}$$

The group of three experiments, at either x_{k-2}, x_{k-1}, and x_{k+1} or x_{k-1}, x_{k+1}, and x_k, depending on whether $y(x_{k-1})$ is greater or less than $y(x_{k+1})$ (see Fig. VI-7*b* and *c*, respectively) is then selected, and a response curve is fitted as discussed above. Since

$$x_k - x_{k+1} = x_{k+1} - x_{k-1} = x_{k-1} - x_{k-2} = \frac{d}{2}$$

the optimum will now be located within the double interval of total length d by the use of a single additional experiment (see also Secs. 4.1 and 4.2), and its estimated location will be (from Sec. 6.1)

$$x^* = x_{k-1} - \frac{d}{4}\frac{y_{k+1} - y_{k-2}}{y_{k+1} - 2y_{k-1} + y_{k-2}} \qquad \text{if } y(x_{k-1}) > y(x_{k+1})$$

or $$x^* = x_{k+1} - \frac{d}{4}\frac{y_k - y_{k-1}}{y_k - 2y_{k+1} + y_{k-1}} \qquad \text{if } y(x_{k-1}) < y(x_{k+1})$$

This modification to the basic search allows the final three points to be closer together, thus affording a much more accurate interpolation.

Search in a Limited Region Once the optimum has been located within a well-defined region, as seen in Sec. 3.2.1, we may accept this as being a sufficiently accurate representation or we may improve the estimation by fitting a curve, but if further experiments are to be used, it may be inefficient in the general case not to utilize one of the most effective search methods developed for this special problem. We should thus often use the unrestricted methods to limit the optimum to a determined region, and then change to one of the specific methods described in Sec. 4.

Continuing the Unrestricted Search On occasion, it may be convenient to use the same unrestricted search routine to avoid writing alternative computer instructions. We should then begin a new sequence of experiments, starting at the best approximation to the optimum already determined. Smith [1962] suggests that a smaller initial step size should be introduced, but as Swann [1964] points out, this may involve several unnecessary function evaluations, especially if the new base point was some distance away from the previous base point [that is, $x_{k-1} - x_0 = (2^{k-1} - 1)s$ is large]. In this case, the final double interval, $\frac{3}{2}d(k)$, might still be appreciably large, and it appears to be more advisable to reduce the initial step size once the optimum has been approached more closely. This approach can be repeated until the desired location is determined within the required degree of accuracy.

Example VI-5: A closer estimation of the optimum

Continuing Example VI-4, we might obtain closer approximations to the location of the optimum by use of the methods of Sec. 3.2.3.

Fitting a surface From Table VI-3, it is seen that the optimum lies in the interval $45 < x < 189$. From the values $y(45)$, $y(93)$, and $y(189)$, we can fit a curve, using the equations of Sec. 6, and find the estimated position of the optimum as given by Eq. (6.1-3). Since

$$\beta_0 = y(45) = 3{,}025$$

$$\beta_1 = \frac{y(93) - y(45)}{93 - 45} = -62$$

$$\beta_{11} = \frac{\dfrac{y(189) - y(45)}{189 - 45} - \dfrac{y(93) - y(45)}{93 - 45}}{189 - 93} = 1$$

the desired location is

$$x^* = \frac{45 + 93}{2} + \frac{62}{2(1)} = 100.0$$

Here, of course, the estimate gives the exact location, since the original function was a quadratic. The fitted curve used in this derivation can be found from Eq. (6.1.1).

Search in a limited region Any suitable method from Sec. 4 can be chosen, since the search is now in a well-defined region. If the number of experiments used is critical, the Fibonacci approach (Sec. 4.2.3)

TABLE VI-5 STAGES IN A CONTINUED ACCELERATED SEARCH

(*a*) *Stage* 2 ($s = 3.0$)					
k	0	1	2	3	
$x_k^{(2)}$	93	96	102	114	
$y(x_k^{(2)})$	49	16	4	196	
(*b*) *Stage* 3 ($s = 0.3$)					
k	0	1	2	3	4
$x_k^{(3)}$	102.0	102.3	101.4	100.2	97.8
$y(x_k^{(3)})$	4.00	5.29	1.96	0.04	4.84
(*c*) *Stage* 4 ($s = 0.03$)					
k	0	1	2	3	4
$x_k^{(4)}$	100.2	100.23	100.14	100.02	99.78
$y(x_k^{(4)})$	0.0400	0.0529	0.0196	0.0004	0.0484
(*d*) *Stage* 5 ($s = 0.03$)					
k	0	1	2		
$x_k^{(5)}$	100.02	100.05	99.96		
$y(x_k^{(5)})$	0.0004	0.0025	0.0016		

TABLE VI-6 A SECOND SEARCH

(a) *Stage* 2 ($s = 0.3$)

k	0	1	2	3	4	5	6
$x_k^{(2)}$	93	93.3	93.9	95.1	97.5	102.3	111.9
$y(x_k^{(2)})$	49.00	44.89	37.21	24.01	6.25	5.29	141.61

(b) *Stage* 3 ($s = 0.03$)

k	0	1	2	3	4	5	6	7
$x_k^{(3)}$	102.30	102.33	102.24	102.12	101.88	101.40	100.44	98.52
$y(x_k^{(3)})$	5.29	5.43	5.02	4.49	3.53	1.96	0.19	2.19

(c) *Stage* 4 ($s = 0.003$)

k	0	1	2	3	4	5	6
$x_k^{(4)}$	100.440	100.443	100.434	100.422	100.398	100.350	100.254
$y(x_k^{(4)})$	0.1936	0.1962	0.1884	0.1781	0.1584	0.1225	0.0645
k	7	8					
$x_k^{(4)}$	100.192	99.678					
$y(x_k^{(4)})$	0.0369	0.1037					

(d) *Stage* 5 ($s = 0.0003$)

k	0	1	2	3	4	5
$x_k^{(5)}$	100.1920	100.1923	100.1914	100.1902	100.1878	100.1830
$y(x_k^{(5)})$	0.03686	0.03698	0.03663	0.03618	0.03527	0.03349
k	6	7	8	9	10	
$x_k^{(5)}$	100.1734	100.1542	100.1158	100.0390	99.8854	
$y(x_k^{(5)})$	0.03007	0.02378	0.01341	0.00152	0.01313	

would prove the most economical,[1] but the modified form (Sec. 4.2.4) might be as suitable in the majority of cases.

Continuing with the unrestricted search routine Since $y(93)$ is the best estimate of the desired minimum, let us repeat the search, using the technique of Example VI-4. If we wish to find the minimum value of y within 0.05 units of y and its location within 0.1 units of x, the various stages of one approach to the search are given in Tables VI-5*a* to *d*. Using various initial step sizes s, the location is $99.96 < x < 100.05$, with the best approximation $y(100.02) = 0.0004$. It will be seen that the initial step size was not reduced after every stage.

An alternative sequence of search, using progressively smaller values of s at each stage, is shown for comparison in Tables VI-6*a* to *d*. We

[1] While the Fibonacci search (Sec. 4.2.3) guarantees the upper limit on the number of experiments used, this does not mean that other search methods cannot on occasion locate the optimum region with a smaller number.

may conclude that there is little advantage in reducing the step size at each stage if the optimum is still a significant distance from the stage base point. Better results are obtained, as deduced in Sec. 3.2.2, when the initial step size s is of the same order of magnitude as the "scale" of the problem. After the fifth stage the optimum will lie in the range $99.88 < x < 100.12$, with $y(100.039) = 0.0015$ as the best approximation. This is still not within the required degree of accuracy.

3.3 Gradient Search with Acceleration

As in the case of direct search methods, there are three stages in gradient search. The first stage is that of selecting a step size or, as we have seen in Sec. 3.2.2, an estimate of the scale. Let us assume that an estimate of the optimal value of the objective function is available. Call this estimate y_e, where $y_e \approx y(x^*)$. If we assume that $y(x)$ is a quadratic function, then by a Taylor series expansion

$$y(x) = y_0 + \left(\frac{dy}{dx}\right)_0 (x - x_0) + \frac{1}{2}\left(\frac{d^2y}{dx^2}\right)_0 (x - x_0)^2 \tag{3.3-1}$$

This series terminates at the second-order terms because of the assumed quadratic behavior of $y(x)$. The derivatives are evaluated at the base point x_0. The scale S is $x^* - x_0$. The first derivatives are assumed here to be easily computable, and indeed form the basis of our computational scheme; the second derivatives will not, however, be generally known. We can express the second derivative in terms of the scale and the estimate of the objective function by setting $x = x^*$ in Eq. (3.3-1) and solving for the second derivative as follows:

$$\left(\frac{d^2y}{dx^2}\right)_0 = \frac{2(y_e - y_0)}{S^2} - \left(\frac{dy}{dx}\right)_0 \frac{2}{S} \tag{3.3-2}$$

Since the first derivative vanishes at the optimum, we have, by differentiating Eq. (3.3-1),

$$\frac{dy}{dx} = \left(\frac{dy}{dx}\right)_0 + \left(\frac{d^2y}{dx^2}\right)_0 S = 0 \tag{3.3-3}$$

Eliminating the second derivative between Eqs. (3.3-2) and (3.3-3), we find that

$$S = \frac{2(y_e - y_0)}{(dy/dx)_0} \tag{3.3-4}$$

Equation (3.3-4) will be exact, provided that the estimate of the optimal value is precise and the objective function is quadratic. This is rarely the case. The expression does, however, yield an estimate of the scale in practical problems, and in fact Fletcher and Reeves [1964] recommend choosing

$$s = |S| \qquad \text{if } 0 < |S| < 1 \tag{3.3-5a}$$

but otherwise take

$$s = 1 \tag{3.3-5b}$$

In other words, the step size s is taken as the minimum of $(1, |S|)$, with $|S|$ given by Eq. (3.3-4). From Eqs. (3.3-4) and (3.3-5a), it will be seen that the step size is chosen to be of the same order of magnitude as the scale if $y_e - y_0$ is small.

The next stage in the search is the determination of both y and dy/dx at the sequence of points x_0; $x_1 = x_0 + s$; $x_2 = x_1 + 2s$; . . . ; $x_k = x_{k-1} + 2^{k-1}s$. Here x_k, the location of the last point in the search, is the first point at which dy/dx changes sign or at which y does not further improve.

Some steps are shown in Fig. VI-8. It will be seen that the derivative has enabled us to reduce our search by one point compared with a direct search, and in general the optimum will now be located within a single interval, in contrast to the previous double interval required when no derivatives are

Fig. VI-8 Gradients in numerical search. (a) $y_k < y_{k-1}$; (b) $y_k > y_{k-1}$.

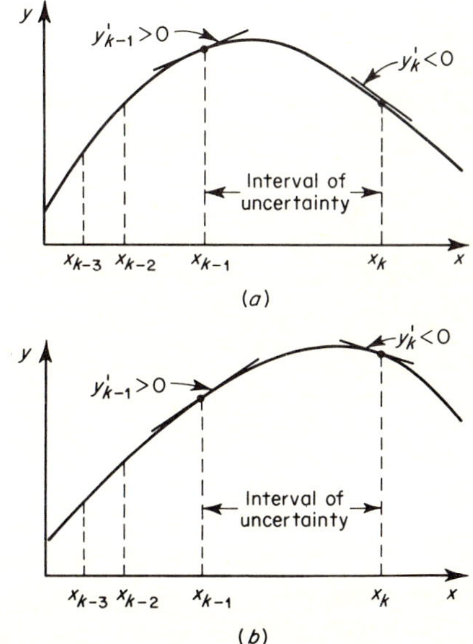

available. The interval of uncertainty is thus

$$d = x_k - x_{k-1} = 2^{k-1}s \tag{3.3-6}$$

The final stage in the calculation is an interpolation between the last two points, x_k and x_{k-1}, evaluated. Since we have the additional information from the derivative, four constants can be determined. We can therefore take the third-degree polynomial,

$$y \approx \beta_0 + \beta_1(x - x_k) + \beta_{11}(x - x_k)(x - x_{k-1}) + \beta_{111}(x - x_k)^2(x - x_{k-1}) \tag{3.3-7}$$

as the approximation to the objective function. The four constants β are to be determined from the known values at $x = x_k$ ($y = y_k$, $dy/dx = y'_k$) and at $x = x_{k-1}$ ($y = y_{k-1}$, $dy/dx = y'_{k-1}$). These constants are $\beta_0 = y_k$, $\beta_1 d = y_k - y_{k-1}$, $\beta_{11}d^2 = y'_k d + y_{k-1} - y_k$, and $\beta_{111}d^3 = (y'_k + y'_{k-1})d + 2y_{k-1} - y_k$, where $d = x_k - x_{k-1}$. Differentiating Eq. (3.3-7) and setting the resultant expression equal to zero yields the estimate x_e^* for the optimum (if $\beta_{111} \neq 0$)

$$x_e^* = x_{k-1} + \frac{A + y'_{k-1}d + B}{2A + (y'_k + y'_{k-1})d}\,d \tag{3.3-8}$$

where $A = 3(y_{k-1} - y_k) + d(y'_{k-1} + y'_k)$ and $B^2 = A^2 - y'_k y'_{k-1} d^2$. If $\beta_{111} = 0$, then

$$x_e^* = x_{k-1} + \frac{d}{2}\left(1 + \frac{y_k - y_{k-1}}{y_k - y_{k-1} - dy'_k}\right) \tag{3.3-9}$$

$y(x_e^*)$ should then be tested. If both y_k and y_{k-1} are worse than $y(x_e^*)$, we may accept this estimate of the optimum. Otherwise, according to whether the gradient at this location is negative or positive, the interpolation would be repeated over the interval x_{k-1} to x_e^* or x_e^* to x_k.

We now have located the optimum at an estimated position within a single interval. This method is essentially that of Fletcher and Reeves [1964], which is based upon the work of Fletcher and Powell [1963] and Davidon [1959].

As with direct search (Sec. 3.2), we may increase the accuracy of the search by repeating the procedure, starting from the estimated optimum and using a smaller step size or by using a technique for restricted systems as described in Sec. 4.

Example VI-6: Accelerated gradient search

Let us assume that the values of y and dy/dx can be represented by

$$y = (x - 100)^2$$

$$\frac{dy}{dx} = 2(x - 100)$$

for the purposes of illustration. If the origin is chosen as the base point, and if the minimum is estimated to have a value of 1.0, we may determine its location as follows: First let us choose a scale S. From Eq. (3.3-4) we have that

$$S = \frac{2(y_e - y_0)}{(dy/dx)_0} = \frac{2(1 - 10{,}000)}{-200} \approx 100$$

where y_0 and $(dy/dx)_0$ are calculated values. Then, from Eq. (3.3-5), $|S| > 1$, so that we may take $s = 1$. The results of the search from the origin, using this initial step size, are given in Table VI-7. At $k = 6$ the

TABLE VI-7 RESULTS FOR SEARCH OF OBJECTIVE FUNCTION DEFINED IN EXAMPLE VI-6

Gradient search ($s = 1.0$)

k	$2^{k-1}s$	x_k	$y(x_k)$	$\left(\frac{dy}{dx}\right)_k$
0	—	0	10,000	−200.0
1	1.0	1.0	9,801	−198.0
2	2.0	3.0	9,409	−194.0
3	4.0	7.0	8,649	−186.0
4	8.0	15.0	7,225	−170.0
5	16.0	31.0	4,761	−138.0
6	32.0	63.0	1,369	−74.0
7	64.0	127.0	729	+54.0
8	128.0	255.0	24,025	+310.0

unchanged sign of the derivative indicates that the minimum still lies in the region $x > 63.0$. At $k = 7$, however, the derivative has changed sign, so that the minimum will be in the region $63.0 < x < 127.0$, with $y(127.0) = 729$ as the best estimate.

Interpolation using Eq. (3.3-8) or (3.3-9) then could allow the location of the minimum to be estimated more accurately. If derivatives had not been used, we should have required a further calculation ($k = 8$).

4 RESTRICTED FUNCTIONS

Any restrictions upon functions of a single variable will resolve themselves into a restriction upon the permitted range of the variable, i.e., into constraints of the form

$$x_L \leq x \leq x^L$$

Let us now develop methods suitable for handling restricted functions. The discussion will be confined to direct search techniques, and it will be assumed that the objective functions exist for all x (although discontinuities may be included), and hence values of the independent variable can be chosen at will. Moreover, in all cases, the function will be assumed to be unimodal in the region of search. This assumption is necessary to permit a logical choice of the promising region in which to continue the search. We shall initially consider the methods based upon regularly spaced search intervals (Sec. 4.1) and then remove this restriction (Sec. 4.2). The situation in which the objective function exists only at certain discrete values of the independent variable is examined in Sec. 5.

4.1 Regularly Spaced Divisions

In this section we shall consider those direct search techniques in which the search region $a \leq x \leq b$ is divided equally into a certain number of intervals.

4.1.1 Preplanned (or simultaneous) uniform search methods

The most obvious search method utilizing regularly sized subintervals is that in which the original interval L_0 $(= b - a)$ is divided into $N + 1$ equal intervals. Here the values of the function $y(x)$ will be calculated at the N intermediate points:

$$x_i = a + \frac{i(b - a)}{N + 1} \tag{4.1-1}$$

where $i = 1, 2, \ldots, N$.†

Application of the ideas expressed in Sec. 2.1, based upon a unimodal function, then allows us to determine the double subinterval containing $y(x^*)$. The position of the extremum will then be known to $\pm 1/(N + 1)$ of the length, i.e., within the length $2L_0/(N + 1)$. Thus, in Fig. VI-9, the optimum location found using three experiments will be in the range $\frac{1}{4} < x < \frac{3}{4}$, and

† Here N is number of experiments, provided none are performed at the boundary of the interval, that is, $y(a)$ or $y(b)$ are not usually determined.

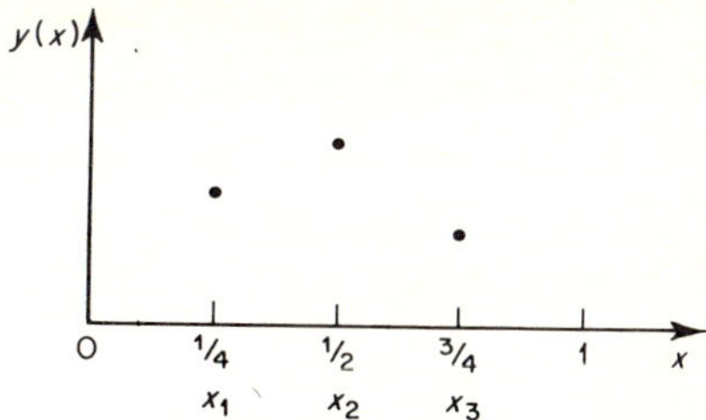

Fig. VI-9 Three-experiment preplanned search.

the optimum value of y, designated as $y(x^*)$, is equal to or greater than $y(\frac{1}{2})$, since this latter value was the largest found by the three experiments.[1]

Example VI-7: Three-experiment preplanned search

Suppose, in a liquid-liquid extraction unit, such as that described in Example VII-7, that the profit function y is dependent upon the total quantity x of solvent used. The function y will then be a unimodal function of x. How accurately can we find the position of the maximum profit if we are allowed only four experiments? (For the purposes of illustration, base any decisions upon the assumption that the optimum will be found at $x = 4.9$, using an initial search region of $0 \leq x \leq 10$.)

Place the four experiments at equal intervals along the range, i.e., at positions $x = 2$, 4, 6, and 8. Calculate or determine the value of y at each of these positions. Compare any two values. Thus,

If $y_4 > y_2$, then the optimum does not lie in $x \leq 2$

If $y_6 > y_8$, then the optimum does not lie in $x \geq 8$

If $y_4 > y_6$, then the optimum does not lie in $x \geq 6$

If $y_6 > y_4$, then the optimum does not lie in $x \leq 4$

The optimum will therefore, in the first case, $y_4 > y_6$, lie in the range $2 < x < 6$, but, in the second case, will be in the range $4 < x < 8$. For the first, the optimum may lie in either $2 < x < 4$ or $4 < x < 6$, but we cannot decide this without further experiments.

The final length of the interval containing the optimum position will be $x = 4$ units, so that the accuracy of the search will be within $\frac{4}{10}$,

[1] Note that the variable x can be transformed so that the region of interest is $0 \leq x' \leq 1$. This transformation is, of course,

$$x' = \frac{x - a}{b - a}$$

that is, 40 percent. This could have been derived from the formula

$$\frac{L_N}{L_0} = \frac{2L_0}{1+N}\frac{1}{L_0} = \frac{2L_0}{(1+4)L_0} = 40 \text{ percent}$$

To obtain the optimum within a greater accuracy than that determined by the first search, the final range may be taken as the starting range and be divided up again into $N+1$ divisions. The procedure can be repeated until the desired accuracy is attained.

4.1.2 *Measure of effectiveness in conducting a search*

The approach presented above, while perfectly feasible, is rather unimaginative. It might prove to become a long and tedious calculation if it were desired to calculate the position of an extreme value to a fine degree of accuracy. Thus, to get this optimum position to within 10^{-3} of the original interval, we should have to carry out at least 1,999 calculations $[2/(1+N) = 10^{-3}]$. Other methods of search may prove, and will be shown to be, much more economical, in the sense of the number of calculations, which, after all, cost money, either in man-hours of hand calculation or experiment or in machine time on a computer.

In order to compare the efficiency of any search technique, it is necessary to have some measure of the efficiency of a search. This is taken as either (1) the fractional double subinterval α, attained in N calculations, or (2) the number of calculations or experiments, N, required to achieve a given maximum uncertainty in the size of final interval, α, containing the optimum.

Thus, if L_N is the length of the final interval in which the extreme value $y(x^*)$ is contained after N experiments, and L_0 is the length of the original interval (a,b), then

$$\alpha = L_N/L_0$$

is the fractional interval containing the desired extreme point.

Hence, in the uniform search method, with preplanned experiments,

$$\alpha = \frac{L_N}{L_0} = \frac{2}{N+1} \tag{4.1-2}$$

where N is the number of experiments. This could be rearranged to give N directly, if the number of experiments were the measure of search efficiency, so that alternatively,

$$N \geq \frac{2}{\alpha} - 1 \qquad \text{for integer } N \tag{4.1-3}$$

Either may be used as the case may arise. Such measures of effectiveness are feasible with one-dimensional searches only.

4.1.3 *Sequential uniform search methods*

We have considered the situation in which all the experiments had to be positioned in advance of the trial. Let us now consider the sequential extension of the uniform search technique, but instead of placing all experiments in advance, we must look at the results from one small set of experiments before planning the next set.

Sequential Search with Two Equally Spaced Experiments One such method of sequential search would be the use of two experiments in each cycle.

Calculate, first, the two values of the function at positions at one-third and two-thirds of the search interval, i.e.,

$$x_1 = a + \tfrac{1}{3}L_0 \qquad \text{and} \qquad x_2 = a + \tfrac{2}{3}L_0$$

We then have the two possibilities shown in Fig. VI-10 (the case of equal values of the function having been shown to be covered by either of these). We shall now proceed as before and choose the double interval, which contains the extreme value as shown. Its length $L_N = L_{2m}$ for $N = 2$ is $\tfrac{2}{3}L_0$, where m is the number of the cycle. This new interval of search may now be treated in exactly the same way, adding two experiments and making a choice of a new subinterval containing the extreme value. This will have the length $L_4 = (\tfrac{2}{3})^2L_0$. With m repetitions of this procedure, the position of the extreme value may be located to within a length L_N

$$L_N = L_{2m} = (\tfrac{2}{3})^m L_0$$

The accuracy α of the search would then be

$$\alpha = \frac{L_N}{L_0} = (\tfrac{2}{3})^m = (\tfrac{2}{3})^{N/2} \tag{4.1-4}$$

Alternatively, the number of calculations required, N, to obtain the optimum position within a length α would be

$$N = 2m \geq 2\,\frac{\log \alpha}{\log \tfrac{2}{3}} \qquad \text{for even } N \tag{4.1-5}$$

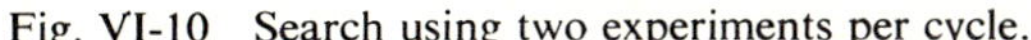

Fig. VI-10 Search using two experiments per cycle.

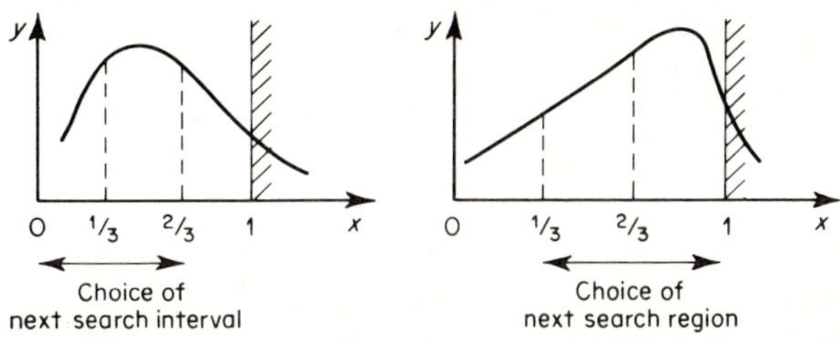

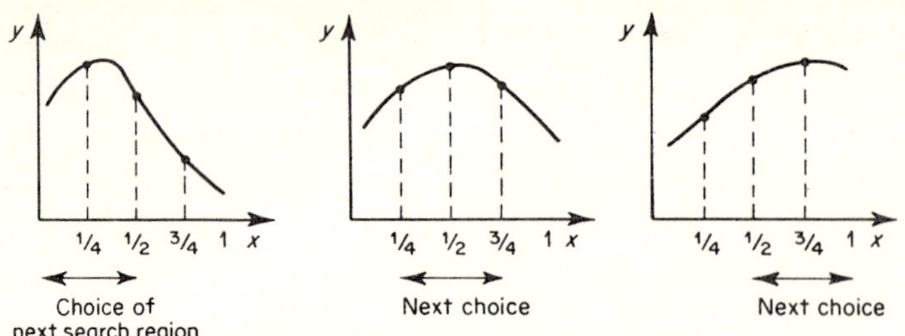

Fig. VI-11 Search using three experiments per cycle.

by rearrangement of the previous equation. Either expression may be used as a measure of the economy of the search procedure.

Sequential Search with Three Equally Spaced Experiments A slightly more economical method, which can be shown to be the best in this type of search, is that in which three experiments are used in each cycle.

First the three values $y(\frac{1}{4})$, $y(\frac{1}{2})$, and $y(\frac{3}{4})$ are determined for the first interval of search ($L_0 = 1$). We must now look for the new subinterval, containing the extreme value. The three possibilities are shown in Fig. VI-11. We see that the new search interval is half the length of the initial search region. Thus, for $N = 3$, it is given by

$$L_3 = \tfrac{1}{2}L_0$$

Treating this new interval in the same way, it will be seen that, while three experiments are still used in the search in the new cycle, the center one of these coincides with one of the experiments of the previous cycle. Thus, in the second cycle, only two new experiments need be made, placed at the quarter and the three-quarter positions in the interval.

Thus, with m iterations, the extreme value will be located to within $(\frac{1}{2})^m$ of the original length, but will involve only $1 + 2m$, rather than $3m$, experiments or calculations. Thus

$$\alpha = \frac{L_N}{L_0} = (\tfrac{1}{2})^m = (\tfrac{1}{2})^{(N-1)/2} \tag{4.1-6}$$

and the number of experiments required will be

$$N = 1 + 2m \geq 1 + 2\frac{\log \alpha}{\log \frac{1}{2}} \qquad \text{for odd } N \tag{4.1-7}$$

A routine such as this can be programmed for the computer without difficulty, since it uses the same basic formula for each cycle of the calculation.

Example VI-8: Comparison of methods using regular intervals

Let us compare the three methods described, the preplanned and the two sequential methods, by computing the number of experiments, N, required to get an accuracy defined by $\alpha \leq 0.01$.

For uniform preplanned search, N is given by Eq. (4.1-3), so that

$$N \geq \frac{2}{\alpha} - 1 = 199$$

and 199 experiments would then be required. For sequential search, using two experiments per cycle,

$$N \geq 2 \log \frac{0.01}{\log 2/3} = 22.7$$

as given by Eq. (4.1-5), so that 24 experiments are required. For modified sequential search

$$N \geq 1 + 2 \log \frac{0.01}{\log 1/2} = 14.3$$

from Eq. (4.1-7), requiring 15 experiments.

Thus the modified sequential is the most efficient of the three methods tested.

4.1.4 *Summary of direct search methods using regularly sized intervals*

We have studied a few representative search methods in which, at any stage, we considered the allocation of calculations or experiments so that the region was divided into equal intervals. Two types of approach were presented:

1. Simultaneous searches, in which all the experiments had to be preplanned before any were carried out.
2. Sequential searches, in which the results of any one set of experiments were examined before the next set or group were positioned. For this it was seen, using a criterion of effectiveness in search, that three equally spaced experiments in each new basic interval was the most economic search which could be achieved, in the sense of the fewest experiments being employed.

While simultaneous searches are sometimes unavoidable, they are not so economic as the sequential methods.

4.2 Direct Search Using Irregularly Sized Intervals

To isolate a subinterval of a certain region in the search for an optimum, at least two experiments are required in that region. In the methods suggested in Sec. 4.1, regularly spaced intervals were used. Let us now see if any advantage can be gained using intervals which are not necessarily uniformly spaced.

4.2.1 *Preplanned search with irregular intervals*

If we specify[1] in advance that N experiments must be planned so that

$$a < x_1 < x_2 < \cdots < x_N < b$$

$y(x)$ being unimodal, with an optimum at x^*, we can say that if $y(x_k)$ is the best experimental value, then the optimum will lie in the double interval

$$x_{k-1} < x^* < x_{k+1} \qquad \text{for } k = 1, 2, \ldots, N$$

Here $x_0 \equiv a$ and $x_{N+1} \equiv b$. The required interval of uncertainty is, then, $\alpha = (x_{k+1} - x_{k-1})/(b - a)$.

Since we do not know in advance where x^* lies, it is equally likely that it can be found in any double interval. All the double intervals used should thus be equal, so that the value of α is the same in all cases. In other words,

$$x_2 - a = x_3 - x_1 = x_4 - x_2 = x_5 - x_3 = x_6 - x_4 = \cdots = b - x_{N-1}$$

Thus $x_3 = x_1 + x_2 - a$; $x_4 = 2x_2 - a$; $x_5 = x_1 + 2x_2 - 2a$; $x_6 = 3x_2 - 2a$; . . . ; or, in general,

$$x_{2k-1} = x_1 + (k - 1)(x_2 - a) \tag{4.2-1a}$$

$$x_{2k} = a + k(x_2 - a) \tag{4.2-1b}$$

for $k = 1, 2, \ldots$. Let us consider separately the situations in which N is even or odd.

An Even Number of Experiments With N even, we may write that $N = 2K$, so that Eqs. (4.2-1) hold for $k = 1, 2, \ldots, K$, where k is an integer. Then

[1] Based on the approach of Jackson [1966a].

the $(N-1)$st interval of uncertainty is $b - x_{N-1}$, and

$$x_2 - a = b - x_{N-1} = b - x_{2K-1} = b - [x_1 + (K-1)(x_2 - a)]$$

from Eq. (4.2-1*a*), so that

$$K(x_2 - a) + (x_1 - a) = b - a$$

Letting $x_1 = x_2 - \delta$, we get

$$(K+1)(x_2 - a) = b - a + \delta$$

or

$$x_2 - a = \frac{b - a + \delta}{K+1} \tag{4.2-2}$$

The interval of uncertainty is then

$$\alpha = \frac{x_2 - a}{b - a} = \frac{1}{1 + N/2}\left(1 + \frac{\delta}{b-a}\right) \tag{4.2-3}$$

When $x_1 \to x_2$, $\delta \to 0$ and $(x_2 - a) \to (b - a)/(1 + N/2)$, and when $x_1 \to a$, $\delta \to x_2 - a$ and $(x_2 - a) \to (b - a)/(N/2)$. The interval of uncertainty is then smallest if x_1 is taken as close to x_2 as possible. From Eqs. (4.2-1) this also implies that $x_3 \to x_4$, $x_5 \to x_6$, . . . , so that pairs of experiments should be equally spaced throughout the original interval $b - a$. The interval of uncertainty reduces to

$$\alpha = \frac{x_2 - a}{b - a} = \frac{2}{2 + N} \tag{4.2-4}$$

if $x_2 - x_1 = \delta$ is sufficiently small. Obviously, we cannot put $x_1 = x_2$, since we should then get identical values of $y(x_1)$ and $y(x_2)$, which yields no information. We therefore wish to choose two values of x that are very close to each other but separated by a small interval δ, which is large enough to distinguish between the results of the two experiments.

The location of the experiments will then be given by Eqs. (4.2-1) and (4.2-2) as

$$x_{2k-1} = a + \frac{k}{1 + N/2}(b - a + \delta) - \delta \tag{4.2-5a}$$

$$x_{2k} = a + \frac{k}{1 + N/2}(b - a + \delta) \tag{4.2-5b}$$

for $k = 1, 2, \ldots, N/2$. Such a paired spacing of experiments throughout the search region is called a *dichotomous* approach.

Example VI-9: Dichotomous search

a. Two experiments With one pair of experiments,[1] these should be placed as close as possible at the center of the search region. We wish to place the x_1 according to Eqs. (4.2-5), so that

$$x_1 = \frac{1}{2} - \frac{\delta}{2}$$

$$x_2 = \frac{1}{2} + \frac{\delta}{2}$$

where δ is a small positive number chosen so that the two experiments give significantly different results. The interval of uncertainty is now $\frac{1}{2} + \delta/2$.

The final region of uncertainty, using a pair of experiments, is in general, from Eq. (4.2-3),

$$\alpha = \frac{L_2}{L_0} = \frac{1}{2} + \frac{\delta}{2L_0} \approx \frac{1}{2}$$

If δ is small, this reduces to $\alpha = \frac{1}{2}$, which is a clear improvement over the corresponding uniform search policy for two experiments, where we found [Eq. (4.1-2)] that

$$\alpha = \frac{L_2}{L_0} = \frac{2}{3}$$

[1] An interesting mini-max problem can be posed if we approach the dichotomous search by a different route. We wish to place two experiments in our initial search region such that $0 < x_1 < x_2 < 1$. We then have one of the two possible intervals of uncertainty, namely, x_2 or $1 - x_1$, and we should like these to be as small as possible, to increase the accuracy of the search. The conservative approach is to assume that the worst will occur in the search in that we should obtain the larger of these two intervals of uncertainty. Thus, knowing only that $x_2 > x_1$, in the placing of these two points in the original range, their location must be chosen in order to reduce the worst interval of uncertainty. This interval can only be x_2 or $1 - x_1$, the larger of the two being chosen. Mathematically, it can be said that the location of the optimum will be found to an accuracy given by the larger of these two intervals, so that

$$\alpha = \max(x_2, 1 - x_1)$$

This will hold for any given pair of x_1, x_2 values. The best procedure is then to choose the values of x_1 and x_2 so that the smallest value of the worst region of uncertainty is found. In other words, we are looking for the procedure which places x_1 and x_2 so as to give the smallest value of α, and thus

$$\alpha^* = \min[\max(x_2, 1 - x_1)]$$

the minimum value of the maximum width. This is called a mini-max problem, a term which will again be encountered in considering game theory (Chap. XII, Sec. 2). The solution is, of course (Wilde [1964]), that $x_1 = x_2 = \frac{1}{2}$.

b. Four experiments Let us compute the accuracy with which an objective function for a given system can be optimized using four experiments if the range is restricted to $0 \leq x \leq 10$ and the search is to be preplanned, using the dichotomous approach.

The four experiments are to be placed as defined by Eqs. (4.2-5), with $K = 2$ and $b - a = 10$, so that

$$x_1 = \frac{10 + \delta}{3} - \delta = 10/3 - 2/3\delta$$

$$x_2 = \frac{10 + \delta}{3} = 10/3 + 1/3\delta$$

$$x_3 = 2/3(10 + \delta) - \delta = 20/3 - 1/3\delta$$

$$x_4 = 2/3(10 + \delta) = 20/3 + 2/3\delta$$

Assume for the purpose of illustration that the maximum lies at $x = 5.9$ (Fig. VI-12) and that $\delta \approx 0$.

Since $y(x_2) > y(x_1)$, we must thus exclude region $x \leq x_1$ if we look for a maximum. Similarly, since $y(x_3) > y(x_4)$, we exclude $x \geq x_4$. The maximum must therefore lie in the range $3.33 < x < 6.67$. The remaining interval in which the optimum lies will have a length of 3.33, or a third of the original search region. This could have been found directly from Eq. (4.2-4) as

$$\alpha = \frac{L_N}{L_0} = \frac{2}{N + 2} = \frac{1}{3}$$

Fig. VI-12 Dichotomous search with four experiments (see Example VI-9).

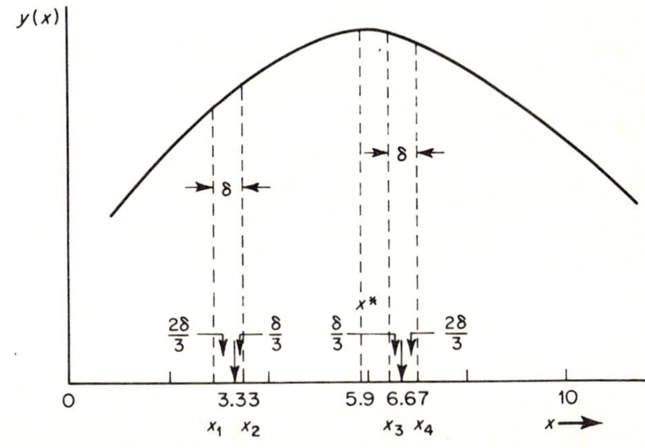

Here we have assumed that the size of the interval between experiments in a pair was small enough not to influence the final answer.

It should be noted that the size of δ may become important relative to the size of the interval of search, as N increases. It may not, then, be convenient to use this method.

An Odd Number of Experiments If N is odd, we can write $N = 2K + 1$, and Eqs. (4.2-1*a*) will be valid for $k = 1, 2, \ldots, K + 1$, and Eqs. (4.2-1*b*) for $k = 1, 2, \ldots, K$, where k is an integer. Then the length of the last double interval, $b - x_{N-1}$, in the set of preplanned experiments, is

$$\begin{aligned} x_2 - a &= b - x_{N-1} \\ &= b - x_{2K} = b - a - K(x_2 - a) \end{aligned}$$

on using Eqs. (4.2-1*b*). Then

$$(K + 1)(x_2 - a) = b - a$$

so that

$$x_2 - a = \frac{b - a}{K + 1}$$

This gives the accuracy of location of the optimum as

$$\alpha = \frac{1}{1 + (N - 1)/2} = \frac{2}{1 + N} \tag{4.2-6}$$

The location of the various experiments[1] is then given by

$$x_{2k-1} = x_{2k-2} + x_1 \tag{4.2-7a}$$

$$x_{2k} = a + \frac{k}{1 + (N - 1)/2}(b - a) \tag{4.2-7b}$$

for $k = 1, 2, \ldots, (N + 1)/2$ and $k = 1, 2, \ldots, (N - 1)/2$, respectively.

A preplanned search with an odd number of experiments would then place the even-numbered experiments at equal intervals over the range $a < x < b$, but since the odd-numbered experiments depend upon the location of x_1, this being placed anywhere in the range $a < x_1 < x_2$, there will be an infinite number of search plans. All, however, will give an interval of uncertainty of $\alpha = 2/(1 + N)$ from Eq. (4.2-6).

Example VI-10: Preplanned search with odd number of experiments

Let us consider the planning of five experiments in the interval $0 \leq x \leq 10$. There will be an infinite number of solutions, all yielding

[1] This can also be expressed as a mini-max problem, since the ratio of the interval of uncertainty of the original search region, α^*, can be written

$$\alpha^* = \min\,[\max\,(x_2 - a, x_3 - x_1, x_4 - x_2, \ldots, x_N - x_{N-2}, b - x_{N-1})]$$

the same interval of uncertainty, namely, from Eq. (4.2-6),

$$\alpha = \frac{2}{1+5} = 0.33$$

Two possible schemes are indicated in Fig. VI-13. In each, the even-numbered experiments are located at

$$x_2 = \tfrac{1}{3}(10) \qquad x_4 = \tfrac{2}{3}(10)$$

using Eq. (4.2-7*b*). If in one scheme we place the first experiment at $x_1 = \frac{1}{6}(10)$, then, from Eq. (4.2-7*a*), we have

$$x_3 = (\tfrac{1}{6} + \tfrac{1}{3}) \times 10 \qquad x_5 = (\tfrac{1}{6} + \tfrac{2}{3}) \times 10$$

while in the second scheme, if we use $x_1 = \frac{1}{10}(10)$, we must place the remaining experiments at

$$x_3 = (\tfrac{1}{10} + \tfrac{1}{3}) \times 10 \qquad x_5 = (\tfrac{1}{10} + \tfrac{2}{3}) \times 10$$

It will be noted from Eq. (4.2-3) that with four experiments we can achieve an accuracy of location of

$$\alpha = \frac{1}{3}\left(1 + \frac{\delta}{10}\right)$$

so that if δ is small, we should not consider using the five-experiment approach.

Fig. VI-13 Preplanned search with five experiments (see Example VI-10).

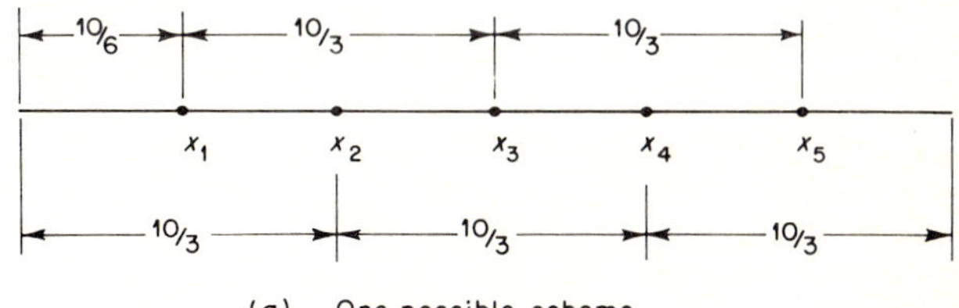

(*a*) One possible scheme

(*b*) A second scheme

It will be seen, by comparing Eqs. (4.2-3) and (4.2-6), that the interval of uncertainty found with an odd number of experiments is exactly that obtained with one experiment less, when δ is negligible. In this situation the dichotomous search with an even number of experiments would be used. If, however, δ is significant compared with the region of search, $b - a$, an odd number of experiments might yield better results.

4.2.2 *Sequential search with irregularly spaced intervals*

Just as in the methods of uniform search, where we found that instead of carrying out our search on a preplanned basis it was better to use a sequential method of operation—performing some experiments, looking at these results before carrying out another set of trials—we can find a better method than the preplanned dichotomous search by using a sequential method, assuming we are allowed to look at the results of previous measurements at any stage of the calculation.

Considering the sequential extension of the preplanned dichotomous search policy, we find that we must carry out a pair of experiments set at the midpoint of the search region, again separated by a distance δ. We shall then find that our new area of search will be either (a,x_2) or (x_1,b), as shown in Fig. VI-14, depending upon whether $y(x_1) > y(x_2)$ or vice versa. Our new interval of search can then be reduced by carrying out a second set of experiments. In this way we may repeat the cycles, carrying out a pair of experiments at each cycle and, assuming that δ is small, successively halving the region of search. We can repeat this procedure until the required degree of accuracy is achieved.

If we neglect δ, the final interval size L_N will be

$$L_N = L_0(\tfrac{1}{2})^K$$

where K is the number of pairs of experiments used, so that we get

$$\alpha = \frac{L_N}{L_0} = (\tfrac{1}{2})^{N/2} \tag{4.2-8a}$$

N being the total number of experiments performed.

Fig. VI-14 Dichotomous sequential search—cycle 1.

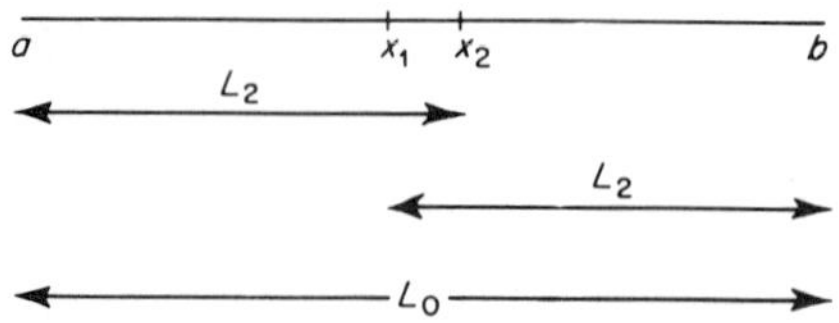

If we consider δ, the separation distance, we get from the first cycle that

$$L_2 = \frac{L_0}{2} + \frac{\delta}{2} = \tfrac{1}{2}(L_0 + \delta)$$

from the second cycle that

$$L_4 = \tfrac{1}{2}(L_2 + \delta) = \tfrac{1}{4}(L_0 + 3\delta)$$

from the third cycle that

$$L_6 = \tfrac{1}{2}(L_4 + \delta) = \tfrac{1}{8}(L_0 + 7\delta)$$

and so on, the $N/2$nd cycle giving

$$L_N = \frac{L_0 + \delta(2^{N/2} - 1)}{2^{N/2}}$$

Hence, if δ is small, Eq. (4.2-8*a*) will hold, but if δ is not small,

$$\begin{aligned} \alpha &= \frac{L_N}{L_0}(\tfrac{1}{2})^{N/2} + [1 - (\tfrac{1}{2})^{N/2}]\frac{\delta}{L_0} \\ &\approx (\tfrac{1}{2})^{N/2} + \frac{\delta}{L_0} \end{aligned} \qquad (4.2\text{-}8b)$$

The sequential dichotomous search method is thus considerably more efficient, if we neglect the effect of δ, than the preplanned uniform dichotomous search, the advantage growing exponentially with the number of experiments performed.

Example VI-11: Number of experiments for a given accuracy

How many experiments will be needed to reduce the final region in which the optimum will lie to within (*a*) 1 percent, (*b*) 0.01 percent, of the original region?

Since, for the sequential dichotomous method, α is defined by Eq. (4.2-8*a*), we then have for

(*a*) $$\alpha \leq 10^{-2}$$

$$N \geq \frac{2 \log 100}{\log 2}$$

giving N as 14. (For a preplanned dichotomous search, $N = 198$.)

(*b*) $$\alpha \leq 10^{-4}$$

so that N is 28. (For a preplanned dichotomous search, $N = 20{,}000$.)

Example VI-12: Sequential dichotomous search

Let us use four experiments to reduce the unknown region containing the optimum, knowing that it is given by a unimodal function in the range $0 \leq x \leq 10$. Use the sequential dichotomous method, and for the purposes of illustration, assume that the maximum lies at $x = 5.9$.

Step 1 Run a pair of experiments at $x = 5$. This allows us to use the unimodality concept to immediately reduce the search interval to five units of x. Here we are neglecting the size of the interval which separates the two experiments. The new interval of search is thus $5 < x < 10$.

Step 2 Carry out another pair of experiments in the optimal region, at its midpoint $x = 7.5$. The result of the experiment, $y_{7.5-} > y_{7.5+}$, allows the region greater than 7.5 to be eliminated. The final length containing the optimum is therefore $5.0 < x < 7.5$, a quarter of the original region of search. Using the expression for α, Eq. (4.2-8a), we should get

$$\alpha = \frac{L_4}{L_0} = (\tfrac{1}{2})^{N/2} = (\tfrac{1}{2})^2 = \tfrac{1}{4}$$

4.2.3 The Fibonacci search technique

In a sequential approach to direct search we must use at least two experiments per cycle to eliminate a region from further consideration. At the end of this cycle the new search region will still contain at least one experiment. Let us consider a technique in which we utilize this experiment and add only one new experiment in each new cycle.

A Search Based on the Fibonacci Sequence of Numbers To be specific, let us examine a region AB of length L_j. Two experiments may be added at points C and D, as indicated in Fig. VI-15, at lengths l_j and l'_j from each end. The optimum will then lie in the region CB of length $L_j - l_j$, or AD of length $L_j - l'_j$, so that, to obtain equal intervals of uncertainty, we should let $l_j = l'_j$. The next search cycle will examine the region AD or CB ($L_{j+1} = L_j - l_j$), depending upon which region has been eliminated. In either case one experiment will still remain within the new region of search, L_{j+1}. Assume that region CB remains and it contains experiment D. Let us use this point, D, and add only one new experiment at E, a length l_{j+1} from B. It is again reasonable to require that $CD = EB = l_{j+1}$. If we define

$$\sigma_j = \frac{l_j}{L_j} \tag{4.2-9}$$

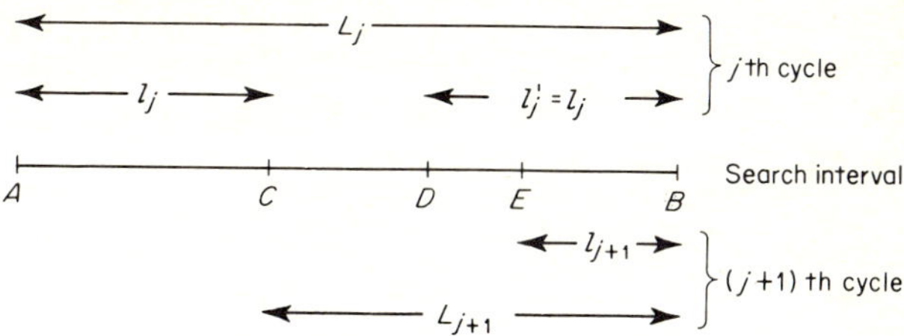

Fig. VI-15 Two cycles of Fibonacci search.

we then have, in the second cycle,

$$\sigma_{j+1} = \frac{l_{j+1}}{L_{j+1}}$$

so that

$$CD = L_{j+1} - l_j = l_{j+1}$$
$$= \sigma_{j+1} L_{j+1}$$

or

$$L_{j+1}(1 - \sigma_{j+1}) = l_j$$

Hence, substituting for L_{j+1} and l_j, we have

$$L_j(1 - \sigma_j)(1 - \sigma_{j+1}) = \sigma_j L_j$$

giving

$$\sigma_j = \frac{1 - \sigma_{j+1}}{2 - \sigma_{j+1}} \tag{4.2-10}$$

This recurrence relation allows σ_j to be found if σ_{j+1} is known. Let us consider the last stage of the search (see also p. 188), in which the final experiment has to be placed in the region L_{N-1}. We know from Example VI-9*a* that the most efficient way to use this is to place it as close as possible to the center of the region in a dichotomous search. Then $\sigma_{N-1} \to \frac{1}{2}$. We may now find all the σ_j from Eq. (4.2-10).

It is convenient to find a general expression for σ_j. If we write

$$\sigma_j = \frac{\beta_j}{\gamma_j}$$

where β_j and γ_j are both integers, then Eq. (4.2-10) becomes

$$\frac{\beta_j}{\gamma_j} = \frac{\gamma_{j+1} - \beta_{j+1}}{2\gamma_{j+1} - \beta_{j+1}}$$

which is satisfied if

$$\beta_j = \gamma_{j+1} - \beta_{j+1}$$

and

$$\gamma_j = 2\gamma_{j+1} - \beta_{j+1}$$

Eliminating γ_j and γ_{j+1} between these two equations, we get the recurrence relations for the β_j as

$$\beta_j = \beta_{j+1} + \beta_{j+2} \tag{4.2-11a}$$

$$\beta_{N-1} = \beta_{N-2} = 1 \tag{4.2-11b}$$

the latter equations being found from the known values of $\sigma_{N-1} = \frac{1}{2}$ and assigning the simplest value of unity.

Let us consider the sequence of integers, the Fibonacci series, given by

$$F_0 = F_1 = 1 \tag{4.2-12a}$$

and for $n > 1$,

$$F_n = F_{n-1} + F_{n-2} \tag{4.2-12b}$$

each number in the series being the sum of the two preceding numbers.[1] (See, for example, Vorob'ev [1966]). Some values of the series are given in Table VI-8.

TABLE VI-8 SOME NUMBERS IN THE FIBONACCI SERIES

n	F_n
0	1
1	1
2	2
3	3
4	5
5	8
6	13
7	21
8	34
9	55
10	89
11	144
12	233
13	377
14	610
15	987
16	1,597
17	2,584
18	4,181
19	6,765
20	10,946

[1] While the search method is a recent discovery, its basis of the Fibonacci number theory goes back to the thirteenth century, to the work of Leonardo of Pisa, in a text of 1202 (Gardner [1969]). It may be shown that the general term in the Fibonacci series is given by the expression

$$F_j = \frac{1}{\sqrt{5}}\left[\left(\frac{1+\sqrt{5}}{2}\right)^{j+1} - \left(\frac{1-\sqrt{5}}{2}\right)^{j+1}\right] \tag{4.2-13}$$

If we write

$$\beta_j \equiv F_{N-(j+1)}$$

then $$\gamma_j = \beta_j + \beta_{j-1} = F_{N-(j+1)} + F_{N-j} = F_{N-(j-1)}$$

from Eq. (4.2-12*b*), so that

$$\sigma_j = \frac{F_{N-(j+1)}}{F_{N-(j-1)}} \qquad \text{for } j = 1, 2, \ldots, N-1$$

We now have all the necessary ingredients for the Fibonacci search technique. This method was developed by Johnson [1956] and Kiefer [1953], being, like the others, based upon the concept of a unimodal function.

Let us now show how the Fibonacci series may be used to obtain an efficient search routine. It will be seen that this is the best method of any we have described, but we shall leave the proof that it is the most efficient of any restricted search to the original articles.

Fibonacci Search In this technique, two experiments are performed in the first search interval (step 1 below), and two of the three intervals will be chosen as the next search region (step 2). If we again place two experiments according to the specified method (step 3), it will be found that one of these coincides with the experiment remaining from the previous search cycle. Thus, in the second cycle, only one new experiment is necessary to determine the third search interval. We may proceed in this manner, adding only one experiment in each cycle.

Step 1 Placing the first two experiments Consider the placing of the two experimental positions x_1 and x_2 ($x_2 > x_1$) within the original search interval L_1, the latter[1] being defined by $a_1 \leq x \leq b_1$. First let us define the distance l_1, such that

$$l_1 = \frac{F_{N-2}}{F_N} L_1 \qquad (4.2\text{-}14)$$

where F_j are the Fibonacci numbers, and N is the total number of experiments finally required. This must be specified in advance. Our two experiments may now be located at the points x_1 and x_2, situated a distance l_1 from each end of the search region, so that

$$x_1 = a_1 + l_1$$

$$x_2 = b_1 - l_1 = a_1 + L_1 - \left(\frac{F_{N-2}}{F_N}\right) L_1 = a_1 + \frac{F_{N-1}}{F_N} L_1$$

[1] As will be seen, it is convenient in this method to label the original search region L_1 rather than L_0.

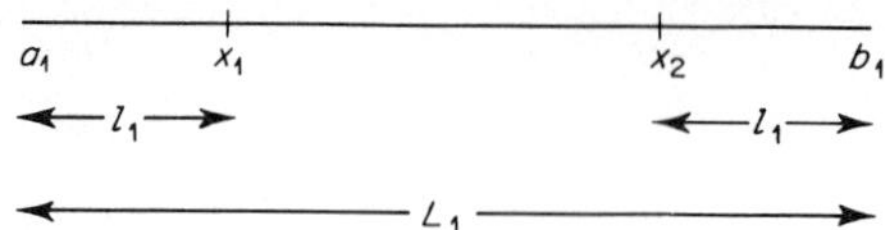

Fig. VI-16 The placing of the first two experiments.

as shown in Fig. VI-16. Since l_1 is always $\leq\frac{1}{2}L_1$,† $x_2 \geq x_1$, and either the subinterval (a_1,x_1) or (x_2,b_1) (both $= l_1$) will be eliminated by examining the results $y(x_1)$ and $y(x_2)$. We shall therefore be able to choose the new reduced interval of uncertainty which will contain the desired optimum point.

Step 2 The new subinterval of search This new length, L_2, after these two first trials, will be less than L_1, and given by

$$L_2 = L_1 - l_1 = b_1 - x_1 \qquad \text{or} \qquad x_2 - a_1$$

On substituting for l_1, we shall get

$$L_2 = L_1 - L_1 \frac{F_{N-2}}{F_N} = L_1\left(1 - \frac{F_{N-2}}{F_N}\right) = L_1 \frac{F_{N-1}}{F_N}$$

since $F_N = F_{N-1} + F_{N-2}$. This is our new interval of search. In general, we could carry on in this manner, placing two experiments in each new subregion as above, so that

$$l_k = \frac{F_{N-(k+1)}}{F_{N-(k-1)}} L_k \tag{4.2-15}$$

Let us now show the advantages of the Fibonacci approach.

Step 3 Two experiments in second interval Let us consider the question of placing two experiments in the new subinterval L_2. These will be placed at the positions x_3' and x_3'', so that they are at a distance l_2 from each end of the interval L_2. The distance l_2 is defined as [Eq. (4.2-15)]

$$l_2 = \frac{F_{N-3}}{F_{N-1}} L_2 \tag{4.2-16}$$

so that we now have the two situations as shown in Fig. VI-17*a* and *b*.

† $$\frac{l_1}{L_1} = \frac{F_{j-2}}{F_j} \leq \frac{1}{2}$$

for $j \geq 2$, as will be seen from Table VI-8.

In the first case, Fig. VI-17*a*, in which the optimum was found to lie in the subinterval (x_1,b_1), we shall get x_3' as

$$x_3' = x_1 + l_2 = x_1 + \frac{F_{N-3}}{F_{N-1}} L_2$$

$$= \left(a_1 + \frac{F_{N-2}}{F_N} L_1\right) + \frac{F_{N-3}}{F_{N-1}}\left(\frac{F_{N-1}}{F_N} L_1\right)$$

$$= a_1 + \frac{F_{N-1}}{F_N} L_1 \equiv x_2$$

it being seen that x_3' coincides with the existing experiment at x_2, since l_2 is the distance between x_1 and x_2, and x_3'' is

$$x_3'' = b_1 - l_2 = x_1 + L_2 - \frac{F_{N-3}}{F_{N-1}} L_2$$

$$= \left(a_1 + \frac{F_{N-2}}{F_N} L_1\right) + \left(\frac{F_{N-1} - F_{N-3}}{F_{N-1}}\right)\frac{F_{N-1}}{F_N} L_1$$

$$= x_1 + \frac{F_{N-2}}{F_N} L_2 \equiv x_3, \text{ say}$$

We have therefore shown that one of the two experiments used in this cycle corresponds with an experiment retained from the previous cycle (here, $x_3' \equiv x_2$), so that, actually, only one new experiment at x_3'' need be carried out in the second cycle. Out of this cycle we shall get a new subinterval for the next search, L_3, as shown in Fig. VI-18.

Fig. VI-17 Alternative experimental locations in the second cycle.

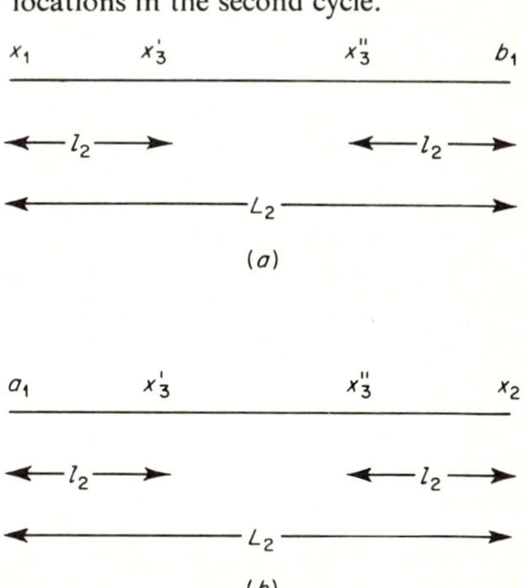

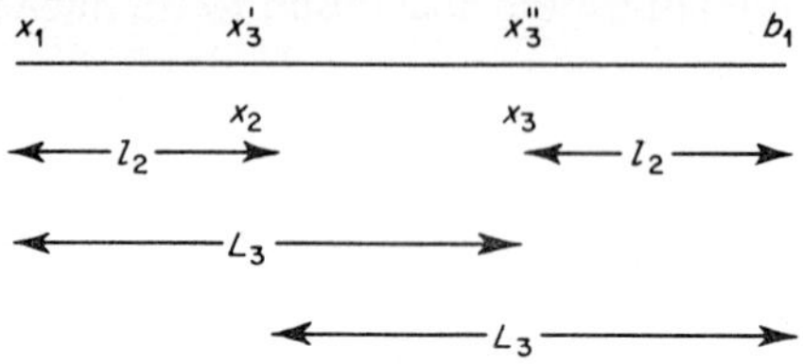

Fig. VI-18 The third-cycle search interval.

In the second possible situation for the second cycle, Fig. VI-17*b*, a similar argument will allow one experiment to be added in order to get the necessary length, L_3, for the next search.

The desired optimum will thus, after the second cycle of search, lie in the length L_3, given by

$$L_3 = L_2 - l_2 = L_2 \frac{F_{N-1} - F_{N-3}}{F_{N-1}} = \frac{F_{N-2}}{F_{N-1}} L_2$$

Accuracy of Search After the first pair of experiments, an interval

$$L_2 = L_1 - l_1 = \frac{F_{N-1}}{F_N} L_1 \tag{4.2-17}$$

remains to be searched. Of course, in the unlikely event of the two experiments yielding the same value of the objective function, the region to be searched is

$$L_2 = L_1 - 2l_1$$

Nevertheless, we must allow for the worst case. Equation (4.2-17) represents the interval of interest after two experiments. On performing a third experiment, the remaining interval of uncertainty, L_3, will be

$$L_3 = L_2 - l_2 = \frac{F_{N-2}}{F_{N-1}} L_2$$

or using Eq. (4.2-17),

$$L_3 = \frac{F_{N-2}}{F_N} L_1$$

is the unsearched interval. Continuing this process, we find, after completion of m of the N predetermined experiments, that

$$L_m = \frac{F_{N-(m-1)}}{F_N} L_1 \tag{4.2-18}$$

Thus, for N experiments, the fractional region of uncertainty α is

$$\alpha = \frac{L_N}{L_1} = \frac{1}{F_N} \tag{4.2-19}$$

This will allow us to determine N, the required number of experiments, for any desired accuracy in locating the extremum location.

Example VI-13: Fibonacci efficiency

Determine the number of experiments required to locate the position of the extreme point to within 1 percent of the original search region by the Fibonacci search technique.

Here we need $\alpha \leq 1/100$, or from Eq. (4.2-19),

$$\frac{1}{F_N} \leq \frac{1}{100}$$

giving $F_N \geq 100$. Looking at Table VI-8 of Fibonacci numbers, we see that this will be satisfied for $N = 11$, since $F_{11} = 144$. A comparison of the number of experiments required by the various methods is shown in Table VI-9 for a given accuracy of at least 1 percent.

The advantages of the Fibonacci method increase with the demand for greater accuracy; e.g., for $\alpha \leq 10^{-4}$, the Fibonacci requires $N = 20$, while the next best, the sequential dichotomous, needs $N = 28$. In general, we can say that the Fibonacci method has the great advantage of few calculations, but suffers from the disadvantage that it requires the calculation of the function $y(x)$ at rather odd values of x.

TABLE VI-9 A COMPARISON OF THE NUMBER OF EXPERIMENTS REQUIRED TO ACHIEVE GIVEN ACCURACY

Search method	α 0.1	0.01	0.001	0.0001
Regular intervals:				
Preplanned	19	199	1,999	19,999
Sequential (two per cycle)	12	24	36	46
Sequential (three per cycle)	9	15	21	29
Irregular intervals:				
Preplanned dichotomous (δ small)	18	198	1,998	19,998
Sequential dichotomous (δ small)	8	14	20	28
Preplanned, odd number	19	199	1,999	19,999
Sequential (three per cycle)	12	21	30	42
Fibonacci	6	11	16	20
Modified Fibonacci	6	11	16	21

The Final Experiment in the Search The last experiment must be placed with some care since the prescription given on p. 181 requires that the final experiment be placed precisely at a point for which an experiment already exists. Indeed, the last point must be placed at

$$\frac{l_{N-1}}{L_{N-1}} = \frac{F_0}{F_2} = \frac{1}{2} \tag{4.2-20}$$

according to Eq. (4.2-15).

After $N - 2$ experiments have been performed, the search interval has been reduced to L_{N-2}, where L_{N-2} is defined by Eq. (4.2-18). On performing the $(N - 1)$st experiment, the interval of search is narrowed to

$$L_{N-1} = L_{N-2} - l_{N-2}$$

in the worst case, as discussed above. Dividing both sides of this equation by L_{N-1} and rearranging,

$$\frac{l_{N-2}}{L_{N-1}} = \frac{L_{N-2}}{L_{N-1}} - 1 = \frac{1}{2} \qquad \text{for all } N \tag{4.2-21}$$

Therefore, after performing $N - 1$ experiments and discarding the appropriate interval at each step (and one experiment), the remaining experiment (the valid one retained from the first $N - 1$) is precisely in the center of the remaining interval to be searched, as found by Eq. (4.2-21). However, the Nth, and final, experiment is also to be placed in the center of the interval, as required by Eq. (4.2-20). This is true no matter how many experiments the searcher has planned; i.e., it is true for all N. This final experiment should not be positioned at precisely the same point as before (obviously), or no new information will be gained. To obtain the accuracy given by Eq. (4.2-19), one must apply the dichotomous concept and place the last experiment very close to the remaining valid experiment, "close" being defined in the same sense as in Sec. 4.2.2. This permits the last experiment to be used to select the location of the optimum to within $\frac{1}{2}L_{N-1}$. If the last experiment is not run in a dichotomous fashion, Eq. (4.2-19) is invalid.

Example VI-14: Placing of the final experiment

Use four experiments to determine the position of the optimum in a certain application in which it may be assumed that the objective function is unimodal and the original search length is taken as unity. (*Note:* The solution is presented in outline form only to indicate the approach and to examine the question of the final experiment.) The solution is presented in Table VI-10. Clearly, the fourth experiment, represented by an unshaded box in row $k = 3$, is to be located at $l_4 = \frac{1}{5}$, which, as

TABLE VI-10 THE FINAL EXPERIMENT IN FIBONACCI SEARCH

Cycle k	$\dfrac{l_k}{L_k}$	l_k	L_k	Position of experiment (□ = New; ■ = Old)
1	2/5	2/5	1	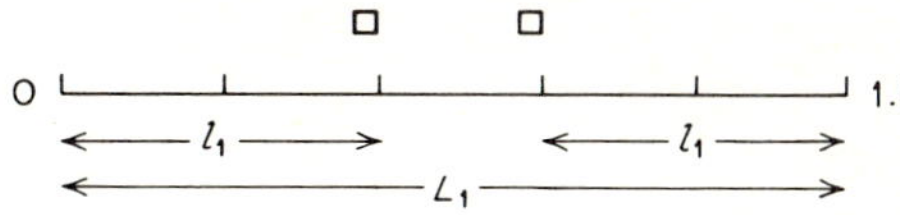
2	1/3	1/5	3/5	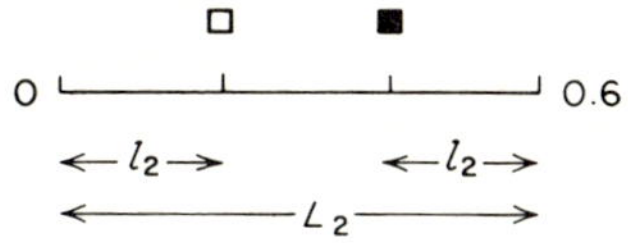
3	1/2	1/5	2/5	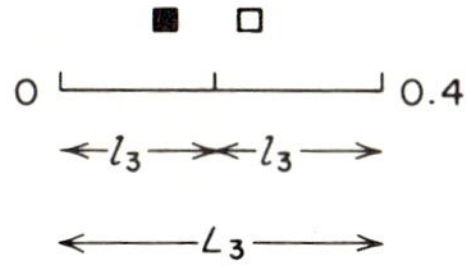
4			1/5	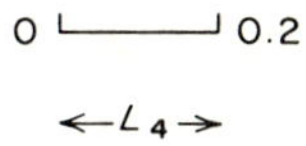

can be seen, is precisely the same point that was found from the valid remaining experiment of the first three cycles. The interval of uncertainty has been reduced to 0.2 in four experiments, in accordance with Eq. (4.2-19).

4.2.4 *Modified Fibonacci search*

Above we considered the general approach to a one-dimensional search, using the Fibonacci technique. There it was necessary to know the final number of experiments required before beginning the calculation. This is defined by the desired accuracy, given by α. In other methods which were considered, such as the preplanned uniform search or the sequential dichotomous search, the search could be started without deciding in advance how accurately it was desired to obtain the optimum value. In other words, the location of the experiments did not depend upon the ultimate number of experiments used. At any later stage in the calculation it would be possible in these latter methods to alter the desired value of α, and this could be accommodated in the search procedure without upsetting the layout of the experimental points. In the Fibonacci method this would not be possible, since the location of the experimental points are at unusual values, determined by the total number of experiments.

If the accuracy or the allowable number of experiments is specified in advance, any of the above techniques may be used, including the very efficient Fibonacci approach.

On the other hand, while all the less efficient methods are applicable if the final number of experiments, or the desired accuracy, is not specified initially, it would be most convenient if we could also use the Fibonacci method. In the Fibonacci method the choice of location of the first experiments is determined by the total number of experiments N. Thus

$$x_1 = a_1 + l_1 \qquad \text{and} \qquad x_2 = b_1 - l_1$$

where this distance l_1 is given by

$$l_1 = L_1 \frac{F_{N-2}}{F_N}$$

It would therefore seem that we could never use this method without first determining N. However, it will be noted that, in reducing the size of the interval at each cycle, we use $l_1, l_2, l_3, \ldots, l_k, \ldots$, as defined by Eq. (4.2-15),

$$l_k = \frac{F_{N-(k+1)}}{F_{N-(k-1)}} L_k = \frac{F_{j-2}}{F_j} L_k$$

Now the ratio F_{j-2}/F_j for large values of j is very near its limit of 0.3820. This may be shown:

1. For the smaller values of j by particular calculation, as seen in Table VI-11. It will be seen that for $j > 4$, this ratio is close to 0.3820.

TABLE VI-11 THE FIBONACCI RATIO F_{j-2}/F_j AT SMALL j

j	3	4	5	6	7	8	9
$\frac{F_{j-2}}{F_j}$	$1/3 =$ 0.3333	$2/5 =$ 0.4000	$3/8 =$ 0.3750	$5/13 =$ 0.3846	$8/21 =$ 0.3810	$13/34 =$ 0.3823	$21/55 =$ 0.3818

2. For large values of j. When j is large, the jth term in the Fibonacci series as given by Eq. (4.2-13) reduces to

$$F_j = \frac{1}{\sqrt{5}}\left(\frac{1+\sqrt{5}}{2}\right)^{j+1} = \frac{1}{\sqrt{5}}(1.6180)^{j+1}$$

We may therefore use this to get the required ratio

$$\frac{F_{j-2}}{F_j} = \frac{1}{(1.6180)^2} = 0.3820$$

In this modified method,[1] we should therefore start the search by placing the points, using the relation

$$l_k = 0.3820L_k$$

for all the initial experiments. We saw that this relation was a reasonable representation of the Fibonacci series for large values of N, which here means those values of N greater than 4. In other words, we may use this relation to calculate the necessary values of l for all but the last five experiments. In the last five, this approach would not give very realistic values of l, and for the last five we can return to the true Fibonacci series.

Thus, if we do not initially know the final number of experiments we want to use, or do not know the desired limits of accuracy, we may start the search using this modified Fibonacci technique. When we decide that not more than five additional experiments are necessary to complete the search, we may then switch to the true Fibonacci sequence; this in the last five experiments will reduce the remaining interval by almost an order of magnitude, since $F_5 = 8$.

The search policy of choosing the new experiments at a length $l_k = 0.3820L_k$ from each end of the interval, L_k (one experiment being the same as a previous one) will be an excellent approximate policy for all except the last few stages. In the last few stages, the search could be switched to the true Fibonacci sequence, but at this stage it is of little importance whether or not the "best" policy is now used, as seen from Table VI-9. This point is discussed by Bellman and Dreyfus [1962, p. 158].

This type of search, using a constant fraction of the new interval, is very convenient for use in digital computers, and gets over the problem that a true Fibonacci method can be used only if the required accuracy is known in advance of the search.

The accuracy achieved by N experiments would then be

$$\alpha = \frac{L_N}{L_1} = \frac{L_N}{L_{N-1}} \frac{L_{N-1}}{L_{N-2}} \cdots \frac{L_2}{L_1} = (0.6180)^{N-1}$$

since $L_{k+1} = L_k - l_k = (1 - 0.3820)L_k$.

[1] Kiefer and Johnson suggested this modification to the Fibonacci method (Bellman [1957]), but the ratio $F_n/F_{n-1} = (1 + \sqrt{5})/2 = 1.6180$ was used much earlier in history. A rectangle formed by the two parts of a line divided in the ratio 1.6180:1 was supposed by the Greeks to have the most pleasing properties, and was widely used in their architecture. It is also found in Euclid's geometry—the division of a segment into two unequal parts so that the ratio of the whole to the larger part equaled the ratio of the larger to the smaller, being known as the golden section, or golden mean. The present search method has also been called "search by golden section" (Wilde [1964]).

Example VI-15: Modified Fibonacci search

Let us locate the minimum value of the function represented by

$$y = (x - 30)^2$$

to within an accuracy of 1 percent if the initial range of search is $0 \leq x \leq 100$.

In the first cycle, irrespective of the accuracy required, the first two experiments will be located at

$$x_1 = 0 + l_1 = 0.3820L_1 = 38.20$$

$$x_2 = 100 - l_1 = 61.80$$

Evaluating the function, we find that

$$y(38.20) < y(61.80)$$

so that, by the assumption of unimodality, the optimum will not lie in the region $x \geq 61.80$. The new region of search, L_2, is then $0 < x < 61.80$.

TABLE VI-12 SUMMARY OF MODIFIED FIBONACCI SEARCH

Cycle k	L_k	l_k	Location and value of kth-cycle experiments: x	$y(x)$	Region containing optimum
0					$0 \leq x \leq 100$
1	100.00	38.20	38.20, 61.80	67.2, 1,011.2	$0 \leq x < 61.80$
2	61.80	23.61	23.61, $\boxed{38.20}$	40.8, 67.2	$0 \leq x < 38.20$
3	38.20	14.59	14.59, $\boxed{23.61}$	237.5, 40.8	$14.59 < x < 38.20$
4	23.61	9.00	$\boxed{23.59}$, 29.20	40.8, 0.6	$23.61 < x < 38.20$
5	14.59	5.57	$\boxed{29.18}$, 32.63	0.6, 6.9	$23.61 < x < 32.63$
6	9.02	3.45	27.06, $\boxed{29.18}$	8.6, 0.6	$27.06 < x < 32.63$
7	5.57	2.13	$\boxed{29.19}$, 30.50	0.6, 0.3	$29.19 < x < 32.63$
8	3.44	1.30	$\boxed{30.49}$, 31.33	0.3, 1.8	$29.19 < x < 31.33$
9	2.14	0.82	30.01, $\boxed{30.51}$	0, 0.3	$29.19 < x < 30.51$
10	1.32	0.50	29.69, $\boxed{30.01}$	0.1, 0	$29.69 < x < 30.51$
11	0.82				

The symbol □ on any x entry signifies an existing experimental location, and corresponding $y(x)$ are taken as the previously calculated value.

In the second cycle, experiments will lie at

$$x_3' = 0 + l_2 = 0.3820L_2 = 23.61$$
$$x_3'' = 61.80 - l_2 = 38.20$$

Thus, only one new experiment at $x_3 = x_3'$ will be used, getting

$$y(23.61) < y(38.20)$$

so that we may eliminate the region $x \geq 38.20$ from further search.

The complete calculation is summarized in Table VI-12, the optimum lying in the range $29.69 < x < 30.51$, with the best approximation $y(30.01) = 0.0001$.

5 OPTIMIZATION OF DISCRETE-VALUED FUNCTIONS

Many functions exist only for discrete values of the independent variable. An excellent example occurs in the determination of the most economical pipe lagging thickness. The total lagging cost may be expressed as the sum of two competing factors, the operating costs and the investment costs. Both of these factors depend on the lagging thickness, which, for all practical purposes, exists only in standard, discrete sizes.

In general, the methods described earlier are not directly suitable for discrete objective functions since the permitted values of the independent variable are unlikely to coincide with the location of experiments required by, say, the Fibonacci method. However, *lattice search problems*, as these discrete cases are often called, can be solved using modifications of most of the search methods already discussed. For example, there are two methods of modifying the lattice problem so that the Fibonacci method is applicable. Once again, it has to be assumed that the functions are unimodal in form.

5.1 Fibonacci Search of Discrete-valued Functions

If the discrete function is defined at K points within the range of interest, the Fibonacci search may be used without modification provided

$$F_N = 1 + K \tag{5.1-1}$$

where F_N is a Fibonacci number. If this relationship is not satisfied, the Fibonacci method requires some modification to achieve an effective search scheme. Let us consider the two problems separately.

5.1.1 Special case

Suppose that, in the range of interest, K cases need to be compared to determine the optimum, where K satisfies Eq. (5.1-1). These cases need not be positioned at equal intervals of the discrete variable, and do not have to be numbered as long as there is some rational way of ordering them to obtain a unimodal function. Once the system has been ordered, it is necessary only to label the various possible cases from 1 to K consecutively. Then the range of interest can be viewed as being $K + 1$ in length. Any given case in this system will then be identified by an associated integer, k, where

$$1 \leq k \leq K + 1$$

Example VI-16: Arranging a discrete-valued objective function

It is required to search for an optimum vessel size which can be obtained from the manufacturer only in the discrete characteristic sizes

2.1, 3.3, 4.6, 5.7, 11.4, 15.2, 19.1

If it is assumed that these sizes are arranged so that the corresponding objective function is unimodal, they will be in the correct order for a search to be carried out. Each case is then numbered from 1 to K, as given in Table VI-13. Thus $k = 2$ corresponds to a vessel size of 3.3. The seven values of k can be viewed as distributed over an interval of length $1 + K$. This is also depicted in Fig. VI-19. Thus the arrangement is complete.

TABLE VI-13 DISCRETE-VALUED OBJECTIVE FUNCTION

k	1	2	3	4	5	6	7
Size	2.1	3.3	4.6	5.7	11.4	15.2	19.1
y	63.1	67.0	73.2	79.3	86.4	81.7	72.4

The reason that a Fibonacci search is possible in the special case of Eq. (5.1-1) is that, for N experiments [N defined by Eq. (5.1-1)], each experiment will correspond to an integral value of K, and hence to a particular case. For example, the positions of the first experiments are defined by Eq. (4.2-14) as

$$l_1 = \frac{F_{N-2}}{F_N}(K + 1)$$

Since Eq. (5.1-1) is presumed valid,

$$l_1 = F_{N-2}$$

which is integral. Thus the experiments are to be performed at $k = l_1$ and

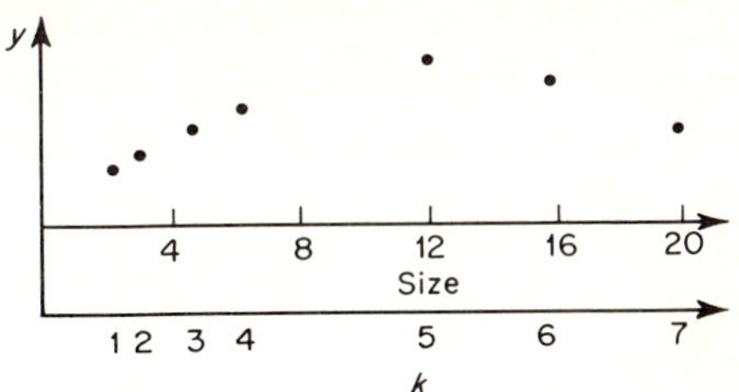

Fig. VI-19 A single-valued function (see Example VI-16).

$k = 1 + K - l_1$. Since l_1 is an integer, these two points correspond to possible discrete cases, as shown in Fig. VI-19, for example.

To prove that a Fibonacci search is generally possible, one must simply show that the point to be searched is given by an integral distance in the interval $1 + K$. As has been shown above, the first experiments are certainly at integral positions. Moreover, since

$$l_j = L_j \frac{F_{N-(j+1)}}{F_{N-(j-1)}} = \frac{F_{N-(j+1)}}{F_N} L_1 = F_{N-(j+1)} \frac{1+K}{F_N}$$

as given by Eqs. (4.2-15) and (4.2-18), then in the special case considered here,

$$l_j = F_{N-(j+1)} \tag{5.1-2}$$

Thus l_j is, clearly, always an integer, and experiments are therefore required which correspond to admissible values of the discrete independent variable.

Example VI-17: Choice of pipe size

Suppose there are four possible pipe sizes, 1, 1¼, 1½, and 2 in., which can be used in a particular service. Let us search to determine the most economic size according to some criteria. Since $K = 4$ (number of pipe sizes), then $F_N = 5$, with Eq. (5.1-1) satisfied by $N = 4$.

The first experiments are to be carried out at the two locations defined by l_1, where

$$l_1 = F_2 = 2$$

[as seen from Eq. (5.1-2), with $j = 1$], l_1 being the distance from each end of the search region. Therefore sizes $k = 2$ and $k = 5 - 2 = 3$ must be tested, which means that 1¼- and 1½-in. pipes are to be considered. Using the assumption of unimodality, and if $y_{k=2} > y_{k=3}$, then the maximum value (if this is the one required) will not be in the range $k \geq 3$. The next experiment is given by

$$l_2 = F_1 = 1$$

and thus the values at $k = 1$ and $k = 2$ must be examined. Since $k = 2$ has been determined in the previous cycle, only the objective function corresponding to the one at $k = 1$ need be considered. The best value is clearly determined by this third experiment at $k = 1$. The complete elimination process is shown in Table VI-14. The best value is found after three experiments.

TABLE VI-14 SEARCH FOR A MAXIMUM

Experiment	0	1	2	3	4	5	
Pipe size, in.		1	$1\frac{1}{4}$	$1\frac{1}{2}$	2		
Cycle							*Comments*
1			[2]	[3]			Assume $y(2) > y(3)$
2		[1]	2				If $y(1) > y(2)$, $y(1)$ is max, and vice versa

New experiment in the cycle [i].

In the preceding example, it was seen that only three experiments were required, although N was 4. The reason for this is clear, since in the last section we proved that the last experiment always corresponds to an existing one at the midpoint of the remaining interval. In the continuous case, this redundancy was interpreted as requiring an experiment at a point slightly displaced from the existing one. However, in a discrete case, this is not possible, since only discrete values of the independent variable are permitted. Moreover, the last experiment is not required to determine the optimum which is defined by $N - 1$ experiments. This can be demonstrated as follows: After $N - 1$ experiments, the length of the interval of uncertainty will be

$$L_{N-1} = L_1 \frac{F_2}{F_N} = 2$$

Thus the interval of uncertainty will be two units, and it will contain only one experiment at its center. Referring to Table VI-14, we can see that on performing the third [and $(N - 1)$st] experiment, the interval to be searched will be from $k = 0$ to $k = 2$ or from $k = 1$ to $k = 3$, depending on whether or not $k = 1$ has a better yield than $k = 2$. Since the adjacent lattice points have already been eliminated, the $k = 1$ (or $k = 2$) will then be the optimum choice. Thus, not only is the last experiment redundant, but the optimum can always be selected with $N - 1$ experiments.

5.1.2 Use of fictitious locations

If the situation arises in which

$$1 + K \neq F_N \tag{5.1-3}$$

then additional, fictitious locations may be added at one end of the sequence to bring the total to a Fibonacci number; i.e.,

$$1 + K + F = F_N \tag{5.1-4}$$

where $F_N > 1 + K > F_{N-1}$, and F is the number of fictitious experiments which have been added. The search should then be conducted as outlined in Sec. 5.1.1.

Example VI-18: Use of fictitious locations

Suppose that, in Example VI-17, six pipe sizes had to be considered, 2½ and 3 in. being added to the list. Since 7 ($= K + 1$) is not a Fibonacci number, we must add one additional fictitious case ($F = 1$) to increase the effective K to 7. Therefore $N = 5$ satisfies Eq. (5.1-4). The progress of the search is depicted in Table VI-15, where it is seen that the fictitious point at $k = 1$ might be compared with the point at $k = 2$. The case corresponding to $k = 2$ will obviously be selected, since for $k = 1$ no possible real case exists.

TABLE VI-15 MAXIMIZATION USING FICTITIOUS LOCATIONS

Experimental location	0	1	2	3	4	5	6	7	8	
Pipe size, in.		F	1	1¼	1½	2	2½	3		
Cycle										*Comment*
1				[3]		[5]				$l_1 = 3$
										$y(3) > y(5)$
2			[2]	3						$l_2 = 2$
										$y(2) > y(3)$
3		⁞1⁞	2							$l_3 = 1$
										$y(1) < y(2)$

F is the location of a fictitious experiment.

[i] is a new experiment.

In this search the optimum is assumed to lie at $y(2)$.

⁞i⁞ is the location of a fictitious experiment which always loses in any comparison.

Thus the optimum was found in three experiments. However, it is not difficult to show that, had $k = 6$ been the optimum, then five experiments $(N - 1)$ would have been required for its location. In general, therefore, we can only say that the optimum will be found with no more than $N - 1$ experiments.

Example VI-19: Sequential search of results from preplanned experiments

Find the minimum value of $y(x)$ within the range $0 < x < 100$ if the location has to be determined to within 5 percent, using a preplanned sequence of experiments.

The objective y is not found directly from an experiment, but the experimental data provide sufficient information for $y(x_j)$ to be calculated. Assume $y(x)$ is given by $y = (x - 61)^2$.

Preplanned experimentation For an accuracy of 5 percent, $\alpha = 0.05$, and since $\alpha = 2/N + 1$, the number of experiments, N, will be

$$N \geq \frac{2}{\alpha} - 1 = 39$$

using Eq. (4.1-2). The preplanned experiments would then be carried out at $x_1 = 2.5$, $x_2 = 5.0$, $x_3 = 7.5, \ldots, x_{39} = 97.5$, the necessary information being determined at each position, allowing $y(x_j)$ to be evaluated when required.

Sequential search It is not necessary, however, to find y at each and every location. Effort can be conserved by using the Fibonacci sequential approach of Sec. 5.1.2.

Here there are 40 intervals; that is, L_1, the range of interest, is $L_1 = K + 1$, K being here the number of preplanned evaluations. Now L_1 is not equal to a Fibonacci number F_N, the next larger being $F_9 = 55$. Place fictitious experimental results at x_k for $k = 40$ to 54 and carry out a Fibonacci search on the modified $L_1 = K + 1 = 55$, with the new value of $K = 54$.

Cycle 1 From Eq. (5.1-2), $l_1 = F_{9-2} = 21$. Evaluate y at $k_1 = l_1$ and $k_2 = L_1 - l_1$, that is, $k_1 = 21$ and $k_2 = 55 - 21 = 34$. We can now find $y(x_{k_1}) = y(52.5) = 72.25$ and $y(x_{k_2}) = y(85) = 576$, so that the new region of search for the minimum will be $0 < k < 34$.

Cycle 2 Place experiments at k_3' and k_3'', where

$$k_3' = 0 + l_2 = 0 + F_{9-3} = 0 + 13 = 13 \equiv k_3$$

$$k_3'' = 34 - l_2 = 34 - 13 = 21 \equiv k_1$$

Thus $y(x_{k_3}) = y(32.5) = 81.23$, and since $y(x_{k_1}) = y(52.5) = 72.25$, we can restrict the next search to the region $13 < k < 34$.

TABLE VI-16 OBJECTIVE FUNCTION VALUES $y(x_{k_j})$ FOR A SEQUENTIAL SEARCH OF PREPLANNED EXPERIMENTS (EXAMPLE VI-9)

Fibonacci experiment			*Cycle*						
j	k_j	x_{k_j}	1	2	3	4	5	6	7
1	21	52.5	72.25	72.25	72.25				
2	34	85.0	576.0						
3	13	32.5		81.23					
4	26	65.0			16.0	16.0	16.0		
5	29	72.5				132.2			
6	24	60.0					1.0	1.0	1.0
7	23	57.5						12.25	
8	25	62.5							2.25
Next region of search			$0 < k < 34$	$13 < k < 34$	$21 < k < 34$	$21 < k < 29$	$21 < k < 26$	$23 < k < 26$	$23 < k < 25$

The full set of calculations is given in Table VI-16, using a large positive penalty value if any fictitious location had been chosen. Note that only eight experiments are actually required, since we find that, after seven cycles of search, $23 < k < 25$, so that $k = 24$ or $x^* = x_{24} = 60$, with $y(x^*) = 1.00$ located within 5 percent. This is quite a reduction over the original 39 preplanned experiments, all of which need not now be examined.

An alternative method has also been proposed by Kiefer [1957] and discussed by Wilde [1964]. However, the method is not more efficient than the modified Fibonacci and will not be discussed here.

6 FINAL STAGE OF THE SEARCH

Any search method proposed in this chapter has as its final result the location of the optimum within a certain interval of uncertainty. The exact position is not defined, except for the case of the discrete function. Given the last few experiments from any of the search techniques, one can improve his knowledge of the optimum by a curve-fitting procedure. The method, a special case of that considered in Chap. III, Sec. 5, involves fitting a polynomial of a certain degree through the known points.

6.1 Polynomial of Second Degree

If there are three points (x_{k-2},x_{k-1},x_k) which have been determined experimentally in the region about the optimum, the curve of $y(x)$ can be approximated as a quadratic function of x as follows:

$$y = \beta_0 + \beta_1(x - x_{k-2}) + \beta_{11}(x - x_{k-2})(x - x_{k-1}) \tag{6.1-1}$$

where
$$\beta_0 = y_{k-2} \equiv y(x_{k-2}) \tag{6.1-2a}$$

$$\beta_1 = \frac{y_{k-1} - y_{k-2}}{x_{k-1} - x_{k-2}} \tag{6.1-2b}$$

$$\beta_{11} = \frac{\dfrac{y_k - y_{k-2}}{x_k - x_{k-2}} - \dfrac{y_{k-1} - y_{k-2}}{x_{k-1} - x_{k-2}}}{x_k - x_{k-1}} \tag{6.1-2c}$$

Here the three known experimental values of the objective function are y_{k-2}, y_{k-1}, and y_k at the points x_{k-2}, x_{k-1}, and x_k, respectively. Of course, if the objective function is quadratic, Eq. (6.1-1) yields the precise relationship between x and y. If the points are closely spaced, the equation tends to be more accurate.

Finally, one can find the optimum by requiring that $y' = 0$, so that the location of the optimum will be

$$x^* = \tfrac{1}{2}(x_{k-2} + x_{k-1}) - \frac{\beta_1}{2\beta_{11}} \tag{6.1-3}$$

This result can be used to estimate the location of the optimum at least to an approximation, whose accuracy depends on the actual functional form $y(x)$ and the interval separating the points. This is the final step in the search. We have an approximate location of the optimum, and the interval of possible error is known. Our task is then complete. An example of its application will be found in Example VI-5.

7 COMPARISON OF THE DIFFERENT METHODS AVAILABLE FOR CALCULATING THE OPTIMUM CONDITIONS FOR ONE-DIMENSIONAL FUNCTIONS

For functions of a single variable which are relatively simple in form and amenable to mathematical treatment, analytical methods would normally be used. While, in general, an optimum can be found at a stationary value, a discontinuity, or a boundary, analytical techniques can be used only for continuous functions. No such limitation is imposed on the numerical methods discussed in this chapter. In addition, an objective function cannot always be expressed in a mathematical form, as is the case when the values of the objective function are obtained by experimentation; where the mathematical function cannot be differentiated easily, if at all; or where the derivative might be available but the resulting equation on setting it to zero cannot be easily solved; in all these cases numerical methods must be utilized. Graphical plotting of values of the function can be used, but the method is generally not efficient because a great number of function evaluations are required to obtain a suitable graph.

Several logical search methods have been discussed. These, whether preplanned or sequential in approach, can locate the desired optimum to within a required degree of uncertainty, and several methods are particularly efficient in the small number of experiments required. These searches, however, can be effectively used only if the function, whether continuous, discontinuous, or discrete, is unimodal. Other functions would have to be considered region by region, in each of which the unimodal assumption was valid.

The preplanned approach to numerical search is very inefficient in terms of the number of experiments required to achieve a given accuracy (Table VI-9) in comparison with any sequential approach. There are, however, situations in which this is unavoidable, but there is no reason why the results of a preplanned study should not be examined in a sequential manner as in Example VI-19.

If the region of search is bounded, a measure of effectiveness of search can be defined. The so-called sequential uniform method, with three experiments per cycle, and the sequential dichotomous method are simple and require relatively few experiments. If experiments are at a premium, however, the most effective method of this group is the Fibonacci technique, which uses only one new experiment in each cycle and guarantees an upper limit on the number of experiments required. This technique might, however, suffer from the disadvantage that the accuracy of the search must be previously stated, but this can be removed by a modification to the basic Fibonacci approach. Even the best method described for such restricted regions may require many iterations and function evaluations if the original search region is large and the required accuracy is high. These methods are best, in a mini-max sense, only if we have no knowledge at all of the form of the function, apart from knowing that it is unimodal.

If we are dealing with relatively "smooth" functions, better, generally more efficient methods are available. These have been discussed under the heading of unbounded search regions. They incorporate acceleration features and depend upon the fitting of polynomial expressions to computed values of the function. Making use of the fact that near the optimum the function can often be approximated by a quadratic, the methods then use these expressions to estimate the position of the optimum. Such methods as Powell's direct search technique [1964] and that of Fletcher and Reeves [1964], based upon Davidon's [1959] gradient method, are particularly useful, being devised originally as one-dimensional searches within multivariable search techniques. Such methods are generally more efficient than the Fibonacci-type approach.

When the gradient of a function can be defined analytically and does not have to be found by additional experimentation, gradient-type search methods are preferred, but otherwise direct search techniques should be utilized.

This discussion has not considered the effect of experimental error upon the accuracy of search, for which purpose the literature[1] should be consulted. It should be noted, however, that the intervals between adjacent experiments at the later stages of search cannot be reduced indefinitely, even without experimental error being present, since the results from these locations must show a significant difference.

In conclusion, it can be stated that while one-dimensional problems may not be so common, an important requirement of many multidimensional search techniques is the provision of a highly effective one-dimensional search routine.

Problems

VI-1 How can the assumption of unimodality help in numerical search? What difference results when a gradient method is used rather than a direct method?

VI-2 The optimum of an unrestricted unimodal function is at $x = -4$.

a. Starting with a base point at $x = 3$ and a step size of $s = 0.1$, bracket the optimum, using an interval-doubling acceleration technique.

b. Repeat (*a*) with $s = 1.0$.

c. Repeat (*a*) with $s = 10.0$.

d. Repeat (*a*) with $s = 100$.

e. Tabulate your results and select, based on both the number of experiments and the length of the final interval of uncertainty, that value of s which is best.

VI-3 The optimum of an unimodal function is at $x = -4$. Using search techniques applicable to unrestricted functions and starting from $x = 3$, show how to locate the optimum to within ± 0.01.

VI-4 The first three experiments of an unrestricted search for a maximum yield the following results:

Experiment	x	$y(x)$
1	1	83
2	2	69
3	4	38

Estimate the scale of the objective function.

VI-5 What are the location and value of the minimum of the unrestricted function

$$y = 6x^4 - 3x + 7$$

Assume that the gradient is easily calculated but that the location cannot easily be determined by solving the algebraic equation. An accuracy of $x = \pm 0.1$ is required.

[1] For example, the work of Robbins and Munro [1951], Wolfowitz [1952], Kiefer and Wolfowitz [1952], Chung [1954], Blum [1954], Dvoretzky [1956], Hodges and Lehman [1956], Sacks [1958], Kesten [1958], Lapidus et al. [1961], and the discussion of Wilde [1964, chap. 6].

VI-6 In solving equations $g(x) = 0$, it is often convenient to seek the minimum of $y = g^2$ by using direct search techniques. Using this approach, find one root of

$$g = 10x^3 + 3x^2 + x + 5 = 0$$

VI-7 Find by direct search the root of

$$g = e^x - x^2 = 0$$

VI-8 Use a gradient method to find the root of the equation given in Prob. VI-7. This should be particularly effective since the objective function value at the optimum is known to be zero.

VI-9 If the output $y(x)$ of a black-box model is given by

$$y = (2x - 9)^2$$

for any value of x in the range $0 \leq x \leq 10$, and it is known that the function is unimodal over this region, locate the smallest value of $y(x)$ to within 5 percent of the initial range if all the experiments must be placed at regular intervals and be carried out at one time.

VI-10 Use a sequential uniform search method to solve Prob. VI-9 if the experiments need not be preplanned. Use two equally spaced experiments per cycle.

VI-11 Repeat Prob. VI-10 using three equally spaced experiments per cycle.

VI-12 It has been shown (Sec. 4.1.3), using regularly sized intervals in a sequential search method in a range $a \leq x \leq b$, that three experiments in each cycle allowed the optimum to be located to finer limits of accuracy than could be achieved using only two experiments per cycle.

Show, by considering other experimental sets or otherwise, that the use of three experiments per cycle gives better results than any other sequential method using uniform-sized search intervals. Derive a general formula, relating the total number of experiments, N, to the final interval of uncertainty, $\alpha = L_N/L_0$.

VI-13 Solve the mini-max problem

$$\alpha^* = \min [\max (x_2, x_3 - x_1, x_4 - x_2, 1 - x_3)]$$

with
$$0 \leq x_1 < x_2 < x_3 < x_4 \leq 1$$

Detail all your arguments. This problem is clearly equivalent to that of best spacing four simultaneous experiments so as to minimize the interval of uncertainty in the worst possible case.

VI-14 In a certain game, a player A can freely choose any value of the variable x_1, while his opponent B can simultaneously freely choose a value of x_2. If A's profit y is given by

$$y = 3 + x_1 x_2$$

which value of x_2 will B select if he wishes to minimize A's gains? A, of course, will attempt to maximize y. Show that this is a mini-max type problem.

VI-15 If the optimum is located at $x = 0.5$ in the interval $0 \leq x \leq 2$, perform a dichotomous search, reducing the interval of uncertainty α to 0.1. Compare the number of experiments required with that predicted by Eq. (4.2-8a), assuming δ to be very small.

VI-16 Repeat Prob. VI-15 assuming $\delta = 0.05$. What effect does this have on the number of experiments?

VI-17 Solve Prob. VI-9 using the dichotomous technique. Here the magnitude of δ is determined by the fact that the computations are to be done using a slide rule.

VI-18 Find by dichotomous search the minimum of

$$y = 10x^2 - 3x + 5$$

subject to the constraint

$$g = x^2 - 5 \leq 0$$

VI-19 Solve Prob. VI-9 using the Fibonacci method.

VI-20 Use the Fibonacci method to seek the position of the maximum value of the function

$$y = 9x - 0.1x^2$$

in the range $0 \leq x \leq 100$ to within an accuracy of 0.1 percent of the original range. Give the results of the calculations for the first four stages only.

VI-21 Show that the Fibonacci series can be represented by the expression

$$F_n = \frac{1}{\sqrt{5}}\left[\left(\frac{1+\sqrt{5}}{2}\right)^{n+1} - \left(\frac{1-\sqrt{5}}{2}\right)^{n+1}\right]$$

(It may be assumed that $x_n = K^n$ is a solution of $K^n = K^{n-1} + K^{n-2}$.)

VI-22 Show that the relationship for F_n in Prob. VI-21 reduces to

$$F_n = 0.4472(1.6180)^{n+1}$$

for large values of n. In this case, what might be the lower range of these n values?

VI-23 Carry out the solution to Prob. VI-9 by the modified Fibonacci method. Present only the first three stages of the solution. How do the results compare with those of Prob. VI-9?

VI-24 In Prob. VI-23, consider in detail the calculation of the last six experiments. Examine only the length of the search interval estimated to remain after the $(N - 6)$th experiment, and assign an arbitrary range for the purpose of allocating the final set of experiments.

VI-25 It has been found that a function, represented by $y = (x - 13)^2$, exists only at certain values of the independent variable, x, namely, at 0.7, 1.1, 7.3, 8.4, 11.7, 13.5, 14.1, 15.3, 18.1, 23.9, 27.4, and 30.1. If the function can be assumed to be unimodal in this range, find its minimum value.

VI-26 Glycerine at 15°C is to be pumped at 1,000 lb/hr from a storage tank at ground level to a vessel situated at a height of 35 ft. The total length of piping is 75 ft, and includes four right-angle bends. Considering only the installed cost of the pipe (depreciated over a 10-year period) and the pumping costs, determine the most suitable diameter of pipe for the plant.

The pressure drop in a pipe may be calculated from the equation

$$h_f = \frac{\Delta p}{\rho} = -\frac{2fLV^2}{D}$$

where the nomenclature has the usual meaning and the friction factor f may be represented by $f = 16/Re$ for viscous flow.

The purchased cost of 1-in. dia. standard-weight black pipe is 24¢/ft length. [The $\frac{8}{10}$ rule applies, with respect to diameter, for pipes less than 1 in. in diameter, and the $\frac{10}{10}$ rule above that size. Standard sizes of pipe (in inches) may be taken as $\frac{1}{4}$, $\frac{3}{8}$, $\frac{1}{2}$, $\frac{3}{4}$, 1, $1\frac{1}{4}$, $1\frac{3}{4}$, 2, 2 $\frac{1}{4}$, $2\frac{1}{2}$,3, ...] Glycerine (15°C): density 1.264 g/ml, viscosity 8 poises. The cost of electricity may be taken as 1¢/kwhr.

VI-27 A multiple-effect evaporator is to be installed to handle a feed of 360,000 lb of 5 percent, by weight, salt solution daily. If the solution has to be concentrated to 30 percent, determine the number of effects which should be used in order to achieve the lowest annual cost.

Data: The initial cost of the evaporator may be taken as $10,500 for each effect installed, together with an additional $3,000 for auxiliary equipment which must be fitted, regardless of the number of effects used. The life of the unit is estimated to be 12 years, and at the end of this time the salvage value is considered to be $1,500 per effect. The evaporator will operate for 320 days annually. The fixed charge (excluding depreciation) may be taken as 12 percent annually, and the annual maintenance costs as 5 percent, both based upon the original capital investment. Steam costs are 84¢/1,000 lb of steam used, and $0.8N$ lb of water is evaporated in the unit for each pound of steam, N being the number of effects installed. All the other costs, except depreciation charges, may be considered as being independent of the number of effects used.

VI-28 Consider Prob. VI-27 for the situation when an alternative investment is also competing for the same available capital. This alternative would realize a 10 percent rate of return on the capital invested. What would be the optimum number of effects chosen for the multiple-effect evaporator on the basis of the marginal rate of return?

VI-29 The final stages of a direct search have yielded the values $y(x_{k-2})$, $y(x_{k-1})$, and $y(x_k)$. If

$$x_k - x_{k-1} \neq x_{k-1} - x_{k-2}$$

fit the quadratic curve

$$y = a + bx + cx^2$$

to the results, and hence obtain a more accurate estimate of the location of the optimum.

VI-30 Prove that Eqs. (6.1-2) provide the required constants in using Eq. (6.1-1) for three-point curve fitting.

VI-31 For the function

$$y = 3x_1^2 - 2x_1$$

which is limited by the constraint

$$g_1 = x_1^2 - 4 \leq 0$$

find the location and value of the global maximum.

CHAPTER VII

MULTIVARIABLE SEARCH—ANALYTICAL METHODS

1 CLASSIFICATION OF THE MULTIVARIABLE PROBLEM

Although many optimization problems can be reduced by an appropriate series of direct eliminations to a problem involving just a single independent variable, important cases do arise in which such elimination is either impossible or so complex as to be prohibitive. It is therefore necessary to develop methods of optimization which can handle more than one independent variable.

In considering the various types of analytical methods of optimization it is convenient to divide the discussion into three main sections, the classification being based upon the absence of restrictions (Sec. 2), the presence of equality restrictions (Sec. 3), and the presence of inequality restrictions (Sec. 4). These restrictions have been described (Chap. II, Sec. 6) and will not be discussed further at this point. A potential alternative approach through dual systems is described in Sec. 5.

2 UNRESTRICTED OBJECTIVE FUNCTIONS

This section will describe the various methods available for the analytical optimization of multivariable functions upon which no constraints have been imposed.

2.1 General Considerations

The maximum (or minimum) of an unrestricted objective function must lie at a stationary point or at a position of discontinuity of the function or its gradient (Chap. IV, Sec. 4). We must therefore search for positions at which the partial derivatives of the function are zero or are discontinuous or at points at which the function itself is discontinuous.

The general procedure for finding the optimum of a differentiable function of several variables is more difficult than for a single variable and requires care in defining the regions in which differentiation of a continuous function may be safely carried out. In order to study a given objective function with the aim of determining its optimum value, the following considerations must be examined:

1. Divide the region into subintervals in which the objective function is defined by a formula possessing continuous first and second derivatives.

2. In each such subinterval the point or points should be found which make the first derivative with respect to all the derivatives equal to zero.

3. Conditions must be checked for each such point to determine if the nature of the stationary value is a maximum, a minimum, or a saddle point by a procedure to be defined in subsequent sections.

4. The value of the objective function should be found at each of the stationary values yielding the desired type of optimum.

5. The best value of the various possible stationary-value optimum objective functions should be found and compared with values of the objective function calculated at the points of discontinuity previously marked by the boundaries of the subintervals.

6. Compare the best results for each subinterval to find the best optimum value for the system.

Since, for most objective functions, it is possible to avoid the presence of discontinuities by the above procedure, let us now consider the problems involved in using analytical methods for continuous, continuously differentiable functions with an unrestricted choice of the independent variables.

2.2 The Character of the Stationary Point

To prove that a certain stationary point is a local maximum (minimum), one must show that there exists a region, albeit small, about the stationary point in which the value of the objective function is less (greater) than or equal to its value at the stationary point. The nature of the stationary point can thus be defined by an exploration of the domain surrounding the point. This exploration can be formalized in definite mathematical terms if the Taylor series expansion exists. The Taylor series in its general form is usually stated as follows: If the function has $j + 1$ derivatives which exist and are continuous in a region about the point $\mathbf{x}^*$ (see, for example, Kaplan [1952]), we can write that

$$y(\mathbf{x}) = y(\mathbf{x}^*) + \frac{\Delta y}{1!} + \frac{\Delta^2 y}{2!} + \cdots + \frac{\Delta^j y}{j!} + R_{j+1} \qquad (2.2\text{-}1)$$

where

$$\Delta y \equiv (x_1 - x_1^*)y_{x_1}^* + (x_2 - x_2^*)y_{x_2}^* + \cdots + (x_n - x_n^*)y_{x_n}^*$$

$$\Delta^2 y \equiv (x_1 - x_1^*)^2 y_{x_1x_1}^* + 2(x_1 - x_1^*)(x_2 - x_2^*)y_{x_1x_2}^* + (x_2 - x_2)^2 y_{x_2x_2}^* + \cdots + (x_n - x_n^*)^2 y_{x_nx_n}^*$$

$$\Delta^k y \equiv \sum \frac{k!}{p_1!p_2!\cdots p_n!}(x_1 - x_1^*)^{p_1}(x_2 - x_2^*)^{p_2}\cdots(x_n - x_n^*)^{p_n} y_{x_1{}^{p_1}x_2{}^{p_2}\cdots x_n{}^{p_n}}^*$$

the summation being taken over all integers p_i such that $\sum_{i=1}^{n} p_i = k$. The remainder in the Taylor's expansion is R_{j+1}. This notation implies that

$$y_{x_1{}^{p_1}x_2{}^{p_2}\cdots x_n{}^{p_n}}^* \equiv \left(\frac{\partial y_k}{(\partial x_1)^{p_1}(\partial x_2)^{p_2}\cdots(\partial x_n)^{p_n}}\right)$$

the asterisk signifying that the partial derivative is evaluated at $\mathbf{x}^*$.

Since this is a relatively complex structure, the first terms in the series will be written out in full for the case of two variables, x_1 and x_2. Thus we get

$$\Delta y \equiv (x_1 - x_1^*)\left(\frac{\partial y}{\partial x_1}\right)^* + (x_2 - x_2^*)\left(\frac{\partial y}{\partial x_2}\right)^*$$

$$\Delta^2 y \equiv (x_1 - x_1^*)^2\left(\frac{\partial^2 y}{\partial x_1{}^2}\right)^* + 2(x_1 - x_1^*)(x_2 - x_2^*)\left(\frac{\partial^2 y}{\partial x_1\,\partial x_2}\right)^* + (x_2 - x_2^*)^2\left(\frac{\partial^2 y}{\partial x_2{}^2}\right)^*$$

and

$$\Delta^3 y \equiv (x_1 - x_1^*)^3\left(\frac{\partial^3 y}{\partial x_1{}^3}\right)^* + 3(x_1 - x_1^*)^2(x_2 - x_2^*)\left(\frac{\partial^3 y}{\partial x_1{}^2\,\partial x_2}\right)^* + 3(x_1 - x_1^*)(x_2 - x_2^*)^2\left(\frac{\partial^3 y}{\partial x_1\partial x_2{}^2}\right)^* + (x_2 - x_2^*)^3\left(\frac{\partial^3 y}{\partial x_2{}^3}\right)^*$$

Based on this Taylor series expansion, some very pertinent conclusions can be deduced. It should be noted that if $y(\mathbf{x})$ is to be a local maximum or minimum, then by an argument parallel to that of Chap. V, Sec. 2, it is necessary for

$$\Delta y = 0 \qquad \text{at } \mathbf{x}^* \tag{2.2-2}$$

i.e., the point $\mathbf{x}^*$ must be a stationary point. Second, if

$$\Delta^2 y < 0 \qquad \text{for all nontrivial } \mathbf{x} - \mathbf{x}^* \tag{2.2-3}$$

then the stationary point is, clearly, a maximum. Equation (2.2-3) is, as before, a sufficient condition for a maximum, but is not necessary. Indeed, $\Delta^2 y$ could vanish and the stationary point still be a maximum. As in Chap.

V, an analysis of the latter situation would similarly require higher-order derivatives and will not be considered further here.

The necessary conditions for a maximum thus require $\Delta y = 0$ and $\Delta^2 y \leq 0$, but if $\Delta y = 0$ and $\Delta^2 y < 0$, this is sufficient. Similarly, for a minimum, the necessary conditions are $\Delta y = 0$ and $\Delta^2 y \geq 0$, but it is sufficient if $\Delta y = 0$ and $\Delta^2 y > 0$.

Our general problem is therefore the evaluation of the stationary point at $\Delta y = 0$ by examination of the sign of $\Delta^2 y$ for all possible $\mathbf{x}$ different from $\mathbf{x}^*$. For simplicity, the discussion will be first limited to a two-variable function to allow the pertinent features to be understood.

2.2.1 *Stationary values in two variables*

If a point $\{x_1^*, x_2^*\}$ is a stationary point of the function $y(x_1, x_2)$, the first-order terms Δy in the Taylor expansion, Eq. (2.2-1), must vanish. This requires that both y_{x_1} and y_{x_2} be zero. We then have, assuming the second derivative is continuous and the third exists, that

$$y(\mathbf{x}) - y(\mathbf{x}^*) = \tfrac{1}{2}\Delta^2 y$$

neglecting the third- and higher-order terms. It is obvious that the stationary point will be a maximum if $\Delta^2 y$ is less than zero. (Conversely, a minimum is found if $\Delta^2 y > 0$.)

It is therefore necessary to examine the conditions for which $\Delta^2 y < 0$ (termed negative-definite[1]). If we expand $\Delta^2 y$ as

$$\Delta^2 y = h_1^2 y^*_{x_1x_1} + 2h_1h_2 y^*_{x_1x_2} + h_2^2 y^*_{x_2x_2} \tag{2.2-4a}$$

where $h_1 = x_1 - x_1^*$ and $h_2 = x_2 - x_2^*$, then, clearly, $\Delta^2 y$ will be negative-definite only if

$$h_1^2 y^*_{x_1x_1} + 2h_1h_2 y^*_{x_1x_2} + h_2^2 y^*_{x_2x_2} < 0 \qquad \text{for nontrivial } h_1 \text{ and } h_2 \tag{2.2-4b}$$

Inequality (2.2-4*b*) is a sufficiency requirement for a stationary point to be a maximum. Let us examine this sufficiency requirement in more detail. If $h_1 = 0$, then $y^*_{x_2x_2}$ must clearly be less than zero if inequality (2.2-4*b*) is to be satisfied[2] for all $h_2 \neq 0$.

Similarly, for $h_2 = 0$ and $h_1 \neq 0$, $y^*_{x_1x_1} < 0$. If $y^*_{x_1x_1} < 0$, we may rewrite the inequality (2.2-4*b*)

$$y^*_{x_1x_1}\left(h_1^2 + 2h_1h_2 \frac{y^*_{x_1x_2}}{y^*_{x_1x_1}} + h_2^2 \frac{y^*_{x_2x_2}}{y^*_{x_1x_1}}\right) < 0$$

[1] If a function $\Delta^2 y < 0$ for all nontrivial h_1 and h_2 (h_1 and h_2 not being simultaneously zero), the function is called *negative-definite*. If $\Delta^2 y > 0$ for these values of h_1 and h_2, the function is called *positive-definite*. If $\Delta^2 y$ can be of either sign, it is *indefinite*. If the function is such that $\Delta^2 y \geq 0$, it is *nonnegative*. Similarly, a *nonpositive* function is such that $\Delta^2 y \leq 0$.

[2] Putting both h_1 and h_2 equal to zero is not permitted, since this would be a trivial case.

and by completing the square of the first two terms we get

$$y^*_{x_1x_1}\left(h_1 + \frac{y^*_{x_1x_2}}{y^*_{x_1x_1}} h_2\right)^2 + h_2^{\,2}\left(y^*_{x_2x_2} - \frac{y^{*2}_{x_1x_2}}{y^*_{x_1x_1}}\right) < 0 \qquad (2.2\text{-}5)$$

[If $y^*_{x_1x_1} = 0$, then inequality (2.2-4*b*) cannot be satisfied. This is demonstrated below.] Since h_2 may take any value, it follows that $\Delta^2 y$ will be negative-definite only if

$$y^*_{x_1x_1} < 0 \qquad (2.2\text{-}6a)$$

$$y^*_{x_2x_2} - \frac{y^{*2}_{x_1x_2}}{y^*_{x_1x_1}} < 0 \qquad \text{or} \qquad y^*_{x_1x_1} y^*_{x_2x_2} > y^{*2}_{x_1x_2} \qquad (2.2\text{-}6b)$$

it being inherent in these conditions that $y^*_{x_2x_2} < 0$.

What happens if one or more of the partial derivatives $y^*_{x_ix_j}$ vanish at the stationary point? Certainly, if

$$y^*_{x_2x_2} = 0$$

it follows that

$$\Delta^2 y = h_1^{\,2} y^*_{x_1x_1} + 2h_1h_2 y^*_{x_1x_2}$$

and if we set $h_1 \neq 0$, then $\Delta^2 y$ can be made to change sign by changing the sign and magnitude of h_2. Therefore inequality (2.2-4*b*) is not satisfied, and the function is indefinite. Similarly, $y^*_{x_1x_1}$ cannot vanish, or else the function $\Delta^2 y$ will also be indefinite.

If $y^*_{x_1x_1} = y^*_{x_2x_2} = 0$, the $\Delta^2 y$ is also indefinite, for in this case

$$\Delta^2 y = 2h_1h_2 y^*_{x_1x_2}$$

which changes sign on reversing the sign of h_1.

If $y^*_{x_1x_2} = y^*_{x_2x_2} = 0$, then

$$\Delta^2 y = y^*_{x_1x_1} h_1^{\,2}$$

and inequality (2.2-4*b*) is not satisfied for all nontrivial h_1 and h_2. The $\Delta^2 y$ is nonpositive for $y^*_{x_1x_1} < 0$, but is not negative-definite, since it vanishes for nontrivial values of h_1 and h_2, namely, for all h_2 when $h_1 = 0$.

Finally, if $y^*_{x_1x_2} = 0$, then $\Delta^2 y$ is negative-definite if both $y^*_{x_1x_1} < 0$ and $y^*_{x_2x_2} < 0$.

It therefore follows that inequalities (2.2-6) are necessary and sufficient conditions for $\Delta^2 y$ to be negative-definite. If $\Delta^2 y$ is negative-definite, the stationary point is certainly a maximum.

However, the stationary point can also be a maximum if the $\Delta^2 y$ is nonpositive. This question is discussed more fully by, for example, Hancock [1960] and Kaplan [1952].

A similar argument can be used for a minimum leading to the sufficient condition $\Delta^2 y > 0$ (positive-definite) if

$$y^*_{x_1x_1} > 0 \tag{2.2-7a}$$

$$y^*_{x_2x_2} - \frac{y^{*2}_{x_1x_2}}{y^*_{x_1x_1}} > 0 \qquad \text{or} \qquad y^*_{x_1x_1} y^*_{x_2x_2} > y^{*2}_{x_1x_2} \tag{2.2-7b}$$

Both of these sufficiency conditions, known as *Lagrange's criteria*, can be summarized in the following manner.

For a minimum: $\quad D_i > 0 \quad$ for $i = 1$ and 2 $\quad$ (2.2-8)

For a maximum: $\quad D_1 < 0, \quad D_2 > 0 \quad$ (2.2-9)

In either case, the D_i are defined[1] as the determinants

$$D_1 = |y^*_{x_1x_1}| \qquad D_2 = \begin{vmatrix} y^*_{x_1x_1} & y^*_{x_1x_2} \\ y^*_{x_2x_1} & y^*_{x_2x_2} \end{vmatrix}$$

It has been noted that the sufficiency condition is the strongest requirement, and in fact other weaker conditions may still define an extreme stationary point. Thus, if

$$y^*_{x_1x_1} y^*_{x_2x_2} = y^{*2}_{x_1x_2}$$

the second-order terms in $\Delta^2 y$ will become

$$\left(h_1 + h_2 \frac{y^*_{x_1x_2}}{y^*_{x_1x_1}}\right)^2 y^*_{x_1x_1}$$

The value of this therefore depends upon the sign of $y^*_{x_1x_1}$, a negative sign indicating a maximum and a positive a minimum. Edelbaum [1962] has noted that extrema can exist in exceptional circumstances (see, for example, Hancock [1960]).

Example VII-1: Minimum values

If the total cost of a process can be represented by the expression

$$y = ax_1 + \frac{b}{x_1x_2} + cx_2 + d$$

where a, b, c, and d are positive coefficients, find the optimum values of x_1 and x_2 to give a minimum cost.

If a minimum cost exists in this continuous function, it must be at a discontinuity or a stationary value. Let us consider the latter. Any

[1] See also Sec. 2.2.3.

stationary point will lie at the necessary condition

$$y^*_{x_1} = a - \frac{b}{x_1^{*2} x_2^*} = 0$$

$$y^*_{x_2} = c - \frac{b}{x_1^* x_2^{*2}} = 0$$

so that its location is given by

$$x_1^* = \left(\frac{bc}{a^2}\right)^{1/3} \qquad x_2^* = \left(\frac{ab}{c^2}\right)^{1/3}$$

For the sufficient conditions we must determine the signs of the determinants

$$D_1 = y^*_{x_1x_1} = \frac{2b}{x_1^{*3} x_2^*}$$

$$D_2 = y^*_{x_1x_1} y^*_{x_2x_2} - y^{*2}_{x_1x_2} = \frac{2b}{x_1^{*3} x_2^*} \frac{2b}{x_1^* x_2^{*3}} - \left(\frac{b}{x_1^{*2} x_2^{*2}}\right)^2 = \frac{3b^2}{x_1^{*4} x_2^{*4}}$$

Since both of these are positive at the stationary value, a minimum will exist there.

2.2.2 *The nature of the stationary point in two dimensions*

If a stationary value has been located, and if sufficiency conditions have shown it to be an optimum, any combination of small changes in x_1 and x_2 can only decrease (for a maximum) or increase (for a minimum) the value of the objective function away from the stationary point. This definite condition can be represented by the elliptical contours of Fig. VII-1*a*.

Some functions may result in the indefinite quadratic conditions depicted by the hyperbolic contours of Fig. VII-1*b*. This stationary point is no longer an optimum "peak," but rather the summit of a pass, or a saddle-shaped region from which location some directions will provide an increase or decrease in the value of the objective function. This type of stationary point is sometimes called a *saddle point*. This situation can be found, for example, when

$$y^*_{x_1x_1} y^*_{x_2x_2} < y^{*2}_{x_1x_2}$$

since it is then possible to get some values of h_1 and h_2 for which $\Delta^2 y$ is positive and others for which it is negative.

It is therefore desirable in seeking optimum values to examine the nature of any stationary values, since these may be true extreme values or saddle points, the latter allowing still more favorable positions to be attained.

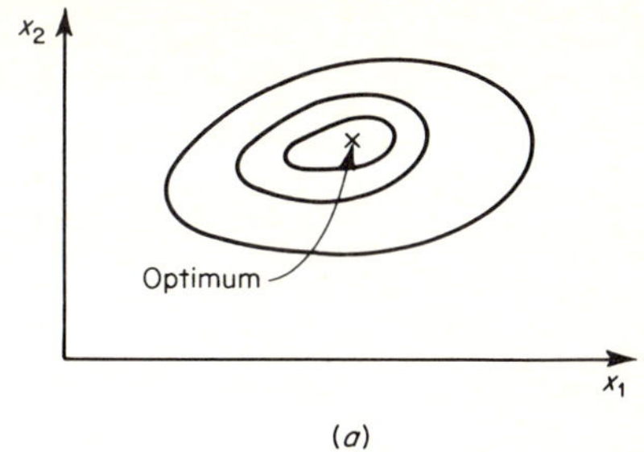

(a)

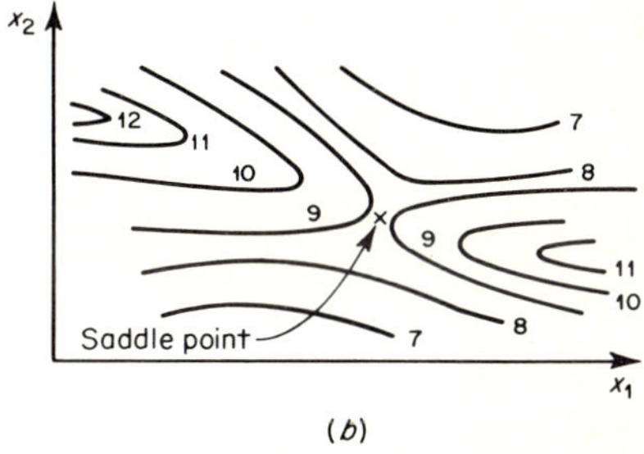

(b)

Fig. VII-1 Stationary values. (a) Extreme-valued stationary point; (b) saddle-point stationary point.

Example VII-2: Stationary values

a. Let us consider the stationary values of the function

$$y = 3x_2^2 + x_1^4 + 2x_2^4$$

Here $y_{x_1} = 4x_1^3$ and $y_{x_2} = 6x_2 + 8x_2^3$, so that a stationary point exists at $x_1^* = x_2^* = 0$. Evaluating the higher derivatives at this point yields

$$y^*_{x_1x_1} = 12(x_1^*)^2 = 0$$
$$y^*_{x_1x_2} = 0$$
$$y^*_{x_2x_2} = 6 + 24(x_2^*)^2 = 6$$

Then the sufficient conditions for an optimum

$$y^*_{x_1x_1} < 0 \qquad \text{or} \qquad y^*_{x_1x_1} > 0$$

and
$$y^*_{x_1x_1}y^*_{x_2x_2} - (y^*_{x_1x_2})^2 > 0$$

are *not* satisfied. If we evaluate $\Delta^2 y$ by Eq. (2.2-4*a*), then

$$\Delta^2 y = +6h_2^2 \geq 0$$

The origin is, in fact, a true minimum, as will be seen in Fig. VII-2 or by examining the objective function further.

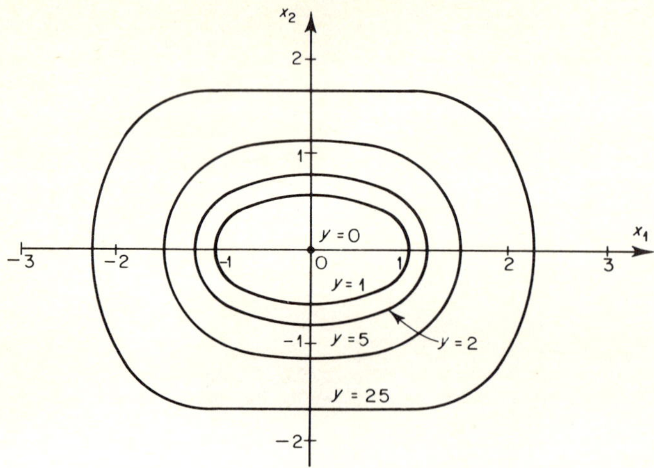

Fig. VII-2 The function $y = 3x_2^2 + x_1^4 + 2x_2^4$ of Example VII-2*a*.

b. On the other hand, the function

$$y = 2x_1^2x_2 - 2x_2^2 + x_1^3$$

is also stationary at the origin and also yields

$$\Delta^2 y = -4h_2^2 \leq 0$$

at that point. However, this function is of a completely different character, for a move along the x_1 axis will either decrease or increase the objective function, depending on the direction of the move, as seen in Fig. VII-3.

Fig.. VII-3 The function $y = 2x_2x_1^2 - 2x_2^2 + x_1^3$ of Example VII-2*b*.

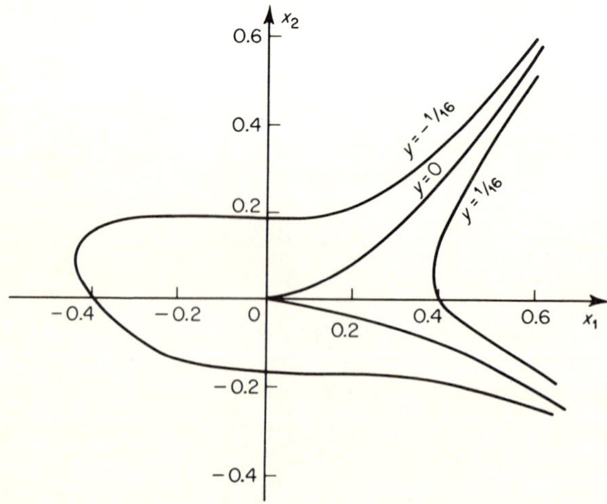

The essential point is that if the sufficiency criteria are not satisfied, further investigation is required to know if a possible optimum exists.

2.2.3 *Necessary and sufficient conditions for an extreme value in the general case*

In the absence of boundaries, it has been seen that optima can be found only at discontinuities or stationary points.

An optimum can exist only at a stationary point if the function and its first derivatives are continuous. We may use the Taylor expansion, Eq. (2.2-1), to obtain the necessary and sufficient conditions for an optimum to exist.

The necessary condition for an optimum is, then, that there is a stationary point, defined by the vanishing of all the first derivatives, so that

$$\Delta y = 0$$

For the stationary value to be an extreme value, the second derivatives, if they do not vanish, must be such that the term $\Delta^2 y$, as defined by Eq. (2.2-1), does not change sign. The sufficient condition for an extreme value is thus

$$\Delta^2 y > 0$$

for a local minimum to exist, and

$$\Delta^2 y < 0$$

for a local maximum.

If neither condition applies and in addition the $\Delta^2 y$ is indefinite, the function is neither a maximum nor a minimum, but is a saddle point. If $\Delta^2 y$ is nonpositive or nonnegative at the stationary point, further investigation of higher derivatives will be necessary to determine the nature of the function.

This Lagrange criterion, the sufficient condition for an extreme value to exist, for $N > 2$, is discussed further by Hancock [1960] and Osgood [1925].

The sufficient condition for a stationary point to be an optimum can in general be investigated by completing the squares of the terms of the second derivative.[1] This has already been used in determining the nature of $\Delta^2 y$ in the two-dimensional problem (Sec. 2.2.1). The general approach may be best illustrated by an example, the process of completing the square being possible for any number of variables for a quadratic function. Each iteration of the procedure gives a perfect square plus a remainder, involving one less variable than before. We thus gather all the terms in x_1, say, and complete the square, leaving a remainder in $x_2, x_3, \ldots, x_n$ only. Next we

[1] Some of the basic theorems which suggest methods of telling whether a real symmetric quadratic form is indefinite, semidefinite, or definite by grouping it into a sum of squares can be found in Turnbull [1960] and Yefimov [1964], among others.

collect all the remaining terms in x_2 and again complete the square. This is continued until all the variables have been included.

Example VII-3: Completing the square

a. If a quadratic function (for example, $\Delta^2 y$) is represented by

$$\Delta^2 y = a_{11}h_1^2 + 2a_{12}h_1h_2 + a_{22}h_2^2$$

the sum of the squares can be computed by adding a multiple of one of the squared variables, h_2, for example, to the first two terms and subtracting a like quantity from the last term. Thus, if $a_{11} \neq 0$,

$$\Delta^2 y = \left(a_{11}h_1^2 + 2a_{12}h_1h_2 + \frac{a_{12}^2}{a_{11}} h_2^2\right) + \left(a_{22}h_2^2 - \frac{a_{12}^2}{a_{11}} h_2^2\right)$$

$$= a_{11}\left(h_1 + \frac{a_{12}}{a_{11}} h_2\right)^2 + \left(a_{22} - \frac{a_{12}^2}{a_{11}}\right)h_2^2$$

This form is particularly useful, as we have seen, in defining the sign of $\Delta^2 y$.

b. If a quadratic has the form

$$\Delta^2 y = a_{11}h_1^2 + 2a_{12}h_1h_2 + a_{22}h_2^2 + 2a_{13}h_1h_3 + a_{33}h_3^2 + 2a_{23}h_2h_3$$

then a similar procedure can be applied to transform this quadratic form into a structure which contains the h_i in squared terms. Of course, such a form is indispensable in attempting to associate a definite sign with $\Delta^2 y$. The procedure is similar to that shown above. Thus, if $a_{11} \neq 0$,

$$\Delta^2 y = \left(a_{11}h_1^2 + 2a_{12}h_1h_2 + 2a_{13}h_1h_3 + \frac{a_{13}^2}{a_{11}} h_3^2 + \frac{a_{12}^2}{a_{11}} h_2^2 + \frac{2a_{13}a_{12}h_2h_3}{a_{11}}\right)$$

$$+ \left(a_{22} - \frac{a_{12}^2}{a_{11}}\right)h_2^2 + \left(a_{33} - \frac{a_{13}^2}{a_{11}}\right)h_3^2 + 2\left(a_{23} - \frac{a_{12}a_{13}}{a_{11}}\right)h_2h_3$$

$$= a_{11}\left(h_1 + \frac{a_{12}}{a_{11}} h_2 + \frac{a_{13}}{a_{11}} h_3\right)^2 + \left(a_{22} - \frac{a_{12}^2}{a_{11}}\right)h_2^2 + \left(a_{33} - \frac{a_{13}^2}{a_{11}}\right)h_3^2$$

$$+ 2\left(a_{23} - \frac{a_{13}a_{12}}{a_{11}}\right)h_2h_3$$

Repeating this procedure and assuming that the coefficient of h_2^2 is not zero, we find that

$$\Delta^2 y = a_{11}\left(h_1 + \frac{a_{12}}{a_{11}} h_2 + \frac{a_{13}}{a_{11}} h_3\right)^2 + \left(a_{22} - \frac{a_{12}^2}{a_{11}}\right)$$

$$\times \left(h_2 + h_3 \frac{a_{23}a_{11} - a_{12}a_{13}}{a_{22}a_{11} - a_{12}^2}\right)^2 + \left[a_{33} - \frac{a_{13}^2}{a_{11}} - \frac{(a_{23}a_{11} - a_{12}a_{13})^2}{a_{22}a_{11}^2 - a_{12}^2 a_{11}}\right]h_3^2 \tag{2.2-10}$$

Thus, if

$$\Delta^2 y = h_1^2 + 5h_2^2 + 6h_3^2 - 4h_1h_2 + 2h_1h_3 - 8h_2h_3$$

then

$$\Delta^2 y = (h_1 - 2h_2 + h_3)^2 + (h_2 - 2h_3)^2 + h_3^2$$

is the required form as a series of squared terms.

The technique of completing the square can evidently be applied to quadratic forms having n variables. These results give rise to the following sufficiency conditions governing the character of a stationary point.

Let us define the determinant D_i as[1]

$$D_i = \begin{vmatrix} y^*_{x_1x_1} & y^*_{x_1x_2} & \cdots & y^*_{x_1x_i} \\ y^*_{x_2x_1} & y^*_{x_2x_2} & \cdots & y^*_{x_2x_i} \\ \cdot & \cdot & \cdot & \cdot \\ y^*_{x_ix_1} & y^*_{x_ix_2} & \cdots & y^*_{x_ix_i} \end{vmatrix}$$

and for n variables, evaluate (see Appendix 1) in turn the n determinants $D_1, D_2, \ldots, D_n$. The following two statements then apply (Edelbaum [1962]):

1. Sufficient conditions for a *minimum* to exist at the stationary point are that *all* the determinants are positive,

$$D_i > 0 \qquad \text{for all } i, i = 1, 2, \ldots, n \tag{2.2-11}$$

2. Sufficient conditions for a *maximum* to exist at the stationary point are that all the *even* determinants are positive and all the *odd* ones are negative; i.e.,

$$D_i < 0 \qquad \text{for all odd } i, i = 1, 3, 5, \ldots \tag{2.2-12a}$$

$$D_i > 0 \qquad \text{for all even } i, i = 2, 4, 6, \ldots \tag{2.2-12b}$$

If neither of these sets of conditions is satisfied, the stationary value may not be an optimum. Weaker conditions, as we have seen, may exist, however. The special case of all D_i being zero is equivalent to $\Delta^2 y = 0$, and higher derivatives must then be tested.

While we shall not give a proof here of these sufficiency criteria, Eqs. (2.2-11) and (2.2-12), for a maximum or a minimum, it is instructive to see their origin when n is 3. For in this case, $\Delta^2 y$ is given by Eq. (2.2-10), provided

$$a_{ji} = y^*_{x_jx_i}$$

[1] The determinant of the second derivatives evaluated at any point $\mathbf{x}$ is sometimes called the *Hessian*.

Examining Eq. (2.2-10), the following inequalities must certainly be satisfied if $\Delta^2 y$ is to be negative-definite for all nontrivial values of the displacements h_i:

$$y^*_{x_1x_1} < 0 \quad y^*_{x_2x_2} - \frac{y^{*2}_{x_1x_2}}{y^*_{x_1x_1}} < 0 \quad y^*_{x_3x_3} - \frac{y^{*2}_{x_1x_3}}{y^*_{x_1x_1}} - \frac{(y^*_{x_2x_3}y^*_{x_1x_1} - y^*_{x_1x_2}y^*_{x_1x_3})^2}{y^*_{x_2x_2}y^{*2}_{x_1x_1} - y^{*2}_{x_1x_2}y^*_{x_1x_1}} < 0$$

The first two inequalities have already been found in Sec. 2.2.1 when considering two variables. If the first inequality is satisfied ($y^*_{x_1x_1} < 0$), the second can be replaced by

$$D_2 \equiv \begin{vmatrix} y^*_{x_1x_1} & y^*_{x_1x_2} \\ y^*_{x_1x_2} & y^*_{x_2x_2} \end{vmatrix} > 0$$

The change in direction of the inequality stems from the fact that we have multiplied both sides by a negative number.

The third inequality can be put into the form suggested by inequality (2.2-12) by first multiplying through by $y^*_{x_1x_1}$, giving

$$y^*_{x_1x_1}y^*_{x_3x_3} - y^{*2}_{x_1x_3} - \frac{(y^*_{x_2x_3}y^*_{x_1x_1} - y^*_{x_1x_2}y^*_{x_1x_3})^2}{y^*_{x_2x_2}y^*_{x_1x_1} - y^{*2}_{x_1x_2}} > 0$$

Now multiply both sides by $y^*_{x_1x_1}y^*_{x_2x_2} - y^{*2}_{x_1x_2}$, a positive quantity. Thus

$$y^{*2}_{x_1x_1}y^*_{x_2x_2}y^*_{x_3x_3} - y^*_{x_1x_1}y^*_{x_3x_3}y^{*2}_{x_1x_2} - y^*_{x_2x_2}y^*_{x_1x_1}y^{*2}_{x_1x_3} - y^{*2}_{x_2x_3}y^{*2}_{x_1x_1} + 2y^*_{x_2x_3}y^*_{x_1x_1}y^*_{x_1x_2}y^*_{x_1x_3} > 0$$

Upon multiplying both sides by $1/y^*_{x_1x_1}$, a negative quantity, the final inequality is immediately seen to be

$$D_3 = \begin{vmatrix} y^*_{x_1x_1} & y^*_{x_1x_2} & y^*_{x_1x_3} \\ y^*_{x_2x_1} & y^*_{x_2x_2} & y^*_{x_2x_3} \\ y^*_{x_3x_1} & y^*_{x_3x_2} & y^*_{x_3x_3} \end{vmatrix} < 0$$

which is consistent with the requirements of inequalities (2.2-12).

Example VII 4: Determinant representation

Find both necessary and sufficient conditions for an optimum to exist in a one- and a two-dimensional function.

One dimension The necessary condition will be defined by $y^*_{x_1} = 0$, and the sufficient condition will be given, using inequalities (2.2-11) and (2.2-12), as

$$D_1 > 0$$

so that

$$y^*_{x_1x_1} > 0$$

for a minimum value, or

$$D_1 < 0$$

so that
$$y^*_{x_1x_1} < 0$$

for a maximum value of the objective function at the stationary point.

Two dimensions The sufficient conditions will again be given by inequalities (2.2-11) and (2.2-12) as

$$D_1 > 0 \qquad \text{and} \qquad D_2 > 0$$

so that
$$y^*_{x_1x_1} > 0 \qquad \text{and} \qquad y^*_{x_1x_1}y^*_{x_2x_2} - (y^*_{x_1x_2})^2 > 0$$

for a minimum value, or

$$D_1 < 0 \qquad \text{and} \qquad D_2 > 0$$

requiring
$$y^*_{x_1x_1} < 0 \qquad \text{and} \qquad y^*_{x_1x_1}y^*_{x_2x_2} - (y^*_{x_1x_2})^2 > 0$$

for a maximum value of the function to exist at the stationary point.

Example VII-5: Nature of the stationary point

Does an optimum exist at the stationary point in the function

$$y = 8x_1 + x_2 + 5x_1^2 - 9x_1x_2 + 2x_2^2$$

The various derivatives are

$$y_{x_1} = 8 + 10x_1 - 9x_2 \qquad y_{x_2} = 1 - 9x_1 + 4x_2$$

$$y_{x_1x_1} = 10 \qquad y_{x_2x_2} = 4 \qquad y_{x_1x_2} = -9$$

The stationary point will lie at $y^*_{x_1} = 0$ and $y^*_{x_2} = 0$, which gives

$$8 + 10x_1^* - 9x_2^* = 0$$

$$1 - 9x_1^* + 4x_2^* = 0$$

or $x_1^* = 1$, $x_2^* = 2$. Since $n = 2$, we must now evaluate the two determinants D_1 and D_2 defined by inequalities (2.2-11) and (2.2-12), getting

$$D_1 = |y^*_{x_1x_1}| = 10$$

$$D_2 = \begin{vmatrix} y^*_{x_1x_1} & y^*_{x_1x_2} \\ y^*_{x_2x_1} & y^*_{x_2x_2} \end{vmatrix} = \begin{vmatrix} 10 & -9 \\ -9 & 4 \end{vmatrix} = (40) - (81) = -41$$

Thus $D_1 > 0$ and $D_2 < 0$, which, from Example VII-4, shows that no optimum exists, but that the stationary value is a saddle point.

2.3 Discontinuities

As in Chap. V, the presence of discontinuities in the objective function can be handled by the insertion of artificial boundary planes, a procedure described in Sec. 2.1. The region within the boundary planes can then be examined using the present analytical approach. The general problem can, however, be left to Sec. 4, where the effect of boundaries is discussed.

2.4 A Simplification of Nomenclature

Up to this stage we have been careful to use asterisks to denote quantities which must be evaluated at the stationary point. This was done to emphasize that such quantities possess numerical values, rather than being functions. For simplicity we shall cease to use this convention, but it must be remembered that this distinction must still be made.

3 PROBLEMS WITH RESTRICTED VARIABLES–EQUALITY CONSTRAINTS

In the developments of the preceding section, the independent variables were altered without restriction. Most realistic problems, however, require solutions subject to a certain number of restrictions, usually arising from physical considerations. Let us direct our attention to the following problem.

Determine the minimum of the function

$$y = y(\mathbf{x}) \tag{3-1}$$

subject to the m subsidiary restrictions on the n variables $x_1 \ldots x_n$ $(n > m)$ when the constraints are of the form

$$g_k(\mathbf{x}) = 0 \qquad \text{for } k = 1, 2, \ldots, m \tag{3-2}$$

In the unrestricted problem, values of the objective function y could be calculated for any $\mathbf{x}$. For every imposed restriction, however, at least one independent variable loses its arbitrariness, and if m is greater than n, the problem is, in general, overdefined, with no solution.

Thus, clearly, the number of independent variables is reduced from n to $n - m$, for if $n - m$ values of the n variables $(x_1, x_2, \ldots, x_n)$ are selected, the remaining m values are determined by the restrictions. In principle, therefore, any m variables, say, $x_1, x_2, \ldots, x_m$, can be eliminated from the objective function and leave an objective function which depends on $n - m$ independent variables, some of which may be restricted in their range of choice. If the eliminations can be accomplished by direct substitution (Sec. 3.1), the unrestricted methods described in Sec. 2, or the methods applicable to inequality restrictions (Sec. 4), can be applied. However, while this elimination is possible in many cases, it is not always convenient, nor even practical. Therefore other methods have to be developed so that the restrictions can be satisfied without actually eliminating variables. Three of these methods, constrained variation, Lagrangian multipliers, and penalty functions, are discussed in some detail in Secs. 3.2, 3.3, and 3.4, respectively.

These methods may be used for any type of problem falling into this class, but as is usual in optimization work, one method may prove to be far superior in use to the others. The choice of the best technique cannot be made without detailed knowledge of the particular task to be accomplished. More powerful methods can be applied if the objective function and the constraints have a special structure. The most notable of these special techniques is that of linear programming, which applies to linear objective functions subjected to linear constraints. This important technique is also applicable to problems restrained by inequalities, and hence discussion of linear programming will be deferred until this more general class of problem is treated (Sec. 4.4).

3.1 Solution by Direct Substitution

One method of solution of problems involving objective functions constrained by equalities is to solve the m restrictions simultaneously for m of the n variables. These may then be substituted into the objective function. This will produce a new objective function,

$$y = y(x_1, x_2, \ldots, x_{n-m})$$

in terms of a reduced number of variables which are no longer subject to the equality restrictions. It may, however, be necessary, as shown in Example VII-8, to restrict the range of the variables by an inequality. The modified problem can therefore be treated by one of the methods of Sec. 2 or Sec. 4.

This approach is often not convenient, however, because the m constraint equations cannot be solved easily for the m variables, or because the resulting solutions become unwieldy. The method of substitution might, however, be used in simpler problems.

Example VII-6: Direct substitution

Find the minimum of the function

$$y = 4x_1^2 + 5x_2^2$$

if it is subject to the restriction that

$$2x_1 + 3x_2 = 6$$

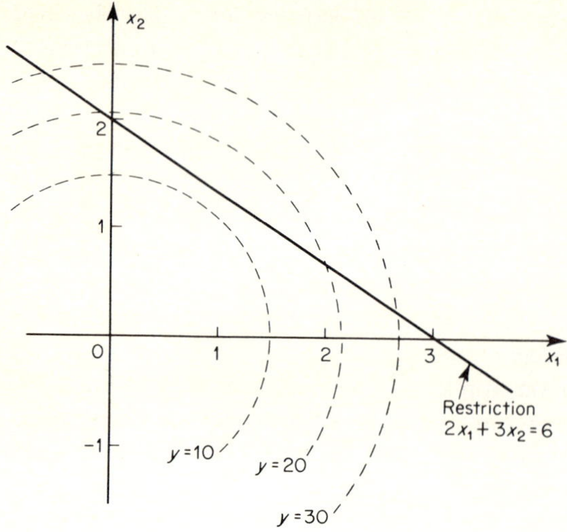

Fig. VII-4 The function $y = 4x_1^2 + 5x_2^2$ of Example VII-6.

From the second equation x_1 can be found and eliminated from the objective function. Thus

$$y = 4\left(\frac{6 - 3x_2}{2}\right)^2 + 5x_2^2 = 36 - 36x_2 + 14x_2^2$$

which can now be treated as an unrestricted function of x_2 only, because the imposed restriction is linear, and both the necessary and sufficient conditions for an optimum determined.

Hence $y_{x_2} = -36 + 28x_2$, which is zero if $x_2 = 1.286$. Also, $y_{x_2x_2} = +28$, which, being positive, indicates that the stationary value at $x_2 = 1.286$ is a minimum. The associated value of x_1 is then 1.071, being completely determined once the choice of an x_2 value has been made.

The minimum value of y is, then, 12.857, which can be compared with the value of zero obtained if an unrestricted choice of x_1 and x_2 can be made. The general nature of the function is indicated in Fig. VII-4.

Example VII-7: Cross-current extraction

Consider the n-stage cross-current extraction process shown in Fig. VII-5. Here a volume flow rate V of solution is contacted in each stage with an immiscible solvent with a view to extracting the solute present in the feed stream entering each stage. Fresh solvent is used in each stage, an amount w_i being added per unit time in the ith stage. The amount of solute in the original solution being fed to the system is denoted by x_0, and that leaving the ith stage is x_i. The amount of solute extracted in the system will then be $V(x_0 - x_n)$. If α is the value of a unit quantity of the extracted solute, and β is the cost of unit quantity of the wash solvent w, determine an expression for the profit from the system if this is given by the value of the extracted solute less the cost of supplying solvent for the extraction.

If each extraction stage is an equilibrium stage, and if z_i is the concentration of the desired solute in the washsolvent exit stream from each stage, we may then relate the values of x_i and z_i at the exit from each stage (these are in equilibrium) by the equilibrium relationship $z_i = f(x_i)$. Using the mass balances for each stage and this equilibrium relation, we can eliminate the w_i from the profit objective function.

We now have an objective function which is a function of n variables. Determine the optimum value of this function for the special case of a fixed feed composition x_0, an n-stage unit, where $n = 4$ and where the equilibrium relation can be expressed by $z_i = mx_i$ for the linear-equilibrium case (Aris, Rudd, and Amundson [1960]).

Since the amount of solute extracted is $\sum_{i=1}^{n} w_i z_i = V(x_0 - x_n)$ and the amount of solvent used is $\sum_{i=1}^{n} w_i$, then the profit p is

$$p = \alpha V(x_0 - x_n) - \beta \sum_{i=1}^{n} w_i \tag{3.1-1}$$

This is, actually, an example of an objective function of $2n$ variables x_i and w_i, of which n variables are restricted, due to the mass balance on each stage, given by

$$\begin{aligned} V(x_{i-1} - x_i) &= w_i z_i \\ &= w_i f(x_i) \end{aligned}$$

after assuming equilibrium between the exit streams from each stage.

The w_i and z_i can then be eliminated from the profit function, giving

$$p = \alpha V(x_0 - x_n) - \beta \sum_{i=1}^{n} \frac{V(x_{i-1} - x_i)}{f(x_i)}$$

Fig. VII-5 Cascade extraction units.

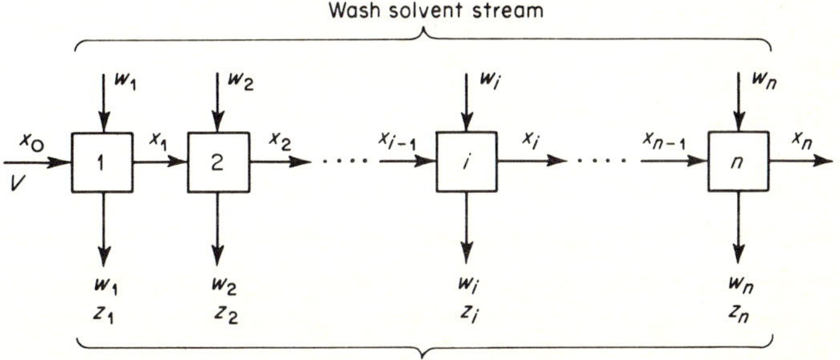

which is a function of only the $n+1$ independent variables x_0, x_1, $x_2, \ldots, x_n$. It is convenient for later use to note that

$$x_0 - x_n = \sum_{i=1}^{n} (x_{i-1} - x_i)$$

and writing $y = p/\alpha V$, to rearrange the objective function into the form

$$y = \frac{p}{\alpha V} = \sum_{i=1}^{n} (x_{i-1} - x_i)\left[1 - \frac{\beta}{\alpha f(x_i)}\right]$$

Let us now consider the special case of four stages and linear equilibrium $z_i = f(x_i) = mx_i$, so that

$$y = (x_0 - x_4) - \sum_{i=1}^{4} \frac{\beta}{\alpha m} \frac{x_{i-1} - x_i}{x_i}$$

If $\beta/\alpha m = A$, we then have that

$$y = x_0 - x_4 + 4A - A\left(\frac{x_0}{x_1} + \frac{x_1}{x_2} + \frac{x_2}{x_3} + \frac{x_3}{x_4}\right) = y(x_1,x_2,x_3,x_4)$$

for a fixed feed x_0. Let us determine the conditions for a maximum value of y to exist.

First, define the various derivatives:

$$y_{x_1} = \frac{Ax_0}{x_1^2} - \frac{A}{x_2} \qquad y_{x_2} = \frac{Ax_1}{x_2^2} - \frac{A}{x_3}$$

$$y_{x_3} = \frac{Ax_2}{x_3^2} - \frac{A}{x_4} \qquad y_{x_4} = \frac{Ax_3}{x_4^2} - 1$$

and

$$y_{x_1x_1} = -\frac{2Ax_0}{x_1^3} \qquad y_{x_1x_2} = \frac{A}{x_2^2} \qquad y_{x_1x_3} = 0 \qquad y_{x_1x_4} = 0$$

$$y_{x_2x_1} = \frac{A}{x_2^2} \qquad y_{x_2x_2} = \frac{-2Ax_1}{x_2^3} \qquad y_{x_2x_3} = \frac{A}{x_3^2} \qquad y_{x_2x_4} = 0$$

$$y_{x_3x_1} = 0 \qquad y_{x_3x_2} = \frac{A}{x_3^2} \qquad y_{x_3x_3} = \frac{-2Ax_2}{x_3^3} \qquad y_{x_3x_4} = \frac{A}{x_4^2}$$

$$y_{x_4x_1} = 0 \qquad y_{x_4x_2} = 0 \qquad y_{x_4x_3} = \frac{A}{x_4^2} \qquad y_{x_4x_4} = \frac{-2Ax_3}{x_4^3}$$

The stationary points, if any, may now be located by setting the first derivatives to zero and solving for the roots. Thus

$$x_0x_2 = x_1^2 \qquad x_1x_3 = x_2^2 \qquad x_2x_4 = x_3^2 \qquad Ax_3 = x_4^2 \tag{3.1-2}$$

which locates the stationary point at

$$x_1 = (Ax_0^4)^{1/5} \qquad x_2 = (A^2x_0^3)^{1/5} \qquad x_3 = (A^3x_0^2)^{1/5} \qquad x_4 = (A^4x_0)^{1/5} \tag{3.1-3}$$

The nature of the stationary point can be found by examining the sufficiency conditions given by inequalities (2.2-12). Let us evaluate

D_1, D_2, D_3, and D_4. Thus

$$D_1 = |y_{x_1x_1}| = -\frac{2Ax_0}{x_1^3}$$

$$D_2 = \begin{vmatrix} y_{x_1x_1} & y_{x_1x_2} \\ y_{x_2x_1} & y_{x_2x_2} \end{vmatrix} = \begin{vmatrix} -\dfrac{2Ax_0}{x_1^3} & \dfrac{A}{x_2^2} \\ \dfrac{A}{x_2^2} & -\dfrac{2Ax_1}{x_2^3} \end{vmatrix}$$

$$= \frac{4A^2x_0x_1}{x_1^3x_2^3} - \frac{A^2}{x_2^4}$$

$$D_3 = \begin{vmatrix} y_{x_1x_1} & y_{x_1x_2} & y_{x_1x_3} \\ y_{x_2x_1} & y_{x_2x_2} & y_{x_2x_3} \\ y_{x_3x_1} & y_{x_3x_2} & y_{x_3x_3} \end{vmatrix} = \begin{vmatrix} -\dfrac{2Ax_0}{x_1^3} & \dfrac{A}{x_2^2} & 0 \\ \dfrac{A}{x_2^2} & -\dfrac{2Ax_1}{x_2^3} & \dfrac{A}{x_3^2} \\ 0 & \dfrac{A}{x_3^2} & -\dfrac{2Ax_2}{x_3^3} \end{vmatrix}$$

$$= \frac{-2Ax_0}{x_1^3}\left(\frac{4A^2x_1x_2}{x_2^3x_3^3} - \frac{A^2}{x_3^4}\right) + \frac{A}{x_2^2}\frac{2A^2x_2}{x_2^2x_3^3}$$

and

$$D_4 = \begin{vmatrix} y_{x_1x_1} & y_{x_1x_2} & y_{x_1x_3} & y_{x_1x_4} \\ y_{x_2x_1} & y_{x_2x_2} & y_{x_2x_3} & y_{x_2x_4} \\ y_{x_3x_1} & y_{x_3x_2} & y_{x_3x_3} & y_{x_3x_4} \\ y_{x_4x_1} & y_{x_4x_2} & y_{x_4x_3} & y_{x_4x_4} \end{vmatrix}$$

$$= \begin{vmatrix} \dfrac{-2Ax_0}{x_1^3} & \dfrac{A}{x_2^2} & 0 & 0 \\ \dfrac{A}{x_2^2} & \dfrac{-2Ax_1}{x_2^3} & \dfrac{A}{x_3^2} & 0 \\ 0 & \dfrac{A}{x_3^2} & \dfrac{-2Ax_2}{x_3^3} & \dfrac{A}{x_4^2} \\ 0 & 0 & \dfrac{A}{x_4^2} & \dfrac{-2Ax_3}{x_4^3} \end{vmatrix}$$

$$= \frac{-2Ax_0}{x_1^3}\left[\frac{-2Ax_1}{x_2^3}\left(\frac{4A^2x_2}{x_3^2x_4^3} - \frac{A^2}{x_4^4}\right) - \frac{A}{x_3^2}\left(\frac{-2A^2}{x_3x_4^3}\right)\right]$$

$$- \frac{A}{x_2^2}\left[\frac{A}{x_2^2}\left(\frac{4A^2x_2}{x_3^2x_4^3} - \frac{A^2}{x_4^4}\right)\right]$$

which, on using Eqs. (3.1-2), reduce to

$$D_1 = \frac{-2Ax_0}{x_1^3} \qquad D_2 = \frac{3A^2}{x_2^4} \qquad D_3 = \frac{-4A^3}{x_2^3x_3^3} \qquad D_4 = \frac{5A^4}{x_2^3x_3^2x_4^3}$$

Thus, at the stationary point, $D_1 < 0$, $D_2 > 0$, $D_3 < 0$, $D_4 > 0$, so that the odd-numbered determinants are negative and the even-numbered are positive, representing sufficient conditions for a maximum, as given by inequalities (2.2-12).

The profit function is therefore a maximum at the point located by Eqs. (3.1-3), with a value of

$$y(\mathbf{x}^*) = x_0 + 4A - 5A\left(\frac{x_0}{A}\right)^{1/5}$$

It is interesting to note that in this special case of linear equilibrium the optimal policy is to allocate equal quantities of each solvent to each stage, of amount

$$w_1 = \cdots = w_i = \cdots = w_n = \frac{V}{m}\left[\left(\frac{x_0}{A}\right)^{1/n+1} - 1\right]$$

Example VII-8: Anomalies in direct substitution

Find the global optimum of the function

$$y = x_2 - x_1^2$$

subject to the restriction that

$$g = 1 - x_1^2 - x_2^2 = 0$$

From the restriction, $x_2^2 = 1 - x_1^2$, so that, using $x_2 = \pm\sqrt{1 - x_1^2}$, x_2 can be eliminated from the objective function.

Using the positive sign, we find that

$$y = \sqrt{1 - x_1^2} - x_1^2$$

which has stationary values at $x_1 = 0$. Examining the other sign, we get

$$y = -\sqrt{1 - x_1^2} - x_1^2$$

which has stationary points at $x_1 = 0$ and $x_1 = \pm 0.866$. These stationary points are valid points, and no note need have been taken of the inherent restriction on x_1, namely, that $x_1^2 \leq 1$; otherwise x_2^2 would be negative.

A study of the second derivatives of y, using the criteria of Chap. V, Sec. 2.3, will reveal the results summarized in Table VII-1. The global maximum thus lies at point 1, while a local maximum lies at point 2, and two equal global minima are found at points 3 and 4.

The danger of eliminating variables without due regard to the resultant imposition of additional restrictions can be illustrated by considering the elimination of the other variable, x_1, in this problem. Here the restriction can be rearranged as $x_1^2 = 1 - x_2^2$, so that the modified objective function becomes

$$y = x_2 - 1 + x_2^2$$

It is important to note that x_2 cannot be freely varied, since values of x_2 larger than 1 yield, from the restriction, imaginary values of x_1. Therefore y is to be maximized, subject to the restriction $-1 \leq x_2 \leq 1$. The

TABLE VII-1 STATIONARY POINTS

Stationary points	1	2	3	4
x_1	0	0	0.866	−0.866
x_2	1	−1	−0.5	−0.5
y	1	−1	−1.25	−1.25
Type	Max	Max	Min	Min

problem has been put into the form discussed in Chap. V, Sec. 3.1, since now boundaries are imposed on the problem. The stationary points of the new y function are then defined by

$$y_{x_2} = 0 = 1 + 2x_2$$

which gives the valid locations of extreme values at $x_2 = -0.5$ and $x_1 = \pm 0.866$. A simple check of the second derivatives will show both of these stationary points to be a minimum. The maximum must then lie on the boundaries. Setting $x_2 = -1$ $(x_1 = 0)$ gives $y = -1$ a local maximum, as shown in the table. The global maximum, however, is found at the boundary $x_2 = 1$ $(x_1 = 0)$, with $y = 1$.

The important point to be noted is that, in eliminating certain variables, using the method of direct substitution, the remaining variables are often limited in the range of permissible values which can be assumed. Thus the problem, while reduced in dimensionality, may still require restrictions to be imposed on the remaining variables. The optimum may therefore lie on a boundary, rather than at a stationary point. This possibility must always be considered.

3.2 Method of Constrained Variation

The method discussed in the preceding section, that of direct substitution, is characterized by a series of steps which fall into two distinct parts, substitution and then differentiation. Thus, in using the method of direct substitution, the search for the stationary points is not started until the objective function has been expressed in terms of *n*-*m* variables, *m* variables having been eliminated. If the constraints are nonlinear, the method of direct substitution can become so complex in the mathematical manipulations as to be hopeless. The method of constrained variation (see, for example, Edelbaum [1962]) in a sense reverses the order of attack and requires differentiation as the first step. Then the resultant equations are linear in the displacements, and *m* of the displacements can be eliminated, using linear algebraic methods. This simplification in the elimination step, i.e., eliminating only the *m* displacements, leaves, as the price, *n* simultaneous equations to be solved, rather than the *n*-*m* encountered using direct substitution.

To understand the general development which follows, it is useful to indicate the salient features of the method, starting with an objective function dependent on only two variables. Suppose the problem is to maximize $y(x_1,x_2)$ subject to the constraint that

$$g(x_1,x_2) = 0 \tag{3.2-1}$$

If the point $\{x_1^*,x_2^*\}$ is to be a maximum, it is necessary for

$$dy = 0 = \frac{\partial y}{\partial x_1} dx_1 + \frac{\partial y}{\partial x_2} dx_2 \tag{3.2-2}$$

to hold for all admissible displacements dx_1, dx_2, and in addition, $g(x_1^*,x_2^*)$ must of course vanish. What is meant by an admissible variation? At the stationary point, $g(x_1^*,x_2^*) = 0$. Any admissible change in dx_1 or dx_2 must still satisfy this equation; i.e., displacements of interest yield

$$g(x_1^* + dx_1, x_2^* + dx_2) = 0 \tag{3.2-3}$$

On expanding $g(x_1,x_2)$ in a Taylor series, and remembering that the variations dx_1 and dx_2 are small, we get

$$dg = \frac{\partial g}{\partial x_1} dx_1 + \frac{\partial g}{\partial x_2} dx_2 = 0 \tag{3.2-4}$$

Equation (3.2-4) must be satisfied by all admissible displacements. This situation is perhaps clarified by considering Fig. VII-6, where the restriction, Eq. (3.2-1), is plotted. For each value of x_1 a value (or values) of x_2 can be determined. Thus values of x_1 and x_2 satisfying Eq. (3.2-1) will lie on the curve shown in Fig. VII-6. Point *A* is not, therefore, an acceptable point, since all admissible values of x_1 and x_2 must lie on the line. Point *B* is

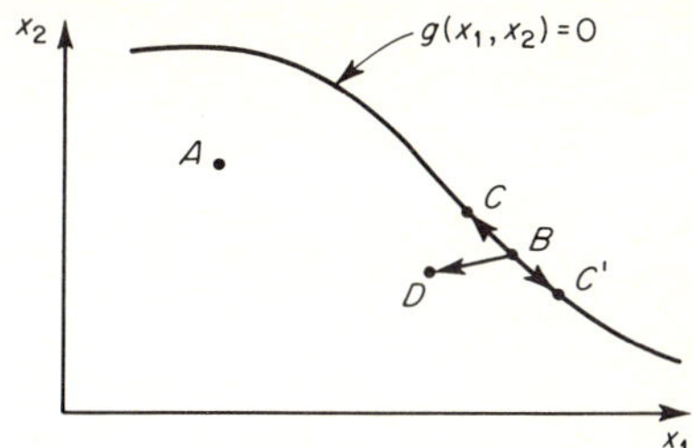

Fig. VII-6 Admissible variations when subject to an equality restriction. Points C and C' are acceptable, whereas A and D are not.

admissible. About point B are shown three displacements, but the displacement to D is not acceptable, since it does not follow the restriction curve. The displacements to C and C' are evidently admissible variations.

Thus a set of displacements not obeying Eq. (3.2-4) gives rise to points which do not satisfy the constraint, Eq. (3.2-1). The constraint thus yields a linear equation in the displacement, Eq. (3.2-4).

If $g_{x_2} \neq 0$, then

$$dx_2 = -\frac{g_{x_1}}{g_{x_2}}\,dx_1$$

This relationship determines the admissible displacement dx_2, given a displacement dx_1. Substituting into Eq. (3.2-2) gives

$$dy = 0 = \left(\frac{\partial y}{\partial x_1} - \frac{\partial g/\partial x_1}{\partial g/\partial x_2}\frac{\partial y}{\partial x_2}\right) dx_1 \tag{3.2-5}$$

Since dy must vanish for all dx_1, and since dx_1 may be arbitrarily chosen (dx_2 then being completely determined), Eq. (3.2-5) yields the relation

$$\frac{\partial y}{\partial x_1}\frac{\partial g}{\partial x_2} - \frac{\partial y}{\partial x_2}\frac{\partial g}{\partial x_1} = 0$$

as a necessary condition for the existence of local maximum or minimum. As in the case of an unrestricted function, sufficiency conditions can also be found (Sec. 3.2-3), but the second-order variations must be included.

3.2.1 Necessary conditions in constrained variation

It is possible to generalize these arguments. The essential point is that each constraint gives rise to a linear equation in the displacements. Thus a number of displacements corresponding to the number of restrictions can be eliminated. In the special case considered above, there was only one restriction, so that one displacement, dx_2, was eliminated. The other displacements are then arbitrary (although they must remain small). These remaining free displacements appear in the differentiated objective function [Eq. (3.2-5)], and their coefficients must vanish by the usual arguments.

The steps shown above may be applied to the general problem. Let $y(\mathbf{x})$ be differentiated so that

$$dy = \sum_{i=1}^{n} y_{x_i}\, dx_i \tag{3.2-6}$$

This differential dy must vanish for all admissible infinitesimal displacements dx_i (for $i = 1, 2, \ldots, n$) about a stationary point. If the constraints

$$g_k = 0 \qquad \text{for } k = 1, 2, \ldots, m \tag{3.2-7}$$

are imposed on the solution, then

$$dg_k = \sum_{i=1}^{n} g_{k_{x_i}}\, dx_i = 0 \tag{3.2-8}$$

where $g_{k_{x_i}}$ are the partial derivatives evaluated at the stationary point. Equation (3.2-8) represents the restrictions to be imposed on a displacement dx_i ($i = 1, 2, \ldots, n$). Only those dx_i which satisfy these equations are of interest here, since any other set will not yield a point simultaneously obeying the imposed constraints. Thus m of the displacements, the first m for definiteness, can be eliminated using Eqs. (3.2-8).

These equations can be rearranged so that

$$\sum_{i=1}^{m} g_{k_{x_i}}\, dx_i = -\sum_{j=m+1}^{n} g_{k_{x_j}}\, dx_j \qquad \text{for } k = 1, 2, \ldots, m \tag{3.2-9}$$

If these equations are linearly independent,[1] then the dx_i ($i = 1, 2, \ldots, m$) can be expressed in terms of the remaining n–m displacements dx_j for $j = m + 1, m + 2, \ldots, n$. If this is done (Sec. 3.2.2), the elimination of the first m displacements from Eq. (3.2-6) leaves

$$dy = \frac{1}{J\left(\dfrac{g_1, \ldots, g_m}{x_1, \ldots, x_m}\right)} \sum_{k=m+1}^{n} J_k\left(\frac{y, g_1, g_2, \ldots, g_m}{x_k, x_1, x_2, \ldots, x_m}\right) dx_k \tag{3.2-10}$$

[1] This linear independence is equivalent to

$$J\left(\frac{g_1, g_2, \ldots, g_m}{x_1, x_2, \ldots, x_m}\right) \neq 0$$

where the notation

$$J_k\left(\frac{y, g_1, g_2, \ldots, g_m}{x_k, x_1, x_2, \ldots, x_m}\right) = \begin{vmatrix} \dfrac{\partial y}{\partial x_k} & \dfrac{\partial y}{\partial x_1} & \dfrac{\partial y}{\partial x_2} \cdots & \dfrac{\partial y}{\partial x_m} \\ \dfrac{\partial g_1}{\partial x_k} & \dfrac{\partial g_1}{\partial x_1} & \dfrac{\partial g_1}{\partial x_2} \cdots & \dfrac{\partial g_1}{\partial x_m} \\ \cdots & \cdots & \cdots & \cdots \\ \dfrac{\partial g_m}{\partial x_k} & \dfrac{\partial g_m}{\partial x_1} & \dfrac{\partial g_m}{\partial x_2} \cdots & \dfrac{\partial g_m}{\partial x_m} \end{vmatrix} \tag{3.2-11}$$

has been introduced. The J_k defined in this way is called a *Jacobian* and represents a very useful shorthand notation found in a variety of applications. It is particularly valuable in the study of constrained variations.

Since dy must vanish for all permissible variations, and since the dx_k appearing in Eq. (3.2-10) are independent, a necessary condition for the existence of a constrained optimum is that

$$J_k\left(\frac{y, g_1, g_2, \ldots, g_m}{x_k, x_1, x_2, \ldots, x_m}\right) = 0 \qquad \text{for } k = m+1, m+2, \ldots, n \tag{3.2-12}$$

There are $n - m$ equations of the type shown by Eq. (3.2-12), and m restrictions, $g_k = 0$. These n equations must be solved simultaneously for the n values of x_i^* $(i = 1, 2, \ldots, n)$. Note that only the displacements dx_1, $dx_2, \ldots, dx_m$ have been eliminated, but the quantities $x_1, x_2, \ldots, x_n$ all appear in Eqs. (3.2-12). This fact distinguishes this method from direct substitution.

Before considering the details in the development of Eqs. (3.2-12), let us consider an example so as to understand more completely its application.

Example VII-9: Constrained variation

Carry out Example VII-6 using constrained variations. Here

$$y = 4x_1^2 + 5x_2^2 \quad \text{and} \quad g_1 = 2x_1 + 3x_2 - 6 = 0 \quad \text{with } n = 2, m = 1$$

Hence Eq. (3.2-12) yields

$$J\left(\frac{y, g_1}{x_2, x_1}\right) = \begin{vmatrix} y_{x_2} & y_{x_1} \\ g_{1_{x_2}} & g_{1_{x_1}} \end{vmatrix} = \begin{vmatrix} 10x_2 & 8x_1 \\ 3 & 2 \end{vmatrix} = 20x_2 - 24x_1 = 0$$

Therefore $$x_1 = \tfrac{5}{6}x_2$$

Substituting in $g_1 = 0$, we get

$$2(\tfrac{5}{6}x_2) + 3x_2 - 6 = 0$$

The solution is, then, that the optimum will lie at $x_1 = 1.071$ and $x_2 = 1.286$ as before.

3.2.2 *Development of the necessary conditions*

Equations (3.2-12) summarize in a convenient and condensed notation the necessary conditions for a constrained function to have an optimum, assuming continuity of the function, the restrictions, and their first derivatives, together with the existence of second derivatives, and assuming linear independence of the gradients of the restrictions. Let us pursue the derivation of this essential result in somewhat more detail.

From the m restrictions, we can write, as in Eq. (3.2-8), that

$$dg_k = \sum_{i=1}^{n} g_{k_{x_i}}\, dx_i = 0$$

which, on being expanded and rearranged as in Eq. (3.2-9), gives

$$\begin{aligned}
g_{1_{x_1}} dx_1 + g_{1_{x_2}} dx_2 + \cdots + g_{1_{x_m}} dx_m &= -g_{1_{x_{m+1}}} dx_{m+1} - \cdots - g_{1_{x_n}} dx_n = d_1 \\
g_{2_{x_1}} dx_1 + g_{2_{x_2}} dx_2 + \cdots + g_{2_{x_m}} dx_m &= -g_{2_{x_{m+1}}} dx_{m+1} - \cdots - g_{2_{x_n}} dx_n = d_2 \\
&\cdots\cdots\cdots\cdots \\
g_{m_{x_1}} dx_1 + g_{m_{x_2}} dx_2 + \cdots + g_{m_{x_m}} dx_m &= -g_{m_{x_{m+1}}} dx_{m+1} - \cdots - g_{m_{x_n}} dx_n = d_m
\end{aligned}$$

Here the m displacements which are now dependent are arranged on the left-hand side, while the arbitrary displacements are placed on the right. Using Cramer's rule (Appendix 1), each of the m dependent displacements can be found from these linear equations in terms of the arbitrary displacements, as expressed by the d_k. Thus we get

$$dx_1 = \frac{\begin{vmatrix} d_1 & g_{1_{x_2}} & g_{1_{x_3}} & \cdots & g_{1_{x_m}} \\ d_2 & g_{2_{x_2}} & g_{2_{x_3}} & \cdots & g_{2_{x_m}} \\ \cdots & \cdots & \cdots & \cdots & \cdots \\ d_m & g_{m_{x_2}} & g_{m_{x_3}} & \cdots & g_{m_{x_m}} \end{vmatrix}}{\begin{vmatrix} g_{1_{x_1}} & g_{1_{x_2}} & \cdots & g_{1_{x_m}} \\ g_{2_{x_1}} & g_{2_{x_2}} & \cdots & g_{2_{x_m}} \\ \cdots & \cdots & \cdots & \cdots \\ g_{m_{x_1}} & g_{m_{x_2}} & \cdots & g_{m_{x_m}} \end{vmatrix}}$$

Since each d_k is the sum of $n - m$ terms, the determinant in the numerator may be split into $n - m$ determinants, thus:

$$-\begin{vmatrix} g_{1_{x_{m+1}}}\,dx_{m+1} & g_{1_{x_2}} & \cdots & g_{1_{x_m}} \\ g_{2_{x_{m+1}}}\,dx_{m+1} & g_{2_{x_2}} & \cdots & g_{2_{x_m}} \\ \cdots & \cdots & \cdots & \cdots \\ g_{m_{x_{m+1}}}\,dx_{m+1} & g_{m_{x_2}} & \cdots & g_{m_{x_m}} \end{vmatrix} - \begin{vmatrix} g_{1_{x_{m+2}}}\,dx_{m+2} & g_{1_{x_2}} & \cdots & g_{1_{x_m}} \\ g_{2_{x_{m+2}}}\,dx_{m+2} & g_{2_{x_2}} & \cdots & g_{2_{x_m}} \\ \cdots & \cdots & \cdots & \cdots \\ g_{m_{x_{m+2}}}\,dx_{m+2} & g_{m_{x_2}} & \cdots & g_{m_{x_m}} \end{vmatrix} - \cdots$$

and since dx_{m+1}, dx_{m+2}, etc,. are common in their respective determinants, they can be factored out, getting

$$dx_1 = -\frac{\left[J\left(\dfrac{g_1, g_2, \ldots, g_m}{x_{m+1}, x_2, x_3, \ldots, x_m}\right)dx_{m+1} + J\left(\dfrac{g_1, g_2, \ldots, g_m}{x_{m+2}, x_2, x_3, \ldots, x_m}\right)dx_{m+2} + \cdots + J\left(\dfrac{g_1, g_2, \ldots, g_m}{x_n, x_2, x_3, \ldots, x_m}\right)dx_n\right]}{J\left(\dfrac{g_1, g_2, \ldots, g_m}{x_1, x_2, \ldots, x_m}\right)}$$

using the convenient Jacobian shorthand notation for these determinants [Eq. (3.2-11)]. Thus, for the kth displacement, dx_k, we shall get

$$dx_k = -\frac{\left[J\left(\dfrac{g_1, g_2, \ldots, g_m}{x_1, \ldots, x_{k-1}, x_{m+1}, x_{k+1}, \ldots, x_m}\right)dx_{m+1} + \cdots + J\left(\dfrac{g_1, g_2, \ldots, g_m}{x_1, x_2, \ldots, x_{k-1}, x_n, x_{k+1}, \ldots, x_m}\right)dx_n\right]}{J\left(\dfrac{g_1, g_2, \ldots, g_m}{x_1, x_2, \ldots, x_m}\right)} \tag{3.2-13}$$

for $k = 1, 2, \ldots, m$

and[1] $$J\left(\frac{g_1, g_2, \ldots, g_m}{x_1, x_2, \ldots, x_m}\right) \neq 0$$

The m restricted displacements dx_k can now be eliminated from Eq. (3.2-6), using Eq. (3.2-13) to obtain dy as a function of the remaining $n - m$ arbitrary

[1] If $J\left(\dfrac{g_1, \ldots, g_m}{x_1, \ldots, x_m}\right) = 0$, we can often select another set of m variables to avoid this situation. An illustration of this is found in Example VII-11.

displacements dx_i. Thus

$$dy = \frac{1}{J\left(\dfrac{g_1, g_2, \ldots, g_m}{x_1, x_2, \ldots, x_m}\right)} \left\{\left[y_{x_{m+1}} J\left(\frac{g_1, g_2, \ldots, g_m}{x_1, x_2, \ldots, x_m}\right)\right.\right.$$

$$- y_{x_1} J\left(\frac{g_1, g_2, \ldots, g_m}{x_{m+1}, x_2, \ldots, x_m}\right) - y_{x_2} J\left(\frac{g_1, g_2, \ldots, g_m}{x_1, x_{m+1}, \ldots x_m}\right) \cdots$$

$$- y_{x_m} J\left(\frac{g_1, g_2, \ldots, g_m}{x_1, x_2, \ldots, x_{m-1}, x_{m+1}}\right)\Bigg] dx_{m+1} + \Bigg[y_{x_{m+2}} J\left(\frac{g_1, g_2, \ldots, g_m}{x_1, x_2, \ldots, x_m}\right)$$

$$- y_{x_1} J\left(\frac{g_1, g_2, \ldots, g_m}{x_{m+2}, x_2, \ldots, x_m}\right) - y_{x_2} J\left(\frac{g_1, g_2, \ldots, g_m}{x_1, x_{m+2}, x_3, \ldots, x_m}\right)$$

$$\left.\left. - \cdots - y_{x_m} J\left(\frac{g_1, g_2, \ldots, g_m}{x_1, x_2, \ldots, x_{m-1}, x_{m+2}}\right)\right] dx_{m+2} + \cdots\right\}$$

Since the dx_j $(j = m + 1, \ldots, n)$ are now independent infinitesimal displacements, a necessary condition for the existence of an optimum at point $\mathbf{x}^*$ is that the coefficients vanish identically.

Thus, for example,

$$\left[y_{x_{m+1}} J\left(\frac{g_1, g_2, \ldots, g_m}{x_1, x_2, \ldots, x_m}\right) - y_{x_1} J\left(\frac{g_1, g_2, \ldots, g_m}{x_{m+1}, x_2, \ldots, x_m}\right)\right.$$

$$\left. - y_{x_2} J\left(\frac{g_1, g_2, \ldots, g_m}{x_1, x_{m+1}, x_3, \ldots, x_m}\right) - \cdots - y_{x_m} J\left(\frac{g_1, g_2, \ldots, g_m}{x_1, x_2, \ldots, x_{m-1}, x_{m+1}}\right)\right] = 0$$

where the derivatives are evaluated at the restricted stationary point.

Using property I of determinants (Appendix 1), this result can be written

$$\left[y_{x_{m+1}} J\left(\frac{g_1, g_2, \ldots, g_m}{x_1, x_2, \ldots, x_m}\right) - y_{x_1} J\left(\frac{g_1, g_2, \ldots, g_m}{x_{m+1}, x_2, \ldots, x_m}\right)\right.$$

$$+ y_{x_2} J\left(\frac{g_1, g_2, \ldots, g_m}{x_{m+1}, x_1, x_3, \ldots, x_m}\right) - \cdots + (-1)^m y_{x_m}$$

$$\left. \times J\left(\frac{g_1, g_2, \ldots, g_m}{x_{m+1}, x_1, x_2, \ldots, x_{m-1}}\right)\right] = 0 \quad (3.2\text{-}14)$$

Note that the x_{m+1} has been transposed in each term after the first, so that it appears first in the Jacobian. Now using property II of determinants, the above expression can be compactly written

$$J\left(\frac{y, g_1, g_2, \ldots, g_m}{x_{m+1}, x_1, x_2, \ldots, x_m}\right) = 0$$

Similar arguments lead to the $n - m$ equations

$$J_k\left(\frac{y, g_1, g_2, \ldots, g_m}{x_k, x_1, x_2, \ldots, x_m}\right) = 0 \qquad \text{for } k = m + 1, m + 2, \ldots, n \quad (3.2\text{-}12)$$

as summarized in Sec. 3.2.1. If these $n - m$ Jacobian functions are set equal to zero, they will be the necessary conditions for an optimum to exist under the m imposed restrictions.

The Necessary Conditions and Constrained Derivatives Equations (3.2-12) have a very interesting and pertinent interpretation, which can be illustrated by the following arguments.

If the first m variables are eliminated from the objective function, using the m restrictions, the objective function y will then depend only on the $n - m$ independent variables, $x_{m+1}, x_{m+2}, \ldots, x_n$. In this case a Taylor expansion in terms of these variables will then become

$$y = y^* + \sum_{j=m+1}^{n} \left(\frac{\partial y}{\partial x_j}\right)_{\underline{g}} h_j + \frac{1}{2!} \sum_{j=m+1}^{n} \sum_{k=m+1}^{n} \left(\frac{\partial^2 y}{\partial x_j \, \partial x_k}\right)_{\underline{g}} h_j h_k + \cdots \quad (3.2\text{-}15)$$

representing admissible variations about an optimum value y^*, since the restrictions have been taken into account.

To clarify this formulation we must expand on the meaning of the partial derivatives. With the m restrictions

$$g_k(\mathbf{x}) = 0 \qquad \text{for } k = 1, 2, \ldots, m$$

it is convenient and proper to visualize varying one of the variables, say, x_{m+3}, while holding $x_{m+1}, x_{m+2}, x_{m+4}, \ldots, x_n$ constant and at the same time permitting $x_1, x_2, \ldots, x_m$ to change so that the m restrictions are satisfied. Thus we may define the constrained derivative as

$$\left(\frac{\partial y}{\partial x_{m+3}}\right)_{\underline{g}} \equiv \left(\frac{\partial y}{\partial x_{m+3}}\right)_{x_{m+1}, x_{m+2}, x_{m+4}, \ldots, x_n, g_1, g_2, \ldots, g_m}$$

Suppose, for example, that the only restriction imposed on a particular problem has the form

$$g_1 = 1 - x_1^2 - x_2^2 - x_3^2 = 0$$

Then the quantity

$$\left(\frac{\partial y}{\partial x_2}\right)_{\underline{g}} \left[\equiv \left(\frac{\partial y}{\partial x_2}\right)_{x_3, g_1}\right]$$

gives the change in y resulting from a change in x_2, holding x_3 constant and at the same time forcing x_1 to change so that the restriction is satisfied. In the particular case cited above this would mean choosing the change dx_1 so that

$$x_1 \, dx_1 + x_2 \, dx_2 = 0$$

Thus $(\partial y/\partial x_j)_{\underline{g}}$ is a derivative determined with respect to one of the $n - m$ free variables, taken so that all the restrictions are satisfied. This quantity must also vanish at an optimum since the $n - m$ quantities h_i appearing in Eq. (3.2-15) are all independent. Therefore the conditions

$$\left(\frac{\partial y}{\partial x_j}\right)_{\underline{g}} = 0 \qquad \text{for } j = m+1, m+2, \ldots, n \tag{3.2-16}$$

are necessary conditions for the existence of an optimum. Indeed, it is not difficult to show that these conditions are precisely identical with those required by Eqs. (3.2-12). Using the chain rule of calculus,

$$\left(\frac{\partial y}{\partial x_j}\right)_{\underline{g}} = \sum_{s=1}^{n} \left(\frac{\partial y}{\partial x_s}\right)_{\underline{x}} \left(\frac{\partial x_s}{\partial x_j}\right)_{\underline{g}} = \sum_{s=1}^{m} \left(\frac{\partial y}{\partial x_s}\right)_{\underline{x}} \left(\frac{\partial x_s}{\partial x_j}\right)_{\underline{g}} + \left(\frac{\partial y}{\partial x_j}\right)_{\underline{x}}$$

$$\text{for } m+1 \le j \le n \tag{3.2-17}$$

Here $(\partial y/\partial x_s)_{\underline{x}}$ is the usual partial derivative in which all x_i are held fixed, except x_s. Thus, for example,

$$\left(\frac{\partial y}{\partial x_1}\right)_{\underline{x}} = \left(\frac{\partial y}{\partial x_1}\right)_{x_2, x_3, \ldots, x_n}$$

The quantity $(\partial x_s/\partial x_j)_{\underline{g}}$ can be computed by differentiating the restrictions so that

$$dg_k = \sum_{p=1}^{n} \left(\frac{\partial g_k}{\partial x_p}\right)_{\underline{x}} dx_p \qquad \text{for } k = 1, 2, \ldots, m$$

Differentiating both sides of this expression gives

$$\left(\frac{\partial g_k}{\partial x_j}\right)_{\underline{g}} = \sum_{p=1}^{n} \left(\frac{\partial g_k}{\partial x_p}\right)_{\underline{x}} \left(\frac{\partial x_p}{\partial x_j}\right)_{\underline{g}} = \sum_{p=1}^{m} \left(\frac{\partial g_k}{\partial x_p}\right)_{\underline{x}} \left(\frac{\partial x_p}{\partial x_j}\right)_{\underline{g}} + \left(\frac{\partial g_k}{\partial x_j}\right)_{\underline{x}} = 0$$

$$\text{for } m+1 \le j \le n,\ k = 1, 2, \ldots, m$$

Since the derivative is at constant g_k, the following m equations are valid.

$$\sum_{p=1}^{m} \left(\frac{\partial g_k}{\partial x_p}\right)_{\underline{x}} \left(\frac{\partial x_p}{\partial x_j}\right)_{\underline{g}} = -\left(\frac{\partial g_k}{\partial x_j}\right)_{\underline{x}} \qquad \text{for } k = 1, 2, \ldots, m \tag{3.2-18}$$

Solving, then, for $(\partial x_p/\partial x_j)_{\underline{g}}$ yields, as before,

$$\left(\frac{\partial x_p}{\partial x_j}\right)_{\underline{g}} = -\frac{J\left(\dfrac{g_1, g_2, \ldots, g_m}{x_1, x_2, \ldots, x_{p-1}, x_j, x_{p+1}, \ldots, x_m}\right)}{J\left(\dfrac{g_1, g_2, \ldots, g_m}{x_1, x_2, x_3, \ldots\ x_m}\right)} \tag{3.2-19}$$

Therefore, substituting this in Eq. (3.2-17) yields

$$\left(\frac{\partial y}{\partial x_j}\right)_{\underline{g}} = -\sum_{s=1}^{m}\left(\frac{\partial y}{\partial x_s}\right)_{\underline{x}} \frac{J\left(\dfrac{g_1, g_2, \ldots, g_m}{x_1, \ldots, x_{p-1}, x_j, x_{p+1}, \ldots, x_m}\right)}{J\left(\dfrac{g_1, g_2, \ldots, g_m}{x_1, x_2, \ldots, x_m}\right)} + \left(\frac{\partial y}{\partial x_j}\right)_{\underline{x}}$$

which can be written

$$\left(\frac{\partial y}{\partial x_j}\right)_{\underline{g}} = \frac{J\left(\dfrac{y, g_1, g_2, \ldots, g_m}{x_j, x_1, x_2, \ldots, x_m}\right)}{J\left(\dfrac{g_1, g_2, \ldots, g_m}{x_1, x_2, \ldots, x_m}\right)} = 0$$

This is, of course, precisely the same result as Eq. (3.2-12).

Thus we see that the method of constrained variations is based upon a derivative taken "along" the restriction so that the restriction is satisfied. This concept will prove invaluable in the following section.

3.2.3 *Sufficient conditions in constrained variation*

Clearly, a sufficient condition for a maximum, from Eq. (3.2-15), is that

$$\Delta^2 y = \sum_{i=m+1}^{n}\sum_{j=m+1}^{n}\left(\frac{\partial^2 y}{\partial x_i\,\partial x_j}\right)_{\underline{g}} h_i h_j < 0 \qquad \text{for all nontrivial } h_i$$

Let us develop the coefficient $\partial^2 y/\partial x_i\,\partial x_j$ in terms of more readily calculable derivatives (Schechter and Beveridge [1966]).† The nature of $\Delta^2 y$ can then be investigated using the same arguments as presented in Sec. 2.2.2. Thus the sufficiency conditions can be examined, for example, by testing the various determinants defined by Eqs. (2.2-11) and (2.2-12). Of course, weaker conditions, as discussed earlier, may also be found.

The problem is thus the expansion of the second derivatives in terms of calculable quantities. Differentiation of Eq. (3.2-17) with respect to x_i $(m + 1 \leq i \leq n)$ yields

$$\left(\frac{\partial^2 y}{\partial x_i\,\partial x_j}\right)_{\underline{g}} = \sum_{s=1}^{m}\frac{\partial}{\partial x_i}\left[\left(\frac{\partial y}{\partial x_s}\right)_{\underline{x}}\right]_{\underline{g}}\left(\frac{\partial x_s}{\partial x_j}\right)_{\underline{g}} + \sum_{s=1}^{m}\left(\frac{\partial y}{\partial x_s}\right)_{\underline{x}}\left(\frac{\partial^2 x_s}{\partial x_i\,\partial x_j}\right)_{\underline{g}} + \frac{\partial}{\partial x_i}\left[\left(\frac{\partial y}{\partial x_j}\right)_{\underline{x}}\right]_{\underline{g}}$$

† A similar approach has been used by Wilde and Beightler [1967]. A slightly different formulation has been suggested by Law and Fariss [1969], being computationally efficient as well as simple with no hinderance from redundant constraints.

Further expansion of the first and last terms on the right-hand side then gives

$$\left(\frac{\partial^2 y}{\partial x_i\,\partial x_j}\right)_{\underline{g}} = \sum_{s=1}^{m}\left(\frac{\partial y}{\partial x_s}\right)_{\underline{x}}\left(\frac{\partial^2 x_s}{\partial x_i\,\partial x_j}\right)_{\underline{g}} + \sum_{s=1}^{m}\sum_{r=1}^{m}\left(\frac{\partial^2 y}{\partial x_r\,\partial x_s}\right)_{\underline{x}}\left(\frac{\partial x_s}{\partial x_i}\right)_{\underline{g}}\left(\frac{\partial x_r}{\partial x_j}\right)_{\underline{g}}$$
$$+ \sum_{s=1}^{m}\left(\frac{\partial^2 y}{\partial x_s\,\partial x_j}\right)_{\underline{x}}\left(\frac{\partial x_s}{\partial x_i}\right)_{\underline{g}} + \sum_{s=1}^{m}\left(\frac{\partial^2 y}{\partial x_s\,\partial x_i}\right)_{\underline{x}}\left(\frac{\partial x_s}{\partial x_j}\right)_{\underline{g}} + \left(\frac{\partial^2 y}{\partial x_i\,\partial x_j}\right)_{\underline{x}} \tag{3.2-20}$$

Equation (3.2-20) is almost in the desired form. Our goal was the evaluation of the second derivative of the objective function "along" the restrictions, for, as noted above, it is these derivatives which determine the character of the restricted stationary point. The $(\partial x_p/\partial x_i)_{\underline{g}}$ are complex, but have already been related to the usual derivatives in Eq. (3.2-19). The quantity which has yet to be evaluated is $(\partial^2 x_s/\partial x_i\,\partial x_j)_{\underline{g}}$.

Let us begin by differentiating Eq. (3.2-18) to obtain

$$\sum_{p=1}^{m}\frac{\partial^2 x_p}{\partial x_j\,\partial x_i}\frac{\partial g_k}{\partial x_p} + \sum_{p=1}^{m}\frac{\partial}{\partial x_i}\left[\left(\frac{\partial g_k}{\partial x_p}\right)_{\underline{x}}\right]_{\underline{g}}\frac{\partial x_p}{\partial x_j} = -\frac{\partial}{\partial x_i}\left[\left(\frac{\partial g_k}{\partial x_j}\right)_{\underline{x}}\right]_{\underline{g}}$$
$$\text{for } k = 1, 2, 3, \ldots, m \tag{3.2-21}$$

Noting again that

$$\frac{\partial}{\partial x_i}[\quad]_{\underline{g}} = \sum_{s=1}^{m}\left(\frac{\partial[\quad]}{\partial x_s}\right)_{\underline{x}}\left(\frac{\partial x_s}{\partial x_i}\right)_{\underline{g}} + \left(\frac{\partial[\quad]}{\partial x_i}\right)_{\underline{x}}$$

where the [ ] could be any function of x_i ($i = 1, 2, \ldots, n$), the application of this result to Eq. (3.2-21) yields

$$\sum_{p=1}^{m}\left(\frac{\partial^2 x_p}{\partial x_j\,\partial x_i}\right)_{\underline{g}}\left(\frac{\partial g_k}{\partial x_p}\right)_{\underline{x}} = -\sum_{p=1}^{m}\sum_{s=1}^{m}\left(\frac{\partial^2 g_k}{\partial x_p\,\partial x_s}\right)_{\underline{x}}\left(\frac{\partial x_s}{\partial x_i}\right)_{\underline{g}}\left(\frac{\partial x_p}{\partial x_j}\right)_{\underline{g}}$$
$$-\sum_{p=1}^{m}\left[\left(\frac{\partial x_p}{\partial x_j}\right)_{\underline{g}}\left(\frac{\partial^2 g_k}{\partial x_p\,\partial x_i}\right)_{\underline{x}} + \left(\frac{\partial x_p}{\partial x_i}\right)_{\underline{g}}\left(\frac{\partial^2 g_k}{\partial x_p\,\partial x_j}\right)_{\underline{x}}\right] - \left(\frac{\partial^2 g_k}{\partial x_i\,\partial x_j}\right)_{\underline{x}}$$
$$\text{for } m+1 \le i, j \le n,\ k = 1, 2, \ldots, m \tag{3.2-22a}$$

To simplify the presentation, let us represent the right-hand side by $d_k^{i,j}$. We may then write that

$$\sum_{p=1}^{m}\left(\frac{\partial^2 x_p}{\partial x_i\,\partial x_j}\right)_{\underline{g}}\left(\frac{\partial g_k}{\partial x_p}\right)_{\underline{x}} = d_k^{i,j} \qquad \text{for } k = 1, 2, \ldots, m \tag{3.2-22b}$$

Note that each $d_k^{i,j}$ can be evaluated by direct partial differentiation, and by use of Eq. (3.2-19). The desired terms $(\partial^2 x_p/\partial x_i\,\partial x_j)_{\underline{g}}$ appear linearly on the left hand of the m equations. Solving as before, using Cramer's rule, we find that

$$\left(\frac{\partial^2 x_p}{\partial x_i\,\partial x_j}\right)_{\underline{g}} = \sum_{k=1}^{m}(-1)^{p+k}\,d_k^{i,j}\,\frac{J\left(\dfrac{g_1, g_2, \ldots, g_{k-1}, g_{k+1}, \ldots, g_m}{x_1, x_2, \ldots, x_{p-1}, x_{p+1}, \ldots, x_m}\right)}{J\left(\dfrac{g_1, g_2, \ldots, g_m}{x_1, x_2, \ldots, x_m}\right)} \tag{3.2-23}$$

Thus our task, while not in a simple, easily discernible form, is complete. Further simplification does not seem possible. The recommended procedure for testing the character of a restricted optimum is to compute the second derivatives $(\partial^2 y/\partial x_i\, \partial x_j)_{\underline{g}}$ by applying the formulas derived above. The calculation is admittedly tedious, and probably prohibitive for cases with more than three restrictions. However, once these second derivatives have been evaluated, the h_i remaining are now arbitrary independent variations, since the effect of restrictions has been removed by the elimination of the m dependent displacements dx_i. The problem of determining the nature of $\Delta^2 y$ can now be handled by precisely the same arguments as were presented in Sec. 2.2.2. Thus the sufficiency conditions can be examined, for example, by testing the various determinants defined by Eqs. (2.2-11) and (2.2-12). Weak conditions, as discussed earlier, may also be found.

These matters may be clarified by the discussion of a problem.

Example VII-10: Necessary and sufficient conditions

Using the illustration in Example VII-6, check that the located stationary value is indeed a minimum.

Here $n = 2$ and $m = 1$, so that Eq. (3.2-20) becomes

$$\left(\frac{\partial^2 y}{\partial x_2^2}\right)_{\underline{g}} = \left(\frac{\partial y}{\partial x_1}\right)_{\underline{x}} \left(\frac{\partial^2 x_1}{\partial x_2^2}\right)_{\underline{g}} + \left(\frac{\partial^2 y}{\partial x_1^2}\right)_{\underline{x}} \left(\frac{\partial x_1}{\partial x_2}\right)_{\underline{g}}^2 + 2\left(\frac{\partial^2 y}{\partial x_1\, \partial x_2}\right)_{\underline{x}} \left(\frac{\partial x_1}{\partial x_2}\right)_{\underline{g}} + \left(\frac{\partial^2 y}{\partial x_2^2}\right)_{\underline{x}} \tag{3.2-24}$$

Clearly,

$$\left(\frac{\partial y}{\partial x_1}\right)_{\underline{x}} = 8x_1 \qquad \left(\frac{\partial^2 y}{\partial x_1^2}\right)_{\underline{x}} = 8 \qquad \left(\frac{\partial y}{\partial x_2}\right)_{\underline{x}} = 10x_2$$

$$\left(\frac{\partial^2 y}{\partial x_2^2}\right)_{\underline{x}} = 10 \qquad \left(\frac{\partial^2 y}{\partial x_1\, \partial x_2}\right)_{\underline{x}} = 0$$

Using Eq. (3.2-19), we find that

$$\left(\frac{\partial x_1}{\partial x_2}\right)_{\underline{g}} = -\frac{J(g/x_2)}{J(g/x_1)} = -\frac{3}{2}$$

Moreover, to evaluate $(\partial^2 x_1/\partial x_2^2)_{\underline{g}}$ the quantity $d_1^{2,2}$ must be computed. For this present example, from Eqs. (3.2-22*a*) and (3.2-22*b*), the $d_1^{2,2}$ is defined as

$$d_1^{2,2} = -\left(\frac{\partial^2 g}{\partial x_1^2}\right)_{\underline{x}} \left(\frac{\partial x_1}{\partial x_2}\right)_{\underline{g}}^2 - 2\left(\frac{\partial x_1}{\partial x_2}\right)_{\underline{g}} \left(\frac{\partial^2 g}{\partial x_1\, \partial x_2}\right)_{\underline{x}} - \left(\frac{\partial^2 g}{\partial x_2^2}\right)_{\underline{x}}$$

Since g is linear in x_1 and x_2, all the second derivatives vanish identically, and so does $d_1^{2,2}$. Thus, from Eq. (3.2-22b),

$$\left(\frac{\partial^2 x_1}{\partial x_2{}^2}\right)_{\underline{g}} = 0$$

as is always the case if the restrictions are linear functions.

Substitution of the numerical values in Eq. (3.2-24) then gives

$$\left(\frac{\partial^2 y}{\partial x_2{}^2}\right)_{\underline{g}} = 28$$

$\Delta^2 y$ is therefore positive, and the stationary point found in Example VII-6 is a minimum. Since this is a problem in two dimensions ($n = 2$) subject to a single restriction ($m = 1$), the variation in the sufficiency condition is in fact only in a single variable (dx_2), which cannot take trivial values. Hence we find that $\Delta^2 y > 0$, and not $\Delta^2 y \geq 0$. It will be noted that $(\partial^2 y/\partial x_2{}^2)_{\underline{g}}$ in this presentation is exactly the same as the $y_{x_2 x_2}$ in Example VII-6, using the method of direct substitution, but the present technique avoids possible problems in the elimination of any variable.

Example VII-11: The vanishing Jacobian denominator

The problem posed in Example VII-8 may be repeated, using the method of constrained variation. Thus we need to find the maximum of

$$y = x_2 - x_1{}^2$$

subject to the constraint

$$g = 1 - x_1{}^2 - x_2{}^2 = 0$$

Here $m = 1$ and $n = 2$, so that the necessary conditions as represented Eq. (3.2-12) reduce to

$$J\left(\frac{y,g}{x_2,x_1}\right) = 0 = \begin{vmatrix} 1 & -2x_1 \\ -2x_2 & -2x_1 \end{vmatrix} = -2x_1(1 + 2x_2)$$

This equation can be satisfied by either $x_1 = 0$ or $x_2 = -\frac{1}{2}$. The stationary points are, accordingly, $\{0,1\}$, $\{0,-1\}$, $\{0.866,-\frac{1}{2}\}$, and $\{-0.866,-\frac{1}{2}\}$. Which, if any, are maxima?

To test, our first step should always be to compute the denominator of Eq. (3.2-16), namely, $J(g/x_1)$ in the present examples. Here

$$J\left(\frac{g}{x_1}\right) = -2x_1$$

Thus, strictly speaking, the second derivatives required in Eq. (3.2-20) cannot be evaluated for $x_1 = 0$. Hence only two of the stationary points appear to be testable. The difficulty can often be resolved by interchanging the roles of the variables, x_1 and x_2. Thus, if we rename x_1 as x_2 and, accordingly, replace x_2 by x_1, then

$$y = x_1 - x_2^2$$

and
$$g = 1 - x_1^2 - x_2^2 = 0$$

Hence
$$J\left(\frac{g}{x_1}\right) = -2x_1$$

with the stationary points now $\{1,0\}$, $\{-1,0\}$, $\{-\tfrac{1}{2},0.866\}$, and $\{-\tfrac{1}{2},-0.866\}$, so that in no case does J vanish.

To examine the character of these stationary points, let us look at $\Delta^2 y$. From Eq. (3.2-15), we have

$$\Delta^2 y = \frac{1}{2!}\left(\frac{\partial^2 y}{\partial x_2^2}\right)_{\underline{g}} h_2^2$$

since $n = 2$ and $m = 1$. The nature of the constrained second derivative $\left(\dfrac{\partial^2 y}{\partial x_2^2}\right)_{\underline{g}}$ can now be found by the use of Eq. (3.2-20). Thus

$$\left(\frac{\partial^2 y}{\partial x_2^2}\right)_{\underline{g}} \quad \left(\frac{\partial y}{\partial x_1}\right)_{\underline{x}} \left(\frac{\partial^2 x_1}{\partial x_2^2}\right)_{\underline{g}} + \left(\frac{\partial^2 y}{\partial x_1^2}\right)_{\underline{x}} \left(\frac{\partial x_1}{\partial x_2}\right)_{\underline{g}}^2 + 2\left(\frac{\partial^2 y}{\partial x_1\,\partial x_2}\right)_{\underline{x}} \times \left(\frac{\partial x_1}{\partial x_2}\right)_{\underline{g}} + \left(\frac{\partial^2 y}{\partial x_2^2}\right)_{\underline{x}}$$

The quantity $(\partial x_1/\partial x_2)_{\underline{g}}$, according to Eq. (3.2-19), is

$$\left(\frac{\partial x_1}{\partial x_2}\right)_{\underline{g}} = -\frac{J(g/x_2)}{J(g/x_1)} = -\frac{x_2}{x_1}$$

and from Eq. (3.2-23),†

$$\left(\frac{\partial^2 x_1}{\partial x_2^2}\right)_{\underline{g}} = d_1^{2,2}\,\frac{1}{J(g/x_1)} = -\frac{d_1^{2,2}}{2x_1}$$

Now, from Eqs. (3.2-22*a*) and (3.2-22*b*),

$$d_1^{2,2} = -\left(\frac{\partial^2 g}{\partial x_1^2}\right)_{\underline{x}}\left(\frac{\partial x_1}{\partial x_2}\right)_{\underline{g}}\left(\frac{\partial x_1}{\partial x_2}\right)_{\underline{g}} - 2\left(\frac{\partial x_1}{\partial x_2}\right)_{\underline{g}}\left(\frac{\partial^2 g}{\partial x_1\,\partial x_2}\right)_{\underline{x}} - \left(\frac{\partial^2 g}{\partial x_2^2}\right)_{x}$$

† With one restriction, $J\left(\dfrac{g_1, g_2, \ldots, g_{k-1}, g_{k+1}, \ldots, g_m}{x_1, x_2, \ldots, x_{p-1}, x_{p+1}, \ldots, x_m}\right) = J\left(\dfrac{0}{0}\right) = 1.$

so that
$$d_1^{2,2} = 2\left[\left(\frac{x_2}{x_1}\right)^2 + 1\right]$$

Since the other derivatives have an obvious meaning, we are now in a position to evaluate $(\partial^2 y/\partial x_2^2)_{\underline{g}}$. Hence from Eq. (3.2-20)

$$\left(\frac{\partial^2 y}{\partial x_2^2}\right)_{\underline{g}} = -\frac{1}{x_1}\left[\left(\frac{x_2}{x_1}\right)^2 + 1\right] - 2$$

which, representing the sign of $\Delta^2 y$, can be used to determine the character of each stationary value. Thus we have

$$\left(\frac{\partial^2 y}{\partial x_2^2}\right)_{\underline{g}} = \begin{cases} -3 & \text{for } \{1,0\}\text{, giving a local maximum} \\ -1 & \text{for } \{-1,0\}\text{, giving a local maximum} \\ +6 & \text{for } \{-\tfrac{1}{2},0.866\}\text{, giving a local minimum} \\ +6 & \text{for } \{-\tfrac{1}{2},-0.866\}\text{, giving a local minimum} \end{cases}$$

Example VII-12: Complex example, using the method of constrained variation

Let us find the minimum of the linear objective function

$$y = 7x_1 - 6x_2 + 4x_3$$

subject to the constraints

$$g_1 = x_1^2 + 2x_2^2 + 3x_3^2 - 1 = 0$$
$$g_2 = 5x_1 + 5x_2 - 3x_3 - 6 = 0$$

Here $m = 2$ and $n = 3$, so that the necessary conditions for a minimum will be given by Eq. (3.2-12) as

$$J\left(\frac{y,g_1,g_2}{x_3,x_1,x_2}\right) = 0$$

For the present case we find that

$$J\left(\frac{y,g_1,g_2}{x_3,x_1,x_2}\right) = \begin{vmatrix} 4 & 7 & -6 \\ 6x_3 & 2x_1 & 4x_2 \\ -3 & 5 & 5 \end{vmatrix} = 4x_1 - 164x_2 - 390x_3 = 0$$

Solving this equation, together with $g_1 = 0$ and $g_2 = 0$, we find two possible solutions to the posed problem:

	x_1	x_2	x_3	y
Solution 1	0.947	0.207	−0.0772	5.08
Solution 2	0.534	0.535	−0.219	−0.346

It appears from the respective objective function values that the first solution is a maximum and the second a minimum. Let us verify the character of these two stationary points. Since $n - m = 1$, we have only a single term contributing to order h^2 in Eq. (3.2-15), and this term, $(\partial^2 y/\partial x_3^2)_{\underline{g}} h_3^2$, determines the character of the stationary point, provided, of course, that it is nonzero. From Eq. (3.2-20) we see that

$$\left(\frac{\partial^2 y}{\partial x_3^2}\right)_{\underline{g}} = 7\left(\frac{\partial^2 x_1}{\partial x_3^2}\right)_{\underline{g}} - 6\left(\frac{\partial^2 x_2}{\partial x_3^2}\right)_{\underline{g}}$$

since $(\partial y/\partial x_1)_{\underline{x}} = 7$, $(\partial y/\partial x_2)_{\underline{x}} = -6$, and the second derivatives of y vanish, because y is a linear function. Expanding Eq. (3.2-23) to suit our particular case yields

$$\left(\frac{\partial^2 x_1}{\partial x_3^2}\right)_{\underline{g}} = \frac{d_1^{3,3} J(g_2/x_2) - d_2^{3,3} J(g_1/x_2)}{J\left(\dfrac{g_1,g_2}{x_1,x_2}\right)} = \frac{5d_1^{3,3} - 4x_2 d_2^{3,3}}{10(x_1 - 2x_2)}$$

and

$$\left(\frac{\partial^2 x_2}{\partial x_3^2}\right)_{\underline{g}} = \frac{-d_1^{3,3} J(g_2/x_1) + d_2^{3,3} J(g_1/x_1)}{J\left(\dfrac{g_1,g_2}{x_1,x_2}\right)} = \frac{-5d_1^{3,3} + 2x_1 d_2^{3,3}}{10(x_1 - 2x_2)}$$

The $d_1^{3,3}$ term represents the right-hand side of Eq. (3.2-22*a*), and for the case considered here, we get

$$d_1^{3,3} = -2\left(\frac{\partial x_1}{\partial x_3}\right)_{\underline{g}}^2 - 4\left(\frac{\partial x_2}{\partial x_3}\right)_{\underline{g}}^2 - 6$$

and

$$d_2^{3,3} = 0$$

Since

$$\left(\frac{\partial x_1}{\partial x_3}\right)_{\underline{g}} = -\frac{J\left(\dfrac{g_1,g_2}{x_3,x_2}\right)}{J\left(\dfrac{g_1,g_2}{x_1,x_2}\right)} = \frac{-6(5x_3 + 2x_2)}{10(x_1 - 2x_2)} = -0.03$$

when evaluated at both solutions, and also

$$\left(\frac{\partial x_2}{\partial x_3}\right)_{\underline{g}} = -\frac{J\left(\dfrac{g_1,g_2}{x_1,x_3}\right)}{J\left(\dfrac{g_1,g_2}{x_1,x_2}\right)} = \frac{6(x_1 + 5x_3)}{10(x_1 - 2x_2)} = 0.63$$

then $d_1^{3,3} = -7.6$ for both our solutions. For solution 2 we find that

$$\left(\frac{\partial^2 x_1}{\partial x_3^2}\right)_{\underline{g}} = 7.1 = -\left(\frac{\partial^2 x_2}{\partial x_3^2}\right)_{\underline{g}}$$

giving $$\Delta^2 y = 92h_3^2$$

which is a sufficient condition definitely establishing solution 2 as a minimum. Similarly, solution 1 could be shown to be a maximum.

We have thus seen that, while somewhat laborious, a sufficiency test is available to determine the character of a stationary point.

It should be noted that the expansion of the Jacobians becomes a considerable chore if the determinants become fourth-order or higher. Thus the method is difficult to apply analytically if m, the number of restrictions, becomes greater than 3. However, in certain cases, the stationary points can be found numerically for a very complex objective function using constrained variations. Such a method has been proposed by Jackson [1964a].

The method of constrained variation is not used much in practice, the Lagrangian multiplier method being more favored. While the two methods are closely related theoretically, the equations resulting from the latter method are usually simpler to solve. However, the constrained variation technique has the advantage of compact notation and allows the nature of the stationary point to be investigated.

3.3 Lagrangian Multipliers

It has been shown that, for restricted problems of optimization, one can often eliminate certain of the variables from the objective function (method of direct substitution) or one can almost always eliminate a number of displacements equal to the number of restrictions (method of constrained variation). Either process yields, as an intermediate result, a variation equation having only arbitrary displacements. At this juncture, the arguments applied to the unrestricted search are equally applicable to restricted problems in this intermediate form. Both methods are characterized by an unnatural distinction between the variables. In the case of direct substitution, certain of the variables are eliminated, thus treating the variables being eliminated as dependent on the remaining variables. But which variables are to be eliminated? Obviously, this choice is made so as to leave the most tractable problem for solution. Nevertheless, a distinction is made between the variables. This is also characteristic of the method of constrained variation, where the first m displacements are determined in terms of the remaining $n - m$ displacements. Thus the methods are not symmetrical with respect to the treatment of variables.

To avoid such distinctions and to treat each variable equally and at the same time preserve the integrity of the constraints, one can use the method of Lagrangian multipliers, in which certain undetermined constants are introduced, seemingly artificially, into the problem (see, for example, Sokolnikoff and Redheffer [1966], Courant [1936], and Mickley, Sherwood, and Reid [1957]). In this method there will be $n + m$ equations to be solved as compared with the n for constrained variation and the $n - m$ encountered in direct substitution. Indeed, the introduction of the arbitrary constants superficially seems to confuse the issue, since one must determine not only the coordinates rendering a certain function stationary, but also the constants. However, the advantage lies in a symmetrical treatment of each variable since no elimination is attempted.

3.3.1 *Necessary conditions*

To begin the development, it is well to note again that a necessary condition for the existence of a local optimum is that the dy, defined as

$$dy = \sum_{i=1}^{n} \frac{\partial y}{\partial x_i} dx_i \tag{3.3-1}$$

must vanish for all admissible variations. The admissible variables are those variations satisfying Eq. (3.2-3), so that we have

$$dg_k = \sum_{i=1}^{n} \frac{\partial g_k}{\partial x_i} dx_i = 0 \qquad \text{for } k = 1, 2, \ldots, m \tag{3.3-2}$$

Moreover, at the restricted stationary point $\mathbf{x}^*$,

$$g_k^* = g_k(\mathbf{x}^*) = 0 \qquad \text{for } k = 1, 2, \ldots, m \tag{3.3-3}$$

Let us now multiply Eq. (3.3-2) by a constant λ_k, of a value to be specified later, and form dF, where

$$dF = dy + \sum_{k=1}^{m} \lambda_k \, dg_k \tag{3.3-4}$$

Thus

$$dF = \left(\frac{\partial y}{\partial x_1} dx_1 + \frac{\partial y}{\partial x_2} dx_2 + \cdots\right) + \lambda_1\left(\frac{\partial g_1}{\partial x_1} dx_1 + \cdots + \frac{\partial g_1}{\partial x_n} dx_n\right) + \cdots + \lambda_m \left(\frac{\partial g_m}{\partial x_1} dx_1 + \cdots + \frac{\partial g_m}{\partial x_n} dx_n\right)$$

Collecting terms in dx_i then gives

$$dF = \left(\frac{\partial y}{\partial x_1} + \lambda_1 \frac{\partial g_1}{\partial x_1} + \cdots + \lambda_m \frac{\partial g_m}{\partial x_1}\right) dx_1 + \cdots + \left(\frac{\partial y}{\partial x_n} + \lambda_1 \frac{\partial g_1}{\partial x_n} + \cdots + \lambda_m \frac{\partial g_m}{dx_n}\right) dx_n$$

or

$$dF = \sum_{i=1}^{n} \left(\frac{\partial y}{\partial x_i} + \sum_{k=1}^{m} \lambda_k \frac{\partial g_k}{\partial x_i}\right) dx_i \tag{3.3-5}$$

This function dF must vanish for all admissible variations, because it is composed of two parts [Eq. (3.3-4)], each of which must vanish for all such permitted variations. Thus

$$dF = 0 \qquad \text{for all admissible variations} \tag{3.3-6}$$

Since there are m constraints, there are $n - m$ displacements which can be arbitrarily chosen. We select values for $\lambda_k (k = 1, 2, \ldots, m)$ so that the coefficients of the first m displacements vanish; that is, the λ_k are defined so that

$$\frac{\partial y}{\partial x_i} + \sum_{k=1}^{m} \lambda_k \frac{\partial g_k}{\partial x_i} = 0 \qquad \text{for } i = 1, 2, \ldots, m \tag{3.3-7a}$$

This will leave only arbitrary displacements in the expression for dF. Since dF must vanish for all values of these arbitrary $n - m$ displacements, the coefficients of these displacements must necessarily vanish, giving

$$\frac{\partial y}{\partial x_i} + \sum_{k=1}^{m} \lambda_k \frac{\partial g_k}{\partial x_i} = 0 \qquad \text{for } i = m + 1, m + 2, \ldots, n \tag{3.3-7b}$$

What has been gained by introducing m undermined multipliers? Essentially, the gain is that each of the x_i $(i = 1, 2, \ldots, n)$ is treated as though it were independent with respect to a modified objective function, which is

$$F = y + \sum_{k=1}^{m} \lambda_k g_k \tag{3.3-8}$$

This new "objective function" F obviously yields Eq. (3.3-4) on calculation of the change resulting from a small displacement. This modified objective function is to be made stationary with respect to all variations, so that

$$\frac{\partial F}{\partial x_i} = 0 \qquad \text{for } i = 1, 2, \ldots, n \tag{3.3-9}$$

and the λ_k are to be selected such that

$$g_k = 0 \qquad \text{for } k = 1, 2, \ldots, m \tag{3.3-10}$$

These conditions provide a system of $n + m$ equations and $n + m$ unknowns, $x_1, x_2, \ldots, x_n$ and $\lambda_1, \lambda_2, \ldots, \lambda_m$.

Example VII-13: Lagrangian multipliers

Let us resolve Example VII-6 using the method of Lagrangian multipliers. Since

$$y = 4x_1^2 + 5x_2^2$$

and

$$g = 2x_1 + 3x_2 - 6 = 0$$

the modified objective function F as defined by Eq. (3.3-8), may now be formed.

$$F = (4x_1^2 + 5x_2^2) + \lambda_1(2x_1 + 3x_2 - 6)$$

Since F must be stationary with respect to variations in both x_1 and x_2 [Eq. (3.3-9)], then

$$\frac{\partial F}{\partial x_1} = 8x_1 + 2\lambda_1 = 0$$

and
$$\frac{\partial F}{\partial x_2} = 10x_2 + 3\lambda_1 = 0$$

Therefore the location of the stationary value in F will be defined by

$$x_1 = -\tfrac{1}{4}\lambda_1$$
$$x_2 = -\tfrac{3}{10}\lambda_1$$

Thus both the x_1 and x_2 depend on λ_1, and the function F is stationary for all λ_1, provided x_1 and x_2 are chosen as above. Let us find the value of λ_1 which satisfies the restriction $g = 0$. Upon substituting for x_1 and x_2 in $g = 0$, we get

$$2(-\tfrac{1}{4}\lambda_1) + 3(-\tfrac{3}{10}\lambda_1) - 6 = 0$$

so that the required value of the Lagrangian multiplier, λ_1, is

$$\lambda_1 = -\tfrac{30}{7}$$

This then yields the desired coordinates of the stationary value as

$$x_1 = 1.071 \qquad x_2 = 1.286$$

as before.

3.3.2 *"Insufficient" conditions*

It is important to note that the modified objective function cannot be tested in the usual way to determine if the stationary point is a maximum or a minimum. This, incidently, is a point missed in some books and technical papers, where it is falsely asserted that sufficiency conditions can be used to test the Lagrangian function. The success of the method of Lagrangian multipliers is, basically, attributable to the fact that Δg_k vanishes for all displacements of interest. However, there is no reason to suppose that $\Delta^2 g_k$ will vanish even if Δg_k does. Therefore, while ΔF is entirely equivalent to Δy, it is not possible to assert that $\Delta^2 F$ equals $\Delta^2 y$. Since $\Delta^2 y$ is to be investigated in the region of the stationary point, the original objective function, and not the fictitious one, F, must be considered in determining the character of the restricted objective function. This point is explored in

Example VII-14, where the method of constrained variation is coupled with that of Lagrangian multipliers to determine whether or not a particular objective function is a maximum, a minimum, or a saddle point. It is generally true that the modified function F can be used only to determine possible locations of any optimum. Another counterexample has been stated by Denn [1965].

Example VII-14: The failure of the modified objective function to determine the character of a stationary point

Let us find the dimensions of a closed cylindrical tank of given volume V which has the minimum surface area A. Since $A = 2\pi R^2 + 2\pi RL$, our objective can be written

$$y = \frac{A}{\pi} = 2R^2 + 2RL$$

which is to be minimized subject to the restriction that $V = \pi R^2 L$ is fixed. Hence the restriction is

$$g_1 = R^2 L - \frac{V}{\pi} = 0$$

Using the method of Lagrangian multipliers, the modified objective function F may now be formed.

$$F = y + \lambda_1 g_1 = 2R^2 + 2RL + \lambda_1\left(R^2 L - \frac{V}{\pi}\right)$$

F is stationary when

$$\frac{\partial F}{\partial R} = 4R + 2L + 2RL\lambda_1 = 0 \qquad (3.3\text{-}11a)$$

and

$$\frac{\partial F}{\partial L} = 2R + R^2\lambda_1 = 0 \qquad (3.3\text{-}11b)$$

Certainly, $R = 0$ and $L = 0$ is a possible solution to these equations, but it contradicts the restriction regarding the fixed size of the volume, except for the trivial case $V = 0$. This solution is thus rejected.[1] Equation (3.3-11*b*) can then be divided by R and rearranged so that

$$R = -\frac{2}{\lambda_1}$$

Therefore

$$L = -\frac{4}{\lambda_1}$$

[1] This example illustrates the general point that the derived stationary values of the objective function F must always be checked to ensure that they do satisfy the system restrictions.

These forms clearly show again that the choice of values for the variables making F stationary depends on λ_1. To properly select λ_1, we substitute for R and L in the restriction, getting

$$g_1 = -\frac{16}{\lambda_1^3} - \frac{V}{\pi} = 0$$

Thus, to satisfy the restriction, we must let λ_1 be

$$\lambda_1 = -\left(\frac{16\pi}{V}\right)^{1/3}$$

which gives, finally, the desired values of L and R as

$$R = \left(\frac{V}{2\pi}\right)^{1/3}$$

$$L = 2\left(\frac{V}{2\pi}\right)^{1/3}$$

The character of this stationary value in the function F can be examined by considering the second derivatives. Thus

$$\frac{\partial^2 F}{\partial R^2} = 4 + 2L\lambda_1$$

$$\frac{\partial^2 F}{\partial R\,\partial L} = 2 + 2R\lambda_1$$

$$\frac{\partial^2 F}{\partial L^2} = 0$$

which, when substituting for L and R values at the stationary point, give

$$\frac{\partial^2 F}{\partial R^2} = 4 + 2\lambda_1\left(-\frac{4}{\lambda_1}\right) = -4$$

$$\frac{\partial^2 F}{\partial R\,\partial L} = 2 + 2\lambda_1\left(-\frac{2}{\lambda_1}\right) = -2$$

$$\frac{\partial^2 F}{\partial L^2} = 0$$

Using the sufficiency conditions of Sec. 2.2.3, we then have that

$$D_1 = \frac{\partial^2 F}{\partial R^2} < 0$$

$$D_2 = \frac{\partial^2 F}{\partial R^2}\frac{\partial^2 F}{\partial L^2} - \left(\frac{\partial^2 F}{\partial R\,\partial L}\right)^2 = -4 < 0$$

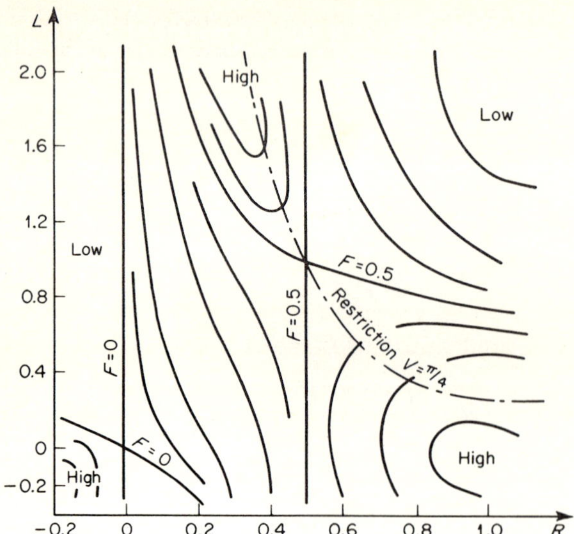

Fig. VII-7 Contours of F for $\lambda = -4$ (see Example VII-14).

which represents a saddle point in F. The form of this F function is shown in Fig. VII-7. No information is therefore available about the character of the stationary value in y when using only the method of Lagrangian multipliers.

The required sufficiency conditions can be found using the method of constrained variation. This requires the investigation of $\Delta^2 y$. For $m = 1$ and $n = 2$, let us treat L as x_1 and R as x_2 in expanding the relevant equations. Thus the important quantity is

$$\left(\frac{\partial^2 y}{\partial R^2}\right)_{\underline{g}} = \left(\frac{\partial y}{\partial L}\right)_{\underline{x}}\left(\frac{\partial^2 L}{\partial R^2}\right)_{\underline{g}} + \left(\frac{\partial^2 y}{\partial L^2}\right)_{\underline{x}}\left(\frac{\partial L}{\partial R}\right)_{\underline{g}}^2 + 2\left(\frac{\partial^2 y}{\partial R\,\partial L}\right)_{\underline{x}}\left(\frac{\partial L}{\partial R}\right)_{g} + \left(\frac{\partial^2 y}{\partial R^2}\right)_{\underline{x}}$$

Certainly,

$$\left(\frac{\partial y}{\partial L}\right)_{\underline{x}} = 2R \qquad \left(\frac{\partial^2 y}{\partial L^2}\right)_{\underline{x}} = 0 \qquad \left(\frac{\partial^2 y}{\partial R\,\partial L}\right)_{\underline{x}} = 2$$

$$\left(\frac{\partial y}{\partial R}\right)_{\underline{x}} = 4R + 2L \qquad \left(\frac{\partial^2 y}{\partial R^2}\right)_{\underline{x}} = 4$$

Moreover, from Eq. (3.2-19), we see that

$$\left(\frac{\partial L}{\partial R}\right)_{\underline{g}} = -\frac{J(g/R)}{J(g/L)} = -\frac{2RL}{R^2} = -\frac{2L}{R}$$

Also, from Eq. (3.2-22), one finds

$$\left(\frac{\partial^2 L}{\partial R^2}\right)_{\underline{g}} = \frac{6L}{R^2}$$

Therefore, substituting for the numerical values, we get

$$\left(\frac{\partial^2 y}{\partial R^2}\right)_{\underline{g}} = 4\left(1 + \frac{L}{R}\right)$$

Then, as $\Delta^2 y = \frac{1}{2}!\,(\partial^2 y/\partial R^2)_{\underline{g}}(dR)^2$, which must be greater than zero for all physically possible values of L and R, the stationary point in y must be a minimum.

3.3.3 Lagrangian multipliers and sequential systems—the so-called discrete maximum principle

The method of Lagrangian multipliers leads to some very interesting results when applied to sequential systems. Let us consider the simplified sequential system depicted in Fig. VII-8. Here we have a system of stages arranged in series with the feed to one stage originating from the previous stage, as, for example, in the cross-current liquid-liquid extraction cascade of Example VII-7. Whatever the particular application, the state of the feed to the ith stage can be specified by the vector $\mathbf{p}_{i-1}$, while the input to this stage is represented by $\mathbf{q}_i$. The product from the ith stage would then be determined by both $\mathbf{p}_{i-1}$ and $\mathbf{q}_i$, the relationship being of the form

$$\mathbf{p}_i = \mathbf{T}_i(\mathbf{p}_{i-1},\mathbf{q}_i) \qquad \text{for } i = 1, 2, \ldots, n,\ \mathbf{p}_0 \equiv \mathbf{q}_0 \qquad (3.3\text{-}12)$$

Alternatively, since these are restrictions on the choice of variables, we can write

$$\mathbf{g}_i = \mathbf{T}_i(\mathbf{p}_{i-1},\mathbf{q}_i) - \mathbf{p}_i = 0 \qquad \text{for } i = 1, 2, \ldots, n,\ \mathbf{p}_0 \equiv \mathbf{q}_0 \quad (3.3\text{-}13)$$

The existence of such a relationship is of course necessary for optimization to proceed. Here the state of the interlinking stream leaving the ith stage

Fig. VII-8 A general sequential system.

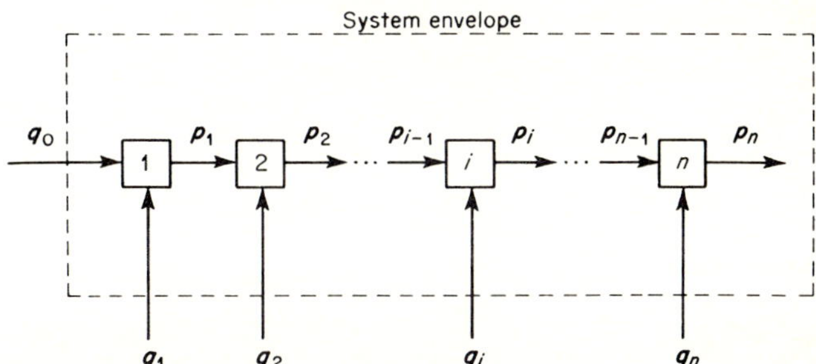

depends solely on the state of the feed to that stage, i.e., the product from the $(i-1)$st stage and the state of the ith input $\mathbf{q}_i$. Sequential systems are characterized by the existence of such a relationship. For such systems Lagrangian multipliers can be a very powerful computational aid (Fan and Wang [1964a–f, 1965], Katz [1962a, b, 1964], Jackson [1964a], Jackson and Horn [1965a, b], Horn and Jackson [1965a, b], and Brosilow and Lasdon [1965]), a closely related approach being that of Denn and Aris [1965a], based upon Green's functions.

To develop this particular application, let us assume that we wish to optimize an objective function y, where

$$y = y(\mathbf{p}_0, \ldots, \mathbf{p}_n; \mathbf{q}_1, \ldots, \mathbf{q}_n) \tag{3.3-14}$$

by an appropriate choice of the input variables $\mathbf{q}_i$ $(i = 1, 2, \ldots, n)$, the system feed $\mathbf{p}_0$ $(\equiv \mathbf{q}_0)$ being specified. Here each vector will contain a set of system variables, $p_{j,i}$ or $q_{j,i}$. As an illustration, however, let us assume that each vector represents a single system variable, and that this variable can be represented by p_i or q_i, as the case may arise (in other words, $Q_i = 1$, $P_i = 1$). Equation (3.3-14) is now a function of $2n$ variables if the feed state q_0 is fixed. Of course, not all these variables are independent, since n restrictions exist, as given by Eqs. (3.3-13).

In principle, n of the $2n$ variables may now be eliminated from the objective function by the use of the restriction relationships. If the state variables p_i $(i = 1, 2, \ldots, n)$ are eliminated, we get

$$y = y(q_0, q_1, \ldots, q_n)$$

a function of only n variables, which are now completely independent, unless additional constraints are imposed. Such an approach of direct substitution was used in Example VII-7, although in general it is too complex to be attempted.

Instead of attempting to reduce the number of variables, let us use the method of Lagrangian multipliers. The condition which must be satisfied if y is to be an optimal value is that the modified objective function F must be stationary relative to arbitrary variations in p_i and q_i. Here

$$F = y + \sum_{i=1}^{n} \lambda_i g_i$$

or

$$F = y + \sum_{i=1}^{n} \lambda_i (T_i - p_i) \tag{3.3-15}$$

where the λ_i are the Lagrangian multipliers. The stationary value in F is found when

$$\frac{\partial F}{\partial p_i} = 0 \quad \text{and} \quad \frac{\partial F}{\partial q_i} = 0 \quad \text{for } i = 1, 2, \ldots, n$$

Thus we require

$$\frac{\partial F}{\partial p_i} = \frac{\partial y}{\partial p_i} - \lambda_i + \lambda_{i+1}\frac{\partial T_{i+1}}{\partial p_i} \qquad \text{for } i = 1, 2, \ldots, n, \text{ with } \lambda_{n+1} \equiv 0 \tag{3.3-16}$$

and

$$\frac{\partial F}{\partial q_i} = \frac{\partial y}{\partial q_i} + \lambda_i \frac{\partial T_i}{\partial q_i} = 0 \qquad \text{for } i = 1, 2, \ldots, n \tag{3.3-17}$$

The solution to this set of $2n$ equations, combined with the n restrictions, Eqs. (3.3-13), defines a set of possible locations for optimal values of y at the restricted stationary points. Whether or not any one of these stationary points is in fact the optimum is a matter for further investigation.

The interesting point to note is that the Lagrangian multipliers play a role which is very similar to that of p_i or q_i. There is a Lagrangian multiplier to be associated with each of the sequential units, and it can be treated, if so desired, as another variable, so that the $3n$ equations found from

$$\frac{\partial F}{\partial p_i} = 0 \qquad \frac{\partial F}{\partial q_i} = 0 \qquad \frac{\partial F}{\partial \lambda_i} = 0$$

will define the required stationary values. The third set, found by differentiation with respect to the Lagrangian multiplier, yields the original n restriction equations,

$$g_i = 0$$

It will be seen that the Lagrangian multipliers are defined starting from the end of the system, with $\lambda_{n+1} \equiv 0$, and in this sense reflect changes from the end of the sequence of stages.[1] On the other hand, p_0 $(= q_0)$ is specified and, in this sense, reflects transformations beginning at the inlet of the sequence. This situation is analogous to the results found in the continuous case, using the method of Lagrangian multipliers. These results, together with the Pontryagin maximum principle, form a very powerful computational tool, as discussed in Chap. IX, Sec. 5.2.2. However, before examining the extent to which the analogy between the continuous and discrete cases persists, it is instructive to consider an example.

[1] This variational method and the dynamic programming technique (Chap. XIII, Sec. 3) use similar computational approaches, both starting at the rear end of the system. Dynamic programming, however, stores information and passes through the calculation only once, while Pontryagin's method, for example, does not store information but carries out multi-iterative cycles through the length of the system. In computation, therefore, the problem of large storage must be weighed against long calculational times due to iteration.

Example VII-15: Cross-current extraction

Consider the cross-current liquid-liquid extraction system of Example VII-7, which is depicted in Fig. VII-9, using the present notation. The objective function can then be reevaluated, getting from Eq. (3.1-1)

$$y = \sum_{i=1}^{n} (p_{i-1} - p_i) - B \sum_{i=1}^{n} q_i$$

where p_i = concentration x_i of solute
q_i = amount w_i of wash solvent added to that stage
B = a constant $(= \beta/\alpha V)$

The function y is now to be maximized subject to the n restrictions imposed by the design relationships at each stage [Eqs. (3.3-13)]. Here the amount of solvent extracted in the ith stage is

$$V(p_{i-1} - p_i) = q_i(mp_i)$$

or

$$g_i = \frac{p_{i-1}}{1 + m'q_i} - p_i = 0 \tag{3.3-18}$$

on writing $m' = m/V$.

The modified objective function F is then

$$F = \sum_{i=1}^{n} (p_{i-1} - p_i) - B \sum_{i=1}^{n} q_i + \sum_{i=1}^{n} \lambda_i \left(\frac{p_{i-1}}{1 + m'q_i} - p_i \right) \tag{3.3-19}$$

The stationary points, if any, can be found from Eqs. (3.3-16) and (3.3-17), getting

$$-\lambda_i + \lambda_{i+1} \left(\frac{1}{1 + m'q_{i+1}} \right) = 0 \qquad \text{for } i \neq n \tag{3.3-20}$$

Fig. VII-9 Cascade of n liquid-liquid extraction units in series using cross-current solvent flow.

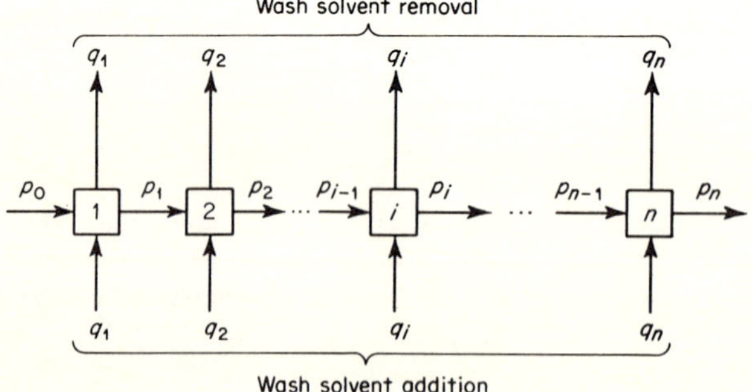

and $$-1 - \lambda_n = 0 \qquad \text{for } i = n \tag{3.3-21}$$

together with

$$-B - \frac{m'\lambda_i p_{i-1}}{(1 + m'q_i)^2} = 0 \qquad \text{for } i = 1, 2, \ldots, n \tag{3.3-22}$$

Equations (3.3-18) to (3.3-22) may now be solved to give the λ_i, p_i, and q_i which define stationary values of F, and thus locate possible optimal values of y. In particular, we are interested in the set $\{q_i\}$, and we can show that these values are equal (for the present problem of linear equilibrium). This can be done by eliminating the q_i, so that, from Eqs. (3.3-18) and (3.3-20), we get

$$\frac{\lambda_i}{\lambda_{i+1}} = \frac{1}{1 + m'q_{i+1}} = \frac{p_{i+1}}{p_i} \tag{3.3-23}$$

giving $$\lambda_1 p_1 = \lambda_2 p_2 = \cdots = \lambda_n p_n = \text{a constant} \tag{3.3-24}$$

and from Eqs. (3.3-18) and (3.3-22), we get

$$-\frac{B}{m'\lambda_{j+1}p_j} = \frac{1}{(1 + m'q_{j+1})^2} = \left(\frac{p_{j+1}}{p_j}\right)^2$$

which, on rearranging, gives

$$-\frac{B}{m'\lambda_{j+1}p_{j+1}} = \frac{p_{j+1}}{p_j}$$

However, from Eq. (3.3-24), the left-hand side of this relation must be a constant; so substitution back into Eq. (3.3-23) gives

$$\frac{p_{i+1}}{p_i} = \frac{1}{1 + m'q_{i+1}} = \text{constant}$$

Thus q_i is a constant, q, say. We must, therefore, in our optimal policy, add equal amounts q of wash solvent in each stage, as was found using direct substitution (Example VII-7) or dynamic programming (Example XIII-6).

Using the recurrence relation, Eq. (3.3-18),

$$p_{n-1} = p_n(1 + m'q)$$

and $$p_n = \frac{p_0}{(1 + m'q)^n}$$

giving $$p_{n-1} = \frac{p_0}{(1 + m'q)^{n-1}}$$

However, from Eq. (3.3-22),

$$p_{n-1} = \frac{-B(1 + m'q)^2}{m'\lambda_N} = \frac{B(1 + m'q)^2}{m'}$$

and equating the two expressions for p_{n-1}, we find

$$(1 + m'q)^{n+1} = \frac{p_0 m'}{B}$$

or in the original notation, the amount of wash solvent in any stage is

$$q = \frac{V}{m}\left[\left(\frac{\alpha p_0 m}{\beta}\right)^{1/(n+1)} - 1\right]$$

as before.

The foregoing example, although quite simple, clearly displays the essential elements of the method. In more complex cases, however, an iterative approach may be required to find the stationary value in F. Fan and Wang [1964b] have presented some such problems, the main difficulty being that the state variables are usually defined starting at one end of the system sequence, while the Lagrangian multipliers are defined from the other end. In the general case, therefore, assumptions as to one or other of these have to be made, and the calculation begun at one end of the sequence, alternative cycles of calculation being made in opposing directions if the original assumptions do not check with the calculated values at the end of each cycle.

The method of Lagrangian multipliers applied to sequential discrete systems yields equations which have analogous counterparts in the continuous case. Indeed, the continuous problem as treated in Chap. IX, Sec. 5.2, is based upon the use of a variable Lagrangian multiplier, so that the discrete form of the equations could be obtained from the continuous result by taking finite differences. For example, Eq. (5.2-38), Chap. IX, is an equation defining the Lagrangian multiplier in the continuous case, and the similarity with Eq. (3.3-16) for the discrete case will be obvious. Other results can likewise be compared. The maximum principle found for the continuous case has, however, no counterpart in the discrete system. A discrete maximum principle for the special case of $y = p_{1,n}$ has been correctly challenged (Horn and Jackson [1965a, b] and Jackson and Horn [1965a, b]). For this objective function and the set of transformation equations (3.3-12), the modified objective function is

$$F = p_{1,n} - \sum_{i=1}^{n} \sum_{j=1}^{s} \lambda_{j,i}(p_{j,i} - T_{j,i})$$

and in analogy with the continuous results of Chap. IX, Sec. 5.2.2, we may define a Hamiltonian to be

$$H_i = \sum_{j=1}^{s} \lambda_{j,i} T_{j,i}$$

If the analogy with the continuous case were complete in every respect, the choice of the input variables to the ith stage, $\mathbf{q}_i$, would have to be made such that H_i is a maximum holding the $\lambda_{j,i}$ fixed if the maximum of $p_{1,n}$ were desired. However, because the modified function F does not necessarily have to be a maximum if the objective function is a maximum (Sec. 3.3.2), no such necessary condition exists for the discrete case. This difficulty is illustrated by the following example.

Example VII-16: Discrete maximum principle—counterexample

To show that the discrete form of the maximum principle is not valid, let us consider the system depicted in Fig. VII-10, in which it is desired to minimize the value of the output $p_{1,2}$ (i.e., component 1 of the vector $\mathbf{p}_2$). The state of each stream in the system is represented by the vector $\mathbf{p}_i$, each of two components $p_{1,i}$ and $p_{2,i}$, while the input to each stage, $\mathbf{q}_i$, is a single-component vector. The transformation of the state variables in each stage is given by the following equations:

$$\mathbf{p}_0 = \begin{Bmatrix} p_{1,0} \\ p_{2,0} \end{Bmatrix} = \begin{Bmatrix} 1 \\ 1 \end{Bmatrix} \tag{3.3-25}$$

$$\mathbf{p}_1 = \begin{Bmatrix} p_{1,1} \\ p_{2,1} \end{Bmatrix} = \mathbf{T}_1(\mathbf{p}_0, q_1) \tag{3.3-26}$$

or

$$p_{1,1} = p_{1,0} - 2q_1 - \frac{q_1^2}{2} \tag{3.3-27}$$

$$p_{2,1} = p_{2,0} + q_1 \tag{3.3-28}$$

$$p_2 = \begin{Bmatrix} p_{1,2} \\ p_{2,2} \end{Bmatrix} = \mathbf{T}_2(\mathbf{p}_1, q_2)$$

Fig. VII-10 Discrete maximum principle—counterexample, see Example VII-16.

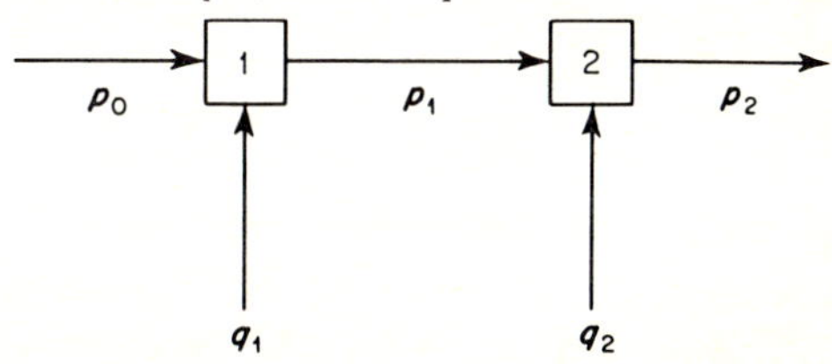

or
$$p_{1,2} = p_{1,1} + p_{2,1}^2 + q_2^2 \qquad (3.3\text{-}29)$$
$$p_{2,2} = p_{2,1} + q_2^2 \qquad (3.3\text{-}30)$$

Since the feed stream $\mathbf{p}_0$ is specified, the minimization of $p_{1,2}$ must be made by an appropriate choice of q_1 and q_2, subject to the restrictions imposed by Eqs. (3.3-27) to (3.3-30). Then, since $y = p_{1,2}$, and by the introduction of four Lagrangian multipliers, one for each restriction in Eqs. (3.3-27) to (3.3-30), we can form the modified function F as

$$F = p_{1,2} - \lambda_{1,1}(p_{1,1} - 1 + 2q_1 + \tfrac{1}{2}q_1^2) - \lambda_{2,1}(p_{2,1} - 1 - q_1)$$
$$- \lambda_{1,2}(p_{1,2} - p_{1,1} - p_{2,1}^2 - q_2^2) - \lambda_{2,2}(p_{2,2} - p_{2,1} - q_2^2)$$

It may then be shown by differentiating in turn with respect to the q_i, $\lambda_{j,i}$, and $p_{j,i}$ that a stationary point in F exists at the location $q_1 = 0$, $q_2 = 0$, $p_{1,1} = 1$, $p_{2,1} = 1$, $p_{1,2} = 2$, $p_{2,2} = 1$, $\lambda_{1,1} = 1$, $\lambda_{2,1} = 2$, $\lambda_{1,2} = 1$, $\lambda_{2,2} = 0$, and that these values also satisfy all the necessary restrictions.

This stationary point in F gives the minimum value of $p_{1,2}$, as can be seen by simply eliminating $p_{1,1}$ and $p_{2,1}$ from Eq. (3.3-29), using Eqs. (3.3-27) and (3.3-28) to obtain

$$p_{1,2} = 2 + \tfrac{1}{2}q_1^2 + q_2^2$$

We therefore see that the stationary point is a minimum of $p_{1,2}$. The discrete Hamiltonian for the first stage is given by

$$H_1 = \lambda_{1,1}(1 - 2q_1 - \tfrac{1}{2}q_1^2) + \lambda_{2,1}(1 + q_1)$$

Since $\lambda_{1,1} = 1$ and $\lambda_{2,1} = 2$, the minimum of H_1 is clearly found by taking q_1 as large as possible (or as small as possible if negative q_1 are permitted). However, the true minimizing solution is for q_1 to vanish. Here we see a failure of the so-called discrete maximum principle, since the value of q_1 required to produce a minimum value of the objective function also is the one which maximizes the Hamiltonian in contradiction with the maximum principle.

It should be noted that Horn and Jackson [1965b] have cited several special cases in which a discrete maximum principle exists. The foregoing illustration, due to Horn and Jackson [1965a], was designed to show that the analogy to Pontryagin's maximum principle fails in the discrete case. The discussion of Sec. 3.3.2 presents the reason for the general nonexistence of a discrete maximum principle. The fact that no discrete maximum principle exists is indeed unfortunate, and detracts from the value of the method. Denn and Aris [1965a] have shown that, for those cases in which the input or control variables $\mathbf{q}_i$ are bounded and a stationary solution does not exist, the $\mathbf{q}_i$ must be selected so as to maximize H_i.

It is essential that great care be taken in the use of Lagrangian multipliers other than for locating constrained stationary points, further tests being generally required to determine the character of a particular stationary value. Any further application of the Lagrangian multiplier, such as in the Kuhn-Tucker method discussed in Sec. 4.3, requires additional proofs, taking any special features of the particular problem into account. The method of Lagrangian multipliers can be used in optimizing very complex systems, but except in the case of sequential systems and some very simple recycle systems, the calculations become almost impossible. Moreover, the final product of such a calculation is still only a stationary point, and the task remains to ensure that it is an optimum.

3.4 Penalty Functions

When the objective function y must be optimized subject to equality restrictions $g_k = 0$, the use of a penalty function to force the solution on to a boundary might be a suitable approach. It finds more use, however, in some of the methods of numerical search.

In the search for a minimum, the basic idea is to form a new function such as

$$F = y + \sum_{k=1}^{m} P_k(g_k)^2 \tag{3.4-1}$$

This new function F can then be treated as in an unrestricted problem and minimized for a succession of increasingly large values of P. For a single restriction, $g_1 = 0$, it has been shown that as $P_1 \rightarrow \infty$, the solution of this modified problem approaches the solution of the original function subject to restriction. Several modifications are possible in the penalty function; thus, for example, we could use

$$F = y + (g_1)^{2P_1} \tag{3.4-2}$$

where P_1 is any large positive integer.

For a maximum we should subtract the penalty function; thus we might use

$$F = y - P_1(g)^2 \tag{3.4-3}$$

Other examples will be found in Sec. 4.4.4 and Chap. VIII, Secs. 4.4.2 and 5.4.1.

Example VII-17: Penalty functions (1)

Find the minimum of the function y, where

$$y = 4x_1^2 + 5x_2^2$$

subject to the restriction

$$2x_1 + 3x_2 = 6$$

using a penalty function method.

Using Eq. (3.4-1), we can form the modified function

$$F = y + P(g)^2$$

where
$$g = 2x_1 + 3x_2 - 6 = 0$$

Hence we have

$$F = 4x_1^2 + 5x_2^2 + P(2x_1 + 3x_2 - 6)^2$$

with stationary values defined by the equations

$$\frac{\partial F}{\partial x_1} = 8x_1 + 4P(2x_1 + 3x_2 - 6) = 0 \tag{3.4-4}$$

$$\frac{\partial F}{\partial x_2} = 10x_2 + 6P(2x_1 + 3x_2 - 6) = 0$$

Thus
$$\frac{8x_1}{4} = -P(g) = \frac{10x_2}{6}$$

so that
$$x_1 = {}^5\!/_6 x_2$$

Upon substituting this in Eq. (3.4-4), we get

$$8({}^5\!/_6 x_2) + 4P({}^5\!/_3 x_2 + 3x_2 - 6) = 0$$

or
$$x_2 = \frac{24P}{{}^{20}\!/_3 + {}^{56}\!/_3 P} = \frac{9}{7 + {}^5\!/_2 P}$$

When P becomes very large, x_2 tends to the value 1.286 and x_1 approaches 1.071 to define the stationary value.

We may now examine the nature of this stationary value. Thus

$$F_{x_1x_1} = 8(1 + P)$$

$$F_{x_1x_2} = 12P$$

$$F_{x_2x_2} = 2(5 + 9P)$$

Hence $D_1 = F_{x_1x_1} = 8(1 + P) > 0$

$$D_2 = F_{x_1x_1}F_{x_2x_2} - (F_{x_1x_2})^2$$
$$= 16[(1 + P)(5 + 9P) - 9P^2] = 16(5 + 14P) > 0$$

thus, by Eqs. (2.2-11), defining a minimum value of y. This method gives the same results as found before, in Example VII-6.

Example VII-18: Penalty functions (2)

Let us repeat the preceding problem using the penalty function defined by Eq. (3.4-2).

The new function is thus

$$F = 4x_1^2 + 5x_2^2 + (2x_1 + 3x_2 - 6)^{2P}$$

whose stationary values will be located by the roots of the equations

$$\frac{\partial F}{\partial x_1} = 8x_1 + 4P(2x_1 + 3x_2 - 6)^{2P-1} \tag{3.4-5}$$

$$\frac{\partial F}{\partial x_2} = 10x_2 + 6P(2x_1 + 3x_1 - 6)^{2P-1}$$

Hence $$\frac{8x_1}{2} = -2P(g)^{2P-1} = \frac{10x_2}{3}$$

so that $$x_1 = {}^5\!/_6 x_2$$

Substitution into Eq. (3.4-5) then gives

$$^{20}\!/_3 x_2 + 4P(^{14}\!/_3 x_2 - 6)^{2P-1} = 0$$

which will be satisfied only as $P \to \infty$ if $^{14}\!/_3 x_2 = 6$ or, as before, $x_2 = 1.286$. Then $x_1 = 1.071$ as before. The nature of $\Delta^2 F$ can also be tested to check that this stationary value is indeed a minimum.

3.5 Special Cases of Equality Restrictions

In certain cases, when a problem falls into a more restricted classification, other analytical methods may be proposed. Let us consider the system with a linear objective function and linear equality restrictions.

If we have a linear objective function, it possesses the special characteristic of having its optimum value on a boundary. This will be seen by considering the one- or two-dimensional problems. Thus, if we have $y = a + bx$, where a and b are positive constants, the maximum value of y will lie at the greatest available value of x, and conversely, the minimum will be found at the smallest x. Accordingly, x must be bounded for an optimum to exist. We need not, therefore, examine the interior of a domain, but only examine the vertices of the system boundaries in our search for an optimum. This simplification allows a simple procedure to be produced for the systematic examination of these boundaries; this has given rise to the specialized optimization study known as *linear programming*. Here the function is limited by linear system boundaries, and it is desired to optimize y, where

$$y = \sum_{i=1}^{n} c_i x_i \tag{3.5-1}$$

subject to the m restrictions ($m < n$)

$$\sum_{i=1}^{n} a_{j,i} x_i = b_j \qquad \text{for } j = 1, 2, \ldots, m \tag{3.5-2}$$

In its simplest form this situation is illustrated by the transportation problem in which ships have to move between ports in such a schedule as to minimize transportation changes. An equivalent problem is the allocation of n people to n jobs in such a manner as to maximize their total value, the value of a given person in a given job being known. Methods of solution for problems having this structure are discussed in Sec. 4.4.

4 INEQUALITY CONSTRAINTS

A more general problem of optimization than previously considered is to find the optimum of an objective function

$$y = y(\mathbf{x}) \tag{4-1}$$

subject to the m constraints

$$g_k(\mathbf{x}) \leq 0 \tag{4-2}$$

for $k = 1, 2, \ldots, m$, where m is no longer necessarily less than n.

A set of constraints such as imposed by these equations will define a region, still of n dimensions, within which all variables must lie to be valid.

The search for an optimum must therefore be in the interior of the region or along the system boundaries (Chap. IV, Sec. 4). Within the interior, an optimum can exist at only one of two possible locations, a stationary value of the function or a discontinuity in the function or its derivatives. Within such a framework the general problem will be initially considered, and then the various simplifications available for special systems, such as those with linear restrictions, will be discussed. The particular case of linear constraints and linear objective function will be investigated in some detail in Sec. 4.4, because of its importance in industrial applications.

4.1 Optimization and Inequality Restrictions

Two main approaches are used in the optimization of problems involving inequality restrictions. The first is the investigation of the system as if it were unrestricted, with an associated check on restrictions to determine whether or not the located optimum is valid (Sec. 4.1.1). (This is the type of approach used in most numerical search methods, as described in Chap. VIII.) The second group of methods involves the conversion of inequality restrictions into equalities by the use of slack variables (Sec. 4.1.2). These new restrictions can now be handled by the methods of Sec. 3.

4.1.1 Unrestricted optimization and inequality restrictions

One method of solving problems with inequality restrictions, and possibly the simplest, is to treat the system as if it were unrestricted, with the subsequent checking of any feasible optimal locations against the imposed inequality restrictions. Thus all stationary points and discontinuities in the objective function could be examined, and the one yielding the global optimum selected. The coordinates of this optimum position could then be checked against the inequality restrictions to see if it is a valid position. If it is not valid, the required optimum may be on the boundary imposed by the conflicting restriction when the inequality is identically zero or at one of the other local optima locations within the bounded region. Here the acceptable or valid bounded region is that region which satisfies all the m restrictions; i.e.,

$$g_j < 0$$

If one or more restrictions is satisfied identically, then

$$g_j = 0$$

and the desired position is said to lie on the boundary.

Example VII-19: Optimum on a boundary

Let us determine the minimum of the function y, where

$$y = 4x_1^2 + 5x_2^2$$

subject to the restriction that

$$x_1 \geq 1$$

which could be rewritten

$$g_1 = 1 - x_1 \leq 0$$

Any optimum will be located within the interior of the region (here for $x_1 > 1$ and any x_2) at a stationary point or at a discontinuity in the function or its derivatives, or it may lie at the boundary defining the valid region. Since no discontinuities are present in this problem, let us seek the stationary values of the unrestricted function y.

Thus the stationary values will be found if

$$\frac{\partial y}{\partial x_1} = 8x_1 = 0$$

$$\frac{\partial y}{\partial x_2} = 10x_2 = 0$$

i.e., at the point $x_1 = 0$, $x_2 = 0$. By testing this position against the restriction $x_1 \geq 1$, it will be seen that the located stationary value does not lie within the valid region. No optimum will therefore be found within the region, and the optimum must thus lie on the system boundary, $x_1 = 1$. Substituting $x_1 = 1$ into the objective function thus gives

$$y = 4 + 5x_2^2$$

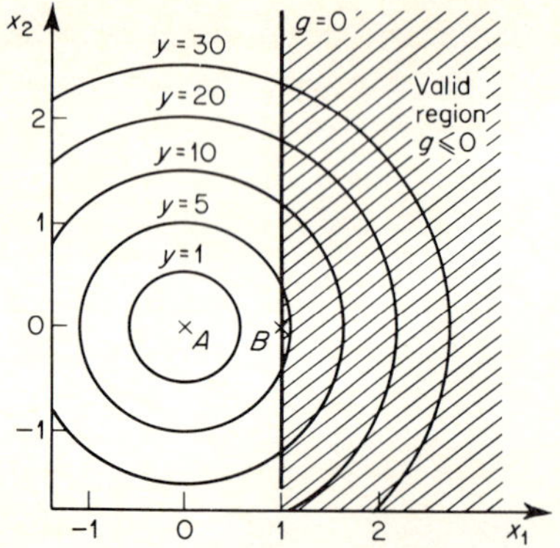

Fig. VII-11 The function $y = 4x_1^2 + 5x_2^2$ subject to $g = 1 - x_1 \leq 0$ (see Examples VII-19 and VII-20). A is unrestricted minimum, and B is restricted minimum.

Since $dy/dx_2 = 10x_2$ and $d^2y/dx_2^2 = 10$, this has a minimum value at $x_2 = 0$. The desired optimum value will therefore lie in the position $x_1 = 1$, $x_2 = 0$, on the system boundary, as shown in Fig. VII-11. It will be noted that the restricted minimum $y(1,0) = 4$ is greater than the unrestricted minimum $y(0,0) = 0$.

4.1.2 *Optimization by modification of the restriction equations—slack variables*

In contrast to the previous method of initially ignoring the restrictions and optimizing an unrestricted function, optimization can also be carried out by converting each inequality relationship into an equality restriction, with subsequent optimization of a function subject to a set of equality constraints.

This modification of the inequality restriction may be carried out by the introduction of a "slack" variable or function, one slack variable being introduced for each inequality constraint in the system. If we now introduce a nonnegative slack function S_i for each restriction, then each inequality restriction may be converted to an equality as follows:

$$\phi_i = g_i + S_i = 0$$

Each function S_i depends upon a single slack variable x_{n+i}, which is an additional variable required solely for the purpose of modifying a restriction.

The simplest relation is of the form

$$S_i = x_{n+i}$$

but an additional restriction is required that $x_{n+i} \geq 0$. This form will be used later, in Sec. 4.4, in the discussion of linear programming. A more suitable form for our purpose might be

$$S_i = x^2_{n+1}$$

which is valid for all values of the slack variable. This function S_i has two properties: it vanishes when the slack variable vanishes (that is, $g_i = 0$) and it is always nonnegative. If the slack function vanishes, the restriction forms a boundary along which the solution to the problem must lie, whereas if $S_i > 0$, the solution will lie away from the boundary (that is, $g_i < 0$), either in the interior of the valid region or at another boundary.

The introduction of slack variables allows the original problem to be transformed into another, subject only to equality restrictions of the form

$$\phi_i = 0$$

although this can be achieved only at the expense of adding more variables to the optimization. The slack variable or function is used to "take up the slack" permitted by the inequality and to yield an equality.

The problem now fits into a category already discussed (Sec. 3), and one of the four methods developed previously can now be applied. An example using the method of direct substitution will be discussed, while the techniques of constrained variation and Lagrangian multipliers will be left to succeeding sections (Secs. 4.2 and 4.3) due to their importance.

Example VII-20: The use of slack variables—a boundary solution

Determine the minimum of y, where

$$y = 4x_1^2 + 5x_2^2$$

subject to the restriction that

$$x_1 \geq 1$$

This can be rewritten

$$g_1 = 1 - x_1 \leq 0$$

Using a slack function $S_1(x_3)$, this inequality constraint can be transformed to

$$\phi_1 = g_1 + S_1 = 1 - x_1 + x_3^2 = 0$$

Using the method of direct substitution (Sec. 3.1), we can eliminate x_1 from the objective function, getting

$$y = 4(1 + x_3^2)^2 + 5x_2^2$$

This may now be minimized freely, since there are no bounds imposed on

x_2 or x_3 due to the original restriction. Hence, in this modified problem,

$$\frac{\partial y}{\partial x_2} = 10x_2 = 0$$

$$\frac{\partial y}{\partial x_3} = 16x_3(1 + x_3^2) = 0$$

for a stationary point to exist. This will thus be found at the point $x_2 = 0$, $x_3 = 0$, so that $x_1 = 1$ as before (Example VII-19). A boundary point in the original system has, however, been found, because the slack variable has vanished.

Let us now investigate the nature of this stationary value of $y(x_2,x_3)$. Thus

$$D_1 = y_{x_2x_2} = 10$$

$$D_2 = \begin{vmatrix} y_{x_2x_2} & y_{x_2x_3} \\ y_{x_3x_2} & y_{x_3x_3} \end{vmatrix} = \begin{vmatrix} 10 & 0 \\ 0 & 16 \end{vmatrix} = 160$$

Since D_1 and D_2 are positive, this stationary point is therefore a minimum of the modified problem and of the original system.

Example VII-21: The use of slack variables—an interior solution

Let us consider a variation of Example VII-20 in which the restriction is altered to

$$x_1 \leq 1 \qquad \text{or} \qquad g_2 = x_1 - 1 \leq 0$$

We should then form

$$\phi_2 = x_1 - 1 + x_4^2 = 0$$

getting

$$y = 4(1 - x_4^2)^2 + 5x_2^2$$

which has stationary values found from

$$y_{x_2} = 10x_2 = 0$$

$$y_{x_4} = -16x_4(1 - x_4^2) = 0$$

The possible locations for an optimum are thus at

$$x_2 = 0, \quad x_4 = 0 \quad \text{and} \quad x_2 = 0, \quad x_4 = \pm 1$$

Also

$$D_1 = 10$$

$$D_2 = -160(1 - 3x_4^2)$$

Here $D_2 < 0$ for $x_4 = 0$ and > 0 for $x_4 = \pm 1$. Thus the solution $x_2 = 0$, $x_4 = 0$, does not define an optimum, but the solution $x_2 = 0$, $x_4 = \pm 1$, defines a minimum lying at a stationary point. From the latter, upon substituting for $x_4^2 = 1$, the desired location is found to be the system origin $x_1 = 0$, $x_2 = 0$, with a minimum y value of zero. This may be compared with the previous problem, in which the reversed restriction ($x_1 \geq 1$) allowed only a restricted minimum of 4 to be attained.

4.2 Slack Variables and Constrained Variations

From the preceding section we have seen how an inequality restriction is converted to an equality before being used in the optimization.

Let us consider the problem of determining the optimum of a function, y, where

$$y = y(\mathbf{x}) \tag{4.2-1}$$

subject to m restrictions

$$g_k(\mathbf{x}) \leq 0 \qquad \text{for } k = 1, 2, \ldots, m \tag{4.2-2}$$

These may be converted to m equality restrictions by the use of m slack functions S_k, each of which is a nonnegative function of a single (slack) variable, x_{n+k}. Thus we have

$$\phi_k = g_k + S_k(x_{n+k}) = 0 \qquad \text{for } k = 1, 2, \ldots, m \tag{4.2-3}$$

The method of constrained variations may now be used to carry out the optimization. From Sec. 3.2, the necessary conditions for a stationary value in y to exist are that

$$J_k\left(\frac{y, \phi_1, \ldots, \phi_m}{x_k, x_1, \ldots, x_m}\right) = 0 \qquad \text{for } k = m+1, m+2, \ldots, n \tag{4.2-4}$$

provided

$$J\left(\frac{\phi_1, \phi_2, \ldots, \phi_m}{x_1, x_2, \ldots, x_m}\right) \neq 0$$

(Sec. 3.2.2). These conditions will, of course, be valid only if all restrictions are included as equalities in Eq. (4.2-4).

Thus it is important to note that these necessary equations need not be satisfied if linear slack functions

$$S_k(x_{n+k}) = x_{n+k}$$

were selected, since we should have to impose the additional restriction that

$$x_{n+k} \geq 0$$

The solution might then lie on the boundary $x_{n+k} = 0$, and such a point need not be a stationary point. However, by choosing the slack function to be

$$S_k(x_{n+k}) = x_{n+k}^2$$

this problem is avoided, since no restrictions on the x_{n+k} need be introduced, and there is therefore no boundary on x_{n+k} at zero. The necessary conditions [Eq. (4.2-4)] will then be valid. For this reason, the quadratic choice is often more convenient (see, for example, Dorn [1963]). An exception is made, however, in the case of linear systems (Sec. 4.4).

Example VII-22: Constrained variations and the slack variable

Find the maximum of y, where

$$y(x_1,x_2) = x_2^2 - 2x_1 - x_1^2$$

subject to the inequality constraint

$$g_1 = x_1^2 + x_2^2 - 1 \leq 0$$

This constraint limits the values of x_1 and x_2 to those falling within the unit circle, as shown in Fig. VII-12. The contours of y are also shown. The region in which the maximum value of y is found will be seen to lie between the contours $y = 1$ and $y = 2$. The precise location of this maximum can be found by defining a slack variable x_3 such that

$$\phi_1 = x_1^2 + x_2^2 - 1 + x_3^2 = 0 \tag{4.2-5}$$

Thus the method of constrained variation requires that

$$J_2\left(\frac{y,\phi_1}{x_2,x_1}\right) = 0 = J_3\left(\frac{y,\phi_1}{x_3,x_1}\right)$$

Now $$J_2\left(\frac{y,\phi_1}{x_2,x_1}\right) = \begin{vmatrix} y_{x_2} & y_{x_1} \\ \phi_{1_{x_2}} & \phi_{1_{x_1}} \end{vmatrix} = \begin{vmatrix} 2x_2 & -2 - 2x_1 \\ 2x_2 & 2x_1 \end{vmatrix} = 0$$

Therefore $$4x_1x_2 + 2x_2(2 + 2x_1) = 0$$

or $$x_2(1 + 2x_1) = 0 \tag{4.2-6}$$

and $$J_3\left(\frac{y,\phi_1}{x_3,x_1}\right) = \begin{vmatrix} y_{x_3} & y_{x_1} \\ \phi_{1_{x_3}} & \phi_{1_{x_1}} \end{vmatrix} = \begin{vmatrix} 0 & -2 - 2x_1 \\ 2x_3 & 2x_1 \end{vmatrix} = 0$$

giving $$4x_3(1 + x_1) = 0 \tag{4.2-7}$$

provided $J(\phi/x_1) \neq 0$, that is, $x_1 \neq 0$.

From Eqs. (4.2-6) and (4.2-7) and the restriction equation (4.2-5), we may locate the stationary points shown in Table VII-2. Since $x_3 = 0$ in every case, it will be seen that each of the restricted stationary points lies on the restriction

$$g_1 = 0$$

represented by the unit circle in Fig. VII-12, and each is, by definition, a boundary point. Thus, for the first stationary value $\{1,0\}$, the only

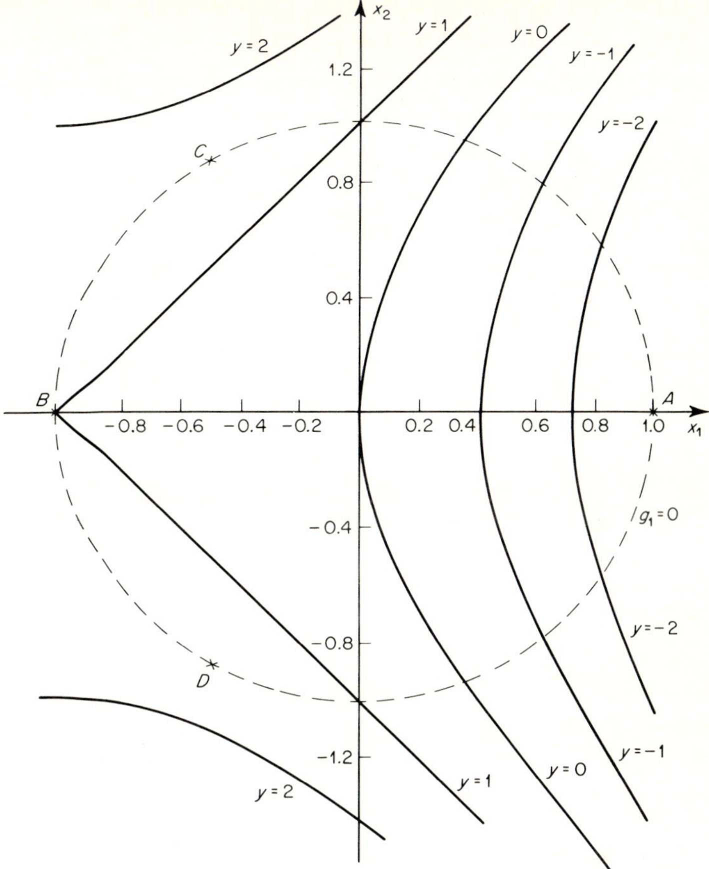

Fig. VII-12 Contour plot for Example VII-22.

permitted changes of dx_1 will be $dx_1 \leq 0$, since otherwise the change $dx_1 > 0$ would allow the point to move out of the acceptable region ($g_1 \leq 0$). In determining the character of these stationary points, we need therefore consider only admissible displacements. These displacements will be limited in direction for any stationary value lying on a boundary, whereas if the latter had fallen within the valid region, any displacements would be feasible.

From the contour plot of Fig. VII-12, the nature of the stationary point may be determined as summarized in Table VII-2, there being a saddle point, a minimum, and two equal maxima.

4.2.1 *The nature of the stationary point*

In the preceding example, the stationary points were investigated graphically. Such graphical methods are quite convenient for investigating the

TABLE VII-2 LOCATION AND NATURE OF THE STATIONARY POINTS IN EXAMPLE VII-22

	Stationary point			
	A	*B*	*C*	*D*
x_1	1.00	−1.00	−0.50	−0.50
x_2	0	0	0.87	−0.87
x_3	0	0	0	0
λ†	2.00	0	−1.00	−1.00
y	−3.00	1.00	1.50	1.50
Type of stationary value	Min	Saddle point	Max	Max

† The Lagrangian multiplier λ is used in Example VII-24.

character of the stationary values if the problem can be defined by one or two independent variables. When, however, the dimensionality of the problem is increased, graphical methods rapidly become impossible, and we must look for help in purely mathematical arguments. In general, these arguments are identical with those already established for the case of restricted problems having equality constraints. Such an analysis will, however, require considerable care if the correct conclusion is to be inferred.

After the necessary conditions for an optimum have been stated and any stationary points determined, the locations of these points may be grouped into two distinct categories, internal points and boundary points. A *boundary point* is defined as a stationary point for which at least one slack variable vanishes. This condition implies, of course that at least one constraint is satisfied as an equality. Thus, in Example VII-22, four stationary points were found, all being boundary points, since the slack variable x_3 vanished in each case. An *internal point*, on the other hand, will be found if all the slack variables are nonzero, so that the stationary point lies within the valid region. The discussion of the nature of these stationary points will differ, depending upon the group under examination. This difference arises in the restrictions imposed by a boundary upon the admissible variations.

In the examination of an internal stationary point, for example, point *C* of Fig. VII-13, any or all of the system variables may be altered by an infinitesimal amount at will. Thus there will be admissible variations dx_i in any variable x_i, no matter which direction is chosen. The sufficiency analysis presented for unrestricted functions (Sec. 2.2.3) will therefore be directly applicable. The nature of any such stationary point (with no zero slack variables) can then be determined by applying the criteria summarized in Eqs. (2.2-11) and (2.2-12).

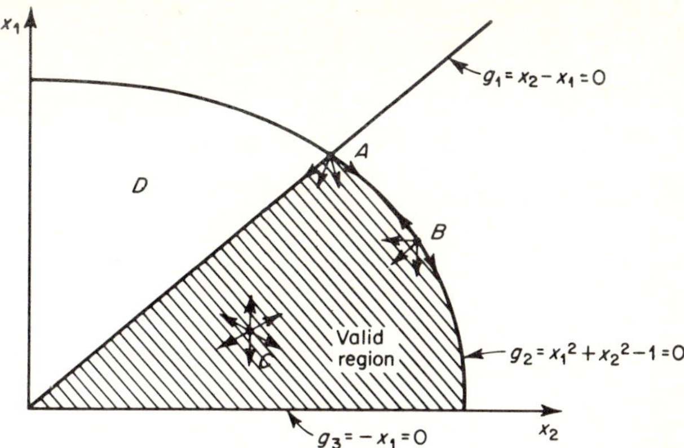

Fig. VII-13 Feasible movements in a restricted problem with a valid region defined by $g_i \leq 0$.

If the stationary point is a boundary point, all possible variations dx_i in x_i are not permitted, since certain of the variations would then fall outside the valid region. Some general variations about a boundary point are depicted in Fig. VII-13. Here the valid region is shown as a shaded region with one internal stationary point, C, and two boundary stationary points, A and B. Clearly, any variation about point C is admissible, but only certain variations about A and B are permitted. Only movement into the shaded region will be allowed. It will be obvious that the general criteria for internal stationary points cannot be used here for testing boundary points, since these criteria depended upon completely arbitrary variations of all variables. Thus we may consider the stationary point A of Fig. VII-14*a*. If the point is an internal stationary point, examination of the sufficiency conditions shows that it is a saddle point. On the other hand, if A is a boundary stationary point, the restriction shown by the dotted line prevents access to the "high" on the right, and A will now be a restricted maximum.

In addition, we cannot use the constrained variation sufficiency conditions on the original problem, since we may use only admissible variations which lie along the boundary. Thus, in Fig. VII-14*b*, the boundary stationary point B is a minimum if the admissible variations about B are confined to the boundary restrictions, but if the variations are permitted over the valid region as shown by arrows, B will be a saddle point.

The sufficiency conditions using constrained variations under inequality constraints may be easily obtained by considering not the original problem in $(x_1, x_2, \ldots, x_n)$, but the transformed problem in $(x_1, x_2, \ldots, x_n, \ldots, x_{n+m})$, which includes the m slack variables x_{n+k} $(k = 1, 2, \ldots, m)$. Since there are no limitations imposed upon the latter problem if the slack functions are of the form

$$S_k = x_{n+k}^2$$

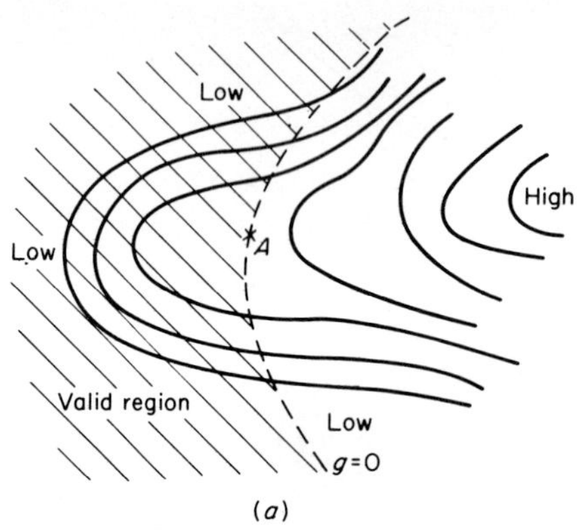

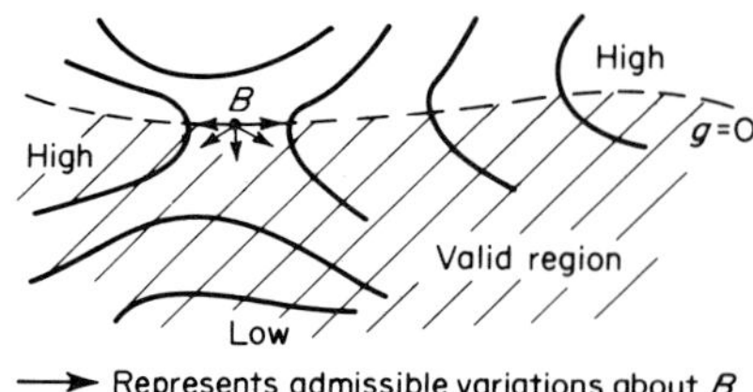

Fig. VII-14 Restricted regions. (*a*) A spur maximum; (*b*) a saddle point.

the methods of Sec. 3.2 can be applied to the transformed problem of determining the nature of y at the stationary points where

$$y = y(x_1, x_2, \ldots, x_{n+m})$$

subject to the m equality restrictions

$$\phi_k = g_k + x_{n+k}^2 = 0 \qquad \text{for } k = 1, 2, \ldots, m$$

This allows us to find and characterize any stationary point.

Example VII-23: Sufficiency conditions and the slack variable

Let us examine the nature of the stationary points found in Example VII-22. Here

$$y = x_2^2 - 2x_1 - x_1^2$$
$$\phi = x_1^2 + x_2^2 - 1 + x_3^2 = 0$$

with stationary values at the four locations given in Table VII-2.

Since $n = 3(x_1, x_2, \text{and } x_3)$ and $m = 1$, we have from Eq. (3.2-15)

$$\Delta^2 y = \frac{1}{2!} \sum_{j=2}^{3} \sum_{k=2}^{3} \left(\frac{\partial^2 y}{\partial x_j \, \partial x_k} \right)_{\underline{\phi}} h_j h_k$$
$$= \frac{1}{2!} \left[\left(\frac{\partial^2 y}{\partial x_2^2} \right)_{\underline{\phi}} h_2^2 + 2 \left(\frac{\partial^2 y}{\partial x_2 \, \partial x_3} \right)_{\underline{\phi}} h_2 h_3 + \left(\frac{\partial^2 y}{\partial x_3^2} \right)_{\underline{\phi}} h_3^2 \right]$$

We must now evaluate the terms $(\partial^2 y/\partial x_j\, \partial x_k)_{\underline{\phi}}$, using Eqs. (3.2-19), (3.2-20), (3.2-22), and (3.2-23). Then

$$\left(\frac{\partial^2 y}{\partial x_2^2}\right)_{\underline{\phi}} = \frac{\partial y}{\partial x_1}\left(\frac{\partial^2 x_1}{\partial x_2^2}\right)_{\underline{\phi}} + \frac{\partial^2 y}{\partial x_1^2}\left(\frac{\partial x_1}{\partial x_2}\right)_{\underline{\phi}}^2 + 2\,\frac{\partial^2 y}{\partial x_1\,\partial x_2}\left(\frac{\partial x_1}{\partial x_2}\right)_{\underline{\phi}} + \frac{\partial^2 y}{\partial x_2^2}.$$

$$\left(\frac{\partial^2 y}{\partial x_2\,\partial x_3}\right)_{\underline{\phi}} = \frac{\partial y}{\partial x_1}\left(\frac{\partial^2 x_1}{\partial x_2\,\partial x_3}\right)_{\underline{\phi}} + \frac{\partial^2 y}{\partial x_1^2}\left(\frac{\partial x_1}{\partial x_2}\right)_{\underline{\phi}}\left(\frac{\partial x_1}{\partial x_3}\right)_{\underline{\phi}} + \frac{\partial^2 y}{\partial x_1\,\partial x_3}\left(\frac{\partial x_1}{\partial x_2}\right)_{\underline{\phi}} + \frac{\partial^2 y}{\partial x_1\,\partial x_2}\left(\frac{\partial x_1}{\partial x_3}\right)_{\underline{\phi}} + \frac{\partial^2 y}{\partial x_2\,\partial x_3}$$

$$\left(\frac{\partial^2 y}{\partial x_3^2}\right)_{\underline{\phi}} = \frac{\partial y}{\partial x_1}\left(\frac{\partial^2 x_1}{\partial x_3^2}\right)_{\underline{\phi}} + \frac{\partial^2 y}{\partial x_1^2}\left(\frac{\partial x_1}{\partial x_3}\right)_{\underline{\phi}}^2 + 2\,\frac{\partial^2 y}{\partial x_1\,\partial x_3}\left(\frac{\partial x_1}{\partial x_3}\right)_{\underline{\phi}} + \frac{\partial^2 y}{\partial x_3^2}$$

Also

$$\left(\frac{\partial x_1}{\partial x_2}\right)_{\underline{\phi}} = -\frac{J(\phi/x_2)}{J(\phi/x_1)} = -\frac{x_2}{x_1}$$

$$\left(\frac{\partial x_1}{\partial x_3}\right)_{\underline{\phi}} = -\frac{J(\phi/x_3)}{J(\phi/x_1)} = -\frac{x_3}{x_1}$$

provided that $J(\phi/x_1) \neq 0$, that is, $x_1 \neq 0$.

$$d_1^{2,2} = -\frac{\partial^2 \phi}{\partial x_1^2}\left(\frac{\partial x_1}{\partial x_2}\right)_{\underline{\phi}}^2 - \frac{\partial^2 \phi}{\partial x_2^2} = -2\left[\left(\frac{x_2}{x_1}\right)^2 + 1\right]$$

$$d_1^{2,3} = -\frac{\partial^2 \phi}{\partial x_1^2}\left(\frac{\partial x_1}{\partial x_2}\right)_{\underline{\phi}}\left(\frac{\partial x_1}{\partial x_3}\right)_{\underline{\phi}} - \frac{\partial^2 \phi}{\partial x_2\,\partial x_3} = -2\,\frac{x_2 x_3}{x_1^2}$$

$$d_1^{3,3} = -\frac{\partial^2 \phi}{\partial x_1^2}\left(\frac{\partial x_1}{\partial x_3}\right)_{\underline{\phi}}^2 - \frac{\partial^2 \phi}{\partial x_3^2} = -2\left[\left(\frac{x_3}{x_1}\right)^2 + 1\right]$$

$$\left(\frac{\partial^2 x_1}{\partial x_2^2}\right)_{\underline{\phi}} = \frac{d_1^{2,2}}{\phi_{x_1}} = -\frac{1}{x_1}\left[\left(\frac{x_2}{x_1}\right)^2 + 1\right]$$

$$\left(\frac{\partial^2 x_1}{\partial x_2\,\partial x_3}\right)_{\underline{\phi}} = \frac{d_1^{2,3}}{\phi_{x_1}} = -\frac{x_2 x_3}{x_1^3}$$

$$\left(\frac{\partial^2 x_1}{\partial x_3^2}\right)_{\underline{\phi}} = \frac{d_1^{3,3}}{\phi_{x_1}} = -\frac{1}{x_1}\left(\frac{x_3^2}{x_1^2} + 1\right)$$

Hence

$$\left(\frac{\partial^2 y}{\partial x_2^2}\right)_{\underline{\phi}} = \frac{2}{x_1}\left[1 + 2x_1 + \left(\frac{x_2}{x_1}\right)^2\right]$$

$$\left(\frac{\partial^2 y}{\partial x_2\,\partial x_3}\right)_{\underline{\phi}} = \frac{2x_2 x_3}{x_1^3}$$

$$\left(\frac{\partial^2 y}{\partial x_3^2}\right)_{\underline{\phi}} = \frac{2}{x_1}\left[1 + x_1 + \left(\frac{x_3}{x_1}\right)^2\right]$$

TABLE VII-3 EVALUATION OF THE SECOND DERIVATIVE FOR EXAMPLE VII-23

	Point			
	A	B	C	D
$\left(\dfrac{\partial^2 y}{\partial x_2^2}\right)_{\underline{\phi}}$	6	2	−12	−12
$\left(\dfrac{\partial^2 y}{\partial x_2\,\partial x_3}\right)_{\underline{\phi}}$	0	0	0	0
$\left(\dfrac{\partial^2 y}{\partial x_3^2}\right)_{\underline{\phi}}$	4	0	−2	−2

Let us evaluate these three terms at each of the four stationary points already located (as summarized in Table VII-2), getting the values shown in Table VII-3. The usual sufficiency equations may now be applied to $\Delta^2 y$, getting a minimum for point A, since

$$D_1 = 6 > 0 \qquad D_2 = (6)(4) - 0^2 = 24 > 0$$

and maxima for points C and D, since

$$D_1 = -12 < 0 \qquad D_2 = (-12)(-1) - 0^2 = 12 > 0$$

For point B, $D_1 = 2 > 0$, and $D_2 = (2)(0) - 0^2 = 0$, which is inconclusive. In fact, $\Delta^2 y$ reduces to $\frac{1}{2}(4h_2^2)$ which is not positive-definite, but nonnegative. This point is thus not a maximum, but as discussed in Sec. 2.2-1, it could be a minimum or a saddle point. As seen in Fig. VII-12, it is actually the latter; otherwise, we would need to examine higher derivatives.

The nature of each stationary point can thus be fully determined.

4.3 Lagrangian Multipliers and a Kuhn-Tucker Theorem

The method described in the preceding section used slack variables to transform the inequalities to equalities, allowing us to find the stationary points by the method of constrained variation. Once the constraints are in this form, the method of Lagrangian multipliers can also be applied. This approach may be preferred, since the value of the multiplier can be of some assistance in determining the character of the stationary point. This aspect of the problem will be developed in some detail.

4.3.1 Necessary conditions

Before examining the question of characterizing the stationary point, let us summarize the techniques of finding them. Recalling that, for equalities, one must seek the stationary values of the function

$$F = y + \sum_{k=1}^{m} \lambda_k \phi_k \tag{4.3-1}$$

if the problem is to determine the optimum value of $y(\mathbf{x})$ subject to m constraints,

$$\phi_k = g_k + S_k(x_{n+k}) = 0 \qquad \text{for } k = 1, 2, \ldots, m \tag{4.3-2}$$

g_k being the original inequality

$$g_k \leq 0$$

and S_k the nonnegative slack function (Sec. 4.1.2), a function of the slack variable x_{n+k}, F is thus a function of $n + m$ variables.

The restricted stationary points, those satisfying Eqs. (4.3-2), can then be found by solving the $n + 2m$ equations

$$\frac{\partial F}{\partial x_j} = 0 \qquad \text{for } j = 1, 2, \ldots, (n + m) \tag{4.3-3}$$

and

$$\frac{\partial F}{\partial \lambda_k} = 0$$

which is equivalent to

$$\phi_k = 0 \qquad \text{for } k = 1, 2, \ldots, m \tag{4.3-4}$$

This is, of course, the same method as was discussed in Sec. 3.3.

Example VII-24: The Lagrangian multiplier and the slack variable

What is the maximum value of y (Examples VII-22 and VII-23), where

$$y = x_2^2 - 2x_1 - x_1^2$$

if the region of search is restricted by

$$g = x_1^2 + x_2^2 - 1 \leq 0$$

or

$$\phi = x_1^2 + x_2^2 - 1 + x_3^2 = 0$$

where x_3 is the slack variable added to convert the constraint to an equality?

Defining

$$\begin{aligned} F &= y + \lambda\phi \\ &= x_2^2 - 2x_1 - x_1^2 + \lambda(x_1^2 + x_2^2 - 1 + x_3^2) \end{aligned}$$

we may form

$$\frac{\partial F}{\partial x_1} = 0 \qquad \frac{\partial F}{\partial x_2} = 0 \qquad \frac{\partial F}{\partial x_3} = 0$$

getting

$$1 + x_1 - \lambda x_1 = 0$$
$$x_2(1 + \lambda) = 0$$
$$x_3\lambda = 0$$

and from $\phi = 0$, we have

$$x_1^2 + x_2^2 + x_3^2 = 1$$

We then have the four possible solutions to this system of four equations and four unknowns, these stationary values being described in Table VII-2.

We shall consider the character of these stationary points in Sec. 4.3.2, but from Example VII-23 we see that there is one minimum (A), two equal maxima (C and D), and one saddle point (B).

Let us now examine this illustration in more detail. First of all, it is clear that a slack variable can appear only in one equation of the set of necessary conditions, and that equation must have the structure

$$x_{n+k}\lambda_k = 0 \tag{4.3-5}$$

This equation arises when differentiating F with respect to the slack variable, and has this form due to the fact that the slack variable appears only once in F and is always a quadratic multiplied by a Lagrangian multiplier. Because of this structure, it is clear that a nonzero value of the Lagrangian multiplier means that the solution is necessarily found on the boundary. Of course, the solution may still lie on the boundary if the Lagrangian multiplier vanishes. One can say, however, that the stationary point can be found only in the interior if the Lagrangian multiplier vanishes. A nonzero Lagrangian multiplier implies that the solution is on the boundary. If we remember these two conclusions, the slack variables need not be introduced. Thus, in future calculations using Lagrangian multipliers, the slack variables will be omitted. This new formulation is the basis for the method of Kuhn and Tucker [1951], some aspects of which will now be discussed.

In setting up the F function in Example VII-24, we thus only need write that

$$F = y + \lambda g$$

and carry out the search for the stationary points as before. This approach may be extended to any number of restrictions.

A second point is also suggested by the results shown in Table VII-2, since here the sign of the Lagrangian multiplier appears to be related to the character of the associated stationary point. Let us examine this in Sec. 4.3.2.

4.3.2 Additional necessary conditions with a single restriction

Let us now look at the results of Example VII-24 shown in Table VII-2. It will be seen that the Lagrangian multiplier is positive for case A, in which the stationary point was a minimum, and is negative when the stationary point is a maximum (cases C and D). These are not chance observations, since it is possible to prove more generally that if $\lambda > 0$, the stationary point cannot be a maximum, and if $\lambda < 0$, it cannot be a minimum. The argument may be presented as follows.

Suppose that there exists a stationary point of F for which $\lambda > 0$ (when using a single restriction). This point $\mathbf{x}_0$ must lie on the boundary, since the Lagrangian multiplier is nonzero. Therefore, since the point is on the boundary,

$$g(\mathbf{x}_0) = 0 \tag{4.3-6}$$

and only displacements which give

$$dg \leq 0 \tag{4.3-7}$$

are admissible. Suppose that there exists a set of displacements dx_i such that

$$dg = \sum_{i=1}^{n} \left(\frac{\partial g}{\partial x_i}\right)_0 dx_i < 0 \tag{4.3-8}$$

The interpretation of this relation will now be discussed using Fig. VII-15. In Fig. VII-15*a* an infinitesimal move away from the boundary stationary point in the direction of one of the arrows will result in a decrease in g (a move from boundary into interior), and therefore each represents a displacement which could be used in Eq. (4.3-8). This is not the case for the stationary point shown in Fig. VII-15*b*. Here there is no possible infinitesimal move from the stationary point into the *interior*, i.e., one for which g decreases. Therefore the statements to follow do not apply to this second point. The term *accessible stationary point* may be defined to mean a stationary point such as A for which there exists at least one set of displacements from a boundary for which the constraining function decreases. An inaccessible stationary point such as B is one for which no such set of displacements exists. At the restricted stationary point

$$\left(\frac{\partial F}{\partial x_i}\right) = 0 \qquad \text{for } i = 1, 2, \ldots, n \tag{4.3-9}$$

where

$$F = y + \lambda g$$

This then gives, for $i = 1, 2, \ldots, n$,

$$\left(\frac{\partial y}{\partial x_i}\right)_0 = -\lambda\left(\frac{\partial g}{\partial x_i}\right)_0 \tag{4.3-10}$$

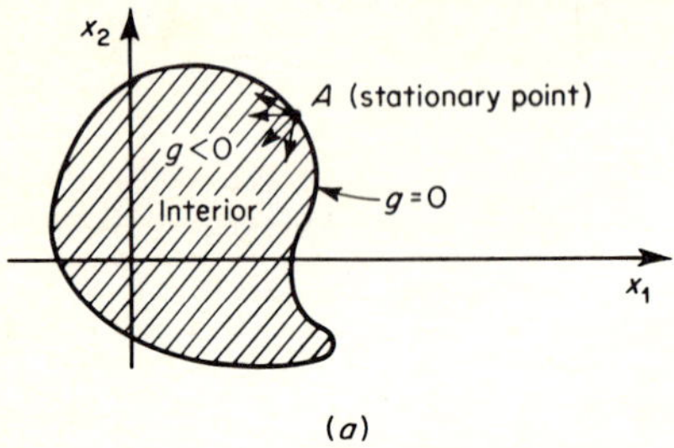

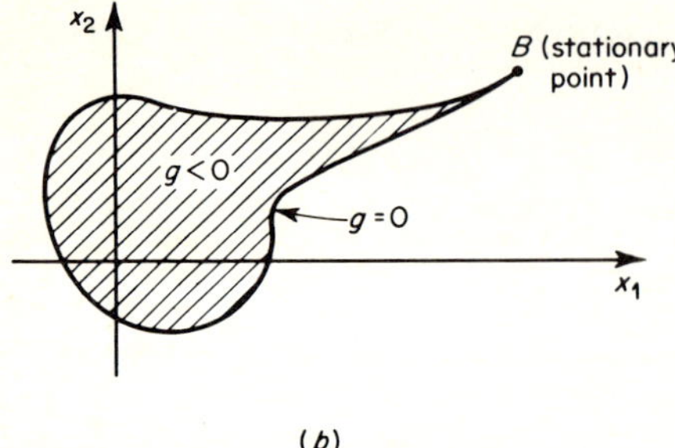

Fig. VII-15 Accessible regions in two dimensions.

Using Eqs. (4.3-10), one can write that

$$\sum_{i=1}^{n} \left(\frac{\partial y}{\partial x_i}\right)_0 dx_i = -\lambda \sum_{i=1}^{n} \left(\frac{\partial g}{\partial x_i}\right)_0 dx_i \tag{4.3-11}$$

or

$$dy = -\lambda\, dg$$

If the restricted stationary point is also an accessible stationary point, there exists a displacement such that

$$dy = -\lambda\, dg > 0 \qquad \text{for positive } \lambda \tag{4.3-12}$$

Consequently, an admissible move dx_i into the interior has resulted in an increase in y. Therefore the stationary point on the boundary could not have been a maximum, since there are points in the interior which yield larger values of y. Therefore, for the accessible restricted point to be a maximum, it is necessary but not sufficient for the Lagrangian multiplier to be nonpositive. Similarly, a necessary condition for an accessible stationary point to be a minimum is that the Lagrangian multiplier be nonnegative. Of course, it should be noted that these are necessary conditions but they are not sufficient. Thus, in Example VII-24, the stationary point associated with $\lambda = 0$ (case B) was neither a maximum nor a minimum.

TABLE VII-4 ADDITIONAL NECESSARY CONDITIONS FOR A SINGLE RESTRICTIONS $g \leq 0$

Sign of Lagrangian multiplier λ	*Stationary point*	
	Location	*Nature*
Nonzero, negative	On boundary	Not a minimum
Nonzero, positive	On boundary	Not a maximum
Zero	On boundary or interior of the valid region	Either a maximum or minimum or neither

If, however, the Lagrangian multiplier is negative-definite (Sec. 2.2.1), the accessible stationary point is clearly a maximum, since y decreases or remains constant for all admissible displacements about the stationary point.[1]

A summary of the conclusions which may be inferred about the nature of a stationary point is presented in Table VII-4, these results being due to Kuhn and Tucker [1951].

Example VII-25: Additional necessary conditions for a constrained minimum

Find the minimum of

$$y = 4(x_1 - a)^2 + 5(x_2 - b)^2$$

for given values of a and b if the choice of x_1 and x_2 is restricted by

$$x_1 + x_2 \geq 1$$

This may be rewritten

$$g_1 = 1 - x_1 - x_2 \leq 0$$

Forming the modified objective function F, where

$$F = y + \lambda_1 g_1$$

[1] These remarks are based upon the convention that

$$g(\mathbf{x}) \leq 0$$

If the restriction was in the form $g \geq 0$, then the sign of the Lagrangian multiplier would have to be reversed in the above discussion. The advantage in using the present convention lies in the parallel between the present relationship of Lagrangian multiplier sign and stationary-point character and that for the determinants D_1 and the nature of the stationary point when considering unrestricted functions (Sec. 2.2.3). Accordingly, any future discussion will continue with negatively arranged inequalities, while of course any inequalities of opposite sign can easily be converted by multiplying through by -1.

we can find the desired stationary points from the following equations:

$$\frac{\partial F}{\partial x_1} = 8(x_1 - a) - \lambda_1 = 0$$

$$\frac{\partial F}{\partial x_2} = 10(x_2 - b) - \lambda_1 = 0$$

and

$$g_1 \leq 0$$

We may now consider the following cases:

Case a $\lambda_1 \neq 0$, so that $g_1 = 0$ and

$$x_1 + x_2 = 1$$

$$x_1 - \frac{\lambda_1}{8} = a$$

$$x_2 - \frac{\lambda_1}{10} = b$$

Fig. VII-16 Solution of Example VII-25. (*a*) For $a + b < 1$ (required minimum is point *A*); (*b*) for $a + b \geq 1$ (required minimum is the global minimum *Q*).

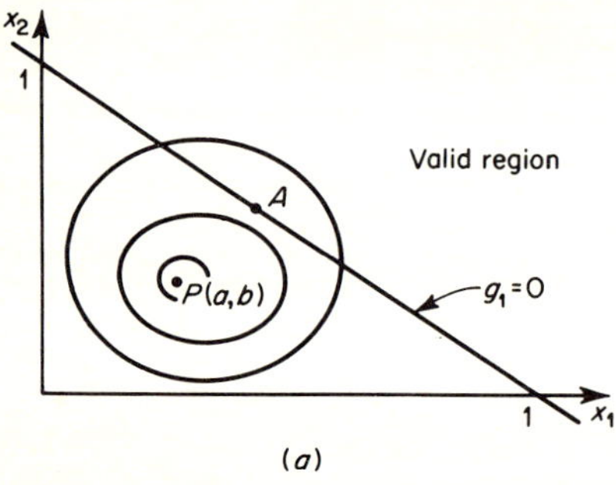

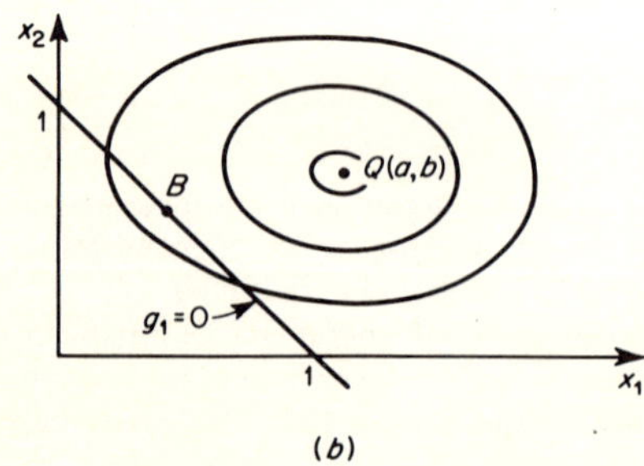

Solving, we get the boundary location $\mathbf{x}_j$, defined by

$$x_{1,j} = \tfrac{1}{9}(5 + 4a - 5b)$$
$$x_{2,j} = \tfrac{1}{9}(4 - 4a + 5b)$$
$$\lambda_1 = \tfrac{40}{9}(1 - a - b)$$

which is valid if $a + b \neq 1$ (otherwise $\lambda_1 = 0$, which would invalidate this solution). When $a + b < 1$, $\lambda_1 > 0$, and the stationary point $j = A$ cannot be a maximum. The present method cannot say, in the general case, that this point is a minimum. This has to be checked by other methods. It will be seen, as shown in Fig. VII-16*a*, that the global minimum $\mathbf{x}_p = \{a,b\}$ is unattainable. Similarly if $a + b > 1$, then $\lambda_1 < 0$, and the boundary point cannot be a minimum. It will be confirmed in Fig. VII-16*b* that the point $j = B$ is not a minimum, and we may conclude that the method does, therefore, allow us to neglect point B since a minimum cannot exist there.

Case b $\lambda_1 = 0$. For this case the point may lie on the boundary or in the interior of the valid region ($x_1 + x_2 \geq 1$). The stationary point which lies at $x_1 = a$; $x_2 = b$ can be attained only if $a + b \geq 1$. No information on the nature of the stationary point can be deduced from the present data. Point Q in Fig VII-16*b* represents a typical location of such a position, but it cannot be determined from the Lagrangian multiplier whether it is a maximum, a minimum, or neither. It is, in fact, the global minimum, attainable if $a + b \geq 1$.

4.3.3 Additional necessary conditions with more than one restriction

If there is more than one constraint, then the rules roughly correspond to those already developed. Our arguments thus far have demonstrated that if there is a single restriction $g \leq 0$, then the stationary point of the function

$$F = y + \lambda g$$

must be such that $\lambda \leq 0$ if y is to be a maximum, assuming that all stationary points on the boundary are accessible. This is a necessary but not sufficient condition. However, if $\lambda < 0$, the accessible stationary point cannot be a minimum.

In the general case, again due to Kuhn and Tucker [1951], the necessary conditions for the existence of an optimum at a certain point are:

1. The modified objective function

$$F = y + \sum_{k=1}^{m} \lambda_k g_k \tag{4.3-13}$$

must be stationary at that point.

2. $$\sum_{k=1}^{m} \lambda_k g_k = 0 \tag{4.3-14}$$

3. Either

$$\lambda_k \leq 0 \qquad \text{for } k = 1, 2, \ldots, m \tag{4.3-15a}$$

if the function is to be a maximum, or

$$\lambda_k \geq 0 \qquad \text{for } k = 1, 2, \ldots, m \tag{4.3-15b}$$

if the function is to be a minimum.

These necessary conditions must be satisfied, provided the boundary near the stationary point is such that the region is accessible. Kuhn and Tucker [1951] define an accessible region as follows: Let l be the number of constraints exactly satisfied at the stationary point [$g_i(\mathbf{x}^*) = 0$ for $i = 1, 2, \ldots, l$]; then the region is accessible if, for any set of displacements $dx_1, dx_2, \ldots, dx_n$, satisfying the inequalities

$$\sum_i \frac{\partial g_k}{\partial x_i} dx_i \leq 0 \qquad \text{for } k = 1, 2, \ldots, l$$

there exists a differentiable arc, tangent to dx_i at $\mathbf{x}^*$, which is contained in the admissible region. This definition rules out the type of region depicted, for example, in Fig. VII-15*b*. An expanded derivation of these Kuhn-Tucker conditions is given by Vajda [1961, chap. 12].

Example VII-26: The optima of a multiconstrained function (1)

Let us determine the minimum values of

$$y = x_1^3 + 2x_2^2 - 10x_1 + 6 + 2x_2^3$$

subject to the restrictions

$$x_1 x_2 \leq 10 \qquad x_1 \geq 0 \qquad x_2 \geq 0$$

or $\quad g_1 = x_1 x_2 - 10 \leq 0 \quad g_2 = -x_1 \leq 0 \quad g_3 = -x_2 \leq 0$

Form the function F so that

$$F = (x_1^3 + 2x_2^2 - 10x_1 + 6 + 2x_2^3) + \lambda_1(x_1 x_2 - 10) + \lambda_2(-x_1) + \lambda_3(-x_2)$$

If F is to be stationary, the following equations must be satisfied:

$$F_{x_1} = 3x_1^2 - 10 + \lambda_1 x_2 - \lambda_2 = 0$$

and $$F_{x_2} = 4x_2 + 6x_2^2 + \lambda_1 x_1 - \lambda_3 = 0$$

Let us first see if there are solutions such that $\lambda_1 \neq 0$, which means that $g_1 = 0$, or

$$x_1 x_2 = 10$$

Since neither x_1 nor x_2 can then vanish, g_2 and g_3 cannot be zero. Hence $\lambda_2 = 0$ and $\lambda_3 = 0$. Our set of equations thus reduces to

$$x_1 x_2 = 10 \tag{4.3-16}$$

$$3x_1^2 - 10 + \lambda_1 x_2 = 0 \tag{4.3-17}$$

$$4x_2 + 6x_2^2 + \lambda_1 x_1 = 0 \tag{4.3-18}$$

TABLE VII-5 NATURE OF THE STATIONARY POINTS OF EXAMPLE VII-26

	Stationary point				
	A	*B*	*C*	*D*	*E*
x_1	3.847	−3.568	0	1.054	−1.054
x_2	2.600	−2.802	0	0	0
λ_1	−13.238	10.063	0	0	0
λ_2	0	0	−10	0	0
λ_3	0	0	0	0	0
g_1	0	0	−10	−10	−10
g_2	−3.847	3.568	0	−1.054	1.054
g_3	−2.600	2.802	0	0	0
Kuhn-Tucker conclusion	Not a minimum	Not a maximum	Not a minimum	None	None
True character	Maximum	Invalid	Maximum	Minimum	Invalid

which can be solved to give the necessary values of x_1, x_2, and λ_1. Thus, from Eqs. (4.3-16) to (4.3-18), $x_1 = 10/x_2$; $\lambda_1 = -x_2(4x_2 + 6x_2^2)/10$, and $3{,}000 - 100x_2^2 - 4x_2^5 - 6x_2^6 = 0$, which can be solved to give the stationary points *A* and *B* of Table VII-5. However, in this case, since $x_2 > 0$, the second relationship shows that $\lambda_1 < 0$. This means, from Eq. (4.3-15*b*), that along $x_1 x_2 = 10$, all admissible stationary points will be precluded from being minimum values. It remains to be shown that any stationary point along this boundary is accessible. The region is clearly accessible, as depicted in Fig. VII-17. Here we see that an arc has been constructed so that it is tangent to a displacement and is contained in a feasible region. Thus stationary points lying along $x_1 x_2 = 10$ can be rejected without further calculation, since we have shown that they can never be minima.

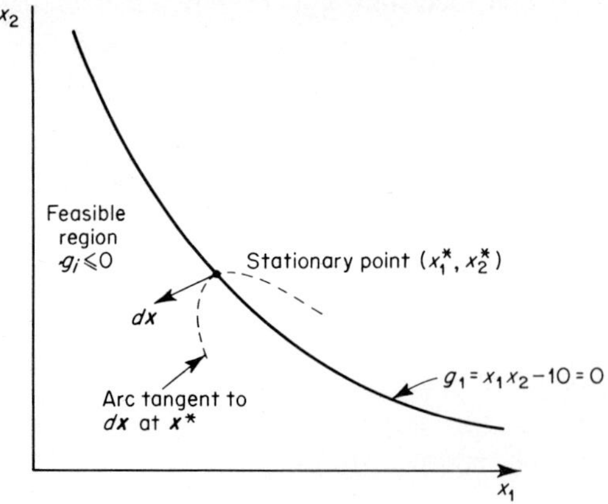

Fig. VII-17 Valid region for Example VII-26.

Let us now test $\lambda_1 = 0$. Then

$$3x_1^2 - 10 - \lambda_2 = 0 \tag{4.3-19}$$

$$4x_2 + 6x_2^2 - \lambda_3 = 0 \tag{4.3-20}$$

If $\lambda_3 \neq 0$, then $g_3 = 0$ or $x_2 = 0$. But from Eq. (4.3-20), $\lambda_3 = 0$, so that our original assumption of $\lambda_3 \neq 0$ is false. Thus $\lambda_3 = 0$ and $x_2 = 0$. If $\lambda_2 \neq 0$, then $g_2 = 0$ and $x_1 = 0$, and from Eq. (4.3-19), $\lambda_2 = -10$. Also, if $\lambda_2 = 0$, $g_2 \leq 0$ and $x_1 \geq 0$, so that, from Eq. (4.3-19), $3x_1^2 = 10$, giving $x_1 = \pm 1.054$. Only the positive root will, however, satisfy the second restriction. The required minimum can lie only at one of these stationary values. The conclusions drawn about the nature of each stationary point are summarized in Table VII-5 (points A to C).

Example VII-27: The optima of a multiconstrained function (2)

Let us investigate the optima of the function

$$y = 3x_1^2 - 2x_1 - 5x_2^2 + 30x_2$$

subject to the restrictions

$$g_1 = 8 - 2x_1 - 3x_2 \leq 0 \tag{4.3-21}$$

$$g_2 = x_2 - 5 \leq 0 \tag{4.3-22}$$

$$g_3 = 3x_1 + 2x_2 - 15 \leq 0 \tag{4.3-23}$$

Forming the modified objective function F, where

$$F = y + \lambda_1 g_1 + \lambda_2 g_2 + \lambda_3 g_3$$

the stationary values will be found from the equations

$$\frac{\partial F}{\partial x_1} = 6x_1 - 2 - 2\lambda_1 + 3\lambda_3 = 0 \tag{4.3-24}$$

$$\frac{\partial F}{\partial x_2} = -10x_2 + 30 - 3\lambda_1 + \lambda_2 + 2\lambda_3 = 0 \tag{4.3-25}$$

We must now search for a set of x_1, x_2, λ_1, λ_2, λ_3, which satisfy these two equations and the three inequality constraints, remembering that if $\lambda_i \neq 0$, then $g_i = 0$.

The search for the valid stationary points of the F function is best organized in terms of the various combinations of zero and nonzero λ_i, as in Table VII-6, showing eight possible combinations, a to h. Each case must be considered separately.

For case a, all three Lagrangian multipliers are nonzero, and therefore $g_i = 0$ for $i = 1$, 2, and 3. Such a solution, however, does not exist, since, as shown in Fig. VII-18, the three restrictions do not intersect at a common point.

In case b, $\lambda_1 = 0$ and λ_2, λ_3 are nonzero. Thus

$$g_2 = g_3 = 0$$

These two restrictions intersect at a single point (point B on Fig. VII-18). The point of intersection is $x_1 = 1.67$ and $x_2 = 5.00$. Solving Eqs.

Fig. VII-18 Contours of y for Example VII-27.

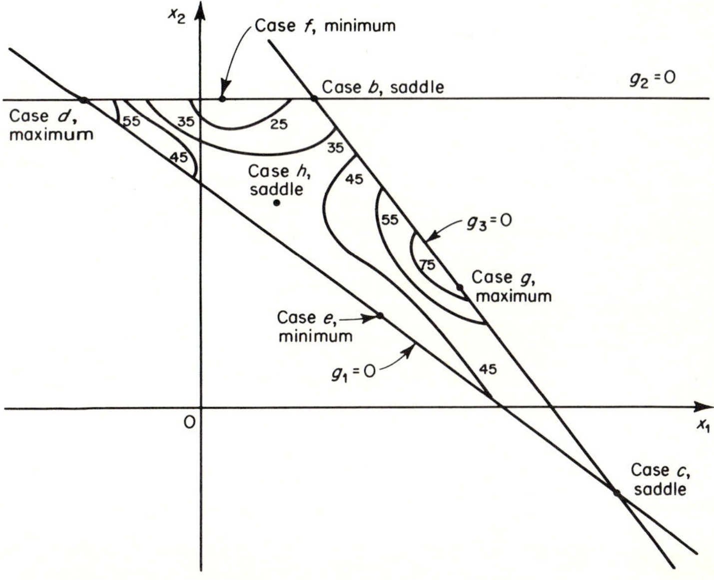

TABLE VII-6 CALCULATION SUMMARY OF STATIONARY-POINT EVALUATION—EXAMPLE VII-27

Case	λ_1	λ_2	λ_3	Equality constraint	Location	Numerical values x_1	x_2	λ_1	λ_2	λ_3	y	Kuhn-Tucker conclusions	True character
a	≠	≠	≠	g_1, g_2, g_3	Invalid			—	—	—	—	—	—
b	0	≠	≠	g_2, g_3	Boundary	1.67	5.00	0	25.33	−2.67	30.0	Not a maximum or minimum	Saddle point
c	≠	0	≠	g_1, g_3	Boundary	5.80	−1.20	12.08	0	−2.88	46.2	Not a maximum or minimum	Saddle point
d	≠	≠	0	g_1, g_2	Boundary	−3.50	5.00	−11.50	−14.50	0	58.7	Not a minimum	Maximum
e	≠	0	0	g_1	Boundary	2.71	0.86	7.01	0	0	38.6	Not a maximum	Minimum
f	0	≠	0	g_2	Boundary	0.33	5.00	0	20.00	0	2.47	Not a maximum	Minimum
g	0	0	≠	g_3	Boundary	3.97	1.55	0	0	−7.27	73.8	Not a minimum	Maximum
h	0	0	0	None	Interior	0.33	3.0	0	0	0	44.7	None	Saddle point

(4.3-24) and (4.3-25) for λ_2 and λ_3, we find the results given in Table VII-6.

Proceeding, similarly, through the other cases, we come to case h. Here the λ_i are all zero, and we have no direct information about the g_i. However, solving Eqs. (4.3-24) and (4.3-25) yields

$$x_1 = 0.33 \qquad \text{and} \qquad x_2 = 3.00$$

and since $g_1 = -1.66$, $g_2 = -2.0$, $g_3 = -8.0$, this is an interior point in the admissible region.

The results of the case studies are summarized in Table VII-6, and also shown in the table are the conclusions regarding the character of each stationary point, which can be inferred by examining the signs of the Lagrangian multipliers. It is seen that certain possibilities can be excluded, but since the conditions are necessary rather than sufficient, positive statements cannot be made.

The true character of each stationary point must finally be determined using the method of constrained variation, together with slack variables.

The preceding example clearly indicates the power of the Kuhn-Tucker theorem in rejecting certain stationary points as possible optima. In most cases the signs of the Lagrangian multipliers are enough to deduce the character of the stationary point, although there is grave danger in accepting such conclusions without continued justification, since the proved conditions are necessary but not sufficient conditions.

Theorems which yield sufficiency conditions have been developed by Kuhn and Tucker [1951] and others for specific cases. One such case, for example, requires a concave form of the objective function. Our presentation has been based upon a general theorem, and is therefore not as restrictive nor as informative.

4.4 Linear Objective Functions and Inequality Restrictions (Linear Programming)

A special case of optimization arises when both the objective function and the constraining restrictions are linear with respect to the independent variables. The problem would then be represented by the set of expressions

$$y = \sum_{i=1}^{n} c_i x_i \tag{4.4-1}$$

subject to the m restrictions that[1]

$$\sum_{i=1}^{n} a_{j,i} x_i \leq b_j \qquad \text{for } j = 1, 2, \ldots, l \tag{4.4-2a}$$

[1] The restriction that the x_i must be nonnegative is not restrictive. Thus, if in practice $x' \geq -10$, the transformation $x = x' + 10$ would allow the feasible values $x \geq 0$ to be used. The nomenclature convention in this section follows common usage.

and $$\sum_{i=1}^{n} a_{j,i}x_i = b_j \qquad \text{for } j = l+1, l+2, \ldots, m \tag{4.4-2b}$$

and $x_i \geq 0$ for $i = 1, 2, \ldots, n$.

Such linear forms might arise from the use of simplified models either directly in their formulation or by linearization of more complex situations (Chap. III, Sec. 2.4.2). This problem has attracted much attention, since the applications are widespread, and very large problems, of several hundred or thousand independent variables, can be handled. This aspect of optimization is commonly known as *linear programming*.

With this type of system it should be noted that the desired extreme value will not only lie at some boundary, but must be situated at an intersection of boundaries.

Example VII-28: Optima at a boundary

Consider the linear function

$$y(x_1,x_2,x_3) = ax_1 + bx_2 + cx_3$$

The partial derivatives are

$$\frac{\partial y}{\partial x_1} = a \qquad \frac{\partial y}{\partial x_2} = b \qquad \frac{\partial y}{\partial x_3} = c$$

There are therefore no values of x_i, which make the partial derivatives vanish—a necessary condition for an internal optimum. Since there are no discontinuities in the function provided a, b, c are constant, the optimum can only lie at a boundary.

Hence, to get an extreme value of this type of function, we must impose boundary restrictions; e.g., the range of independent variables may be given as

$$0 \leq x_1 \leq 1 \qquad 0 \leq x_2 \leq 1 \qquad 0 \leq x_3 \leq 1$$

We should therefore expect the extreme value to lie at the intersections of the constraining planes. This may be illustrated for the one-dimensional case in Fig. VII-19, where $y = ax + b$ is a maximum at $x = x^L$ and a minimum at $x = x_L$ if $a > 0$.

If it is known in advance that the optima must lie on one or more of the boundary restrictions, the search can be confined to these boundaries. Special methods have thus arisen to handle these boundary investigations.

This is a special case of the general constrained problem, since in the previous case it was not known whether or not the desired optimum would lie within the interior of a region or at a boundary.

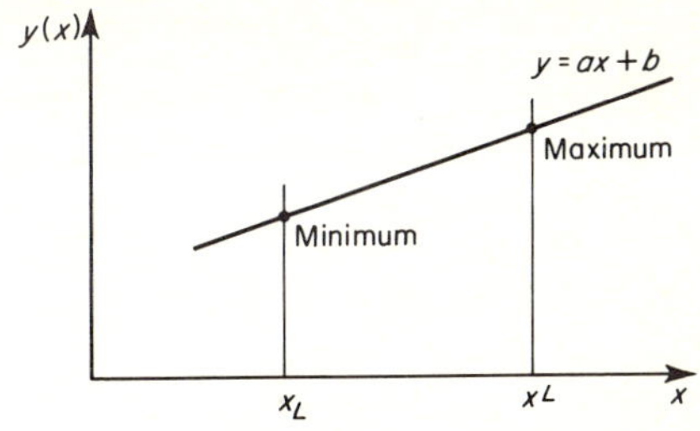

Fig. VII-19 A linear objective function.

4.4.1 *Graphical solution*

We shall first consider a graphical solution of a maximization problem in two dimensions to illustrate the nature of the problem.

Example VII-29: Two-dimensional linear function

Suppose it is required to choose values of x_1 and x_2 to maximize the function

$$y = 2x_1 - x_2 \tag{4.4-3}$$

x_1 and x_2 being subject to the constraints

$$x_1 \geq 0 \qquad x_2 \geq 0 \tag{4.4-4a}$$

$$-3x_1 + 2x_2 \leq 2 \tag{4.4-4b}$$

$$2x_1 - 4x_2 \leq 3 \tag{4.4-4c}$$

$$x_1 + x_2 \leq 6 \tag{4.4-4d}$$

The permitted region in the (x_1,x_2) plane, the area bounded by the lines corresponding to these constraints (as equalities), is shown in Fig. VII-20. The function

$$y = 2x_1 - x_2$$

is a family of parallel lines which increase in the direction indicated by the arrow, and it is clear that this function takes its largest value in the permitted region at the vertex P of the polygon forming its boundary. This is the point [found from the intersection of Eqs. (4.4-4*c*) and (4.4-4*d*)], where $x_1 = 4.5$, $x_2 = 1.5$.

This simple graphical procedure is always possible for two variables, but even in three variables it is very difficult to construct models of the polyhedra representing the permitted regions, and with more than three variables no geometrical construction is possible.

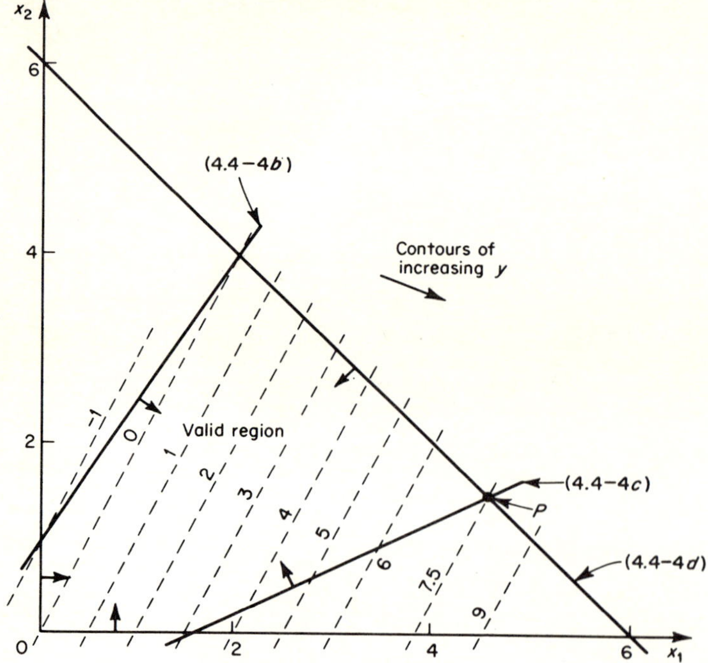

Fig. VII-20 A contour plot for Example VII-29 (arrows indicate valid side of the restriction).

It is therefore necessary to find a formal method. This will be illustrated below for the case of two variables (although the method is generally applicable), so that the analytical results may be followed graphically.

4.4.2 *Logical methods of solving the linear problem*

It has been seen that the linear problem can be easily solved graphically if there are only two independent variables.

Solution is possible by several methods (see, for example, Metzger [1958]), all of which are essentially direct search techniques in which the boundaries are systematically examined. Here attention will be focused only upon the best-known approach, that called the *Simplex method*, first developed in 1947 by Dantzig [1949; 1951a, p. 330], the mathematical origin and subsequent developments being traced in Dantzig [1963].

4.4.3 *The Simplex method*

In general, the Simplex method is an iterative process, in which the calculation is begun with a set of values for the independent variables that satisfy the restrictions, and in which each iterative cycle seeks another set of values that improve the desired objective function. This procedure is continued until the best objective function attainable is found. It is, essentially, a trial-and-error solution, but the procedure is so formulated that a systematic mathematical treatment is followed in order to reduce the labor involved. Before describing the Simplex method, let us carry out a solution algebraically.

Algebraic Solution Let us now indicate the algebraic basis to the Simplex method: at each stage in the calculation a generalized statement will be put forward, and then illustrated in terms of the simplified two-dimensional problem given in Example VII-29.

Step 1 General problem with equality restrictions We know from Sec. 4.1 that an equality can be made to include the case of an inequality restriction by the use of slack variables. For example, if a restriction takes the form

$$a_{j,1}x_1 + a_{j,2}x_2 + \cdots + a_{j,n}x_n \leq b_j$$

a slack variable can be introduced so that an equality is obtained. Thus, if x_{n+j} is added, with the understanding that it is nonnegative, the above expression becomes

$$a_{j,1}x_1 + a_{j,2}x_2 + \cdots + a_{j,n}x_n + x_{n+j} = b_j$$

If the original problem involved n variables and m constraints, of which l were inequalities, we should add l slack variables, and the problem would now be one of optimizing

$$y = \sum_{i=1}^{N} c_i x_i$$

subject to $$\sum_{i=1}^{N} a_{j,i}x_i = b_j \qquad \text{for } j = 1, 2, \ldots, m$$

a problem in N variables and m restrictions, some of the c_i and $a_{j,i}$ coefficients being zero or unity, with $N = n + l$.

Example VII-30: Conversion to equalities

The two-dimensional problem of Example VII-29 will now be carried forward in steps throughout the section to illustrate the approach. The three constraints are

$$-3x_1 + 2x_2 \leq 2 \qquad (4.4\text{-}4b)$$

$$2x_1 - 4x_2 \leq 3 \qquad (4.4\text{-}4c)$$

$$x_1 + x_2 \leq 6 \qquad (4.4\text{-}4d)$$

First introduce three additional slack variables, x_3, x_4, x_5, so that the restrictions can be written in the form of equalities

$$-3x_1 + 2x_2 + x_3 = 2 \qquad (4.4\text{-}5a)$$

$$2x_1 - 4x_2 + x_4 = 3 \qquad (4.4\text{-}5b)$$

$$x_1 + x_2 + x_5 = 6 \qquad (4.4\text{-}5c)$$

and the objective function can be written

$$y = 2x_1 - x_2 + 0(x_3 + x_4 + x_5) \qquad (4.4\text{-}6)$$

Our problem is now to find the nonnegative values of x_i, for $i = 1, 2, \ldots, 5$, in order to maximize y, subject to the set of restrictions given by Eqs. (4.4-5).

We therefore wish to optimize

$$y = \sum_{i=1}^{N} c_i x_i \qquad (4.4\text{-}7)$$

subject to the m constraints

$$\sum_{i=1}^{N} a_{j,i} x_i = b_j \qquad \text{for } j = 1, 2, \ldots, m \qquad (4.4\text{-}8)$$

and the additional restriction that all x_i are nonnegative. We purposely have introduced the convention of denoting the dimensionality of the system by N rather than by n variables, so that slack variables can be included where required.

Step 2 Choice of a basic solution—a valid boundary location We have already seen (Sec. 3.2) that in any linear problem involving m restrictions, when the functions involve N variables, $N - m$ of the variables are thus independent and can be chosen arbitrarily, the other m variables being no longer independent, but completely defined by the restrictions in terms of the $N - m$ independent variables.

Let us choose any $N - m$ variables so that they are zero (the reason for the choice will be seen later); then the other m variables will be completely determined. Such a choice of variables, in the language of linear programming, is termed a *basic solution* of the constraining equations. If these m variables are also nonnegative, the set of N variables is then a permissible or *feasible* solution to the problem, satisfying all the constraints. The m determined variables are called the *basis* of the basic solution.

It will be shown below that the procedure for determining the optimum value of the objective function (which must always lie at a boundary) is, first, to find any feasible basic solution to the problem and then try to improve this solution, getting in turn a sequence of other basic solutions which yield improved values of the objective function. This procedure will be shown to be equivalent to moving along the boundary-restriction lines from vertex to vertex formed by the boundary planes. The search for an optimum can be begun at any vertex, defined by a basic solution, and it does not matter which vertex or basic solution is used as a starting point as long as it is feasible. The example below shows a basic solution lying at the origin of the original problem, and the first basic solution can be evaluated on inspection. Later in the text we shall consider the problem of finding a first basic solution when its values are not obvious.

Example VII-30a: A basic solution representing a vertex

For the simple system given by Eqs. (4.4-5) it is possible to find the first basic solution by inspection. Here we set $N - m$ (that is, $5 - 3 = 2$) variables to zero, and get nonnegative values for the remaining m (that is, 3) variables. Thus $x_1 = 0$, $x_2 = 0$, and $x_3 = 2$, $x_4 = 3$, $x_5 = 6$ provides such a solution. Another possible set might be $x_1 = 2$, $x_2 = 4$, $x_3 = 0$, $x_4 = 15$, $x_5 = 0$. In these, x_3, x_4, x_5 form the basis for the first solution, and x_1, x_2, x_4 the basis for the second.

Referring to the graphical representation of this example (Fig. VII-20), it will be seen that the first solution corresponds to the origin, a vertex, and the second to the intersection of restrictions [Eqs. (4.4-4*b*) and (4.4-4*d*)], another vertex of the permitted region. Let us use the first set of x_i above as the first basic solution to the problem.

It is, in fact, a general property of the system that if we determine such a set of x_i (with m nonnegative values and $N - m = n$† zero values), this solution represents a point at a vertex of the n-dimensional permitted region.

We have said that an extreme value can lie only at particular points within the region or at a boundary. With a linear function, the optimum will always lie on a boundary. It is easily shown that since we have linear restriction equations, the extreme value can normally lie only at the intersection of two or more restrictive relations. Thus, in two dimensions, we may have a situation as shown in Fig. VII-20. Since y increases in the direction shown, the extreme value will lie along (c), for example, but since it increases along it, the best value as far as (c) is concerned is at one end of its valid region, i.e., at the intersection with (d). Only in the special case of the y lines being parallel to the restriction (c) should we get the same value of y along the (c) line. In this case we should be able to account for this result by noting that at each end of (c) we should find the same value of y. In other words, we do

† Because of the introduction of slack variables, the original dimensionality n of the system is obscure. For clarification, let us recall that l slack variables have been introduced and that N is the total number of variables, $n + l$. Due to a possible mixture of restrictions totaling m, there are $N - m$ degrees of freedom ($= n'$, say) so that $n' \leq n$. To avoid complexity we shall use n for n' here.

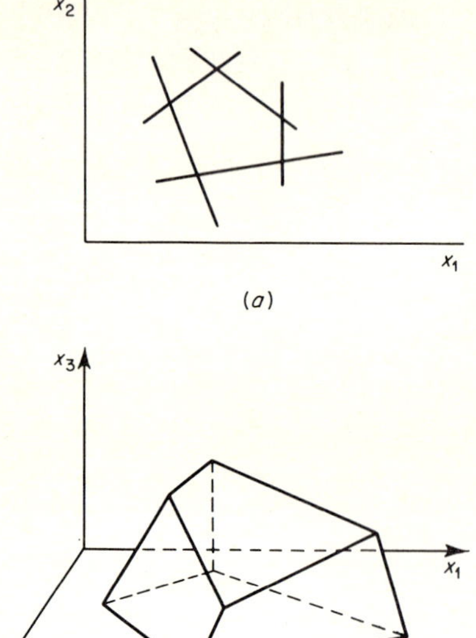

Fig. VII-21 The structure of vertices. (*a*) Polygon in two dimensions; (*b*) polyhedron in three dimensions.

not need to search all the boundaries when we are dealing with linear restrictions; we need only look at the vertices of the valid region.

It will be seen that in a two-dimensional problem the vertex is defined by the intersection of two linear restrictions; in three dimensions by three linear restrictions (Fig. VII-21); and in n-dimensions, the vertex is given by the intersection of n surfaces. It was the purpose of obtaining valid solutions, "basic" solutions lying at a vertex, that we originally choose $N - m$ values of x_i to be zero. Thus n slack variables are zero, allowing $N - m = n$ linear equality restrictions to intersect. The other m variables were taken as non-negative. Once the new N-dimensional problem has been set up, no further distinction is made between the original system variables and the slack variables.

Step 3 Rearrangement of restrictions and objective function At this stage it is convenient to rearrange the restrictions and the objective function so that the objective function and the m variables of the basis are expressed only in terms of the remaining $N - m$ variables, which were set equal to zero in the first basic solution.

This can be done by solving the m constraint equations for the m variables of the basis.

Example VII-30b: Rearrangement in seeking an improved solution

This rearrangement is easily carried out in Example VII-30, since only one variable of the basis appears in each of Eqs. (4.4-5). Thus we get

$$x_3 = 2 + 3x_1 - 2x_2 \tag{4.4-9a}$$

$$x_4 = 3 - 2x_1 + 4x_2 \tag{4.4-9b}$$

$$x_5 = 6 - x_1 - x_2 \tag{4.4-9c}$$

and

$$y = 0 + 2x_1 - x_2 \tag{4.4-10}$$

where the left-hand side of the first three equations represents the variables in the basis of the first basic solution.

Step 4 Changes in x_i to improve the objective function in seeking a new vertex It will be seen from the preceding step that since the nonbasis variables are zero, the variables in the basis still have their values, as given by the first basic solution. Let us now alter one of the nonbasis variables in order to try to improve the objective function and yet satisfy the restrictions. The choice of the most suitable variable is found by considering the objective function as written in step 3. If it is desirable to increase the objective function, we shall look for a variable which, when it increases in value (it can only increase since it starts at zero and cannot be negative), will improve the objective function. This will be the case for any variable in the objective function which has a positive coefficient. If several variables have positive coefficients, we could choose the variable with the largest coefficient, so that the goal is attained most rapidly, although there is no guarantee that this will be so. Conversely, if we wish to find the smallest value of the objective function, we should look for the most negative coefficient.

After deciding upon the nonbasic variable to be increased from its present value of zero, we must consider how much to increase it. This is necessary since it is essential to ensure that all the restrictions are still satisfied. This is carried out by checking that any increase in a nonbasic variable does not invalidate one of the restrictions by making a basis variable negative. It will be seen that if the nonbasis variable has a positive coefficient in the new form of the rearranged restriction equation, there will be no limit to its increase, but if the coefficient is negative, the increase will be limited by the ratio of the basis-variable value to this coefficient.

At the end of the change we shall have a new basis, a new basic solution, and a new value of the objective function.

Example VII-30c: Change in basis and the attainment of a new vertex

In the expression for the objective function, Eq. (4.4-10),

$$y = 0 + 2x_1 - x_2$$

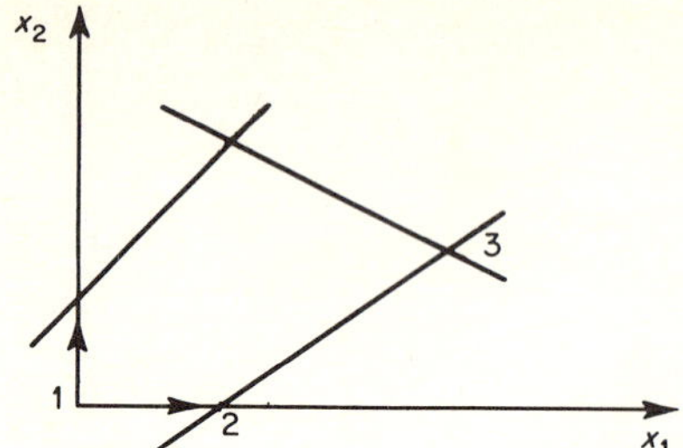

Fig. VII-22 Possible moves from a vertex (see Example VII-30*c*).

any increase in x_2 will decrease y, so that this variation offers no improvement. On the other hand, increasing x_1 will continually increase y, there being no limit on the increase so far. From the first restriction, as given by Eq. (4.4-9*a*), x_3 will be greater than zero for all increases in x_1; but from the second and third restrictions, Eqs. (4.4-9*b*) and (4.4-9*c*), $x_4 = 0$ when x_1 reaches 1.5 and $x_5 = 0$ when x_1 reaches 6. The second restriction therefore imposes a limit of $x_1 = 1.5$ on the choice of the new x_1. In this range of increase of x_1, x_3, x_4, and x_5 still have nonnegative values. During this increase, x_2 is held constant at zero.

At the end of this change of x_1, we have a new basic solution,

$$x_1 = 1.5 \qquad x_2 = 0 \qquad x_3 = 6.5 \qquad x_4 = 0 \qquad x_5 = 4.5$$

Let us call this the second basic solution.

It will be seen that the new basis is x_1, x_3, and x_5, with x_1 introduced and x_4 removed from the previous basis.

Graphically, it will be seen that (Fig. VII-22) the alternative choices of x_1 and x_2 correspond to the two possible routes of traveling from our starting vertex to one of the two adjacent vertices. The second basic solution corresponds to the vertex numbered 2 in the figure, the change from vertex 1 to vertex 2 having been along the boundary of the valid region.

In general, we shall have a choice of $N - m$ directions of movement from each vertex, these directions corresponding to the choice of an x_i from the nonbasis variables, and a movement along a restriction toward a new vertex; thus in two dimensions we have two choices of movement when all the restrictions are equalities.

Step 5 Repetition of the above procedure in seeking other vertices Having reached an adjacent vertex, represented by a new basic solution, we may repeat the same procedure, steps 3 and 4, until a final basic solution (and vertex) is found from which we can no longer move without worsening the final value of the objective function. This point will then be the location of the desired optimum value of the objective function. While illustrated for the two-dimensional case, the same algebraic procedure will lead us in the required direction each time from a vertex, and it is not limited to cases which have a geometric representation in two or three dimensions.

Example VII-30d: Continuation of the search

We may continue with the foregoing illustration, repeating steps 3 and 4 until the desired optimum value of the objective function is reached.

Cycle 2, step 3 Rearrange Eqs. (4.4-9) to give an equivalent set expressing the variables of the new basis in terms of the remaining $N - m$ variables (x_2,x_4), which at present are zero. Thus, from Eq. (4.4-9*b*), we have

$$x_1 = \tfrac{1}{2}(3 - x_4 + 4x_2)$$

Substituting into the remaining pair of equations and the objective function, we get

$$x_1 = 1.5 + 2x_2 - 0.5x_4 \qquad (4.4\text{-}11a)$$

$$x_3 = 6.5 + 4x_2 - 1.5x_4 \qquad (4.4\text{-}11b)$$

$$x_5 = 4.5 - 3x_2 + 0.5x_4 \qquad (4.4\text{-}11c)$$

$$y = 3 + 3x_2 - x_4 \qquad (4.4\text{-}12)$$

Cycle 2, step 4 Examining Eq. (4.4-12) for y, we see that an increase in x_4 will decrease the value of y, whereas y could be further increased by increasing the value of x_2 from zero. From Eqs. (4.4-11), this increase in x_2 is limited to a final value of 1.5, since, then, x_5 will be zero and a further increase will make it negative. Setting $x_2 = 1.5$, we thus obtain a third basic solution corresponding to the third vertex (marked 3) of the geometrical representation (Fig. VII-22). Hence basic solution 3 is

$$x_1 = 4.5 \qquad x_2 = 1.5 \qquad x_3 = 12.5 \qquad x_4 = 0 \qquad x_5 = 0$$

The new basis is now x_1, x_2, and x_3, with x_2 added and x_5 removed from the basis.

Cycle 3, step 3 Rearranging Eqs. (4.4-11) and (4.4-12) to express y and the new basis in terms of the new nonbasis variables (x_4 and x_5), we get, in the same way as before, from the third restriction,

$$x_2 = \tfrac{1}{3}(4.5 + 0.5x_4 - x_5)$$

so that

$$x_1 = 4.5 - 0.167x_4 - 0.667x_5 \qquad (4.4\text{-}13a)$$

$$x_2 = 1.5 + 0.167x_4 - 0.333x_5 \qquad (4.4\text{-}13b)$$

$$x_3 = 12.5 - 0.833x_4 - 1.33x_5 \qquad (4.4\text{-}13c)$$

$$y = 7.5 - 0.50x_4 - x_5 \qquad (4.4\text{-}14)$$

Cycle 3, step 4 No permitted change of x_4 or x_5 will improve y; hence this is the required location of the optimum value of y. The required solution to the original problem is therefore

$$y^* = 7.5 \quad \text{at} \quad x_1^* = 4.5 \quad \text{and} \quad x_2^* = 1.5$$

with the slack variables taking the values

$$x_3 = 12.5 \qquad x_4 = 0 \qquad x_5 = 0$$

The latter show merely how far the solution is from any particular constraint; thus $x_4, x_5 = 0$ mean that the second and third restrictions limit the required solution.

In practice, the algebraic procedure which has just been illustrated is not used in this form. The Simplex method presents a shorthand method of attaining the same procedure as the algebraic method, but does it in such a way, using a set of rules, that there is no need to consider the algebraic transformation, but only requires following an equivalent fixed procedure. Let us now develop this procedure and the rules of transformation.

The general proof of these rules can be derived using matrix theory, but for the present it is sufficient to compare the transformation rule at any stage with the corresponding step in the algebraic method; see, for example, Amundson [1966]. This present approach will, however, provide sufficient understanding of the Simplex method.

General Shorthand Procedure for the Above Algebraic Method—the Simplex Method

Step 1 As before, use slack variables to get all the m restrictions in the form of equalities.

Step 2 As before, choose an initial basic solution, of which m of the variables are nonnegative (the basis) and $N - m$ of the variables are zero. This determines the first vertex of the polyhedral valid region.

Step 3 Write the basis and the objective function in terms of the other $N - m$ nonbasis variables. This is most conveniently carried out by representing the equations by a tabulation of the coefficients of the variables in the form of an array or tableau. This corresponds to step 3 of the algebraic procedure. The use of the array is best explained in terms of a specific example.

Example VII-31: The Simplex representation

Repeating Example VII-30, Eqs. (4.4-9) and (4.4-10) can be rearranged as

$$x_3 - 3x_1 + 2x_2 = 2 \tag{4.4-15a}$$

$$x_4 + 2x_1 - 4x_2 = 3 \tag{4.4-15b}$$

$$x_5 + x_1 + x_2 = 6 \tag{4.4-15c}$$

$$y - 2x_1 + x_2 = 0 \tag{4.4-16}$$

and can be represented by the following array:

ARRAY 1

		x_1	x_2
x_3	2	−3	2
x_4	3	2	−4
x_5	6	1	1
y	0	−2	1

Here each element in the array represents the coefficient of the appropriate variable in the above equations. Thus, for example, the first row in the array corresponds to the first restriction equation, and so on, and the last row represents the objective function given by Eq. (4.4-16). It will be noted that the two variables x_1 and x_2, the nonbasis variables, are at present zero—this is a general property—and that the present values of the basis variables (in the left-hand column) and y are given by the first column of coefficients.

Step 4 In the algebraic procedure it was seen that, once the basis and the objective function have been written in terms of the nonbasis variables (equivalent to the present step 3), the *next stage* was the choice of a nonbasis variable to be altered, the choice of a size increase for this variable, and the calculation of the new basic solution. This procedure represents the movement from one vertex to another. Let us now develop the corresponding shorthand method for this stage in the calculation. In the shorthand method we wish to move from one array to another, since an array represents a basic solution, or vertex, the movement being carried out according to a set of rules of transformation which correspond to the algebraic transformations.

Step 4a Choice of variable to be added to the basis In the algebraic procedure, the process of deciding how to improve the value of the objective function was carried out by determining which nonbasis variable (in terms of which the objective function has already been expressed) could be altered in order to improve the objective function. This will be the variable which will enter the basis of the new basic solution. In seeking a maximum value of the objective function, we looked for a negative coefficient for the nonbasis variable (negative in the present arrangement of the equation) and for a minimum, a positive coefficient.

We must thus look for a negative element in the last line of our array during the search for a maximum, whereas for a minimum, we should look for a positive entry. It will be obvious by comparison with the algebraic procedure that an increase in the corresponding variable (which heads the column containing the chosen coefficient of the final row) from its value of zero will therefore increase (decrease) the objective function. In general, particularly with a large number of variables, there may be several negative elements in the final row of the array. In such a case it might be wise to choose the one with the largest magnitude and to introduce the corresponding variable into the basis; there is, however, no guarantee that this will lead to the final answer in the smallest number of cycles of calculation.

Step 4b Choice of variable to be removed from the basis It has been seen in the algebraic procedure that we moved from one basic solution to another in our search of the vertices for the optimum. In every case, the basis, while changing its components, still has only m variables in it. Thus, if we introduce a new variable into the basis, we must also take one out. By comparison with the algebraic procedure, we see that the variable which is removed is the variable in the basis whose restriction equation limits the increase of the variable being added to the basis.

It is easily seen that the limiting constraint which limits the extent to which the new basis variable (step 4*a*) can be increased is that for which the ratio of the element in the first column of coefficients to the element in the altered nonbasis-variable column takes the smallest positive ratio.

Example VII-31a: Choice of variables to be interchanged

Let us examine Array 1 of Example VII-31.

Step 4a We wish to maximize y; hence we shall choose the most negative coefficient in the last row of the array

		x_1	
y	0	−2	1
		↑	

We shall therefore choose the x_1 column and transfer x_1 into the basis for the new basic solution.

Step 4b Consider the ratios of the elements of the first column to those in the x_1 column, that is, 2:−3, 3:2, 6:1. Take the smallest positive ratio, that for the x_4 row. We shall therefore take x_4 out of the basis. This choice shows that the x_4 restriction limits the increase in size of x_1 in the search for a new basic solution. Our new basic solution will therefore contain the basis x_1, x_3, x_5 and the nonbasis variables x_2 and x_4, as was previously determined.

Step 4c Procedure for transforming the array Having determined the variables to be added and removed from the basis, it is seen from the algebraic method that new equations, representing the new basic solution, were obtained by solving the original equations for the nonbasis variable being transferred to the basis and substituting for this so as to express the new basis in terms of the other variables. By following through the transformations in detail, it is not difficult to see that the algebraic procedure can be reduced to a set of rules. These rules for the equivalent transformation of our first array into another array (corresponding to the new basic solution) are given below.

For convenience, let us draw a line through the row of coefficients in the original array corresponding to the basis variable being removed, and through the column of coefficients corresponding to the nonbasis variable being added to the basis. Call this row and column the *pivotal row* and *pivotal column*, respectively, any coefficient being represented by e_r or e_c, respectively. Let the coefficient at the intersection of the lines be called the *pivotal element*, represented by e_p.

Let us now consider the rules for transformation. All the calculations involve only coefficients in the *old* array, and the answer is placed in the corresponding position in the *new* array for any calculation based upon a particular location in the old array. No entries in the new array will affect any other entry, the rules applying only to locations not previously filled.

Step 4d Rules of transformation

1. Take the left-hand column of variables (not coefficients but the x_i) and the top row of variables, and write these in the corresponding positions in the new array, but interchange the positions of two of them—the variable to be added to the basis (from the top row) and the variable to be removed (from the column)—using steps 4*a* and 4*b*.
2. Calculate the reciprocal of the pivotal element ($= 1/e_p$) and place in the corresponding position in the new array.
3. For each coefficient in the pivotal row (not the pivot), calculate the value e_r/e_p and place in the appropriate location in the new array.
4. For each coefficient in the pivotal column (not the pivot), calculate the value $-e_c/e_p$ and place in the appropriate location in the new array.
5. For any unfilled location, calculate the value

$$e - \frac{e_r e_c}{e_p}$$

where the e_r is the value in the pivotal row in the same column as the element e, and e_c is the value in the pivotal column in the same row as e, and place in the corresponding position in the new array (Fig. VII-23).

Step 5 Repeat step 4, getting a new array in each cycle, until the required optimum objective function is attained. This is obtained when a new variable cannot be found to add to the basis (step 4*a*).

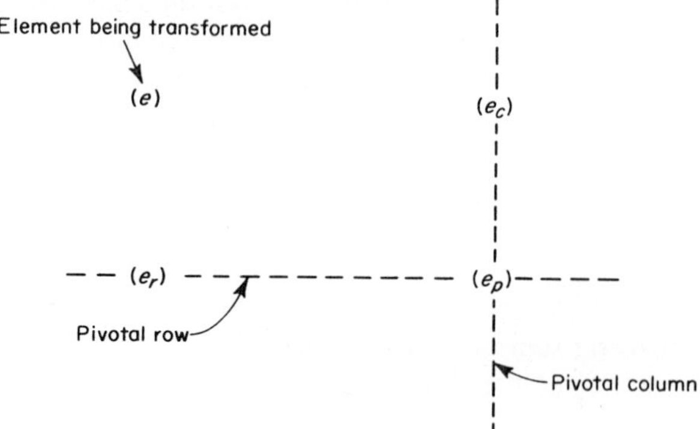

Fig. VII-23 Transformation of nonpivotal column-row elements.

Example VII-31b: Array transformation

Let us now complete the transformation of the first array shown in Example VII-31, according to the rules of step 4.

Steps 4*a* and *b* of Cycle 1 have been discussed in Example VII-31*a*, with variables x_1 and x_4 interchanged.

Step 4c Draw lines through the x_1 column and the x_4 row of coefficients. Then the term 2 at their intersection is the pivotal element.

Step 4d Construct the new array according to the rules:

2. New pivot element is $1/2 = 0.5$.
3. New pivotal row is 3/2 and $-4/2$, that is, 1.5 and -2.
4. New pivotal column is $-(-3/2)$, gap, $-(1/2)$, $-(-2/2)$, that is, 1.5, gap, -0.5, 1.
5. Other values: for example, the top left-hand coefficient has the new value $2 - (-3)(3/2) = 6.5$.

This procedure gives the new array:

ARRAY 2

		x_4	x_2
x_3	6.5	1.5	-4.0
x_1	1.5	0.5	-2.0
x_5	4.5	-0.5	3.0
y	3	1	-3.0

Let us now consider the same sequence of operations acting on array 2.

Cycle 2 The only negative element in the last row is -3; so bring x_2 into the next basis. The smallest positive ratio is that in the x_5 row; thus remove x_5 from the basis. Apply rules to get the third array:

ARRAY 3

		x_4	x_5
x_3	12.50	0.833	1.333
x_1	4.50	0.167	0.667
x_2	1.50	−0.167	0.333
y	7.5	0.50	1.0

All the elements of the last row are positive, so that no further change of basis can lead to an improvement of y. The final result is, thus, that the maximum $y^* = 7.5$ obtained at $x_1^* = 4.5$ and $x_2^* = 1.5$, which checks with the results found by other methods (Sec. 4.4.1).

Example VII-32: Simplex illustration

Let us consider the problem of maximizing y, where

$$y = 2x_1 - 3x_2 + 3x_3$$

is subject to the restrictions that $x_i \geq 0$ and

$$\begin{aligned} -2x_1 + 3x_2 &\leq 2 \\ x_1 - 2x_2 - 4x_3 &\leq 5 \\ 2x_1 + x_2 + x_3 &\leq 6 \\ x_1 + x_2 + 5x_3 &\leq 10 \end{aligned}$$

The solution using the Simplex procedure is as follows: Use slack variables to get the equalities

$$\begin{aligned} -2x_1 + 3x_2 \qquad + x_4 \qquad\qquad\qquad &= 2 \\ x_1 - 2x_2 - 4x_3 \qquad + x_5 \qquad\qquad &= 5 \\ 2x_1 + x_2 + x_3 \qquad\qquad + x_6 \qquad &= 6 \\ x_1 + x_2 + 5x_3 \qquad\qquad\qquad + x_7 &= 10 \end{aligned}$$

Then a first basic solution can be written $x_1 = 0$, $x_2 = 0$, and $x_3 = 0$, $x_4 = 2$, $x_5 = 5$, $x_6 = 6$, and $x_7 = 10$. The successive arrays can now be formed.

ARRAY 1

(Interchange x_3 and x_7 to get array 2.)

		x_1	x_2	x_3
x_4	2	−2	3	0
x_5	5	1	−2	−4
x_6	6	2	1	1
→ x_7	10	1	1	5
y	0	−2	3	−3
				↑

ARRAY 2

(Interchange x_3 and x_7 to get array 3.)

		x_1	x_2	x_7
x_4	2	−2.0	−3.0	0.0
x_5	13	1.8	−1.2	0.8
→ x_6	4	1.8	0.8	−0.2
x_3	2	0.2	0.2	0.2
y	6	−1.4	3.6	0.6
		↑		

ARRAY 3

		x_6	x_2	x_7
x_4	6.444	−1.111	3.889	−0.222
x_5	9.000	−1.000	−2.000	1.000
x_1	2.222	0.556	0.444	−0.111
x_3	1.556	−0.111	0.111	0.222
y	9.111	0.778	4.222	0.444

Since all elements of the last row are positive, no further change of basis can lead to an improvement of the objective function y. The final result is thus $y^* = 9.111$ at $x_1^* = 2.222$, $x_2^* = 0$, and $x_3^* = 1.556$.

It will be seen that the use of the algebraic transformation, or its equivalent Simplex representation, allows the boundaries to be systematically searched for the optimum by examining each vertex in turn. Since the objective function is linear, no optimum can be found unless restrictions are imposed upon the system, and once this is done, it is then known that the optimum will lie at these boundary constraints only. Indeed, it is not necessary to examine all the boundary planes, but only their intersection at the vertices.

A vertex can be found only at the intersection of n boundaries (in the n-dimensional problem), and this requires n variables to be zero. That this is indeed the case may be seen in the previous three-dimensional problem, where the first basic solution (which defines a vertex) has x_1, x_2, and x_3 set at zero, and the second vertex tested has x_1, x_2, and x_7 at zero, and the third and final vertex has x_2, x_6, and x_7 set at zero. In each case the setting of a slack variable equal to zero implies that the point under discussion is lying on a boundary. Thus, at the final vertex, $x_2 = 0$ puts the point on the x_1-axis boundary, while $x_6 = 0$ puts it on the restriction

$$2x_1 + x_2 + x_3 = 0$$

and $x_7 = 0$ puts it on

$$x_1 + x_2 + 5x_3 = 10$$

The final vertex thus lies at the intersection of these three restrictions.

Calculation checks can be introduced at each step in the manipulation. These are discussed, for example, by Vajda [1960] and Gass [1964].

Example VII-33: A mixture of inequality and equality restrictions

Let us maximize the linear function

$$y = x_1 + 3x_2 - x_3$$

subject to the two linear constraints

$$x_1 + 2x_2 + x_3 = 4$$
$$2x_1 + x_2 \leq 5$$

Adding a slack variable, x_4, to the second restriction, we have

$$2x_1 + x_2 + x_4 = 5$$

The first array may then be written as follows:

ARRAY 1

		x_1	x_2
→ x_3	4	1	2
x_4	5	2	1
y	−4	−2	−5
			↑

which transforms to the final array

ARRAY 2

		x_1	x_3
x_2	2	0.5	0.5
x_4	3	1.5	−0.5
y	6	0.5	2.5

The solution is then $y^* = 6$, where $x_1^* = 0$, $x_2^* = 2$, $x_3^* = 0$, the approach being no different from that for unmixed systems.

4.4.4 *Apparent difficulties arising in the Simplex method*

The general procedure in setting up the tables for the Simplex method of solving the problem with a linear objective function and linear restrictions and for carrying out the transformations from one array to another is normally straightforward. Several situations may arise, however, in which the procedure requires clarification or the results need an amplified interpretation (see, for example, Vajda [1956, 1960]).

The difficulties which might arise in the Simplex method are as follows:

1. In the choice of a *first basic solution*, none may be obvious or none may exist (see below).
2. Having constructed the first array, the variable to be introduced into the basis must be selected. This is normally carried out by selecting the most negative (for a maximum) or the most positive coefficient (for a minimum). Obviously, there can be no further improvements in the objective function if only coefficients of the opposite sign remain, but what happens if some of the coefficients are equal to each other or zero, no other choice remaining? (This yields a *multiple solution;* see page 312.)
3. The variable to be removed from the basis is chosen using the smallest positive ratio of the coefficients in the first column and the chosen column (corresponding to the variable being added to the basis). What should be done if there is no positive ratio or only zero values? What should be done if more than one identical value exists for the smallest positive ratio? (These yield *infinite* and *degenerate solutions*, respectively; see pages 316 to 318.)

Each situation will now be discussed in more detail.

Choice of First Basic Solution In the procedure outlined in Sec. 4.4.3, the problem could be started using any set of values of the variables which formed a basic solution. This basic solution with m nonvanishing positive variables (basis) and $N - m$ vanishing variables corresponds to a vertex of the figure formed by the m restrictions, and thus must contain $N - m$ variables which are zero, and m nonnegative variables (where N is the total number of system and slack variables).

A solution can be found by inspection In the simple problems introduced earlier, where the system of equations originated from inequalities and the right-hand side had the same sign as the slack variable, for example,

$$3x_1 - 8x_2 \leq 12$$

or

$$3x_1 - 8x_2 + x_3 = 12$$

so that the origin is one of the vertices, the basis is easily determined by setting the nonslack variables equal to zero, allowing the slack variables to take the values of the positive constants in the equations. Thus part of the first basic solution is found to be, here, $x_1 = 0$, $x_2 = 0$, $x_3 = 12$.

This type of feasible basic solution can be often found by inspection, as was seen in Example VII-30.

A solution exists, but is not obvious In many cases the location of the first vertex (the first basic solution) is not obvious. This is generally the case when the valid region defined by the restrictions does not include the origin as one of the vertices. Vajda [1956], for example, presents a systematic procedure to overcome this difficulty.

Let us introduce, as before, the slack variables, as necessary to obtain equality restrictions. These equations may now be rearranged so that the constants on the right-hand side of the equations are nonnegative. Normally, we shall find in most equations that there is one variable (usually, a slack variable) which does not appear in any other equation. Examine the coefficient of this variable in the rearranged equation. If all the coefficients are positive, our first feasible solution can be written, these variables forming the basis, as in the previous section.

Example VII-34: An easily found first solution

Find a valid boundary intersection satisfying the restrictions

$$\begin{aligned} -2x_1 + 3x_2 + x_3 \qquad\qquad &= 2 \\ 5x_1 + 2x_2 \qquad + x_4 \qquad &= 8 \\ 7x_1 + x_2 \qquad\qquad + x_5 &= 6 \end{aligned}$$

The variables x_3, x_4, x_5, each appear in only one equation; the right-hand sides are positive in each equation; and the coefficients of these x_i are positive. The first basic solution can then be written

$$x_1 = 0 \qquad x_2 = 0 \qquad x_3 = 2 \qquad x_4 = 8 \qquad x_5 = 6$$

the last three forming the basis. This represents one valid boundary intersection.

It may be found, however, that some of these variables do not have positive coefficients. (If we tried to form a basic solution, using them, we should find that some of the variables had negative values, which is not permitted.)

For these particular equations, let us adopt the following procedure, using *artificial* variables. First, add an artificial variable to the left-hand side of the equation; second, add a new term ($\pm P$ multiplied by the sum of all artificial variables) to the objective function, P being a large positive penalty number, larger than any other coefficient in the problem. The sign in front of this penalty term is chosen so that the presence of the penalty term will force the solution away from the desired optimum value unless the artificial variables are all zero (see also Sec. 3.4). Thus, for a maximum, use $-P$, and for a minimum, use $+P$. In this way, the artificial variable may be used under the normal rules in order to get a feasible solution, which will be reached when all the artificial variables have been found to be zero. The procedure will be obvious upon considering an example. Another method of obtaining the first basic and feasible solution is given by Vajda [1961, p. 9].

Example VII-35: Finding a basis which is not obvious

Find the maximum value of y, where

$$y = 6x_1 + 5x_2$$

subject to the restrictions

$$2x_1 + 5x_2 \leq 20 \tag{4.4-17a}$$

$$-5x_1 - x_2 \leq -5 \tag{4.4-17b}$$

$$-3x_1 - 11x_2 \leq -33 \tag{4.4-17c}$$

Graphically, this may be shown to be the triangle away from the origin, as in Fig. VII-24. If we now introduce the slack variables as before,

$$2x_1 + 5x_2 + x_3 = 20 \tag{4.4-18a}$$

$$-5x_1 - x_2 + x_4 = -5 \tag{4.4-18b}$$

$$-3x_1 - 11x_2 + x_5 = -33 \tag{4.4-18c}$$

it is easily seen that the origin is not a feasible point for a vertex of the permitted region, since this would require that two of the slack variables be negative; i.e., if $x_1 = 0$, $x_2 = 0$, then $x_3 = 20$, $x_4 = -5$, $x_5 = -33$. To handle the problem, rearrange all equations so that the right-hand sides are positive; this will then change Eqs. (4.4-18*b*) and (4.4-18*c*) to

$$5x_1 + x_2 - x_4 = 5$$

$$3x_1 + 11x_2 - x_5 = 33$$

We can then find in each equation a variable which does not appear in any other equation—here x_3, x_4, and x_5. Let us check their coefficients. The only negative coefficients belong to x_4 and x_5; so artificial variables

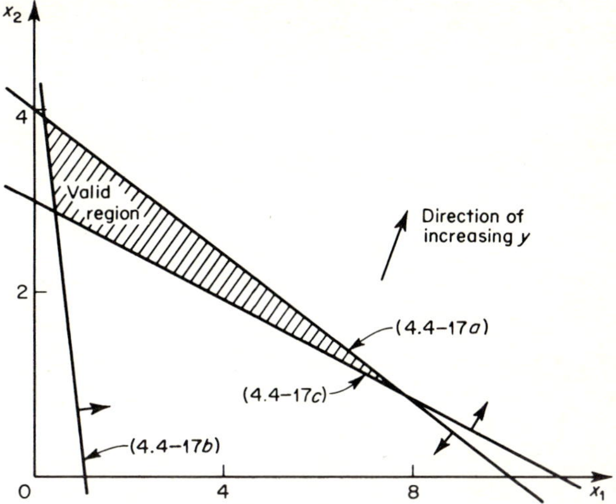

Fig. VII-24 The system for Example VII-35 (arrows indicate valid side of the restriction).

may be added as follows:

$$5x_1 + \quad x_2 - x_4 + x_{100} = 5$$
$$3x_1 + 11x_2 - x_5 + x_{101} = 33$$

Having done this, we may choose our first basic solution as

$$x_1 = 0 \qquad x_2 = 0 \qquad x_3 = 20 \qquad x_4 = 0$$
$$x_5 = 0 \qquad x_{100} = 5 \qquad x_{101} = 33$$

where x_{100}, x_{101}, and x_3 form the basis.[1] Since we do not wish the final solution to contain the artificial variables, we must add a penalty function to the objective function to ensure that the final solution includes x_{100} and x_{101} only as zeros.

Let us write the modified objective function as

$$y = 6x_1 + 5x_2 - P(x_{100} + x_{101})$$

or in terms of the nonbasis variables

$$y = 6x_1 + 5x_2 - P(38 - 8x_1 - 12x_2 + x_4 + x_5)$$

where P is a very large number, larger than any other in the calculation. Thus, if there is a solution to the original problem, the large coefficient P will prevent the appearance of the artificial variables in the optimal

[1] It may be suggested that the present problem is irrelevant since the original problem does not contain x_{100} and x_{101}. However, we can ensure that it is omitted from our final solution and that the remaining variables have precisely their optimal values by adding a penalty term to the objective function y.

solution. We may now write our first array:

ARRAY 1*a*

		x_1	x_2	x_4	x_5
x_3	20	2	5	0	0
x_{100}	5	5	1	-1	0
→ x_{101}	33	3	11	0	-1
y	0	-6	-5	0	0
(P)	-38	-8	-12	1	1
			↑		

It should be noted that this array does not represent a feasible basic solution, but is only a step in achieving a feasible vertex. The latter will be determined after we get rid of the artificial variables. In the array we have set out the objective function in two rows. This has been done for convenience, since the P terms will be much larger than any other terms in y when P is made to take any very large value which dominates the expression. In any decision, as long as x_{100} and x_{101} are not zero, we shall consider only these terms in the P row of y. Apply the usual rules for transforming this array. We shall therefore interchange x_2 and x_{101}, getting

ARRAY 1*b*

		x_1	x_{101}	x_4	x_5
x_3	5	0.636	-0.455	0	0.455
→ x_{100}	2	4.727	-0.091	-1.000	0.091
x_2	3	0.273	0.091	0	-0.091
y	15	-4.636	0.455	0	-0.455
(P)	-2	-4.727	1.091	1.000	-0.091
		↑			

Since x_{101} is zero, we can drop this column. Interchanging x_1 and x_{100}, we get:

ARRAY 1*c*

		x_{100}	x_4	x_5
→ x_3	4.731	-0.135	0.135	0.442
x_1	0.423	0.212	-0.212	0.019
x_2	2.885	-0.058	0.058	-0.096
y	16.962	0.981	-0.981	-0.365
(P)	0	1.000	-1.000	0
			↑	

The artificial variable x_{100} has now become nonbasic and zero. This means that we can now ignore both x_{100} and the P penalty function in the objective function, since the relations represented in this array are those in the original problem. We thus have found a basic solution which is feasible. We can now finish the problem in the usual manner, upon dropping the x_{100} column and the P row from the last array. We then have the final array, after interchanging x_3 and x_4:

ARRAY 2

		x_3	x_5
x_4	35.141	7.428	3.286
x_1	7.857	1.571	0.714
x_2	0.857	−0.429	−0.286
y	51.429	7.285	2.857

The optimum value of y is thus 51.429, found at $x_1^* = 7.857$, $x_2^* = 0.857$, the intersection of restrictions (4.4-17*a*) and (4.4-17*c*).

No feasible solution exists In some cases no feasible solution exists because the restrictions are contradictory, and thus exclude all possible valid regions. Such a situation will be detected if the preceding method is used to attempt to get a first feasible basic solution, and if the artificial variables can never be removed from the basis and made zero.

Example VII-36: No feasible basis

Find the maximum of y, where

$$y = 6x_1 + 5x_2$$

subject to the restrictions that

$$-2x_1 - 5x_2 \leq -20 \tag{4.4-19a}$$
$$5x_1 + x_2 \leq 5 \tag{4.4-19b}$$
$$3x_1 + 11x_2 \leq 33 \tag{4.4-19c}$$

Applying the previous rules, we get, in turn,

$$-2x_1 - 5x_2 + x_3 = -20 \tag{4.4-20a}$$
$$5x_1 + x_2 + x_4 = 5 \tag{4.4-20b}$$
$$3x_1 + 11x_2 + x_5 = 33 \tag{4.4-20c}$$

$$2x_1 + 5x_2 - x_3 = 20 \tag{4.4-21a}$$
$$5x_1 + x_2 + x_4 = 5 \tag{4.4-21b}$$
$$3x_1 + 11x_2 + x_5 = 33 \tag{4.4-21c}$$

$$2x_1 + 5x_2 - x_3 + x_{100} = 20 \tag{4.4-22a}$$
$$5x_1 + x_2 + x_4 = 5 \tag{4.4-22b}$$
$$3x_1 + 11x_2 + x_5 = 33 \tag{4.4-22c}$$

and
$$y = 6x_1 + 5x_2 - P(x_{100})$$
$$= 6x_1 + 5x_2 - P(20 - 2x_1 - 5x_2 + x_3)$$

The succession of arrays is then as follows:

ARRAY 1*a*

		x_1	x_2	x_3
x_{100}	20	2	5	−1
x_4	5	5	1	0
→ x_5	33	3	11	0
y	0	−6	−5	0
(P)	−20	−2	−5	1
			↑	

ARRAY 1*b*

		x_1	x_5	x_3
x_{100}	5	0.636	−0.455	−1
→ x_4	2	4.727	−0.091	0
x_2	3	0.273	0.091	0
y	15	−4.636	0.455	0
(P)	−5	−0.636	0.455	1
		↑		

ARRAY 1*c*

		x_4	x_5	x_3
x_{100}	4.731	−0.135	−0.442	−1
x_1	0.423	0.212	−0.019	0
x_2	2.885	−0.058	0.096	0
y	16.962	−0.981	0.446	0
(P)	−4.731	0.135	0.442	1

Since the overriding coefficients of P in the objective function are all positive, we seem to have obtained the optimal array. However, an artificial variable is still nonzero in the basis. This is the final answer to the modified problem, and it shows that there is *no* possible answer to the original problem. This is indicated by our inability to lose the artificial variable. The reason for the nonexistence of a feasible region can be seen graphically in Fig. VII-25.

Choice of Nonbasis Variable Column—Zero Entry in Last Row—Multiple Solutions Having constructed the first array, the variable to be introduced

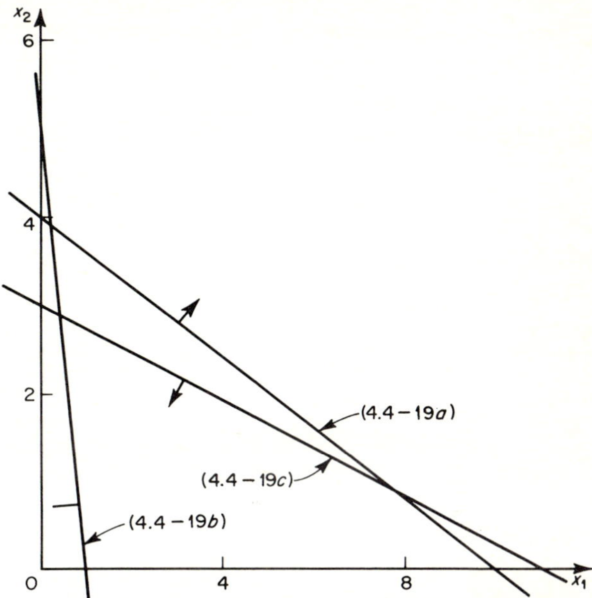

Fig. VII-25 Mutually exclusive restrictions in Example VII-36 (arrows indicate valid side of the restriction).

into the basis must be chosen. Normally, this is done by choosing the most negative or most positive entry in the last row of the array, depending upon whether a maximum or minimum is being sought. When no more negative (or no more positive) coefficients remain in this row, we have found the maximum (or minimum) value of the objective function for which we are searching. Let us now consider the problem of what is meant by a zero coefficient appearing in the last row of the final array. If it appears earlier, of course, we can ignore it in favor of a negative or positive coefficient.

Example VII-37: A multiple solution

Maximize y, where

$$y = 6x_1 + x_2$$

subject to

$$3x_1 + 5x_2 \leq 13 \qquad (4.4\text{-}23a)$$

$$6x_1 + x_2 \leq 12 \qquad (4.4\text{-}23b)$$

$$4x_1 + 3x_2 \leq 12 \qquad (4.4\text{-}23c)$$

It will be noted that the right-hand side of the objective function is the same as the left-hand side of the second equality and that in the geometric representation of Fig. VII-26 the y family of curves

$$6x_1 + x_2 = K$$

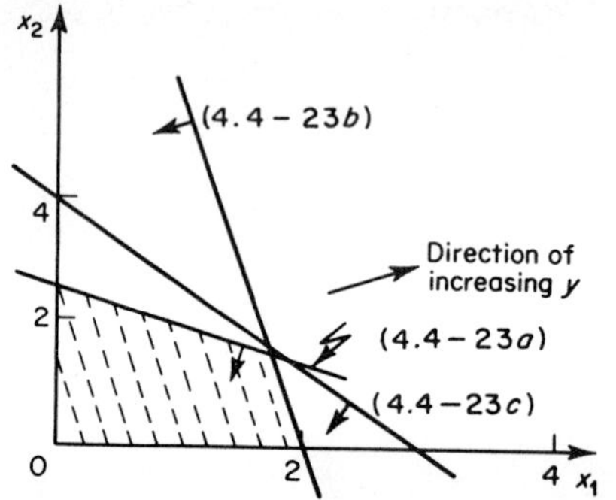

Fig. VII-26 Objective function contours parallel to the limiting restriction (see Example VII-37).

is thus parallel to that restriction. It would therefore be possible to find a value of K so that y lies along the boundary. If this were the optimal value of y, all the points along the boundary between the two vertices would be optimal.

Finding $x_1 = 0$, $x_2 = 0$, $x_3 = 13$, $x_4 = 12$, and $x_5 = 12$ as a first basic solution, the first array can be constructed, and the solution carried out as before, getting:

ARRAY 1

		x_1	x_2
x_3	13	3	5
→ x_4	12	6	1
x_5	12	4	3
y	0	−6	−1
		↑	

ARRAY 2

		x_4	x_2
→ x_3	7.000	−0.500	4.500
x_1	2.000	0.167	0.167
x_5	4.000	−0.667	2.333
y	12.000	1.000	0
			↑

The point $x_1 = 2$, $x_2 = 0$, thus yields a maximum value of y. As shown in the two-dimensional example, the last row contains a zero coefficient, and all the other coefficients are positive, so that the value of y cannot be increased further. We may therefore conclude that no further change can be made to alter profitably the value of the objective function. This is generally true. However, since the x_i term which has the zero coefficient does not appear in the expression for the optimal objective function, no harm can be done by increasing this variable from its value of zero. Let us do this in the present example.

We may increase the variable x_2 without altering the optimal value of $y = 12$. Applying the usual rules of transformation, we get the new array:

ARRAY 3

		x_4	x_3
x_2	1.556	−0.111	0.222
x_1	1.741	0.148	−0.037
x_5	1.371	−0.407	−0.518
y	12.000	1.000	0

The location $x_1 = 1.741$, $x_2 = 1.556$, thus also satisfies the problem, yielding the same optimal value of y as in the preceding array. If we now carried on as before and changed x_3, we should get back to the preceding array.

We thus have the situation in which two basic solutions exist which yield the same optimal value of y (note that if here we had required a minimum y, no problem would have arisen). Recalling that each array represents a vertex of the valid region, we have now shown in this special case that two vertices have the same optimal value of y, so that the line joining these two points must also have the same value of y. In fact, we shall have an infinite number of solutions to the problem, since every point on the intermediate line will satisfy the restrictions on the problem and give the optimum value of the objective function. Thus, for example, Table VII-7 shows the basic solutions holding at the two

TABLE VII-7 POSSIBLE MULTIPLE SOLUTIONS IN EXAMPLE VII-37

	Solution 1	2	3	4
x_1	2.000	1.741	1.870	1.974
x_2	0	1.556	0.778	0.156
x_3	7.000	0	3.500	6.300
x_4	0	0	0	0
x_5	4.000	1.371	2.685	3.737
$y(\mathbf{x})$	12.00	12.00	12.00	12.00

vertices (columns 1 and 2), and in addition two other possible solutions (columns 3 and 4) from the infinite number existing on the line joining the two vertices, all the answers giving the maximum value of $y = 12$.

For the general problem, when a zero coefficient is found in the final row, this means that another vertex of the valid region can be found which will also yield the same value of the objective function. Since the objective function has the same optimum value at two adjacent vertices, the line joining them must also yield the same value, since we have a linear function. We can therefore choose any point on this line as our final solution. In other words, if more than one basic solution (vertex) exists, with the same value of the objective function, an infinite number of feasible (although not basic) solutions exist, each giving the optimal value of y. Geometrically, it may be seen from Fig. VII-26 and the original statement of the problem that the line representing one of the restrictions is parallel to the line representing the objective function, thus yielding more than one optimal location. It should be noted, however, that if in this illustration a minimum had been required, no difficulty would have been encountered.

Choice of a Pivot Having chosen the variable to be added to the basis, the Simplex method then looks for the smallest positive ratio of the coefficients in the first column to the x_i column. In some cases there is no positive ratio, or only zero values; and in others, more than one identical smallest positive ratio may be found. In addition, some coefficients in the x_i column may be zero. Let us consider these cases.

Zero coefficient in column (infinite values in the solution) If a zero coefficient is found in any column, this indicates that this row does not restrict the range of values which this column variable may take, since the latter does not appear in the restriction. A zero value could therefore never be used as a pivot. If all the other coefficients produced negative ratios, a row with the zero coefficient indicates that the value of x in that column would be able to increase to infinity, correspondingly increasing or decreasing the objective function.

Example VII-38: An infinite solution (1)

If the array is given as:

		x_4	x_5
x_1	1	−0.33	−0.33
x_2	4	0	0.67
x_3	9	−1	−1
y	3	−0.67	0.33
		↑	

and if we wish to maximize y, we should choose the x_4 column. There are no positive ratios, except the one with zero in the denominator. The final solution to this problem is that x_4, and hence y, may increase to infinity without limit.

No positive ratio (infinite values in the solution) It may be found that no positive ratios exist. If we examine the meaning of each row in the array, we shall see that this means that there are no restrictions upon the upper value of the x in the column being considered. In other words, this x may go to infinity, so that the objective function will be unbounded.

Example VII-39: An infinite solution (2)

Maximize y, where

$$y = 2x_1 - x_2$$

subject to

$$-3x_1 + 2x_2 \leq 2$$
$$2x_1 - 4x_2 \leq 3$$
$$x_1 + x_2 \geq 6$$

We may write the restrictions, after adding slack and artificial variables, as

$$-3x_1 + 2x_2 + x_3 = 2$$
$$2x_1 - 4x_2 + x_4 = 3$$
$$x_1 + x_2 - x_5 + x_{100} = 6$$

with the modified objective function as

$$\begin{aligned} y &= 2x_1 - x_2 - P(x_{100}) \\ &= 2x_1 - x_2 - P(6 - x_1 - x_2 + x_5) \end{aligned}$$

The arrays are, respectively, as follows:

ARRAY 1*a*

			x_1	x_2	x_5
	x_3	2	−3	2	0
→	x_4	3	2	−4	0
	x_{100}	6	1	1	−1
	y	0	−2	1	0
	(P)	−6	−1	−1	1
			↑		

ARRAY 1*b*

			x_4	x_2	x_5
	x_3	6.5	1.5	−4.0	0
	x_1	1.5	0.5	−2.0	0
→	x_{100}	4.5	−0.5	3	−1
	y	3.0	1.0	−3	0
	(P)	−4.5	0.5	−3	1
				↑	

ARRAY 1*c*

		x_4	x_5
x_3	12.5	0.83	−1.33
x_1	4.5	0.17	−0.67
x_2	1.5	−0.17	−0.33
y	7.5	0.5	−1.00

Since x_{100} has now left the basis and is zero, we may drop the x_{100} column and the P row. This array is, then, our *first feasible basic solution.* Applying the usual rules, we wish to take x_5 into the basis, since its coefficient in the last row is negative and we wish to find the maximum of y. However, we cannot find a positive ratio to determine the pivot. We can certainly increase x_5, and we can do this without limit, since there is no danger of any variable becoming negative. Looking at the third restriction, we see that x_1, x_2, and x_3 can reach infinite values as x_5 goes to infinity. The objective function will, similarly, have no upper limit. Our failure to find a pivot in the x_5 column is an indication that there is no finite bound to the limit of y.

In general, we may conclude that a failure to find a pivot in the Simplex table means that the variable being introduced into the basis may take an unbounded range of values and allow the objective function to go to infinite values.

An alternative way of discovering that the value of objective function is unbounded by the use of an artificial restriction is described by Vajda [1956, p. 64].

Identical values of the smallest positive ratio (possible degeneracy) In carrying out the Simplex rules, it may be found that more than one row satisfies the condition that we choose the smallest positive ratio between the first column and the column containing the variable being introduced into the basis.

Example VII-40: A zero basis variable

Minimize y, where

$$y = x_2 - 3x_1$$

subject to the constraints that

$$4x_1 - x_2 \leq 8 \qquad (4.4\text{-}24a)$$

$$x_1 - x_2 \leq 2 \qquad (4.4\text{-}24b)$$

$$x_1 + x_2 \leq 6 \qquad (4.4\text{-}24c)$$

Proceeding in the usual way, the first array is as follows:

ARRAY 1

		x_1	x_2	
x_3	8	4	−1	← First choice
x_4	2	1	−1	← Second choice
x_5	6	1	1	
y	0	3	−1	
		↑		

Taking x_1 into the basis, it will be seen that both the x_3 and the x_4 rows give the same value of 2 for the smallest positive ratio.

If we consider either of the two indicated rows (x_3 and x_4) as the desired pivot, we may carry out the transformation of the array. The same final array will be obtained for most problems.

Approach a Taking the x_3 row as our first choice for the pivotal row, we get the next array:

ARRAY 2*a*

		x_3	x_2
x_1	2	0.25	−0.25
x_4	0	−0.25	−0.75
x_5	4	−0.25	1.25
y	−6	−0.75	−0.25

which is optimal, so that the basic solution is $x_1^* = 2$, $x_2^* = 0$, with $y^*_{\min} = -6$.

Approach b Alternatively, we could have used the x_4 row in the first array as our pivotal row, and we should get:

ARRAY 2*b*

		x_4	x_2
→ x_3	0	−4	3
x_1	2	1	−1
x_5	4	−1	2
y	−6	−3	2
			↑

ARRAY 3*b*

		x_4	x_3
x_2	0	−1.33	0.33
x_1	2	−0.33	0.33
x_5	4	1.67	−0.67
y	−6	−0.33	−0.67

which is optimal, producing the same optimal solution as by the first pivot route, that is, $x_1^* = 2$, $x_2^* = 0$, with $y_{\min}^* = -6$.

It will be seen that *one of the basic variables* has a value of zero. It will also be noted that this value of zero (in particular, x_4 or x_3 in the first array, by each route) is due to the equality of the ratios in the first array. In the two final arrays three variables are zero, which conforms with the fact that three straight lines pass through the same point. It will also be seen that the objective function does not alter between the two last arrays, since a zero basis variable was interchanged for a zero nonbasis variable. Geometrically, we may interpret this system as shown in Fig. VII-27. Here the x_1 axis and restrictions (a) and (b) pass through the same point ($x_1 = 2$, $x_2 = 0$). At a normal vertex of the valid region, only two lines meet, and correspondingly only two variables are zero. Here three lines intersect.

In general, this situation will arise if $n + 1$ restrictions intersect at one vertex, instead of the necessary n intersections.

Normally, it follows from the choice of the new basic variable that the objective function to be maximized (or minimized) will increase (or decrease) at each step, and a new basic feasible solution is obtained. There are only a finite number of feasible basic solutions, and since the linear objective function increases (decreases) at each step, we can never go back to one we

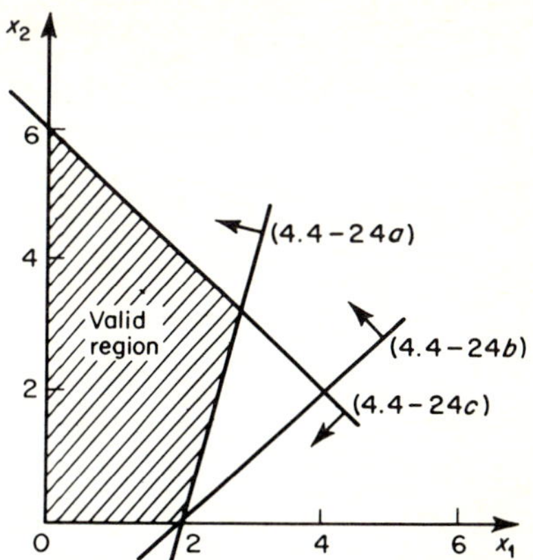

Fig. VII-27 Superfluous restrictions (see Example VII-40) (arrows indicate valid side of the restriction).

have already examined. The process must therefore terminate in a finite number of steps.

In our present case, however, the objective function does not change between steps; so, in principle, it is possible that we might take an infinite number of steps to reach an optimum. In other words, a case may exist in which the Simplex method gets into a cycle, producing the same sequence of arrays time and time again, repeating the same set of basic variables in each loop, without breaking out to attain the required solution. A rule equivalent to a small perturbation of the variables has been postulated to handle such a multiplicity of choices.[1] This in the last problem would ensure that the three lines did not pass exactly through the same vertex, one line being shifted slightly before starting the calculation, and replacing it in its correct position after the solution is complete.

Even without a rule, however, a loop is not normally found in practical applications of linear programming.[2] In the present example we do not get into a loop because the final array has already been reached.

[1] Discussed by Dantzig in chap. 21 of Koopmans [1951] and by Charnes [1952]. See, for example, Vajda [1956, p. 66] for details.

[2] In fact, only three examples are known in the literature, Beale [1955b], Hoffman [1953], and Gass [1964, p. 119]. Gass' example is the minimization of

$$y = -\tfrac{3}{4}x_1 + 150x_2 - \tfrac{1}{50}x_3 + 6x_4$$

subject to $x_i \geq 0$, and

$$\tfrac{1}{4}x_1 - 60x_2 - \tfrac{1}{25}x_3 + 9x_4 + x_5 = 0$$

$$\tfrac{1}{2}x_1 - 90x_2 - \tfrac{1}{50}x_3 + 3x_4 + x_6 = 0$$

$$x_3 + x_7 = 1$$

This situation which has been described is referred to as *degeneracy*. If we ignore all these variables which have the value zero, then in a case where one of the basic variables has this zero value, we obtain more equations than we have variables; such a system can be satisfied only in exceptional circumstances, and these, in algebra, are described as degenerate (Dantzig [1963]).

4.4.5 General comments on linear programming

While it is possible to solve these linear problems by several techniques, the Simplex method (Secs. 4.4.3 and 4.4.4) is a particularly convenient method. Essentially, it starts with any feasible solution lying at a vertex formed by the intersection of boundaries, and the direction of improvement is determined by computing the constrained derivatives. While this might be done by evaluating the Jacobians, here the linearity allows matrix inversion and multiplication to accomplish the task more easily, resulting in the simple transformation rules described above. If inspection of the signs of these constrained derivatives (as in the last row of any array) show that they are all nonnegative (in the original equations), the Kuhn-Tucker conditions for a global minimum are satisfied (conversely, for a maximum). If this is not the case, then one of the independent variables (one not in the basis) with a negative constrained derivative is altered, being increased in value until some dependent (basis) variable goes to zero. This repeated procedure thus determines in each cycle a new solution which satisfies the restrictions and shows an improvement over the previously calculated values of the objective function.

For problems involving up to about two hundred restrictions, linear programming is feasible in a single computer calculation. However, it is not directly practical for very large problems, such as would describe company operations of even moderate size. To get large problems on a computer, one must use the Dantzig and Wolfe [1960] decomposition principle to split the original problem into pieces of convenient size (Chap. XIII, Sec. 5.1.2). Dynamic programming (Chap. XIII, Sec. 4) might also be used to handle large linear systems (Roberts [1964b, chap. 5]). Some typical applications of linear programming are presented in Table VII-8, Riley and Gass [1958], Gass [1964], and Dantzig [1963], among others, presenting detailed bibliographies.

In certain cases, it might prove advantageous to consider not the original problem, but an associated dual system (Sec. 5.1). The special cases in which the variables might have upper limits or be confined to integer values are noted below.

It is also possible, as described by Wilde and Beightler [1967], Beightler and Wilde [1965], and Chung [1963], to derive the necessary information from the final optimal arrays for a sensitivity analysis of the optimal region.

TABLE VII-8 SOME TYPICAL APPLICATIONS OF LINEAR PROGRAMMING

General texts

Dantzig [1963], Charnes and Cooper [1961], Gass [1964], Vajda [1960, 1961, 1956].

Bibliography of applications

Riley and Gass [1958], Gass [1964], Dantzig [1963], Vajda [1961], Rohde [1957].

Applications[1]

(a) Allocation of resources among activities (blending, production scheduling, etc.)

Eisemann [1957], Charnes, Cooper, and Mellon [1952, 1954], Deam [1959], Bowman [1956], Symonds [1955, 1956], Sears [1963], Fenech and Acrivos [1956], Gaddum, Hoffman, and Sokolowsky [1954], Paull [1957], Prager [1956], Jacobs [1954], Manne [1956], Schwan and Wilkinson [1956], Ferguson and Dantzig [1956], Charnes and Cooper [1953, 1955, 1957], Smith [1965], Amey [1964], Gilmore and Gomory [1961, 1963, 1965].

(b) Transportation and distribution from r sources to s sinks, network flows, traveling-salesman problems, and personnel assignment

Houthakker [1955], Hitchcock [1941], Schwan and Wilkinson [1956], Symonds [1955], Flood [1954, 1956], Dantzig [1951b, 1959], Dantzig, Fulkerson, and Johnson [1954], Orden [1956], Prager [1955], Ford and Fulkerson [1956], Williams [1962, 1963, 1965], Gilmore and Gomory [1964], Balas [1965], Balinski and Quandt [1964].

(c) Technical problems

Battersby [1958], Dantzig, Johnson, and White [1958], Koenigsberg [1961].

(d) Management decisions

Rosander [1956], Stanley, Honig, and Gainen [1954], Charnes, Cooper, and Farr [1953], Gunther [1955], Charnes, Cooper, and Ferguson [1954–1955], Hobbs [1966].

[1] Only a few classifications have been attempted. Gass [1964], for example, has classified applications into agricultural, contracts, industrial (chemical, coal, iron and steel, airlines, communications, paper, petroleum, railroad), economic analysis, military, personnel assignment, production scheduling, inventory, planning, structural design, traffic analysis, transportation and network analysis, and the routing of traveling salesmen.

A similar problem was considered by Shetty [1961], who used an optimal solution to derive the solution for different parameter values. Barnett [1962] examined the stability of the linear programming solution, determining how the parameter values $a_{i,j}$ may vary without affecting the choice of the optimal basis.

Upper Bounded Variables In a bounded-variable problem, some or all of the variables are limited, not only by the nonnegativity restrictions on the lower side, but also by upper limits. These upper bounds, such as $x_j \leq x_j^L$, present no fundamental difficulties, since they can be handled by the Simplex method in the normal manner. The only problem is a computational one, since the presence of these additional constraints will greatly enlarge the basis and the Simplex arrays. Various workers[1] have proposed techniques which modify the Simplex decision and transformation rules rather than enlarge the array.

Integer Linear Programming The integer linear programming problem is one in which some or all of the variables must assume integer values.[2] When all the variables must be integers, the method of Gomory [1958] is widely used, the linear programming problem being first solved without taking the integer restriction into account. If all the variables turn out to be integers, the problem is solved, but if this is not the case, a further constraint is added. Gomory's method has been republished (Graves and Wolfe [1963, p. 269]), and the method is discussed, for example, by Vajda [1961, p. 197], Hadley [1964, p. 272], and Dantzig [1963, p. 521]. As Gomory [1965] has noted, the rounding off of a linear programming solution to the nearest set of feasible integers does not, in general, provide a solution to this problem.

If the restriction on integral values applies only to some specified variables, a method due to Land and Doig [1960] can be used in which once again the usual Simplex approach provides an initial optimal solution. This method can of course be used when all the variables must be integers, while Gomory's method can be modified as discussed by Dantzig [1963, chap. 26] to handle the mixed situation. Harris [1964] has also considered this problem.

Some Special Forms of Linear Programming The algorithms described above can be used to solve any linear programming problem, but their use is not advocated in all cases. On occasions special features allow easier solutions to be obtained by different computational procedures. One such case is the so-called *transportation problem*, which has the structure

$$\sum_j x_{i,j} = a_i \qquad \sum_i x_{i,j} = b_j \qquad x_{i,j} \geq 0 \qquad \min \sum_i \sum_j c_{i,j} x_{i,j}$$

Others which are special forms of this type are termed the assignment and warehousing problems. These and others are discussed, for example, by Vajda [1961], Hadley [1962], and Dantzig [1963].

[1] The work of Dantzig [1954] is reported, for example, in Dantzig [1963, chap. 18], while that of Charnes and Lemke [1954] is described by Chung [1963, chap. 10].

[2] Such additional restrictions can arise naturally, when, for example, a manufacturer dispatches an integral number of batches of product annually, or might arise by the transformation of certain nonlinear systems. The latter have been discussed, for example, by Dantzig [1960] in his "fixed-charge" problem. Dantzig [1963, p. 535] has also systematically reviewed these latter problems. One well-known problem in this category is that of minimizing the traveling salesman's route between a certain number of cities (Dantzig, Fulkerson, and Johnson [1954], Dantzig [1959], and Flood [1956]).

4.5 Some Special Cases in Nonlinear Programming

4.5.1 *Quadratic programming*

The special form of the nonlinear programming problem

$$y(\mathbf{x}) = \sum_{i=1}^{n} c_i x_i + \sum_{i=1}^{n} \sum_{j=1}^{n} c_{i,j} x_i x_j \tag{4.5-1}$$

subject to the m linear constraints

$$\sum_{i=1}^{n} a_{i,j} x_i \leq b_j \qquad \text{for } j = 1, 2, \ldots, m \tag{4.5-2}$$

is referred to as a *quadratic programming* problem. The main difference here, compared with linear programming, lies in the possibility of an extreme value existing whether or not constraints are present, the extreme value being found if the quadratic form is positive- or negative-definite (see, for example, Debreu [1952]).

A number of techniques for the solution of this system have been proposed,[1] most of them varying only in computational detail. The best-known approach is that of Wolfe [1959], based on the theorem of Kuhn and Tucker [1951]. This has the advantage that it reduces the task of solving the quadratic problem to a form allowing application of the linear programming Simplex method. We shall not discuss this special problem further, but might also note that theorems similar to those in linear programming exist for quadratic duals. Vajda [1961, p. 242], for example, has discussed the work of Dorn [1958, 1960] in this field. Boot [1964, chap. 8] has considered the question of sensitivity analysis.

4.5.2 *Geometric programming*

Another special form of nonlinear programming is that termed *geometric programming* (see Sec. 5.3.2 and Duffin and Peterson [1964], Duffin, Peterson, and Zener [1967], and Eben and Ferron [1968]).

5 DUALS IN OPTIMIZATION

We have been concerned with the optimization of an objective function, $y(\mathbf{x})$, by selecting a suitable set of variables, $\mathbf{x}$, which satisfy all the imposed

[1] Early work includes that of Beale [1955a, 1959], Barankin and Dorfman [1955], Markowitz [1956], Frank and Wolfe [1956], Hildreth [1957], Wolfe [1959], and Houthakker [1960]. A recent text by Boot [1964] presents a comprehensive discussion of most methods, while Vajda [1961, p. 223] describes the work of Beale, and Hadley [1964, chap. 7] and Dantzig [1963, chap. 24] discuss the work of Wolfe in some detail.

constraints. We might thus denote the solution to a minimization problem as

$$M = \min_{\mathbf{x}} y(\mathbf{x}) \tag{5-1}$$

where $\mathbf{x}$ must satisfy all imposed restrictions. Let us suppose that there exists a second function $W(\mathbf{v})$ which has the same value, M, at its maximum, so that

$$M = \max_{\mathbf{v}} W(\mathbf{v}) \tag{5-2}$$

with $\mathbf{v}$ satisfying certain restrictions. Then we can state that

$$y(\mathbf{x}) \text{ for all permitted } \mathbf{x} \geq M \geq W(\mathbf{v}) \text{ for all permitted } \mathbf{v} \tag{5-3}$$

Therefore $y(\mathbf{x})$ provides an upper bound on the particular value of M, and $W(\mathbf{v})$ is a lower bound. The two problems defined by Eqs. (5-1) and (5-2) are dual problems, with, in general, one specified as the primal problem and the other its dual. Here the objective function $W(\mathbf{v})$ is defined as the dual of the primal problem.

The concept of duality has not played a prominent role in optimization since duals cannot always be constructed. If, however, the dual to an optimization problem can be formulated, important computational advantages can often be gained. For example, upper and lower limits to the true value of the optimal objective function are given immediately by approximate solutions from inequalities (5-3). Such bounds can play an important part in terminating numerical search routines. In some cases the dual might be a simpler problem than the primal. Important examples of this advantage are found in the field of linear programming. Since the concept of duality is potentially valuable, we shall consider its formulation at this point. Unfortunately, our remarks are pertinent only to a rather restricted group of cases, for example, in linear problems or those minimization problems having convex objective functions.

5.1 Duals in Linear Systems

The linear problem is one of those cases for which it is possible to construct a dual system. Hadley [1962, chap. 8], Vajda [1961], and Dantzig [1963], among others, have noted some of the useful computational applications in linear programming.

Let us denote as the primal problem (P) the linear problem of minimizing y, where

$$y = \sum_{i=1}^{n} c_i x_i \tag{5.1-1}$$

subject to the m constraints

$$d_j - \sum_{i=1}^{n} b_{j,i}x_i \leq 0 \qquad \text{for } j = 1, 2, \ldots, m \tag{5.1-2}$$

In addition, we shall restrict our attention to nonnegative values of x_i, so that

$$x_i \geq 0 \tag{5.1-3}$$

This primal problem can be expressed in Lagrangian form, which is of course valid even in the case of linear problems. Thus

$$F = \sum_{i=1}^{n} c_i x_i + \sum_{j=1}^{m} \lambda_j \left(d_j - \sum_{i=1}^{n} b_{j,i}x_i \right) - \sum_{i=1}^{n} \alpha_i x_i \tag{5.1-4}$$

Here both λ_j and α_i are Lagrangian multipliers which are restricted by the Kuhn-Tucker conditions (see Sec. 4.3), so that

$$\alpha_i \geq 0 \qquad \text{for } i = 1, 2, \ldots, n \tag{5.1-5}$$

and

$$\lambda_j \geq 0 \qquad \text{for } j = 1, 2, \ldots, m \tag{5.1-6}$$

Our aim is to construct a second problem (D) dual to the first, or primal, problem defined above. Consistent with this goal, we note, first, that at the minimum, either inequality (5.1-2) must be satisfied identically or the λ_j associated with that inequality must vanish. Similarly, either $x_i = 0$ or $\alpha_i = 0$. These conditions are discussed in connection with the use of slack variables, in Sec. 4.3.1. If, therefore, $\mathbf{x}^*$ is the admissible point which makes y a minimum, it follows from Eq. (5.1-4) that

$$F = \sum_{i=1}^{n} c_i x_i^* = y(\mathbf{x}^*) \tag{5.1-7}$$

F then reduces to a value which is the solution to the primal problem (P).

A second important result in seeking a dual is obtained by noting that F must be stationary at the optimum. Thus

$$\frac{\partial F}{\partial x_k} = 0 = c_k - \sum_{j=1}^{m} \lambda_j b_{j,k} - \alpha_k \tag{5.1-8}$$

Eliminating c_k between Eqs. (5.1-4) and (5.1-8), we find that

$$F = \sum_{j=1}^{m} d_j \lambda_j^* \tag{5.1-9}$$

Here the asterisk has been appended to the λ_j to denote the fact that the particular values of λ_j satisfy Eq. (5.1-8) as well as (5.1-6). It is interesting to note that if we define

$$W(\boldsymbol{\lambda}) = \sum_{j=1}^{m} d_j \lambda_j \tag{5.1-10}$$

$$W(\boldsymbol{\lambda}^*) = y(\mathbf{x}^*)$$

when $\boldsymbol{\lambda}$ satisfies the required constraints. This expression is particularly interesting when we recall that one property of a dual is that its maximum value must equal the minimum of the primal problem. Thus $W(\boldsymbol{\lambda})$ is equal to $y(\mathbf{x}^*)$ for a certain set of the Lagrangian multipliers, namely, those satisfying the Kuhn-Tucker conditions and Eq. (5.1-8). A dual must also have the following property: it has a maximum value at the point where its objective function equals the minimum value of the primal problem. How can we suitably restrict $\boldsymbol{\lambda}$ so that, for all admissible $\boldsymbol{\lambda}$, $W(\boldsymbol{\lambda}) \leq y(\mathbf{x}^*)$? We recognize that the inequality

$$\sum_{j=1}^{m} \lambda_j b_{j,k} - c_k \leq 0 \tag{5.1-11}$$

contains Eq. (5.1-8), since the α_k are positive. Thus, if the $\boldsymbol{\lambda}$ satisfy Eq. (5.1-8), they will certainly satisfy inequality (5.1-11), although the converse is not true. Imposing inequality (5.1-11) thus gives a new freedom to $\boldsymbol{\lambda}$. Nevertheless, we can show that $W(\boldsymbol{\lambda})$ is never greater than $y(\mathbf{x}^*)$. The demonstration is as follows: Multiply inequality (5.1-11) by $\mathbf{x}_k^*$ and sum from $k = 1$ to n. Thus, since

$$\sum_{j=1}^{m} \sum_{k=1}^{n} \lambda_j x_k^* b_{j,k} \leq \sum_{k=1}^{n} c_k x_k^* = y(\mathbf{x}^*) \tag{5.1-12}$$

multiplying inequality (5.1-2) by λ_j and summing on j, we find

$$\sum_{j=1}^{m} \sum_{k=1}^{n} \lambda_j x_k^* b_{j,k} \geq \sum_{j=1}^{m} d_j \lambda_j = W(\boldsymbol{\lambda}) \tag{5.1-13}$$

since we shall restrict our attention to nonnegative values of λ_j. Therefore we see that

$$y(\mathbf{x}^*) \geq W(\boldsymbol{\lambda}) \tag{5.1-14}$$

with the equality sign holding for the set $\boldsymbol{\lambda}^*$. We then have the necessary ingredients for $W(\boldsymbol{\lambda})$ to be the dual of P, since, when $W(\boldsymbol{\lambda})$ takes its maximum value, this will equal the minimum value of $y(\mathbf{x})$.

To summarize, if we define two problems

(P) Minimize $y = \sum_{i=1}^{n} c_i x_i$

subject to $x_i \geq 0$ for $i = 1, 2, \ldots, n$

and $d_j - \sum_{i=1}^{n} b_{j,i} x_i \leq 0$ for $j = 1, 2, \ldots, m$

(D) Maximize $W = \sum_{j=1}^{m} \lambda_j d_j$

subject to $\lambda_j \geq 0$ for $j = 1, 2, \ldots, m$

and $\sum_{j=1}^{m} \lambda_j b_{j,i} - c_i \leq 0$ for $i = 1, 2, \ldots, n$

then we find that

$$W(\boldsymbol{\lambda}) \text{ for all admissible } \boldsymbol{\lambda} \leq M \leq y(\mathbf{x}) \text{ for all admissible } \mathbf{x}$$

and
$$W(\boldsymbol{\lambda}^*) = M = y(\mathbf{x}^*)$$

The two problems must therefore be dual.

Example VII-41: Linear duals

Let us consider the problem P of minimizing

$$y = x_1 + x_2$$

with nonnegative values of x_1 and x_2, satisfying the constraints

$$x_1 - 2x_2 \geq -4$$
$$x_1 + 2x_2 \geq 2$$
$$-x_1 \geq -2$$

The dual system D can now be formulated. From inequality (5.1-11), the following two restrictions for this particular case can be written

$$\lambda_1 + \lambda_2 - \lambda_3 \leq 1$$
$$-2\lambda_1 + 2\lambda_2 \leq 1$$

The dual objective function to be maximized will take, from Eq. (5.1-10), the form

$$W = -4\lambda_1 + 2\lambda_2 - 2\lambda_3$$

Let us first select a few feasible, but not necessarily optimal, solutions. Some cases are given in Table VII-9. It will be seen that the largest value of $W(\boldsymbol{\lambda})$ determined is -3, whereas the smallest value of $y(\mathbf{x})$ is 1.5. These two values thus bound the true solution, as expected from a primal-dual relationship, giving

$$-3 \leq y(\mathbf{x}^*) \leq 1.5$$

TABLE VII-9 SOME FEASIBLE SOLUTIONS OF THE PRIMAL AND DUAL PROBLEMS (EXAMPLE VII-41)

Primal	x_1	1	2	2	0.5
	x_2	1	3	1	1.0
	$y(\mathbf{x})$	2	5	3	1.5
Dual	λ_1	1	0	6	
	λ_2	0	0.5	1	
	λ_3	1	2	9	
	$W(\boldsymbol{\lambda})$	−6	−3	−40	

By making further guesses at the values of $\mathbf{x}$ and $\boldsymbol{\lambda}$, these limits could be narrowed. The optimal solutions, however, can easily be found in this case by the Simplex method, yielding $y(\mathbf{x}^*) = 1.0$ at $x_1^* = 0$, $x_2^* = 1$, and $W(\boldsymbol{\lambda}^*) = 1.0$ at $\lambda_1^* = 0$, $\lambda_2^* = 0.5$, and $\lambda_3^* = 0$. $y(\mathbf{x}^*)$ of course equals $W(\boldsymbol{\lambda}^*)$.

This example illustrates the nature of the dual problem. A number of details, however, must be considered before making full use of the primal-dual relationship. It has been seen that our demonstration of Eq. (5.1-8), using Lagrangian multipliers, required the assumption of an optimum at a stationary point. If the optimum to the primal problem does not exist (for example, lying at infinity), the dual system will not have a feasible solution (Vajda [1961, p. 42] and others). If one problem, dual or primal, has a finite solution, then so must the other.

It is of importance to note that our statement of the primal problem included restrictions in the form of inequalities. What differences are introduced if one or more of the restrictions become strict equalities? If one of the inequalities is replaced by an equality, the corresponding λ is no longer sign-limited. That is, it need not be nonnegative as required by inequality (5.1-6). This difference is easily understood if we recall that the λ's in the dual problem are Lagrangian multipliers and the non-negative character of the Lagrangian multipliers arose because of the Kuhn-Tucker conditions, which do not apply in the case of equality restrictions. Thus an equality restriction in the primal problem, yielding a variable of unrestricted sign in the dual problem, is reversible in the following sense: a variable of unrestricted sign in the primal problem will yield an equality restriction in the dual. This fact can again be best seen from the Lagrangian formulation. If x_p is unrestricted, a Lagrangian multiplier, α_p, will not appear in Eq. (5.1-8). Equation (5.1-11) cannot, therefore, be reduced to an inequality, as was done in our previous analysis.

Apart from the use of duals to bound an estimate of the optimum solution, a dual system might prove easier to solve than the primal if in the latter it is more difficult to obtain a first feasible basic solution. In general, it is more efficient to solve the dual problem by the Simplex method if the linear programming problem has many constraints but few variables, since this requires a large basis matrix. It can also be shown that the optimal conditions in the primal can be derived from the optimal solution to the dual. Let us illustrate these points.

Example VII-42: An easier dual system

Consider the problem of minimizing

$$y = 3x_1 + 5x_2$$

subject to the restrictions that

$$x_1 + 3x_2 \geq 14 \tag{5.1-15a}$$

$$2x_1 - x_2 \geq 2 \tag{5.1-15b}$$

$$x_1 - 4x_2 \leq 2 \tag{5.1-15c}$$

$$x_1 + x_2 \leq 20 \tag{5.1-15d}$$

Adding slack variables x_3, x_4, x_5, and x_6 (Sec. 4.4.3) and artificial variables x_{100} and x_{101} (Sec. 4.4.4), we may write these constraints as

$$x_1 + 3x_2 - x_3 + x_{100} = 14$$

$$2x_1 - x_2 - x_4 + x_{101} = 2$$

$$x_1 - 4x_2 + x_5 = 2$$

$$x_1 + x_2 + x_6 = 20$$

to get the first array in the Simplex technique:

ARRAY 1*a*

			x_1	x_2	x_3	x_4
	x_{100}	14	1	3	−1	0
→	x_{101}	2	2	−1	0	−1
	x_5	2	1	−4	0	0
	x_6	20	1	1	0	0
	y	0	−3	−5	0	0
	(P)	16	3	2	−1	−1
			↑			

Applying the usual rules of Simplex array transformation then gives:

ARRAY 1*b*

			x_2	x_3	x_4
→	x_{100}	13.0	3.5	−1.0	0.5
	x_1	1.0	−0.5	0	−0.5
	x_5	1.0	−3.5	0	0.5
	x_6	19.0	3.5	0	0.5
	y	3.0	−6.5	0	−1.5
	(P)	13.0	3.5	−1.0	0.5
			↑		

ARRAY 1c

		x_3	x_4
x_2	3.714	−0.286	0.143
x_1	2.857	−0.143	−0.429
x_5	14.000	−1.000	1.000
x_6	13.429	0.429	0.286
y	27.143	−1.857	−0.571

This will be the optimal solution, so that

$$y^*(2.857,3.714) = 27.143$$

Let us now consider the optimization of the dual problem. Before this dual can be stated we must rewrite inequalities (5.1-15c) and (5.1-15d) as

$$-x_1 + 4x_2 \geq -2$$
$$-x_1 - x_2 \geq -20$$

The dual can then be written as the maximization of W, where

$$W = 14\lambda_1 + 2\lambda_2 - 2\lambda_3 - 20\lambda_4$$

subject to $\lambda_i \geq 0$ and

$$\lambda_1 + 2\lambda_2 - \lambda_3 - \lambda_4 \leq 3$$
$$3\lambda_1 - \lambda_2 + 4\lambda_3 - \lambda_4 \leq 5$$

Adding slack variables λ_5 and λ_6, the stages in the solution are:

ARRAY 1

			λ_1	λ_2	λ_3	λ_4
	λ_5	3	1	2	−1	−1
→	λ_6	5	3	−1	4	−1
	W	0	−14	−2	2	20
			↑			

ARRAY 2

			λ_6	λ_2	λ_3	λ_4
→	λ_5	1.333	−0.333	2.333	−2.333	−0.667
	λ_1	1.667	0.333	−0.333	1.333	−1.333
	W	23.333	4.667	−6.667	20.667	15.333
				↑		

ARRAY 3

		λ_6	λ_5	λ_3	λ_4
λ_2	0.571	−0.143	0.429	−1.000	−0.286
λ_1	1.857	0.286	0.143	1.000	−0.429
W	27.143	3.714	2.857	14.000	13.429

In this case the dual problem required fewer calculations, since the first basic solution was feasible. It will be seen from the optimal arrays for the primal and the dual that not only are the optimal values of y and W equal, but there is a correspondence between elements in each array and the subscripts of the main and slack variables are interchanged between the problems. Thus values of the optimal λ_j can be found from the y row of the x_j columns of the primal, but with an opposite sign. Similarly, the optimal x_i could have been found from the W row corresponding to the λ_i. The other elements in the two final arrays have a similar correspondence, with a changed sign and in the mirror-image position.

The computational advantages of duals in linear systems have been discussed, for example, by Hadley [1962, chap. 8], Vajda [1961, chap. 5], and Dantzig [1963, chap. 11]. This dual alternative to the Simplex method has been called the *Dual*, or *Dual Simplex*, method, being due to Lemke [1954]. Computationally, similar variants of the Simplex approach are due to Beale [1954] and Dantzig, Ford, and Fulkerson [1956].

Once a problem has been solved, any further constraints to be added to the system can be accommodated by adding a further variable to the dual and starting from the optimal solution to the original dual problem, with the new variable initially zero. Duals can also be used to test the sensitivity of the solution to changes in values of the system coefficients in the objective function and restrictions (see, for example, Chung [1963, chap. 7]). Dual linear programming has also found a use, for instance, in the formulation of approximating functions, using the Chebyshev criterion (Leung and Quon [1966]).

5.2 Theory of Legendre Transformations

The understanding of duality in certain nonlinear systems can best be achieved by first considering Legendre transformations. Important in many aspects of physics and chemistry (especially thermodynamics), Legendre transformations allow us to resolve the following general problem: Given a function $y(\mathbf{x})$, where the $\mathbf{x}$ are independent variables, find an equivalent representation $y(\mathbf{d})$ in which the partial derivatives ($d_i = \partial y/\partial x_i$) of the original function are used as the independent variables. In other words, we want to transform the problem so that the new independent variables are given by the original partial derivatives. At first thought the problem seems trivial. Thus, for a single variable, the function $y(x_1)$ can be differentiated, the resultant expression solved so that x_1 can be expressed in terms of d_1 ($= y_{x_1}$), and finally, x_1 eliminated from $y(x_1)$.

Example VII-43: A simple transformation

Suppose that

$$y(x_1) = x_1^2 + 3$$

Then, since $d_1 = dy/dx_1 = 2x_1$, we can write that $x_1 = \frac{1}{2}d_1$ and get

$$y(d_1) = 0.25d_1^2 + 3$$

It will be observed, however, that by accepting this approach to the stated problem, we have made a certain sacrifice. This can be seen by noting that the transformation cannot be reversed. Thus, given a function $y(d_1)$, we cannot recover the function $y(x_1)$, since a constant of integration arises in solving the differential equation.

Example VII-44: The impossibility of inverting the simple transform

Suppose that we are given the relationship

$$y(d_1) = 0.25d_1^2 + 3$$

and we wish to find $y(x_1)$, where $d_1 = y_{x_1}$. We must, obviously, solve the differential equation

$$y = 0.25\left(\frac{dy}{dx_1}\right)^2 + 3$$

Rearranging gives

$$\frac{dy}{2\sqrt{y-3}} = dx_1$$

which, on integration, yields

$$y = 3 + (x_1 + C)^2$$

C being a constant of integration which cannot be evaluated, since no other conditions are to be satisfied.

The obvious solution to the stated problem of finding a representation in terms of the partial derivatives is thus not acceptable, since the transformation is irreversible. A satisfactory solution can, however, be found by noting that a given curve might be represented equally well as a family of tangents, as shown in Fig. VII-28. Clearly, by being able to construct the tangent lines, the curve $y(x_1)$ can also be found. These tangent lines are determined by two numbers, the slope d_1 and the intercept of the tangent with the y axis. The latter depends on the slope and can therefore be represented as $\psi(d_1)$. A knowledge of $\psi(d_1)$ is equivalent to a knowledge of $y(x_1)$. The Legendre transformation is the mathematical manipulation giving the

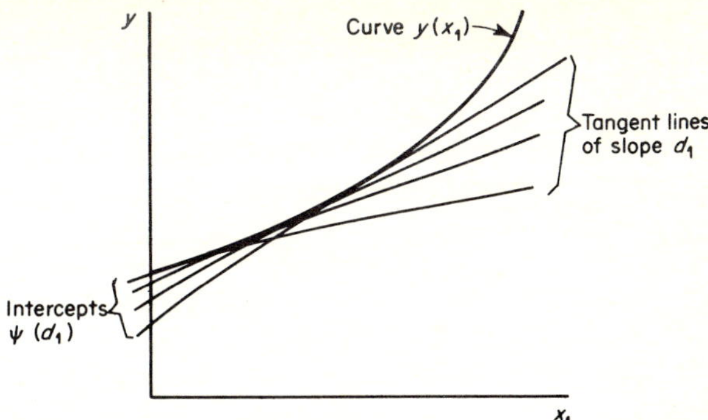

Fig. VII-28 Representation of a curve by a family of tangents.

relationship between $\psi(d_1)$ and $y(x_1)$. For the simple case of a single variable, this construction can easily be seen. From Fig. VII-29 we see that

$$d_1 = \frac{y - \psi}{x_1}$$

or

$$\psi = y - d_1 x_1 \tag{5.2-1}$$

The x_1 can then be eliminated from this equation by forming $d_1 = y_{x_1}$ and solving for x_1 as before.

Example VII-45: The Legendre transformation

Suppose that we are given the function

$$y = x_1^2 + 3$$

Fig. VII-29 Construction for the Legendre transformation.

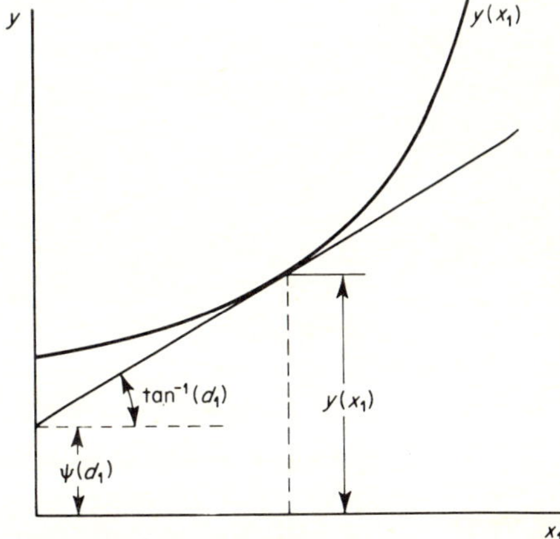

Then, from Eq. (5.2-1), we have

$$\psi = x_1^2 + 3 - d_1 x_1$$

Since $d_1 = y_{x_1} = 2x_1$, the intercept can be written in terms of d_1 as follows:

$$\psi(d_1) = 0.25d_1^2 + 3 - 0.5d_1^2 = 3 - 0.25d_1^2$$

Given this locus of intercepts, $\psi(d_1)$, the original curve $y(x_1)$ could be reconstructed by geometric means. We can thus conclude that $y(x_1)$ and $\psi(d_1)$ are equivalent representations of the same curve.

The function $\psi(d_1)$ is called the *Legendre transformation* of the function $y(x_1)$. This transformation exists when the determinant of the second derivatives (the Hessian, Sec. 2.2.3) does not vanish (Lanczos [1966]). We have, moreover, indicated how the curve can be constructed geometrically, given the function $\psi(d_1)$. We must now obtain the reverse transformation so as to find $y(x_1)$, given $\psi(d_1)$.

The inverse transformation is easily found by considering the total derivative of Eq. (5.2-1) as follows:

$$d(\psi) = dy - x_1\, d(d_1) - d_1\, dx_1$$

or since y depends on x_1

$$d(\psi) = \frac{\partial y}{\partial x_1} dx_1 - x_1\, d(d_1) - d_1\, dx_1 = d_1\, dx_1 - x_1\, d(d_1) - d_1\, dx_1$$

$$= -x_1\, d(d_1)$$

Therefore

$$x_1 = -\frac{\partial \psi}{\partial d_1} \tag{5.2-2}$$

This result reveals a striking analogy with the generating equation for d_1, since

$$d_1 = \frac{\partial y}{\partial x_1}$$

Thus, apart from a sign, d_1 plays the same role in ψ as x_1 does in y. The $y(x_1)$ can now be found by solving Eq. (5.2-2) for d_1 in terms of x_1 and eliminating d_1 from Eq. (5.2-1).

Example VII-46: The inverse transformation

If, as in Example VII-45,

$$\psi = 3 - 0.25d_1^2$$

then, by Eq. (5.2-2), we find that

$$x_1 = 0.5d_1$$

Substituting this into Eq. (5.2-1) then gives

$$y = \psi + d_1x_1 = 3 - 0.25d_1^2 + d_1x_1 = 3 - x_1^2 + 2x_1^2 = 3 + x_1^2$$

This is, of course, our original function of Example VII-45.

These ideas and mathematical results can be extended to n dimensions. Thus $\psi(\mathbf{d})$ will be the intercept of a plane which is tangent to a surface $y(\mathbf{x})$ at a point at which the slope of the plane is $\mathbf{d}$ $(d_i = \partial y/\partial x_i)$. By an analogous reasoning, we can show that the intercept ψ is given as a function of the slopes of the tangent planes as

$$\psi = y - \sum_{i=1}^{n} d_i x_i \qquad (5.2\text{-}3)$$

By using the n equations $d_i = \partial y/\partial x_i$, the x_i can in principle be eliminated, leaving ψ as a function of the derivatives. The reverse transformation can then be made by using Eq. (5.2-3), together with

$$x_i = -\frac{\partial \psi}{\partial d_i} \qquad (5.2\text{-}4)$$

So far we have discussed the Legendre transformation without showing its relationship to the theory of duals. Let us now do this.

5.3 Duals in Convex Systems

Let us consider the more general problem of finding dual systems. We shall confine our attention to minimization problems in convex systems, but will find (Sec. 5.3.1) that duals cannot be constructed for all such problems, or indeed prove useful in every case. One particular field of application will be examined further in Sec. 5.3.2.

5.3.1 Duals and a general convex function

Before examining convex objective functions and duality, let us recall, from Chap. IV, Sec. 2.3, the mathematical definition of such functions. A convex function is, by definition, one which satisfies the inequality

$$(1 - \theta)y(\mathbf{x}_1) + \theta y(\mathbf{x}_2) \geq y[(1 - \theta)\mathbf{x}_1 + \theta\mathbf{x}_2] \qquad (5.3\text{-}1)$$

for all $0 \leq \theta \leq 1$ and all points $\mathbf{x}_1$ and $\mathbf{x}_2$. A somewhat weaker inequality has also been developed (p. 111), yielding

$$y(\mathbf{x}_2) \geq y(\mathbf{x}_1) + \sum_{i=1}^{n} d_{i,1}(x_{i,2} - x_{i,1}) \qquad (5.3\text{-}2)$$

The existence of the Legendre transformation requires a further restriction on the class of objective functions under consideration. Thus, while linear functions satisfy inequality (5.3-1), and are therefore convex, it was found in Sec. 5.2 that the Legendre transformation cannot exist for a linear function, since the Hessian vanishes. Not all convex functions, therefore, possess Legendre transformations, and these cannot be considered further. The particular case of linear functions has already been discussed in Sec. 5.1. Let us now suppose that the Legendre transformation of $y(\mathbf{x})$ exists. Inequality (5.3-2), on substitution from Eq. (5.2-3), then reduces to

$$y(\mathbf{x}_2) \geq \psi(\mathbf{d}_1) + \sum_{i=1}^{n} d_{i,1}x_{i,2} \tag{5.3-3}$$

If we select $\mathbf{x}_2$ to be the optimum $\mathbf{x}^*$ and at the same time restrict our attention to those values of $\mathbf{d}$ satisfying the restriction

$$\sum_{i=1}^{n} x_i^* \, d_i = 0 \tag{5.3-4}$$

we then have

$$M = y(\mathbf{x}^*) \geq \psi(\mathbf{d}) \tag{5.3-5}$$

Since, at the minimum, $\mathbf{d} = 0$, this latter restriction includes the optimum, and from Eq. (5.2-3) we then see that

$$M = y(\mathbf{x}^*) = \psi(0) \tag{5.3-6}$$

In view of inequality (5.3-5), we see that the maximum value that ψ can attain is in fact M, and that this is found with $\mathbf{d} = 0$. The function $\psi(\mathbf{d})$, with the $\mathbf{d}$ suitably restricted in the way noted above, is therefore a dual to the original problem of minimizing $y(\mathbf{x})$, since the maximum of the dual objective function equals the minimum of the primal. These results may seem rather abstract, but will become clarified by the following example. We may note, in particular, that the general dual-system specification depends upon knowledge of the optimal solution of the primal problem, as required by Eq. (5.3-4).

Example VII-47: The dual of a convex function

Let us find the dual to the primal problem of minimizing the objective function

$$y = 2 - 5x_1 - 12x_2 + \tfrac{1}{2}x_1^2 + 2x_2^2 \tag{5.3-7}$$

Then

$$d_1 = \frac{\partial y}{\partial x_1} = -5 + x_1 \tag{5.3-8a}$$

$$d_2 = \frac{\partial y}{\partial x_2} = -12 + 4x_2 \tag{5.3-8b}$$

The Legendre transformation then becomes, after substituting from Eqs. (5.3-7), (5.3-8a), and (5.3-8b) into (5.2-3),

$$\psi = -28.5 - 5d_1 - 3d_2 - 0.5d_1^2 - 0.125d_2^2 \tag{5.3-9}$$

This function is dual to the y of Eq. (5.3-7) only if the d_1 and d_2 are suitably restricted. In other words, feasible values of d_1 and d_2 must satisfy Eq. (5.3-4), giving here

$$x_1^* d_1 + x_2^* d_2 = 0 \tag{5.3-10}$$

where $\mathbf{x}^*$ is the optimal solution to the primal problem. Any choices of $\mathbf{d}$ which do not satisfy Eq. (5.3-10) will not yield a dual. The construction of the dual system cannot, therefore, be completed, in the general case, unless we can determine the optimal solution $\mathbf{x}^*$ of the primal problem. From Eqs. (5.3-8a) and (5.3-8b) we see[1] that the stationary points in y exist when $\partial y/\partial x_i = 0$, so that $x_1^* = 5$ and $x_2^* = 3$. All admissible values of d_1 and d_2 must thus satisfy Eq. (5.3-10) in the form

$$5d_1 + 3d_2 = 0 \tag{5.3-11}$$

Thus, substituting for d_1 from this, we get

$$\psi = -28.5 - 0.305d_2^2 \tag{5.3-12}$$

which is a maximum when $d_2 = 0$. Hence $d_1 = 0$. Thus, for all feasible $\mathbf{d}$, the Legendre transformation is $\psi \leq -28.5$, with the strict equality holding only when $\mathbf{d} = 0$. The minimum value of y is thus -28.5, as could be verified by substitution of $\mathbf{x}^*$ in Eq. (5.3-7).

It can be easily seen that $\psi(\mathbf{d})$ is not a dual unless the condition $\sum_{i=1}^{n} d_i x_i^* = 0$ is satisfied. By selecting values of $\mathbf{d}$ which do not satisfy Eq. (5.3-11), larger values of $\psi(\mathbf{d})$ can be found. Thus, if $d_1 = -5$, $d_2 = -3$, then $\psi = -8.875$. This is larger than the minimum value of y, and thus does not provide a lower bound for the primal problem.

We have seen that the duals developed here can be applied only in convex-minimization (or concave-maximization) problems when $\sum_{i=1}^{n} d_i x_i^* = 0$. Thus the solution to the primal is required before the dual can be constructed. The value of this is therefore questionable. There are, however, special cases

[1] In case of possible confusion, it should be noted that the specification of a stationary value in y by letting $d_i = \partial y/\partial x_i = 0$ does not mean that all the feasible d_i in the dual are zero. It will be remembered that the primal and dual systems share a common solution only when this is the optimal solution. The dual variables d_i need therefore be zero only at the optimal solution.

in which the restriction can be satisfied without first solving the original problem. Let us now consider such a case.

5.3.2 *Geometric programming—a successful application of the dual of convex functions*

There is a significant application in the class of problems discussed by Zener [1961], in which the objective function has the separable form

$$y = \sum_{i=1}^{n} y_i \tag{5.3-13}$$

with further restrictions that each term in the series has the form[1]

$$y_i = c_i t_1^{a_{i,1}} t_2^{a_{i,2}} \cdots t_k^{a_{i,k}} \qquad \text{for } i = 1, 2, \ldots, n \tag{5.3-14}$$

where the c_i are positive, and the independent variables t_j are constrained by

$$0 < t_j < \infty \tag{5.3-15}$$

From the results of Sec. 5.3.1, we can find a dual for Zener's problem. This dual is the basis for the technique often called *geometric programming*. By the use of this, we shall find that the dual has a potential usefulness, since its construction does not require the solution to the primal. Let us demonstrate this. Since the c_i and $\mathbf{t}$ are all positive, a variable x_i can be defined as

$$x_i = \sum_{j=1}^{k} a_{i,j} \ln t_j \tag{5.3-16}$$

so that

$$y(\mathbf{x}) = \sum_{i=1}^{n} c_i e^{x_i} \tag{5.3-17}$$

The derivatives of $y(\mathbf{x})$ are then

$$d_i \equiv \frac{\partial y}{\partial x_i} = c_i\, e^{x_i} \tag{5.3-18a}$$

or

$$x_i = \ln \frac{d_i}{c_i} \tag{5.3-18b}$$

The Legendre transformation of $y(\mathbf{x})$ can then be written

$$\psi(\mathbf{d}) = y(\mathbf{x}) - \sum_{i=1}^{n} d_i x_i = \sum_{i=1}^{n} d_i - \sum_{i=1}^{n} d_i \ln \frac{d_i}{c_i} \tag{5.3-19}$$

[1] We have used t to denote the independent variables because of a later intended change of variable which will bring the resulting expressions more nearly in agreement with the usual notation. Before proceeding, we should note that optimization problems such as that posed here are frequently encountered in engineering design, where the total cost is often the sum of subsystem costs, and the latter are in many instances expressed as power functions of the design variables $t_1, t_2, \ldots, t_k$. Alternatively, the power function may provide a good fit over some limited range of operation.

As seen in Sec. 5.3.1, the $\psi(\mathbf{d})$ is a dual, provided that $y(\mathbf{x})$ is a convex function and has a Legendre transformation and the **d** satisfy

$$\sum_{i=1}^{n} x_i^* d_i = 0 \tag{5.3-20}$$

Here $y(\mathbf{x})$ is a convex function of **x**, and it remains only to select the **d** so that

$$\sum_{i=1}^{n} x_i d_i = \sum_{i=1}^{n} \sum_{j=1}^{k} a_{i,j} (\ln t_j) d_i = 0$$

It is possible to satisfy this condition without knowing t^* by demanding that the **d** satisfy the more restrictive condition

$$\sum_{i=1}^{n} a_{i,j} d_i = 0 \qquad \text{for } j = 1, 2, \ldots, k \tag{5.3-21}\dagger$$

This second, more restrictive condition will also include the optimum, and hence the **d**, which gives ψ its maximum value. To see that this is indeed the case, let us examine the stationary points of $y(\mathbf{t})$. Differentiating Eq. (5.3-13), we see that

$$\frac{dy}{dt_j} = \sum_{i=1}^{n} \frac{dy_i}{dt_j} = \sum_{i=1}^{n} \frac{a_{i,j} y_i}{t_j}$$

or at the optimum,

$$\sum_{i=1}^{n} a_{i,j} y_i^* = 0$$

Since in this case $y_i = d_i$ [Eq. (5.3-18)], we see that the **d** evaluated at the optimum also satisfy Eq. (5.3-21). The new and more stringent restriction on governing admissible choices of **d** does, therefore, also contain the optimum. If the **d** are selected in accordance with Eq. (5.3-21), then Eq. (5.3-20) will automatically be satisfied. This feature of the dual in Zener's problem thus allows us to select suitable **d** using Eq. (5.3-21) without actually having the solution to the original problem. A number of solutions of the k equations (5.3-21) are possible if $n > k$. A useful form of dual system can thus be established.

Duffin [1962a, 1962b] has shown that Eq. (5.3-19) can be simplified considerably by normalizing the vector **d**. Since each component of **d** is positive, a positive quantity θ can be defined as the sum of the components of **d** as

$$\theta = \sum_{i=1}^{n} d_i \tag{5.3-22}$$

† If all the $a_{i,j}$ are of the same sign, it is impossible to find a set of **d** which give a feasible solution to this equation, since the **d** are restricted to take on only positive values.

A normalized vector $\boldsymbol{\delta}$ can then be defined as having components

$$\delta_i = \frac{d_i}{\theta} \tag{5.3-23}$$

Hence we have that

$$\sum_{i=1}^{n} \delta_i = 1 \tag{5.3-24}$$

which becomes a restriction on the δ_i. Equation (5.3-19) then gives

$$\psi(\boldsymbol{\delta},\theta) = \theta - \theta \ln \theta - \theta \sum_{i=1}^{n} \ln \left(\frac{\delta_i}{c_i}\right)^{\delta_i} \tag{5.3-25}$$

Since ψ provides a lower limit to the original problem, we should now like to make ψ as large as possible, so that the lower limit is set as near to the true solution as can be attained with assumed values of **d**. We shall therefore choose the θ so that $d\psi/d\theta$ vanishes. This derivative is, for constant $\boldsymbol{\delta}$,

$$\frac{d\psi}{d\theta} = -\ln \theta - \sum_{i=1}^{n} \ln \left(\frac{\delta_i}{c_i}\right)^{\delta_i} = 0$$

or

$$\theta = \left(\frac{c_1}{\delta_1}\right)^{\delta_1} \left(\frac{c_2}{\delta_2}\right)^{\delta_2} \cdots \left(\frac{c_n}{\delta_n}\right)^{\delta_n}$$

$$= \prod_{i=1}^{n} \left(\frac{c_i}{\delta_i}\right)^{\delta_i} \tag{5.3-26}$$

Substituting this best value of θ in Eq. (5.3-25), we then find that

$$\psi(\boldsymbol{\delta}) = \prod_{i=1}^{n} \left(\frac{c_i}{\delta_i}\right)^{\delta_1} \tag{5.3-27}$$

with the $\boldsymbol{\delta}$ satisfying Eqs. (5.3-24) and

$$\sum_{i=1}^{n} a_{i,j}\delta_i = 0 \qquad \text{for } j = 1, 2, \ldots, k \tag{5.3-28}$$

This is the desired dual to the original problem of minimizing y.

In using the dual, a set of $\boldsymbol{\delta}$ satisfying the restrictions, Eqs. (5.3-24) and (5.3-28) must be selected, and a value of ψ computed from Eq. (5.3-27). This value will be less than or equal to the true solution to the original problem. We should not forget, however, that the solution of Zener's problem, using a dual, is in general as difficult, if not more so, as solving the primal problem. The main advantage is found, however, in the fact that any feasible primal and dual solutions will bound the desired optimal solution.

Example VII-48: The dual in Zener's problem

Let us suppose that the objective function y is of the form

$$y = t_1^2 + 2t_2 + \frac{3}{t_1 t_2} + 2t_1 t_2 \tag{5.3-29}$$

and that it is desired to find those positive values of t_i which maximize y.

From Eqs. (5.3-13) and (5.3-14) the $a_{i,j}$ and c_i coefficients will be $a_{1,1} = 2$, $a_{1,2} = 0$, $a_{2,1} = 0$, $a_{2,2} = 1$, $a_{3,1} = -1$, $a_{3,2} = -1$, $a_{4,1} = 1$, $a_{4,2} = 1$; $c_1 = 1$, $c_2 = 2$, $c_3 = 3$, $c_4 = 2$. The dual function is therefore, from Eqs. (5.3-27) and (5.3-28),

$$\psi = \left(\frac{1}{\delta_1}\right)^{\delta_1} \left(\frac{2}{\delta_2}\right)^{\delta_2} \left(\frac{3}{\delta_3}\right)^{\delta_2} \left(\frac{2}{\delta_4}\right)^{\delta_4} \tag{5.3-30}$$

subject to the restrictions

$$\delta_1 + \delta_2 + \delta_3 + \delta_4 = 1 \tag{5.3-31a}$$

$$2\delta_1 \qquad - \delta_3 + \delta_4 = 0 \tag{5.3-31b}$$

$$\delta_2 - \delta_3 + \delta_4 = 0 \tag{5.3-31c}$$

Any value of $\boldsymbol{\delta}$ which satisfies these restrictions will provide a lower limit to the optimal solution of the original problem. Thus, for example,

$$\boldsymbol{\delta}_a \equiv \{\delta_1 = 1/15,\ \delta_2 = 2/15,\ \delta_3 = 7/15,\ \delta_4 = 1/3\}$$

and $$\boldsymbol{\delta}_b \equiv \{\delta_1 = 1/10,\ \delta_2 = 1/5,\ \delta_3 = 9/20,\ \delta_4 = 1/4\}$$

both satisfy the restrictions, as can easily be verified, with $\psi(\boldsymbol{\delta}_a) = 7.44$ and $\psi(\boldsymbol{\delta}_b) = 7.89$. Since the optimal solution is larger than or equal to all values of ψ computed using feasible values of $\boldsymbol{\delta}$, the larger of the two calculated values of ψ will provide a lower limit. Thus

$$\min y = \max \psi \geq 7.89$$

The solution can be given an upper bound by considering the primal problem and assuming a set of feasible values of t_i. Thus, taking $t_1 = 1$, $t_2 = 1.5$, we get

$$y(1,1.5) = 9$$

The desired solution is thus bounded as

$$9 \geq \min y = \max \psi \geq 7.89$$

We see here how the dual has allowed us to limit the desired solution within a range of values without having to solve the problem. If this range is narrow enough for all practical purposes, the solution is entirely unnecessary, and one of the assumed sets could be adopted as the solution. For example, if the data prove very inaccurate, it may not be desirable to determine the precise minimum, and the values $t_1 = 1$, $t_2 = 1.5$ may be acceptable, since we know that the true optimal operating conditions cannot produce a value of y less than 7.89. We might then stop the optimization. If, however, a better set of conditions is required to produce narrower limits, these can be established by numerical search procedures.

The exact solution can of course be found here. By solving the restrictions, δ_1, δ_2, and δ_3 can be expressed in terms of δ_4. Since the restrictions are linear, Cramer's rule (Appendix 1) can be used to get

$$\delta_1 = 0.2 - 0.4\delta_4 \tag{5.3-32a}$$

$$\delta_2 = 0.4 - 0.8\delta_4 \tag{5.3-32b}$$

$$\delta_3 = 0.4 + 0.2\delta_4 \tag{5.3-32c}$$

Since all the components must be positive, then $0 < \delta_4 < 0.5$. We could now write ψ as a function of δ_4. Since the logarithm of ψ is a monotonic function of ψ, the maximum of ψ can be found by maximizing the logarithm of ψ. Thus we need

$$\frac{d(\ln \psi)}{d\delta_4} = -(1 + \ln \delta_1)\frac{d\delta_1}{d\delta_4} - \left(1 + \ln \frac{\delta_2}{2}\right)\frac{d\delta_2}{d\delta_4} - \left(1 + \ln \frac{\delta_3}{3}\right)\frac{d\delta_3}{d\delta_4} - \left(1 + \ln \frac{\delta_4}{2}\right) = 0 \tag{5.3-33}$$

derived from Eq. (5.3-30). From Eqs. (5.3-32) we see that

$$\frac{d\delta_1}{d\delta_4} = -0.4 \qquad \frac{d\delta_2}{d\delta_4} = -0.8 \qquad \frac{d\delta_3}{d\delta_4} = 0.2$$

so that Eq. (5.3-33) reduces to

$$\frac{\delta_1^2(\delta_2/2)^4}{\delta_3/3(\delta_4/2)^5} = 1$$

The simultaneous solution of this expression and Eqs. 5.3-32 yield

$$\boldsymbol{\delta}^* = \{\delta_1 = 0.1137,\ \delta_2 = 0.2274,\ \delta_3 = 0.4431,\ \delta_4 = 0.2158\}$$

as the maximizing set, with max $\psi = 7.92$.

This maximum should be the same as the minimum of y. The necessary conditions for a minimum in y are

$$\frac{\partial y}{\partial t_1} = 2t_1 - \frac{3}{t_1^2 t_2} + 2t_2 = 0$$

$$\frac{\partial y}{\partial t_2} = 2 - \frac{3}{t_1 t_2^2} + 2t_1 = 0$$

Solving, we get $t_1^* = 0.9490$, $t_2^* = 0.9006$, with min $y = 7.92$.

There may often be little computational advantage in solving the dual system rather than the primal. In this particular illustration, for example, it was simpler to solve for **t**, rather than to maximize the dual. In some cases, however, the dual will be the easier problem. Thus, if the number of terms, n, in the objective function is one more than the number of independent variables, k, then the restrictions given by Eqs. (5.3-24) and (5.3-28) provide an immediate solution to the primal problem, since these restrictions are linear. This might prove an important computational advantage.

Example VII-49: A unique set of dual variables

Minimize y, where

$$y = t_1^2 + 2t_2 + \frac{3}{t_1 t_2} \tag{5.3-34}$$

Then $a_{1,1} = 2$, $a_{1,2} = 0$, $a_{2,1} = 0$, $a_{2,2} = 1$, $a_{3,1} = -1$, $a_{3,2} = -1$, $c_1 = 1$, $c_2 = 2$, $c_3 = 3$, and

$$\psi = \left(\frac{1}{\delta_1}\right)^{\delta_1}\left(\frac{2}{\delta_2}\right)^{\delta_2}\left(\frac{3}{\delta_3}\right)^{\delta_3} \tag{5.3-35}$$

where the $\boldsymbol{\delta}$ are subject to the restrictions

$$\delta_1 + \delta_2 + \delta_3 = 1 \tag{5.3-36a}$$

$$2\delta_1 \qquad - \delta_3 = 0 \tag{5.3-36b}$$

$$\delta_2 - \delta_3 = 0 \tag{5.3-36c}$$

Since there are three unknowns and three restrictions, the $\boldsymbol{\delta}$ are uniquely defined by Eqs. (5.3-36). Thus $\boldsymbol{\delta}^* \equiv \{\delta_1 = 0.2, \delta_2 = 0.4, \delta_3 = 0.4\}$. This situation may be contrasted with the preceding illustration in which there were more unknowns than restrictions.

The maximum value of ψ and the minimum value of y are then

$$\psi = (5)^{0.2}(5)^{0.4}(7.5)^{0.4} = 5.88$$

We can also find the corresponding set of **t** from these results. Since, from Eq. (5.3-23), we have $d_i = \theta\delta_i$, then

$$y = \sum_{i=1}^{n} d_i = \theta$$

At the minimum we thus get that

$$\min y = \theta = 5.88 \tag{5.3-37}$$

Substitution for θ then gives, from Eq. (5.3-23),

$$d_1 = 5.88(0.2) = 1.176$$

$$d_2 = 5.88(0.4) = 2.352$$

$$d_3 = 5.88(0.4) = 2.352$$

From Eqs. (5.3-16) and (5.3-18*a*) we can now write that

$$d_i = c_i t_1^{a_{i,1}} t_2^{a_{i,2}} \cdots t_k^{a_{i,k}}$$

or as here, $d_1 = t_1^2$, $d_2 = 2t_2$, $d_3 = 3/t_1 t_2$. These equations can be solved for t to give the optimal conditions as $t_1^* = 1.084$ and $t_2^* = 1.176$.

In this example the dual turns out to be a far easier problem than the primal. It can also be shown, for this case in which $n = k + 1$, that the solution of the dual can be transformed into knowledge of the primal problem. Thus, if $\boldsymbol{\delta}^*$ is the vector satisfying the restrictions and maximizing ψ, we know that

$$\psi(\boldsymbol{\delta}^*) = y(\mathbf{t}^*) = \theta \tag{5.3-38}$$

It will be remembered that y and ψ are equivalent only under optimal conditions, and Eq. (5.3-38) is thus valid only for this situation. Moreover,

$$y = \sum_{i=1}^{n} d_i = \sum_{i=1}^{n} y_i$$

Hence $\delta_i \theta = d_i$ gives the portion of the total objective function which is associated with the ith subsystem. If y is an overall cost which is to be minimized, it can be seen from the solution of the dual which of the subsystems are making the major cost contribution to the total operation. The largest values of δ_i are associated with the largest costs, and this aspect of the dual solution is valuable to the extent of recognizing the important cost terms without further calculation.

We should emphasize that the main value of a dual system lies in the estimation of the approximate solution to the primal problem by obtaining feasible, but not necessarily optimal, solutions. This allows bounds to be placed upon the range of values expected. This aspect could find valuable application in numerical optimization methods. In certain cases the dual system may also represent a simpler problem than the original, and thus allow the optimal solution to be determined more easily. The intrinsic weakness of the dual in the minimization of convex systems as developed is that the optimal solution to the general primal problem must be known before the specifications of the dual system can be completed.

6 SOME GENERAL CONCLUSIONS FOR CHAPTER VII

It will be clear that the general procedure for finding the greatest or least value of a function is much more difficult with several variables than it is with one. The analytical methods proposed can handle only continuous functions with continuous derivatives. If necessary, the region under analysis must be restricted by artificial boundaries, so that each region contains a function which can be defined by a single formula and possesses continuous derivatives.

When no restrictions are present and the derivatives and the function are continuous, the location of the possible optima can be determined by solving the n simultaneous equations formed by setting the first derivatives equal to zero, and the nature of these stationary points can in principle then be examined using the second derivatives. If such an analysis can be carried out, it is obviously possible to discuss the system in a more general manner than is possible with purely numerical methods, since analytical methods define all the stationary points in a system. In many cases, however, the search for the roots of the first-derivative equations is a formidable one which must often be carried out by numerical methods. The question should then be asked if the search should not, in fact, be attempted by a purely numerical technique, in which the function, and possibly its gradient, are evaluated at a particular location, rather than the location of the roots being sought after the gradients have been set equal to zero.

Several techniques are possible when equality constraints are present, the method of Lagrangian multipliers finding the most application. This approach can, however, define only potential locations of optima, in the general case. If sufficiency conditions are required, recourse must be made to the method of constrained variations. Another disadvantage of the Lagrangian multipliers method is that it expands the dimensionality of the problem by introducing a new variable for each constraint. There is, actually, no need to distinguish between problems containing equalities or inequality restrictions, since the latter, as in the linear programming techniques, can always be converted to the former class by the use of slack variables. While the two classes were separated for ease of analysis, this distinction is not normally necessary, and problems containing mixed constraints can be handled.

In some situations it may be found easier to modify the stated problem by considering an equivalent or dual mathematical statement of the problem.

All these methods can determine only the conditions for local optima to exist. The values of the objective function must then be determined at each of these positions and compared with each other and with values of the function at any artificial boundaries imposed at discontinuities. This is the only way of finding a global result. This solution may not be the absolute optimum for the system if the latter is unattainable, due to the imposition of restrictions, and it may be generally concluded that a restricted optimum has a value which equals, or is worse than, the value of the unrestricted, or absolute, optimum. As we have seen, sufficiency is often very difficult to prove, especially with a large number of variables. In practice, the physical interpretation of the mathematical model may give sufficient information about the character of a stationary point.

An essential point in analytical argument is that these methods concentrate not on the value of the objective function, but on the necessary conditions for an optimum to exist. A search must first be made for the solution yielding this location, and the function must then be compared for each of these locations, so that even with analytical methods some direct search is necessary. The advantage of the analytical technique lies in the closely defined locations for search in comparison with the free search of a region in numerical work. The field of optimization known as linear programming provides an excellent illustration of this point. However, if the roots of the equations defining the locations of the stationary points are not easily obtained analytically, and numerical methods have to be employed, it might prove better to use numerical techniques on the objective function from the start of the problem.

Problems

VII-1 Find out if any stationary values exist for the function

$$y = 3x_1^2 + x_2^3 + 3x_1x_2 + 3x_2^2$$

and state the necessary conditions and sufficient conditions for these to be maximum or minimum values.

VII-2 Investigate the stationary points of

a. $y = x_1^3 - 3x_1x_2^2$

b. $y = 3x_1^2 + 4x_1 + x_2^2 + x_1x_2 + 2x_3 + 2x_3^2 + x_2x_3 - 2x_2$

VII-3 Is the following function positive-definite?

$$x_1^2 + 10x_2^2 - x_3^2 + 3x_1x_2 + 15x_2x_3$$

VII-4 Find the specification of an open-topped rectangular tank whose total wall and base area is to be 108 ft², if a maximum volume is required. Use the method of direct substitution.

VII-5 Find the maximum and minimum distances from the origin to the curve

$$5x_1^2 + 6x_1x_2 + 5x_2^2 = 8$$

Use the method of constrained variation.

VII-6 Find the global minimum and maximum of the function

$$y = x_2 - x_1^2$$

if it is subject to the restriction that

$$g = 1 - x_1^2 - x_2^2 = 0$$

(Use a penalty function method.)

VII-7 What should be the proportions of a cone-roofed tank to obtain the greatest volume of storage for a fixed surface area? Take the diameter and the height of the lower cylindrical section to be $2R$ and L, respectively, and the slant height of the upper conical section to be S. Use the method of direct substitution.

VII-8 Repeat Prob. VII-7 using the penalty function method.

VII-9 Carry out Prob. VII-7 by the method of Lagrangian multipliers.

VII-10 Attempt the solution of Prob. VII-7 using constrained variations.

VII-11 Rework Prob. VII-6 using Lagrangian multipliers.

VII-12 Rework Prob. VII-5 using Lagrangian multipliers.

VII-13 Find values of x_1 and x_2 which minimize the function

$$y = x_1^2 + 4x_2^2 - 4x_1$$

subject to the restriction that

$$2x_2 - x_1 = 12$$

VII-14 Show that the stationary points of

$$y = x_1^2 + x_2^2$$

subject to

$$g = x_1 + x_2^2 - 5 = 0$$

consists of two minima and one maximum. Then show that

$$F = y + \lambda g$$

has one saddle point and two minima for unconstrained variations about the three stationary points.

VII-15 The energy balance about the ith heat exchanger shown in the sketch is given by

$$T_i = \frac{T_{i-1} + t_i A_i U}{1 + A_i U}$$

where U and t_i are both given quantities, and A_i is the area.

a. Develop a set of necessary conditions which are to be satisfied if the outlet temperature T_n is to be maximized by varying the individual areas A_i, subject to the restriction that the total area be fixed at a value A_t.

b. Discuss a numerical technique which you would use to solve this problem.

c. Start to make calculations for the special case

$$N = 3$$
$$U = 0.1 \text{ ft}^{-2}$$
$$t_1 = 150°\text{F}$$
$$t_2 = 200°\text{F}$$
$$t_3 = 250°\text{F}$$
$$A_t = 100 \text{ ft}^2$$
$$T_0 = 100°\text{F}$$

Complete at least one iteration.

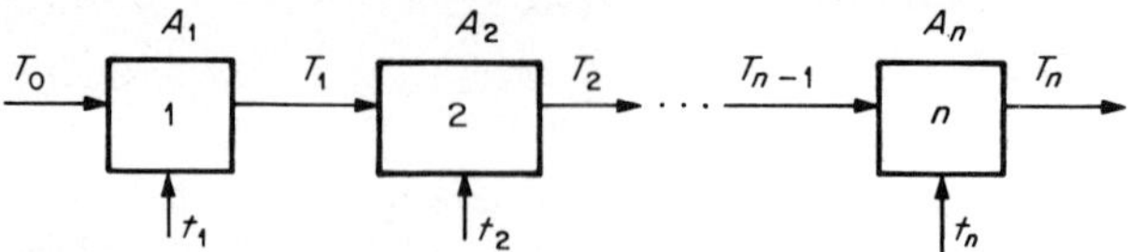

TRAIN OF HEAT EXCHANGERS

VII-16 Is the discrete maximum principle valid for the system shown below, where it is desired to maximize $p_{1,2}$ by an appropriate choice of q_1 and q_2?

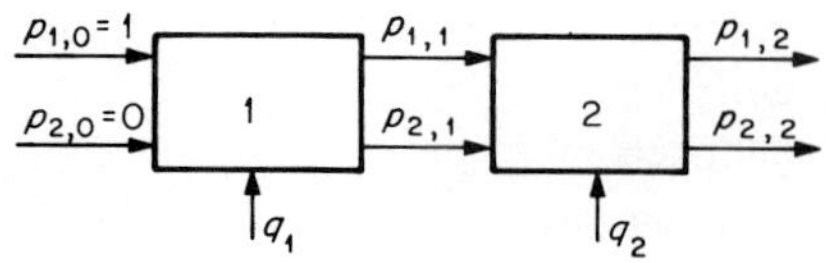

Stage 1: $p_{1,1} = q_1^2 + 1 - 2q_1$

$p_{2,1} = 2q_1^2$

Stage 2: $p_{1,2} = -p_{2,1} + p_{1,1} - q_2^2$

$p_{2,2} = p_{2,1} + q_2$

VII-17 Minimize the function

$$y = (x - 3)^2$$

subject to the inequality restriction that

$$x \geq 1$$

using slack variables.

VII-18 Repeat Prob. VII-17, but with the restriction that $x \leq 1$.

VII-19 Find the minimum of the function $y = 8x_1^2 + 10x_2^2$ if it is subject to the restriction that $4x_1 + 6x_2 \geq 3$.

VII-20 What is the minimum of the function

$$y = 4x_1^2 - 2x_1 + 3x_1x_2 + 5x_2^2 - 4x_2$$

subject to $x_1 + x_2 \geq 1$? Do not use Lagrangian multipliers.

VII-21 Find the positions of possible minima of the objective function

$$y = x_1 + 3x_2 - 6x_3^2$$

subject to the constraints

$$x_1^2 + x_2^2 + x_3^2 \leq 1$$

VII-22 Find the maximum of

$$y = 10x_1^2 - 4x_1x_2 + 3x_2^2 + 5x_2x_3$$

subject to

$$x_1 + 2x_2 \leq 3$$
$$x_2 - x_3 \geq 2$$
$$x_1 \geq 1$$

but not using Lagrangian multipliers.

VII-23 Repeat Prob. VII-20 using the method of Lagrangian multipliers, and examine the Kuhn-Tucker conditions.

VII-24 Repeat Prob. VII-22 using the method of Lagrangian multipliers, and examine the Kuhn-Tucker conditions.

VII-25 Minimize the function

$$y = 3x_1 + x_2$$

subject to the restriction

$$3x_1 + x_2 \geq 14$$
$$5x_1 - 2x_2 \geq 4$$
$$x_1 \geq 0 \qquad x_2 \geq 0$$

but not using a linear programming technique.

VII-26 Find the global maximum of the function

$$y = 3x_1^2 - 2x_1 - 5x_2^2 + 30x_2 + x_3^2$$

subject to the restrictions that

$$g_1 = 8 - 2x_1 - 3x_2 \leq 0$$
$$g_2 = 2x_2 - 10 \leq 0$$
$$g_3 = 3x_1 + 2x_2 + x_3 - 15 \leq 0$$
$$g_4 = x_3^2 - 1 \leq 0$$

VII-27 A chemical manufacturer finds that there is a high demand for two of his products, A and B. How much should he produce of each? Obviously, as much as he is capable. However, he is faced with the problem that only limited supplies of his two raw materials, α and β, are available. Both raw materials are necessary in the production of his products. What should be the production rate of each of his products under the limitation of a limited raw-material supply if he wishes to make a maximum profit?

Let x_i = rate of production of product i (A or B)
$a_{j,i}$ = rate of consumption of raw material j (α or β) per unit rate of production of i
c_i = net profit from sale of unit quantity of i
b_j = maximum available rate of supply of j
p = total profit from operation per unit time

$a_{j,i}$, c_i, b_j are known, x_i are the independent variables, p is the objective function. Set up a mathematical model for the system. (Based on Fenech and Acrivos [1956].)

VII-28 Find the solution to Prob. VII-27 using a graphical approach, if

$$a_{\alpha,A} = 1 \qquad b_\alpha = 3$$
$$a_{\alpha,B} = 1 \qquad b_\beta = 4$$
$$a_{\beta,A} = 2 \qquad c_A = 3$$
$$a_{\beta,B} = 1 \qquad c_B = 2$$

VII-29 Find the solution to Prob. VII-28 using an algebraic approach.

VII-30 Find the solution to Prob. VII-28 using the Simplex method of linear programming.

VII-31 Find the maximum value of the function y, where

$$y = 2x_1 + x_2$$

if the independent variables are subjected to the constraints

$$x_1 + x_2 \leq 14$$
$$x_1 - 2x_2 \leq 4$$
$$x_i \geq 0$$

VII-32 Determine the nonnegative values of the independent variables x_i which maximize the function

$$y = 3x_1 + 5x_2 + 4x_3$$

subject to the restrictions that

$$2x_1 + 3x_2 \leq 8$$
$$5x_1 + 2x_2 \leq 10$$
$$3x_1 + 2x_2 + 4x_3 \leq 15$$

VII-33 Determine by the Simplex method the minimum value of the function

$$y = 2x_1 + x_2 + x_3$$

subject to the restrictions $x_i \geq 0$ and

$$3x_1 + 5x_2 + 2x_3 \geq 16$$
$$4x_1 - 2x_2 + x_3 \geq 3$$

VII-34 Maximize

$$y = x_1 - x_2$$

subject to $x_i \geq 0$ and

$$-2x_1 + x_2 \leq 2$$
$$-x_1 + 2x_2 \geq 8$$
$$x_1 + x_2 \leq 5$$

(Vajda [1956]—an example of no feasible solution existing.)

VII-35 A manufacturer requires an alloy consisting of 40 percent tin, the remainder being made up of lead and zinc in equal proportions. This alloy can be made up by mixing a number of available tin-lead-zinc alloys, the properties and costs of which are tabulated below. Find the cost of the cheapest blend and the amount of each type of alloy which should be purchased per pound of alloy produced.

		Available alloy				
		1	2	3	4	5
Analysis:	Percent lead	10	10	40	60	30
	Percent zinc	10	30	50	30	30
	Percent tin	80	60	10	10	40
Cost:	\$/lb metal	4.1	4.3	5.8	6.0	7.6

Dantzig [1963].

VII-36 *a.* Find the minimum of

$$y = 4x_1 - 3x_2$$

subject to the constraints

$$x_1 - 3x_2 \leq 4$$
$$2x_1 + 4x_2 \leq 15$$
$$-x_1 + x_2 \leq 6$$

with $x_i \geq 0$.

b. Construct the dual of this problem and find the solution of the dual problem.

c. Find for both the primal and the dual problem several feasible points, and show that the objective function of the primal problem is greater than the objective function of the dual.

VII-37 *a*. Find the minimum of the function

$$y = 4x_1 - 3x_2 + x_3$$

subject to the restrictions

$$3x_1 + 6x_2 + x_3 \leq 12$$
$$-2x_1 - 6x_2 + x_3 \leq 4$$

with $x_i \geq 0$.

b. Construct the dual for this problem and solve.

c. Find for both the dual and the primal problem several feasible points, and show that the objective functions provide limits about this solution.

d. Write the dual for this problem, with y to be maximized rather than minimized.

VII-38 Repeat Prob. VII-36, except maximize the objective function rather than minimizing it.

VII-39 Prove, using Lagrangian multipliers, that equality constraints in the primal problem lead to independent variables in the duals which may be both positive and negative.

VII-40 What is the Legendre transformation of

$$y = x_1^2 - 2x_1x_2 + 2x_2^2$$

Show how to recover this function, starting with the transformed form.

VII-41 If the Legendre transformation of a function of x is given by

$$\psi = -6d_1^2 + 2d_1d_2 - d_2^2 + 7$$

find the function. Here $d_i = \partial y/\partial x_i$.

VII-42 Find the dual of

$$y = \frac{2}{t_1t_2} + \frac{3}{t_2} + 4t_1t_2^2 + 6t_2^2$$

Solve both the dual and the primal problems with $t_i \geq 0$.

VII-43 Find the dual of

$$y = \frac{4}{t_1t_2} + 3t_2^{-2} + 4t_1t_2^2$$

and solve both the primal and the dual problems with $t_i \geq 0$.

VII-44 Maximize P with respect to x_i when it is subject to the restrictions that

$$\begin{aligned} 11x_1 + 3x_2 - 2x_3 &\geq P \\ 4x_1 + 5x_2 + x_3 &\geq P \\ 6x_1 + 10x_2 + 5x_3 &\geq P \\ x_1 + 8x_2 + 12x_3 &\geq P \end{aligned}$$

and

$$\sum_{i=1}^{3} x_i = 1 \qquad x_i \geq 0$$

(*Hint:* Add slack variables, maximize $y = P$, and eliminate P from restrictions. See also Example XII-2.)

VII-45 Consider the dual system to Prob. VII-44: Minimize L with respect to λ_j, subject to

$$\begin{aligned} 11\lambda_1 + 4\lambda_2 + 6\lambda_3 + \lambda_4 &\leq L \\ 3\lambda_1 + 5\lambda_2 + 10\lambda_3 + 8\lambda_4 &\leq L \\ -2\lambda_1 + \lambda_2 + 5\lambda_3 + 12\lambda_4 &\leq L \end{aligned}$$

$$\sum_{1}^{4} \lambda_j = 1 \qquad \lambda_j \geq 0$$

CHAPTER VIII

MULTIVARIABLE FUNCTIONS—NUMERICAL METHODS

1 INTRODUCTION

We have already seen (Chap. III, Sec. 1) that two types of models may be proposed for any system, and we have examined the interrelation between them. While analytical methods have been developed for handling optimization problems with objective functions simulated from analytical models, these methods cannot be applied in most practical cases, owing to the complexity of the required mathematical manipulations or to the discontinuous nature of the functional relationships. Moreover, black-box models cannot be directly investigated in this manner. Numerical search techniques must therefore be found to handle the n-dimensional problem, and the remarks of Chap. VI, Sec. 1, will also apply here. In this chapter search methods will be discussed as if all the objective functions are derived from black-box models. This is consistent with the interchangeability of models, as discussed in Chap. III.

It is unfortunate that the methods which were found to be so powerful in one-dimensional search (for example, the Fibonacci technique of Chap. VI, Sec. 4.2.3) cannot be simply extended to handle the n-dimensional system. We shall therefore find an entirely different philosophy being applied to multidimensional direct search, although the general principles of search (Sec. 1.3) are similar, irrespective of the dimension. Just as the analytical methods become more complex by the inclusion of more than one variable (as will be seen by a comparison of Chaps. V and VII), so do the numerical search routines. In the problems of one-dimensional numerical search, a measure of the search efficiency was defined for unimodal objective functions, this measure providing a means of "optimizing" the search technique. As will be recalled, for a sequential search in a restricted region, the Fibonacci method was shown to be preferable when the number of experiments was of major concern. No such measure of search efficiency has been proposed for the n-dimensional search techniques, and even for unimodal functions a "best" search technique cannot be defined. In the general problem, no one technique can thus be singled out as the most efficient, and each type must be examined in relation to the problem under study. We can only note that certain methods are well tested and usually effective. However, by the acquisition of some experience in applying the various methods in common use, it is possible to develop a preference for one or other method in a particular task.

The general organization of the search for an extreme value can be summarized as a sequential search involving the successive calculation of new values of the objective function and the comparison of these values with the best value which has been so far obtained. With this iterative procedure, the basic difference in the methods proposed in this chapter lies in the philosophy dictating the choice of the next location for objective function evaluation. In the general case study method (Sec. 2) this choice may take place in an arbitrary manner, perhaps randomly, and possibly guided by intuition. If this choice is not made effectively, the resulting search will be most inefficient as regards the number of experiments required to reach the desired optimum region.

In any logical method, the choice of the new location for evaluation of the objective function is of major importance. The procedure for making such a choice is often described in terms of search strategy. In general, once a feasible base point is chosen, a set of exploratory experiments is carried out in the vicinity of this base point. This exploration is made, initially, to study the behavior of the objective function in the neighborhood of the starting location, thus providing information about those directions of movement which might yield favorable results. This step might, for illustration, be likened to the action of a blind man standing on a rough hillside, in which he feels the surrounding ground with his foot in order to ascertain the gradient of the hillslopes. The data from this step may be left in their original form, either qualitative or quantitative, or be used to fit an approximating surface to represent the local conditions around the base point.

Thus one important feature of any search method is the choice of a *direction of movement.* If the direction can be chosen only from a finite number of choices, the direction of movement will be in a favorable direction only, and not necessarily lying along the line of greatest improvement. Thus, for example, the direction might be limited to directions parallel to the axes of the independent variables. If, on the other hand, the choice is unrestricted, the direction along the line of steepest gradient may be taken. Other directions, however, might be taken, based upon the results of earlier moves, if acceleration of the search is being attempted.

The second essential feature of any multidimensional search is the *distance of movement* along the chosen direction. Only at this stage does possible improvement occur in the values of the objective function, all previous work being required to lay the basis for efficient movement. The move in this selected direction can be a single step, or alternatively, a series of steps, in which the objective function is tested at each step, the movement in this direction being continued as long as the objective function improves.

Using the final location of the previous move as the new base, the cycle of exploratory experiments, choice of direction for movement, and movement are repeated as long as is required. In each cycle of the search the value of the objective function will improve, or at least remain constant. Eventually, no further improvement can be obtained. Further optimization might, however, be attempted by reducing the step size in the search. The search for the desired optimum will then end if it has been located within the required degree of accuracy or if changes in its value have been found to fall within some preselected fraction of its value. The basic structure of exploration movement will thus allow the search to move in the direction of the region containing at least a local optimum. If constraints are present, an alternative routine might be necessary to allow movement to continue in valid regions, and in any case a special procedure will always be necessary to define the local optimum and stop the search.

In summary, we note that the methods applicable for numerical search of one-dimensional objective functions cannot be directly extended to the n-dimensional case. Even more perplexing is the complete lack of a measure of the efficiency of search. Most investigators use a particular objective function, or set of objective functions, to compare various proposed search methods. This procedure is not entirely satisfactory, since there is an element of chance involved, and one method may perform better for one particular problem and yet may suffer when applied to another. General conclusions are thus difficult. We have seen, too, in the n-dimensional search, that two essential features characterize the various available methods: one important part of any search method is the choice of a direction of movement to a better region, and another is the choice of a suitable distance of movement. We shall use these two features to serve as a basis for organizing the various proposed search methods into appropriate groups within this chapter.

1.1 Local and Global Optima

Only in special cases, discussed in Chap. IV, Sec. 4.4, will an extreme value, found starting from a particular initial base, be known to be a global optimum. If a function is not known to contain a single optimum, an accepted procedure is to begin the search at a number of initial base points widely separated; and if the same value is found for all tested cases, this value can be termed global with some degree of confidence. If several optima are found, the best, of course, is selected. It should be understood, however, that the finding of an absolute optimum is a real problem, which is only partially resolved by the means suggested above. This particular aspect of numerical search in face of possible multiple optima is a subject of continuing research. As yet, however, a rigorous method of establishing the global optimum has not been proposed, except in certain restricted classes of problems, such as the minimization of a convex function in a convex region, whereas in general there is no guarantee of achieving the desired point (Chap. XI, Sec. 5.3.2).

1.2 Preplanned versus Sequential

As was found in one dimension, experiments can be either simultaneously carried out or planned sequentially. Generally, the former approach is most inefficient, and would be used only if it were unavoidable. This preplanned search over the whole region might lead us to the subregion containing the required extreme value, once a network of experiments has been performed. The search in the particular subregion could be carried out by a sequential approach such as that described in Sec. 1.3. Another approach would be the fitting of a polynomial to the network covering the whole region. However, since this would probably involve fitting a polynomial of high degree, it is most probable that a technique based on sequential methods would be simpler.

If the search is not forced to be preplanned, by far the best approach lies in the full use of sequential methods, the techniques described below being of this type. Here a few experiments are carried out in a local region and analyzed before deciding upon the location of the next cycle of experiments. By such an iterative procedure in both direction and step size, the optimum can be approached, usually a much smaller number of experiments being required.

1.3 General Principles of Sequential Numerical Search

The principles of sequential search in one dimension or in n dimensions are, basically, the same, and can be summarized in the following six steps:

1. Select a set of feasible values for the independent variables which satisfy any restrictions. This is the initial base point.
2. Evaluate the objective function at this base point.
3. Choose, by one of the methods to be described, a second feasible location.
4. Evaluate the objective function at the second location.
5. Compare the value of the objective function at the second location with that at the base point.
6. If the second point is better, move to this point, which is now taken as the new base point, and continue the search. If the initial base point

is better, the search is continued in some alternative direction, perhaps with a smaller step size, or stopped.

Basically, the different search methods vary only in step 3. In case study methods (Sec. 2) the choice of the new direction and distance of search is usually more or less random, or at least dependent very strongly on the intuition of the optimizer, and the choice of variable is not confined to the system variables, but may include the system itself. The logical methods are discussed in the three following sections: unrestricted (Sec. 3), equality restrictions (Sec. 4), and inequality restrictions (Sec. 5), the search being made for an optimum lying at either a stationary point, a discontinuity, or a boundary (Chap. IV, Sec. 4).

The various techniques are grouped according to the methods used to determine the direction and distance of the next experiment from the base point (step 3). Thus, for the unconstrained problem, direct searches (Sec. 3.1) might follow directions initially parallel to the coordinate axes or repeatedly form new geometric figures, while the gradient techniques (Sec. 3.2) require an evaluation of the first derivatives to locate a better point along a favorable direction of movement.

1.4 The Direction of Search

As has been noted, one of the important characteristics of a particular search technique is the criterion applied to determine the direction of search from a base point. For our purposes, the direction of move, given by a vector $\boldsymbol{\xi}$, will be represented by n numbers $\{m_1, m_2, \ldots, m_n\}$ such that any move will be made along a line drawn parallel to the line drawn from the origin to the point $\{m_1, m_2, \ldots, m_n\}$, as shown in Fig. VIII-1. Clearly, in a two-dimensional space the vector $\{1,0\}$ represents a direction parallel to the positive x_1 axis, while $\{0,-1\}$ is the direction parallel to the x_2 axis in the negative direction, and $\{1/\sqrt{2},1/\sqrt{2}\}$ is a direction parallel to a 45° line in the direction of increasing x_1 and x_2. Thus the direction $\boldsymbol{\xi}$ in a two-dimensional space can be defined by the pair of numbers $\{m_1,m_2\}$, with n numbers obviously required in the general case.

Most of the computational schemes to be developed in this chapter seek to improve the objective function by moving a certain distance, denoted here as s, from a feasible base point in a particular direction $\boldsymbol{\xi}$. The coordinates $\mathbf{x}_{j+1}$ of the new point are given as

$$\mathbf{x}_{j+1} = \mathbf{x}_j + s\boldsymbol{\xi} \tag{1.4-1a}$$

where $\mathbf{x}_j$ is the base point. In component form these equations are

$$x_{i,j+1} = x_{i,j} + sm_i \tag{1.4-1b}$$

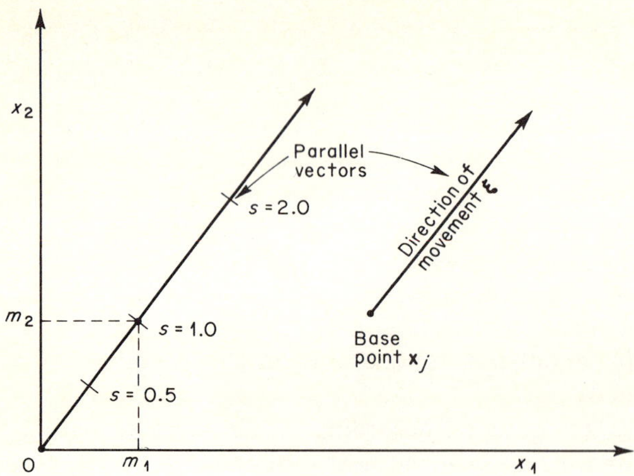

Fig. VIII-1 Direction of movement.

Here we have assumed that ξ is a unit vector, i.e.,

$$\xi \cdot \xi = \sum_{i=1}^{n} m_i^2 = 1 \tag{1.4-2}$$

To understand these arguments, let us note that if a line is drawn from the origin through the point $\{m_1, m_2, \ldots, m_n\}$, then any point on that line can be found by an appropriate choice of the parameter s in the equation

$$x_i = sm_i \qquad \text{for } i = 1, 2, \ldots, n \tag{1.4-3}$$

This is the parametric representation of a straight line which passes through the origin. If s is 0, then the point determined is the origin; if s is unity, the point determined is $\{m_1, m_2, \ldots, m_n\}$; and if s is bounded between 0 and 1, then a point in the interval between the origin and $\{m_1, m_2, \ldots, m_n\}$ is found.

Let us clarify these considerations by referring to the geometric representation of Fig. VIII-1, in which a straight line is shown in two-dimensional space. Clearly, any point satisfying Eq. (1.4-3) lies along this line, since all such points will yield a line of the same slope when connected with the origin. Since the lines pass through the origin, the slope is then x_2/x_1 for a two-dimensional surface. All points satisfying Eq. (1.4-3) will yield the same slope, since the slope is given by

$$\frac{x_2}{x_1} = \frac{sm_2}{sm_1} = \frac{m_2}{m_1}$$

which is independent of s. However, s determines the distance from the origin, since the distance is

$$\sqrt{x_1^2 + x_2^2} = \sqrt{s^2 m_1^2 + s^2 m_2^2} = s$$

provided Eq. (1.4-2) is satisfied.

Equations (1.4-3) are the parametric representation of a certain straight line whose direction is defined by the n numbers m_i. If we write Eq. (1.4-1) as

$$s\xi_i = x_{i,j+1} - x_{i,j} = sm_i$$

then, by defining a new origin for the coordinate system at $\mathbf{x}_j$, we find the same form as Eq. (1.4-3). Therefore the two lines must be parallel but displaced, since one passes through $\mathbf{x}_j$ and the other through the origin. Here the origin has been translated, but the coordinates have not been rotated.

Our conclusion is, then, that Eq. (1.4-1) represents a movement parallel to the direction $\boldsymbol{\xi}$, and the distance of movement is characterized by the parameter s. Any specific search method must provide a means of choosing a "good" direction $\boldsymbol{\xi}$ and distance s. By "good" we mean an eventual convergence to the close vicinity of optimum, provided the search is repeated a sufficient number of times.

1.5 Final Stages in a Search

Irrespective of the actual direct search technique being used and whether or not the search is preplanned or sequential in nature, the end result is the definition of the desired optimum within a given degree of accuracy. When the region of the optimum has been attained by some method to be described (Secs. 3 to 5), the continued application of the search procedure, even with progressively reduced step size, is inherently self-defeating, due to lack of definition in the linear terms of the gradients. It is then desirable to fit some quadratic approximation to the objective response to enable the optimum to be located more precisely. This is discussed in Sec. 7. Some particularly important points regarding convergence of a search and the development of stopping criteria, together with the difficulties of scaling, are discussed in Chap. XI, Sec. 5.3.2.

2 CASE STUDY METHODS

All numerical search methods are essentially case study methods in which one solution to the system is compared with another, but it is usually accepted that the description is applied to those techniques in which choices are not governed by systematic logic.

For some optimization problems, the case study method is the only possible approach, for example, in the comparison of different arrangements of units comprising a system flowsheet, or of different selections of equipment to carry out a specific duty. In such cases, for which there is generally no common set of system variables, but only a common objective function, one is forced to use the case study approach. The technique is, in fact, the method of search often employed by the experienced designer. Several searches (perhaps randomly distributed) are carried out, and these results, usually the best for each subsystem, such as a reactor type, are then compared and studied carefully to see what guides to a better result can be found. Clearly, the case study method is not readily usable when there are more than two or three variables, because it becomes exceedingly difficult to visualize the interactions and their contributions to the objective functions. While the general philosophy underlying the case study approach is similar to that discussed in the search strategy of Sec. 1.3, the choice of new locations or systems might be carried out in a less logical manner than might be desired.

2.1 Graphical Search

A typical case study approach is represented by one of the graphical search methods.

For the case of two variables, the following method is quite frequently used. The general approach is to fix one variable, x_1, say, and plot the objective function against the free variable x_2. The optimal value of x_2 for that given value of x_1 may then be determined. Another value of x_1 is then selected, and the process repeated. The dotted lines of Fig. VIII-2 depict curves of constant x_1, showing the objective function $y(x_1,x_2)$ as a function of x_2. The locus of the various restricted minima is represented by the solid line, and the solution to the optimization problem is clearly found at the minimum along this locus.

Fig. VIII-2 Graphical solution of two-dimensional problem.

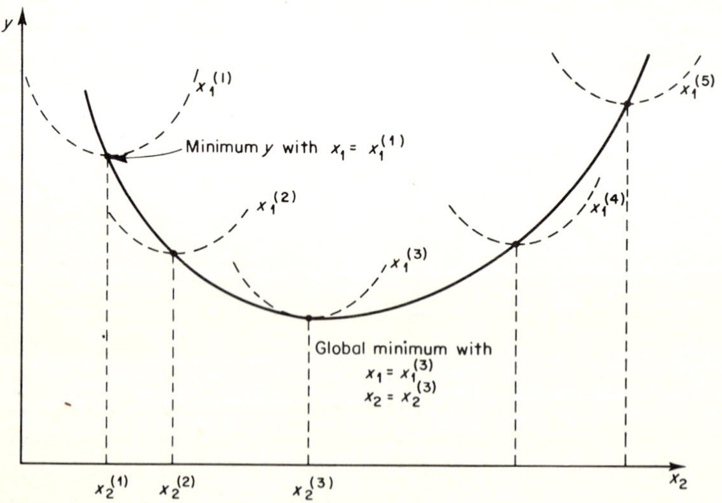

This scheme is certainly not recommended for more than two variables, and even for two variables the method is not really efficient in the number of computations required to achieve an acceptable estimate of the optimum. This is true, since there is no real guide provided in the selection of the fixed values of x_1 necessary to produce the result.

3 NUMERICAL SEARCH OF UNRESTRICTED FUNCTIONS

Let us initially consider the problem of locating the optimum of a function $y(\mathbf{x})$ when both the function and the system variables are unrestricted. Two separate types of search method may be distinguished, those in which only the function is evaluated and those in which both the function and its first derivatives (the gradient) are evaluated about any given point. These two groups are classified as direct and gradient methods, discussed in Secs. 3.1 and 3.2, respectively.

3.1 Direct Search Methods

Direct search methods require only the evaluation of the objective funct at a particular location. This is in contrast to the gradient techniques, which require both function and gradient values to be found at any position, the gradient being estimated by local exploration, using several experiments, if it cannot be found directly from an analytical expression. It may therefore be found that the direct "gradient-free" search is more efficient in terms of computation time, although it might not move in the "best" direction each time, because it does not need a local exploration to define each direction of movement. Typical directions might be chosen to lie parallel to the coordinate axes (Sec. 3.1.1) or in some pattern move such as a regular geometric figure (Sec. 3.1.2). Not only might the choice of direction be defined by such simple rules, but so might the length of move along that direction. Thus this may be of a single fixed step length or be found by seeking the optimum in a one-dimensional search along this direction. Methods of accelerating these choices in direction and distance are also examined (Sec. 3.1.3).

3.1.1 Univariant search parallel to an axis

The simplest direct search is that in which only one variable at a time is changed, being also called the *one-at-a-time*, or *sectioning*, method. By keeping $n - 1$ of the n variables fixed at some level—those of the base point—the remaining variable can be altered over its range, and a search conducted by any suitable technique described in Chaps. V or VI, since the search is now one-dimensional. When the greatest improvement in the

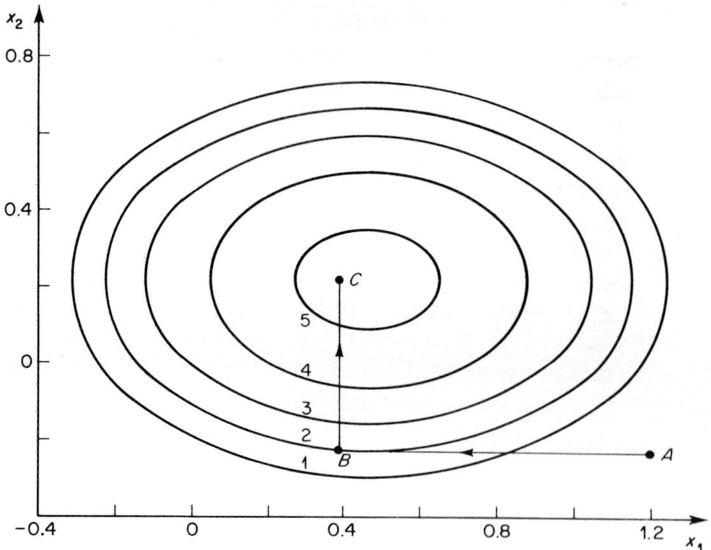

Fig. VIII-3 Univariant search.

objective function has been found within a given degree of accuracy, the corresponding location is determined and retained as the new base point. The search may then be continued in a new direction. This direction—one of a possible $n - 1$, since one direction has just been searched—may be chosen systematically by taking each coordinate direction in turn. After all n directions have been searched sequentially, the first stage is complete, and the entire process of sequential optimization should then be repeated. The calculation would cease when no further improvement was observed in the objective function after several stages of calculation.

To make this general description of the method of Friedman and Savage [1947] more lucid, let us consider the two-dimensional optimization problem depicted in Fig. VIII-3. An initial guess at the optimum conditions, the initial base point $\mathbf{x}_1$, is shown here as point A. The x_2 is fixed, and a one-dimensional search commences along x_1 to find an approximation to the restricted optimum at B for the fixed value of x_2. This approximation may be found to an accuracy determined by the initial step size or by reducing the step length to closer tolerances. This unidimensional search will initially be carried out by a method of Chap. VI, Sec. 3, until an optimal region is defined, whereupon the methods of Chap. VI, Sec. 4, might be applied. Continuing the search from point B in a direction of fixed x_1, as shown in Fig. VIII-3, the search proceeds until a reasonable approximation to the new restricted optimum is found, point C, say. This stagewise procedure of alternately searching in directions x_1 and x_2 is repeated until no further progress is possible.

The direction of search $\boldsymbol{\xi}_j$ is thus always parallel to one of the coordinate axes, and the distance moved s is determined as the distance yielding an optimum along $\boldsymbol{\xi}_j$. It will be seen, however, that it is not too important in any one-dimensional-stage search to find the local optimum. It is, in fact, better to find an approximation rapidly and then to move to the next stage. Any disadvantage in not achieving each local optimum is outweighed by the relative advantage of rapid movement.

The univariant search is a very simple and easily applied method, although not particularly rapid in its convergence to a solution. The choice of the direction and distance of any move in the n-dimensional problems can be summarized as follows.

Step 1 Assuming that a base point $\mathbf{x}_1$ has been selected initially, the direction of movement is chosen in the jth cycle in accordance with

$$\mathbf{x}_{j+1} = \mathbf{x}_j + s_j\boldsymbol{\xi}_j$$

$$\text{where} \quad \boldsymbol{\xi}_j = \begin{cases} (1, 0, 0, \ldots, 0) & \text{for } j = 1, n+1, 2n+1, \ldots \\ (0, 1, 0, \ldots, 0) & \text{for } j = 2, n+2, 2n+2, \ldots \\ (0, 0, 1, 0, \ldots, 0) & \text{for } j = 3, n+3, 2n+3, \ldots \\ \cdots\cdots\cdots\cdots\cdots\cdots & \\ (0, 0, 0, \ldots, 1) & \text{for } j = n, 2n, 3n, \ldots \end{cases}$$

Step 2 The distance of the move s_j is selected so that the optimal, or almost optimal, value of the objective function is found along the direction $\boldsymbol{\xi}_j$. This is a problem in one-dimensional search and is handled by the methods of Chap. VI.

The first two steps define the selection of the distance and direction of any movement.

Step 3 From steps 1 and 2 we have moved from base point $\mathbf{x}_j$ to a new position, $\mathbf{x}_{j+1}$, which is taken as the new base point. Advancing j by one increment, steps 1 and 2 are repeated cyclically, until there is no significant change in the objective function, or until any change is within a preselected value.

The general method described above has the disadvantage of tending to oscillate with steadily decreasing progress toward the optimum. It is essential, therefore, to stop the computation at some point short of the optimum. Moreover, when "sharp" ridges are present, the method may not even converge. For example, in Fig. VIII-4, the first search along AB locates the local optimum C. If the one-dimensional search routine employed in conjunction with the univariant uses steps of finite length, a second search along DE which passes through C may also give C as the local

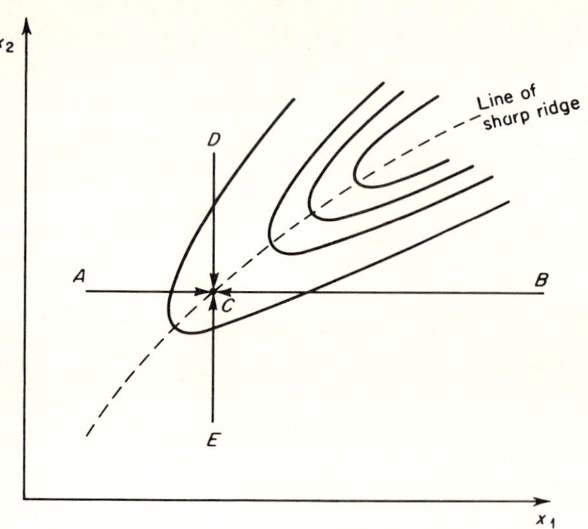

Fig. VIII-4 Failure of univariant search on a sharp ridge.

optimum. Thus the search comes to a halt, centered on the point C, although this is certainly not the global optimum. This simple illustration shows a failure in the search technique, when the contours of the system come to a sharp point, as on a narrow ridge. "Sharpness" is a relative term which need not imply a discontinuity. A sharp ridge will be found if the ridge is narrow compared with the step size of the search (Friedman and Savage [1947] and Wilde [1964]).

It will also be found that if the main ridges of the objective function do not lie parallel to a coordinate axis, the search will be slow. Since it is frequently found with a finite step size in n-dimensional functions that it is impossible to improve the function value unless two or more variables are altered simultaneously, the univariant method is not recommended for general use. It has been introduced here, however, as a basis for the development of more refined techniques, such as that of Hooke-Jeeves (Sec. 3.1.3).

Example VIII-1: Univariant search

The contours for the objective function

$$y = 3 + 6x_1 + 7x_2 - 7x_1^2 + 2x_1x_2 - 16x_2^2$$

are shown in Fig. VIII-3. Let us seek the maximum of this function, using the univariant method, starting at the point $\{1.2, -0.2\}$—point A. The x_1 direction is searched first, holding x_2 constant (-0.2 in the present cycle). Thus we move along the line $x_2 = -0.2$, until the maximum value of y is found. Here the maximum is at point B. Actually, point

B is here determined by letting $x_2 = -0.2$ in the expression for y and subsequently maximizing with respect to x_1. Thus

$$y(x_1,-0.2) = 0.96 + 5.4x_1 - 7x_1^2$$

The coordinate maximizing this particular objective function is, clearly,

$$x_1 = 0.3857$$

yielding $$y = 2.00$$

This first cycle of the search yields point B as the new improved position. In the next cycle, x_1 is now fixed, and x_2 varied, so as to find the maximum of y along the line $x_1 = 0.3857$. In this second cycle, point C is located as the improved position.

The process is continued until consecutive improvements become sufficiently small so as to make further experimentation unjustifiable.

3.1.2 *Pattern search*

The basic ideas underlying a direct search have now been presented. It will be seen, however, that the only method described so far, that based upon the univariant search, shows two weaknesses. First, the search may not converge to the true optimum, and second, the solution may oscillate as it nears the optimum. Some modification to the basic approach is thus required to circumvent these problems.

Instead of following directions parallel to the coordinate axes, a more effective method might be to permit the direction of search to change in some favorable way while fixing the distance of the move. This is in contrast to the univariant method, in which the directions were fixed without regard for the experimental results, but in which the distance of move was selected in an optimal sense.

A good illustration of this is found in the Sequential Simplex method of Spendley, Hext, and Himsworth [1962]. This method takes a regular geometric figure (known as a simplex) as a basis. Thus, in two dimensions, we should choose an equilateral triangle. Experiments are located such that the objective function is evaluated at the points formed by the vertices of the geometric figure. One vertex is then rejected as being inferior in value to the others. The general direction of search may then be taken in a direction away from this worst point, being chosen so that the movement passes through the center of gravity of the remaining points. A new point is then selected along this direction so as to preserve the geometric shape of the figure, and the function is evaluated anew at this point. The method proceeds by this process of vertex rejection and regeneration until the figure "straddles" the optimum, when no subsequent moves would lead to further improvement, since the last few geometric figures are essentially repeated. The two-dimensional problem will be presented first, followed by a generalization of the various steps within the search.

The Two-dimensional Simplex The maximization of an unconstrained two-dimensional objective function of the two variables x_1 and x_2 will be considered. Three points are initially chosen so that they are equidistant from each other, thus forming the equilateral triangle shown in Fig. VIII-5 by points 1 to 3. The value of the objective function is then calculated for the values of x_1 and x_2 corresponding to these three vertices. In general, the values of y at the three points will not be equal, and in the example illustrated, the value of y at point 2, y_2, is the worst of the three.

The technique then chooses a direction of movement to obtain new points, where better values of the objective function will be found. The procedure is now to reject the point at which y is worst (here point 2) and to replace it by a point obtained by reflecting the rejected point in the opposite side of the original equilateral triangle, giving point 4, as shown in Fig. VIII-5. By taking this mirror image of the worst point, together with the two other points, a new equilateral triangle is formed, which can then be used as the basis for fresh moves. Thus the value of y will be calculated at point 4, and a comparison made of the values of y at points 1, 3, and 4. The procedure is repeated, taking the point to be rejected as that at which the worst value of the objective function is obtained and replacing it by the reflected point in the opposite side of the triangle. Let us call this *rule 1.* This then produces a new triangle to act as the next basis. We thus see that the selection of the

Fig. VIII-5 Basic experimental layout of the two-dimensional simplex; triangle 1-2-3 is the starting figure, and a new point 4 is found in the first search cycle.

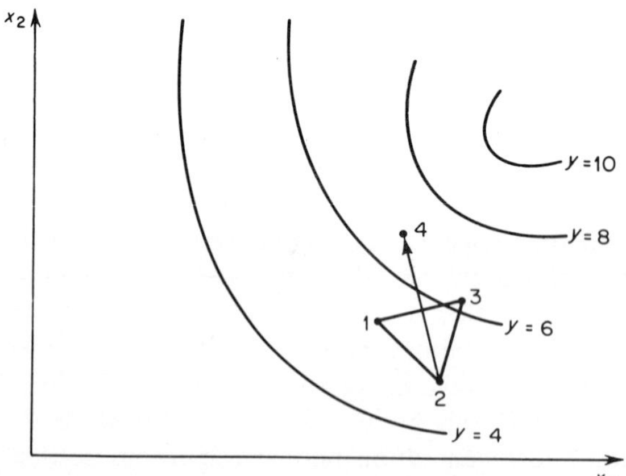

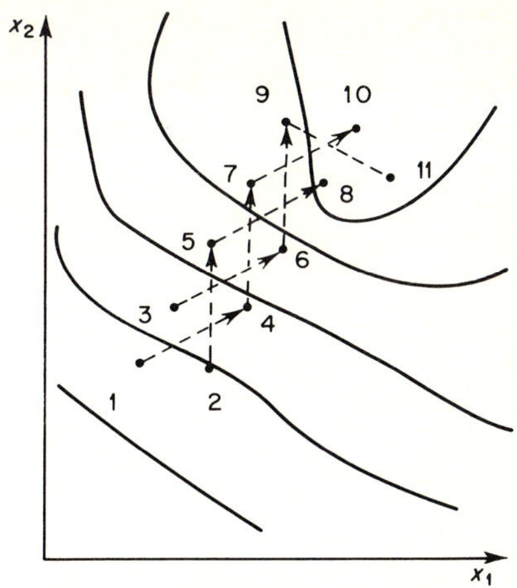

Fig. VIII-6 Progression of the simplex toward the optimum.

new direction of movement is always away from the poorest result; so our move will lie in a favorable direction. As long as the surface y is sloping and reasonably planar over the area of the triangle, every move leads to a region in which more favorable values of y will be found. Continuing these repetitive operations, the sequence of points shown in Fig. VIII-6 will be obtained. The process is seen to lead to a path which zigzags in the general direction of high ground, oscillating about the line of steepest ascent (page 407).

Several difficulties may arise in the application of this rule. The first of these is encountered when one of the triangles straddles a ridge formed by the function y, as shown by the triangle ABC in Fig. VIII-7. In this case a closed cycle of operations is entered. Thus, if y is worst at A, the first rule rejects point A and reflects it in the side BC to give point D, getting the new triangle BCD. In the new triangle BCD, y has its worst value at point D, so that the application of the same rule would cause point D to be rejected and replaced by the original point A, and the original triangle ABC would be reentered. At this point in the calculation, the method simply oscillates between points A and D, and no further moves can be made to approach the true optimum. The process of optimization has been stranded on the ridge. This trouble can easily be avoided, however, by postulating *rule 2*, that no return can be made to points which have just been left. If the first rule resulted in this, as in the above illustration, instead of rejecting the point at which y has the worst value, the point at which y has the second-worst value should be rejected. The effect of this on the behavior of the process in the neighborhood of a rising ridge is shown in Fig. VIII-8. Here the arrows drawn as continuous lines represent steps in which the lowest values of y are

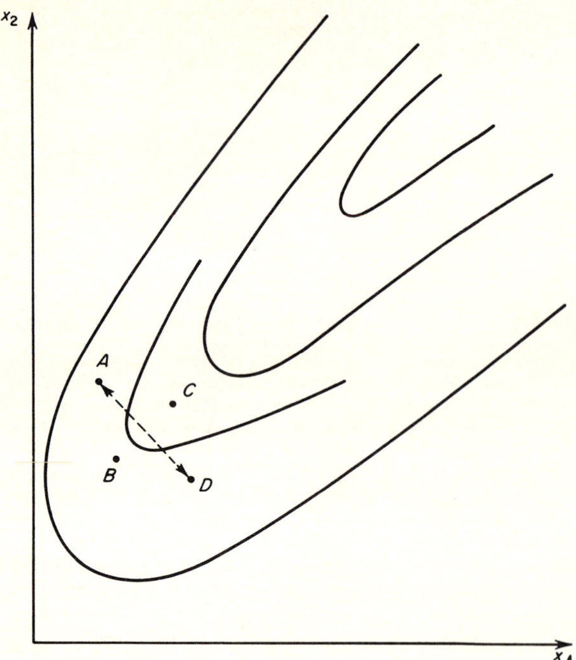

Fig. VIII-7 Failure of rule 1 on a ridge.

Fig. VIII-8 Simplex moves along a sharp ridge (→ moves by rule 1, --➤ moves by rule 2).

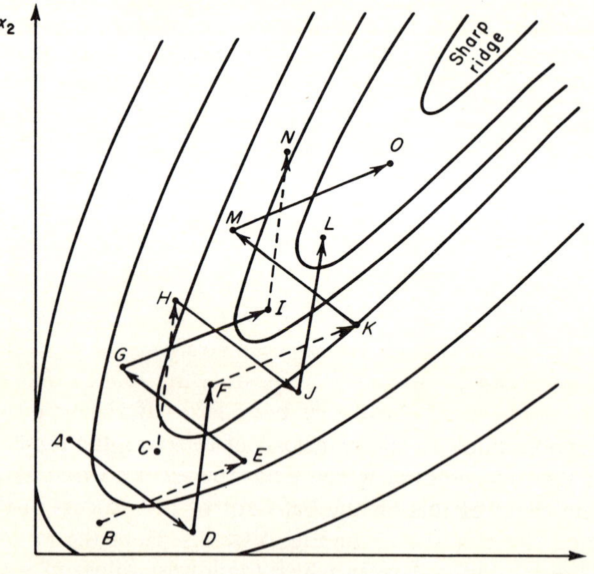

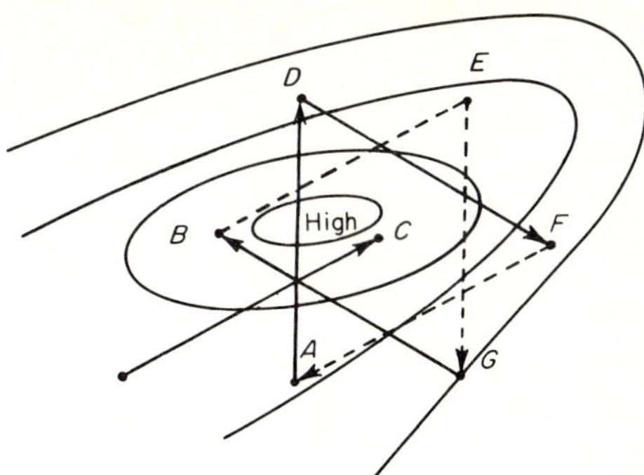

Fig. VIII-9 The approach to the final region. C is the best approximation to the optimum; a closed region is defined by the sequence of triangles ABC, BCD, . . . , CGA, ABC.

rejected in turn, while the arrows drawn as dotted lines represent the application of the second rule, in which the second smallest value of the objective function is rejected. It is seen that the sequence of operations formed from these two rules allows the search to move up the ridge, instead of becoming stranded.

Using these two rules, the optimization can be carried out until the region containing the desired optimum is attained. In this region the last equilateral triangle formed will straddle the summit, or at least lie within a distance of the order of its own size from the summit. At this stage it is not possible for a set of three points at the vertices of the last equilateral triangle of this size to get nearer to the optimum position, and the pattern begins to cycle and repeat itself, as shown in Fig. VIII-9. When such cycling is found, the vertex on which the whole pattern pivots is the best approximation to the optimum position that can be obtained without reducing the size of the triangular pattern. In general, the final steps close to the desired optimum are characterized by one vertex of the simplex remaining at the same point for a long time. The remedy is, clearly, a reduction in size of the simplex. The introduction of *rule 3* allows this to take place, the rule stating that if the best-valued vertex remains unchanged for more than M iterations, then the simplex size is reduced by, for example, halving the distance of all other vertices from that vertex. The next stage can then start. The size of M depends on the number of variables, Spendley suggesting that

$$M = 1.65n + 0.05n^2$$

The search can finally be stopped when the simplex size is small enough to locate the optimum adequately.

The optimum position has therefore been obtained to within a degree of accuracy determined by the size of the side of the equilateral triangle. The similarity to the final interval of uncertainty in one-dimensional searches will be obvious. If it is desired to obtain the location of the optimum to a closer tolerance, the size of the triangle could be reduced, and the search started again, or alternatively, a surface could be fitted to the final points obtained in the last cycle (Sec. 7).

The method may from its simplicity have some advantages over other numerical search techniques, especially the gradient methods. Instead of seeking directional accuracy in movement, the emphasis has been placed upon rapidity of choice and movement in a favorable direction, only one new experiment being required in each cycle. Any disadvantages in not moving along a more direct route to the optimum are compensated by the speed of calculation and movement.

Let us now generalize the approach outlined above so that it can be extended to n dimensions. First, the general property of these simplex figures will be formulated. This will then be used to obtain a suitable starting simplex, and the process of rejection and regeneration of points will be discussed mathematically.

The Simplex in n Dimensions We have seen in two dimensions that the simplex form is the equilateral triangle. For three variables, the equilateral pattern is simply extended by adding one point, so that there are four points lying at the vertices of a regular tetrahedron (Fig. VIII-10*b*). Here again, the distance between any two points is the same. The generalization of these figures will be an obvious procedure.

Let us now express these concepts in mathematical terms. In two dimensions, we may write the length of any one side of the equilateral triangle in terms of the coordinates of the two points concerned. Thus, if the coordinates of points 1 and 2 of Fig. VIII-10*a* are $\mathbf{x}_1 = \{x_{1,1},x_{2,1}\}$ and $\mathbf{x}_2 = \{x_{1,2},x_{2,2}\}$, respectively, the square of the distance between 1 and 2 is then

$$(\mathbf{x}_1 - \mathbf{x}_2)^2 = (x_{1,1} - x_{1,2})^2 + (x_{2,1} - x_{2,2})^2 = a^2$$

Proceeding, similarly, for other combinations of pairs of points, the general expression for the square of the length of the side between the points j and k may be written

$$(x_{1,j} - x_{1,k})^2 + (x_{2,j} - x_{2,k})^2 = \sum_{i=1}^{2} (x_{i,j} - x_{i,k})^2 = a^2 \qquad \text{for } j \neq k \quad (3.1\text{-}1)$$

This will hold for the combinations (j,k) of (1,2), (2,3), and (3,1), so that there will be three equations to define the figure, one for each choice of two vertices from the set of three, that is, $C_2^3 = 3!/2!1! = 3$.† Thus, for a

† Here C_a^b denotes the number of ways in which a group of a objects can be chosen from an original selection of b objects irrespective of the order in which they are arranged. This number is given by (Margenau and Murphy [1961])

$$C_a^b = \frac{b(b-1)\cdots(2)1}{a(a-1)\cdots(2)(1)(b-a)(b-a-1)\cdots(2)(1)} = \frac{b!}{a!(b-a)!}$$

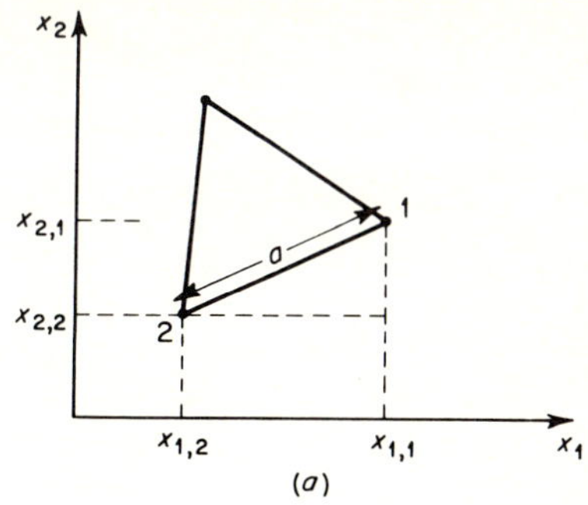

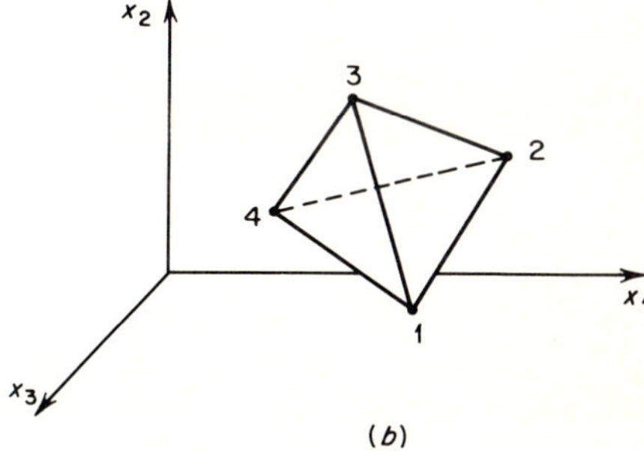

Fig. VIII-10 Simplexes. (*a*) Equilateral triangle for two dimensions; (*b*) triangular pyramid (tetrahedron) for three dimensions.

specified point $\mathbf{x}_1$, Eq. (3.1-1) would represent the locus of points $\mathbf{x}_2$ as a circle of radius a with the center at point $\mathbf{x}_1$. Taking any point $\mathbf{x}_2$ on that circle, Eq. (3.1-1) may be used to define a point $\mathbf{x}_3$. Since this will also be on the two circles of radius a, centered, respectively, on $\mathbf{x}_1$ and $\mathbf{x}_2$, only two possible locations for point $\mathbf{x}_3$ exist, as shown in Fig. VIII-11. The definition of a triangle by the use of Eqs. (3.1-1) is therefore not unique, even if two of the points are specified. This latter condition is necessary, however, since, when applying rule 1 of Simplex transformation, we hold two points constant and reject the third. It is therefore necessary for two valid positions to exist for point $\mathbf{x}_3$. We may look upon the Eqs. (3.1-1) as not, in fact, determining the simplex vertices, but rather checking that the specified points form a simplex.

Similarly, for three dimensions, we have the generalization of the square of the distance between the jth and kth points,

$$\sum_{i=1}^{3} (x_{i,j} - x_{i,k})^2 = a^2$$

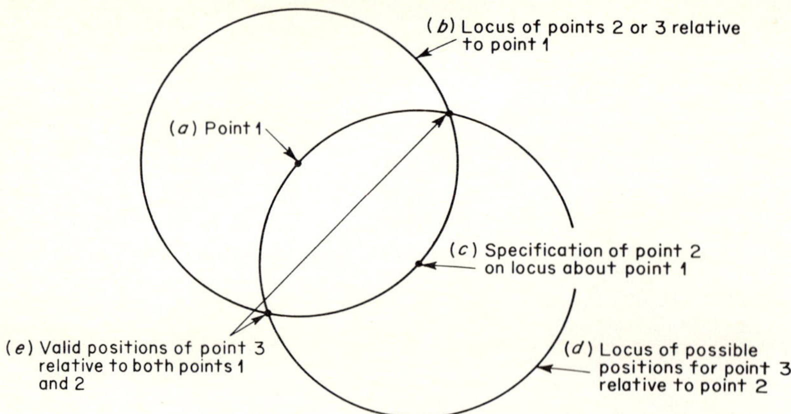

Fig. VIII-11 Construction of a two-dimensional simplex; the letters *a* to *e* represent the sequence of steps in the construction. (*a*) Point 1; (*b*) locus of points 2 or 3 relative to point 1; (*c*) specification of point 2 on locus about point 1; (*d*) locus of possible positions for point 3 relative to point 2; (*e*) valid positions of point 3 relative to both points 1 and 2.

Since the figure has six edges, six of these equations are required to define the simplex—the number of ways of choosing two different vertices from the set of four; that is, $C_2^4 = 4!/2!\,2! = 6$.

Extending this to n dimensions, the general simplex will have $n + 1$ vertices, with the equation

$$\sum_{i=1}^{n} (x_{i,j} - x_{i,k})^2 = a^2 \tag{3.1-2}$$

defining the distance a between any two points j and k. Since there are $n + 1$ points, any two different vertices can be chosen in C_2^{n+1} ways, so that

$$\frac{(n+1)!}{2!(n-1)!} = \frac{(n+1)n}{2}$$

equations must be used to define the simplex figure in n dimensions.

Starting Simplex Having defined the necessary relations for any simplex, an initial figure must be obtained in order to begin the search. Once this has been found, the successive simplex figures will be determined by changing only one vertex in each cycle. It does not particularly matter which simplex is used to start the calculation, and any set of points can be checked, using the above equations to determine if, in fact, they form the desired type of figure.

The set of $n + 1$ points listed in Table VIII-1 is proposed as the starting simplex for a problem in n variables, when the search is started at the origin. Of course, by coordinate transformation, the latest base point can always

TABLE VIII-1 COORDINATES FOR A SET OF SIMPLEX VERTICES

	n coordinates of each point						
Point j	$\zeta_{1,j}$	$\zeta_{2,j}$	$\zeta_{3,j}$	$\zeta_{4,j}$	$\cdots$	$\zeta_{n-1,j}$	$\zeta_{n,j}$
1	0	0	0	0	$\cdots$	0	0
2	p	q	q	q	$\cdots$	q	q
3	q	p	q	q	$\cdots$	q	q
$\cdots$	$\cdots$	$\cdots$	$\cdots$	$\cdots$	$\cdots$	$\cdots$	$\cdots$
n	q	q	q	q	$\cdots$	p	q
$n+1$	q	q	q	q	$\cdots$	q	p

where

$$p = \frac{a}{n\sqrt{2}}(\sqrt{n+1} + n - 1)$$

$$q = \frac{a}{n\sqrt{2}}(\sqrt{n+1} - 1)$$

be taken as the origin for any new search cycle. We shall therefore refer to the magnitude of the vertex coordinates $\mathbf{x}_j$ by their position $\boldsymbol{\zeta}_j$ relative to a base point $\mathbf{x}_1$. Then the coordinates of any vertex, j, will be given by

$$\mathbf{x}_j = \boldsymbol{\zeta}_j + \mathbf{x}_1 \qquad \text{for } j = 1, 2, 3, \ldots, n+1$$

It will be seen that, apart from the first row of zeros (the origin) and a diagonal of p terms, the table is composed only of q terms.

Other sets of points may be postulated, but this is one of the simplest. Let us now determine a suitable set of values of p and q so that they form a simplex. The conditions for these points to form the vertices of a simplex [Eqs. (3.1-2)] may be applied in turn to each pair of points. Thus, for points 1 and j, the square of the distance between them is

$$(0 - p)^2 + (n - 1)(0 - q)^2 = a^2$$

giving

$$p^2 + (n - 1)q^2 = a^2 \qquad (3.1\text{-}3)$$

n equations in all. Similarly, between points 2 and 3 (and between any other pair of points which do not include the first point, C_2^n sets in all), the equation is

$$2(p - q)^2 + (n - 2)(q - q)^2 = a^2$$

or

$$2(p - q)^2 = a^2 \qquad (3.1\text{-}4)$$

Values of p and q must therefore be chosen to satisfy both Eqs. (3.1-3) and

(3.1-4). Solving for p and q then gives

$$p = \frac{a}{n\sqrt{2}}(n - 1 + \sqrt{n+1}) \tag{3.1-5a}$$

$$q = \frac{a}{n\sqrt{2}}(-1 + \sqrt{n+1}) \tag{3.1-5b}$$

Here a is the size of the edge of the simplex. The general technique can therefore be started using this set of points as the initial simplex.

Example VIII-2: Starting simplex in two dimensions

Let us choose a starting simplex for the two-dimensional case. Here $n = 2$, so that

$$p = \frac{a(1 + \sqrt{3})}{2\sqrt{2}} = 0.9657a$$

$$q = \frac{a(\sqrt{3} - 1)}{2\sqrt{2}} = 0.2587a$$

Using the coordinate arrangement of Table VIII-1 for $n = 2$, the relative coordinates of the required points may be written {0,0}, {0.9657a,0.2587a}, and {0.2587a,0.9657a}. Here a, the side of the triangle, may be taken arbitrarily as any reasonable value. This value corresponds to the size of the step in the search, so that the smaller it is, the more accurate is the search, but the more slowly it will progress. It may be checked that, in fact, these are the coordinates of the vertices of an equilateral triangle. If it were desired to start the search at a base point {3.1,6.7}, then the coordinates of the starting simplex of side a would be

Vertex j	$x_{1,j}$	$x_{2,j}$
1	3.1	6.7
2	$3.1 + 0.9657a$	$6.7 + 0.2587a$
3	$3.1 + 0.2587a$	$6.7 + 0.9657a$

as shown in Fig. VIII-12.

Generation of a New Simplex In the course of the pattern search, using the Sequential Simplex method, one vertex is rejected and a new one chosen. The new simplex in each cycle is therefore made up of the remaining vertices of the original simplex, together with the new vertex. This new vertex is

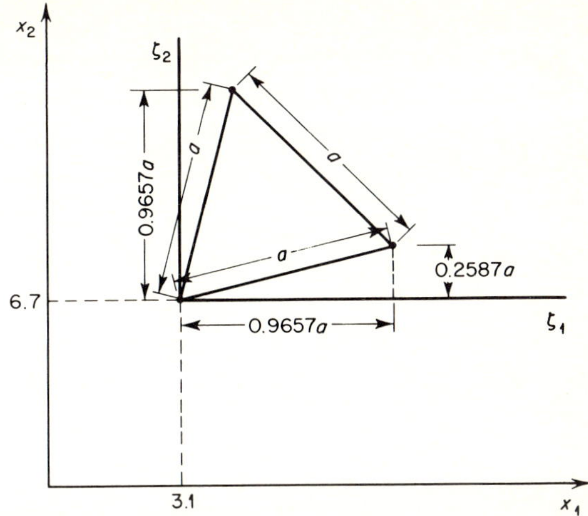

Fig. VIII-12 A possible simplex for starting the search.

chosen as the mirror image of the rejected vertex reflected in the plane formed by the remaining vertices of the original simplex. This will be considered first in two and three dimensions.

In two dimensions, referring to Fig. VIII-13, if it is desired to reflect one vertex R of an equilateral triangle in the opposite face, a line is drawn from R perpendicular to the line joining 1 and 2, cutting it at the point F. The line RF is projected its own length to obtain the required new vertex N. The point F is therefore the midpoint, or center of gravity, of both the lines 1-2 and RF. The coordinates of the required point N can then be found by determining the coordinates $\mathbf{x}_F$ of F in terms of the four points as follows: Since F is the midpoint of the line 1-2, then

$$x_{1,F} = \tfrac{1}{2}(x_{1,1} + x_{1,2}) \qquad x_{2,F} = \tfrac{1}{2}(x_{2,1} + x_{2,2})$$

or in general, $\quad x_{i,F} = \tfrac{1}{2}(x_{i,1} + x_{i,2}) \qquad \text{for } i = 1, 2$

Similarly, F is the midpoint of the line RN, so that

$$x_{i,F} = \tfrac{1}{2}(x_{i,R} + x_{i,N}) \qquad \text{for } i = 1, 2 \tag{3.1-6}$$

Combining the two expressions gives

$$\frac{x_{i,R} + x_{i,N}}{2} = \frac{x_{i,1} + x_{i,2}}{2} \tag{3.1-7}$$

thus defining the required value $x_{i,N}$. Hence

$$x_{i,N} = \frac{2(x_{i,1} + x_{i,2})}{2} - x_{i,R} \tag{3.1-8}$$

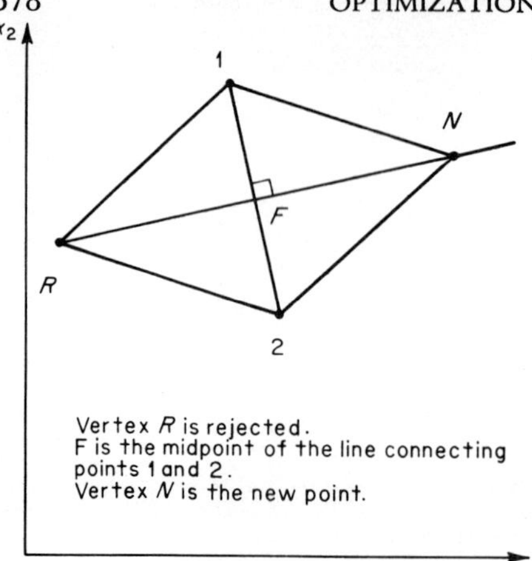

Fig. VIII-13 Generation of a new vertex.

This may be repeated for three dimensions. Considering the triangular pyramid shown in Fig. VIII-14, the new vertex N is obtained by joining the rejected vertex R to the center of gravity F of the plane formed by the remaining vertices 1, 2, and 3 of the original simplex and producing the line RF its own length to N. Proceeding in the same manner, getting the coordinates of point F in two ways, we have that F lies at the center of gravity

Fig. VIII-14 Regeneration of simplex. x_F is the centroid of x_1, x_2, and x_3.

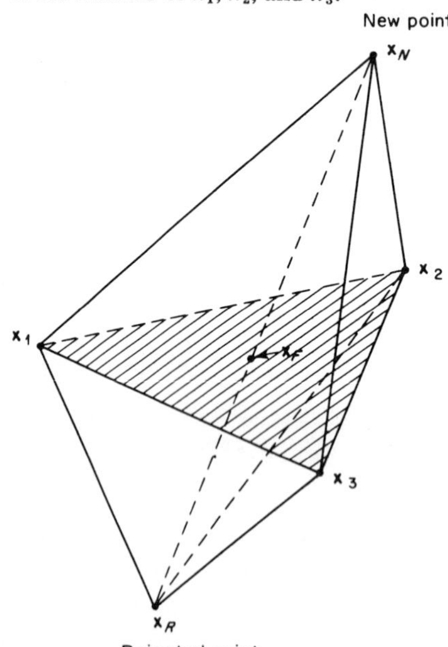

of all vertices, excluding R, so that, by taking moments,

$$(x_{i,F} - x_{i,1}) + (x_{i,F} - x_{i,2}) + (x_{i,F} - x_{i,3}) = 0$$

or

$$x_{i,F} = \tfrac{1}{3}(x_{i,1} + x_{i,2} + x_{i,3}) = \frac{1}{3}\sum_{j=1}^{3} x_{i,j} \tag{3.1-9}$$

and also lies at the center of gravity of RN when its coordinates are expressed by Eq. (3.1-6). Equating (3.1-6) and (3.1-9), we get the coordinates $x_{i,N}$ of the new point N as

$$\frac{x_{i,R} + x_{i,N}}{2} = \frac{1}{3}\sum_{j=1}^{3} x_{i,j} \qquad \text{for } i = 1, 2, 3 \tag{3.1-10}$$

Alternatively, we can write the right-hand side as

$$\frac{1}{3}\left(\sum_{j=1}^{4} x_{i,j} - x_{i,R}\right)$$

so that the numerator, in general, is the sum of all the coordinates, excluding the rejected point.

Generalizing, the coordinates of the new vertex N for the n-dimensional case will therefore be given by

$$\frac{x_{i,R} + x_{i,N}}{2} = x_{i,F} = \frac{\sum_{j=1}^{n+1} x_{i,j} - x_{i,R}}{n}$$

Hence the required coordinate is

$$x_{i,N} = \left[\frac{2}{n}\left(\sum_{j=1}^{n+1} x_{i,j} - x_{i,R}\right)\right] - x_{i,R} \qquad \text{for } i = 1, 2, \ldots, n \tag{3.1-11}$$

It can be checked that this equation will in fact produce a new point N which, with the n unchanged points in the original simplex, will form a new regular simplex figure.

All the necessary equations have now been developed to allow the Sequential Simplex method to be used, applying the same simple rules of rejection, regeneration, and cessation as presented earlier in Sec. 3.1.2 for two dimensions.

Example VIII-3: Sequential Simplex search

Locate the maximum of the function represented by

$$y = 100 - (10 - x_1)^2 - (5 - x_2)^2$$

to within 1 percent accuracy.

Stage 1 The first simplex in the two-dimensional case is given by the coordinate arrangements in Table VIII-1.

p and q are given by Eq. (3.1-5) and Example VIII-2 as

$$p = 0.9657a$$
$$q = 0.2587a$$

a being the step size. If this is chosen arbitrarily to be $a = 2$ and the search is started at the origin, the coordinates and corresponding values of the objective function are as given by points 1 to 3 in Table VIII-2.

The search can now proceed according to rule 1 or 2, as explained above. Thus, in the initial simplex, we shall eliminate point 1, since it has the smallest value of y, and generate a new point, $\mathbf{x}_4$, using Eq. (3.1-11). Thus

$$\frac{x_{i,4} + x_{i,1}}{2} = \frac{x_{i,2} + x_{i,3}}{2}$$

giving

$$x_{i,4} = x_{i,2} + x_{i,3} - x_{i,1}$$

Hence $\mathbf{x}_4$ is

$$x_{1,4} = 1.9314 + 0.5174 - 0 = 2.4488$$
$$x_{2,4} = 0.5174 + 1.9314 - 0 = 2.4488$$

and

$$y(\mathbf{x}_4) = 26.5$$

as entered in line 4 of Table VIII-2. We have thus rejected point 1 from the original triangle 1-2-3 and formed a new figure 2-3-4. We may proceed in this manner, using rule 2 only when the application of rule 1 would force the search to return to a previously rejected point.

It will be seen from the table that points 18 and 19 are exactly those given by points 10 and 12. We are now in a closed cycle about a pivot 13, as shown in Fig. VIII-9. This final closed search cycle can be detected by the number of times M the pivot appears in the final set of triangles, point 13 having appeared more than six times. Normally, we should stop the search on the fourth occurrence of point 13. This point can therefore be taken as the best approximation to the optimum value, so that the optimum will be in the region of

$$x_1 = 10.1744 \pm a = 10.2 \pm 2$$
$$x_2 = 4.5184 \pm a = 4.5 \pm 2$$

with a best value of $y(x_{13}) = 99.74$.

This step size has not located the final region within the desired accuracy because we purposely used a large step size to allow the optimum to be rapidly defined. Let us now conduct a new search, using a smaller step size of $a = 0.1$.

TABLE VIII-2 FIRST SEARCH STAGE IN EXAMPLE VIII-3 (Using a Step Size of $a = 2$)

Point j	$x_{1,j}$	$x_{2,j}$	y_j	*Point rejected*	*Rule for rejection*	*Points in simplex*		
1	0	0	−25					
2	1.9314	0.5174	14.9	Starting simplex		1	2	3
3	0.5174	1.9314	0.6					
4	2.4488	2.4488	26.5	1	1	4	2	3
5	3.8628	1.0348	46.6	3	1	4	2	5
6	4.3802	2.9662	64.3	2	1	4	6	5
7	5.7942	1.5522	70.3	4	1	7	6	5
8	6.3116	3.4836	84.1	5	1	7	6	8
9	7.7256	2.0696	79.4	6	1	7	9	8
10	8.2430	4.0010	95.9	7	1	10	9	8
11	6.8290	5.4150	89.7	9	1	10	11	8
12	8.7604	5.9324	97.60	8	1	10	11	12
13	10.1744	4.5184	99.74	11	1	10	13	12
14	10.6918	6.4498	97.42	10	1	14	13	12
15	12.1054	5.0358	95.57	12	2	14	13	15
16	11.5884	3.1044	93.88	14	2	16	13	15
17	9.6574	2.5870	94.06	15	2	16	13	17
18	8.2434	4.0010	95.92	16	1	18	13	17
19	8.7604	5.9324	97.60	17	1	18	13	19

Stage 2 With $a = 0.1$; we can form the initial simplex in the new stage either by retaining the best previous point and generating two more vertices by moving distances a toward the remaining vertices in the final simplex of the previous stage or by using $p = 0.0966$ and $q = 0.0259$ in Table VIII-1. By the latter we shall find the ζ_i variations from the new base point so that the coordinates and objective function value can be calculated as

$$\mathbf{x}_1 = \begin{Bmatrix} x_{1,1} \\ x_{2,1} \end{Bmatrix} = \begin{Bmatrix} x_{1,0} + 0 \\ x_{2,0} + 0 \end{Bmatrix} = \begin{Bmatrix} 10.1744 \\ 4.5184 \end{Bmatrix}$$

$$\mathbf{x}_2 = \begin{Bmatrix} x_{1,2} \\ x_{2,2} \end{Bmatrix} = \begin{Bmatrix} x_{1,0} + p \\ x_{2,0} + q \end{Bmatrix} = \begin{Bmatrix} 10.2709 \\ 4.5442 \end{Bmatrix}$$

$$\mathbf{x}_3 = \begin{Bmatrix} x_{1,3} \\ x_{2,3} \end{Bmatrix} = \begin{Bmatrix} x_{1,0} + q \\ x_{2,0} + p \end{Bmatrix} = \begin{Bmatrix} 10.2002 \\ 4.6149 \end{Bmatrix}$$

to give points 1 to 3 of Table VIII-3. Here $\mathbf{x}_0 \equiv \mathbf{x}_{13}$ of stage 1.

It is finally found in this stage that point 13 (again by coincidence) appears six times in the search, so that it forms the final pivot. The

TABLE VIII-3 SECOND SEARCH STAGE OF EXAMPLE VIII-3 (Using a Step Size of $a = 0.1$)

Point j	$x_{1,j}$	$x_{2,j}$	y_j	*Point rejected*	*Rejection rule*	*Points in simplex*		
1	10.1744	4.5184	99.738					
2	10.2709	4.5442	99.719	Starting simplex		1	2	3
3	10.2002	4.6149	99.811					
4	10.2967	4.6407	99.783	1	1	4	2	3
5	10.2260	4.7114	99.866	2	1	4	5	3
6	10.1295	4.6856	99.884	4	1	6	5	3
7	10.1553	4.7821	99.929	3	1	6	5	7
8	10.0588	4.7563	99.932	5	1	6	8	7
9	10.0846	4.8528	99.972	6	1	9	8	7
10	9.9881	4.8270	99.970	7	1	9	8	10
11	10.0135	4.9235	99.994	8	1	9	11	10
12	10.1100	4.9493	99.985	10	1	9	11	12
13	10.0389	5.0200	99.998	9	1	13	11	12
14	9.9424	4.9942	99.997	12	1	13	11	14
15	9.9678	5.0907	99.991	11	1	13	15	14
16	10.0643	5.1165	99.982	14	2	13	15	16
17	10.1354	5.0458	99.980	15	2	13	17	16
18	10.1100	4.9493	99.985	16	2	13	17	18

optimum thus will lie in the final region

$$x_1^* = 10.0389 \pm a = 10.03 \pm 0.10$$
$$x_2^* = 5.0200 \pm a = 5.02 \pm 0.10$$

within 2 and 4 percent, respectively, point 13 giving the best approximation to the optimum, with $y = 99.998$.

Further stages can now be carried out with smaller step sizes a. The exact solution is, of course, that

$$y^* = 100.00 \qquad \text{at } x_1^* = 10.0,\ x_2^* = 5.0$$

Comments on the Sequential Simplex Technique This method can be particularly recommended since it is conceptually simple and easy to program and requires only one additional experiment for each move.

Experimental errors in the determination of the objective function at any location may cause some moves to go downhill instead of uphill, but these false moves are rapidly corrected and simply slow down the average rate of climb. When errors are present, an additional rule, *rule 1a*, can be introduced to take care of observations which are too old. This rule states that if a value of a vertex has occurred in the last $n + 1$ successive simplexes and is not then eliminated by rule 1, no move should be made at all; the result should be discarded and replaced by a new observation at the same point.

If the point is a genuine optimum, the repeat observation will also tend to be high, but if it is not, the result is likely to be rejected in due course.

Himsworth [1962] states that the Simplex method is always more efficient than the others, the more so if the number of variables is large and the experimental error large. Brooks and Mickley [1961] have shown that the most efficient experimental or calculational design for estimating the slope of a plane, i.e., the design giving the most information per observation in the presence of experimental error, is the regular simplex. This is the figure used in this Sequential Simplex method.

The method has the advantage that all the calculations involved are trivial, calling for no mathematical or statistical knowledge, and each move is completely determined by the previous results, so that judgment or extrapolation is not required. It is not necessary to maintain a large storage space in the computer, since it requires only knowledge of the last simplex and a count of the number of times each vertex in this last simplex has appeared, and it is easy to add a fresh or extra variable at any time on any face of the original simplex to complete a new simplex in order to accommodate a further variable in the investigation. Thus a new variable, added to a two-dimensional problem, can be taken into account by adding one point to the existing equilateral triangle and forming a regular tetrahedron. This expansion feature seems to be unique, but the converse reduction by one variable is not so straightforward, and the system of simplex figures must be restarted at one of the final vertices. Another advantage is that the technique will follow an optimum which moves with time, and thus can be used in "evolutionary operation" optimization (Spendley et al. [1962]; see also Chap. XI, Sec. 6.6).

It will be realized that the Simplex method lacks any form of acceleration, since the basic form can only maintain its size, or contract. This is satisfactory, unless flatter contour regions are subsequently encountered. The simplex should then ideally be increased in size to make faster progress. Modifications might also be made to alter the regularity of the figure to take advantage of faster movement in some directions than others. Modifications to the basic Simplex method to include such magnification and distortion have been proposed by Campey and Nickols [1961], Nelder and Mead [1964, 1965], and Box [1965] (see also Sec. 5.2.1).

3.1.3 Acceleration methods in direct search

The methods of direct search presented so far use either a fixed pattern of search or a fixed direction, the choice of the direction and distance being independent of past successes and failures. Acceleration methods are those search schemes which rely to some extent on previous successful results. There are acceleration methods for both nongradient and gradient searches, but we shall confine our attention in this section to the direct search methods.

The acceleration can be associated with a better choice of distance (see below), a better choice of direction (page 390), or a combination of both distance and direction (page 396), all being based upon knowledge of past and present results.

Acceleration in Distance—the Hooke and Jeeves Method This method, due to Hooke and Jeeves [1961], is designed to follow a ridge, so that it does not suffer from the main disadvantages of the basic univariant approach. The method is based on a local exploration in the directions parallel to the coordinate axes at points a given distance δ from the base point. A larger move of distance s in a direction to be defined is then attempted. If this move is successful, and if a new local exploration indicates that the direction is a good one, a further move is made with a much larger step size. Thus, although the technique starts cautiously, with short steps, the step length increases with repeated success, it being assumed that previous moves are worth repeating. If the objective function did not improve along a given direction, smaller steps would be used again, and if this were not successful, a new stage and a new pattern would be started. As an introduction to the method, let us first consider the details of a two-dimensional search.

The first stage starts with a local exploration at an arbitrary base point $\mathbf{x}_0^{(0)}(=\mathbf{x}^{(0)}$, with components $\{x_1^{(0)},x_2^{(0)}\})$, both x_1 and x_2 being changed as in a univariant search, but now using only fixed step sizes of $\pm\delta$. We should therefore initially test $y(x_{1,0}^{(0)}+\alpha,\ x_{2,0}^{(0)})$ against $y(\mathbf{x}_0^{(0)})$, where $\alpha = \delta, -\delta$, in turn, stopping the search when the first function improvement is located, and calling this point $\mathbf{x}_1^{(0)}$. Thus, in Fig. VIII-15, we should test $+\delta$, getting a worse value, and then $-\delta$, finding an improvement, so that $\mathbf{x}_1^{(0)}$ is here $\{x_{1,0}^{(0)}-\delta,\ x_{2,0}^{(0)}\}$.† This procedure is repeated for the x_2 direction, starting from the most recently established base point $\mathbf{x}_1^{(0)}$, testing first $y(x_{1,1}^{(0)},\ x_{2,1}^{(0)}+\delta)$, and if this is not an improvement, then $y(x_{1,1}^{(0)},\ x_{2,1}^{(0)}-\delta)$. The result of this direction evaluation is a new base point, $\mathbf{x}_2^{(0)}$, which, for the case shown in Fig. VIII-15, is $\{x_{1,0}^{(0)}-\delta,\ x_{2,0}^{(0)}+\delta\}$. In some cases, it may be found that no improvement is obtained at either $+\delta$ or $-\delta$, as for example in Fig. VIII-16*a*, where both $y(x_{1,0}^{(0)}-\delta,\ x_{2,0}^{(0)}+\delta)$ and $y(x_{1,0}^{(0)}-\delta,\ x_{2,0}^{(0)}-\delta)$ are inferior in value to $y(x_{1,0}^{(0)}-\delta,\ x_{2,0}^{(0)})$, and we should in this case

† The notation adopted here defines a vector $\mathbf{x}_l^{(k)}$, where

$$\mathbf{x}_l^{(k)} = \{x_{1,l}^{(k)}, x_{2,l}^{(k)}, \ldots, x_{n,l}^{(k)}\} \qquad (3.1\text{-}12)$$

so as to represent the location of a point in n-dimensional space. The superscript k refers to the kth stage in the search, while the subscript l indicates the lth cycle of search within the kth stage. If $\mathbf{x}_l^{(k)}$ denotes the sequence of base points used in the search as we approach the optimum at $\mathbf{x}^*$, we should expect (for a maximum) that

$$y(\mathbf{x}^*) \geq \cdots \geq y(\mathbf{x}_l^{(k)}) \geq y(\mathbf{x}_{l-1}^{(k)}) \geq \cdots \geq y(\mathbf{x}_1^{(k)}) \geq \cdots \geq y(\mathbf{x}_1^{(0)})$$

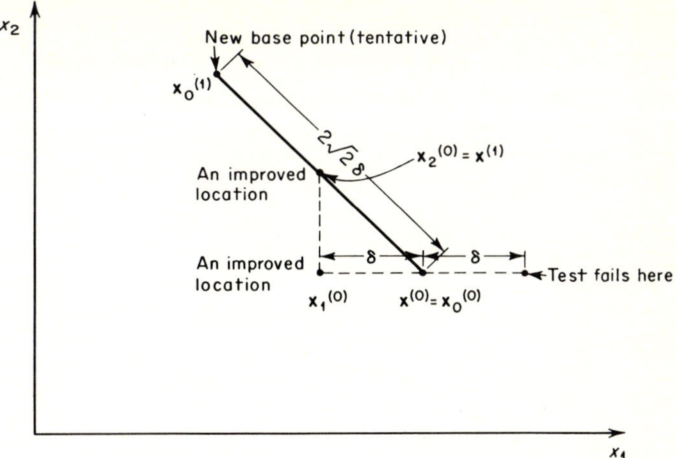

Fig. VIII-15 Local exploration in the Hooke-Jeeves method.

Fig. VIII-16 Failure of the local exploration in the x_2 direction (case *a*) and in both directions (case *b*).

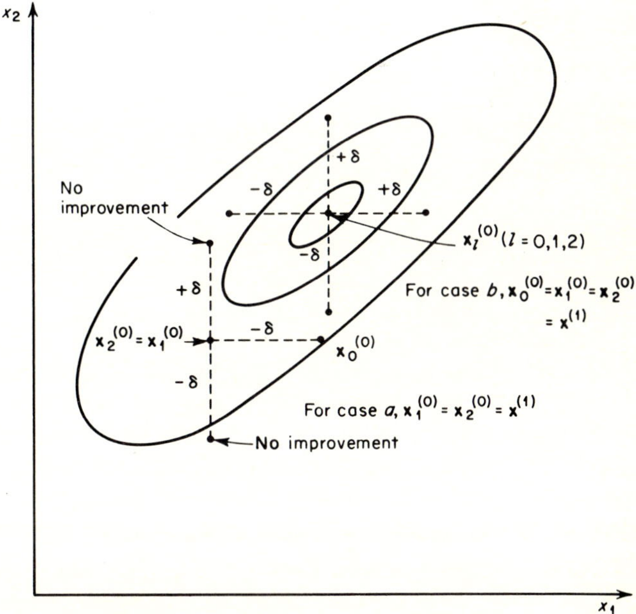

retain $\mathbf{x}_2^{(0)}$ at the latter position, $\mathbf{x}_1^{(0)}$. Again it might be found that, as in Fig. VIII-16*b*, the original base point proved better than all the tested points —here, $\mathbf{x}_2^{(0)} = \mathbf{x}_1^{(0)} = \mathbf{x}^{(0)}$. The appropriate action to be taken in the latter situation will be discussed later.

Let us first assume that at least one of the two directions x_1 or x_2 gives an improved objective function, so that the final temporary base point for the first stage, $\mathbf{x}_2^{(0)}$, is different from the initial base point $\mathbf{x}^{(0)}$. $\mathbf{x}_2^{(0)}$ is called the final temporary base point because each coordinate, here x_1 and x_2, has been searched once. This completes the first part of the stage.

Let $\mathbf{x}_2^{(0)} \equiv \mathbf{x}^{(1)}$, the starting point for the first-stage move. An accelerated move is now made based upon the first-stage results. The direction of this accelerated move is represented by a line drawn from $\mathbf{x}^{(0)}$ and passing through $\mathbf{x}^{(1)}$ (previously called $\mathbf{x}_2^{(0)}$) to a point $\mathbf{x}_0^{(1)}$. The latter point is obtained by defining the length of the accelerated move to be double the distance from $\mathbf{x}^{(0)}$ to $\mathbf{x}^{(1)}$. Thus

$$\begin{aligned} \mathbf{x}_0^{(1)} &= \mathbf{x}^{(0)} + 2(\mathbf{x}^{(1)} - \mathbf{x}^{(0)}) \\ &= 2\mathbf{x}^{(1)} - \mathbf{x}^{(0)} \end{aligned} \tag{3.2-13}$$

Then, if $\mathbf{x}^{(0)} \equiv \{x_{1,0}^{(0)}, x_{2,0}^{(0)}\}$ and $\mathbf{x}^{(1)} \equiv \{x_1^{(1)}, x_2^{(1)}\} \equiv \mathbf{x}_2^{(0)} \equiv \{x_{1,2}^{(0)}, x_{2,2}^{(0)}\}$, we have

$$\mathbf{x}_0^{(1)} \equiv \{x_{1,0}^{(1)}, x_{2,0}^{(1)}\}$$

where $\qquad x_{1,0}^{(1)} = 2x_{1,2}^{(0)} - x_{1,0}^{(0)} \qquad$ and $\qquad x_{2,0}^{(1)} = 2x_{2,2}^{(0)} - x_{2,0}^{(0)}$

It will be seen that this length of this move will vary, depending upon past successes and failures. Thus, in Fig. VIII-15, the acceleration distance from $\mathbf{x}^{(1)}$ is $\sqrt{2}\,\delta$, while in Fig. VIII-16*a*, it is only δ.

Before moving to this new point, $\mathbf{x}_0^{(1)}$, we must exercise some caution by evaluating $y(\mathbf{x}_0^{(1)})$. If this objective function is indeed an improvement over the best value in hand, $y(\mathbf{x}_2^{(0)})$, then we shall accept $\mathbf{x}_0^{(1)}$. However, if there is no improvement, we shall let $\mathbf{x}_0^{(1)} = \mathbf{x}_2^{(0)}$, and begin a new local exploration, as described above.

Let us assume that the new point $\mathbf{x}_0^{(1)}$ is an improvement. We then place the new base point $\mathbf{x}^{(1)}$ at $\mathbf{x}_2^{(0)}$, as shown in Fig. VIII-17. Stage 2 is then begun, and temporary base points $\mathbf{x}_1^{(1)}$ and $\mathbf{x}_2^{(1)}$ are established about $\mathbf{x}_0^{(1)}$ by a sequential local exploration at distances $\pm\delta$ along the coordinate directions, as in stage 1. If the direction of movement remains the same, the distance of the acceleration move is now increased, since it equals twice the distance between $\mathbf{x}_2^{(1)}$ ($\equiv \mathbf{x}^{(2)}$) and $\mathbf{x}^{(1)}$ [in other words, $2(\mathbf{x}^{(2)} - \mathbf{x}^{(1)})$, as before]. For the example illustrated in Fig. VIII-17, it can be checked that the move from the best point located in the second stage is $2\sqrt{2}\,\delta$ (the total distance measured from base point being $4\sqrt{2}\,\delta$), which is twice the magnitude of the previous accelerated move. If the point reached is not an improvement over the best so far attained, the location of $\mathbf{x}_0^{(j)}$ is then fixed at $\mathbf{x}^{(j)}$,

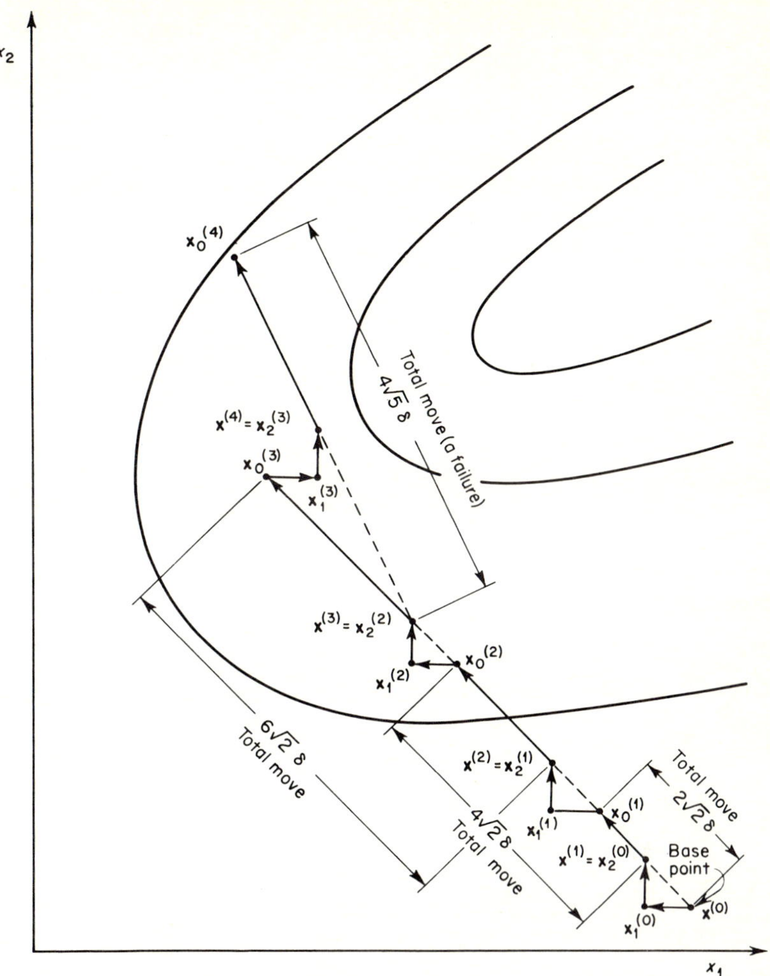

Fig. VIII-17 A Hooke-Jeeves search.

which in effect cancels the accelerated move, and a new local search is conducted. This procedure is recommended by Klingman and Himmelblau [1964], instead of the alternative of reducing the accelerated distance and again testing. They claim that the search proceeds more slowly near the optimum if the unsuccessful accelerated move is reduced, rather than completely rejected.

Thus, continuing the case depicted in Fig. VIII-17, it will be seen that the accelerated move from $\mathbf{x}^{(4)}$ to $\mathbf{x}_0^{(4)}$ fails, and therefore one would conduct a local search about point $\mathbf{x}^{(4)}$. All previous successes and failures are then forgotten, and the search again starts cautiously.

If a local search fails to find a better point, the interval δ is reduced, and a new local search is conducted. The entire process terminates when, on successive reductions of the interval, the local searches fail to find an enhancement of the objective function.

Let us now summarize the entire procedure for n dimensions.

Step 1 Select an arbitrary base point $\mathbf{x}^{(0)}$ and a distance δ. Set $k = 0$.

Step 2 Sequentially explore the locations

$$\mathbf{x}_l^{(k)} = \mathbf{x}_{l-1}^{(k)} \pm \delta \boldsymbol{\xi}_l$$

in the lth directional cycle (i.e., along x_l direction) of the kth stage in the search. Here $\boldsymbol{\xi}_l$ is the direction vector defined as

$$\boldsymbol{\xi}_l = \{m_{1,l}, m_{2,l}, \ldots, m_{n,l}\} \tag{3.1-14}$$

where $m_{i,l} = 0$ for $i \neq l$, and $= 1$ for $i = l$; $1 \leq i$, $l \leq n$. We then move the temporary base point $\mathbf{x}_l^{(k)}$ to each position for which y is improved. Thus, we first go to

$$\mathbf{x}_l^{(k)} = \mathbf{x}_{l-1}^{(k)} + \delta \boldsymbol{\xi}_l$$

and, if in looking for a maximum, we find $y(\mathbf{x}_l^{(k)}) > y(\mathbf{x}_{l-1}^{(k)})$, we should accept the point and put $l = l + 1$, but if not, we should move to

$$\mathbf{x}_l^{(k)} = \mathbf{x}_{l-1}^{(k)} - \delta \boldsymbol{\xi}_l$$

If $y(\mathbf{x}_l^{(k)}) > y(\mathbf{x}_{l-1}^{(k)})$, we should accept this point and put $l = l + 1$, but if not, we should return to

$$\mathbf{x}_l^{(k)} = \mathbf{x}_{l-1}^{(k)}$$

and put $l = l + 1$. This will be the situation when neither of the previous two cases holds.

Once this local search is performed for all n dimensions in turn, the exploratory part of any stage is complete, and the information accumulated is used to direct a large move.

Step 3 Let $\mathbf{x}^{(k+1)} \equiv \mathbf{x}_n^{(k)}$. If $\mathbf{x}^{(k+1)} \neq \mathbf{x}^{(k)}$, then move and test a new location, $\mathbf{x}_0^{(k+1)}$, where

$$\mathbf{x}_0^{(k+1)} = 2\mathbf{x}^{(k+1)} - \mathbf{x}^{(k)}$$

Two cases arise:

1. If $y(\mathbf{x}_0^{(k+1)})$ is an improvement over the previous best value, a new search can be begun, using step 2, with $k = k + 1$.
2. If $y(\mathbf{x}_0^{(k+1)})$ is not an improvement, than all previous moves are forgotten, the base point being established at the best point as yet found, namely, $\mathbf{x}_n^{(k)} = \mathbf{x}^{(k+1)}$. A new local exploration is conducted, beginning again at step 2, with k set equal to zero. Essentially, all past successes and failures are forgotten.

If, after a stage, it is found that $\mathbf{x}^{(k+1)} = \mathbf{x}^{(k)}$, corresponding to no move at all, the step size should be reduced, and step 2 repeated. If this point is the optimum, the step size will continue to decrease until the search is stopped at some small specified step length.

This is a convenient search scheme, being simple to use and yet having acceleration features. Klingman and Himmelblau [1964] have tested it rather extensively and found it to be reliable in most cases. It is not, however, the best direct search method available.

Example VIII-4: Hooke-Jeeves method of search

To illustrate the method of Hooke-Jeeves, let us resolve the problem posed in Example VIII-1, starting from the base point $\mathbf{x}^{(0)} = \{1.2, -0.2\}$ and using a value of $\delta = 0.1$. We shall trace the progress of the search in Fig. VIII-18. The result of the first local exploration yields point b. The accelerated move is then to point c. A test of this point will show that it is better than b. Therefore this improved point is designated $\mathbf{x}_0^{(1)}$, and a local search is then initiated about this point. $\mathbf{x}^{(2)}$ (point d) is found. The accelerated move to $\mathbf{x}_0^{(2)}$ succeeds. $\mathbf{x}^{(3)}$ is the improved point resulting from the improved search about point $\mathbf{x}_0^{(2)}$, but the accelerated move to g fails. The entire accelerated move is then abandoned, and a local search about $\mathbf{x}^{(3)}$ reveals that h is an improvement. This is the best approximation to the optimum which can be found, using the present value of $\delta = 0.1$. This interval must be reduced to refine the answer further.

Fig. VIII-18 Hooke-Jeeves search for Example VIII-4.

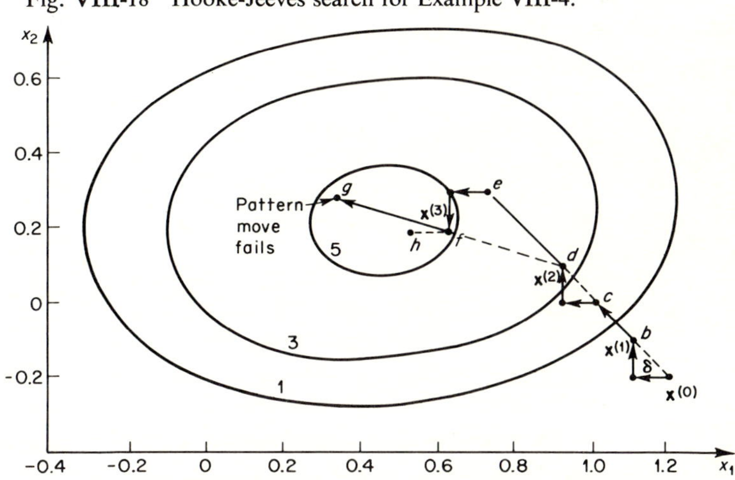

Acceleration in Direction—the Powell method Mugele [1962] has modified the univariant search so that a better direction is selected when the one-at-a-time choice fails. The essential idea of the method is the fitting of a quadratic approximation between the points defined by $n + 1$ experiments. This surface is then used to move toward the optimum; and once movement is again possible by the one-at-a-time technique, the latter is again introduced. Although this method would seem to be reliable, the convergence rate will probably not be as fast as the Sequential Simplex method for most problems, and the two methods seem to be of comparable complexity in application. (Methods using the gradient to speed up the univariant method are mentioned in Sec. 3.2.3.)

The main techniques for accelerating the choice of direction in a direct search carry out a set of n search cycles and then use the resulting information to define the acceleration direction. The simplest is the two-dimensional application due to Forsythe and Motzkin [1951]. Thus, as in Fig. VIII-19, if A is the optimum from a previous stage in the univariant search, and B and C are the respective optima from the subsequent searches along x_1 and x_2 in the next stage of the search, the acceleration direction is along a line from A produced through C. More generally, this approach is a member of the group of acceleration methods developed to find within a finite number of experiments the exact optimum (the minimum for definiteness) of a function

Fig. VIII-19 An acceleration in direction.

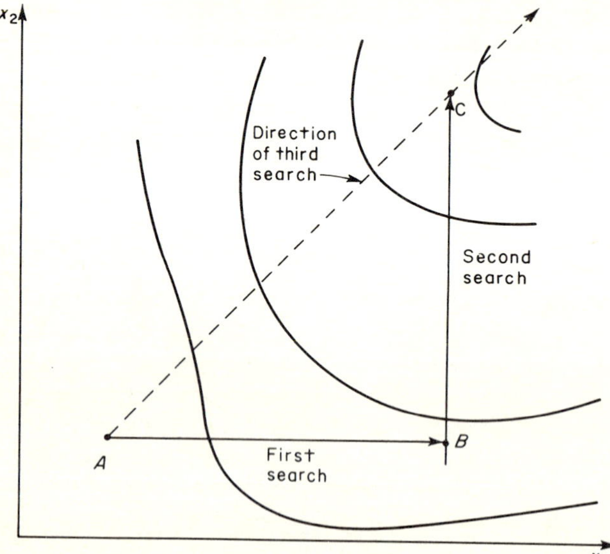

having the form

$$y(\mathbf{x}) = y(\mathbf{x}^*) + \frac{1}{2}\sum_{i=1}^{n}\sum_{j=1}^{n} a_{i,j}(x_i - x_i^*)(x_j - x_j^*) \qquad (3.1\text{-}15)$$

where the $a_{i,j}$ are constants which satisfy $a_{i,j} = a_{j,i}$ and $\sum_{i=1}^{n}\sum_{j=1}^{n} a_{i,j}(x_i - x_i^*) \times (x_j - x_j^*) > 0$ for all nontrivial $x_i - x_i^*$. This has considerable merit because, near the minimum, most functions which are continuous and have continuous first and second derivatives will display this functional dependence. To see this, consider a Taylor expansion of y about the minimum $\mathbf{x}^*$, so that (Chap. VII, Sec. 2.2)

$$y(\mathbf{x}) = y(\mathbf{x}^*) + \sum_{i=1}^{n} \frac{\partial y}{\partial x_i}(x_i - x_i^*) + \frac{1}{2}\sum_{i=1}^{n}\sum_{j=1}^{n} \frac{\partial^2 y}{\partial x_i\,\partial x_j}(x_i - x_i^*)(x_j - x_j^*)$$

Since $y(\mathbf{x}^*)$ is the minimum, the first derivatives will vanish, and if we identify

$$a_{i,j} \equiv \frac{\partial^2 y}{\partial x_i\,\partial x_j} \equiv a_{j,i}$$

the Taylor expansion reduces to Eq. (3.1-15). Moreover, the additional requirement that the $a_{i,j}$ be such that the second term on the right be positive-definite is also a sufficient condition for a minimum (Chap. VII, Sec. 2.2.3).

A search technique which finds the minimum of y defined by Eq. (3.1-15) in a finite number of one-dimensional searches will have considerable merit for a general optimization problem, since it should converge rapidly once the optimum is approached. The basis for these methods can be seen in the geometric construction shown in Fig. VIII-20. Here any point, $\mathbf{x}_0$, is taken as the initial base point, and a search is made in an arbitrary direction (shown here as parallel to the x_1 axis) until a minimum, $\mathbf{x}_1$, is found. A second point, $\mathbf{x}_2$, not on the line through $\mathbf{x}_0$ and $\mathbf{x}_1$, is now taken and a search carried out for a minimum, $\mathbf{x}_3$, on the line parallel to that through $\mathbf{x}_0$ and $\mathbf{x}_1$. If the contours are ellipsoidal, as defined by Eq. (3.1-15), a line connecting points $\mathbf{x}_1$ and $\mathbf{x}_3$ will pass exactly through the true minimum (see Sec. 3.2.2 and Prob. VIII-17). A one-dimensional search along this connecting line will then reveal the desired optimum. We can therefore find the optimum without further iteration, provided the objective function has the quadratic form given by Eq. (3.1-15). If the objective function is not quadratic, then of course convergence is not assured in a finite number of iterations, or even after a very large number of tries. Since most problems of interest are not expressed in terms of quadratic objective functions, we might inquire as to the value of quadratic convergence. Most functions, however, approach quadratic behavior in the neighborhood of their optimum, and since quadratic functions can have very elongated elliptical contours, quadratically convergent methods are capable of coping with such shapes. In essence, it

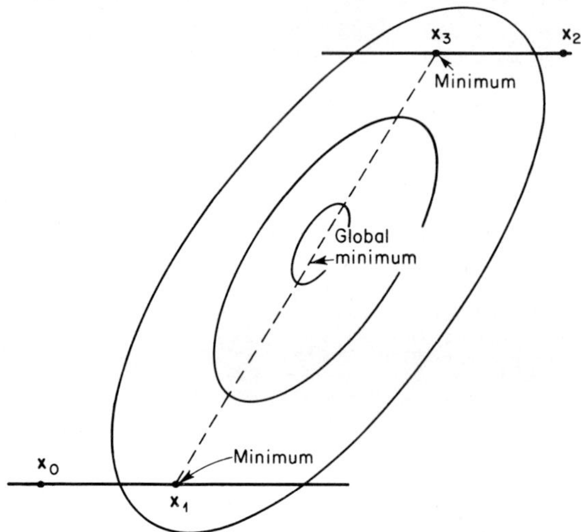

Fig. VIII-20 Principle of quadratic convergence.

seems very reasonable to expect a quadratically convergent procedure to be effective in searching nonquadratic objective functions.

One of the most widely accepted direct search methods, that developed by Powell [1964], is based on the concept of quadratic convergence depicted in Fig. VIII-20. Powell has, however, made some very important modifications to facilitate convergence of his method when applied to nonquadratic objective functions. Because of these modifications, Powell's method is not perfectly quadratically convergent, but his basic method of generating the directions of search are guided by the desire to achieve such convergence where practical. This method is a very powerful one, and we shall detail it here.

To fully appreciate Powell's method, let us first restrict our attention to two-dimensional problems and try to envisage a search scheme based on the construction shown in Fig. VIII-20. One possible quadratically convergent two-dimensional search method might start with two not necessarily conjugate search directions $\boldsymbol{\xi}_1^{(0)}$ and $\boldsymbol{\xi}_2^{(0)}$ (see page 423). In the first stage these could be directions parallel to the coordinate axes. We then carry out a univariant search along these, starting from an arbitrarily chosen base point $\mathbf{x}_0^{(0)}$ and ending at the improved point $\mathbf{x}_2^{(0)}$. Just as in the Forsythe-Motzkin method, we define a favorable search direction as being $\boldsymbol{\mu}^{(0)} = \mathbf{x}_2^{(0)} - \mathbf{x}_0^{(0)}$. Here $\boldsymbol{\mu}^{(0)}$ is a direction from the base point to the final point of the univariant search. This is very similar to the situation depicted in Fig. VIII-19, except that the point $\mathbf{x}_0^{(0)}$ is not the optimum point of a previous search, and hence we have no assurance that $\boldsymbol{\mu}$ is a direction which will pass through the optimum even if the function were quadratic. Let us now search along $\boldsymbol{\mu}^{(0)}$ reaching the improved point $\mathbf{x}_0^{(1)}$, the base point for the second stage of the search. Let us use the directions $\boldsymbol{\xi}_1^{(1)} = \boldsymbol{\xi}_2^{(0)}$ and $\boldsymbol{\xi}_2^{(1)} = \boldsymbol{\mu}^{(0)}$ in our second stage. If $\mathbf{x}_2^{(1)}$ is the optimum found after searching both of these directions, then direction $\boldsymbol{\mu}^{(1)} = \mathbf{x}_2^{(1)} - \mathbf{x}_0^{(1)}$ will pass through the optimum of a quadratic function since there have been two searches along the direction $\boldsymbol{\mu}^{(0)}$ in this stage as indicated in Fig. VIII-20.

Of course, if the function is not quadratic, then $\boldsymbol{\mu}^{(1)}$ need not pass through the optimum but would still probably be a favorable search direction. The iterations could then be repeated.

In the case of three dimensions, it can be shown that the optimum of a quadratic function can be found after no more than three stages of the above procedure. Higher dimensionalities require correspondingly more stages, but nevertheless, the number of searches required remains finite. The proofs of these assertions are not given here, but the basic theorems required for the proofs are obtained in the solution of Probs. VIII-53 and VIII-54. The practical application of the method outlined above is seriously hampered by the fact that, when $\mathbf{x}_1$ and $\mathbf{x}_0$ are by chance very close to one another, as would be the case if points A and B of Fig. VIII-19 were near each other, the new search direction, the third direction of search in Fig. VIII-19, is almost colinear with the second, and in the next stage of our search we should be using two directions which are colinear. The search would not, therefore, proceed in the x_1 direction. This is a serious defect when optimizing nonquadratic objective functions.

Powell [1964] has modified the scheme presented above to avoid these difficulties and yet retain as nearly as possible the feature of quadratic convergence. The Powell method can be summarized as follows for n-dimensional search:

Step 1 Starting with the best previous value position $\mathbf{x}_0^{(k)}$ and a series of linearly independent directions of search, $\boldsymbol{\xi}_1^{(k)}, \boldsymbol{\xi}_2^{(k)}, \ldots, \boldsymbol{\xi}_n^{(k)}$ (the property of linear independence simply implies that any arbitrary vector in the n-dimensional space can be expressed as a linear combination of the n vectors listed here), begin the search by finding the position of the optimum along the line passing through $\mathbf{x}_0^{(k)}$ which is parallel to $\boldsymbol{\xi}_1^{(k)}$. At this optimum point $\mathbf{x}_1^{(k)}$, begin a second search from this new point in the $\boldsymbol{\xi}_2^{(k)}$ direction. Continue this procedure until all n search directions have been explored.

Step 2 We now find the particular point $\mathbf{x}_m^{(k)}$ for which the greatest improvement of the objective function over its previous value is realized. The point $\mathbf{x}_m^{(k)}$ therefore yields the largest change, Δ, of any of the n moves, where $\Delta = |y(\mathbf{x}_m^{(k)}) - y(\mathbf{x}_{m-1}^{(k)})|$. Also determine the vector $\boldsymbol{\mu} = \mathbf{x}_n^{(k)} - \mathbf{x}_0^{(k)}$.

Step 3 Determine $y(2\mathbf{x}_n^{(k)} - \mathbf{x}_0^{(k)}) = y_t^{(k)}$.

Step 4 If, in seeking a minimum,

$$y_t^{(k)} \geq y_0^{(k)} \tag{3.1-16a}$$

and/or

$$(y_0^{(k)} - 2y_n^{(k)} + y_t^{(k)})(y_0^{(k)} - y_n^{(k)} - \Delta)^2 \geq \frac{\Delta(y_0^{(k)} - y_t^{(k)})^2}{2} \quad (3.1\text{-}16b)$$

then $\boldsymbol{\mu}$ is not a good direction to introduce into our search, and we begin the search again, starting at the last point and using the same directions, that is, $\mathbf{x}_0^{(k+1)} = \mathbf{x}_n^{(k)}$ and $\boldsymbol{\xi}_i^{(k+1)} = \boldsymbol{\xi}_i^{(k)}$ for $i = 1, 2, \ldots, n$. Step 1 is then repeated. If neither of these inequalities is satisfied, search along the direction $\boldsymbol{\mu}$ until the minimum is found. This point is defined as $\mathbf{x}_0^{(k+1)}$ and the new search directions for the $(k+1)$ stage are $\boldsymbol{\xi}_i^{(k+1)} = \boldsymbol{\xi}_i^{(k)}$, $i = 1, 2, \ldots, m-1$; $\boldsymbol{\xi}_i^{(k+1)} = \boldsymbol{\xi}_{i+1}^{(k)}$, $i = m, \ldots, n-1$; and $\boldsymbol{\xi}_n^{(k+1)} = \boldsymbol{\mu}$. We then repeat the entire process, starting with step 1.

The steps detailed above show that the Powell method determines a new search direction in n dimensions, using the constructional procedure depicted in Fig. VIII-19, but questions the value of using this new direction rather than blindly introducing it into the list of search directions while rejecting another. There are two inequalities to be tested. The first considers the value of a tentative value of the objective function y_t found by exploring ahead along the new search direction. Here Powell tests the objective function at the point $\mathbf{x}_t = 2\mathbf{x}_n - \mathbf{x}_0$, which is a point along the direction $\boldsymbol{\mu}$, located the same distance from $\mathbf{x}_n$ as the original base point $\mathbf{x}_0$. If this tentative feeler in the new direction does not produce a better value of the objective function than the original base point, the new direction is rejected without further tests as being a poor prospect.

The second inequality is used to determine if the function may not be rising sharply (when seeking a minimum) in the move from the point $\mathbf{x}_n$ to the tentative point $2\mathbf{x}_n - \mathbf{x}_0$ after dropping sharply from $\mathbf{x}_0$ to $\mathbf{x}_n$. In other words, the test is used to determine if the direction $\boldsymbol{\mu}$ is pointing across a deep valley, and if it is, we reject it as a reasonable direction of search, retaining the previous search directions for use in the new stage.

Finally, we note that the two inequalities of step 4 are to be reversed in the search for a maximum.

Example VIII-5: Application of the Powell search method

Let us seek the minimum of the objective function

$$y = 10(x_1 + x_2 - 5)^2 + (x_1 - x_2)^2$$

We shall begin our search at the point $\mathbf{x}_0^{(0)} = \{0,0\}$, using the search directions $\boldsymbol{\xi}_1^{(0)} = \{1,0\}$ and $\boldsymbol{\xi}_2^{(0)} = \{0,1\}$.

Stage 1 Starting from the base point, we perform a one-dimensional search in the $\boldsymbol{\xi}_1^{(0)}$ direction. The distance of movement is determined so as to minimize

$$y = 10(s - 5)^2 + s^2$$

giving $s = 4.545$. Thus $\mathbf{x}_1^{(0)} = \{4.545,0\}$, and the value of the objective function is reduced from 250 to 22.727. Searching in the $\boldsymbol{\xi}_2^{(0)}$ direction, we find $\mathbf{x}_2^{(0)} = \{4.545,0.826\}$, with $y_2^{(0)} = y(\mathbf{x}_2^{(0)}) = 15.214$.

At this point in the search we must decide whether or not to use the direction $\boldsymbol{\mu}^{(0)} = \mathbf{x}_2^{(0)} - \mathbf{x}_0^{(0)}$ in the next stage. Examining the first of inequalities (3.1-16), we see that a test of the point $\mathbf{x}_t = 2\mathbf{x}_2^{(0)} - \mathbf{x}_0^{(0)}$ must be performed. In this case $\mathbf{x}_t = \{2(4.545) - 0, 2(0.826) - 0\} = \{9.091,1.653\}$, and $y_t^{(0)}$ is seen to be 385.24. Since $y_t^{(0)} > y_2^{(0)}(y_2^{(0)} = 15.214)$, the direction $\boldsymbol{\mu}^{(0)}$ is not used in the subsequent calculations.

Stage 2 The directions in this stage are the same as in the preceding stage since $\boldsymbol{\mu}^{(0)}$ is not introduced into the calculations. Successive searches yield $\mathbf{x}_1^{(1)} = \{3.869,0.826\}$ with $y_1^{(1)} = 10.185$ and $\mathbf{x}_2^{(1)} = \{3.869,1.380\}$ with $y_2^{(1)} = 6.818$. We must again test point $\mathbf{x}_t = \{2(3.869) - 4.545, 2(1.380) - 0.826\} = \{3.193,1.933\}$. Here $y_t = 1.747$, and since $y_t^{(1)} < y_2^{(1)}$, the first of inequalities (3.1-16) is not satisfied. To test the second, we compute

$$(y_0^{(1)} - 2y_2^{(1)} + y_t^{(1)})(y_0^{(1)} - y_2^{(1)} - \Delta)^2$$
$$= [15.214 - 2(6.818) + 1.747][15.214 - 6.818 - 5.029]^2 = 37.7$$

Here Δ is the largest change in the objective function in a given direction and is $|y_0^{(1)} - y_1^{(1)}| = 5.029$. The right-hand side of the second inequality is

$$\tfrac{1}{2}\Delta(y_0^{(1)} - y_t^{(1)})^2 = 456.0$$

The second inequality is thus not satisfied, and the direction $\boldsymbol{\mu}^{(1)} = \{3.869 - 4.545, 1.380 - 0.826\} = \{-0.676,0.553\}$ is to be introduced into the search.

Before moving to the third stage, a search is required along the new direction $\boldsymbol{\mu}^{(1)}$. This search yields the base point for the third stage as $\mathbf{x}_0^{(2)} = \{2.500,2.500\}$ with $y_0^{(2)} = 0.00$.

Stage 3 The search directions for the third stage are $\boldsymbol{\xi}_1^{(2)} = \boldsymbol{\xi}_2^{(1)} = \{0,1\}$ and $\boldsymbol{\xi}_2^{(2)} = \boldsymbol{\mu}^{(1)} = \{-0.676,0.553\}$. Searches along these two directions will reveal that no movement is possible since the global minimum was found in the previous stage. Of course, this is not surprising since the objective function is quadratic.

Other methods which are quadratically convergent and do not require gradients have been reported by Shah et al. [1964][1], Powell [1962], and Smith [1962]. All these quadratically convergent direct search methods, once a direction has been chosen, proceed to determine the optimum along

[1] Their method of parallel tangents, Partan, being also described by Wilde [1964], quoting from the original reports of Buehler et al. [1961a, b, 1962].

that direction by a one-dimensional search. However, Powell's modifications appear to be significant and yield, in general, the best of these search techniques. The best method for a particular application will, however, depend upon the specific situation, Fletcher [1965] having critically reviewed the nongradient techniques.

Acceleration in Both Direction and Distance—the Rosenbrock Method The search technique of Rosenbrock [1960] gives, after an initial exploration, an acceleration in both the direction and distance of movement.

We shall first consider the two-dimensional problem and then generalize the results to n dimensions. The essential aim of the Rosenbrock method is to align one of the search directions in the most favorable direction of search, the most favorable being selected from information gleaned from previous experiments. A second direction of search is then to be selected, so that it is orthogonal to the first and at the same time oriented in a favorable direction. This process is continued until n directions of search have been generated. The Rosenbrock method can therefore be conveniently divided into two levels of computation. There are, at the upper level, stages in which the new directions of search are generated; and at the lower level, each stage consists of sequential searches along fixed directions of search. A stage is therefore characterized by the fact that the directions of search are fixed, with these directions being changed only in the transition from one stage to another.

Let us begin by assuming that a step size for each direction of search has been selected; i.e., there are n step sizes, $s_1, s_2, \ldots, s_n$, which are known. Initially, let us choose these directions of search to correspond with the coordinate axes. A stage comprised of a series of cycles can now be completed. This starts from an arbitrary base point, for example, point A of Fig. VIII-21, and, searching for a maximum, looks first in the direction $\boldsymbol{\xi}_1^{(1)} = \{1,0\}$ at point B, a distance s_1 from the base point A.† For this illustration $y(B) < y(A)$, so that the step is a failure. In the event of a failure, the distance of movement s_1 is replaced in the next cycle by $-\beta s_1$, with $0 < \beta < 1$, Rosenbrock recommending $\beta = \frac{1}{2}$ as a reasonable value. The first stage is continued by searching in the $\boldsymbol{\xi}_2^{(1)}$ direction $\{0,1\}$ at point C, a distance s_2 from A. Success. For such a success replace s_2 by αs_2 in the next cycle of search, α being greater than 1, Rosenbrock recommending $\alpha = 3$. Now try again in the $\boldsymbol{\xi}_1^{(1)}$ direction getting point D, located a distance s_1 from the last successful point, in this case point C. Remember, however, that s_1 is now $-\beta$ times the original choice, since the last move in the $\boldsymbol{\xi}_1^{(1)}$ direction was

† The notation is explained more fully by Eq. (3.1-18), the vector $\boldsymbol{\xi}_j^{(k)}$ representing the direction along which the jth exploration move is aligned in the kth stage of the search, with $k \geq 1$ and $1 \leq j \leq n$, the n directions being placed orthogonally and realigned after each stage.

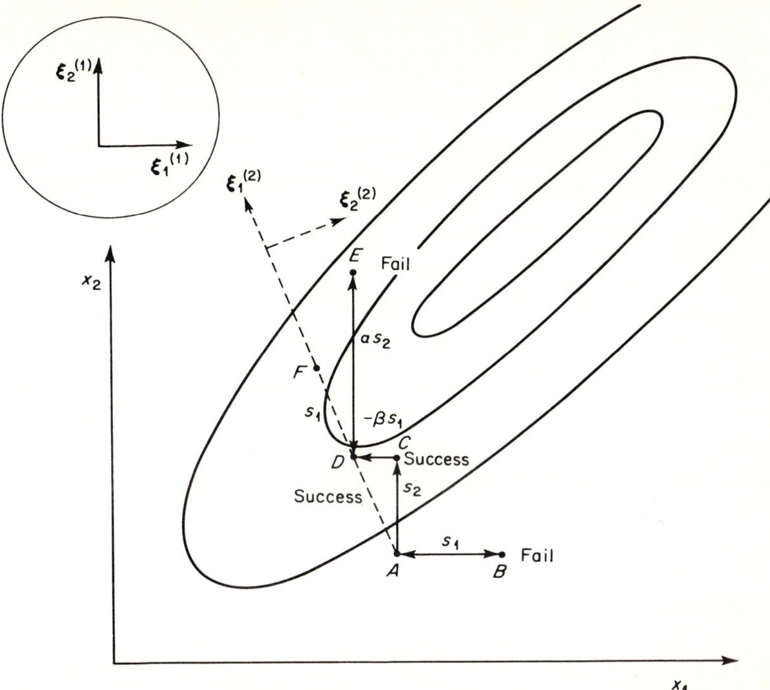

Fig. VIII-21 Two-dimensional Rosenbrock search.

a failure. Testing the objective function, we find that $y(D) > y(C)$. Let us move again in the direction $\xi_2^{(1)}$ and test point E at a distance αs_2 from D. Here the test fails, since $y(E) < y(D)$, and D is retained as the best point in the stage. So far the method is similar to the univariant technique (Sec. 3.1.1), although there is no attempt to find the optimum along a given path. We have now had at least one success and one failure in each of the directions $\xi_1^{(1)}$ and $\xi_2^{(1)}$. At this juncture Rosenbrock recommends starting a second stage and changing the direction of search to $\xi_1^{(2)}$ and $\xi_2^{(2)}$ in such a way that eventually, after several changes of direction, the new ξ_1 will tend to lie along the direction of fastest advance, and ξ_2 will be perpendicular to it. The set of calculations for a given set of directions is called a stage; and at the end of a stage, which occurs when there has been at least one success and one failure in each direction, the directions are changed.

Before noting the way in which the directions are altered, it should be pointed out that the series of trials in a certain direction must ultimately succeed, since the step size for that direction becomes so small after repeated failures that it causes no change in objective function. Thus, by defining success as being a step in which the new objective function is better than or equal to the value at the base point, we are bound to succeed eventually.

Having completed the first stage, let us define a set of directions for the second stage. First, define the following functions for use in choosing the new direction $\boldsymbol{\xi}_1^{(2)}$.

$$\mathbf{A}_1^{(1)} = d_1\boldsymbol{\xi}_1^{(1)} + d_2\boldsymbol{\xi}_2^{(1)} \tag{3.1-17a}$$

or written out fully,[1]

$$A_{i,1}^{(1)} = d_1 m_{i,1}^{(1)} + d_2 m_{i,2}^{(1)} \qquad \text{for } i = 1, 2 \tag{3.1-17b}$$

Here d_1 is the algebraic sum of all the successful moves in the $\boldsymbol{\xi}_1^{(1)}$ direction, and d_2 is the same sum for the $\boldsymbol{\xi}_2^{(1)}$ direction. The $A_{i,j}^{(1)}$ are thus defined in terms of the directional components and the magnitudes of the successful moves of the first stage. Hence, for the case depicted in Fig. VIII-21,

$$d_1 = -\beta s_1$$

$$d_2 = s_2$$

corresponding to the single successful moves from C to D and A to C in the $\boldsymbol{\xi}_1^{(1)}$ and $\boldsymbol{\xi}_2^{(1)}$ directions, respectively. Then, since

$$\boldsymbol{\xi}_1^{(1)} = \{1,0\} \qquad \boldsymbol{\xi}_2^{(1)} = \{0,1\}$$

we get

$$A_{1,1}^{(1)} = -\beta s_1 \qquad \text{and} \qquad A_{2,1}^{(1)} = s_2$$

or

$$\mathbf{A}_1^{(1)} = \{-\beta s_1, s_2\}$$

The necessary directional components for the second stage of the search can now be written

$$m_{i,1}^{(2)} = \frac{A_{i,1}^{(1)}}{\sqrt{(A_{1,1}^{(1)})^2 + (A_{2,1}^{(1)})^2}} \qquad \text{for } i = 1, 2 \tag{3.1-19}$$

These define the new direction

$$\boldsymbol{\xi}_1^{(2)} = \{m_{1,1}^{(2)}, m_{2,1}^{(2)}\}$$

We must now find a second direction of search, which is perpendicular (orthogonal) to $\boldsymbol{\xi}_1^{(2)}$. Let us define

$$\mathbf{A}_2^{(1)} = d_2\boldsymbol{\xi}_2^{(1)} \tag{3.1-20}$$

[1] In this notation the vector $\mathbf{x}_l^{(k)}$ represents the local base point after the lth cycle of search in the kth stage, as given by Eq. (3.1-12); $\boldsymbol{\xi}_l^{(k)}$ is the directional vector in the kth stage such that

$$\boldsymbol{\xi}_l^{(k)} = \{m_{1,l}^{(k)}, \ldots, m_{n,l}^{(k)}\} \tag{3.1-18}$$

similar to the definition of Eq. (3.1-14), but now, since the direction component $m_{i,l}$ may change in magnitude, it is necessary to add a superscript k to denote the stage; and

$$\mathbf{A}_l^{(k)} = \{A_{1,l}^{(k)}, \ldots, A_{n,l}^{(k)}\}$$

where the $\mathbf{A}_l^{(k)}$ are given by Eqs. (3.1-22) [for $l = 1$, being also given by (3.1-17)].

For the situation in Fig. VIII-21, we find that

$$A_{1,2}^{(1)} = 0 \qquad \text{and} \qquad A_{2,2}^{(1)} = s_2 \qquad \text{or} \qquad \mathbf{A}_2^{(1)} = \{0,s_2\}$$

We may now define the new direction $\boldsymbol{\xi}_2^{(2)}$ so that

$$\boldsymbol{\xi}_1^{(2)} \cdot \boldsymbol{\xi}_2^{(2)} = 0$$

a necessary condition for the two directions to be orthogonal. The direction $\boldsymbol{\xi}_1^{(2)}$ has been normalized so that

$$\boldsymbol{\xi}_1^{(2)} \cdot \boldsymbol{\xi}_1^{(2)} = \sum_{i=1}^{n} m_{i,1}^{(2)} m_{i,1}^{(2)} = 1$$

This is an important property, since the vector

$$\boldsymbol{\xi}_2^{(2)} = \frac{\mathbf{A}_2^{(1)} - (\boldsymbol{\xi}_1^{(2)} \cdot \mathbf{A}_2^{(1)})\boldsymbol{\xi}_1^{(2)}}{|\mathbf{A}_2^{(1)} - (\boldsymbol{\xi}_1^{(2)} \cdot \mathbf{A}_2^{(1)})\boldsymbol{\xi}_1^{(2)}|}$$

satisfies the orthogonality condition and is normalized as well. These properties can be verified by first forming $\boldsymbol{\xi}_2^{(2)} \cdot \boldsymbol{\xi}_1^{(2)}$ as follows:

$$\boldsymbol{\xi}_2^{(2)} \cdot \boldsymbol{\xi}_1^{(2)} = \frac{\boldsymbol{\xi}_1^{(2)} \cdot \mathbf{A}_2^{(1)} - (\boldsymbol{\xi}_1^{(2)} \cdot \mathbf{A}_2^{(1)})\boldsymbol{\xi}_1^{(2)} \cdot \boldsymbol{\xi}_1^{(2)}}{|\mathbf{A}_2^{(1)} - (\boldsymbol{\xi}_1^{(2)} \cdot \mathbf{A}_2^{(1)})\boldsymbol{\xi}_1^{(2)}|} = 0$$

Moreover,

$$\boldsymbol{\xi}_2^{(2)} \cdot \boldsymbol{\xi}_2^{(2)} = 1$$

because the quantity $|\mathbf{A}_1^{(2)} - (\boldsymbol{\xi}_1^{(2)} \cdot \mathbf{A}_2^{(1)})\boldsymbol{\xi}_1^{(2)}|$ represents the magnitude of this vector. We can, for clarity, write this expression in component form

$$\boldsymbol{\xi}_2^{(2)} = \{m_{1,2}^{(2)}, m_{2,2}^{(2)}\} = \{BD_1^{(1)}, BD_2^{(1)}\} \tag{3.1-21}$$

where

$$B = [(D_1^{(1)})^2 + (D_2^{(1)})^2]^{-1/2}$$

and

$$D_i^{(1)} = A_{i,2}^{(1)} - m_{i,1}^{(2)}[A_{1,2}^{(1)} m_{1,1}^{(2)} + A_{2,2}^{(1)} m_{2,1}^{(2)}] \qquad \text{for } i = 1, 2$$

This method of forming the orthogonal set of directions is called the *Gram-Schmidt orthogonalization* (Birkhoff and MacLane [1953]). This completes the first stage.

There are several essential features to be observed. First of all, the vector $\boldsymbol{\xi}_1^{(2)}$, a new direction of search in the second stage, corresponds to a line connecting the base point at the beginning of the first stage to the terminal point in that stage, which is a vector sum of all the successful moves. Therefore $\boldsymbol{\xi}_1^{(2)}$ now points in the direction showing the greatest promise based on past results (direction ADF in Fig. VIII-21). The second direction, $\boldsymbol{\xi}_2^{(2)}$, is orthogonal to $\boldsymbol{\xi}_1^{(2)}$and, in effect, points in the next most successful direction, which is perpendicular to the first.

The analytical evaluation of these vectors for the particular case shown in Fig. VIII-21 is straightforward. Thus $\boldsymbol{\xi}_1^{(2)}$, as determined by Eqs. (3.1-19), is given, if $s_1 = s_2$, by

$$m_{1,1}^{(2)} = \frac{-\beta s_1}{\sqrt{(\beta s_1)^2 + (s_2)^2}} = \frac{-\beta}{\sqrt{\beta^2 + 1}}$$

and $$m_{2,1}^{(2)} = \frac{s_2}{\sqrt{(\beta s_1)^2 + (s_2)^2}} = \frac{1}{\sqrt{\beta^2 + 1}}$$

Similarly, from Eqs. (3.1-21) we find that $\boldsymbol{\xi}_2^{(2)}$ is given by

$$m_{1,2}^{(2)} = \frac{1}{\sqrt{\beta^2 + 1}}$$

and $$m_{2,2}^{(2)} = \frac{\beta}{\sqrt{\beta^2 + 1}}$$

It will be seen that this new direction, $\boldsymbol{\xi}_2^{(2)}$, is perpendicular to the direction $\boldsymbol{\xi}_1^{(2)}$, by showing that

$$m_{1,1}^{(2)} m_{1,2}^{(2)} + m_{2,1}^{(2)} m_{2,2}^{(2)} = 0$$

or in vector notation,

$$\boldsymbol{\xi}_1^{(2)} \cdot \boldsymbol{\xi}_2^{(2)} = 0$$

This construction gives us a new set of directions along which we can search in stage 2, as illustrated in Fig. VIII-21. Let us move from point D in a direction $\boldsymbol{\xi}_1^{(2)}$, a distance s_1 to point F. This is a failure, since the objective function decreases. Again, at D, we test along the direction $\boldsymbol{\xi}_2^{(2)}$, a distance s_2 (not shown in the figure). This test will also fail. We are still at point D. Now move a distance $-\beta s_1$ from point D and in direction $\boldsymbol{\xi}_1^{(2)}$. Again a failure. Thus we have two failures in our trials in direction $\boldsymbol{\xi}_1^{(2)}$, and our base point is still D. Now move a distance $-\beta s_2$ in direction $\boldsymbol{\xi}_2^{(1)}$. The test again fails. Now move in direction $\boldsymbol{\xi}_1^{(2)}$ a distance $\beta^2 s_1$. This test might be fruitful, depending on the value of β, since the move is along the line connecting D to F, but only a fraction (β^2) of the distance. This would give us a new base point and a successful try in the $\boldsymbol{\xi}_1^{(2)}$ direction. This computational procedure would continue until a successful try in the $\boldsymbol{\xi}_2^{(2)}$ direction was found. Then new directions of search would again be selected, and stage 3 initiated. The whole process is continued for either a fixed number of moves or until some predetermined lower limit on $d_1^2 + d_2^2$ is achieved. This latter criterion would indicate that the distances moved in the stage just completed are sufficiently small to suppose that an optimum has been found.

Example VIII-6: The Rosenbrock search technique

We shall detail three stages in the Rosenbrock search technique by seeking the minimum of the function

$$y = 25(x_2 - x_1^2)^2 + (1 - x_1)^2$$

The contour lines of this banana-shaped function are shown in Fig. VIII-22. The function given here is a modification of one proposed by Rosenbrock for testing various search techniques.

The base point chosen arbitrarily is point A {2,13} in Fig. VIII-22. The first point to be searched in the first cycle is found by moving a distance s_1 along a direction parallel to the x_1 axis. As can be seen in Table VIII-4A, the distance s_1 has been chosen to be 0.5, thus giving the

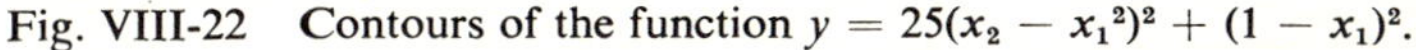

Fig. VIII-22 Contours of the function $y = 25(x_2 - x_1^2)^2 + (1 - x_1)^2$.

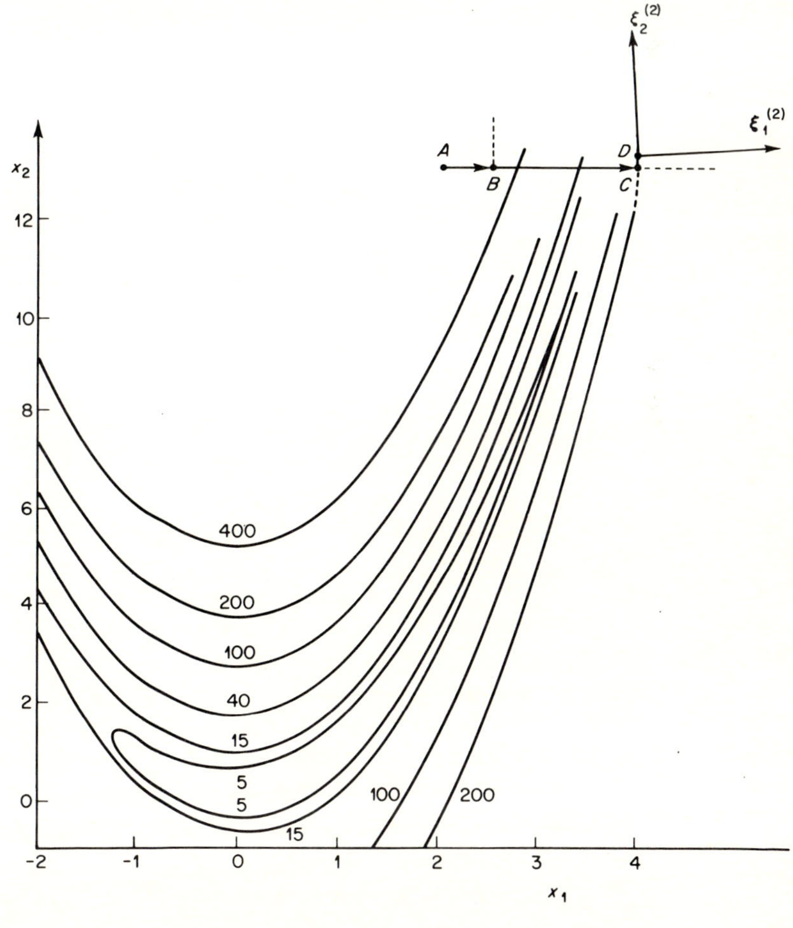

TABLE VIII-4A ROSENBROCK SEARCH

First stage					Action			
Cycle	x_1	x_2	y	*Result of test*	d_1	d_2	s_1	s_2
Base point	2	13	2,026		0	0	0.5	0.5
First	2.5	13	1,141	Success	0.5	0	1.5	0.5
Second	2.5	13.5	1,316	Failure	0.5	0	1.5	−0.25
Third	4	13	234	Success	2.0	0	4.5	−0.25
Fourth	4	12.75	273	Failure	2.0	0	4.5	0.125
Fifth	8.5	13	87,820	Failure	2.0	0	−2.25	0.125
Sixth	4	13.125	216	Success	2.0	0.125	−2.25	0.375

first point to be tested as {2.5,13}, which is point B in Fig. VIII-22. As can be seen, the move was a success, since the objective function decreased, and the following action is taken: (1) The quantity d_1 is made equal to 0.5, since this was the distance moved along the first direction of search; and (2) the s_1 is multiplied by a factor α (here $\alpha = 3$), since the move was a success.

The results of each of the successive cycles is shown in Table VIII-4A. Note that the value of s_2 is given as −0.25 after the second search, which in this case was unsuccessful. We are using the value of $\beta = \frac{1}{2}$, as recommended by Rosenbrock.

The search terminates when there has been at least one success and one failure along each of the directions of search. This rule is essential, because the next step, that of selecting new search directions, is not feasible unless all the d's are nonzero.

Let us now turn our attention to the selection of new coordinates for search. During the first stage, we used

$$\xi_1^{(1)} = \{m_{1,1}^{(1)}, m_{2,1}^{(1)}\} = \{1,0\} \qquad \text{and} \qquad \xi_2^{(1)} = \{m_{1,2}^{(1)}, m_{2,2}^{(1)}\} = \{0,1\}$$

From Eq. (3.1-17) we find

$$\mathbf{A}_1^{(1)} = \{d_1 m_{1,1}^{(1)} + d_2 m_{1,2}^{(1)},\ d_1 m_{2,1}^{(1)} + d_2 m_{2,2}^{(1)}\} = \{2.0, 0.125\}$$

The normalized components[1] of the direction are given as

$$m_{1,1}^{(2)} = \frac{A_{1,1}^{(1)}}{\sqrt{(A_{1,1}^{(1)})^2 + (A_{2,1}^{(1)})^2}} = 0.9981$$

$$m_{2,1}^{(2)} = \frac{A_{2,1}^{(1)}}{\sqrt{(A_{1,1}^{(1)})^2 + (A_{2,1}^{(1)})^2}} = 0.0624$$

[1] These components are normalized because

$$\xi_1^{(2)} \cdot \xi_1^{(2)} = m_{1,1}^{(2)} \cdot m_{1,1}^{(2)} + m_{2,1}^{(2)} \cdot m_{2,1}^{(2)} = (0.9981)^2 + (0.0624)^2 = 1.0$$

The components of the first new direction of search are now formed. This direction is heavily weighted in those directions yielding the greatest movement in the preceding stage. For the present example, the search along the x_1 direction was by far the more fruitful, and consequently, the new direction of search almost coincides with the x_1 axis. The last direction of search can be constructed perpendicular to the first, and here it is unique, so that we do not need to use the Gram-Schmidt method. Let us use the equations given in the text in order to familiarize ourselves with their meaning.

Using Eq. (3.1-20), we see that

$$\mathbf{A}_2^{(1)} = \{A_{1,2}^{(1)}, A_{2,2}^{(1)}\} = \{0, 0.125\}$$

Therefore $\quad D_1^{(1)} = -0.9981[0.125(0.0624)] = -0.00778$

$$D_2^{(1)} = 0.125 - 0.0624[0.125(0.0624)] = 0.125$$

giving $\quad m_{1,2}^{(2)} = \dfrac{D_1^{(1)}}{\sqrt{(D_1^{(1)})^2 + (D_2^{(1)})^2}} = -0.0624$

$$m_{2,2}^{(2)} = \frac{D_2^{(1)}}{\sqrt{(D_1^{(1)})^2 + (D_2^{(1)})^2}} = 0.9981$$

This new direction is easily seen to be orthogonal to the first. We have now generated our two new directions for search. Let us begin the second stage. The results of this stage are not shown on Fig. VIII-22, but are summarized in Table VIII-4B.

Most of this table is self-explanatory; however, it is instructive to see the origin of the point tested in the first cycle of this second stage. Thus the point to be tested is

$$x_{1,1} = 4 + 0.5(0.9981) = 4.499$$

and $\quad x_{2,1} = 13.125 + 0.5(0.0624) = 13.156$

TABLE VIII-4B ROSENBROCK SEARCH (Continued)

Second stage					*Action*			
Cycle	x_1	x_2	y	*Result of test*	d_1	d_2	s_1	s_2
Base point	4.000	13.125	216		0	0	0.5	0.5
First	4.499	13.156	1,267	Failure	0	0	−0.25	0.5
Second	3.9769	13.624	122	Success	0	0.5	−0.25	1.5
Third	3.720	13.608	8.66	Success	−0.25	0.5	−0.75	1.5
Fourth	3.627	15.105	103	Failure	−0.25	0.5	−0.75	−0.75

Finally, let us construct the set of search directions to be used in the third stage.

$$\mathbf{A}_1^{(2)} = \{d_1 m_{1,1}^{(2)} + d_2 m_{1,2}^{(2)}, d_1 m_{2,1}^{(2)} + d_2 m_{2,2}^{(2)}\}$$
$$= \{-0.25(0.9981) + 0.5(-0.0624), -0.25(0.0624) + 0.5(0.9981)\}$$
$$= \{-0.281, 0.485\}$$

Therefore $$m_{1,1}^{(3)} = \frac{-0.281}{\sqrt{(0.281)^2 + (0.485)^2}} = -0.502$$

and $$m_{2,1}^{(3)} = \frac{0.483}{\sqrt{(0.281)^2 + (0.485)^2}} = 0.865$$

Continuing,

$$\mathbf{A}_2^{(2)} = \{d_2 m_{1,2}^{(2)}, d_2 m_{2,2}^{(2)}\} = \{-0.031, 0.499\}$$
$$D_1^{(2)} = A_{1,2}^{(2)} - m_{1,1}^{(3)}(m_{1,1}^{(3)} A_{1,2}^{(2)} + m_{2,1}^{(3)} A_{2,2}^{(2)}) = 0.193$$
$$D_2^{(2)} = A_{2,2}^{(2)} - m_{2,1}^{(3)}(m_{1,1}^{(3)} A_{1,2}^{(2)} + m_{2,1}^{(3)} A_{2,2}^{(2)}) = 0.112$$

giving for the components of the second search direction

$$m_{1,2}^{(3)} = 0.865$$
$$m_{2,2}^{(3)} = 0.502$$

The search now continues through the third stage, but we have gone far enough to understand the method. Let us consider the general problem.

Let us summarize the Rosenbrock method for the n-dimensional case. Consider stage k.

Step 1 Select a set of step sizes $s_1, s_2, \ldots, s_n$ and a set of directions $\boldsymbol{\xi}_1^{(k)}, \boldsymbol{\xi}_2^{(k)}, \ldots, \boldsymbol{\xi}_n^{(k)}$. One usually selects the directions as being parallel to the axes when $k = 1$, although this is not necessary.

Step 2 Sequentially search parallel to each of the n directions in turn, adopting the new point if the move is successful (objective function is greater than or equal to the value at the base point when looking for a maximum) and retaining the base point if the move is unsuccessful. If a move in a given direction is successful, a move of length α times the distance last moved should be made when the direction is next searched in its turn. If the previous move in a given direction is unsuccessful, then the next time that particular direction is to be searched, the move should be of a distance $-\beta$ times the previous distance. Here $\alpha > 1$ and $0 < \beta < 1$, with, usually, $\alpha = 3$, $\beta = 0.5$.

Step 3 After searching in each direction sequentially until at least one move has been successful and one move has failed in each of the directions, a new set of directions can be computed from the following equations:

$$\mathbf{A}_l^{(k)} = \sum_{j=l}^{n} d_j \boldsymbol{\xi}_j^{(k)} \qquad \text{for } 1 \le l \le n \tag{3.1-22}$$

In particular, $\mathbf{A}_1^{(k)}$ defines the direction $\boldsymbol{\xi}_1^{(k+1)}$, and it will be seen that this direction of the n to be chosen is the most greatly influenced by past successful moves, and so is most likely to indicate the new direction of search. As the orthogonal directions are chosen in turn, there is less and less chance of finding the optimum along them.

Equation (3.1-22) can be written out in full as

$$A_{i,l}^{(k)} = \sum_{j=l}^{n} d_j m_{i,j}^{(k)} \qquad \text{for } i = 1, 2, \ldots, n$$

Having defined these vectors, we must now determine the required directions. Thus the principal direction $\boldsymbol{\xi}_1^{(k+1)}$ is given by

$$\boldsymbol{\xi}_1^{(k+1)} = \frac{\mathbf{D}_1^{(k)}}{|\mathbf{D}_1^{(k)}|} \tag{3.1-23}$$

with

$$\mathbf{D}_1^{(k)} = \mathbf{A}_1^{(k)}$$

giving

$$m_{i,1}^{(k+1)} = \frac{D_{i,1}^{(k)}}{\sqrt{\sum_{j=1}^{n} D_{j,1}^{(k)} D_{j,1}^{(k)}}} \qquad \text{for } i = 1, 2, \ldots, n$$

The full set of $\mathbf{D}_i^{(k)}$ are defined as

$$\mathbf{D}_1^{(k)} = \mathbf{A}_1^{(k)}$$

$$\mathbf{D}_2^{(k)} = \mathbf{A}_2^{(k)} - (\mathbf{A}_2^{(k)} \cdot \boldsymbol{\xi}_1^{(k+1)})\boldsymbol{\xi}_1^{(k+1)}$$

$$\mathbf{D}_3^{(k)} = \mathbf{A}_3^{(k)} - (\mathbf{A}_3^{(k)} \cdot \boldsymbol{\xi}_1^{(k+1)})\boldsymbol{\xi}_1^{(k+1)} - (\mathbf{A}_3^{(k)} \cdot \boldsymbol{\xi}_2^{(k+1)})\boldsymbol{\xi}_2^{(k+1)}$$

$$\cdots\cdots\cdots\cdots\cdots\cdots\cdots\cdots\cdots$$

$$\mathbf{D}_n^{(k)} = \mathbf{A}_n^{(k)} - \sum_{l=1}^{n-1} (\mathbf{A}_n^{(k)} \cdot \boldsymbol{\xi}_l^{(k+1)})\boldsymbol{\xi}_l^{(k+1)}$$

and the directions of search are defined by

$$\boldsymbol{\xi}_j^{(k+1)} = \frac{\mathbf{D}_j^{(k)}}{|\mathbf{D}_j^{(k)}|} \qquad \text{for } j = 1, 2, \ldots, n$$

This procedure is then repeated through a succession of stages until the optimum is reached.

Rosenbrock's method represents an attempt to align the direction of search along the axes of a valley, based on results of past successes in local search. It overcomes all the objections to univariant methods; however, it can be slow in starting, especially if one of the steps is too large and must be reduced several times. This can add wasteful experiments, since, for each halving of the step length, the function must be evaluated once along all the other $n - 1$ directions. The method possesses no satisfactory convergence criterion.

Rosenbrock's method has been modified by Davies, Swann, and Campey, so that in a given cycle, the search along a given direction is continued, until the optimum along that direction is found in a manner similar to univariant search. Details are given by Swann [1964], a comparison of its effectiveness being presented by Fletcher [1965]. It is said to be superior, in general, to either Hooke-Jeeves' or Rosenbrock's methods.

3.1.4 Comment on direct search methods

While gradient methods are generally more effective than direct search methods, there are situations, for example, when the function gradient is not analytically defined, in which the direct search methods are more appropriate. These methods vary from the elementary univariant form to the sophisticated Powell and Rosenbrock techniques.

The univariant method should rarely be used due to its tendency to stick on ridges and its slow progress, especially when the ridges are not parallel to the coordinate axes. This is a fault of all methods which change one variable at a time or which are constrained to limited directions of movement. Another simple form—the Sequential Simplex method—is preferable for general use. Simple to program, it has the advantage of needing only the smallest number of points to start and only one new experiment for each movement. Its main disadvantages lie in the necessity to scale the problem beforehand so that unit changes in each variable are of equal interest and in its inability to accelerate. The modified forms of Campey and Nichols, of Nelder and Mead, and of Box are more efficient, but lose some of the basic simplicity of the Simplex approach.

The other methods discussed include acceleration features and are more elegant. An extension of the univariant method, the Hooke-Jeeves method, works well on well-behaved, smooth functions, but is not highly recommended. The Rosenbrock approach overcomes all the disadvantages of the univariant method, and its coordinate rotation feature allows it to follow ridges closely. While it has no satisfactory convergence criterion and is sometimes slow to get started, particularly when one of the steps is too large and has to be reduced several times, it is effective and had found widespread application. The methods exhibiting quadratic convergence are said to be superior to the others. Powell's development is more economical than Smith's original method and possesses good convergence in most applications.

Each of the methods therefore has some basic strengths and some weaknesses, and indeed a universally effective method of numerical search with or without knowledge of the gradients has yet to be proposed.

3.2 Gradient Methods

The derivatives of an objective function with respect to each of the n variables are collectively called the *gradient of the objective function.* The gradient is a vector at a point in n-dimensional space. The greatest rate of improvement of the function will be found by moving along the gradient. This is called the *direction of steepest ascent*, if we are seeking a maximum. Unfortunately, the direction of steepest ascent is a local property rather than a global one, since a move starting in the direction of steepest ascent will not necessarily end up aligned along the new direction of steepest ascent associated with the new point, as indicated in Fig. VIII-23. In other words, the direction of steepest ascent generally varies from point to point, and the locus of infinitely small moves along the line of steepest ascent will generally be a curved line. This important aspect will be pursued in greater detail (Sec. 3.2.1). However, for the purposes of introduction, it is sufficient to note that the direction of steepest ascent has been used as the basis of many possible search techniques. These methods, called *gradient methods*, can generally be divided into two categories. One group of gradient search methods follows the direction of steepest ascent as closely as possible (Sec. 3.2.1), while the second class of methods uses the gradient to guide the search (Sec. 3.2.2), but the moves are not necessarily in the direction of the steepest ascent.

Fig. VIII-23 A sequence of steepest ascent moves.

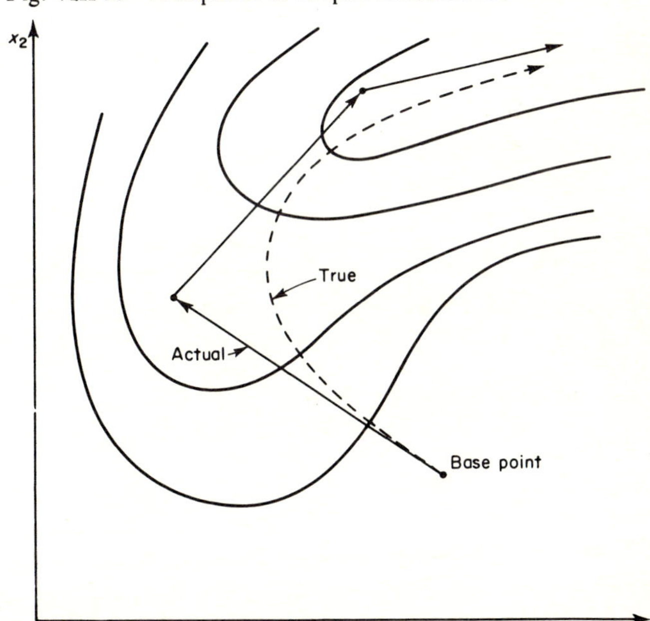

3.2.1 Search in the direction of the steepest ascent

The basic method of steepest ascent (descent in minimization) is a cyclic search technique, which does not follow the continuous line of steepest ascent, but approximates it by a succession of straight lines (Fig. VIII-23). Each line corresponds to a stage in the search in which the local function gradient is initially determined, and a set of moves then made along the resulting direction of steepest ascent, searching one-dimensionally for the approximate location of the restricted optimum. This point serves as the base for the next stage. Before considering the method, let us examine in turn the meaning of the direction of steepest ascent and the question of its evaluation.

Two-dimensional Gradients and Contour Tangents Let us consider an objective function $y(x_1,x_2)$ having continuous partial derivatives. The objective function has a well-defined value for each point $\{x_1,x_2\}$, such as point P of Fig. VIII-24. For small changes about P in any direction whatsoever, the change in y is given by

$$dy = \left(\frac{\partial y}{\partial x_1}\right)_P dx_1 + \left(\frac{\partial y}{\partial x_2}\right)_P dx_2 \tag{3.2-1}$$

where the subscript P denotes that the partial derivatives are evaluated at P and do not change, regardless of the relative values of the small displacements from P, dx_1, and dx_2. Suppose that we select a special set of values for dx_1 and dx_2 and that we are forced to move along the line with slope α

Fig. VIII-24 Directional derivative.

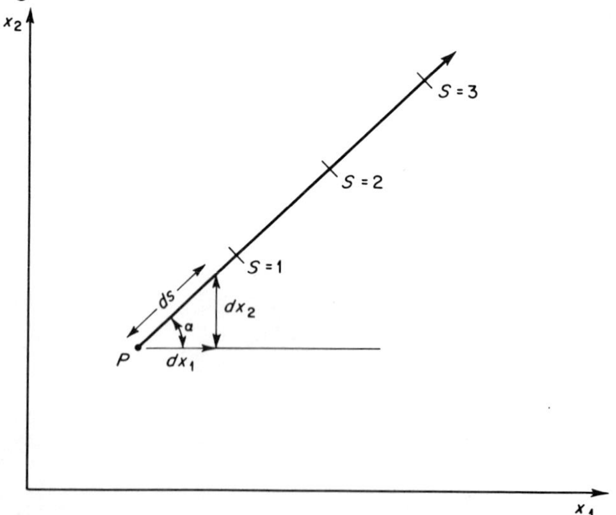

which passes through P, as shown in Fig. VIII-24. Let s be a measure of distance along this line. Then

$$\frac{dy}{ds} = \left(\frac{\partial y}{\partial x_1}\right)_P \left(\frac{dx_1}{ds}\right) + \left(\frac{\partial y}{\partial x_2}\right)_P \left(\frac{dx_2}{ds}\right) \tag{3.2-2}$$

and since, by construction (Fig. VIII-24), $dx_2/dx_1 = \tan\alpha$, $dx_1/ds = \cos\alpha$, and $dx_2/ds = \sin\alpha$, the directional derivative defined by Eq. (3.2-2) becomes

$$\frac{dy}{ds} = \left(\frac{\partial y}{\partial x_1}\right)_P \cos\alpha + \left(\frac{\partial y}{\partial x_2}\right)_P \sin\alpha \tag{3.2-3}$$

This expression can be interpreted as the change in y resulting from a move along a line having a slope α in the x_1, x_2 plane. As α changes with P fixed, the directional derivative changes, and there will be a certain value of α for which the directional derivative vanishes. This condition then represents a move along a constant value of y, and it is thus a move tangent to the contour y_P (y evaluated at P). If dy/ds is to vanish, then from Eq. (3.2-3),

$$\tan\alpha_1 = \frac{-(\partial y/\partial x_1)_P}{(\partial y/\partial x_2)_P} \tag{3.2-4}$$

This expression determines the contour tangent at the point P. Of course, there are two values of α_1 which satisfy this equation, depending on which direction along the contour the move is made. As will be seen from Fig. VIII-25, α_1 is one solution; the other is $\alpha_1 + \pi$. Between these two values of α there must be at least one maximum of dy/ds. This maximum is located along a direction such that dy/ds is stationary with respect to variations in α.

Fig. VIII-25 The contour tangent.

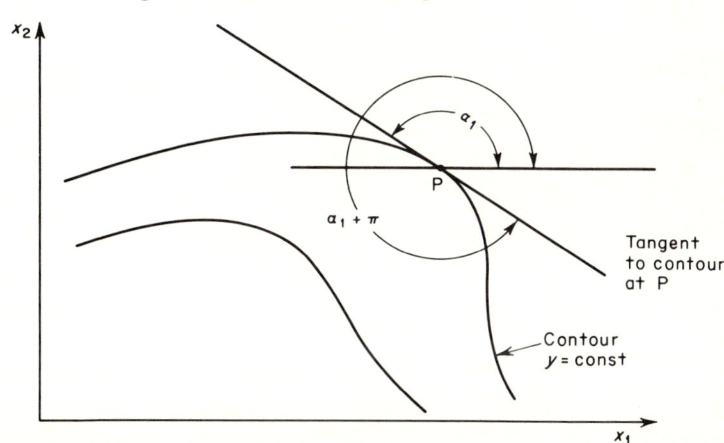

The stationary condition $d(dy/ds)/d\alpha = 0$ then leads to

$$\tan \alpha^* = \frac{(\partial y/\partial x_2)_P}{(\partial y/\partial x_1)_P} \tag{3.2-5}$$

where α^* is the angle making dy/ds stationary. Therefore

$$\left(\frac{dy}{ds}\right)_{\text{stationary}} = \cos \alpha^* \left[\left(\frac{\partial y}{\partial x_1}\right)_P + \left(\frac{\partial y}{\partial x_2}\right)_P \frac{(\partial y/\partial x_2)_P}{(\partial y/\partial x_1)_P}\right] \tag{3.2-6}$$

Since, from Eq. (3.2-5),

$$\cos \alpha^* = \frac{(\partial y/\partial x_1)_P}{\pm\sqrt{(\partial y/\partial x_1)_P{}^2 + (\partial y/\partial x_2)_P{}^2}}$$

Eq. (3.2-6) reduces to

$$\left(\frac{dy}{ds}\right)_{\text{stationary}} = \pm\sqrt{\left(\frac{\partial y}{\partial x_1}\right)_P^2 + \left(\frac{\partial y}{\partial x_2}\right)_P^2} \tag{3.2-7}$$

Clearly, the positive sign makes $(dy/ds)_{\text{stationary}}$ a maximum, and the negative gives the minimum value. Therefore one of the angles defined by Eq. (3.2-5) makes the directional derivative a maximum. It can be seen that this direction is perpendicular to the contour tangent, since the product of the gradients is

$$\tan \alpha_1 \tan \alpha^* = -1$$

a condition for two lines in the x_1, x_2 plane to be perpendicular. The greatest change in y therefore comes when a move is made perpendicular to the contour $y = \text{constant}$.

The Direction of Steepest Ascent in n Dimensions An alternative method of determining the direction of steepest ascent (or descent) has been given by Kelley [1962], but variations of this approach can also be found in most books which discuss vectors, for example, Hildebrand [1956]. This approach demonstrates an analytical, rather than geometric, argument behind the direction of steepest ascent, and this viewpoint should enhance our understanding of this search method.

The problem is to take a small step ds in such a way that the objective function increases (decreases) as much as possible. Here

$$(ds)^2 = \sum_{i=1}^{n} (dx_i)^2 \tag{3.2-8}$$

Assuming y to be differentiable, the change in y associated with a set of displacements dx_i is then given by

$$dy = \sum_{i=1}^{n} \left(\frac{\partial y}{\partial x_i}\right) dx_i$$

or

$$\frac{dy}{ds} = \sum_{i=1}^{n} \frac{\partial y}{\partial x_i} \frac{dx_i}{ds} \tag{3.2-9}$$

Now one particular set of displacements will make dy/ds as large (small) as possible. This is the direction of steepest ascent (descent). Posed another way, we can see that the problem is one of optimization with a constraint. We must maximize (minimize) dy/ds subject to the condition that ds be given by Eq. (3.2-8). Clearly, the correct set of dx_i/ds is the set which finds the stationary points of the modified objective function

$$F = \sum_{i=1}^{n} \frac{\partial y}{\partial x_i} \frac{dx_i}{ds} + \lambda\left[1 - \sum_{i=1}^{n} \left(\frac{dx_i}{ds}\right)^2\right]$$

where λ is a Lagrangian multiplier. Differentiating with respect to dx_j/ds, we find that the necessary condition is

$$\frac{\partial y}{\partial x_j} - 2\lambda \frac{dx_j}{ds} = 0$$

or

$$\frac{dx_j}{ds} = \frac{1}{2\lambda} \frac{\partial y}{\partial x_j} \qquad \text{for } j = 1, 2, \ldots, n \tag{3.2-10}$$

This important result tells us that the direction of the steepest ascent (descent) is proportional to the partial derivatives y, or in vector terminology, the direction of the steepest ascent (descent) is a vector parallel to the gradient of the function.

If at some point $\mathbf{x}_0$, the gradient has the n values $(\partial y/\partial x_i)_0$, then a line drawn from this point in the direction of steepest gradient can be represented in parametric form (Sec. 1.4) as

$$x_j - x_{j,0} = \frac{1}{2\lambda} \left(\frac{\partial y}{\partial x_j}\right)_0 s = m_{j,0} s \qquad \text{for } j = 1, 2, \ldots, n \tag{3.2-11}$$

where $m_{j,0}$ denotes a direction evaluated at the point $\mathbf{x}_0$. This expression then defines the direction of the steepest gradient at the point $\mathbf{x}_0$.

As yet we have not indicated the direction of maximum increase (or decrease), but only the line of steepest gradient. Let us therefore investigate this aspect further. Substituting Eq. (3.2-10) into the constraint represented by Eq. (3.2-8) yields

$$1 = \sum_{j=1}^{n} \left(\frac{dx_j}{ds}\right)^2 = \frac{1}{4\lambda^2} \sum_{j=1}^{n} \left(\frac{\partial y}{\partial x_j}\right)^2$$

Solving for λ then yields

$$2\lambda = \pm \sqrt{\sum_{j=1}^{n} \left(\frac{\partial y}{\partial x_j}\right)^2}$$

Moreover, from Eq. (3.2-9), we see that

$$\frac{dy}{ds} = \sum_{j=1}^{n} \frac{1}{2\lambda}\left(\frac{\partial y}{\partial x_j}\right)^2 = \pm\sqrt{\sum_{j=1}^{n}\left(\frac{\partial y}{\partial x_j}\right)^2} \tag{3.2-12}$$

The positive sign therefore gives the direction of the greatest increase in y, whereas the negative sign gives the direction of the greatest decrease in y. Thus, if a move to larger values of y is desired, then, by Eq. (3.2-11), the direction $\boldsymbol{\xi}$ is defined by

$$m_j = \frac{\partial y/\partial x_j}{\sqrt{\sum_{j=1}^{n}\left(\frac{\partial y}{\partial x_j}\right)^2}} \tag{3.2-13}$$

On the other hand, the direction leading to smaller values of y is defined by

$$m_j = \frac{-\partial y/\partial x_j}{\sqrt{\sum_{j=1}^{n}\left(\frac{\partial y}{\partial x_j}\right)^2}} \tag{3.2-14}$$

Example VIII-7: The contour tangent and direction of steepest ascent using a two-dimensional linear objective function

Determine the directions of steepest ascent and of the contour tangent for a linear objective function in two dimensions.

Let us first determine the direction of the contour tangent for the linear objective function

$$y = a_1x_1 + a_2x_2$$

using Eq. (3.2-4). Thus the slope of the contour tangent, $\tan \alpha_1$, is given by

$$\tan \alpha_1 = -\frac{a_1}{a_2}$$

The tangent directions are therefore independent of position, which is true only for linear contours, as we have here. It is easily shown that any line having the slope $-a_1/a_2$ is a line of constant y. Such a line is, in general,

$$x_2 = \left(-\frac{a_1}{a_2}\right)x_1 + b$$

where b is the value of the intercept. Substituting into the objective function, we get

$$y = a_1x_1 + a_2\left[\left(-\frac{a_1}{a_2}\right)x_1 + b\right] = a_2b = \text{constant}$$

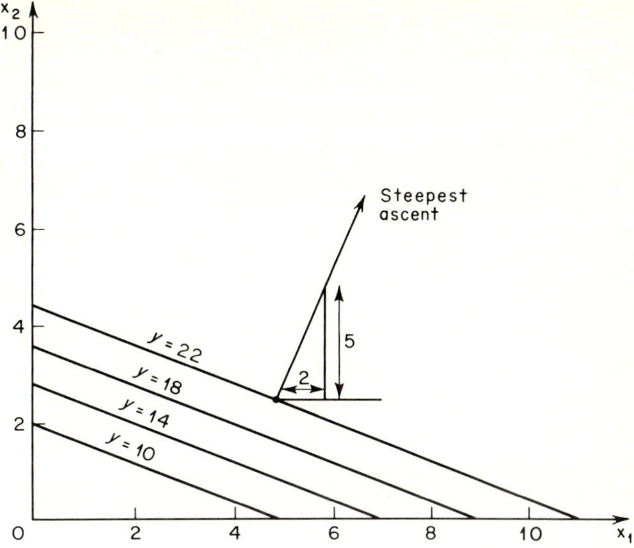

Fig. VIII-26 Steepest ascent and a linear function.

For the specific case of $a_1 = 2$ and $a_2 = 5$, the contours are shown in Fig. VIII-26.

What about the direction of steepest ascent? From Eq. (3.2-13), we see that

$$m_1 = \frac{\partial y/\partial x_1}{\sqrt{(\partial y/\partial x_1)^2 + (\partial y/\partial x_2)^2}} = \frac{a_1}{\sqrt{a_1^2 + a_2^2}} = \frac{5}{\sqrt{29}}$$

$$m_2 = \frac{\partial y/\partial x_2}{\sqrt{(\partial y/\partial x_1)^2 + (\partial y/\partial x_2)^2}} = \frac{a_2}{\sqrt{a_1^2 + a_2^2}} = \frac{2}{\sqrt{29}}$$

so that the direction of greatest increase is $\boldsymbol{\xi}$, or in parametric form, from Eq. (3.2-11),

$$x_1 - x_{1,0} = \frac{a_1 s}{\sqrt{a_1^2 + a_2^2}} = \frac{5s}{\sqrt{29}} \tag{3.2-15a}$$

$$x_2 - x_{2,0} = \frac{a_2 s}{\sqrt{a_1^2 + a_2^2}} = \frac{2s}{\sqrt{29}} \tag{3.2-15b}$$

Here we see that the m_1 and m_2 are independent of x_1 and x_2, indicating that the direction of greatest increase is a global, rather than local, property. This situation is a special case, however, due to linearity, and in general, the direction of the gradient does depend on position.

If we take any point, A, in Fig. VIII-26, and construct the line having the parametric form given by Eq. (3.2-15), then this line will be perpendicular to the contours and point in the direction of the greatest increase in y. Moreover, any vector parallel to that constructed at A will also be in the direction of greatest increase for the linear case considered here.

Finally, it should be noted that the direction of greatest decrease is defined by

$$m_1 = -\frac{5}{\sqrt{29}} \qquad m_2 = -\frac{2}{\sqrt{29}}$$

It should be stressed that, in the general case, the contour tangents and directions of steepest descent (or ascent) depend on position. This fact is an essential ingredient in understanding the weaknesses of some direct search methods based on steepest ascent.

Example VIII-8: The direction of steepest descent is a local property

Let us evaluate the gradient of the function $y = 3x_1^2 + 2x_2^2 + 4x_1x_2$ and show that it is a local property. In particular, we shall find the gradient at a point on the $y = 10$ contour, and also at the intersection of the resulting line of steepest descent with the $y = 5$ contour. The gradient at any point is then defined by the two numbers $\partial y/\partial x_1 = 6x_1 + 4x_2$ and $\partial y/\partial x_2 = 4x_2 + 4x_1$. These have the values

$$\frac{\partial y}{\partial x_1} = -10.92 \qquad \frac{\partial y}{\partial x_2} = -7.28$$

at point $A\{-1.82, 0\}$, as shown in Fig. VIII-27. From Eq. (3.2-14), the direction of steepest descent is seen to be $\boldsymbol{\xi}_A$, where

$$m_{1,A} = \frac{10.92}{\sqrt{(10.92)^2 + (7.28)^2}} = 0.833$$

$$m_{2,A} = \frac{7.28}{\sqrt{(10.92)^2 + (7.28)^2}} = 0.555$$

The parametric form of this direction of steepest descent will then be, from Eq. (3.2-11),

$$x_1 + 1.82 = 0.833s$$
$$x_2 = 0.555s$$

This line is shown in Fig. VIII-27, drawn from point A and intersecting the $y = 5$ contour when s satisfies

$$5 = 3(0.833s - 1.82)^2 + 2(0.555s)^2 + 4(0.833s - 1.82)0.555s$$

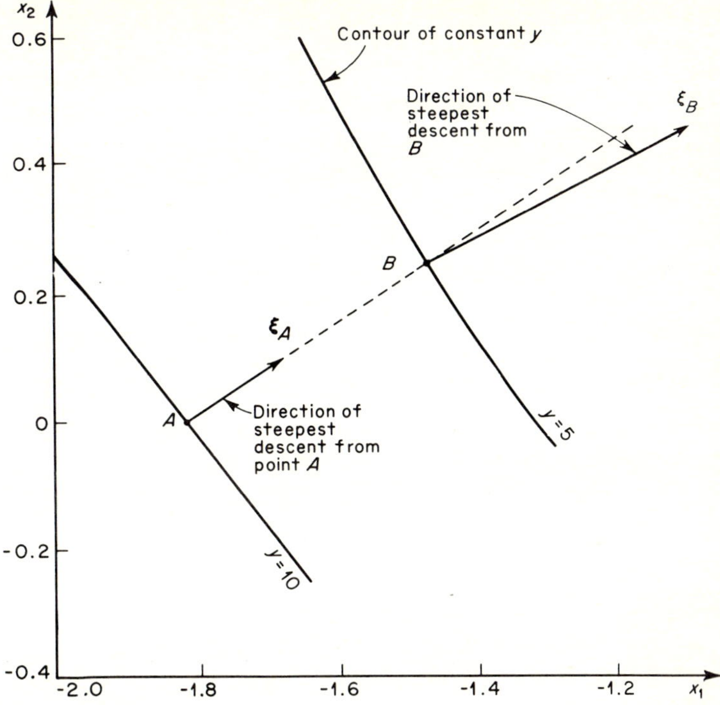

Fig. VIII-27 Steepest descent is a local property.

yielding the two values of 0.44 and 2.44 for s. The location of the nearest intersection B is then

$$x_1 = (0.833s - 1.82)_{s=0.44} = -1.454$$
$$x_2 = (0.555s)_{s=0.44} = 0.244$$

The gradients at this new point, B, are

$$\left(\frac{\partial y}{\partial x_1}\right)_B = 6(-1.44) + 4(0.244) = -7.75$$

and
$$\left(\frac{\partial y}{\partial x_2}\right)_B = 4(0.244) - 4(1.44) = -4.84$$

The direction of steepest descent at point B is thus defined by ξ_B, where, from Eq. (3.2-14),

$$m_{1,B} = 0.849$$
$$m_{2,B} = 0.528$$

It will be seen that this direction is different from that determined at point A, clearly indicating that the direction of steepest gradient is a local property. The two directions are shown in Fig. VIII-27.

Evaluation of the Gradient In some cases it is possible to find an analytical expression for the gradient by differentiating the objective function. In the majority of problems, however, the objective function is too complex to be differentiated or is black-box in nature. In such cases the gradient can be determined by a series of experiments placed in some pattern so that the coefficients of a polynomial surface representing the objective function can be estimated. A detailed discussion of the problems associated with surface fitting has been presented in Chap. III, Sec. 5. The simplest representation, suitable for all but the final stages of search, is that using a plane surface with a linear equation

$$\Delta y = \beta_1 \zeta_1 + \beta_2 \zeta_2 + \cdots + \beta_n \zeta_n \tag{3.2-16}$$

where Δy = value of objective function relative to that at base point $\mathbf{x}_0$
ζ_i = variations $x_i - x_{i,0}$ about base point
β_i = experimentally determined coefficients

This expression is, of course, valid only in a region close to this base point. Nevertheless, an estimate of the direction of steepest ascent can be obtained by noting that

$$\frac{\partial y}{\partial x_j} = \frac{\partial(\Delta y)}{\partial x_j} = \beta_j \qquad \text{for } j = 1, 2, \ldots, n \tag{3.2-17}$$

so that the direction of steepest ascent, $\boldsymbol{\xi}_1$ is then defined by the n numbers,

$$m_j = \frac{\beta_j}{\sqrt{\sum_{i=1}^{n} \beta_i^2}} \qquad \text{for } j = 1, 2, \ldots, n \tag{3.2-18}$$

The direction of steepest ascent can therefore be determined by a series of experiments which provide estimates of the β_i. If a statistically designed set of experiments is used, estimates of error in the β_i can also be obtained (Box [1956]).

The Distance of Movement in the Method of Steepest Ascent (Descent) At the given base point, with the direction of greatest improvement determined, the next problem is to decide on the distance of the move. Several methods are available. One choice might be to advance along the line of steepest ascent from the base point, starting with a given step size, using

$$\mathbf{x}_{j+1} = \mathbf{x}_j + \boldsymbol{\xi}_j s \tag{3.2-19}$$

The unrestricted one-dimensional search would continue, using a method from Chap. VI, until an approximation to the optimum is found. This can be adopted as the new base point, and the procedure repeated. This forms the basis of the method of steepest ascent (or descent). The search is stopped

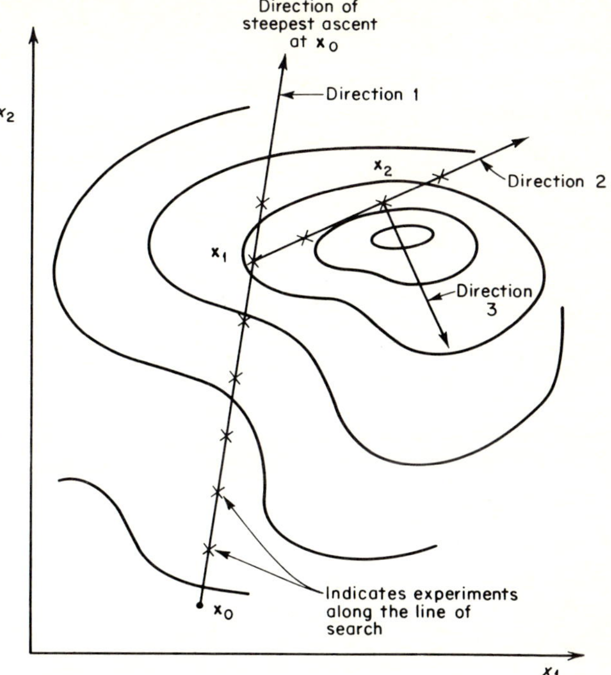

Fig. VIII-28 Search based on direction of steepest ascent.

when the step size falls below an agreed minimum. The results of a few stages of such a search are depicted in Fig. VIII-28 for the two-dimensional case. The search begins at base point $\mathbf{x}_0$, and the direction of steepest ascent determined. Then a series of experiments along this direction can be carried out to locate the restricted optimum at $\mathbf{x}_1$ to within the approximation of a step length. A new local exploration about $\mathbf{x}_1$ then fixes the new direction of steepest ascent, and the calculation is repeated.

It is obvious that the movements along the direction of steepest gradient in each stage can be made in various ways, yielding modifications of the search technique, one of the earliest being due to Cauchy [1847]. Apart from that illustrated, we might have moved only a single step of predetermined length, or we might have moved to the restricted optimum itself. In the latter situation, it should be noted that successive directions of motion will be orthogonal to each other since, having found one stage optimum, the next direction of greatest improvement will be at right angles to it.

The method of steepest ascent may appear to be the "ultimate" in search techniques, since each one-dimensional search starts in the "best" direction. However, owing to the fact that the direction of steepest ascent is a local property, in many cases, especially those with large interactions between

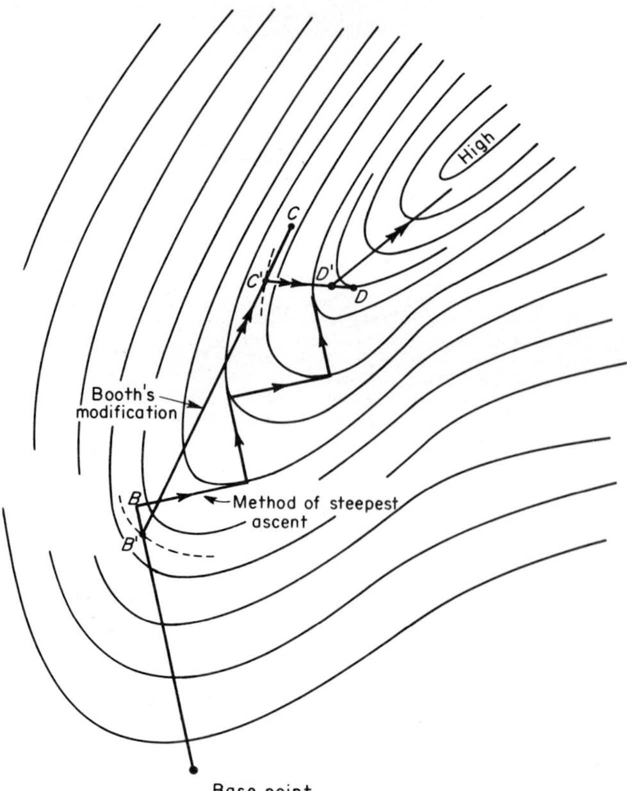

Fig. VIII-29 A comparison of the method of steepest ascent with Booth's modification.

variables, the method is not really effective. Consider the situation depicted in Fig. VIII-29, in which the contours form a long and narrow ridge. The method of steepest ascent will "zigzag" up the ridge in much the same way as the one-at-a-time method. Since the derivatives of the objective function must be evaluated at each new base point, $\mathbf{x}_0$, $\mathbf{x}_1$, . . . , a fairly large number of experiments will evidently be required to find the optimum.

To circumvent this difficulty, Booth [1955] has suggested that the method be modified by moving only 90 percent of the distance to the optimum value along the lines of steepest ascent during four or five moves and then taking a full step on the fifth move. This procedure has the advantage of following the ridge more closely, as shown in Fig. VIII-29. Thus, in starting from point A along the direction of steepest ascent, point B is found as the optimum. Instead of accepting B as our next base point, we shall choose B', where $\overrightarrow{AB'} = 0.9(\overrightarrow{AB})$. Similarly, in the next search cycle, C' is taken as the new base point, where $\overrightarrow{B'C'} = 0.9(\overrightarrow{B'C})$. The net effect is that long narrow ridges are traversed with fewer experiments than by the basic approach.

Other modifications of the method of steepest ascent have been reported by Levenberg [1944] and Marquardt [1959]. However, we shall not discuss these modifications further, because the basic principles underlying all those variations have already been presented and since better search techniques are always possible if the gradients are known.

Example VIII-9: Steepest descent applied to a Rosenbrock's function

Rosenbrock [1960] used the following test problem for the comparison of various search techniques. It is similar to the function shown in Fig. VIII-22.

Minimize y, where

$$y = 100(x_2 - x_1^2)^2 + (1 - x_1)^2 \tag{3.2-20}$$

Let us use the method of steepest descent to find the optimum, the distance of movement being that which optimizes the objective function along the line of search.

Stage 1 Let point $\mathbf{x}_0 = \{2,2\}$ be the initial base point. The first derivatives evaluated at the base point are found to be

$$\left(\frac{\partial y}{\partial x_1}\right)_0 = -400(x_2 x_1 - x_1^3) - 2(1 - x_1) = 1{,}602 \tag{3.2-21a}$$

and

$$\left(\frac{\partial y}{\partial x_2}\right)_0 = 200(x_2 - x_1^2) = -400 \tag{3.2-21b}$$

The direction of steepest descent at the point $\mathbf{x}_0$ is then given by $\boldsymbol{\xi}_0$, with components

$$m_{1,0} = \frac{-1{,}602}{\sqrt{(1{,}602)^2 + (400)^2}} = -0.9702$$

$$m_{2,0} = \frac{400}{\sqrt{(1{,}602)^2 + (400)^2}} = +0.2422$$

The parametric representation of the straight line in the direction of the steepest descent is then given by

$$x_1 = 2 - 0.9702s$$

$$x_2 = 2 + 0.2422s$$

Thus the objective function becomes

$$y = 100[2 + 0.2422s - (2 - 0.9702s)^2]^2 + [1 - (2 - 0.9702s)]^2$$

and the value of s can now be selected so that y is minimized along this direction. Therefore, by using a one-dimensional search technique, we find $s = +0.555(\pm 0.003)$ and $y = 0.213$ at this new point compared with $y = 401$ at the base point. The new coordinates are

$$x_{1,1} = 2 - 0.9702(0.555) = 1.4615$$

$$x_{2,1} = 2 + 0.2422(0.555) = 2.1345$$

Stage 2 The gradient at this new base point has components

$$\frac{\partial y}{\partial x_1} = 0.922 \qquad \text{and} \qquad \frac{\partial y}{\partial x_2} = 0$$

Note that these values are quite different from those of the previous point, showing again that the gradient is a local property. We now search a second time in the direction parallel to the new gradient, defined by

$$m_{1,2} = -1.0 \qquad \text{and} \qquad m_{2,2} = 0.0$$

Thus we want to search along a line defined parametrically as

$$x_1 = 1.4615 - s$$

$$x_2 = 2.1345$$

to find the minimum of the function

$$y = 100[2.1345 - (1.4615 - s)^2]^2 + (0.4615 - s)^2$$

The distance of movement is quite small, for here we find $s = 0.0010$, giving $y = 0.212$. Indeed, the steepest descent search will proceed exceedingly slowly to the optimum, for this function has been designed by Rosenbrock to challenge the method of steepest descent. The function is very similar in shape to that shown in Fig. VIII-22. We shall not carry the search further.

Even with such modifications, there may be a significant amount of oscillation, since the direction of the steepest gradient does not always point toward the desired optimum. This can be seen by minimizing the function $y = 16x_1^2 + x_2^2$. This produces an oscillating search path, since the successive directions of steepest descent do not pass through the minimum. If, however, we define $x_3 = 4x_1$, then the function becomes $y = x_3^2 + x_2^2$, and a move from any point along the direction of steepest descent will pass through the minimum. The problem of oscillation may therefore be one of scaling (Chap. XI, Sec. 5.3.2). The methods of steepest gradient will work better if the function can be transformed to one with nearly circular contours. This can be sometimes achieved, but it is probably better to consider other gradient methods, such as those now discussed.

3.2.2 *Another gradient method—conjugate directions*

In Sec. 3.2.1 we have studied those gradient search techniques in which the direction of movement is made from a base point along the direction of the steepest ascent. Most of these steepest ascent methods tend to yield an oscillatory approach to the local optimum, particularly when the contours of the objective function are elongated. There are other methods of search which utilize the gradients in fixing the direction of movement, but do not necessarily move in the direction of steepest ascent. Those to be discussed here have the property of quadratic convergence described earlier (see page 392) in that the minimum $y(\mathbf{x}^*)$ of the objective function

$$y = y(\mathbf{x}^*) + \frac{1}{2}\sum_{i=1}^{n}\sum_{j=1}^{n} a_{i,j}(x_i - x_i^*)(x_j - x_j^*) \qquad (3.2\text{-}22)$$

is found within some finite number of iterations, usually n. Here the $a_{i,j}$ are symmetric ($a_{i,j} = a_{j,i}$) and are positive-definite in the sense that the term

$$\frac{1}{2}\sum_{i=1}^{n}\sum_{j=1}^{n} a_{i,j}h_ih_j$$

is greater than zero for all nontrivial $h_i(= x_i - x_i^*)$. This is a sufficient condition for the stationary point of y to be a minimum (Chap. VII, Sec. 2.2.3).

In these methods, searches are performed in a certain set of directions, called *conjugate directions*, chosen to ensure that the optimum of the quadratic function is found in a finite number of iterations. A geometric interpretation of such a set of directions has already been given in discussing nongradient acceleration techniques (see page 391). At this point we shall introduce a mathematical argument to supplement our previous discussion.

Common Characteristics of Conjugate Direction Methods Once a direction $\boldsymbol{\xi}^{(k)} = \{m_1^{(k)}, m_2^{(k)}, \ldots, m_n^{(k)}\}$ in the kth stage of the iteration has been selected, all the methods to be discussed here require a move of length $s^{(k)}$ along a direction such that the then restricted objective function is optimized. Specifically, if $\mathbf{x}^{(k)}$ is the base point after k stages and if $\boldsymbol{\xi}^{(k)}$ is the direction of move, then $s^{(k)}$ is selected so that $y(\mathbf{x}^{(k)} + s^{(k)}\boldsymbol{\xi}^{(k)})$ is optimized in the $(k+1)$th stage. It is necessary, therefore, to choose $s^{(k)}$ such that

$$\frac{dy^{(k+1)}}{ds^{(k)}} = 0$$

where $y^{(k+1)} = y(\mathbf{x}^{(k+1)}) = y(\mathbf{x}^{(k)} + s^{(k)}\boldsymbol{\xi}^{(k)})$.

Now, for any set of displacements about the point $\mathbf{x}^{(k+1)}$,

$$dy = \sum_{i=1}^{n} y_{x_i}^{(k+1)}\, dx_i$$

where the partial derivatives are evaluated at $\mathbf{x}^{(k+1)}$. For a displacement along the direction of movement from point $\mathbf{x}^{(k)}$ to $\mathbf{x}^{(k+1)}$, we then have

$$\frac{dy}{ds^{(k)}} = \sum_{i=1}^{n} y_{x_i}^{(k+1)} \frac{dx_i}{ds^{(k)}} = \sum_{i=1}^{n} y_{x_i}^{(k+1)} m_i^{(k)}$$

If we set

$$\sum_{i=1}^{n} y_{x_i}^{(k+1)} m_i^{(k)} = 0 \tag{3.2-23}$$

then the $s^{(k)}$ will be chosen so that y is optimized along this direction. This is a very important result and will be used several times in the arguments to follow. It states that the search along a given direction will continue until the point is found at which the gradient of the objective is perpendicular (orthogonal) to the direction of search.

Equation (3.2-23) can be used to find an explicit expression for $s^{(k)}$. Thus, by differentiating Eq. (3.2-22) and using the symmetry property of the $a_{i,j}$, we find that

$$y_{x_i} = \sum_{r=1}^{n} a_{i,r}(x_r - x_r^*)$$

Therefore

$$y_{x_i}^{(k+1)} = \sum_{r=1}^{n} a_{i,r}(x_r^{(k+1)} - x_r^*) \tag{3.2-24}$$

or

$$y_{x_i}^{(k+1)} = \sum_{r=1}^{n} a_{i,r}(x_r^{(k)} + s^{(k)} m_r^{(k)} - x_r^*)$$

Rearrangement then yields

$$y_{x_i}^{(k+1)} = \sum_{r=1}^{n} a_{i,r}(x_r^{(k)} - x_r^*) + s^{(k)} \sum_{r=1}^{n} a_{i,r} m_r^{(k)} \tag{3.2-25}$$

After substituting this result into Eq. (3.2-23), we find that

$$s^{(k)} = -\frac{\sum_{i=1}^{n}\sum_{r=1}^{n} a_{i,r}(x_r^{(k)} - x_r^*) m_i^{(k)}}{\sum_{i=1}^{n}\sum_{r=1}^{n} a_{i,r} m_i^{(k)} m_r^{(k)}} \tag{3.2-26}$$

which reduces, with the aid of Eq. (3.2-24), to

$$s^{(k)} = -\frac{\sum_{i=1}^{n} y_{x_i}^{(k)} m_i^{(k)}}{\sum_{i=1}^{n}\sum_{r=1}^{n} a_{i,r} m_i^{(k)} m_r^{(k)}} \tag{3.2-27}$$

The methods to be discussed all select $s^{(k)}$ so as to satisfy Eq. (3.2-27) and differ only in the choice of a direction of search.

Conjugate Directions How do we choose a set of conjugate directions for use in search? In any practical problem the objective function will not have the form given by Eq. (3.2-22), or if it does, this fact will be unknown to the optimizer, since there would be no need to search when the objective function is stated in this form. However, in the arguments to follow, it will be supposed that the quadratic form of the objective function is the proper one and that it is known. This supposition will be removed in the subsequent section.

Starting with the base point at the $(j + 1)$st stage, we can write

$$x_i^{(j+2)} = x_i^{(j+1)} + s^{(j+1)} m_i^{(j+1)} \tag{3.2-28}$$

where $s^{(j+1)}$ is selected as prescribed in Eq. (3.2-27).

Since

$$x_i^{(j+3)} = x_i^{(j+2)} + s^{(j+2)} m_i^{(j+2)}$$

then

$$x_i^{(j+3)} = x_i^{(j+1)} + s^{(j+1)} m_i^{(j+1)} + s^{(j+2)} m_i^{(j+2)}$$

Continuing, we can write, in general, that

$$x_i^{(n)} = x_i^{(j+1)} + \sum_{r=j+1}^{n-1} s^{(r)} m_i^{(r)} \qquad \text{for any } j \text{ in } 0 \leq j \leq n-1 \tag{3.2-29}$$

Moreover, the expression for the gradient, Eq. (3.2-24), can be written

$$y_{x_i}^{(n)} = y_{x_i}^{(j+1)} + \sum_{r=1}^{n} \sum_{l=j+1}^{n-1} a_{i,r} s^{(l)} m_r^{(l)} \tag{3.2-30}$$

If we multiply both sides of this expression by $m_i^{(j)}$ and sum on i, we then get the equation

$$\sum_{i=1}^{n} y_{x_i}^{(n)} m_i^{(j)} = \sum_{i=1}^{n} y_{x_i}^{(j+1)} m_i^{(j)} + \sum_{l=j+1}^{n-1} \sum_{r=1}^{n} \sum_{i=1}^{n} a_{i,r} s^{(l)} m_r^{(l)} m_i^{(j)} \tag{3.2-31}$$

A conjugate set of directions is defined as a set which satisfies the equations

$$\sum_{r=1}^{n} \sum_{i=1}^{n} a_{r,i} m_r^{(l)} m_i^{(j)} = 0 \qquad \text{for } l \neq j. \tag{3.2-32}$$

It should be noted that this expression does not define a unique set of directions. There are, in fact, an infinite number of conjugate directions corresponding to a given set of $a_{i,j}$.

If we admit only conjugate directions for search, then Eq. (3.2-31) reduces to

$$\sum_{i=1}^{n} y_{x_i}^{(n)} m_i^{(j)} = 0 \tag{3.2-33}$$

The first term on the right-hand side vanished because $s^{(j)}$ was chosen to satisfy Eq. (3.2-23), and the second term vanished because the directions are conjugate.

Since Eq. (3.2-33) is satisfied for all j in the interval $0 \leq j \leq n - 1$, the only possible solution is

$$y_{x_i}^{(n)} = 0$$

However, for the quadratic expression under consideration, the gradients can vanish only at the minimum, and therefore $\mathbf{x}^{(n)}$ is the solution.

Actually, the conclusion that the $y_{x_i}^{(n)}$ must vanish for Eq. (3.2-33) to be satisfied is contingent on the hypothesis that the $\boldsymbol{\xi}^{(j)}$ form a linear independent set of n vectors. However, the n vector directions $\boldsymbol{\xi}^{(0)}, \boldsymbol{\xi}^{(1)}, \ldots, \boldsymbol{\xi}^{(n-1)}$ do form a linear independent set if they are conjugate directions and if the quadratic term is positive-definite, as is assumed here. This proof has been given by Beckman [1960].

The situation can be summarized as follows: Given any conjugate set of directions (and there are an infinite number of such sets), then, by searching each of these directions in turn, we can find the minimum of a quadratic function in n or less stages. Fewer than n stages might be used if some of the distances for movement vanish.

Example VIII-10: Conjugate directions

Let us first consider the minimization of the objective function

$$y = 3x_1^2 + 2x_1x_2 + x_2^2 + y(\mathbf{x}^*) \tag{3.2-34}$$

The optimum is clearly at the origin. The search for this minimum will be started at the point {5,5}, and a method of conjugate gradients will be used.

Clearly, this objective function is of the form

$$y = \frac{1}{2}\sum_{i=1}^{2}\sum_{j=1}^{2} a_{i,j}x_i x_j + y(\mathbf{x}^*)$$

with $a_{1,1} = 6$, $a_{1,2} = 2$, $a_{2,1} = 2$, and $a_{2,2} = 2$. Since there are two independent variables x_1 and x_2, then two stages of iteration will be required to find the optimum, using conjugate directions. We shall therefore need to select two directions of search. Those will be fixed by the requirement that they are conjugate directions in the sense of Eq. (3.2-32). If the first direction $\boldsymbol{\xi}^{(0)}$ of movement is $\{m_1^{(0)}, m_2^{(0)}\}$ and the second is $\{m_1^{(1)}, m_2^{(1)}\}$, Eq. (3.2-32) requires that they be selected so that

$$\sum_{r=1}^{2}\sum_{i=1}^{2} a_{r,i}m_r^{(0)}m_i^{(1)} = 6m_1^{(0)}m_1^{(1)} + 2m_1^{(0)}m_2^{(1)} + 2m_2^{(0)}m_1^{(1)} + 2m_2^{(0)}m_2^{(1)} = 0$$

Of course, there are many combinations of directions which will satisfy this expression. No unique set of directions is defined. For example, the following sets of directions are all conjugate directions:

$$(A) \qquad \xi^{(0)}\{m_1^{(0)} = 1, m_2^{(0)} = 0\} \qquad \xi^{(1)}\left\{m_1^{(1)} = \frac{1}{\sqrt{10}}, m_2^{(1)} = \frac{-3}{\sqrt{10}}\right\}$$

$$(B) \qquad \xi^{(0)}\{m_1^{(0)} = 0, m_2^{(0)} = 1\} \qquad \xi^{(1)}\left\{m_1^{(1)} = \frac{-1}{\sqrt{2}}, m_2^{(1)} = \frac{1}{\sqrt{2}}\right\}$$

$$(C) \qquad \xi^{(0)}\left\{m_1^{(0)} = \frac{1}{\sqrt{2}}, m_2^{(0)} = \frac{1}{\sqrt{2}}\right\} \qquad \xi^{(1)}\left\{m_1^{(1)} = \frac{1}{\sqrt{5}}, m_2^{(1)} = \frac{-2}{\sqrt{5}}\right\}$$

Note that, for consistency, these sets have also been selected so that $(m_1)^2 + (m_2)^2 = 1$; that is, they each form a unit direction.

Any one of these three sets will suffice as a basis for search. Select the first set, set A. Then, from the base point $\{5,5\}$, our line of search is initially defined by the parametric equations

$$x_1^{(1)} = 5 + m_1^{(0)}s^{(0)} = 5 + s^{(0)} \qquad (3.2\text{-}35a)$$

$$x_2^{(1)} = 5 + m_2^{(0)}s^{(0)} = 5 \qquad (3.2\text{-}35b)$$

Our first iteration is therefore along a line parallel to the x_1 axis. All methods using conjugate directions now require that $s^{(0)}$ be selected so that y, when evaluated along the line of search, is a minimum (or a maximum as the case may be). Thus we write that

$$y = 3(5 + s^{(0)})^2 + 10(5 + s^{(0)}) + 25 + y(\mathbf{x}^*)$$

so that the optimum value of y will be given when

$$\frac{dy}{ds^{(0)}} = 6(5 + s^{(0)}) + 10 = 0 \qquad (3.2\text{-}36)$$

at a stationary point in y.†

Solving Eq. (3.2-36) gives the appropriate choice of $s^{(0)}$ as

$$s^{(0)} = -20/3$$

The base point, found after the first stage in the search, is then, from Eqs. (3.2-35),

$$x_1^{(1)} = -\frac{5}{3} \qquad x_2^{(1)} = 5$$

† It may be noted that Eq. (3.2-23) gives precisely this result, since we can write that

$$y_{x_1}^{(1)} m_1^{(0)} + y_{x_2}^{(1)} m_2^{(0)} = 0$$

giving in this case

$$y_{x_1}^{(1)} = 6x_1^{(1)} + 2x_2^{(1)} = 6(5 + s^{(0)}) + 10 = 0$$

Thus the two results are clearly identical, which of course must be the case, since both were obtained by differentiating the objective function with respect to the parameter $s^{(0)}$.

Taking the second direction $\boldsymbol{\xi}^{(1)}$ of the set A, the search will seek a new point $\mathbf{x}^{(2)}$ such that

$$x_1^{(2)} = -\frac{5}{3} + \frac{s^{(1)}}{\sqrt{10}} \tag{3.2-37a}$$

$$x_2^{(2)} = 5 - \frac{3s^{(1)}}{\sqrt{10}} \tag{3.2-37b}$$

and the objective function becomes

$$y = 3\left(-\frac{5}{3} + \frac{s^{(1)}}{\sqrt{10}}\right)^2 + 2\left(5 - \frac{3s^{(1)}}{\sqrt{10}}\right)\left(-\frac{5}{3} + \frac{s^{(1)}}{\sqrt{10}}\right) + \left(5 - \frac{3s^{(1)}}{\sqrt{10}}\right)^2 + y(\mathbf{x}^*)$$

The value of $s^{(1)}$ minimizing y is easily seen to be

$$s^{(1)} = \frac{5\sqrt{10}}{3}$$

so that the final base point $\mathbf{x}^{(2)}$, from Eq. (3.2-37), is

$$x_1^{(2)} = 0 \qquad x_2^{(2)} = 0$$

We have thus found the precise solution in only two stages, using the conjugate directions. Any other set (such as B or C) of conjugate directions would have served equally well in locating the minimum within two stages.

The Generation of Conjugate Directions In the preceding section we saw that the search for the optimum of a quadratic function can be completed in a definite number of iterations, provided a set of conjugate directions is used. This is a very important property because quadratic behavior is observed near the optimum in all functions which are expandable in a Taylor series.

In the general search problem the coefficients $a_{i,j}$ are not known, and it is not possible to compute the conjugate directions in advance, as was done in Example VIII-10. Various proposals have been made to generate a set of such directions as seen in Sec .3.1.3. These utilize certain geometric properties of elliptical surfaces to generate conjugate directions rather than using the gradient to construct such directions. Fletcher and Powell [1963] have modified a method proposed by Davidon [1959] to generate a set of conjugate directions using the gradient, and Fletcher and Reeves [1964] have adapted a technique suggested by Hestenes and Stiefel [1952] in connection with the solution of linear equations to find a set of conjugate directions. This latter method for generating conjugate directions has been described by Beckman [1960].

All these methods are very similar, the essential problem being the generation of a set of conjugate directions when the form of the quadratic is unknown. This, at first thought, will be quite difficult, since the definition of the conjugate directions uses the coefficients of the quadratic terms in the objective function. It is possible, however, to generate sets of conjugate directions based on a knowledge of the gradients. One method of doing this is the method of Fletcher and Reeves [1964]. This approach appears to be efficient in search, as well as one easily programmed for the digital computer. We shall therefore describe only this method, since it would be redundant to examine each similar available method in detail, even though they may be as good or better.

A set of conjugate directions can be developed sequentially as follows: A base point $\mathbf{x}^{(0)}$ is selected, and the gradients $y_{x_1}^{(0)}, y_{x_2}^{(0)}, \ldots$ determined. The first direction of search will be along the path of steepest descent,[1] Let

$$m_i^{(0)} = -y_{x_i}^{(0)} \tag{3.2-38}$$

Move in this direction until the objective function is minimized. Therefore, $s^{(0)}$ is given by Eq. (3.2-27). We have now located a new base point $\mathbf{x}^{(1)}$.

We now need to define a new direction of search which is constructed so that

$$m_i^{(1)} = -y_{x_i}^{(1)} + \beta^{(0)} m_i^{(0)} \tag{3.2-39}$$

The form of Eq. (3.2-39) arises because a value of $\beta^{(0)}$ can be selected so that the two directions are conjugate. Other forms, of course, are possible, but this particular equation has some very convenient properties from the computational viewpoint. If $\boldsymbol{\xi}^{(1)}$ and $\boldsymbol{\xi}^{(0)}$ are to be conjugate directions, then

$$\sum_{i=1}^{n} \sum_{r=1}^{n} a_{r,i} m_i^{(1)} m_r^{(0)} = 0 = -\sum_{i=1}^{n} \sum_{r=1}^{n} y_{x_i}^{(1)} a_{r,i} m_r^{(0)} + \beta^{(0)} \sum_{i=1}^{n} \sum_{r=1}^{n} a_{r,i} m_r^{(0)} m_i^{(0)}$$

and hence we must choose $\beta^{(0)}$ as

$$\beta^{(0)} = \frac{\sum_{i=1}^{n} \sum_{r=1}^{n} y_{x_i}^{(1)} a_{r,i} m_r^{(0)}}{\sum_{i=1}^{n} \sum_{r=1}^{n} a_{r,i} m_r^{(0)} m_i^{(0)}} \tag{3.2-40}$$

This expression is not, however, in a convenient form, since we are assuming that the $a_{r,s}$ are unknown. The new direction cannot, therefore, as yet be

[1] In the remaining work we shall not demand that

$$\sum_{i=1}^{n} m_i^{(l)} m_i^{(l)} = 1$$

since this would complicate the formulas. The directions can, however, easily be normalized if this seems desirable.

determined. We must write $\beta^{(0)}$ in terms of the gradients. From Eq. (3.2-25), it is seen that

$$y_{x_i}^{(1)} = \sum_{r=1}^{n} a_{i,r}(x_r^{(0)} - x_r^*) + s^{(0)} \sum_{r=1}^{n} a_{i,r} m_r^{(0)}$$

or

$$y_{x_i}^{(1)} = y_{x_i}^{(0)} + s^{(0)} \sum_{r=1}^{n} a_{i,r} m_r^{(0)} \tag{3.2-41}$$

Substituting this result into Eq. (3.2-40), we find that

$$\beta^{(0)} = \frac{\sum_{i=1}^{n} y_{x_i}^{(1)}(y_{x_i}^{(1)} - y_{x_i}^{(0)})}{s^{(0)} \sum_{i=1}^{n} \sum_{r=1}^{n} a_{r,i} m_i^{(0)} m_r^{(0)}}$$

This expression can be simplified using Eqs. (3.2-23), (3.2-27), and (3.2-38), getting

$$\beta^{(0)} = \frac{\sum_{i=1}^{n} y_{x_i}^{(1)} y_{x_i}^{(1)}}{\sum_{i=1}^{n} y_{x_i}^{(0)} y_{x_i}^{(0)}} \tag{3.2-42}$$

This result is indeed useful, since a second direction conjugate to the first can now be computed using only gradients at the new and old base points.

The next move can now be made along the second direction, and a third base point found. We are now ready to make the third move. A third direction conjugate to both the first and second directions is required. How should we generate this third direction of search? Again we can use a linear combination of the gradient and the two previous directions

$$m_i^{(2)} = -y_{x_i}^{(2)} + \alpha^{(0)} m_i^{(0)} + \beta^{(1)} m_i^{(1)} \tag{3.2-43}$$

Here we have two undetermined constants $\alpha^{(0)}$ and $\beta^{(1)}$, which must be selected so that both $m_i^{(0)}$ and $m_i^{(1)}$ are conjugate to $m_i^{(2)}$. Therefore, using

$$\sum_{i=1}^{n} \sum_{r=1}^{n} a_{r,i} m_i^{(2)} m_r^{(0)} = 0$$

Eq. (3.2-43) yields

$$-\sum_{i=1}^{n} \sum_{r=1}^{n} a_{r,i} y_{x_i}^{(2)} m_r^{(0)} + \alpha^{(0)} \sum_{i=1}^{n} \sum_{r=1}^{n} a_{r,i} m_r^{(0)} m_i^{(0)} + \beta^{(1)} \sum_{i=1}^{n} \sum_{r=1}^{n} a_{r,i} m_r^{(0)} m_i^{(1)} = 0$$

The last term on the left-hand side vanishes because the $m_r^{(0)}$ and $m_i^{(1)}$ define conjugate directions. Therefore

$$\alpha^{(0)} = \frac{\sum_{i=1}^{a} \sum_{r=1}^{n} a_{r,i} y_{x_i}^{(2)} m_r^{(0)}}{\sum_{i=1}^{n} \sum_{r=1}^{n} a_{r,i} m_r^{(0)} m_i^{(0)}} \tag{3.2-44}$$

The convenience of this method of generating conjugate directions is that the numerator of this expression vanishes.[1] We therefore need not include $m_i^{(0)}$ in the construction of $m_i^{(2)}$ because $\alpha^{(0)} = 0$. Thus Eq. (3.2-43) reduces to

$$m_i^{(2)} = -y_{x_i}^{(2)} + \beta^{(1)} m_i^{(1)} \tag{3.2-48}$$

The similarity with Eq. (3.2-39) will be seen. We must now determine $\beta^{(1)}$. If $m_i^{(2)}$ is to be conjugate to $m_i^{(1)}$, then

$$\beta^{(1)} = \frac{\sum_{i=1}^{n} \sum_{r=1}^{n} a_{r,i} m_r^{(1)} y_{x_i}^{(2)}}{\sum_{i=1}^{n} \sum_{r=1}^{n} a_{r,i} m_i^{(1)} m_r^{(1)}}$$

From Eqs. (3.2-23), (3.2-27), (3.2-39), (3.2-46) (remembering that $D = 0$) and (3.2-47), we then get

$$\beta^{(1)} = \frac{\sum_{i=1}^{n} y_{x_i}^{(2)} y_{x_i}^{(2)}}{\sum_{i=1}^{n} y_{x_i}^{(1)} y_{x_i}^{(1)}} \tag{3.2-49}$$

[1] Let us show that this numerator does in fact vanish. Let

$$D = \sum_{i=1}^{n} \sum_{r=1}^{n} a_{r,i} y_{x_i}^{(2)} m_r^{(0)} \tag{3.2-45}$$

Substituting for $\sum_{r=1}^{n} m_r^{(0)} a_{r,i}$ from Eq. (3.2-41), we get

$$D = \sum_{i=1}^{n} y_{x_i}^{(2)} \frac{y_{x_i}^{(1)} - y_{x_i}^{(0)}}{s^{(0)}}$$

Now $y_{x_i}^{(0)}$ and $y_{x_i}^{(1)}$ can be eliminated, using Eqs. (3.2-38) and (3.2-39), respectively. Thus

$$D = \sum_{i=1}^{n} y_{x_i}^{(2)} \frac{m_i^{(0)} + \beta^{(0)} m_i^{(0)} - m_i^{(1)}}{s^{(0)}}$$

or

$$D = \frac{1 + \beta^{(0)}}{s^{(0)}} \sum_{i=1}^{n} y_{x_i}^{(2)} m_i^{(0)} \tag{3.2-46}$$

since $\sum_{i=1}^{n} y_{x_i}^{(2)} m_i^{(1)} = 0$ by Eq. (3.2-23). Finally, D can be shown to vanish, since

$$y_{x_i}^{(2)} = y_{x_i}^{(1)} + s^{(1)} \sum_{r=1}^{n} a_{r,i} m_r^{(1)} \tag{3.2-47}$$

giving

$$D = (1 + \beta^{(0)}) \left(\sum_{i=1}^{n} y_{x_i}^{(1)} m_i^{(0)} + s^{(1)} \sum_{i=1}^{n} \sum_{r=1}^{n} a_{r,i} m_r^{(1)} m_i^{(0)} \right)$$

which is zero, since both terms on the right-hand side vanish.

Continuing this process, it can be shown (see Prob. VIII-55) that a series of conjugate directions can be generated by applying the recursion relationship

$$m_i^{(k)} = -y_{x_i}^{(k)} + \beta^{(k-1)} m_i^{(k-1)} \qquad (3.2\text{-}50)$$

where

$$\beta^{(k-1)} = \frac{\sum_{i=1}^{n} y_{x_i}^{(k)} y_{x_i}^{(k)}}{\sum_{i=1}^{n} y_{x_i}^{(k-1)} y_{x_i}^{(k-1)}}$$

We have therefore established a method of generating a set of conjugate directions based on the gradients of the function. A rigorous proof is presented by Beckman [1960].

The Search Method Based on Conjugate Gradients The Fletcher and Reeves [1964] method selects the direction of search at a given point in accordance with Eq. (3.2-50). Their search method is therefore guaranteed to locate the optimum of any quadratic function in at most n stages, provided the numerical errors are negligible, since Eq. (3.2-50) generates conjugate directions. For functions which are not quadratic, one cannot, of course, guarantee convergence in n steps, and a test for convergence is certainly required. The directions of movement are related to some quadratic approximation, and depending on the type of objective function, the convergence to the vicinity of the minimum may be fast or slow. However, once near the optimum, the method converges rapidly, because most functions tend to behave quadratically in this region.

Finally, Fletcher and Reeves recommend discarding all previous experience after each $n + 1$ stages, because the successive directions tend to become very nearly parallel and the new base points become very closely spaced. Of course, as long as the restarts are not more frequent than n, the process is still quadratically convergent. This method converged more rapidly than Rosenbrock's method when tested on several trial functions by Fletcher and Reeves [1964]. However, if the gradients need to be evaluated experimentally, there may not be any advantage in using a conjugate direction method based on gradients or the method of steepest ascent. If, however, an analytical expression is available for the gradient of the objective function, the methods of conjugate direction look very promising.

Example VIII-11: Minimization by the Fletcher-Reeves method

Let us try Rosenbrock's function using conjugate directions starting from $\{2,2\}$. We note that, since the first search is along the gradient, just as the method of steepest descent, our first trial will correspond to the first move in Example VIII-9. There

$$y_{x_1}^{(0)} = \frac{\partial y}{\partial x_1} = 1{,}602 \qquad \text{and} \qquad y_{x_2}^{(0)} = \frac{\partial y}{\partial x_2} = -400$$

for the point {2,2}. The first move led to the point $x_1^{(1)} = 1.461$ and $x_2^{(1)} = 2.135$. The components of the gradient, evaluated at this point, are as follows (again taken from Example VIII-9):

$$y_{x_1}^{(1)} = 0.922 \qquad \text{and} \qquad y_{x_2}^{(1)} = 0$$

Therefore, from Eq. (3.2-50),

$$\beta^{(0)} = \frac{(y_{x_1}^{(1)})^2 + (y_{x_2}^{(1)})^2}{(y_{x_1}^{(0)})^2 + (y_{x_2}^{(0)})^2} = 3.4 \times 10^{-7}$$

From Eq. (3.2-50) our second search direction, conjugate to the first, is given by

$$m_1^{(1)} = -0.922 + 3.4 \times 10^{-7}(-1.602 \times 10^3) = -0.922$$

$$m_2^{(1)} = 3.4 \times 10^{-7}(4 \times 10^2) \approx 0$$

Therefore we need to select $s^{(1)}$ so that

$$y = 100[2.135 - (1.461 - 0.922s^{(1)})^2]^2 + (0.461 - 0.922s^{(1)})^2$$

is a minimum. The value $s^{(1)}$ minimizing y along this second direction of search is very small, being about 0.0006. The second point is

$$x_1^{(2)} = 1.461 + (0.0006)(-0.922) = 1.460$$

$$x_2^{(2)} = 2.135$$

and the components of the gradient at this new position are

$$y_{x_1}^{(2)} = -0.783 \qquad y_{x_2}^{(2)} = 0.584$$

giving

$$\beta^{(1)} = \frac{(0.783)^2 + (0.584)^2}{(0.922)^2} = 1.123$$

The third direction of search is therefore

$$m_1^{(3)} = 0.783 + 1.123(-0.922) = -0.252$$

$$m_2^{(3)} = -0.584 + 1.123(0) = -0.584$$

and $s^{(2)}$ is to be determined so as to minimize

$$y = 100[2.135 - 0.584s^{(2)} - (1.460 - 0.252s^{(2)})^2]^2 + (-0.460 + 0.252s^{(2)})^2$$

In this step $s^{(2)} = 0.032$, and the new point is {1.452,2.116}.

Since this is the fourth stage, having searched along direction $\xi^{(0)}$, $\xi^{(1)}$, and $\xi^{(2)}$, we now forget the preceding searches, and again search along the gradient as required in the Fletcher-Reeves method. After each $n + 1$ stages (here $n = 2$), we begin anew.

3.2.3 Other gradient methods

Wilde [1963] has proposed a method in which the contour tangents are used to eliminate unfavorable (as illustrated in Fig. VIII-30) regions, rather than searching for more favorable positions in a fashion similar to that for one-dimensional search. This is essentially a gradient method, since the contour tangent is found by constructing a perpendicular to the gradient. The method is dependent on the objective function being "strongly unimodal." This latter condition is rather restrictive (see Chap. IV, Sec. 2.3), and unless the mathematical behavior of the objective function is well known, the method cannot be used.

Two methods for improving the univariant direct search technique (Sec. 3.1.1) may be noted. Instead of considering each direction in turn, the first method (Storey [1962]) chooses that axial direction along which the function

Fig. VIII-30 Failure of contour tangent method.

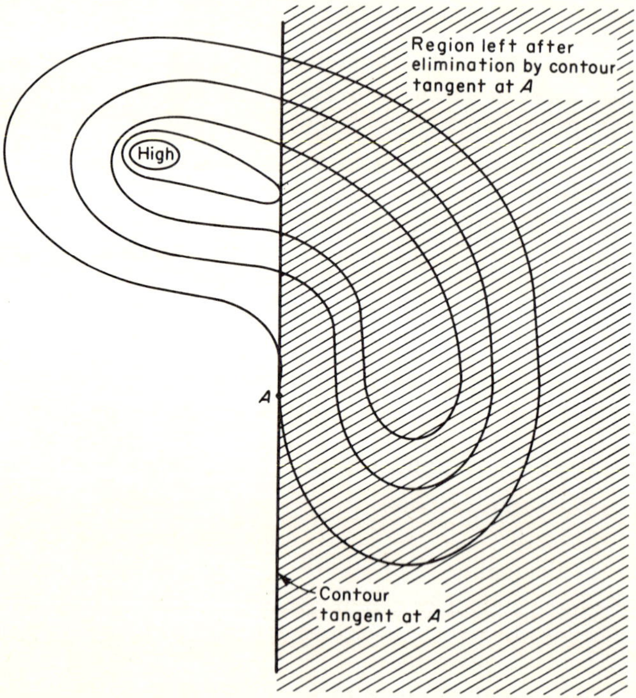

improves most rapidly by comparing the derivatives y_{x_i} in each direction. The other approach (suggested by Synge [1944]) chooses that axial direction along which

$$\frac{(y_{x_i})^2}{y_{x_i x_i}}$$

is the greatest. This is rather more convergent than the former modification, but of course requires the calculation of second derivatives. It is worth noting that this basic approach is more restrictive than in other gradient searches, since the choice of direction is restricted to those parallel to the coordinate axes.

3.2.4 Comments on gradient methods

The gradient methods appear to converge more rapidly than the direct search methods, provided that the number of experiments required to establish the gradient at a point is not excessive. This assertion has been established by numerical experiments (Fletcher and Reeves [1964] and Box [1966]), rather than by formal mathematical arguments. It seems reasonable. It is therefore recommended that a gradient method be adopted, provided that suitable analytical expressions are available for defining the gradient. The methods exhibiting quadratic convergence then appear to represent a good choice, the method of Fletcher and Reeves (Sec. 3.2.2) being particularly simple to program. The best method, however, appears to be that of Davidon [1959] as extended by Fletcher and Powell [1963] (page 426). Box [1966] has suggested that this method is apparently more successful than that of Fletcher and Reeves because, in the latter, past information is periodically discarded. The method of steepest ascent does not seem to be a profitable approach compared with the more convergent gradient techniques.

4 NUMERICAL OPTIMIZATION UNDER EQUALITY CONSTRAINTS

So far we have been concerned with unrestricted functions. Let us now consider the effect of constraints upon function search. As we have seen in Chap. VII, these restrictions fall into the two categories of equality and inequality constraints, a distinction which, it was noted, is largely arbitrary, since any inequality can be written in terms of an equality by the use of slack variables (Chap. VII, Sec. 4.1). Let us examine equality restrictions here and leave the problem of inequalities to Sec. 5. The discussion of these restrictions in which the independent variables exist only in discrete values will be left to Sec. 6. It might be noted at this stage that search methods for handling constraints are not in such an advanced stage of development as the unrestricted search techniques.

With unrestricted and inequality-constrained optimization, search is carried out in an n-dimensional space. Equality restrictions, on the other hand, impose stricter limitations, with the result that the search is confined to a part of the space.

Restrictions may arise from many sources, but in an existing system analyzed, using a black-box approach, many of the constraints will be automatically satisfied. An example of this is the equality restriction implicit in the conservation of mass. Some restrictions cannot, however, be so absorbed into the problem, and must be accounted for in any optimization search.

In rare circumstances the method of direct substitution (Chap. VII, Sec. 3.1) can be used, and the restricted problem converted to an unrestricted one. This might be a feasible proposition if one of the constraints, $g_k(\mathbf{x}) = 0$, can be arranged to give, for example, x_n in terms of the other variables, since then the number of constraints and independent variables can each be reduced by one, rendering the problem easier and smaller. With more complex restrictions, such an arrangement is in general not possible, and the restrictions must be satisfied by an iteration scheme so as to determine x_n numerically (by Newton's method, say, Chap. III, Sec. 4.1.2) at each stage of the search. This latter scheme would require an excessive number of experiments, and is not recommended.

The numerical techniques applicable to searches subject to equality constraints can be conveniently divided into two classes for the purposes of discussion. In the first, the original objective function is examined, with a simultaneous check of the feasibility of any new search locations.

The problem is a difficult one, and the techniques considered are all subject to criticism. They are, however, the best available; better search methods are still needed to adequately solve the more complex problems. Once again, the search methods may be examined under the headings of direct search and gradient methods (Secs. 4.2 and 4.3, respectively), although the former do not seem feasible. A major problem might be the determination of a first feasible base point (Sec. 4.1). The second group of methods utilize a modified objective function, which includes the constraints as an integral part, and therefore permits an unrestricted search (Sec. 4.4).

4.1 An Initial Solution

In the unrestricted problem we did not have to give special consideration to the choice of a starting location for the search. The presence of equality restrictions now forces us to choose only those points which satisfy the equations $g_k(\mathbf{x}) = 0$. Such a feasible point may, in fact, be difficult to find, and it might be necessary to conduct a special search, using some optimization technique, before starting the main optimization.

One method of finding a feasible point is to determine a value **x** which minimizes the sum of the squares

$$S = \sum_{k=1}^{m} g_k^2 \tag{4.1-1}$$

The problem of finding a valid search location is therefore converted to an unrestricted search problem of minimizing S. Any appropriate numerical or analytical method may be used. If a valid solution exists, then clearly, the minimum will exist at $S = 0$. Otherwise the constraints are mutually exclusive, and no solution exists.

The problem of finding a first feasible solution does not arise if we use penalty functions (Sec. 4.4). An illustration of this in linear systems has been discussed in Chap. VII, Sec. 4.4.4. Having obtained a valid starting position, let us now examine the various search methods available for use under equality constraints.

4.2 Direct Search Methods

It is unlikely that the direct search methods, such as the Sequential Simplex method of Sec. 3.1.3, which search in a restricted range of directions, can be used for handling problems involving equality constraints, since they cannot move along a boundary. These methods might, however, be used with profit when the objective function is modified, as in Sec. 4.4, to form an unrestricted problem.

4.3 Gradient Methods

The methods applicable to the unmodified restricted problem require the evaluation of the function gradient, the basic problem being, of course, the improvement of the objective function while still satisfying all the constraints. Several methods have been proposed, but none have been tested extensively. Only one approach, that of the projection gradient, will be described.

4.3.1 Projected gradient methods

The methods of constrained variation were developed in Chap. VII, Sec. 3.2, where variations of the objective function are taken in such a way that the constraints are automatically satisfied. Thus, for an infinitesimal displacement, the m equations [Eqs. (3.2-8), Chap. VII]

$$dg_k = \sum_{i=1}^{n} g_{k_{x_i}}\, dx_i = 0 \tag{4.3-1}$$

must be satisfied for $k = 1, 2, \ldots, m$. The search for an optimum cannot therefore freely proceed in the most desirable direction (steepest ascent for very small distances of movement), but must always be restricted to movements satisfying the constraints. Let us now, following Kelley [1962], develop a method of search which again uses constrained variation, but now treats each of the variables symmetrically, rather than expressing the variations of the first m variables in terms of the last $n - m$, as was done in Chap. VII.

As in the development of the steepest ascent, let us again seek the direction which makes the change in the objective function the greatest. The rate of change of the objective function along a direction is

$$\frac{dy}{ds} = \sum_{i=1}^{n} y_{x_i} \frac{dx_i}{ds} \tag{4.3-2}$$

The direction cosines dx_i/ds must obey the relationship (page 410)

$$\sum_{i=1}^{n} \left(\frac{dx_i}{ds}\right)^2 = 1 \tag{4.3-3}$$

and satisfy $dg_k = 0$ for $k = 1, 2, \ldots, m$ in the form

$$\frac{dg_k}{ds} = \sum_{i=1}^{n} g_{kx_i} \frac{dx_i}{ds} = 0 \tag{4.3-4}$$

This is a problem in optimization most easily approached using the method of Lagrangian multipliers. The desired values of dx_i/ds are those which make the modified objective function

$$F = \frac{dy}{ds} + \lambda_0\left[1 - \sum_{i=1}^{n}\left(\frac{dx_i}{ds}\right)^2\right] + \sum_{k=1}^{m} \lambda_k \sum_{i=1}^{n} g_{kx_i} \frac{dx_i}{ds}$$

take on a stationary value. Therefore the necessary condition

$$\frac{\partial F}{\partial(dx_l/ds)} = 0$$

gives

$$y_{x_l} - 2\lambda_0 \frac{dx_l}{ds} + \sum_{k=1}^{m} \lambda_k g_{k_{x_l}} = 0 \tag{4.3-5}$$

Rearranging Eq. (4.3-5) then gives

$$\frac{dx_l}{ds} = \frac{1}{2\lambda_0}\left(y_{x_l} + \sum_{k=1}^{m} \lambda_k g_{k_{x_l}}\right) \qquad \text{for } l = 1, 2, \ldots, n \tag{4.3-6}$$

This result gives the directions of movement yielding the greatest change in y subject to the condition that the m equality constraints be satisfied.

Substituting Eq. (4.3-6) into the constraint equation (4.3-4), we find that the λ_k are defined by

$$\sum_{i=1}^{n}\sum_{l=1}^{m}\lambda_l g_{l_{x_i}} g_{k_{x_i}} = -\sum_{i=1}^{n} g_{k_{x_i}} y_{x_i} \qquad \text{for } k = 1, 2, \ldots, m \tag{4.3-7}$$

Similarly by eliminating dx_i/ds from Eq. (4.3-3), using Eqs. (4.3-6) and (4.3-7), we can obtain the following equation for λ_0:

$$2\lambda_0 = \pm\sqrt{\sum_{i=1}^{n}\left(y_{x_i}{}^2 + y_{x_i}\sum_{k=1}^{m}\lambda_k g_{k_{x_i}}\right)} \tag{4.3-8}$$

The rate of change in the objective function is easily computed as

$$\frac{dy}{ds} = \pm\sqrt{\sum_{i=1}^{n}\left(y_{x_i}{}^2 + y_{x_i}\sum_{k=1}^{m}\lambda_k g_{k_{x_i}}\right)} \tag{4.3-9}$$

using Eqs. (4.3-6) to (4.3-8) in Eq. (4.3-2). The positive sign is selected if the greatest feasible increase in the objective function is desired, as is the case in the search for a maximum.

We have now assembled all the ingredients to develop a computational method. Before discussing the calculation, it is interesting to examine the significance of these results in a geometrical context.

The geometrical features of the mathematical results are best depicted by restricting our attention to a three-dimensional problem having a single restriction. For this case, Eq. (4.3-7) is

$$\lambda_1\sum_{i=1}^{3} g_{1_{x_i}}^2 = -\sum_{i=1}^{3} g_{1_{x_i}} y_{x_i} \tag{4.3-10}$$

Equation (4.3-9), giving the maximum feasible rate of change, here reduces to

$$\left(\frac{dy}{ds}\right)_{\text{constrained}} = \sqrt{\sum_{i=1}^{3} y_{x_i}{}^2}\sqrt{1 - \frac{\left(\sum_{i=1}^{3} g_{1_{x_i}} y_{x_i}\right)^2}{\sum_{i=1}^{3} g_{1x_i}^2 \sum_{i=1}^{3} y_{x_i}{}^2}} \tag{4.3-11}$$

which can be written

$$\left(\frac{dy}{ds}\right)_{\text{constrained}} = \sqrt{\sum_{i=1}^{3} y_{x_i}{}^2 f} \tag{4.3-12}$$

From Eq. (3.2-12) we know that the greatest rate of increase is given by

$$\left(\frac{dy}{ds}\right)_{\text{unconstrained}} = \sqrt{\sum_{i=1}^{3} y_{x_i}{}^2} \tag{4.3-13}$$

Clearly, the presence of the constraint $g_1(x_1,x_2,x_3) = 0$ is to reduce the greatest possible rate of increase by some factor f which is bounded $0 \leq f \leq 1$. This factor can be shown[1] to be the cosine of the angle between the gradient and its projection on to the tangent plane. The best possible differential move in the presence of constraints is therefore along the projection of the steepest ascent on the constraining surface.

Given a feasible point $\mathbf{x}^{(p)}$, a new point $\mathbf{x}^{(p+1)}$ can then be defined as

$$x_i^{(p+1)} = x_i^{(p)} + \frac{1}{2\lambda_0}\left(y_{x_i} + \sum_{k=1}^{m} \lambda_k g_{k_{x_i}}\right) s \tag{4.3-16}$$

[1] Let us show that

$$\left(\frac{dy}{ds}\right)_{\text{constrained}} = \left(\frac{dy}{ds}\right)_{\text{steepest ascent}} \cos \alpha \tag{4.3-14}$$

Since the normal to the surface of constant g is defined by the direction ξ', where

$$m_i' = \frac{g_{x_i}}{\sqrt{\sum_{j=1}^{3} (g_{x_j})^2}}$$

(as was seen from our work on the method of steepest gradient, the gradient is perpendicular to a contour surface), the gradient of g evaluated at a point where g vanishes must necessarily be normal to the surface, $g = 0$, as shown in Fig. VIII-31. Then, certainly, the projection of the gradient on to the constraint plane is given by

$$\left(\frac{dy}{ds}\right)_{\text{projection}} = \left(\frac{dy}{ds}\right)_{\text{steepest ascent}} \cos \alpha \tag{4.3-15}$$

If f can be shown to be identically $\cos \alpha$, then

$$\left(\frac{dy}{ds}\right)_{\text{constrained}} \equiv \left(\frac{dy}{ds}\right)_{\text{projection}}$$

and the relationship given in Eq. (4.3-14) will hold.

From Fig. VIII-31, we see that

$$\cos(90 - \alpha) = \frac{y_{x_1}g_{x_1} + y_{x_2}g_{x_2} + y_{x_3}g_{x_3}}{\sqrt{y_{x_1}^2 + y_{x_2}^2 + y_{x_3}^2}\sqrt{g_{x_1}^2 + g_{x_2}^2 + g_{x_3}^2}}$$

since the cosine of the angle β, between two vectors, is given by

$$\cos \beta = \frac{\mathbf{a} \cdot \mathbf{b}}{|\mathbf{a}|\,|\mathbf{b}|}$$

Moreover,

$$\cos(90 - \alpha) = \cos 90 \cos \alpha + \sin 90 \sin \alpha = \sin \alpha$$

and

$$\cos \alpha = \sqrt{1 - \sin^2 \alpha}$$

so that

$$\cos \alpha = \sqrt{1 - \cos^2(90 - \alpha)} = f$$

from the definition of f in Eqs. (4.3-11) and (4.3-12). Thus f and $\cos \alpha$ are the same, and the best constrained move in the presence of constraints is the projection of the steepest ascent gradient on to the constraining surface.

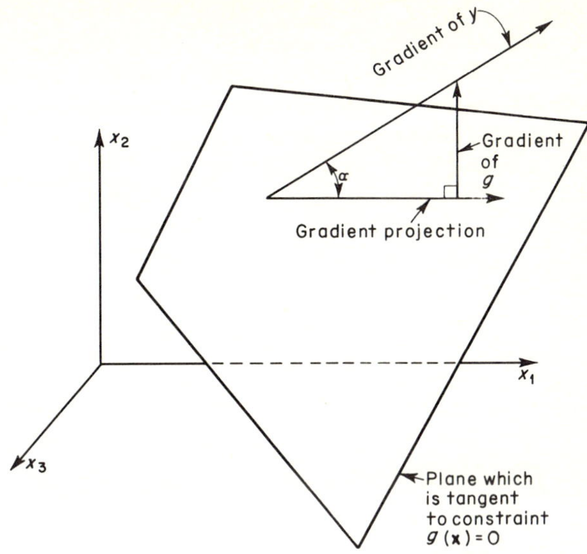

Fig. VIII-31 Geometry of gradient projection.

This will be an improved position for a sufficiently small distance of movement, s. This expression is derived from Eq. (4.3-8) and is analogous to the movements in steepest ascent. Equations (4.3-7) and (4.3-8) form the basis of a computational scheme. The m Lagrangian multipliers are first found[1] from the m linear algebraic equations (4.3-7), while Eq. (4.3-8) gives λ_0. The new base point in the search can then be computed from Eq. (4.3-16). These two steps form the basis of the cyclic procedure described by Kelley [1962].

The procedure should be effective if the constraints are linear, since large values of s can be selected without fear of violating the constraints. This is the basis of the gradient projection method due to Rosen [1960]. If, however, the constraints are nonlinear, very small steps must be taken, and a calculation to bring the new point back on to the constraining surfaces must be included in each cycle. This step is necessary because the new point lies along the tangent surface when computed as suggested above and will not in the case of curving surfaces lie on the true surface. Thus we see that the

[1] Difficulties might arise if some of the planes defined by Eqs. (4.3-7) are nearly parallel. Mathematically, there will be a solution if the $m \times m$ matrix of elements

$$a_{k,j} = \sum_{i=1}^{n} g_{j_{x_i}} g_{k_{x_i}}$$

is nonsingular. This is the Gram determinant criterion for linear independence of Eqs. (4.3-7), and is equivalent to a similar requirement on the Jacobians in constrained variation (Chap. VII, Sec. 3.2.2; Kelley [1962]).

convergence rate of this method will depend strongly upon the curvature of the constraining surfaces. These same criticisms must apply to the methods of Rosen [1961] and Fiacco, Smith, and Blackwell [1960], which differ only in the way in which the step size s is selected and in the method of returning to the surface of the constraint.

Example VIII-12: Minimization by the gradient projection method

Find the minimum of the function

$$y = x_1^2 - 2x_1x_2 + x_2^4$$

subject to the constraint

$$g = x_1 + x_2 - 5 = 0$$

Let us start our computation at the feasible point {5,0}. Here $y_{x_1} = 10$, $y_{x_2} = -10$, $g_{x_1} = 1$, $g_{x_2} = 1$. Therefore, from Eq. (4.3-7), we find that $\lambda_1 = 0$. Equation (4.3-8) then gives

$$2\lambda_0 = -14.14$$

where the negative sign has been selected since a minimum is sought. The new point defined by Eq. (4.3-16) is

$$x_i^{(1)} = x_i^{(0)} + \frac{s}{2\lambda_0}(y_{x_i} + \lambda_1 g_{x_i})$$

in the case of the single constraint, as is studied here.

Then, with $s = 1$,

$$x_1^{(1)} = 5 - 10/14.14 = 4.293$$

$$x_2^{(1)} = 10/14.14 = 0.707$$

Using these coordinates, we find that the objective function $y(\mathbf{x}^{(1)})$ has the value 12.61, which is an improvement over the previous value $y(\mathbf{x}^{(0)}) = 25$.

The second stage of the search is a repeat of the first. With $y_{x_1} = 7.17$ and $y_{x_2} = -7.17$, λ_1 can again be shown to be zero. This Lagrangian multiplier vanishes because any move along the gradient will produce equal and opposite changes in x_1 and x_2, since the components of the gradient are equal in magnitude and opposite in sign. If such a change is made, the new point will satisfy the constraint, since any increase in x_1 will be compensated by a decrease in x_2. Thus, if the gradient of the objective function is tangent to the constraint, the move is along the gradient. This is the significance of the Lagrangian multiplier vanishing.

In the second stage

$$2\lambda_0 = -10.28$$

Thus $x_1^{(2)} = 3.59$ and $x_2^{(2)} = 1.41$

giving $y(\mathbf{x}^{(2)}) = 6.72$

The third stage is determined by the components of the gradient $y_{x_1} = 4.36$ and $y_{x_2} = 4.03$. Since these are no longer the negatives of one another, we should not expect the Lagrangian multiplier to be zero for this third stage of the calculation. The constraint now forces the search away from moves strictly in the direction of steepest ascent. From Eq. (4.3-7), $\lambda_1 = -4.20$. Also

$$2\lambda_0 = \sqrt{(4.36)^2 + (-4.20)(4.36) + (4.03)^2 + (-4.20)(4.03)} = -0.112$$

This gives $x_1^{(3)} = 2.12$ and $x_2^{(3)} = 2.88$

A test will show that the objective function is not improved by this move. The distance of movement $s = 1$ is too large. This distance must be reduced, and the calculation continued. It should be noted that $2\lambda_0$ is a measure of the magnitude of the constrained derivative, since we know from the method of Lagrangian multipliers (Chap. VII, Sec. 3.3.1) that the derivative

$$y_{x_i} + \sum_k \lambda_k g_{k_{x_i}} = 0$$

is a necessary condition for the existence of a minimum. Thus, when λ_0 becomes very small, the point is near the optimum. The Lagrangian multiplier is then in some sense a measure of the distance of a point from the optimum, and if this quantity becomes small, s should be reduced appropriately. Note that in the present example the absolute value of λ_0 has become progressively smaller, reaching a value of 0.112 at the last stage. Our step size should be reduced to, roughly, the same proportions.

An important point to note is that the gradient projection method has been applied in this example to a problem restricted by linear constraints, and no difficulty was encountered in using a large step size ($s = 1$). If, however, the constraints are nonlinear, the new point generated using this method will in general be nonfeasible, and some corrective action must be made before making the next move.

4.3.2 *A modified gradient projection approach*

The methods described by Kelley [1962] can be modified so that the new base point will remain on the constraints, rather than on the tangent surface. Again we can determine $x_i^{(p+1)}$ from Eq. (4.3-16), leaving the values of λ_k undetermined, and use a value of s selected *a priori*. The values of λ_k might then be selected so that the function S, defined by

$$S = \sum_{k=1}^{m} (g_k)^2 = S(\lambda_1, \lambda_2, \ldots, \lambda_m)$$

is a minimum. This can now be minimized by an m-dimensional unrestricted search. Once the λ_k minimizing S have been found, and the new location determined, the entire cycle would again be repeated. The distance of movement, s, should probably be progressively reduced during search, but this method has not been fully tested (in fact, none of these search methods have, according to Wolfe [1963]) and has not, apparently, been previously published. No proven algorithm for computing s is therefore given, nor is there any guarantee of convergence. This procedure is somewhat similar to the approach of Zoutendijk [1960] under inequality restrictions (page 466).

4.4 Objective Functions Modified to Give an Unrestricted Form

Optimization in the presence of equality constraints may also be carried out by using methods which modify the objective function by incorporating the constraints. We shall recall that, in obtaining analytical solutions to such problems, it was found (Chap. VII, Sec. 3) that a restricted problem could be so modified that techniques for unrestricted optimization could be used in its solution. Let us consider the methods of Lagrangian multipliers (Sec. 4.4.1) and penalty functions (Sec. 4.4.2) in connection with numerical optimization.

4.4.1 *Lagrangian multipliers in numerical optimization*

One possible technique for modifying the constrained problem might be to use Lagrangian multipliers (Chap. VII, Sec. 3.3). In this method a modified objective function

$$F = y + \sum_{k=1}^{m} \lambda_k g_k \tag{4.4-1}$$

is searched for stationary values. In the general case, however, this function F cannot be investigated by search methods,[1] since purely numerical methods are unlikely to locate all types of stationary point (including the saddle points). It will be recalled that a restricted optimum of y might indeed be at such a saddle point (Example VII-14). A more feasible approach is that of penalty functions, as discussed in the next section.

[1] The use of Lagrangian multipliers is possible, of course, when g is strictly convex and y is strictly convex or concave for minimization and maximization, respectively (see, for example, Wolfe [1963]). The optimal value of y will then lie at an optimal value of F, and it is only necessary to find the unrestricted optimum in F by some numerical search.

4.4.2 *Penalty functions in numerical optimization*

The incorporation of the constraints into a modified objective function by the use of penalty factors has been discussed in Chap. VII, Sec. 3.4. One such form of modified function is

$$F = y + \sum_{l=1}^{m} P_l g_l^2 \tag{4.4-2}$$

where the P_l are large positive-valued constants if y is to be minimized and large negative quantities for these y which are to be maximized.[1]

As noted in Chap. VII, Sec. 3.4, the problem should now be treated as an unrestricted one, and as the P_l are increased, the g_l are forced to zero at the optimum. Although the idea of a penalty function seems to have been conceived some years ago (Courant [1943]), there has been very little reported computational experience with regard to its application. It is clear, upon considering Example VIII-13, that the choice of P_l is critical, and it should probably be progressively increased during the course of the search. The reason for this is that large values of P_l effectively produce a sharp ridge in the contours of F, and most search techniques are troubled by the existence of such a ridge. It is therefore advantageous to search with smaller values of P_l so far as the number of experiments is concerned, whereas values of P_l which are not large enough result in a failure to satisfy the constraints with sufficient accuracy. The choice of P_l is therefore an optimization problem.

Kelley [1962] has considered the problem of estimating the size of P_l when it is practical to form the partial derivatives of both the constraints and the objective function. We suppose that there are preassigned tolerances, and at the end of each stage the values of P_l are to be adjusted, depending on the magnitude of the deviations of g_l from their required zero value. Since the derivatives of Eq. (4.4-2) must vanish, then

$$\frac{\partial y}{\partial x_i} + 2\sum_{l=1}^{m} P_l g_l \frac{\partial g_l}{\partial x_i} = 0 \qquad \text{for } i = 1, 2, \ldots, n \tag{4.4-3}$$

This result can also be written

$$2\sum_{i=1}^{n}\sum_{l=1}^{m} P_l g_l \frac{\partial g_l}{\partial x_i}\frac{\partial g_k}{\partial x_i} = -\sum_{i=1}^{n} \frac{\partial y}{\partial x_i}\frac{\partial g_k}{\partial x_i} \qquad \text{for } k = 1, 2, \ldots, m \tag{4.4-4}$$

by multiplying Eq. (4.4-3) by $\partial g_k/\partial x_i$ and summing on i. We have put the necessary condition for F to be an optimum in this form because we now have a structure quite similar to that found in gradient projection and valid near

[1] Fiacco and McCormick [1966] have shown that the form $F = y + P\sum_{i=1}^{m} g_i^2$ produces convergence in the minimization of a convex function as $P \to \infty$. See also Rubin and Ungar [1957].

the true optimum (where the linear approximation is valid). Thus Eq. (4.3-7),

$$\sum_{i=1}^{n} \sum_{l=1}^{m} \lambda_l \frac{\partial g_l}{\partial x_i} \frac{\partial g_k}{\partial x_i} = - \sum_{i=1}^{n} \frac{\partial y}{\partial x_i} \frac{\partial g_k}{\partial x_i} \tag{4.3-7}$$

derived in our study of the gradient projection method is to be satisfied at the optimum, as was shown in Sec. 4.3.1. Comparing this condition with that given by Eq. (4.4-4), we see that, near the optimum, the violations are given by

$$g_k \approx \frac{\lambda_k}{2P_k} \tag{4.4-5}$$

and the violations are seen to vary inversely with the magnitude of the penalty, P_k.

Based on this result, Kelley has suggested, but not tested, the possibility of increasing or decreasing the current value of P_k at the end of each stage, $P_k^{(i)}$, based on the equation

$$P_k^{(i)} = \left| \frac{\lambda_k(\text{estimated})}{\epsilon_k^{(i)}} \right| \tag{4.4-6}$$

where λ_k(estimated) are values of the Lagrangian multipliers, found by solving Eqs. (4.3-7), the derivatives of the objective function and the constraints being evaluated at the most recent estimate of the optimal solution, and the $\epsilon_k^{(i)}$ are the preassigned tolerances for the kth constraint in the ith stage which define the acceptable deviations of g_k from zero.

A solution should never be accepted unless some check is performed to ensure that the constraints are indeed satisfied within the desired accuracy. The modified penalty function of Eq. (4.4-2) or equivalent forms (Chap. VII, Sec. 3.4) may be searched, using any method for unrestricted search.

The advantage of this type of approach for handling constraints, compared with those methods which search the objective function subject to restrictions, lies in the avoidance of iterative corrections necessary to follow the constraints. In addition, it is not necessary to worry about obtaining a feasible initial solution.

Example VIII-13: Penalty functions

Let us consider the search for the minimum of

$$y = (x_1 - 3)^2 + (x_2 - 2)^2$$

subject to the restriction that

$$g = x_1 + x_2 - 4 = 0$$

(The solution is at the point {2.5,1.5}.)

Using the concept of the penalty function, we form the modified objective function

$$F = y + Pg^2 = (x_1 - 3)^2 + (x_2 - 2)^2 + P(x_1 + x_2 - 4)^2$$

and now use a suitable unrestricted search technique to determine the optimum of F. For sufficiently large values of P, this will yield a solution which satisfies the imposed constraint with a prescribed accuracy. Care, however, must be taken in the selection of an initial value of P.

Using the univariant method, for the sake of simplicity, we find as a necessary condition for the existence of a local optimum along the x_1 direction the expression

$$F_{x_1} = 2(x_1 - 3) + 2P(x_1 + x_2 - 4) = 0$$

or

$$x_1 = \frac{3 + 4P - x_2 P}{1 + P}$$

For a starting point {0,6}, chosen arbitrarily, and a value $P = 100$, we therefore find that x_1 is, approximately, -1.9. This is the position of the minimum, here a minimum along the line $x_2 = 6$. The next search, as is customary in the univariant technique, is along a line of constant x_1, and in the x_2 direction the necessary condition for the existence of a minimum is

$$F_{x_2} = 2(x_2 - 2) + 2P(x_1 + x_2 - 4) = 0$$

giving

$$x_2 = \frac{2 + 4P - Px_1}{1 + P}$$

so that $x_2 = 5.93$. Continuing, we find, in turn, the points {0,6}, {−1.95,6}, {−1.95,5.93}, {−1.88,5.93}, {−1.88,5.84}, ... The search is now oscillatory about a narrow ridge created by the large value of P. The contours of F for various values of P are shown in Fig. VIII-32, where it is seen that the ridge becomes sharpened as the value of P is increased. Indeed, larger values of P would eventually cause the search to become stranded at the point {−2,6} if it started at {0,6}. It seems very reasonable, therefore, to start with a relatively small value of P, so that y and Pg^2 terms comprising F will have the same order of influence, and once the region of the optimum is achieved, then again to increase the value of P so as to retain the two contributions to F at the same order of magnitude. Thus the search scheme would be to increase P in successive stages.

Hence, starting from {0,6} with $P = 1$, we get in the first cycle

$$x_1 = \frac{3 + 4 - x_2}{1 + 1} = \frac{7 - x_2}{2}$$

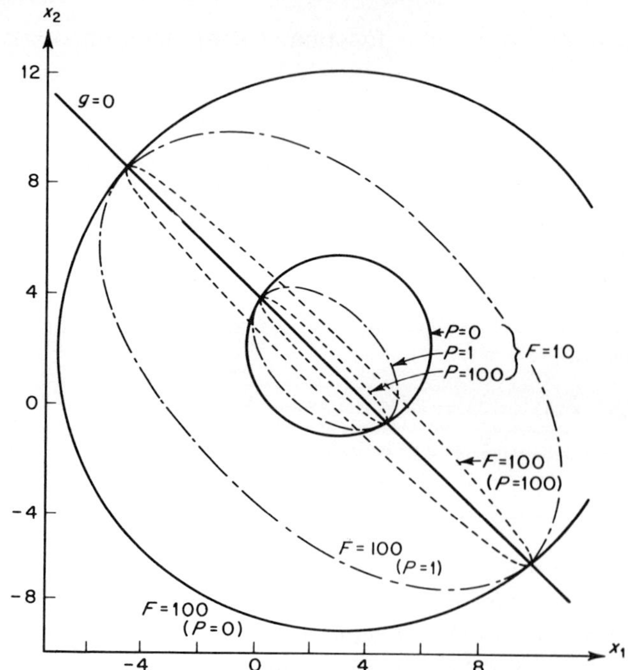

Fig. VIII-32 Contour ridges formed by the penalty factor *P*.

so that we get {0.5,6}. The next cycle produces a new point {0.5,2.75}. Continuing in this way, we shall approach the point {2.66,1.66}. Since this is a relatively wide ridge, fewer iterations are required in locating this minimum. Then, gradually increasing *P*, we should in turn reach $\{2.54,1.54\}_{P=5}$, $\{2.52,1.52\}_{P=10}$, and thus eventually reach the desired location.

Example VIII-14: Choice of a sequence of penalties

We have seen that the choice of the penalty factor in successive searches is critical in governing the rate of convergence. If the penalty is too large, the search becomes stranded on a ridge far from the optimum. If it is not increased rapidly enough, the search is slow. For the problem posed in Example VIII-13, Eq. (4.3-7) reduces to

$$\lambda_1\left[\left(\frac{\partial g_1}{\partial x_1}\right)^2 + \left(\frac{\partial g_1}{\partial x_2}\right)^2\right] = -\left(\frac{\partial y}{\partial x_1}\frac{\partial g_1}{\partial x_1} + \frac{\partial y}{\partial x_2}\frac{\partial g_1}{\partial x_2}\right)$$

or

$$\lambda_1 = 5 - x_1 - x_2$$

Let us select a sequence of tolerances; $\epsilon_1^{(1)} = 10$, $\epsilon_1^{(2)} = 1$, $\epsilon_1^{(3)} = 0.1, \ldots$. These tolerances will determine approximately the extent to which the constraint is violated in a particular iteration. Since Eq. (4.4-6) is an estimate, the deviation of the constraint from zero may be more or less than ϵ but will be of the same order of magnitude or less. Thus

$$|x_1 + x_2 - 4| \leq \text{Order } (\epsilon_1)$$

in a particular iteration.

Starting at the base point $\mathbf{x}^{(0)} = \{5,6\}$, we obtain

$$P_1^{(1)} = \left|\frac{\lambda_1}{\epsilon_1^{(1)}}\right| = \frac{6}{10} = 0.6$$

On minimizing

$$F_1 = y + P_1^{(1)} g_1^{\,2} = (x_1 - 3)^2 + (x_2 - 2)^2 + P_1^{(1)}(x_1 + x_2 - 4)^2$$

by an unconstrained technique, we find $\mathbf{x}^{(1)} = \{2.727,1.727\}$. Note that $|x_1 + x_2 - 4| = 0.454$, which is less than the order of magnitude of $\epsilon_1^{(1)}$.

Starting from point $\mathbf{x}^{(1)}$, we use a second penalty

$$P_1^{(2)} = \left|\frac{\lambda_1}{\epsilon_1^{(2)}}\right| = \frac{0.546}{1.0} = 0.546$$

With this value of the penalty, we find $\mathbf{x}^{(2)} = \{2.739,1.739\}$ so that $|x_1 + x_2 - 4| = 0.478$ is the same order of magnitude as the preassigned tolerance.

A third stage with

$$P_1^{(3)} = \left|\frac{\lambda_1}{\epsilon_1^{(3)}}\right| = 5.22$$

yields $\mathbf{x}^{(3)} = \{2.544,1.544\}$. The $|x_1 + x_2 - 4| = 0.087$ is again a reduction from the previous stage.

Thus we see that Kelley's method of selecting the penalties gives a logical structure for their choice. By choosing values of ϵ rather than P, one is fixing the error allowed in the constraint violation.

A second method of utilizing penalty functions is due to Paviani [1969]. The method uses the unconstrained search technique of Nelder and Mead [1964] and combines all restrictions into a single tolerance. New points are checked to ensure both that they improve the objective function and that they satisfy the tolerance. The tolerance is allowed to be large at the start of the search, but it is steadily decreased as the search progresses. This is a very attractive idea since, in effect, equality constraints are treated as though they were a single inequality; thus

$$-\phi(\mathbf{x}) + \sum_{k=1}^{m} g_k^{\,2}(\mathbf{x}) \leq 0$$

Here $\phi(\mathbf{x})$ is the positive tolerance, and Paviani conducts his search in such a way that $\phi(\mathbf{x})$ decreases during the search. At the end, when $\phi(\mathbf{x})$ is nearly zero, the equality constraints can be seen to be very nearly satisfied.

Stocker [1969] has tested this method and compared it with a number of others. The findings reported by Stocker are given in Appendix 2.

5 NUMERICAL OPTIMIZATION UNDER INEQUALITY CONSTRAINTS

A more general type of problem is that in which the objective function $y(\mathbf{x})$ is to be optimized subject to m inequality constraints

$$g_k(\mathbf{x}) \leq 0 \qquad \text{for } k = 1, 2, \ldots, m \tag{5-1}$$

A number of methods have been developed to solve this constrained problem, but this aspect of optimization is still under active investigation.

The problem may, in general, be tackled by one of two approaches.

First, the search for an optimum constrained by inequalities may be carried out using the numerical search methods already developed in Sec. 3. An additional routine, however, must be added to ensure that each new point is a feasible one and to direct the search in a feasible direction should a nonfeasible point be found. The problem of finding a feasible initial point is examined in Sec. 5.1. A search of the objective function, using any of the unrestricted techniques, may result in a move from this feasible point out of the valid region to some point contradicting one or more of the constraints. The methods of choosing a new direction of movement from an invalid point so as to return within the feasible region or on to the boundaries can be divided into two broad classes, which can be called *nongradient* (Sec. 5.2) and *gradient methods* (Sec. 5.3). Some of the techniques for moving out of an invalid region require only local values of the objective function and the constraints. These methods, therefore, are more likely to be used in conjunction with direct search methods, whereas those routines which require local values of the gradient to move to a feasible point are usually applied in connection with gradient methods. This is, of course, not a firm rule, but would seem to be a reasonable one. While methods originally designed for use without restrictions might be satisfactory in the present type of problem, they may not prove particularly efficient in comparison with methods specifically designed to handle constrained search.

The second approach depends upon the problem being converted into the form of an unrestricted system (Sec. 5.4), and the methods of Sec. 3 used directly to seek an optimum.

5.1 A Feasible Starting Point

As in the work on equality constraints (Sec. 4.1), the first task in the process of constrained optimization is to find a feasible base point from which to begin the search. One such approach is to choose any seemingly suitable point, check all the constraints, and, if any are contradicted, to minimize the function

$$S = \sum_{k=1}^{m} g_k \tag{5.1-1}$$

the summation being taken only over those constraints which are violated. A more formal presentation of this procedure is to minimize

$$S = \sum_{k=1}^{m} H(g_k)g_k \tag{5.1-2}$$

where $H(g_k)$ is the Heavyside function, defined as

$$H(g_k) = \begin{cases} 1 & \text{if } g_k \geq 0 \\ 0 & \text{if } g_k < 0 \end{cases}$$

so that the summation S is indeed only over the violated constraints. When S vanishes, we have found a feasible point. Of course, it is not certain that a solution will exist. Since all minimization techniques find local rather than global optima, obtaining a nonzero minimum of S will not necessarily imply the nonexistence of a feasible point. Three approaches might be considered:

1. An unrestricted search of S might be made (unrestricted in the sense that the inequalities are not tested).
2. S is minimized, constrained not to violate inequalities which are already satisfied.
3. If l of the inequalities are satisfied, a search for the minimum of S might be made, subject only to these and the $(l + 1)$st inequality. This procedure is repeated until all the inequalities are satisfied. Finally, $S = \sum g_k$, where g_k are the remaining equalities, is minimized subject to all the inequalities. This last step can be omitted if Carroll's method (Sec. 5.4.1) is being used.

Box et al. [1969] recommend the first approach, since unconstrained search is easier, and note that any optimization techniques utilized must not be defeated by having independent variables present upon which S does not currently depend. Fiacco, Smith, and Blackwell [1960] and Fiacco [1961] have also discussed the problem of getting a first feasible location, basing their approach upon the optimization technique in use in the main search, in particular the method of Carroll [1961].

The method of centers (Hensgen [1966]) replaces the usual objective function by a modification

$$F = \left[\prod_{k=1}^{m} (b_k - g_k)\right] y^a$$

for maximization. Initially the power a is kept small to obtain feasible points, and it is thereafter increased as the search progresses. Normally $b_k = 0$, but if a constraint is violated, the b_k can be temporarily increased to make the constraint approach positive, a being then reduced to force the search into the valid region with little or no regard for the objective contours.

Example VIII-15: A first feasible point

Maximize

$$y(x_1,x_2) = [(1 + x_1)^2 + x_2^2]^{-1}$$

subject to the constraints

$$g_1(x_1,x_2) = 4 - x_1^2 - x_2^2 \leq 0$$

$$g_2(x_1,x_2) = x_1^2 + x_2^2 - 16 \leq 0$$

Before we can start a direct search for the maximum of y, we must find an initial feasible base point. This may be done by minimizing

$$S = \sum_k g_k$$

the summation being taken only over those constraints violated. Starting at $\mathbf{x}_1 = \{5,4\}$, we find that $g_1 = -37.00$ and $g_2 = 25.00$. We shall therefore minimize $S = g_2$, the other constraint not being required, since it is satisfied. Details of the calculation, using the Sequential Simplex method, are presented in Table VIII-5. $\mathbf{x}_9$ is the first feasible point which can be used to start the search for the maximum of y.

TABLE VIII-5 SEARCH FOR A FEASIBLE POINT
(Sequential Simplex method: $a = 1.0$, $p = 0.9657$, $q = 0.2587$)

Point	x_1	x_2	g_1	$g_2 \equiv S$	*Point rejected*	*Rule*	*New simplex*
1	5.0000	4.0000	−37.00	25.00			
2	5.9657	4.2587	−49.73	37.73	Starting simplex		1, 2, 3
3	5.2587	4.9657	−48.31	36.31			
4	4.2930	4.7070	−36.58	24.58	2	1	1, 4, 3
5	4.0343	3.7413	−26.27	14.27	3	1	1, 4, 5
6	3.3273	4.4483	−26.86	14.86	1	1	6, 4, 5
7	3.0686	3.4826	−17.54	5.54	4	1	6, 7, 5
8	3.7756	2.7756	−17.96	5.96	6	1	8, 7, 5
9	2.8099	2.5159	−10.23	−1.77	5	1	8, 7, 9

5.2 Nongradient Moves and Nonfeasible Regions

In using any of the direct search methods under constraint, it is necessary to test the feasibility of each new point. To indicate the underlying difficulties, Fig. VIII-33*a* shows the situation in which a search, using the Hooke-Jeeves method, can continue toward the optimum with feasible moves in the valid region. The initial pattern move arriving at point *A* finds that a continued move to *B* would be unsuccessful, since *B* violates $g_1 = 0$. An exploration may be begun about *A*, whence it is found that *C* is a feasible improvement but point *D*, a further improvement, is invalid. The search could, however, continue with a pattern move from *C* in the usual manner (Sec. 3.1.3). In this search the Hooke-Jeeves approach is used unmodified, except in that each possible search location, *A*, *B*, . . . , is tested for validity. In general, such a procedure might be followed, and we may go ahead as if we were examining an unrestricted system.

This approach cannot, however, always cope with constraints. Thus, if the system is as shown in Fig. VIII-33*b*, the Hooke-Jeeves type of move will stop at *P*, and the only two directions of improvement do not yield feasible results. The reduction of the step size would only result in the cessation of

Fig. VIII-33 Hooke-Jeeves method under constraint with successes and failures.

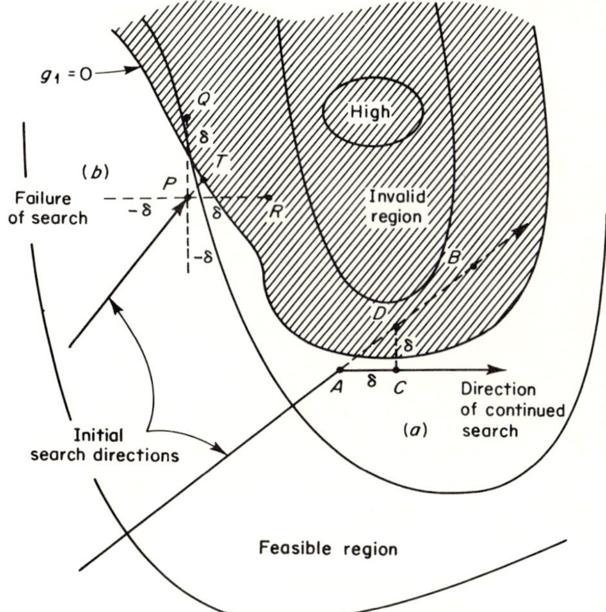

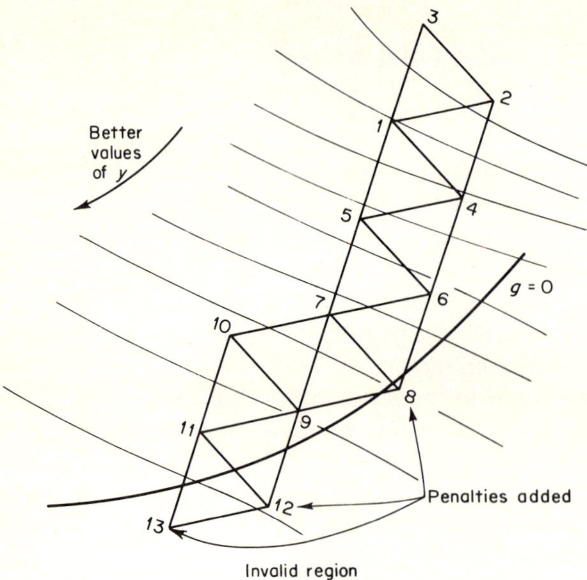

Fig. VIII-34 Unsuccessful Sequential Simplex search.

the search at T on the constraint, $g = 0$, a location which, however, proves not to be the optimum. Singer [1962] has described the *maze method* of avoiding such false optima when using the univariant technique, but it will not be described, since a better method is available. The Sequential Simplex method, while convenient for unrestricted systems, can also get into trouble against constraints. Retaining the usual rejection and regeneration rules (Sec. 3.1.2), the presence of constraints might be taken into account by testing the validity of each new location and replacing the objective function value by a large negative penalty number for maximization (positive for minimization) if a violation is found. This will force the search back into the valid region. Since the rejection of this invalid point would cause the search to return to the previous simplex, we should instead reject the second worst point, as in rule 2. A typical set[1] of moves is shown in Fig. VIII-34. Even with these simple contours the search will come to a false halt (at $\mathbf{x}_{11}$), and little further improvement would result from a decrease in the step size. It is also possible for the search to move in the "wrong" direction. Difficulties can also arise in finding a suitable initial simplex.

One direct search method which appears to cope successfully with constraints is a modification of the Sequential Simplex technique. Let us consider it.

[1] In the search shown, points 8 and 12 are successfully avoided by the use of penalty terms, but the search is forced to halt at $\mathbf{x}_{11}$. Any further decrease in step size will tend finally to stop the search where $g = 0$ crosses the line from $\mathbf{x}_{11}$ to $\mathbf{x}_{13}$.

5.2.1 The Complex Method of Box

A modification of the Sequential Simplex technique has been proposed by Box [1965] to avoid such problems as an inability to construct a first simplex. Called the *Complex* method, it handles constraints by the use of a flexible figure of more than $n + 1$ vertices, which can expand or contract in any or all directions and can extend round "corners." This method is straightforward and easy to program for convex regions. Vertices are rejected and generated, as in the Simplex method, but no attempt is made to preserve a regular figure in which each vertex is equidistant from all other points.

The Complex method requires the use of $k \geq n + 1$ vertices, each of which must satisfy all the imposed constraints. These vertices may be initially found by starting at one point which satisfies all m constraints. This initial point might be found as in Sec. 5.1. The method not only allows for the inclusion of the region restrictions $g_k(\mathbf{x}) \leq 0$, but also includes limits on the x_i variables in the form

$$x_{iL} \leq x_i \leq x_i^L$$

These limits might be inferred from the $g_k \leq 0$ restrictions, and would then represent a rectangular space within which the search must take place (Rosen [1961]).

The additional $k - 1$ points in the first complex figure can now be found, based on the relations

$$x_{i,j} = x_{iL} + r_{i,j}(x_i^L - x_{iL}) \qquad \text{for } i = 1, 2, \ldots, n \tag{5.2-1}$$

where $r_{i,j}$ are random numbers generated in the range 0 to 1. The points so generated will lie within the rectangular search region, no further scaling having to be considered. These points, however, will not necessarily satisfy all the implicit restrictions. If a constraint violation occurs, the trial point is moved halfway in toward the centroid of the vertices already determined (*rule 1*).[1] There is always such a centroid, since we start with a feasible point. We shall certainly ultimately get a satisfactory new vertex point if the region is convex, the retraction process being repeated until a feasible point is found. This process can be repeated until the complete set of k vertices is defined.

If bounds are available on each of the independent variables, this procedure automatically adjusts to the order of magnitude of each variable, and no further scaling is required. When no bounds are present and the method does not require these to be imposed, scaling must be considered so that an

[1] Guin [1968] has given a modification of these rules which permits continued movement even though the region of search may not be convex. Convergence is not assured, however.

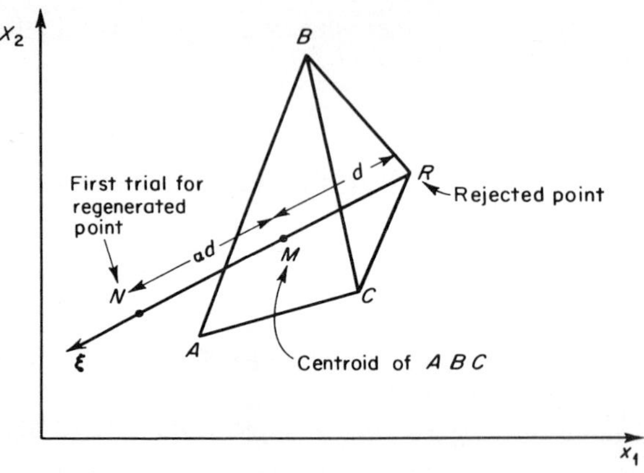

Fig. VIII-35 A "complex."

adequate coverage of the search region is ensured in the initial stages. The use of this variable sized figure avoids the difficulty of constructing a regular simplex which satisfies all the constraints and which is of a reasonable size.

The objective function is evaluated as before at each vertex, and the vertex yielding the poorest value of the objective function is rejected and replaced by a point which is located a distance $\alpha(\alpha \geq 1)$ times as far from the centroid of the remaining points as the distance of this rejected vertex, but in a direction defined by a vector pointing from the rejected point to the centroid. Call this *rule 2*. This procedure is best illustrated by a two-dimensional example, as depicted in Fig. VIII-35. Point R has the worst value and is to be rejected. The centroid of points A, B, and C is designated as M. If the distance between the point R and the centroid M is d, then point R will be replaced by a point N, which is a distance αd from the centroid M, in a direction defined by the line drawn from R through M. The coordinates of N are then given by

$$x_{i,N} = \alpha(x_{i,M} - x_{i,R}) + x_{i,M} = (1 + \alpha)x_{i,M} - \alpha x_{i,R} \qquad (5.2\text{-}2a)$$

where the centroid M is given by

$$x_{i,M} = \frac{1}{k-1}\left(\sum_{j=1}^{k} x_{i,j} - x_{i,R}\right) \qquad (5.2\text{-}2b)$$

This may be compared with Eq. (3.1-11) of the Sequential Simplex method. If this new point N is a feasible point, the function is evaluated there, and the process of vertex rejection and regeneration is repeated.

If, however, the new point has the worst value in the set of vertices in the new figure, its location is moved halfway in toward the previous centroid. Call this *rule 3*. Thus the point N' is

$$x_{i,N'} = \tfrac{1}{2}(x_{i,N} + x_{i,M})$$

If a constraint on the feasible region, $g_k \leq 0$, is at any time violated, the vertex is again moved halfway in toward the centroid. If this constraint is still violated, this retreat is continued until a valid point is attained. Call this *rule 4*. For a convex region, the vertex will ultimately satisfy the restriction, although difficulties might arise in nonconvex regions, especially if the centroid is found outside the feasible region. In addition, if a regenerated point does not satisfy one of the explicit constraints on x_i, that variable is reset at a suitable small distance inside the appropriate boundary to yield a feasible point (*rule* 5).

Progress will continue with repeated rejection and regeneration until the complex is reduced essentially to the centroid. Values of α greater than unity tend to cause a continued enlargement of the complex, and thus compensate for moves halfway toward the centroid. Progress is faster when $\mathbf{x}_j$ is far from the optimum, but the speed of convergence does not seem to be particularly sensitive to α and k, Box recommending that α be about 1.3 and k be about $2n$ (although less are used if n is larger, say, than 5). If insufficient vertices are used, however, such as $k = n + 1$, the complex tends to collapse and flatten along the first constraint located.

While the method ceases when five consecutive function evaluations give equal values of y to within the accuracy of the calculation, this usually means that the technique stops only when the optimum is located exactly and the complex has collapsed into its centroid.

During a search, the complex rolls over and over, normally expanding. When, however, a boundary is reached, the figure contracts and flattens itself against this restriction. It can then roll along the boundary and leave it if the contours change. It can accommodate more than one boundary and can turn corners.

A similar approach has been suggested by Nelder and Mead [1964, 1965].

Example VIII-16: Minimization by the Complex method

Find the minimum of the function

$$y = (x_1 - 0.5)^2 + (x_2 - 1.0)^2$$

subject to the restriction that

$$g = x_1^2 + 2x_2^2 - 4 \leq 0$$

The restrictions on this problem are, then, that

$$-2 \leq x_1 \leq 2$$

$$-\sqrt{2} \leq x_2 \leq \sqrt{2}$$

$$g \leq 0$$

The restrictions on the x_i correspond to the rectangular region surrounding $g \leq 0$ and allow scaling to take place. Let us now choose $k = 2n = 4$ vertices for the first complex, starting the search at the origin. Using random numbers and Eq. (5.2-1), the second vertex might be found to be $\{-1.600, 1.131\}$, as shown in Table VIII-6. The constraint $g \leq 0$ is, however, violated, and applying rule 1, point 3 is found midway between $\mathbf{x}_1$ and $\mathbf{x}_2$. This is a feasible second vertex. Similarly, points $\mathbf{x}_4$ and $\mathbf{x}_5$ are generated to complete the first figure.

The objective function is evaluated at each vertex, and the highest value is rejected. This corresponds to $\mathbf{x}_5$, the new figure 1,3,4,6 being formed once point $\mathbf{x}_6$ is determined, using Eq. (5.2-2). Applying the appropriate rules, the result of a few cycles of search is presented in Table VIII-6.

5.3 Gradient Methods

In this section we shall discuss those search methods which depend upon the gradient of the constraints or the gradient of the objective function or both. The gradient of the constraint, like that of the objective function, is the direction of the greatest rate of change of the constraint. *Hemstitching* is one gradient method (Sec. 5.3.1) which, when any constraint is violated, satisfies these as rapidly as possible, regardless of objective function improvement or deterioration. It therefore relies entirely on the constraint gradients to direct its moves when constraints are encountered. Other, more sophisticated methods attempt to avoid leaving the feasible region, while at the same time seeking an improvement of the objective function. These more complex methods usually require a linearization of the operative constraints (Sec. 5.3.2) and are perhaps best understood by considering the special case of linear inequalities (Sec. 5.3.3) before examining the more difficult problem of nonlinear inequalities (Sec. 5.3.4).

Before proceeding, we should note that proofs of convergence for these methods exist only in the maximization (minimization) of concave (convex) functions in convex regions. In all other circumstances great caution has to be exercised in the interpretation of the results.

5.3.1 A hemstitching approach

The methods of hemstitching proposed by Roberts and Lyvers [1961] and deflected ascent due to Singer [1962], both use the gradients of the violated constraints to return to the feasible region as rapidly as possible. If, for example, the first j constraints were violated, the next move in the search would be given by

$$\mathbf{x}_{p+1} = \mathbf{x}_p + \xi_p s_p \tag{5.3-1}$$

in which the direction $\boldsymbol{\xi}_p$ is defined by

$$m_{i,p} = -\sum_{k=1}^{j} \frac{\partial g_k/\partial x_i}{\sqrt{\sum_{l=1}^{n} \left(\frac{\partial g_k}{\partial x_l}\right)^2}} \tag{5.3-2}$$

the distance of the move being selected in advance. Several steps along this direction, which is perpendicular to the constraints,[1] could be taken before a new direction is again selected. The process is repeated until the feasible region is entered. Of course, these steps will not necessarily improve the objective function, since $y(\mathbf{x})$ plays no role in fixing the direction of movement.

Example VIII-17: Hemstitching

Roberts and Lyvers [1961] consider the problem of maximizing the function

$$y = 5x_1^3 - 3x_1^2x_2 + x_2^2$$

subject to the restrictions

$$g_1 = 1 + \tfrac{2}{3}x_1 - x_2 \leq 0$$

and $$g_2 = (x_1 - 1)^2 + (x_2 - 2)^2 - 4 \leq 0$$

starting at the base point $\mathbf{x}_0$ {2,2.5}. At this position $g_1 = -0.166 < 0$ and $g_2 = -2.75 < 0$, so that $\mathbf{x}_0$ is a feasible experimental location. If the next move (selected by some method which is not relevant in this discussion) reaches the point {1.0,0.5}, then $g_1 = 1.17 > 0$. The first restriction is therefore violated, and we must move back to a feasible region. This move is then defined by the direction ξ_1, where the $m_{i,1}$ are defined by Eqs. (5.3-2) as

$$m_{1,1} = \frac{-\tfrac{2}{3}}{\sqrt{\tfrac{4}{9} + 1}} = -0.555 \qquad \text{and} \qquad m_{2,1} = 0.832$$

This direction is shown in Fig. VIII-36. It will be seen that the second move is perpendicular to the violated constraints, and thus represents a move back to the feasible region without regard to the enhancement of the objective function.

The new base point $\mathbf{x}_2$ in the feasible region can then be found as

$$x_{i,2} = x_{i,1} + m_{i,1}s_1$$

[1] Here a perpendicular to the constraint surface $g_l(\mathbf{x}) = K$ is constructed, where K is the positive constant found by evaluating g_l at the invalid point $\mathbf{x}_p$. This direction is of course different from that determined by a perpendicular to the constraint $g(\mathbf{x}) = 0$, since the gradient of a function is a local property.

TABLE VIII-6 A COMPLEX-METHOD SOLUTION (Example VIII-16)

k	r_1	r_2	*Centroid M*			*Vertex k*				*Calculation notes*		
			$x_{1,M}$	$x_{2,M}$	y_M	$x_{1,k}$	$x_{2,k}$	g	y_k	*Point rejected*	*Rule*	*Resulting figure*
1						0	0	−4	1.2500	1st vertex		1
2	0.1	0.9				−1.600	1.131	1.118		2d vertex		1, 2
3			0	0	1.2500	−0.800	0.566	−2.719	1.8784	2	1	1, 3
4	0.6	0.7				0.400	0.566	−3.199	0.1984	3d vertex		1, 3, 4
5	0.5	0.2				0	−0.848	−2.562	3.6651	4th vertex		1, 3, 4, 5
6			−0.133	0.377	0.7888	−0.306	1.970			5	2	1, 3, 4, 6
7						−0.306	1.414	0.094		6	5	1, 3, 4, 7
8						−0.220	0.896	−2.346	0.5292	7	4	1, 3, 4, 8
9			0.060	0.487	0.4568	1.178	0.384	−2.317	0.8391	3	2	1, 9, 4, 8
10			0.453	0.615	0.1504	1.042	1.415			1	2	10, 9, 4, 8
11						1.042	1.414	1.086		10	5	11, 9, 4, 8
12						0.748	1.015	−1.380	0.0617	11	4	12, 9, 4, 8

13		0.309	0.826	0.0668	−0.821	1.401	0.600		9	2	12, 13, 4, 8
14					−0.256	1.114	−1.452	0.5845	13	4	12, 14, 4, 8
15					0.027	0.970	−2.117	0.2246	14	3	12, 15, 4, 8
16		0.392	0.850	0.0342	1.188	0.790	−1.340	0.5174	8	2	12, 15, 4, 16
17					0.790	0.820	−2.031	0.1165	16	3	12, 15, 4, 17
18		0.646	0.800	0.0613	1.451	0.579	−1.224	1.0816	15	2	12, 18, 4, 17
19					1.049	0.690	−1.947	0.3975	18	3	12, 19, 4, 17
20					0.848	0.745	−2.171	0.1861	19	3	12, 20, 4, 17
21		0.795	0.860	0.1066	1.309	1.242	0.799		4	2	12, 20, 21, 17
22					1.052	1.051	−0.684	0.3073	21	4	12, 20, 22, 17
23					0.924	0.956	−1.318	0.1817	22	3	12, 20, 23, 17
24		0.821	0.930	0.1079	0.786	1.171	−0.640	0.1110	20	2	12, 24, 23, 17
25		0.775	1.002	0.0756	0.581	1.062	−1.407	0.0104	23	2	12, 24, 25, 17
26		0.705	1.083	0.0489	0.595	1.425	0.415		17	2	12, 24, 25, 26
27					0.650	1.254	−0.432	0.0870	26	4	12, 24, 25, 27

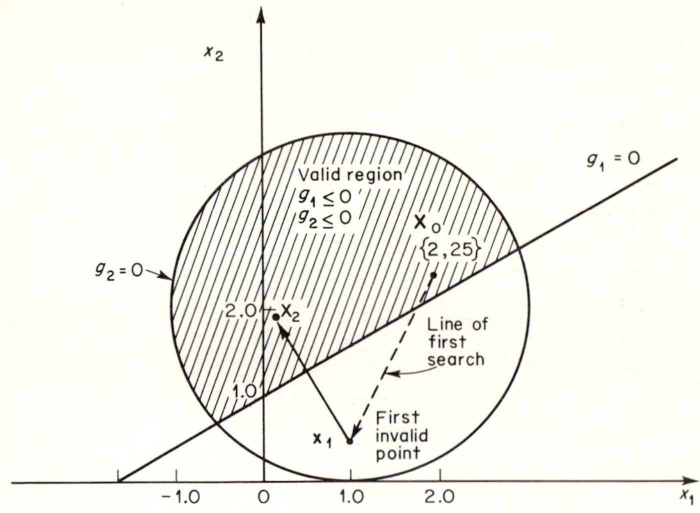

Fig. VIII-36 Hemstitching—return from an invalid region.

so that

$$x_{1,2} = 1.0 - 0.555s_1$$

$$x_{2,2} = 0.5 + 0.832s_1$$

Here the step size s_1 could be selected so that a feasible point would be found. Thus s_1 could be picked so that

$$g_1 = 1 + \tfrac{2}{3}(1.0 - 0.555s_1) - (0.5 + 0.832s_1) < 0$$

Once the search has been moved back to some feasible point, $\mathbf{x}_2$, the optimum can again be sought by conventional methods. The name of the method arises from the zigzag pattern formed along the constraint by the successive use of gradient moves with reference to the objective function and the constraints, respectively.

5.3.2 *Operative constraints*

It is clear that the move back into a feasible region by the hemstitching method will not necessarily result in an improvement of the objective function. It is preferable to use more sophisticated methods of constrained search, in which the objective function, if at all possible, is to be enhanced at every iteration.

Before delving into the details of the various computational methods, it is convenient to distinguish between "operative" and "inoperative" boundaries. After the pth iteration, the computations will begin from a known feasible point, $\mathbf{x}_p$. Associated with this point are m' operative boundaries or constraints, for which the inequalities (5-1) are identically satisfied, or very nearly so. In other words, there are m' inequalities

$$[g_k(\mathbf{x}_p)]^2 \le \delta^2 \qquad \text{with } k = 1, 2, \ldots, m' \tag{5.3-3}$$

Here δ is a small preassigned number. We may say that the point $\mathbf{x}_p$ lies on or near these m' boundaries, and that such boundaries are operative because they restrict any small movement, whereas inoperative or more distant boundaries can restrict only larger moves. We must therefore take greater care in accounting for operative boundaries because of the immediate dangers they pose. Inoperative boundaries are obviously found at the remaining $m - m'$ constraints, but these will not affect the immediate search.

5.3.3 *Linear inequality constraints*

A number of gradient methods have been proposed to solve the general nonlinear problem. Methods include the *cutting plane* technique of Kelley [1960], which has been modified by Wolfe [1961], the method of feasible directions due to Zoutendijk [1960], the gradient projection method of Rosen [1960, 1961], and those of Glass and Cooper [1965] and Hartley [1960]. As yet comparative tests have not been reported on the effectiveness of the various methods, and we shall not recommend one method to the exclusion of others. Since we cannot adequately describe all the available approaches, we shall consider three of them, selected for their apparent effectiveness and their diversity in philosophy. We shall first consider the method of Glass and Cooper, primarily because of its simplicity, and then the methods of Zoutendijk and Rosen.

Method of Glass and Cooper The method of Glass and Cooper [1965] moves, essentially, in the direction of steepest ascent in the search for a maximum, until a boundary is encountered. When all constraints are inoperative, the gradient of the objective function determines the direction of movement as

$$m_i = \frac{\pm y_{x_i}}{d} \qquad \text{for } i = 1, 2, \ldots, n \tag{5.3-4}$$

where

$$d = \sum_{i=1}^{n} |y_{x_i}|$$

the positive sign being selected if a maximum is sought. A move $\mathbf{x}_{j+1} = \mathbf{x}_j + s\boldsymbol{\xi}$ is then made, with a preselected value of s, and if successful, s is increased (by a factor of 2, say) and a second move is made in the same direction. This strategy is continued until an unsuccessful move is made. Such a move can occur in one of two ways: the objective function can be found to be worse, or a constraint can be violated.

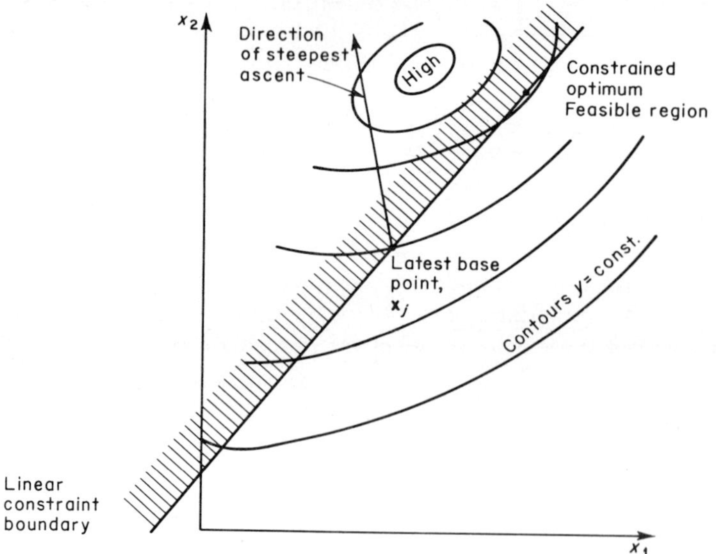

Fig. VIII-37 Steepest ascent leads to invalid region.

If the objective function worsens, the gradient of the objective function is evaluated at the last successful point, a new direction of search is defined, and moves of length s are made along this new direction. If any move is found to be unsuccessful, owing to the violation of one or more constraints, a new base point is established in the feasible region, but near one or more of the constraint boundaries at a predetermined distance δ in Eq. (5.3-3). It is then recommended that new rules be adopted in determining the next moves, since we can no longer expect to follow the path of steepest ascent. The size of s is also reduced. The problem facing us can perhaps be best understood by referring to Fig. VIII-37, in which the direction of steepest ascent leads to the nonfeasible region. We must therefore define an alternative route to the optimum, one which remains in the feasible region. If the components of the gradient evaluated at $\mathbf{x}_j$ are y_{x_i}, then, for a small move in the direction $\boldsymbol{\xi}_j$, the change in the objective function can be approximated as

$$\Delta y = s \sum_{i=1}^{n} y_{x_i} m_{i,j} \tag{5.3-5}$$

This small move must not only enhance the value of y to the greatest extent for a predetermined value of s, but must also take care not to violate a constraint. These constraints can be written

$$\sum_{i=1}^{n} b_{k,i} x_i - d_k \leq 0 \qquad \text{for } k = 1, 2, \ldots, m \tag{5.3-6}$$

in the linear case. If the point $\mathbf{x}_j$ is located so that m' of the m constraints are operative, then for these m' constraints the inequalities of Eq. (5.3-6) must be equalities. In making any move we must take great care with these, since they pose an immediate danger. The direction of movement is therefore restricted so that these constraints will not be violated. Thus, if the new point is to be $\mathbf{x}_{j+1}$, we are required to satisfy

$$\sum_{i=1}^{n} b_{k,i}x_{i,j+1} - d_k \leq 0$$

In addition, $x_{i,j+1} = x_{i,j} + sm_{i,j}$, which then gives

$$\sum_{i=1}^{n} b_{k,i}x_{i,j} - d_k + s\sum_{i=1}^{n} b_{k,i}m_{i,j} \leq 0$$

For the operative constraints at $\mathbf{x}_j$,

$$\sum_{i=1}^{n} b_{k,i}x_{i,j} - d_k \approx 0$$

and we must therefore select the new direction of movement so as to satisfy the operative constraints. The movement will, hopefully, be small enough so that the other constraints will not be violated. All the constraints must therefore be checked before finally accepting the new point. For the operative constraints we require that

$$\sum_{i=1}^{n} b_{k,i}m_{i,j} \leq 0 \tag{5.3-7}$$

In addition, we shall require that the $m_{i,j}$ shall satisfy the linear restriction[1]

$$-1 \leq m_{i,j} \leq 1 \tag{5.3-8}$$

This equation permits us to control the distance of movement by selecting values of s, since the $m_{i,j}$ are forced to remain within prescribed limits.

We now wish to select the $m_{i,j}$ so as to optimize the Δy as defined by Eq. (5.3-5), while satisfying the inequalities (5.3-7) and (5.3-8). This is a linear problem which can be solved using the Simplex method of linear programming, provided we make the following transformation:

$$m_{i,j} = t_{i,j} - 1 \tag{5.3-9}$$

Here the new variables $t_{i,j}$ have the property of being nonnegative as contrasted with the $m_{i,j}$, which can be negative.

[1] It might seem more reasonable to insist that the $m_{i,j}$ be normalized as in the steepest ascent or gradient projection, where we require $\sum_{i=1}^{n} m_{i,j}^2 = 1$. This would mean, however, that the linear restriction of Eq. (5.3-8) would be replaced by a quadratic. The present "distance" s is therefore actually proportional to the distance due to the adopted definition of the m_i.

Substituting Eq. (5.3-9) into (5.3-5), (5.3-7), and (5.3-8), our problem can be redefined as the selection of the n values of $t_{i,j}$ which optimize $\sum_{i=1}^{n} y_{xi}(t_{i,j} - 1)$, subject to the constraints

$$\sum_{i=1}^{n} b_{k,i}t_{i,j} - \sum_{i=1}^{n} b_{k,i} \leq 0 \tag{5.3-10}$$

$$t_{i,j} - 2 \leq 0 \tag{5.3-11a}$$

and

$$t_{i,j} \geq 0 \tag{5.3-11b}$$

for $i = 1, 2, \ldots, n$.

The solution of this linear problem defines a new direction of search which will satisfy the operative constraints. The new point found by moving a distance s along this direction should be tested to see if it both satisfies the constraints and enhances the objective function. If either test fails, the value of s should be reduced and another point nearer $\mathbf{x}_j$ tested. An improved point will be found for some sufficiently small s. If the gradient should vanish at an interior point or if the $t_{i,j}$ all vanish at a boundary, a local optimum has been found.

Example VIII-18: A Glass-Cooper search

Let us seek the minimum of

$$y = -\ln(1 + x_1) - 2\ln(1 + x_2)$$

subject to the restrictions that $x_1 + x_2 - 2 \leq 0$ and $x_i \geq 0$ (Dorn [1963]). This function and the feasible region are shown in Fig. VIII-38.

Stage 1 Taking $\mathbf{x}_0 = \{\frac{1}{2},\frac{1}{2}\}$ as a first feasible base point and letting $s = \frac{1}{2}$, the direction of the first move may be computed from the gradients $\partial y/\partial x_1 = -\frac{2}{3}$ and $\partial y/\partial x_2 = -\frac{4}{3}$, there being no operable constraints. From Eq. (5.3-4), using the negative sign in minimization, we get

$$m_{1,0} = \frac{\frac{2}{3}}{\frac{2}{3} + \frac{4}{3}} = \frac{1}{3} \qquad m_{2,0} = \frac{\frac{4}{3}}{\frac{2}{3} + \frac{4}{3}} = \frac{2}{3}$$

The new point is given by

$$\mathbf{x}_1 = \{x_{1,0} + sm_{1,0}, x_{2,0} + sm_{2,0}\} = \{\tfrac{2}{3},\tfrac{5}{6}\}$$

as shown in Fig. VIII-38. It is then checked that the constraints are satisfied and that the objective function is enhanced. Both tests are satisfied, so that we shall accept point $\mathbf{x}_1$. Continuing this successful move, we shall again move in the same direction from $\mathbf{x}_1$, with s now

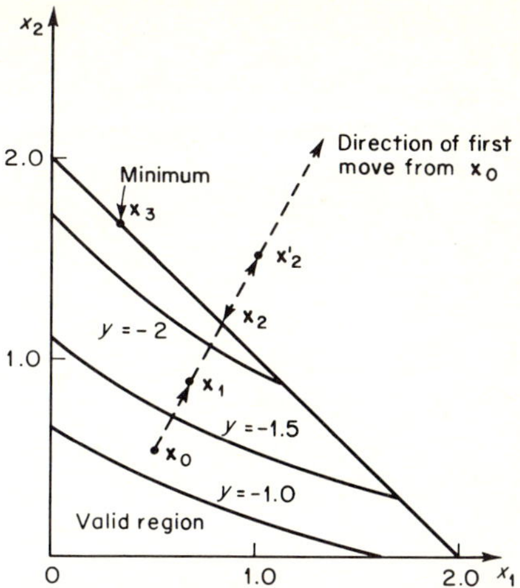

Fig. VIII-38 The Glass and Cooper method; Dorn's example,

$$y = -\ln(1 + x_1) - 2\ln(1 + x_2).$$

doubled to a value of 1, yielding the new point

$$\mathbf{x}_2 = \{x_{1,1} + sm_{1,0}, x_{2,0} + sm_{2,0}\} = \{1, 9/6\}$$

Since this point violates a constraint, we must retreat toward $\mathbf{x}_1$ until the feasible region is found. This requires s to be selected so that

$$x_{1,2} + x_{2,2} - 2 = 0$$

or

$$x_{1,1} + x_{2,1} + s(m_{1,0} + m_{2,0}) = 2$$

Solving, we get $s = 1/2$ and $\mathbf{x}_2 = \{5/6, 7/6\}$, with $y(\mathbf{x}_2) = -2.15$, an improvement over $y(\mathbf{x}_1)$.

Stage 2 We are now on the boundary since $x_{1,2} + x_{2,2} - 2 = 0$, and this constraint is therefore operative. We should thus like to move from this point $\mathbf{x}_2$ so as to avoid violating the constraints. Equation (5.3-7) applied here gives $m_{1,2} + m_{2,2} \leq 0$, while Eq. (5.3-8) requires $-1 \leq m_{i,2} \leq 1$. The $m_{i,2}$ satisfying these inequalities must be selected so that the Δy of Eq. (5.3-5) is as small as possible. At point $\mathbf{x}_2$

$$\Delta y = -s(6/11 m_{1,2} + 12/13 m_{2,2})$$

The solution of this linear problem for a given value of s is found by inspection, so that we need not apply the Simplex method after making the transformations suggested by Eq. (5.3-9). The solution is $m_{1,2} = -1$ and $m_{2,2} = 1$. These components now direct a move along the constraint. We may take a smaller value of $s = 1/2$, say, since the last move was unsuccessful, and move to a new point, $\mathbf{x}_3$, given by

$$\mathbf{x}_3 = \{x_{1,2} + sm_{1,2}, x_{2,2} + sm_{2,2}\} = \{\tfrac{1}{3}, \tfrac{5}{3}\}$$

which is still on the boundary, being linear. In fact, this is the optimal solution, but since we do not know this, we must evaluate $y(\mathbf{x}_3)$ and attempt a new stage of search.

Stage 3 At point $\mathbf{x}_3$

$$\Delta y = s(-\tfrac{3}{4}m_{1,3} - \tfrac{3}{4}m_{2,3})$$

is to be minimized subject to the condition that the operative constraint is not violated. Thus

$$m_{1,3} + m_{2,3} \leq 0$$

and

$$-1 \leq m_{i,3} \leq 1$$

No direction can be found which improves the objective function. We have therefore found a local minimum. In this example it is the global result.

The Zoutendijk Method The method of Glass and Cooper used two different schemes for computing the direction of movement, depending on whether the base point is an interior or boundary point. The method of Zoutendijk [1960], called *feasible directions*, uses, however, only a single computational method for finding the direction of movement at every point, this being the same as that proposed by Glass and Cooper for movements near the boundary. The method of feasible directions also differs from Glass and Cooper in the selection of a distance of movement. In determining the direction of movement, Zoutendijk optimizes the Δy defined by Eq. (5.3-5), with the $m_{i,j}$ subject to restrictions (5.3-7) and (5.3-8). Once the direction of movement is found, a one-dimensional search is conducted along the direction and an optimum found, if one exists. The search stops, however, when constraint boundaries are encountered. We therefore need to determine which is the smaller, the distance measured from the base point to the optimum along the line of search or the distance to the first violated constraint. Since the point $\mathbf{x}_j$ is feasible, we know that

$$\sum_{i=1}^{n} b_{k,i}x_{i,j} - d_k \leq 0 \qquad \text{for } k = 1, 2, \ldots, m$$

In moving from $\mathbf{x}_j$ along the direction of search, we have

$$\sum_{i=1}^{n} b_{k,i}x_{i,j} - d_k + s\sum_{i=1}^{n} b_{k,i}m_{i,j} \leq 0$$

Which of these m constraints are in danger of being violated as s increases? Certainly, only those for which $\sum_{i=1}^{n} b_{k,i}m_{i,j} > 0$ can possibly be violated.

Assume that there are m' ($m' \le m$) such constraints. The greatest allowable distance of movement s_k, without violating a constraint, is found by solving the equation

$$s_k = \frac{d_k - \sum_{i=1}^{n} b_{k,i}x_{i,j}}{\sum_{i=1}^{n} b_{k,i}m_{i,j}} \tag{5.3-12}$$

for the m' inequalities for which

$$\sum_{i=1}^{n} b_{k,i}m_{i,j} > 0$$

The distance of movement is selected as

$$s = \min\,(s^*, s_1, s_2, \ldots, s_{m'}) \tag{5.3-13}$$

Here s^* is the distance to the optimum along the line of search. A new base point is thus located, and the procedure repeated.

Example VIII-19: The Zoutendijk method

Let us repeat Example VIII-18 using the method of feasible directions.

Stage 1 Starting with the feasible base point $\{\frac{1}{2},\frac{1}{2}\}$, we want to find the values of $m_{1,0}$ and $m_{2,0}$ which minimize

$$\Delta y = s\left(m_{1,0}\frac{\partial y}{\partial x_1} + m_{2,0}\frac{\partial y}{\partial x_2}\right) = s(-\tfrac{2}{3}m_{1,0} - \tfrac{4}{3}m_{2,0})$$

subject to the restrictions $-1 \le m_{i,0} \le 1$, there being no operative constraints. The solution to this problem is $m_{1,0} = 1$ and $m_{2,0} = 1$. Thus $\mathbf{x}_1 = \{\frac{1}{2} + s, \frac{1}{2} + s\}$, where s is to be selected applying Eq. (5.3-13). To find s^* we need to optimize

$$y(\mathbf{x}_0 + s\boldsymbol{\xi}_0) = -\ln\,(\tfrac{3}{2} + s) - 2\ln\,(\tfrac{3}{2} + s)$$

Since

$$\frac{dy}{ds} = \frac{-3}{\frac{3}{2} + s}$$

there is no stationary point, and s^* should be as large as is permitted. Let us consider each of the constraints in turn. For the first constraint, $x_1 + x_2 - 2 \le 0$, we see that $\sum_{i=1}^{n} b_{1,i}m_{i,0} > 0$, so that this constraint poses a danger. Indeed, a move from the base point along the direction $\{1,1\}$ intersects this constraint, as shown in Fig. VIII-39. The distance

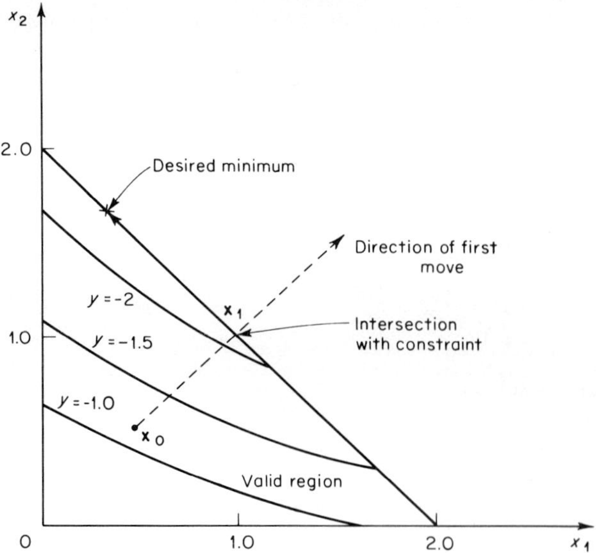

Fig. VIII-39 The Zoutendijk method.

to this point of intersection is found by Eq. (5.3-12) as

$$s_1 = \frac{2 - 1(\frac{1}{2}) - 1(\frac{1}{2})}{1 + 1} = \frac{1}{2}$$

For the second constraint we have $-x_1 \leq 0$, but $\sum_{i=1}^{n} b_{2,i} m_{i,0} < 0$, so that this constraint cannot be violated by the move. Similarly, the third constraint, $-x_2 < 0$, will not be violated by the move. Applying Eq. (5.3-13), we find that $s = \frac{1}{2}$ and the new point $\mathbf{x}_1 = \{1,1\}$.

Stage 2 To generate our second direction of search, we need to solve the linear problem given by Eq. (5.3-8) as

$$\Delta y = s\left(m_{1,1}\frac{\partial y}{\partial x_1} + m_{2,1}\frac{\partial y}{\partial x_2}\right) = s(-\tfrac{1}{2}m_{1,1} - m_{2,1})$$

subject to the constraint $m_{1,1} + m_{2,1} \leq 0$ and $-1 \leq m_{i,1} \leq 1$. The first constraint stems here from Eq. (5.3-7), since we have an operative constraint. The solution to this linear problem is $m_{1,1} = -1$ and $m_{2,1} = 1$. Having specified the directions, we must now find the distance of the second move.

Let us first find s^*. Since $\mathbf{x}_2 = \{1 - s, 1 + s\}$, we must minimize

$$y(\mathbf{x}_2) = -\ln(2 - s) - 2\ln(2 + s)$$

The minimum is found at $s^* = 2/3$. Let us now check if any of the constraints are violated before this optimum is reached. The second constraint, $-x_1 \leq 0$, is the only one for which $\sum_{i=1}^{n} b_{2,i}m_{i,1} > 0$. In this case

$$s_2 = \frac{1}{1} = 1$$

Therefore $s = \min(2/3,1) = 2/3$, and the new point is $\mathbf{x}_2 = \{1/3, 5/3\}$. This is the solution, but one more attempted move is required to establish it.

Stage 3 To find our third direction of search, we must solve the linear problem

$$\Delta y = s\left(m_{1,2}\frac{\partial y}{\partial x_1} + m_{2,2}\frac{\partial y}{\partial x_2}\right) = s(-3/4 m_{1,2} - 3/4 m_{2,2})$$

subject to the constraints $m_{1,2} + m_{2,2} \leq 0$ and $-1 \leq m_{i,2} \leq l$. The solution to this problem is $m_{1,2} = m_{2,2} = 0$, so that, with $\mathbf{x}_2$, we are at a local optimum.

The Rosen Method The methods of Zoutendijk and of Glass and Cooper are similar. Near the boundary each requires the solution of a linear problem to define the new direction of search. The third method to be discussed employs a different strategy, in that the search is initially along the boundaries. Just as for equality constraints, a move is directed by the projection of the gradient of the objective function. This method, called the *gradient projection method*, developed by Rosen [1960, 1961], uses certain rules of computation, starting from a base point $\mathbf{x}_p$, with m' operative constraints, l of which $(l \leq m')$ are to be associated with $\mathbf{x}_p$. These rules are:

1. The components of the gradient, $\partial y/\partial x_i$, are evaluated at the point $\mathbf{x}_p$.
2. Calculate the projection of the gradient on to the intersection of all of the l planes associated with the point $\mathbf{x}_p$. (Initially, let l be equal to m'. In case l is zero, the gradient projection is the gradient itself.) This projection determines a direction in space which is constrained to lie along the l associated planes. This is the direction derived in Sec. 4.3.1 as

$$m_i = \frac{\pm\left(y_{x_i} + \sum_{k=1}^{l} \lambda_k g_{k_{x_i}}\right)}{\sqrt{\sum_{j=1}^{n}\left(y_{x_j} + \sum_{k=1}^{l} \lambda_k g_{k_{x_j}}\right)^2}} \tag{5.3-14}$$

with the λ_k determined by the l equations

$$\sum_{i=1}^{n}\sum_{j=1}^{l} \lambda_j g_{j_{x_i}} g_{k_{x_i}} = -\sum_{i=1}^{n} g_{k_{x_i}} y_{x_i} \tag{5.3-15}$$

3. If at least some of the m_i are different from zero, initiate a search in the direction of the gradient projection until a constraint boundary is encountered. Call this point, $\mathbf{x}_t$, a temporary value.
   a. If $y(\mathbf{x}_t)$ is an improvement when compared with $y(\mathbf{x}_p)$, the iteration is complete and $\mathbf{x}_{p+1}$ is set equal to $\mathbf{x}_t$. Step 1 is started again, with the newly encountered constraint included in the list of associated planes.
   b. If $y(\mathbf{x}_t)$ is not an improvement, find the point along the line connecting $\mathbf{x}_p$ and $\mathbf{x}_t$ which yields the best value of y. This is a one-dimensional search. Once $\mathbf{x}_{p+1}$ is found, we return to step 1.

4. If the projection along the l planes vanishes, there are no components of the gradient in the l planes. This implies that the gradient is perpendicular to all these planes. In other words, we have

$$\frac{\partial y}{\partial x_i} + \sum_{j=1}^{l} \lambda_j \frac{\partial g_j}{\partial x_i} = 0$$

Two courses of action are required in this event (see Wolfe [1963]):
   a. If all $\lambda_j \geq 0$, the point $\mathbf{x}_p$ satisfies all the Kuhn-Tucker conditions (Chap. VII, Sec. 4.3) for a minimum. Similarly, if all $\lambda_j \leq 0$, the point satisfies all the Kuhn-Tucker conditions for a maximum. The point is then accepted as a local optimum.
   b. If the λ_j do not all have the same sign, define a new set of planes to be associated with $\mathbf{x}_p$ by deleting from the set of l some single plane for which $\lambda_j < 0$ (If a minimum is sought) and return to step 2.

Example VIII-20: Gradient projection

Let us repeat Example VIII-18 using the method of gradient projection, starting from the feasible base point $\{\frac{1}{2},\frac{1}{2}\}$.

Stage 1 For this first feasible point $m' = 0$, and therefore the number of constraints associated with this base point must also be zero; that is, $l = 0$. The first move is in the direction of gradient given by

$$m_{1,0} = \frac{-\partial y/\partial x_1}{\sqrt{(\partial y/\partial x_1)^2 + (\partial y/\partial x_2)^2}} = \frac{1}{\sqrt{5}}$$

$$m_{2,0} = \frac{-\partial y/\partial x_2}{\sqrt{(\partial y/\partial x_1)^2 + (\partial y/\partial x_2)^2}} = \frac{2}{\sqrt{5}}$$

The new point $\mathbf{x}_1$ is to be found by moving along this direction until the first boundary is encountered. From Fig. VIII-40, we see that the constraint $x_1 + x_2 - 2 = 0$ is encountered first. This is determined by solving $x_{1,t} + x_{2,t} - 2 = 0$, with $\mathbf{x}_t = \{\frac{1}{2} + s/\sqrt{5}, \frac{1}{2} + 2s/\sqrt{5}\}$. Since $s = \sqrt{5}/3$, the new point on the boundary is $\mathbf{x}_t = \{\frac{5}{6},\frac{7}{6}\}$. A test

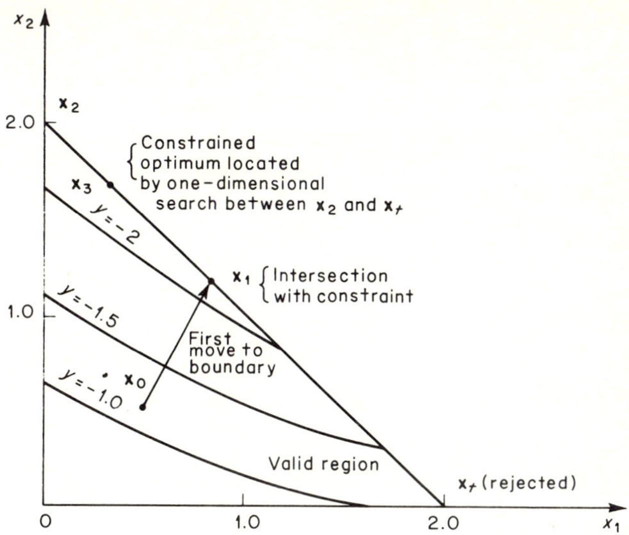

Fig. VIII-40 The Rosen method.

shows that $y(\mathbf{x}_t) = -2.16$, which is less than $y(\mathbf{x}_0) = -1.22$. Therefore we put $\mathbf{x}_1 = \mathbf{x}_t$, and the search begins in a new direction.

Stage 2 At this new point, $m' = 1$, since we are on a boundary (satisfying a single inequality as an identity). We also associate this constraint with the new point, as required by step 3*a*. We now begin again with step 1, as required. The gradient at this new point is $\{-0.546, -0.920\}$, and since $l = 1$, we seek the direction given by the projection of the gradient on to this constraint. This direction is found by solving Eqs. (5.3-14) and (5.3-15). Equation (5.3-15) gives, for $g_1 = x_1 + x_2 - 2$,

$$\lambda_1\left[\left(\frac{\partial g_1}{\partial x_1}\right)^2 + \left(\frac{\partial g_1}{\partial x_2}\right)^2\right] = -\frac{\partial g_1}{\partial x_1}\frac{\partial y}{\partial x_1} - \frac{\partial g_1}{\partial x_2}\frac{\partial y}{\partial x_2}$$

or

$$\lambda_1 = 0.733$$

The $m_{i,1}$ are then found from Eq. 5.3-14 as

$$m_{1,1} = -\frac{-0.546 + 0.733}{\sqrt{(0.733 - 0.546)^2 + (-0.920 + 0.733)^2}} = -1$$

$$m_{2,1} = -\frac{-0.920 + 0.733}{\sqrt{(0.733 - 0.546)^2 + (-0.920 + 0.733)^2}} = 1$$

where the negative sign in Eq. (5.3-14) has been selected because we seek a minimum. The next move is in this direction, until a new constraint is encountered. From Fig. VIII-40 we see that a move from point $\mathbf{x}_1$ will encounter the constraint $x_1 = 0$, and we have a temporary point $\mathbf{x}_t = \{0,2\}$. The objective function at this point is $y(\mathbf{x}_t) = -2.2$, so that we accept it as point $\mathbf{x}_2$.

Stage 3 The new point $\mathbf{x}_2$ now has two constraints associated with it $(x_1 + x_2 - 2 = 0$ and $-x_1 = 0)$, so that $l = 2$, since, according to step 3*a* we incorporate a newly encountered constraint in our list of associated planes. We therefore seek the projection along both planes. Equation (5.3-15) gives

$$\lambda_1\left[\left(\frac{\partial g_1}{\partial x_1}\right)^2 + \left(\frac{\partial g_1}{\partial x_2}\right)^2\right] + \lambda_2\left(\frac{\partial g_2}{\partial x_1}\frac{\partial g_1}{\partial x_1} + \frac{\partial g_2}{\partial x_2}\frac{\partial g_1}{\partial x_2}\right) = -\frac{\partial g_1}{\partial x_1}\frac{\partial y}{\partial x_1} - \frac{\partial g_1}{\partial x_2}\frac{\partial y}{\partial x_2}$$

$$\lambda_1\left(\frac{\partial g_1}{\partial x_1}\frac{\partial g_2}{\partial x_1} + \frac{\partial g_1}{\partial x_2}\frac{\partial g_2}{\partial x_2}\right) + \lambda_2\left[\left(\frac{\partial g_2}{\partial x_1}\right)^2 + \left(\frac{\partial g_2}{\partial x_2}\right)^2\right] = -\frac{\partial g_2}{\partial x_1}\frac{\partial y}{\partial x_1} - \frac{\partial g_2}{\partial x_2}\frac{\partial y}{\partial x_2}$$

Since

$$\frac{\partial y}{\partial x_1} = -1 \qquad \frac{\partial y}{\partial x_2} = -\frac{2}{3} \qquad \frac{\partial g_1}{\partial x_1} = 1 \qquad \frac{\partial g_1}{\partial x_2} = 1 \qquad \frac{\partial g_2}{\partial x_1} = -1 \qquad \frac{\partial g_2}{\partial x_2} = 0$$

we find that $\lambda_1 = \frac{2}{3}$ and $\lambda_2 = -\frac{1}{3}$.

From Eq. (5.3-14), $m_{1,2} = m_{2,2} = 0$. There is therefore no projection. Referring to step 4, we see that this is not a solution, since, for a minimum, we require both Lagrangian multipliers to be nonnegative. Therefore we must remove one of the constraints from their association. Since $\lambda_2 < 0$, we relax $-x_1 = 0$ and only require $x_1 + x_2 - 2 = 0$. Again, solving Eq. (5.3-15) with $l = 1$, we find $\lambda_1 = \frac{5}{6}$; $m_{1,2} = 1$, $m_{2,2} = -1$. Following this direction we find the intersection with the constraint $-x_2 = 0$, and we fix a temporary point there. However, a test of this point $\{2,0\}$ gives $y = -1.1$, which is not an improvement over $\mathbf{x}_2 = \{0,2\}$. Therefore we seek the minimum value along the constraint $x_1 + x_2 - 2 = 0$. Since the directions of move are $m_{1,2} = 1$ and $m_{2,2} = -1$, we have

$$\mathbf{x} = \{x_{1,2} + s, x_{2,2} - s\} = \{s, 2 - s\}$$

or

$$y = -\ln s - 2\ln(3 - s)$$

The minimum with respect to s is given when $s = \frac{1}{3}$. Therefore we establish point $\mathbf{x}_3$ at $\{\frac{1}{3},\frac{5}{3}\}$.

Stage 4 At point $\mathbf{x}_3$ we have one constraint associated with the point $(x_1 + x_2 - 2 = 0)$, so that for $l = 1$, we find from Eq. (5.3-14)

$$\lambda_1 = \tfrac{3}{4} \qquad \text{and} \qquad m_{1,3} = m_{2,3} = 0$$

This is therefore a possible local minimum, since the Lagrangian multiplier is positive and the constrained derivatives vanish.

5.3.4 *Nonlinear inequality constraints*

Most optimization problems of practical importance include one or more inequality constraints, some of which are usually nonlinear. The inventors of the three methods detailed in the preceding section have all extended their respective methods to include nonlinear constraints. The usual procedure in this generalization is to make a linear approximation of the constraints as opposed to altering the strategy of determining a new move. Thus, for example, if $g_l \leq 0$ is an operative constraint, the move will be planned based on the linear approximation

$$\sum_{i=1}^{n} \frac{\partial g_l}{\partial x_i} m_i \leq 0 \tag{5.3-16}$$

rather than on the constraint itself. If the operative constraints do not possess large curvature in the vicinity of the base point, this procedure of linearization will work, but if the constraints form curving boundaries, difficulties can possibly arise, as shown in Fig. VIII-41. Here the base point $\mathbf{x}_0$ is located on a constraint boundary. Both the linearized version of the

Fig. VIII-41 Failure of linearization near a curving boundary.

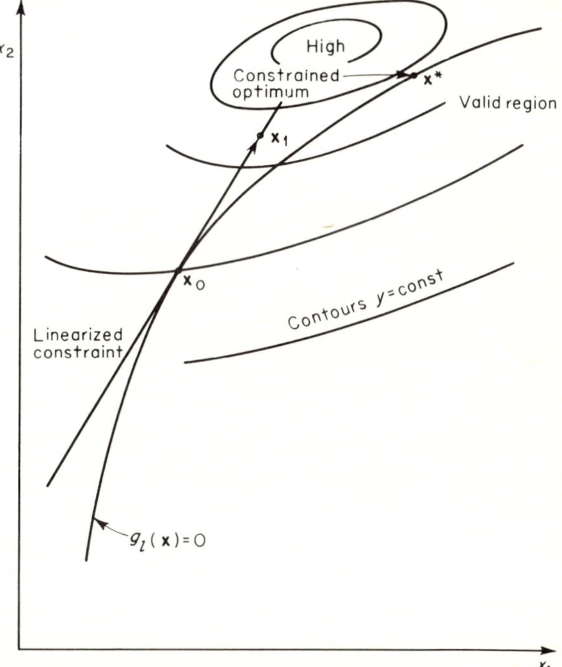

constraint, shown by the straight line, and the true constraint boundary are shown passing through $\mathbf{x}_0$. It is assumed that the desired constrained optimum is at the point $\mathbf{x}^*$. Any of the strategies of Sec. 5.3.3 used in conjunction with the linearized constraints will produce a move directed along the linearized boundary to point $\mathbf{x}_1$, say, where $\mathbf{x}_1$ is, however, invalid. The difficulty behind this approach is, obviously, that a move which satisfies a linearized form of the constraints does not necessarily result in a move which satisfies the nonlinear inequalities.

All these methods try, in general, to end all moves within the feasible region while still obtaining some enhancement of the objective function. Two alternative philosophies might be considered to ensure that a feasible point is attained. On the one hand, we can modify the direction of the move, as in the Glass and Cooper method, so as to stop at a feasible point. Alternatively, we can return to the feasible region after a nonfeasible point is found, the gradient projection method being an example of this strategy. Because of its simplicity and apparent effectiveness, we shall detail the method of Glass and Cooper here.

Suppose that a move from a point on or near a constraint boundary leads to a nonfeasible point because the inequality satisfied is in the form of inequality (5.3-16) rather than the real constraint $g_l \leq 0$. Glass and Cooper then try to force the search into the feasible region by modifying the linear inequality (5.3-16) as follows:

$$\sum_{i=1}^{n} \frac{\partial g_l}{\partial x_i} m_i \leq -T_l \tag{5.3-17}$$

where T_l is a positive constant. A linear programming problem is again solved to determine the direction of the move using inequality (5.3-17) rather than (5.3-16). These new m_i will give a direction pointing more to the feasible region than before. If T_l is sufficiently large, the new direction will yield a point in the feasible region.

Two points should be considered with respect to the choice of T_l. First, the T_l should be as small as possible and yet serve to force the search back into the feasible region so that the search direction can be permitted sufficient freedom to point to an enhanced value of the objective function. Second, we can select a value of T_l so large as to make it impossible to find a set of m_i which will satisfy inequality (5.3-17). Certainly, since $-1 \leq m_i \leq 1$, the smallest value of the left-hand side of the inequality is found by taking m_i to be of opposite sign to $\partial g_l/\partial x_i$ and having an absolute value of unity. The left-hand side can thus be no smaller than $-\sum_{i=1}^{n} |\partial g_l/\partial x_i|$. Therefore any T_l exceeding $\sum_{i=1}^{n} |\partial g_l/\partial x_i|$ will result in the linear programming problem not having a solution. In fact, it is not desirable to have the value of T_l exceed that which would be required to force the direction of search to be perpendicular to the constraints. Since the gradients of the constraints represent a direction perpendicular to the constraints, the m_i also give a perpendicular direction if

$$m_i = -D \frac{\partial g_l}{\partial x_i} \qquad \text{for } i = 1, 2, \ldots, n \tag{5.3-18}$$

where D is a constant of proportionality. If the m_i are perpendicular to the constraint, then clearly, the maximum desirable value of T_l is given by

$$(T_l)_{\max} = D \sum_{i=1}^{n} \left(\frac{\partial g_l}{\partial x_i}\right)^2 \tag{5.3-19}$$

Since the m_i can be no larger than unity, the value of D will be given by

$$D = \frac{1}{\max\limits_j \left|\dfrac{\partial g_l}{\partial x_j}\right|} \tag{5.3-20}$$

at least one of the m_i having the maximum value of unity. The maximum desirable T_l is then given by

$$T_l = \frac{1}{\max\limits_j \left|\dfrac{\partial g_l}{\partial x_j}\right|} \sum_{i=1}^{n} \left(\frac{\partial g_l}{\partial x_i}\right)^2 \tag{5.3-21}$$

It should be clearly understood that solutions with larger values of T_l are possible, but the present value is all that is required to force the direction of the move to be perpendicular to the constraint.

In determining the direction of a new move, we should not want to use $(T_l)_{\max}$ in inequality (5.3-17) on the first violation of a constraint. On discovering that a move leads to a nonfeasible point, we should return to the original point and determine the best search direction, imposing inequality (5.3-17), with $T_l = (1/N)(T_l)_{\max}$, where N is some arbitrary integer in place of inequality (5.3-16). The effect of the factor $1/N$ is to impose only a fraction of the maximum in the first trial. If the new point is again nonfeasible, T_l is increased by decreasing N by unity and recomputing the fraction $1/N$ of $(T_l)_{\max}$. This procedure is continued until a feasible move is made.

If several constraints are violated by a particular move, each is modified as in inequality (5.3-17), with a separate value of T_l associated with each violation. The T_l are to be increased until the individual constraints are satisfied by the trial move, at which point the T_l are left unchanged. No move is made unless to a feasible point.

Example VIII-21: The move from a constraint

Let us assume, in the optimization of the objective function

$$y = x_1 + 10x_2$$

subject to the inequality constraints $x_i \geq 0$ and

$$g = x_1^2 + x_2^2 - 1 \leq 0$$

that the base point is at $\{1,0\}$. Then the Glass and Cooper method requires the new direction of search to be selected, so that

$$\Delta y = \sum_{i=1}^{n} m_i \frac{\partial y}{\partial x_i} = m_1 \frac{\partial y}{\partial x_1} + m_2 \frac{\partial y}{\partial x_2} = m_1 + 10m_2$$

is to be maximized subject to inequality (5.3-16); thus

$$2m_1 \leq 0 \qquad \text{and} \qquad -1 \leq m_i \leq 1$$

The solution to this linear programming problem is $m_1 = 0$ and $m_2 = +1$. Assume that our step size s is $\frac{1}{2}$, so that the new point is $\{1 + 0 \times \frac{1}{2}, 0 + 1 \times \frac{1}{2}\} = \{1,\frac{1}{2}\}$, a nonfeasible point. We must modify the inequality arising from linearizing the constraints. First, let us determine $(T)_{\max}$. Since $\partial g/\partial x_1 = 2$ and $\partial g/\partial x_2 = 0$ at $\{1,0\}$, then $\max_j |\partial g/\partial x_j| = 2$, and from Eq. (5.3-21) we see that $(T)_{\max} = \frac{1}{2}(4) = 2$. Now let us arbitrarily choose $N = 5$.

The new linear programming problem is then to find the maximum of

$$\Delta y = 10m_2 + m_1$$

subject to
$$2m_1 \leq \frac{-(T)_{\max}}{N} = -\frac{2}{5}$$

together with $-1 \leq m_i \leq 1$. The solution to this problem is $m_1 = -\frac{1}{5}$ and $m_2 = 1$. This direction is orientated more toward the feasible region than to the previous search direction. The new point is $\{1 - \frac{1}{5} \times \frac{1}{2}, 0 + \frac{1}{2} \times 1\} = \{0.9,0.5\}$.

This new point is also in the nonfeasible region, so that we must again solve the linear programming problem, this time with $N = 4$. Again maximize

$$\Delta y = 10m_2 + m_1$$

subject to
$$2m_1 \leq -\tfrac{1}{2} \qquad \text{and} \qquad -1 \leq m_i \leq 1$$

The solution is $m_1 = -\frac{1}{4}$ and $m_2 = 1$. The new point is $\{0.875,0.5\}$, which is again nonfeasible.

Setting $N = 3$ and continuing, we find the next point as $\{0.833,0.5\}$. This point is feasible, and our search may continue.

In addition to the references cited above, Hanson [1963] and Klingman and Himmelblau [1964] have presented an empirical search routine in which the direction of search is chosen, after a boundary is violated, as the weighted vector sum of the boundary gradient and the direction of steepest ascent.

5.3.5 *Some general remarks regarding gradient methods for search with inequality constraints*

All the gradient methods which aim at enhancing the objective function at every step while confining all movements to the feasible region operate using the linearized equations, but before any tentative move is accepted, the true objective function and all the constraints are tested to ensure that an improved feasible point has indeed been found. If not, some corrective action is immediately taken. Although we cannot recommend any method as being generally better than any of the others because a definitive series of testing has not been reported, two points are clear. All these methods will be much more effective when used with linear constraints than with nonlinear ones, and they all essentially reduce to the method of steepest ascent when there are no operative boundary conditions present. This latter fact is somewhat disturbing, since we have already seen that unrestricted search methods which follow the gradient tend to oscillate in many situations (Sec. 3.2.1), and therefore these methods would all suffer from the same difficulty when an interior point is being sought. In view of these criticisms, better search methods for inequality constraints are certain to be developed in the future.

5.4 Modified Objective Functions

The numerical search for constrained optima is often facilitated by a modification of the objective function and constraints. The most common method combines the objective function and constraints in a single modified function by the use of penalty factors (Sec. 5.4.1). In a sense, the penalty function method alters both the objective function and the constraints. There are also techniques which modify the constraints by making a change of variable, and these are discussed in Sec. 5.4.2.

5.4.1 *Penalty functions*

Penalty functions offer a reasonable method of converting a restricted problem into one which is unrestricted by including the inequality constraints in a modified objective function. There are two basic approaches.

First, we may form, after Kelley [1962], the modified objective function

$$F = y + \sum_{k=1}^{m} P_k g_k^2 H(g_k)$$

where $H(g_k)$ is the Heavyside unit step function defined so that

$$H(g_k) = \begin{cases} 1 & \text{for } g_k \geq 0 \\ 0 & \text{for } g_k < 0 \end{cases}$$

and the P_k are large penalty constants which are positive if seeking a minimum and negative for maximization. In this form, we may treat F as an unrestricted function and may then use any appropriate search routine.

Example VIII-22: Minimization using penalty functions

Find the minimum of the function

$$y = x_1^2 - 2x_1x_2 + x_2^4$$

subject to the constraint

$$g = -x_1^2 - x_2^2 + 4 \leq 0$$

The penalized objective function has the form

$$F = x_1^2 - 2x_1x_2 + x_2^4 + 100(g)^2 H(g)$$

where we have selected P to be 100. It is essential to select P large enough so that the search is forced to remain in or very near a feasible region. If the search terminates at a nonfeasible point, P should be increased and the search continued. The example shown here is of interest since the optimum lies at a boundary.

To be explicit, we must now select an unconstrained search technique and seek the minimum of the function F. As an example, let us use an univariant search technique (Sec. 3.1.1), starting from the base point {2.000,0.909}. At this point $F = 1.05$, and we seek to improve this value by searching in the x_1 direction. We can use any one-dimensional search scheme which seems desirable.[1]

Here we use a one-dimensional direct search, first establishing limits and then making a dichotomous search to narrow the interval of uncertainty. To an accuracy of ± 0.05, $x_1 = 1.78$ is found to be the optimum, with $F = 0.6172$. It should be noted that this point,

[1] It should be noted that a gradient search is possible because, although dH/dx and $(dg/dx)H$ are both discontinuous at the system boundary, $g^2(dH/dx)$ and $gH(dg/dx)$ are continuous.

If the penalty function had been

$$F' = y + \sum_k P_k g_k H(g)$$

the gradients of the objective function would not have been continuous at the boundary of the feasible region, and use of an unrestricted gradient technique would be extremely hazardous.

{1.78,0.909}, is somewhat outside the feasible region. It could be forced nearer to the acceptable region by increasing P. We shall, however, continue our search from this point and concern ourselves with the feasibility of a point only after no further improvement is possible. Thus, holding $x_1 = 1.78$, we search in the x_2 direction to find the minimum at $x_2 = 1.0$. The next search should again be in the x_1 direction; however, we shall not continue. It is important only to note that the search has been of the unrestricted type. At the end of the search we should determine if the optimum found is feasible; if it is not, P should be increased and the search continued.

An alternative penalty function formulation which introduces a natural optimum within the feasible region has been proposed by Carroll [1961]. A sequence of the resulting surfaces are then optimized, where these surfaces have the property that they tend to the true value of the objective function. This approach has been modified and formalized by Fiacco and McCormick[1] [1964a] for minimization of convex functions, and in a more general form by Strong [1965]. A search of the modified objective function

$$F = y - P \sum_{k=1}^{m} \frac{1}{g_k} \tag{5.4-1}$$

is carried out, with P a positive penalty factor in minimization and a negative factor for maximization. A typical application has been reported by Brander et al. [1964]. Requiring a feasible starting point, the initial search is carried out with a large value of P, P_1 say, using an appropriate unconstrained search method. Those of Powell (page 393) and Davidon [as extended by Fletcher and Powell (page 426)] are powerful techniques for use here. The resulting optimum $\mathbf{x}^{(1)}$ will of course be a function of the magnitude of the penalty term P, and it cannot lie at a point where any of the g_k vanish, since g_k^{-1} would become infinite. Thus, as any constraint is approached, the penalty term forces F toward ∞ in minimization. An artificial optimum is thus introduced within the feasible region, F approximating to y away from the constraint. We now have a feasible point, $\mathbf{x}^{(1)} = \mathbf{f}(P_1)$, which cannot, however, be the optimum of the original problem if any of the constraints have to be satisfied as equalities. A better approximation to the true optimum can now be found by the use of a smaller value of P, P_2 say, in the next search stage. The new function F can then be optimized starting

[1] Carroll, in his "created response surface technique," suggested using

$$F = y - P \sum_{k=1}^{m} \frac{W_k}{g_k}$$

where the W_k are weighting factors, but these seem unnecessary. Fiacco and McCormick [1964b] discuss an appropriate computational algorithm.

from $\mathbf{x}^{(1)}$ to get a new approximation, $\mathbf{x}^{(2)} = \mathbf{f}(P_2)$. This solution will be allowed to approach the constraints more closely if the optimum does in fact lie at a boundary. Thus this procedure, using a large value of P_j initially to delineate a region of search, will gradually reduce the region as P_j is progressively reduced in size. As these penalty constants in each stage become smaller, forming a monotonic-decreasing sequence, $P_1 > P_2 > \cdots$, which approaches zero, the corresponding optimum for each stage, $\mathbf{x}^{(j)} = \mathbf{f}(P_j)$, will form a sequence which tends to the desired optimum. It has been suggested that the search should stop when 20 iterations fail to produce a change in the objective function value. It will be seen that the method essentially replaces a constrained optimization by a sequence of unrestricted optimizations. Since the search is restricted to feasible points only, and the function is modified rapidly as it approaches the boundaries, with the original function only slightly altered or unaltered within the valid region away from the boundaries, it introduces a sharp valley or ridge just inside the constraints, which forces the search away from the boundaries and avoids function evaluation at invalid locations.

If the initial choice of P_1 is too small, the first-stage optimization will lead to the original function optimum, which might be difficult to find and would require more calculations than needed with larger values of P_1 and several more stages with $P_2, P_3, \ldots$. A small value of P_1 gives rise to a highly asymmetric surface, and if a quadratically convergent unrestricted optimization method is used, the optimum will not be located rapidly. On the other hand, if large values of P_1 are chosen, the optima from the first few stages of calculation will depend almost entirely upon the constraints, and all lie near the center of the valid region. Although suggestions have been put forward, no entirely satisfactory procedure exists for the selection of P_1.†

It has been found in practice that the rate of reduction of P_j to P_{j+1} is not a critical factor in determining the amount of work to be done to locate the final optimum, since in general, for large rates of decrease of P_j, fewer surfaces are generated with fewer optima, but each is more difficult to find compared with the situation for slower rates with more optima, but each located more rapidly.

An alternative form (for minimization) for F,

$$F = y + P_j \sum_{k=1}^{m} \frac{1}{g_k^2} \tag{5.4-2}$$

† Fiacco and McCormick [1964b] have suggested several possibilities. One is the following:

$$P_1 = \frac{\sum_{l=1}^{m} \sum_{i=1}^{n} y_{x_i}(g_{l_{x_i}}/g_l^2)}{\sum_{l=1}^{m} \sum_{k=1}^{m} \sum_{i=1}^{n} (g_{l_{x_i}} g_{k_{x_i}}/g_l^2 g_k^2)}$$

in which all functions are evaluated at the initial feasible point.

has been suggested by Box et al. [1969] as being more likely to lead to the global result for problems in which two or more local optima exist.

The above use of penalty factors cannot accommodate strict equalities, but if we wish to optimize a function subject to $g_k = 0$ for $i = 1, 2, \ldots, l$ and to $g_i \leq 0$ for $i = l + 1, l + 2, \ldots, m$, we could form the modified function

$$F = y - P_j \sum_{i=l+1}^{m} \left(\frac{1}{g_i}\right) + \frac{1}{P_j^{1/2}} \sum_{i=1}^{l} (g_i)^2 \tag{5.4-3}$$

after Sec. 4.4.2 and Fiacco and McCormick [1964a, 1964b, 1966] and McCormick, Mylander and Fiacco [1965].

Example VIII-23: The Carroll-Fiacco penalty function

Maximize the function x subject to the constraints

$$g_1 = -x \leq 0$$

$$g_2 = x - 1 \leq 0$$

(Carroll [1961]).

Rather than displaying a step-by-step calculation of the search, as has been done in preceding examples, let us consider the effect of the Carroll-Fiacco penalty function by noting the effect of the parameter P as shown in Fig. VIII-42. The penalized objective function is drastically reduced as the independent variable nears the boundary. Of course, one sees

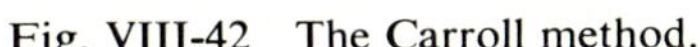

Fig. VIII-42 The Carroll method.

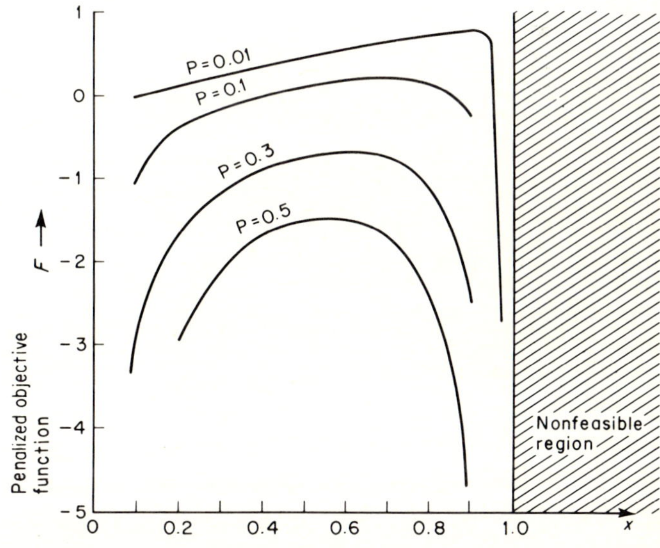

clearly that nonfeasible points cannot be admitted for comparison, because the large negative contributions to the objective function can suddenly become large enhancing positive contributions and confound the search.

Note that as P is decreased, the true optimum is more nearly achieved. It is this property which one hopes to use as a means of avoiding constraints and yet approaching the optimum.

5.4.2 *Other modifications*

Several other approaches can be made in dealing with constrained problems.

Removal of Constraints Since unrestricted problems are easier to handle than constrained problems, it is advantageous to modify the system to include as many restrictions as possible in the objective function (Box [1966]).

One approach is to use transformations such as

$$x_i = |y_i| \qquad x_i = y_i^2 \qquad x_i = e^{y_i}$$

all of which keep $x_i \geq 0$, or

$$x_i = \sin^2 y_i$$

$$x_i = \frac{e^{y_i}}{e^{y_i} + e^{-y_i}}$$

which are equivalent to the restriction that $0 \leq x_i \leq 1$. If the restriction is of the form $x_{iL} \leq x_i \leq x_i^L$, then the transformation

$$x_i = x_{iL} + (x_i^L - x_{iL}) \sin^2 y_i$$

will allow an unrestricted search in y_i to be made, all the x_i being restricted automatically to feasible values. Again, a restriction of the type $0 \leq x_1 \leq x_2$ can be handled by transforming the variables to

$$x_1 = y_1^2 \qquad x_2 = y_1^2 + y_2^2$$

so that an unrestricted optimization can take place in y_i space.

Example VIII-24: Restriction removal

Let us consider the post office parcel problem of Rosenbrock [1960], as discussed by Box [1966]. Here we must maximize

$$y = x_1 x_2 x_3 \tag{5.4-4}$$

subject to $x_i \geq 0$ and

$$0 \leq x_i \leq 42 \qquad i = 1, 2, 3 \tag{5.4-5}$$

$$0 \leq x_1 + 2x_2 + 2x_3 \leq 72 \tag{5.4-6}$$

Equation (5.4-6) implies that x_2 and x_3 are less than 36, so that Eq. (5.4-5) is redundant when $i = 2$ or 3. It also implies that $x_1 \leq 72$ which is weaker then $x_1 \leq 42$, so that the latter will be retained. Since the $x_i \geq 0$, we do not need to write $0 \leq x_1 + 2x_2 + 2x_3$. The problem restrictions are now

$$0 \leq x_1 \leq 42 \qquad x_2 \geq 0 \qquad x_3 \geq 0 \qquad x_1 + 2x_2 + 2x_3 \leq 72$$

The latter is equivalent to

$$x_2 \leq 36 - 0.5x_1 \qquad \text{for given value of } x_1$$

$$x_3 \leq 36 - 0.5x_1 - 0.5x_2 \qquad \text{for given values of } x_1 \text{ and } x_2$$

so that the restrictions become

$$0 \leq x_1 \leq 42$$

$$0 \leq x_2 \leq 36 - 0.5x_1$$

$$0 \leq x_3 \leq 36 - x_2 - 0.5x_1$$

If we define $y_1 = x_1/42$, $y_2 = x_2/(36 - 0.5x_1)$, $y_3 = x_3/(36 - 0.5x_1 - x_2)$, then $x_1 = 42y_1$, $x_2 = (36 - 21y_1)y_2$, $x_3 = (36 - 21y_1)(1 - y_2)y_3$, and $0 \leq y_i \leq 1$ for all i. The latter allows us to write $y_i = \sin^2 z_i$, with no restriction on the z_i.

The original problem, Eq. (5.4-4), can now be defined in terms of z and treated by an unrestricted optimization technique.

Slack variables As we have seen in Chap. VII, Sec. 4.1.2, slack variables may be added to change any inequality restrictions to equalities. This implies, however, that the number of system variables will increase, which expands the problem, a trend which should be opposed where possible. In addition, the search techniques available for searching in the face of equality constraints appear to be less effective than those which can be used with inequality constraints.

5.5 Comment upon Constrained Optimization

Since constraints are often of vital importance, the search methods originally devised for unrestricted systems may not work effectively under constraint merely by adding some penalty or alternative routine when constraints are violated. Thus, for example, in direct search methods, a penalty term might be added outside the search region boundaries whenever a constraint is violated, creating deep valleys or ridges, so that invalid points appear unattractive. Such deep valleys or ridges present at all the boundaries create problems in moving along a boundary, and only a slow zigzag progress along the feature will be possible. This is also true with gradient methods.

Of the direct search methods, Rosenbrock's technique has been used widely to solve constrained problems. It has found extensive application, mainly due to its relatively early publication, but other methods (especially the Complex method and the penalty function method of Carroll) are often preferable. The powerful conjugate gradient methods are not suitable in this context, since the solution might lie on a boundary, while the basis for their effectiveness is the existence of an optimum at a stationary point, so that the Taylor expansion reduces to a quadratic form.

When the problem is unmodified and still subject to constraints, the best method available appears to be the Complex routine of Box. The gradient projection methods are also effective, but their progress depends upon successive approximations by linearizing the system, and this can be slow if the constraints are nonlinear.

When the problem can be modified so as to include the constraints within a modified objective function, using, for example, penalty functions, the more efficient unrestricted methods can be brought into action. Carroll's method of adding penalties is now in fairly widespread use and, in conjunction with some unrestricted method such as Powell's or Davidon's, rapidly locates the approximate location of the optimum. It is not, however, particularly precise when this lies in a sharp corner. In this situation the Complex technique can be used to refine the result from Carroll's method, and the combination is better than either Rosenbrock's or Carroll's methods when these optima correspond to several variables lying along constraints (Box et al. [1969]). This feature of the Complex method is valuable, since it is the only one of the three methods which does not modify the objective function near the boundaries. Away from the boundaries the Complex method is slower than Carroll's when the latter is used with a quadratically convergent method. Since unrestricted methods of optimization are generally better, the possibility of replacing a constrained optimization by an equivalent problem without constraints should always be considered.

One important feature of constrained optimization is the difficulty of showing that a local optimum is in fact a global value. In certain cases, such as in the minimization of a convex function in a convex region, this correspondence will be obvious. In general, however, our policy has been to start from different base points, and if all the searches lead to the same solution, it is likely that this is the global optimum. This procedure can be used freely in unconstrained problems. With restrictions, however, it is not so easy to obtain valid alternative starting points which satisfy all the constraints and are significantly different. The method of using random numbers to generate the first figure in the Complex method seems to be a reasonable feature in striving for a global value.

6 DISCRETE-VALUED VARIABLES

When the objective function exists only at certain discrete values of the independent variables or is dependent on a mixture of continuous and discrete-valued variables, the search is then limited to the nodes of a lattice or grid which may be imagined to be constructed in an n-dimensional space. Of course, not all the nodes need be filled with accessible states. For example, a certain type of distillation column might be supplied as standard units with column diameter D available in multiples of 6 in. If the smaller columns ($D < 36$ in.) have 12-, 14-, 15-, and 16-in. plate spacings, while the larger columns have possible plate spacings of 12, 14, 16, and 18 in., then the accessible nodes are as shown in Fig. VIII-43. If the cost of making a particular separation in one of these columns can be shown by contours added to Fig. VIII-43, the problem of optimization might then be to find the particular arrangement giving the least cost. This class of problem, restricted by discrete-valued variables, has not received much attention, although it arises frequently in plant design.

6.1 Stranding of the Search

One possible method of solution might seem to be the univariant search technique (Sec. 3.1.1). We might start at point f of Fig. VIII-43 and, using

Fig. VIII-43 Accessible nodes in a discrete-valued function.

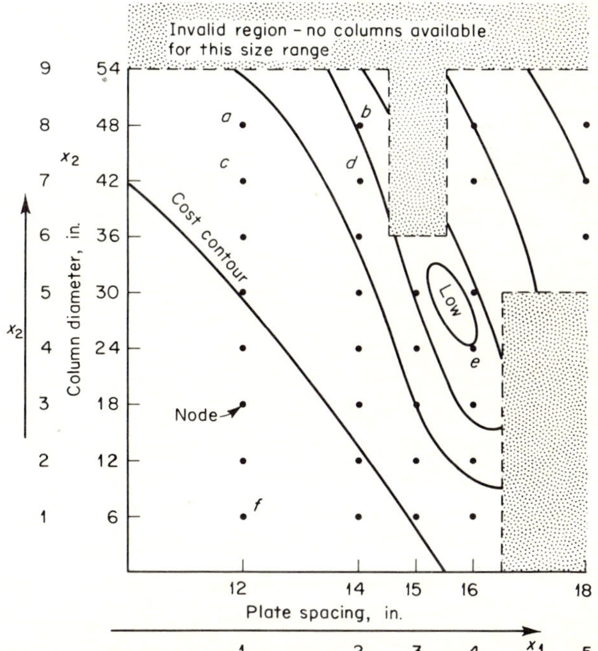

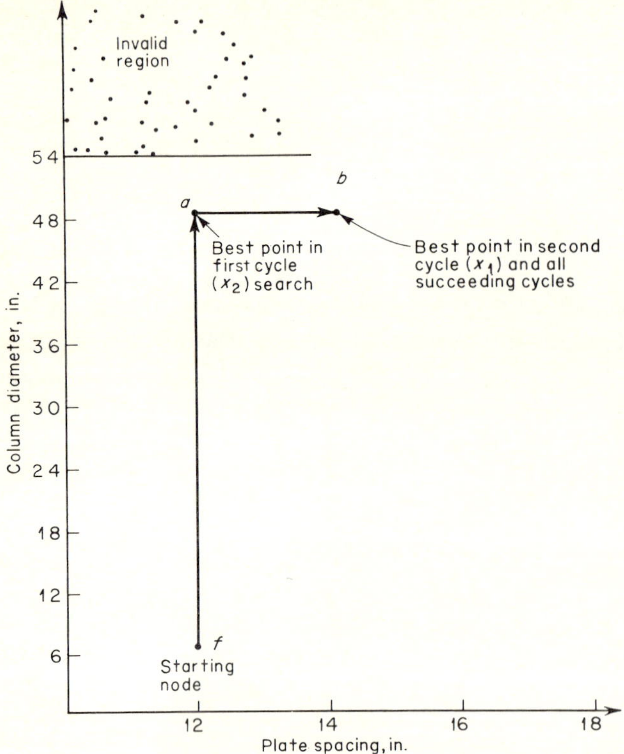

Fig. VIII-44 The location of a local optimum by univariant search (detail from Fig. VIII-43).

fixed step sizes to correspond to node distances, proceed in the x_2 direction to point a, as shown in Fig. VIII-44. The next cycle in x_1 would then reach point b, which is clearly better, but the search now stops. We have reached a local optimum. Starting from the same original base point, but moving first in the x_1 and then in the x_2 direction, we should not reach the same position, but attain the global optimum, point e. The univariant method would thus seem to be as satisfactory for use with discrete-valued variables as it is for continuous variables. However, when the basic or accelerated one-at-a-time methods are used in which some or all the variables take on only discrete values, the problems of becoming stranded on a ridge may be aggravated, since the step length cannot be reduced below a given value. Thus, in Fig. VIII-45, search about point $\{x_{1,1}, x_{2,1}\}$ shows no improvement, and the step size cannot be altered. If some of the variables are continuous, perturbations in their values might, however, allow the search to proceed along the crest of the ridge.

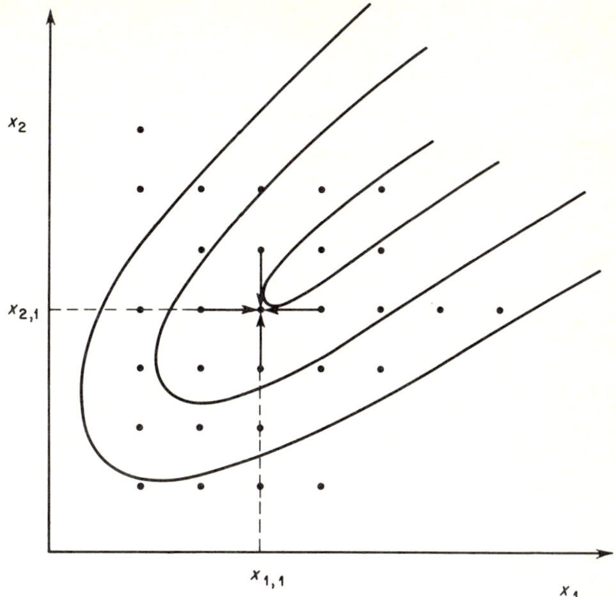

Fig. VIII-45 Discrete-valued variables and stranding of ridge search.

One approach might be that of treating the variables as continuous and, once the optimum has been determined, of finding the nearest valid set of discrete-valued variables. However, this is not always practicable, since the objective function may not be well behaved. Such an approach is essentially used in linear systems (Sec. 6.3). A more suitable technique is that presented below.

6.2 A Modification of the Complex Method

The problem of optimization when discrete-valued variables are found might be tackled by a modification of the Complex search method (Sec. 5.2.1). A figure of $2n$ feasible vertices would be selected at feasible nodes, and the objective function evaluated at each. The poorest vertex would then be discarded, being replaced in the $2n$ figure by a vertex which is selected in a fashion similar to that recommended for the continuous case, but now including the additional condition that the new choice must lie at a node.

It is convenient to replace the actual magnitudes of the discrete values of the independent variables by a sequence of integers. Thus, in Fig. VIII-43, the valid values of x_1, 12, 14, . . . , 18 in., might be labeled 1, 2, . . . , 5. Each node, the jth say, can then be defined by $\mathbf{v}_j$, representing a set of n integers, $\{v_{1,j}, v_{2,j}, \ldots, v_{n,j}\}$. In the Complex method, we choose k vertices, all of which lie at feasible nodes. The complex is thus represented by the k points, $\mathbf{v}_1, \mathbf{v}_2, \ldots, \mathbf{v}_k$. Let us suppose, in a maximization problem, that the smallest value of the objective function is found at the lth vertex ($1 \leq l \leq k$). The centroid M of the remaining vertices can then be defined as $\mathbf{v}_M$, where

$$v_{j,M} = \frac{1}{k-1} \sum_{i=1}^{k} (v_{j,i} - v_{j,l}) \qquad \text{for } j = 1, 2, \ldots, n \tag{6.2-1}$$

The $v_{j,M}$ need not be integers. The new point to be added to the figure is now found as follows: First, let $\boldsymbol{\xi}$ be the direction from the rejected vertex $\mathbf{v}_l$ to the centroid $\mathbf{v}_M$, as defined by the n numbers (not necessarily integers)

$$m_j = v_{j,M} - v_{j,l} \tag{6.2-2}$$

If α_j is now defined as

$$\alpha_j = \text{nearest† integer value of } \{\alpha m_j\} = I\{\alpha m_j\} \tag{6.2-3}$$

where α is some scale factor, then the new vertex, $\mathbf{v}_l'$, which replaces the old vertex, $\mathbf{v}_l$, is given by

$$v_{j,l}' = v_{j,l} + \alpha_j \tag{6.2-4}$$

If this point, $\mathbf{v}_l'$, is not feasible, then, following the rules of the Complex method, we move along $\boldsymbol{\xi}$, halfway in toward $\mathbf{v}_M$, of course accepting only solutions which lie on nodes. Similarly, if $\mathbf{v}_l'$ produced a worse value of the objective function than $\mathbf{v}_l$, we should use the same rule to contract the complex figure. In any one cycle, this contraction rule is applied until a feasible point is attained, even if this should coincide with another vertex. The final objective is to contract the figure until all vertices lie at a given node—the desired optimum.

Example VIII-25: Discrete-valued variables and the Complex methods

The minimum cost of a distillation column design, valid only at the nodes shown in Fig. VIII-43, is to be found. Design costs are shown by contours in that figure.

Four points, $\mathbf{v}_1, \mathbf{v}_2, \mathbf{v}_3, \mathbf{v}_4$, are first chosen, taking, for example,

$$v_{1,1} = 1, \quad v_{2,1} = 2 \qquad v_{1,2} = 1, \quad v_{2,2} = 3$$
$$v_{1,3} = 2, \quad v_{2,3} = 2 \qquad v_{1,4} = 2, \quad v_{2,4} = 3$$

as shown in Fig. VIII-46*a*.

† Thus, if $\alpha m_j = 2.5$, $\alpha_j = 3$, but if $\alpha m_j = 2.4$, $\alpha_j = 2$.

First cycle (complex 1-2-3-4): The vertex $\mathbf{v}_1 = \{1,2\}$ has the largest value of the objective function in this complex and will be rejected. Then $l = 1$, and the centroid M is given by Eq. (6.2-1) as

$$v_{1,M} = \frac{1}{k-1}(v_{1,2} + v_{1,3} + v_{1,4}) = \tfrac{1}{3}(5)$$

$$v_{2,M} = \frac{1}{k-1}(v_{2,2} + v_{2,3} + v_{2,4}) = \tfrac{1}{3}(8)$$

Then the direction of movement, $\boldsymbol{\xi}$ is given from Eq. (6.2-2) as

$$m_1 = \tfrac{5}{3} - 1 = \tfrac{2}{3}$$
$$m_2 = \tfrac{8}{3} - 2 = \tfrac{2}{3}$$

The new vertex $\mathbf{v}'_1$ is then, from Eqs. (6.2-3) and (6.2-4), given as

$$v'_{1,1} = 1 + I\{2.5 \times \tfrac{2}{3}\} = 1 + 2 = 3$$
$$v'_{2,1} = 2 + I\{2.5 \times \tfrac{2}{3}\} = 2 + 2 = 4$$

if we take the scale factor α to be 2.5. The new figure, after the first cycle, is thus as shown in Fig. VIII-46*b*, with $\mathbf{v}_5 = \mathbf{v}'_1$, since the tentative point is both feasible and an improvement.

Second cycle (Fig. VIII-46*b*, complex 2-3-4-5): Point 2 is now the largest in value, and is thus to be rejected. The centroid M is then

$$v_{1,M} = \tfrac{1}{3}(v_{1,5} + v_{1,3} + v_{1,4}) = \tfrac{7}{3}$$
$$v_{2,M} = \tfrac{1}{3}(v_{2,5} + v_{2,3} + v_{2,4}) = \tfrac{9}{3}$$

The direction of movement $\boldsymbol{\xi}$ is given by

$$m_1 = \tfrac{7}{3} - 1 = \tfrac{4}{3}$$

$$m_2 = 3 - 3 = 0$$

Thus
$$I\{m_1\alpha\} = I\{\tfrac{10}{3}\} = 3$$

$$I\{m_2\alpha\} = I\{0\} = 0$$

to give the new point $\mathbf{v}'_2 = \mathbf{v}_6$ as

$$v'_{1,2} = 1 + 3 = 4 = v_{1,6}$$

$$v'_{2,2} = 3 + 0 = 3 = v_{2,6}$$

This point {4,3} exists and is feasible, so that the new figure will be that shown in Fig. VIII-46*c*.

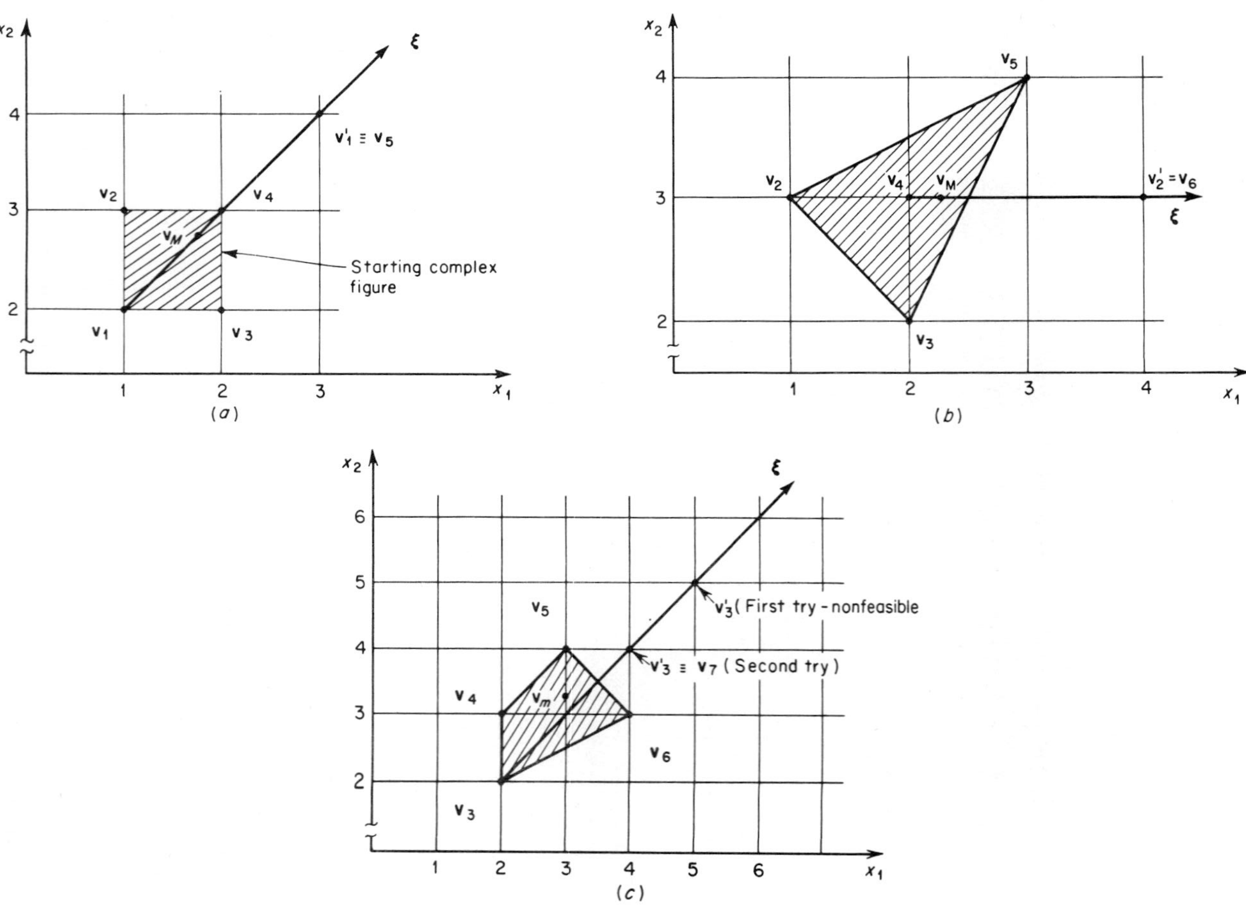

Starting complex figure
v'1 ≡ v5
(a)
v'2 = v6
(b)
v'3(First try - nonfeasible
v'3 ≡ v7 (Second try)
(c)

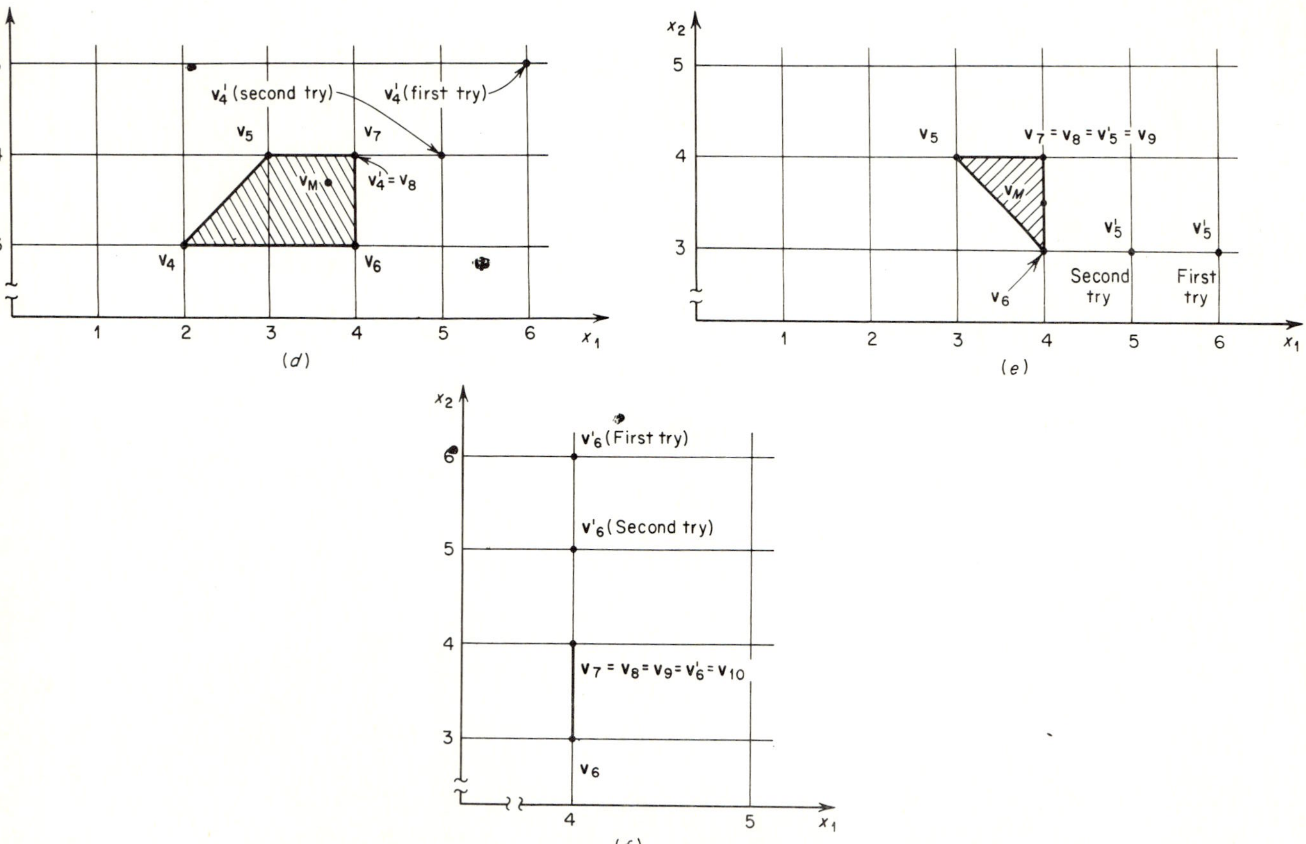

Fig. VIII-46 A discrete-valued search. (a) First cycle; (b) second cycle; (c) third cycle; (d) fourth cycle; (e) fifth cycle; (f) sixth cycle.

Third cycle (Fig. VIII-46*c*, complex 3-4-5-6): Point 3 being the largest, it is rejected, and the new vertex v_3' is generated.

$$\mathbf{v}_M: \quad v_{1,M} = \tfrac{1}{3}(2 + 3 + 4) = 3$$

$$v_{2,M} = \tfrac{1}{3}(3 + 4 + 3) = \tfrac{10}{3}$$

$$\boldsymbol{\xi}: \quad m_1 = 3 - 2 = 1$$

$$m_2 = \tfrac{10}{3} - 2 = \tfrac{4}{3}$$

$$\mathbf{v}_3': \quad v_{1,3}' = 2 + I(2.5) = 5$$

$$v_{2,3}' = 2 + I(\tfrac{10}{3}) = 5$$

This point {5,5} does not exist, so that we should withdraw back along the direction $\boldsymbol{\xi}$ until a valid point is found. Now $\mathbf{v}_3'$ is given by

$$v_1'^{(3)} = 2 + I(\alpha)$$

$$v_2'^{(3)} = 2 + I\left(\frac{\alpha 4}{3}\right)$$

so a reduction of α from its previous value of 2.5 will lead us to nodes, if they exist, which lie closer to the centroid $\mathbf{v}_M$. For the purposes of this example, we shall reduce the value of α by one-half, in the event that the new point generated fails, either because it is nonfeasible or because it does not enhance the objective function. On reducing α by one-half to 1.75, we find the new point, {4,4}, which is acceptable, as shown in Fig. VIII-46*c*.

Fourth cycle (Fig. VIII-46*d*, complex 4-5-6-7): Rejecting point 4, we get the new vertex from

$$\mathbf{v}_M: \quad v_{1,M} = \tfrac{1}{3}(3 + 4 + 4) = \tfrac{11}{3}$$

$$v_{2,M} = \tfrac{1}{3}(3 + 4 + 4) = \tfrac{11}{3}$$

$$\boldsymbol{\xi}: \quad m_1 = \tfrac{11}{3} - 2 = \tfrac{5}{3}$$

$$m_2 = \tfrac{11}{3} - 3 = \tfrac{2}{3}$$

$$\mathbf{v}_4': \quad v_{1,4}' = 2 + I(\tfrac{25}{6}) = 6$$

$$v_{2,4}' = 3 + I(\tfrac{5}{3}) = 5$$

This point is not valid; so the complex figure is contracted by moving in along $\boldsymbol{\xi}$. Try $\alpha = 1.75$, getting {5,4}, which is not valid. Reduce α again to $\alpha = 1.4$, giving {4,4}, which is valid ($= \mathbf{v}_8$). As shown in Fig. VIII-46*d*, two vertices now lie at one node.

Fifth cycle (Fig. VIII-46*e*, complex 5-6-7-8): Reject point 5, and get the new vertex from

$\mathbf{v}_M$:
$$v_{1,M} = \tfrac{1}{3}(4 + 4 + 4) = 4$$
$$v_{1,M} = \tfrac{1}{3}(3 + 4 + 4) = {}^{11}\!/_3$$

$\boldsymbol{\xi}$:
$$m_1 = 4 - 3 = 1$$
$$m_2 = {}^{11}\!/_3 - 4 = -\tfrac{1}{3}$$

$\mathbf{v}'_5$:
$$v'_{1,5} = 3 + I(2.5) = 6$$
$$v'_{2,5} = 4 + I\left(\frac{2.5}{3}\right) = 3$$

Point {6,3} does not exist. Try $\alpha = 1.75$, which gives the invalid location {5,3}. Further reduction to $\alpha = 1.4$ gives the valid position $\mathbf{v}_9${4,4}. Let us accept this point. We now have contracted the figure to two nodes, as seen in Fig. VIII-46*e*.

Sixth cycle (Fig. VIII-46*f*, complex 6-7-8-9): Reject vertex 6. The new vertex is found from

$\mathbf{v}_M$:
$$v_{1,M} = \tfrac{1}{3}(4 + 4 + 4) = 4$$
$$v_{2,M} = \tfrac{1}{3}(4 + 4 + 4) = 4$$

$\boldsymbol{\xi}$:
$$m_1 = 4 - 4 = 0$$
$$m_2 = 4 - 3 = 1$$

$\mathbf{v}'_6$:
$$v'_{1,6} = 4 + I(0) = 4$$
$$v'_{2,6} = 3 + I(2.5) = 6$$

Point {4,6} has a worse value of objective function than the rejected point; so we shall have the new point halfway inwards. Try $\alpha = 1.75$, getting the point {4,5}, which is still inferior. Try $\alpha = 1.4$, getting point {4,4}, which is valid ($\equiv \mathbf{v}_{10}$).

We have now forced the complex figure to contract on to a single node. This is the desired optimum.

This illustration shows all the basic rules required in manipulating the complex figure, although in this situation with discrete-valued variables it is necessary to ensure that all vertices lie at feasible node points. It is interesting to note that a starting figure, such as *a-b-c-d* in Fig. VIII-43, would seek the local optimum *b*, while the global optimum in fact lies at point *e*. This situation could arise with any numerical search technique, and is not confined to the present discrete-valued problem.

It will be obvious from this example that no special technique is required to handle discrete, as opposed to continuous, variables as long as the method forces the solution to lie at feasible node points.

6.3 Linear Discrete-valued Systems

If the function is linear, but exists only in discrete values, or if the function can be linearized in a local region, integer linear programming might be used when the system is bounded, methods for the situation in which all variables have to be integers having been developed, for example, by Gomory [1958]. The case of mixed discrete- and continuous-variables programming, when only some of the variables in the optimal solution must be integer-valued, has been discussed, for example, by Dakin [1965]. The integer system problem will of course satisfy the continuous problem, but not the converse.

7 FINAL STAGES IN A SEARCH

Irrespective of the actual approach to a numerical search problem—preplanned or sequential, direct search or gradient search—the end result is the definition of the desired optimum to within a given degree of accuracy. Most of the numerical optimization techniques described in previous sections are, essentially, based upon linear approximations of the objective function. Accordingly, when the region containing the optimum has been attained, these methods will run into difficulty, unless special methods are introduced to account for the quadratic terms in the surface representation (Sec. 7.1). It is also essential to check whether or not the result represents a local or a global optimum (Sec. 7.2), and in many cases a sensitivity study of the region surrounding the optimum is a necessary adjunct to optimization (Sec. 7.3).

7.1 Final Steps in the Locating of the Optimum

When a near-optimum (or near-stationary) value of the objective function has been achieved, it is found that most techniques converge rather slowly. Thus, while the accuracy of search may be improved by decreasing the step size, it is common for the search to oscillate about the region containing the optimum. The reason for this can be best visualized in the two-dimensional case when the search is for a mountain peak and the gradient of ascent decreases as the summit is approached. The subsequent lack of definition in the gradient thus prevents the search, whether it is a gradient or nongradient method, being guided quickly to a conclusion. Instead, it oscillates or cycles around the true optimum. Such difficulties arise even if the objective function can be expressed mathematically, since in numerical optimization most search methods (even if they do not explicitly require such a formulation) depend upon the fitting of a plane surface approximation

$$y = \beta_0 + \sum_{i=1}^{n} \beta_i \zeta_i$$

to a general response function. This approximation essentially neglects second- and higher-order terms in the function (Chap. III, Sec. 2.4.2) and is valid only when these terms have a negligible effect. When they are significant, the plane approximation remains valid only over smaller and smaller regions in the neighborhood of the starting point, and for a set of experimental locations situated at finite distances from this base point, the quadratic and higher terms in the objective function

$$y = \beta_0 + \sum_{i=1}^{n} \beta_i \zeta_i + \sum_{i=1}^{n} \sum_{j=1}^{n} \beta_{i,j} \zeta_i \zeta_j + \cdots$$

may bias the calculated coefficients β_i of any linear approximation. This approximation may therefore depart significantly from the true tangent plane, which would give the correct direction or steepest line of ascent. While unimportant far from the optimum, this effect is particularly important on approaching the final stationary value, where the true linear coefficients tend to zero, and are therefore smaller than, or of the same magnitude as, the second-order terms. As a consequence, the search procedure is likely to give rise to an oscillating path, which makes only slow progress in the direction of the true optimum. A similar effect is encountered in searching ridges.

The final stages of the search are best carried out by fitting a higher-order polynomial surface to the final group of experimental results. Since a plane surface is inherently used in the main search, the final stages can therefore be carried out using a second-order surface. Analytical methods can then be used to determine the nature of the optimum, if one exists, and to pinpoint its location and value. The fitting of such a quadratic approximation has been dealt with in some detail in Chap. III, and need not be discussed again. The minimum number of experiments required to fit an n-dimensional surface is given in Table III-4, and the appropriate equations defining the linear and quadratic coefficients, β_i and $\beta_{i,j}$, are given by Eqs. (5.4-1) to (5.4-3), Chap. III. A typical illustration of such curve fittings is given by Example III-18. It will also be recalled that the placing of experiments might be in an orthogonal design which eases the calculation of the various coefficients. Such a calculation was described in Example III-19.

We have been discussing, in the foregoing paragraphs, methods applicable to most types of search. The use of linear approximations is obvious in the gradient-type methods, but its application in the direct search technique might not be so obvious. However, if we examine a typical method, such as the Sequential Simplex approach of Spendley et al. [1962] (Sec. 3.1.2), we shall see that the $n + 1$ vertices defining a simplex figure will also define a linear surface, and the rejection of the worst or second-worst vertex will define a gradient along that plane surface. Continued application of the direct search procedure, with progressively reduced step size, would thus be

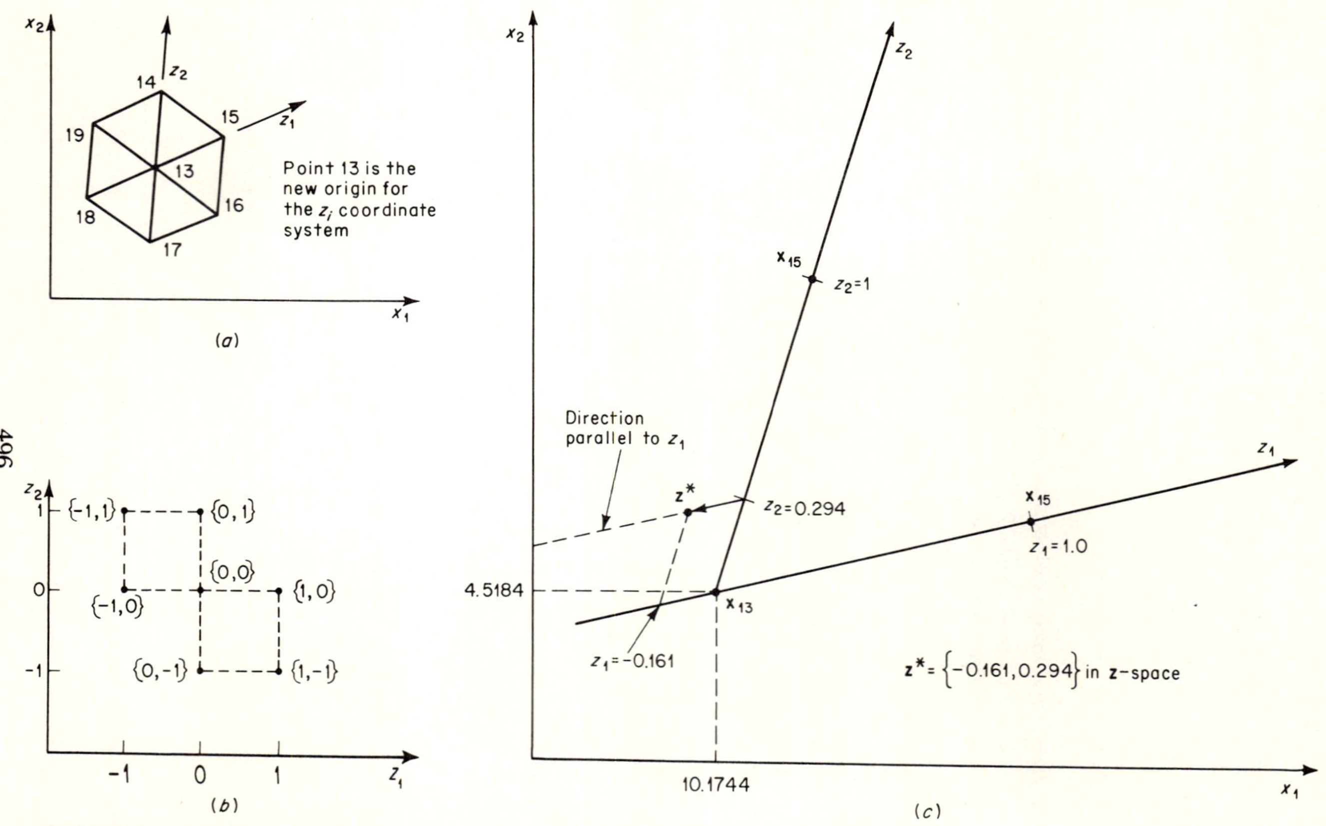

Fig. VIII-47 Fitting of final results of Sequential Simplex search. (*a*) New coordinate system; (*b*) arrangement of experiments in z space; (*c*) relationship of point z^* to x coordinates.

expected to run into the same difficulties as are found in the use of gradient methods.

Let us consider the fitting of a quadratic surface to the data produced in the final stages of a Simplex cycle. The following illustration does not require any additional experiments to be carried out—an advantage, if these have to be kept to a minimum.

Example VIII-26: Quadratic approximation in the Simplex method

Let us fit a second-order surface to the data produced in the final closed loop of stage 1 in Example VIII-3. The form of this loop is shown in Fig. VIII-47*a*. It will be seen that, by aligning new axes z_1 and z_2 along the adjacent sides of one equilateral triangle and letting the sides of the triangle be of unit length, we shall obtain the simple experimental design shown in Fig. VIII-47*b*. From this and Table VIII-2 we may draw up the data shown in Table VIII-7.

Following the approach of Chap. III, Sec. 5.4, we may calculate from this information the coefficients $\tilde{C}_{i,j}$, $\tilde{C}_{i,j,k}$, $\tilde{C}_{i,j,k,l}$, $\tilde{D}_i$, and $\tilde{D}_{i,j}$, using the definitions of Eqs. (5.4.3). Chap. III. The β_i and $\beta_{i,j}$ can then be found by solving Eqs. (5.4.1) and (5.4-2), Chap. III. This gives

$$\Delta y = \beta_1 z_1 + \beta_2 z_2 + \beta_{1,1} z_1^2 + \beta_{1,2} z_1 z_2 + \beta_{2,2} z_2^2$$

where

$$\beta_1 = -0.176 \qquad \beta_2 = 1.680 \qquad \beta_{1,1} = -3.998$$
$$\beta_{1,2} = -3.999 \qquad \beta_{2,2} = -3.999$$

The location of the optimum can then be determined, since

$$\frac{\partial(\Delta y)}{\partial z_1} = \beta_1 + 2\beta_{1,1} z_1 + \beta_{1,2} z_2 = 0$$

$$\frac{\partial(\Delta y)}{\partial z_2} = \beta_2 + \beta_{1,2} z_1 + 2\beta_{2,2} z_2 = 0$$

TABLE VIII-7 THE FITTING OF A SURFACE (Example VIII-26)

Simplex point	*Data from Table* VIII-2 x_1	x_2	y	*Relative coordinates* z_1	z_2	*Relative objective function* Δy
13	10.1744	4.5184	99.74	0	0	0
14	10.6918	6.4498	97.42	0	1	−2.32
15	12.1054	5.0358	95.57	1	0	−4.17
16	11.5884	3.1044	93.88	1	−1	−5.86
17	9.6574	2.5870	94.06	0	−1	−5.68
18	8.2434	4.0010	95.92	−1	0	−3.82
19	8.7604	5.9324	97.60	−1	1	−2.14

giving, by Cramer's rule,

$$z_1^* = \frac{-2\beta_1\beta_{2,2} + \beta_2\beta_{1,2}}{4\beta_{1,1}\beta_{2,2} - \beta_{1,2}^2} = -0.161$$

$$z_2^* = \frac{-2\beta_2\beta_{1,1} + \beta_1\beta_{1,2}}{4\beta_{1,1}\beta_{2,2} - \beta_{1,2}^2} = 0.294$$

It remains only to translate this location back into the original coordinate system. This is a relatively trivial task if we recognize that z_1 is a linear measure of distance along a direction $\boldsymbol{\xi}_1 = \mathbf{x}_{15} - \mathbf{x}_{13}$, and z_1 is a measure of distance along a direction $\boldsymbol{\xi}_2 = \mathbf{x}_{14} - \mathbf{x}_{13}$. Thus the final point is, obviously, given by the sum of two moves, first along the z_2 direction and then along the z_2 direction, as seen in Fig. VIII-47c, is

$$\mathbf{x}^* = \mathbf{x}_{13} + \boldsymbol{\xi}_1 z_1^* + \boldsymbol{\xi}_2 z_2^*$$

or

$$\mathbf{x}^* = \{10.1744,4.5184\} - 0.161\{1.9310,0.5174\} + 0.294\{0.5174,1.9310\}$$

giving $$\mathbf{x}^* = \{10.01562,5.00281\}$$

which is very nearly the true optimum at {10.0,5.0}.

It will be seen that, with no extra experimentation, this has a much improved value over the results of cycle 1 in Example VIII-3. A new cycle of search might be started, centered on this approximation to the optimum.

Spendley et al. [1962] propose a somewhat similar method for fitting a quadratic surface, but based upon a single simplex, preferably the final one. It requires additional experiments, however. Thus, if the $n + 1$ vertices of the simplex are given in z space by $\mathbf{V}_{0,0}, \mathbf{V}_{1,1}, \ldots, \mathbf{V}_{n,n}$, it is necessary to find the response at a further $\frac{1}{2}n(n + 1)$ points. These may be conveniently taken at the midpoints of each edge of the simplex, so that the locations are

$$\mathbf{V}_{i,j} = \tfrac{1}{2}(\mathbf{V}_{i,i} + \mathbf{V}_{j,j}) \qquad \text{for } i \neq j$$

If the original points of the simplex are used to define a set of oblique axes with coordinates z_i, these vertices may be taken as

$$\begin{array}{lccccc}
\mathbf{V}_{0,0}: & 0 & 0 & 0 & \cdots & 0 \\
\mathbf{V}_{1,1}: & 1 & 0 & 0 & \cdots & 0 \\
\mathbf{V}_{2,2}: & 0 & 1 & 0 & \cdots & 0 \\
\cdots & \cdots & \cdots & \cdots & \cdots & \cdots \\
\mathbf{V}_{n,n}: & 0 & 0 & 0 & \cdots & 1
\end{array}$$

so that the additional points are

$$\begin{array}{llllll} \mathbf{V}_{0,1}: & \frac{1}{2} & 0 & 0 & \cdots & 0 \\ \mathbf{V}_{1,2}: & \frac{1}{2} & \frac{1}{2} & 0 & \cdots & 0 \\ \cdot & \cdot & \cdot & \cdot & \cdot & \cdot \end{array}$$

If the response at a vertex $\mathbf{V}_{i,j}$ is $y_{i,j}$, the desired quadratic approximation will be

$$y = \beta_0 + \sum_{i=1}^{n} \beta_i z_i + \sum_{i=1}^{n} \sum_{j=1}^{n} \beta_{i,j} z_i z_j$$

where $\beta_0 = y_{0,0}$

$$\beta_i = 4y_{0,i} - y_{i,i} - 3y_{0,0}$$

$$\beta_{i,j} = 2(y_{i,j} + y_{0,0} - y_{0,i} - y_{0,j})$$

for $i, j = 1, 2, \ldots, n$. Nelder and Mead [1964, 1965] discuss the problems of accurately locating the stationary point of this quadratic approximation. It is necessary to steer a course between two hazards. From one viewpoint the simplex can become so small that $\beta_{i,j}$ consists largely of roundoff errors in calculating the $y_{i,j}$. On the other hand, the simplex can become so large that the quadratic approximation is poor, and the $\beta_{i,j}$ are correspondingly biased.

Further second-order designs based on the simplex are discussed by Box and Behnken [1960] and Box and Hunter [1957].

7.2 The Global Optimum

As discussed earlier (Sec. 1.1), all numerical search methods seek local optima. Thus, once a particular local optimum has been located by an appropriate search technique, it is important to check that other, better optima do not exist. There is no vigorous method for this search, except in certain restricted classes of problem. One can only begin the search procedure at a number of different initial base points. Zellnik et al. [1962], Brooks [1958], and Anderson [1953] have suggested a random scanning method to achieve this.

7.3 Sensitivity Analysis

After the search for an optimum has been completed, it is usually desirable to examine the nature of the result. This examination will ensure, for example, that the solution is in fact optimal, and not a saddle point or a false "stationary" value, such as might be obtained by being stranded on a ridge. In addition, it is most important to know the sensitivity of the objective function to small changes in the independent variables.

We may best accomplish these ends by fitting a polynomial of higher order than the first degree to the available experimental data adjacent to the optimum. Thus we might use a quadratic approximation about a given base point, getting

$$\Delta y = \sum_{i=1}^{n} \beta_i \zeta_i + \sum_{i=1}^{n} \sum_{j=1}^{n} \beta_{i,j} \zeta_i \zeta_j \tag{7.3-1}$$

Here Δy and ζ_i are the respective changes in value of the objective function and distance in moving from the base point. If the β_i vanish or nearly vanish, indicating that the optimum lies at or near our base point, we may use the canonical analysis approach (Box [1954]) to redefine the coordinate system relative to the optimum and obtain a quadratic approximation having a particularly simple structure. A sensitivity analysis could then be carried out by examining the nature of the surface in this neighborhood represented by these quadratic terms.

Although most reliable search methods will inherently avoid a saddle point, care must be taken that a fitted quadratic representation does not indicate a false result. Thus, for example, a univariant search might strand itself on a ridge near a saddle point. As shown in Fig. VIII-48, a surface fitted to the final set of points would indicate an optimum location—a fallacious result introduced by the inflexible nature of the search technique

Fig. VIII-48 False optimum on a ridge.

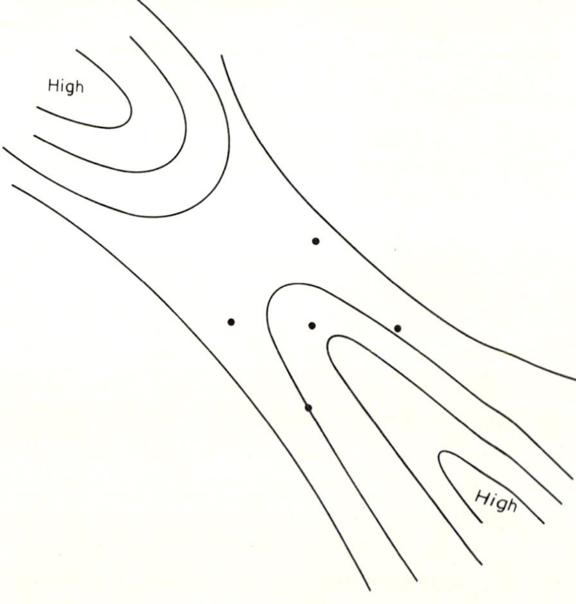

employed. It is also possible for the convergence, or stopping, criteria to stop a search far from the desired optimum, especially if the search region has "flat" sections. It is conceivable, therefore, that the search can halt in the region of all stationary points (even though these are saddle points) or low-gradient regions. Some of the problems involved in the choice of suitable criteria for stopping a search are given in Chap. XI, Sec. 5.3.2.

8 RECOMMENDED TECHNIQUES FOR NUMERICAL OPTIMIZATION

Since most systems of practical interest cannot be optimized by analytical methods, numerical search techniques must be applied. These methods are characterized by one essential idea, namely, to find a new feasible point which is better than the existing one. These techniques are applied iteratively until no further improvement is possible or until the optimum has been located to within the desired accuracy.

The distinguishing characteristic of a particular method is the mode of selecting a direction and distance of movement. Such methods have been broadly divided into two classes, namely, gradient and nongradient techniques, depending on whether or not the function gradient is required to define the direction of movement. Some general comments are made in Secs. 3.1.4, 3.2.4, 4.3, 5, 5.3, and 5.5.

If the derivatives of the objective function are not given in explicit terms, a direct search (nongradient) method is usually selected, since additional work in local experimentation would be required to determine the gradient. A simple method for programming, and one which seems to be reliable, is that known as the Sequential Simplex method (Sec. 3.1.2). While not converging as rapidly as some of the other direct search techniques, the simplex has the real advantage of simplicity, requiring only one new experiment in each move. An extension of the basic ideas in the Complex method (Sec. 5.2.1) allows inequality constraints to be effectively handled.

These two direct search methods are recommended for general use, although the Rosenbrock technique (Sec. 3.1.3) and some of the search techniques exhibiting quadratic convergence, such as that of Powell [1964] (Sec. 3.1.3), are more powerful. These latter methods require a one-dimensional search, in addition to the determination of the next direction of motion. Moreover, it is difficult to see how to extend these methods to constrained optimization, except by problem modification, in which the constraints are included in some manner within an unrestricted function. The Carroll methods provides an illustration of this approach, using penalty functions. This possibility of using an equivalent unrestricted problem should always be considered.

If function gradients are easily evaluated, for example, from an analytical expression, some very powerful search techniques can be used. Such gradient techniques should always be used, where possible, in preference to direct search methods. For the unrestricted problem, the methods exhibiting quadratic convergence have a clear advantage over the method of steepest ascent. Of several such methods, that of Davidon [1959], as extended by Fletcher and Powell [1963], seems to be the best. Unfortunately, the methods giving quadratic convergence have not as yet been extended to include constraints, except by modifying the problem as in the Carroll method. For this situation, methods similar to the steepest ascent technique are of value, an initial linearization of the objective function and constraints being followed by search moves. Such an approach is used by the gradient projection techniques (Secs. 4.3.1 and 5.3.3).

In general, most methods, whether gradient or direct search, will probably show the same oscillatory approach to the optimum as exhibited by the unrestricted method of steepest ascent. One solution to this problem is the fitting of a quadratic approximation to the experimental data in the region of the optimum. None of these methods guarantee convergence to the global optimum in the general case, and it is essential that the nature of any optimal location be examined with care, and the possible existence of multiple local optima checked. The success of any search method might depend upon the choice of stopping criteria. These can be placed upon the variables x_i or upon the objective function, a full discussion of the problems involved being presented in Chap. XI, Sec. 5.3.2.

In conclusion, it should be remembered that there is no criterion for defining the effectiveness of a particular numerical optimization technique, so that no one method can be singled out as being better than another. The above recommendations are based upon published trials on specific problems, such as those of Swann [1964], Fletcher and Reeves [1964], Box [1965, 1966], Fletcher [1965], and Stocker [1969] (see Appendix 2).

Problems

VIII-1 Search for the minimum of the objective function

$$y = 6x_1^2 + 3x_2^2$$

using the univariant approach. Terminate the search after each of the coordinate directions has been searched three times.

VIII-2 In performing a numerical search of the objective function

$$y = x_1^2 + x_1x_2 + 16x_2^2 + x_3^2 - x_1x_2x_3$$

suppose that a one-dimensional search in the direction

$$\xi = \left\{\frac{1}{\sqrt{3}}, \frac{-1}{\sqrt{3}}, \frac{-1}{\sqrt{3}}\right\}$$

is required, starting from the base point $\{1,-2,-2\}$. Determine the position of the minimum along this direction and the value of the objective function at the minimum point.

VIII-3 Find a direction orthogonal to the direction

$$\xi = \left\{\frac{1}{\sqrt{3}}, \frac{-1}{\sqrt{3}}, \frac{-1}{\sqrt{3}}\right\}$$

VIII-4 Find a direction orthogonal to the direction

$$\xi = \{2/3, -1/3, -2/3\}$$

VIII-5 Determine a simplex figure in a three-dimensional space such that the distance between vertices is 0.2 unit and one vertex is at the point $\{-1,2,-2\}$.

VIII-6 Starting with the simplex figure generated in Prob. VIII-5, carry out eight cycles of vertex rejection and regeneration in searching for a minimum of the objective function

$$y = x_1^2 + 3x_2^2 + 5x_3^2$$

using the Sequential Simplex method.

VIII-7 Repeat Prob. VIII-6, but simulate the presence of experimental error by adding or subtracting an amount equal to ± 10 percent of the value of the objective function at each vertex, evaluated with the sign of the error to be determined by the flip of a coin. Remember the additional Simplex rule introduced in the face of experimental error.

VIII-8 Determine a simplex figure in a four-dimensional space such that the distance between vertices is 0.5 unit and one vertex is at the point $\{1,-2,4,2\}$.

VIII-9 Using the simplex figure generated in Prob. VIII-8, carry out eight cycles of vertex rejection and regeneration in searching for a minimum of the objective function

$$y = 2x_1^2 + x_2^2 + 4x_3^2 + x_4^2$$

using the Sequential Simplex method.

VIII-10 Carry out three stages of a Hooke-Jeeves search of the objective function given in Prob. VIII-6. Use $\delta = 0.5$, starting from the base point $\{2,-1,1\}$. A stage consists of a local exploration, together with an accelerated move.

VIII-11 Repeat Prob. VIII-10 with $\delta = 1.0$.

VIII-12 Using the Gram-Schmidt procedure, develop an orthogonal set of vectors, basing your development on the following set of linearly independent vectors:

$$\mathbf{v}_1 = \{1,2,1\} \qquad \mathbf{v}_2 = \{-1,3,1\} \qquad \mathbf{v}_3 = \{0,5,1\}$$

VIII-13 Repeat Prob. VII-12, using the following set of vectors:

$$\mathbf{v}_1 = \{1,2,1\} \qquad \mathbf{v}_2 = \{-1,3,1\} \qquad \mathbf{v}_3 = \{2,4,2\}$$

These vectors are not linearly independent. What difficulties do you encounter? What effect does this result have on the Rosenbrock method?

VIII-14 A completed stage of the Rosenbrock method gives the following results in the search for a minimum:

y	*Result*	x_1	x_2	d_1	d_2	s_1	s_2
55		3	7	0	0	$\sqrt{2}$	$\sqrt{2}$
103	Failure	4	8	0	0	$-1/\sqrt{2}$	$\sqrt{2}$
38	Success	2	8	0	$\sqrt{2}$	$-1/\sqrt{2}$	$3\sqrt{2}$
29	Success	1.5	7.5	$-1/\sqrt{2}$	$\sqrt{2}$	$-3/\sqrt{2}$	$3\sqrt{2}$
35	Failure	1.5	10.5	$-1/\sqrt{2}$	$\sqrt{2}$	$-3/\sqrt{2}$	$\sqrt{2}$

(*a*) Determine the directions used in this stage of the search.

(*b*) Calculate the new directions of search.

VIII-15 Using the Rosenbrock search technique, seek the minimum of the objective function

$$y = 6x_1^2 + x_2^2$$

using $s_1 = s_2 = 1$, $\beta = \frac{1}{2}$, and $\alpha = 3$, until no further movement is possible.

VIII-16 Given the objective function

$$y = 3x_1^2 - 6x_1x_2 + 4x_2^2$$

find:

(*a*) The minimum along the line

$$x_1 = 8 + 2s$$
$$x_2 = 4 + 3s$$

(*b*) The minimum along the parallel line

$$x_1 = -3 + 2s$$
$$x_2 = 4 + 3s$$

(*c*) The minimum value along a line connecting the two points found in (*a*) and (*b*).

(*d*) The global minimum of the objective function. The results found in parts (*c*) and (*d*) should be identical, as shown in Prob. VIII-17.

VIII-17 Prove that if the objective function has the form

$$y = a_{11}x_1^2 + 2a_{12}x_1x_2 + a_{22}x_2^2$$

with $a_{11} > 0$ and $a_{11}a_{22} - a_{12}^2 > 0$, then a line joining the minima found along any two parallel lines will pass through the global minimum. How does your solution differ if the inequality restrictions are relaxed?

VIII-18 Given the objective function

$$y = 3x_1^2 - 6x_1x_2 + 4x_2^2$$

find:

(*a*) The minimum point along the line

$$x_1 = 2 + s$$
$$x_2 = -2$$

(*b*) The minimum point along the line x_1 = constant, starting from the minimum on the line used in (*a*).

(*c*) The minimum point along a line x_2 = constant, starting from the minimum on the line used in (*b*).

(*d*) The minimum point along a line connecting the two minima found in parts (*b*) and (*c*).

Is the minimum found in part (*d*) in fact the global minimum?

VIII-19 Prove that if the objective function has the form

$$y = a_{11}x_1^2 + 2a_{12}x_1x_2 + a_{22}x_2^2$$

with $a_{11} > 0$ and $a_{11}a_{22} - a_{12}^2 > 0$, then a line connecting the first and the last minima found by three cycles of search along perpendicular lines passes through the global minimum of the objective function.

VIII-20 Find the minimum of the objective function

$$y = x_1^2 + 3x_2^2 + 5x_3^2$$

starting from the point $\{1,6,-7\}$ and using the Powell method. Analytical techniques may be used to locate the one-dimensional optima.

VIII-21 Using the Powell method, conduct three stages of search seeking the minimum of the objective function

$$y = x_1^2 + 3x_1^2x_2^2 - 2x_1x_2x_3 + 5x_3^2$$

VIII-22 Write the equation defining the direction of the contour tangent through the point {1,1} for a function

$$y = 3x_1^2 + 4x_2^2$$

VIII-23 What is the direction of steepest ascent through the point {1,1} for the objective function

$$y = 3x_1^2 + 4x_2^2$$

Show that the contour tangent and this direction are orthogonal.

VIII-24 Carry out three cycles of the method of steepest descent in seeking the minimum of the function

$$y = 3x_1^2 + 4x_2^2$$

starting at the point {1,1}.

VIII-25 Carry out three cycles of the method of steepest descent in seeking the minimum of the function

$$y = x_1^2 + 3x_1^2x_2^2 - 2x_1x_2x_3 + 5x_3^2$$

starting at the point {−2,3,−4}.

VIII-26 Repeat Prob. VIII-24 using Booth's modification of the method of steepest descent.

VIII-27 Repeat Prob. VIII-25 using Booth's modification of the method of steepest descent.

VIII-28 Given the objective function

$$y = 5x_1^2 + 3x_2^2 + x_3^2 - x_1x_2$$

generate three different sets of conjugate directions.

VIII-29 Using one of the sets of conjugate directions found in Prob. VIII-28, search for the minimum of the objective function defined in that problem.

VIII-30 Search for the minimum of the objective function given in Prob. VIII-28 using the Fletcher-Reeves method.

VIII-31 Carry out three cycles of the Fletcher-Reeves method in seeking the minimum of the objective function defined in Prob. VIII-25, starting at the point {−2,3,−4}

VIII-32 Find the projection of the gradient of the function

$$y = 5x_1 - 3x_2 + 6x_3$$

onto the x_1x_2 plane. Prepare a sketch showing both the gradient and its projection.

VIII-33 Repeat Prob. VIII-32, except project the gradient of the function

$$y = x_1^2 + 3x_2^2 + 2x_3^2$$

evaluated at the point {1,1,−1} onto the x_1x_2 plane.

VIII-34 Devise a search scheme, based on the gradient projection, to find the minimum of the function

$$y = x_1^2 + x_2^2 + 2x_3^2 - 2x_1x_2$$

subject to the constraints

$$g_1 = 2x_1 + x_2 - 4 = 0$$

and

$$g_2 = 5x_1 - x_3 - 8 = 0$$

Carry out a few iterations of your search method.

VIII-35 Solve Prob. VII-34 using penalty functions forming an objective function

$$F = y + P(g_1^2 + g_2^2)$$

Optimize first with $P = 1$ and then with $P = 10$. Use the Powell search technique.

VIII-36 Using the Sequential Simplex method in conjunction with penalty functions, find the minimum of

$$y = x_1^2 + 2x_2^2 + x_3^2$$

subject to $$g_1 = x_1 + 2x_2 + x_3 - 6 = 0$$

VIII-37 What are the next three cycles of search in Example VIII-16 (Table VIII-6) using the Complex approach?

VIII-38 Define a suitable search region and a feasible initial base point for the Complex method of search in minimizing a function subject to the restrictions that $x_i \geq 0$ and

$$x_1 + x_2 + x_3 \leq 6$$
$$5x_1 - x_2 + x_3 \leq 4$$
$$x_1 + 3x_2 + 2x_3 \geq 1$$

VIII-39 Set up a Complex method of search and carry out the first 10 cycles of search in minimizing the function

$$y = 4x_1 + x_2 + 2x_3$$

subject to the restrictions of Prob. VIII-38.

VIII-40 Use the Complex method to determine the minimum of the function

$$y = x_1^2 - 2x_1x_2 + x_2^4$$

subject to the restriction that

$$x_1^2 + 2x_2^2 - 4 \leq 0$$

Set up only the first figure, and carry out three subsequent cycles of vertex rejection and regeneration.

VIII-41 Find the minimum of

$$y = x_1^2 + 2x_2^2$$

subject to $$3x_1 + x_2 \geq 2$$
$$x_1 + 2x_2 \geq 3$$

using a penalty function method.

VIII-42 Find the maximum of

$$y = x_1^2 + x_2^2$$

subject to $$x_1 + x_2 = 5$$
$$x_1 \geq -3 \quad x_2 \geq 4$$

using a penalty function approach.

VIII-43 Find one feasible point in the region

$$g_1 = 1 + \tfrac{2}{3}x_1 - x_2 \leq 0$$
$$g_2 = (x_1 - 1)^2 + (x_2 - 2)^2 - 4 \leq 0$$

VIII-44 What is the search region defined by the restriction

$$g = x_1 + x_2^2 - 5 \leq 0$$

Set up the first figure of search and carry out five cycles of vertex regeneration using the Complex method to minimize the function

$$y = x_1^2 - 2x_1x_2 + x_2^4$$

VIII-45 Maximize

$$y = x_1 x_2 x_3$$

subject to the constraints

$$x_i \geq 0 \qquad \text{and} \qquad 0 \leq x_1 \leq 42$$

together with

$$0 \leq x_1 + 2x_2 + 2x_3 \leq 72$$

after a suitable transformation of the variables to obtain an unrestricted problem.

VIII-46 Find the maximum of the objective function

$$y = x_1 x_2$$

subject to the constraints

$$x_2 - x_1 - 1 \leq 0$$
$$2x_2 + x_1 - 6 \leq 0$$
$$x_1 \geq 0$$

Use the Glass-Cooper method. Sketch the feasible region on a sheet of graph paper and indicate a few contour lines. Follow the progress of your search by denoting the various points on the plot as you find them.

VIII-47 Repeat Prob. VIII-46 using the method of feasible directions.

VIII-48 Repeat Prob. VIII-46 using the gradient projection method.

VIII-49 Solve Prob. VIII-41 using the method of feasible directions.

VIII-50 Solve Prob. VIII-41 using the method of Glass and Cooper.

VIII-51 Solve Prob. VIII-41 using the gradient projection method.

VIII-52 Using the Glass-Cooper method, find a move in the feasible region, starting from the point {6,0}, if the objective function is given by

$$y = 10x_2 + x_1$$

subject to the constraints

$$g_1 = x_2^2 + 6 - x_1 \leq 0$$

and

$$g_2 = 15 - x_1 \leq 0$$

VIII-53 Given an objective function to be minimized having the form

$$y = \sum_{i=1}^{n} \sum_{j=1}^{n} a_{i,j} x_i x_j + \sum_{j=1}^{n} b_j x_j + c$$

in which $a_{i,j} = a_{j,i}$ prove that the optimum $\mathbf{x}^* = \mathbf{x}_0 + \sum_{i=1}^{n} \alpha_i \boldsymbol{\xi}_i$ is found by searching along each of the n conjugate directions only once. (HINT: Show that each of the α_i are independent of the others. Searching in each of the directions only once will then yield the absolute minimum.)

VIII-54 If the objective function to be minimized has the form

$$y = \sum_{i=1}^{n} \sum_{j=1}^{n} a_{i,j} x_i x_j + \sum_{i=1}^{n} b_i x_i + c$$

and if $\mathbf{x}_0$ and $\mathbf{x}_1$ are the minima found in searching along the direction $\boldsymbol{\xi}$ starting from two different base points, prove that $\mathbf{x}_1 - \mathbf{x}_0$ is conjugate to $\boldsymbol{\xi}$ (see Fig. VIII-20).

VIII-55 Prove that Eq. (3.2-50) is generally true. Use the methods of algebraic continuation. Assume that the first $k - 1$ directions have been selected in accordance with Eq. (3.2-50) and that they are all conjugate to one another. Then show that the new direction is conjugate to the first $k - 1$ directions.

CHAPTER IX

FUNCTIONS OF A CONTINUOUS VARIABLE—ANALYTICAL METHODS

1 INTRODUCTION TO THE VARIATIONAL CALCULUS

In our previous work the objective function was optimized by an appropriate numerical choice of a set of variables which may or may not be subject to certain constraints. Thus the best operating temperature, optimal shape, or most economic arrangement was determined, the solution of a given problem being represented by a single numerical value associated with each variable. The solution to all problems in optimization cannot, however, always be so simply represented. Consider, for example, the task of determining reactor temperature as a function of length in order to maximize the production of some product, A_2, in a tubular reactor of fixed length, θ. The exit concentration c_2 is obtained by integrating the local rate of production along the length of the reactor. Then, the steady-state solution is given by

$$y = c_2 - c_{2,0} = \int_0^\theta a_2 r(c_1, c_2, \ldots, c_n, T)\, dt \tag{1-1}$$

where r = rate of production of A_2
T = temperature
c_i = concentration of substance A_i
t = distance along reactor of length θ

Suppose that A_2 is an intermediate product in the reaction scheme

$$A_1 \rightarrow A_2 \rightarrow A_3$$

Here the rate at which each reactant changes will be a function of the concentration of the reactants and of the temperature. It is clear that a specific temperature, say, 300°C, cannot be the solution of this problem, since one must specify an appropriate temperature for each axial position. The solution to this problem must thus be a function of length, $T(t)$. Contrast this situation with those which were encountered in earlier chapters, where a single number was associated with each variable. The optimization of the objective function defined by Eq. (1-1) is a quite different problem from those previously studied, falling into that branch of mathematics known as the *calculus of variations* (Forsyth [1960] and Weinstock [1952]). It is this subject we want to develop in this chapter. Optimization problems relating to systems for which certain of the variables are operating continuously very often give rise to the structure shown above, and thus lead to problems which have to be treated using the techniques of variational calculus.

The fundamental problem of variational calculus is to select the function $x(t)$ which makes the y defined by the integral

$$y = \int_0^\theta Y[t,x(t),x_t(t)]\,dt \tag{1-2}$$

as large, say, as possible. Here Y, the integrand, is in general both an implicit and an explicit function of t, since it depends on t directly and also through the function $x(t)$ and its derivative $x_t(t)$.

This fundamental problem must be modified in a number of ways if a large spectrum of applied problems is to be considered. First of all, various types of boundary conditions must be included in the study (Sec. 4), and certain types of restrictions must also be imposed (Sec. 5). The development of restrictions leads to formulations in terms of Lagrangian multipliers, the global restrictions giving rise to multipliers which are constants, whereas for local restrictions the Lagrangian multipliers depend on position. Those variable Lagrangian multipliers are commonly called *adjoint functions* in the literature, and they take on even greater importance within the framework of Pontryagin's maximum principle, which is introduced in Sec. 5.2.2. The problem with local restrictions can also be studied using the principle of optimality, but for the continuous cases discussed here, the method of Lagrangian multipliers, together with Pontryagin's maximum principle, yields identical results. This important point is discussed in Sec. 5.2.3. The importance of boundary conditions arises again in considering more complex recycle systems in Sec. 5.3.

2 THE FUNDAMENTAL PROBLEM—A SINGLE INDEPENDENT FUNCTION

The fundamental problem is stated mathematically by Eq. (1-2), where y is an objective function, often called a *functional*, for the continuous case. The name functional arises from the fact that y provides a "measure" for the function; i.e., given any function $x(t)$ which is continuous and has continuous derivatives, a number y can be associated with this function through Eq. (1-2). The problem is to find the function $x(t)$ which optimizes the objective function (functional).

Example IX-1: Formulation of a variational problem

To indicate the nature of the problem let us consider a simple example. Suppose that it is desired to determine the curve of minimum length

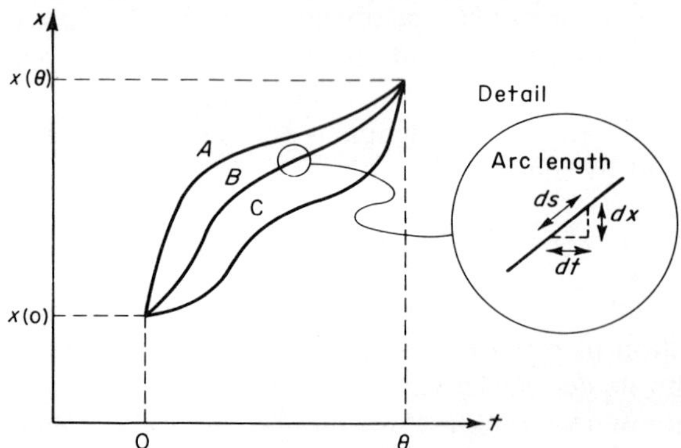

Fig. IX-1 Alternative paths connecting two points.

connecting two points. Several possible curves connecting the points $\{0,x(0)\}$ and $\{\theta,x(\theta)\}$ are shown in Fig. IX-1. Since an element of arc length ds can be written

$$ds = \sqrt{dx^2 + dt^2} = \sqrt{x_t^{\,2} + 1}\, dt$$

the length y of the curve connecting the two points will be

$$y = \int_0^\theta \sqrt{1 + x_t^{\,2}}\, dt \tag{2-1}$$

The function $x(t)$ must now be selected so as to minimize y. This is, clearly, a problem falling into the realm of the calculus of variations, since the integrand depends on an unspecified function which is to be selected so as to optimize the integral. Certainly, curves A, B, and C shown in Fig. IX-1 have different lengths, and if the problem has a solution, one particular curve will yield a shorter length than any other possible curve.

Comparing Eq. (2-1) with the general structure given in Eq. (1-2), we have

$$Y[x(t),x_t(t),t] \equiv \sqrt{1 + x_t^{\,2}}$$

Let us now examine the conditions which $x(t)$ must satisfy to ensure that y is as small as possible. It will be shown in the work to follow that a necessary but not sufficient condition is that $x(t)$ must satisfy the differential equation[1]

$$\frac{\partial Y}{\partial x} - \frac{d}{dt}\left(\frac{\partial Y}{\partial x_t}\right) = 0$$

provided the function $x(t)$ and its first derivative $x_t(t)$ are both continuous in the interval $0 \leq t \leq \theta$. This equation is the Euler-Lagrange equation.

To show that this equation is a necessary condition to be imposed on $x(t)$ if y is to be a minimum or a maximum, it is necessary to introduce the concept of a variation.

2.1 The Variation and Its Properties

Suppose $x^*(t)$ is indeed the function of t which gives y the smallest possible value, and $x(t)$ is a second function of t which is at most infinitesimally different from x^* at every point t within the open interval $(0,\theta)$. Let us define the variation of $x(t)$ as

$$\delta x = x(t) - x^*(t) \tag{2.1-1}$$

Here δ is called the variation symbol. The variation of a function should be understood to represent an infinitesimal change in both the function and its derivative at a given value of t, the change being arbitrary. Thus, although $x(t)$ can be changed by changing t, the variation symbol implies that t is fixed, and the function itself is altered as from curve A to curve B in Fig. IX-2.

[1] The partial derivatives have the following meaning:

$$\frac{\partial Y}{\partial x} \equiv \left(\frac{\partial Y}{\partial x}\right)_{t,x_t} \quad \text{and} \quad \frac{\partial Y}{\partial x_t} \equiv \left(\frac{\partial Y}{\partial x_t}\right)_{t,x}$$

Therefore, in differentiating Y with respect to x, one must treat both t and x_t as independent variables. Of course, the total change in Y is given by

$$dY = \frac{\partial Y}{\partial x}\,dx + \frac{\partial Y}{\partial x_t}\,dx_t + \frac{\partial Y}{\partial t}\,dt$$

Fig. IX-2 The variation shown graphically.

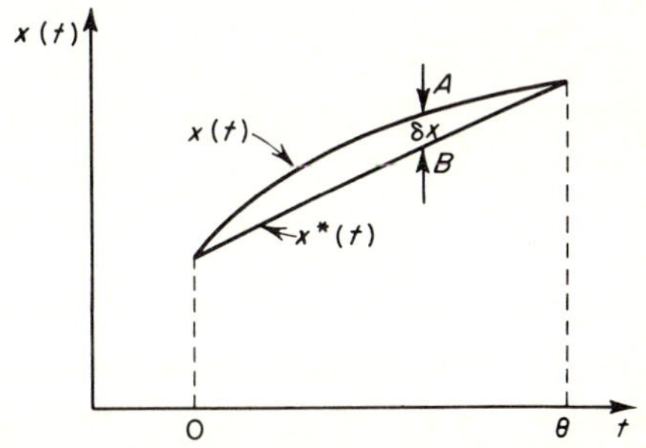

Of course, if the change is to be infinitesimally small, A and B must be close together at every point t.

The variation thus differs from the usual differentiation. The latter gives a measure of the change of a function resulting from a specified change in the independent variable (in this case, t). It is well to keep this difference in mind, since both the variation and differentiation operations will often appear in the same equation. Equation (2.1-1) can be represented also as

$$\delta x(t) = x(t) - x^*(t) = \epsilon\phi(t) \tag{2.1-2}$$

where $\phi(t)$ is an arbitrary but continuous and differentiable function of t, and the variable parameter ϵ tends to zero.

At this point it is useful to develop several properties of the variation operator. Since the variation operator imparts an infinitesimal change to a function at fixed t, then

$$\delta t = 0 \tag{2.1-3}$$

so that the independent variable does not participate in the variation process. Furthermore, it should be noted that the variation operator is interchangeable with both the operations of differentiation and integration. These two valuable properties can easily be demonstrated in the following way: First compute the derivative of a variation

$$\frac{d}{dt}(\delta x) = \frac{d}{dt}[\epsilon\phi(t)] = \epsilon\frac{d\phi}{dt}$$

Next take the variation of a derivative

$$\delta\left(\frac{dx}{dt}\right) = \frac{dx}{dt} - \frac{dx^*}{dt} = \frac{d}{dt}(x - x^*) = \epsilon\frac{d\phi}{dt}$$

Since both computations yield the same result, then

$$\frac{d}{dt}(\delta x) = \delta\left(\frac{dx}{dt}\right) \tag{2.1-4}$$

which demonstrates that the operations of differentiation and variation commute. Integration and variation also commute, since

$$\delta\int_0^\theta x(t)\,dt = \int_0^\theta x(t)\,dt - \int_0^\theta x^*(t)\,dt = \int_0^\theta [x(t) - x^*(t)]\,dt = \int_0^\theta \delta x(t)\,dt$$

giving

$$\delta\int_0^\theta x(t)\,dt = \int_0^\theta \delta x(t)\,dt \tag{2.1-5}$$

and the assertion that the two operators are commutative has been verified.

The change in the value of a functional y resulting from a change or variation in the functional form of $x(t)$ can thus be found as follows:

$$\delta y = y(x) - y(x^*) = \int_0^\theta \{Y[t,x(t),x_t(t)] - Y[t,x^*(t),x_t^*(t)]\}\,dt \quad (2.1\text{-}6)$$

The necessary conditions which must apply for an optimal policy $x^*(t)$ to be found can now be derived from this expression.

2.2 The Necessary and Sufficient Conditions

Let us now return to our original task of determining the function $x(t)$ which optimizes an objective of the form

$$y = \int_0^\theta Y[t,x(t),x_t(t)]\,dt$$

The concept of a variation will prove invaluable in this search. Let us recall the principles which guide the search for a local minimum (or a maximum) of a function of one or more independent variables (Chap. VII, Sec. 2.2). In this search the function is studied in an infinitesimal region about the specified point to determine if the function is the same (to the first order) at all neighboring points. This condition, defining a stationary value at the point where all the first derivatives with respect to all the independent variables vanish, is necessary for the function to be a minimum at the point in question. However, the vanishing of the first derivative is not a sufficient condition to ensure that the stationary point is a minimum, and the second derivatives at least must be examined to establish the character of the stationary point.

A similar situation is found in the calculus of continuous variations. Indeed, a *stationary point* is defined as one for which

$$\delta y = 0 \quad (2.2\text{-}1)$$

for sufficiently small ϵ.

To establish the nature of the stationary point, one must examine higher-order terms in ϵ. To make this point more definite, let us expand the integrand of the first integral on the right-hand side of Eq. (2.1-6) into a Taylor series, as follows:

$$Y[t,x(t),x_t(t)] = Y(t,x^*,x_t^*) + \frac{\partial Y}{\partial x}(x - x^*) + \frac{\partial Y}{\partial x_t}(x_t - x_t^*)$$

$$+ \frac{1}{2}\frac{\partial^2 Y}{\partial x^2}(x - x^*)^2 + \frac{\partial^2 Y}{\partial x\,\partial x_t}(x - x^*)(x_t - x_t^*) + \frac{1}{2}\frac{\partial^2 Y}{\partial x_t^2}(x_t - x_t^*)^2 + \cdots$$

where the t is held fixed and the partial derivatives are evaluated at $x^*(t)$, as is usual in the Taylor expression (Chap. VII, Sec. 2.2). A change in Y

resulting from changes in the function $x(t)$ at each t can then be computed. Substituting this expression into Eq. (2.1-6) and using the alternative definition of the variation as expressed by Eq. (2.1-2), we find that

$$y(x) - y(x^*) = \epsilon \int_0^\theta \left(\phi \frac{\partial Y}{\partial x} + \phi_t \frac{\partial Y}{\partial x_t} \right) dt$$
$$+ \frac{\epsilon^2}{2} \int_0^\theta \left(\phi^2 \frac{\partial^2 Y}{\partial x^2} + 2\phi\phi_t \frac{\partial^2 Y}{\partial x \, \partial x_t} + \phi_t^2 \frac{\partial^2 Y}{\partial x_t^2} \right) dt + \cdots \qquad (2.2\text{-}2)$$

The arguments which can now be put forward are indeed quite analogous to those used in both Chaps. V and VII, where functions of one or more variables were optimized. As in those two chapters, we can find necessary conditions for an optimum to exist by examining terms of first order, and the sufficient conditions stem from investigating second-order changes.

2.2.1 Necessary conditions: the Euler-Lagrange equations

The necessary conditions for an optimum to exist can be developed as follows: Suppose that ϵ is so small that all terms in the Taylor expression of $y(x) - y(x^*)$ can be neglected, except the first. Thus Eq. (2.2-2) for sufficiently small ϵ becomes

$$y(x) - y(x^*) \equiv \delta y = \epsilon \int_0^\theta \left(\phi \frac{\partial Y}{\partial x} + \phi_t \frac{\partial Y}{\partial x_t} \right) dt$$

If $y(x^*)$ is to be an optimum, δy must vanish since ϵ can be either positive or negative, $y(x^*)$ can therefore be a minimum or maximum only if

$$\int_0^\theta \left(\phi \frac{\partial Y}{\partial x} + \phi_t \frac{\partial Y}{\partial x_t} \right) dt = 0 \qquad (2.2\text{-}3)$$

This is analogous to demanding that the first derivatives vanish in the ordinary calculus of maxima and minima.

The second term in the integrand of Eq. (2.2-3) can be integrated by parts, giving

$$\int_0^\theta \phi_t \frac{\partial Y}{\partial x_t} dt = \phi \frac{\partial Y}{\partial x_t} \bigg|_0^\theta - \int_0^\theta \phi \frac{d}{dt} \left(\frac{\partial Y}{\partial x_t} \right) dt \qquad (2.2\text{-}4)$$

If it is assumed that the boundary conditions are specified so that the function $x(t)$ is fixed at the limits of integration, the function $\phi(t)$ must vanish on the boundaries. The first term on the right-hand side will then vanish. Of course, if other boundary conditions are imposed, or if the boundary conditions are free (unspecified), further consideration must be given to the evaluation of the first term (Sec. 4). For the moment assume that the function is specified on the boundary and no variation is permitted at the end points. Using Eq. (2.2-4), Eq. (2.2-3) becomes

$$\delta y = \epsilon \int_0^\theta \left[\frac{\partial Y}{\partial x} - \frac{d}{dt}\left(\frac{\partial Y}{\partial x_t}\right)\right] \phi \, dt = 0 \qquad (2.2\text{-}5)$$

The integration by parts was an essential step in the analysis of the problem. This allowed the integral to be rearranged so as to have the arbitrary[1] function $\phi(t)$ multiplied by a grouping of terms which are dependent on the true solution but not on the choice of $\phi(t)$. Since $\phi(t)$ is arbitrary, and if Eq. (2.2-5) is to be satisfied for all admissible $\phi(t)$, then

$$\frac{\partial Y}{\partial x} - \frac{d}{dt}\left(\frac{\partial Y}{\partial x_t}\right) = 0 \qquad (2.2\text{-}6)$$

Indeed, suppose that Eq. (2.2-6) is not satisfied for all t in the interval $(0,\theta)$. This would mean that

$$\frac{\partial Y}{\partial x} - \frac{d}{dt}\left(\frac{\partial Y}{\partial x_t}\right) \neq 0$$

for some values of t in the range of integration. Clearly, it would be possible to construct a function ϕ (since ϕ is arbitrary) which has the same sign as this term. Therefore the integrand of Eq. (2.2-5) would be positive everywhere that it is not zero. If this is so, Eq. (2.2-5) would not be satisfied, for the integral would be positive. Thus Eq. (2.2-6) must be true for all t in the range of integration if Eq. (2.2-5) is to be true for *all* ϕ. This differential equation is the Euler-Lagrange equation, which must be satisfied if the objective function of the type given by Eq. (1-2) is to be an optimum.

Example IX-1a: The necessary conditions

The Euler-Lagrange equation may now be applied to the problem of determining the shortest distance between two points (Example IX-1). Since

$$\frac{\partial Y}{\partial x} = \frac{\partial}{\partial x}(1 + x_t^2)^{1/2} = 0$$

and

$$\frac{\partial Y}{\partial x_t} = \frac{\partial}{\partial x_t}(1 + x_t^2)^{1/2} = x_t(1 + x_t^2)^{-1/2}$$

[1] The function $\phi(t)$ is not completely arbitrary, for there are three restrictions which must be imposed on this function: the function must be continuous, have a continuous first derivative, and vanish at the end points to be an admissible function.

the Euler-Lagrange equation in this case becomes

$$-\frac{d}{dt}\left[\frac{x_t}{(1+x_t^2)^{1/2}}\right]=0$$

The necessary condition for a minimum is therefore

$$x_t(1+x_t^2)^{-1/2}=\text{constant}$$

which implies that x_t is a constant. Thus, since the curve must pass through the end points, the curve connecting two points and giving the shortest path has a constant slope; i.e., the curve is a straight line. Of course, to ensure that a minimum has indeed been found, further investigation is required. This will be considered later (Sec. 2.2.2).

If the integrand Y of a variational integral does not depend explicitly on t, the Euler-Lagrange equation can be written in the special form

$$\frac{d}{dt}\left(Y-x_t\frac{\partial Y}{\partial x_t}\right)=0 \tag{2.2-7}$$

To demonstrate this identity, this expression may be differentiated to obtain

$$\frac{dY}{dt}-x_{tt}\frac{\partial Y}{\partial x_t}-x_t\frac{d}{dt}\left(\frac{\partial Y}{\partial x_t}\right)=0 \tag{2.2-8}$$

Since Y is presumed not to depend on t explicitly (but only through x and x_t), we can write

$$\frac{dY}{dt}=\frac{\partial Y}{\partial x}x_t+\frac{\partial Y}{\partial x_t}x_{tt}$$

which, when substituted into Eq. (2.2-8), gives

$$\frac{\partial Y}{\partial x}-\frac{d}{dt}\left(\frac{\partial Y}{\partial x_t}\right)=0$$

This is the Euler-Lagrange equation, and the proof is complete.

Fig. IX-3 Surface of revolution.

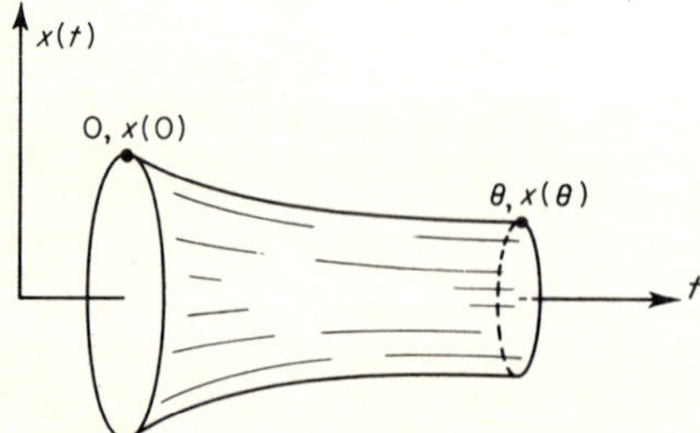

Example IX-2: Minimum surface of revolution

Let us determine the function $x(t)$ which, when rotated about the t axis, passes through the points $\{0,x(0)\}$ and $\{\theta,x(\theta)\}$ giving the minimum surface of revolution. This surface of revolution is depicted in Fig. IX-3. It is not difficult to show that the surface area y of the surface of revolution is given by

$$y = 2\pi \int_0^\theta x(1 + x_t^2)^{1/2}\, dt$$

The $x(t)$ must satisfy the Euler-Lagrange equation if a minimum is to exist. Since the integrand does not depend on t explicitly, the form of the Euler-Lagrange equation expressed by Eq. (2.2-7) is applicable. The necessary condition for an optimum value of the surface area to be obtained is thus

$$\frac{d}{dt}\left[x(1 + x_t^2)^{1/2} - x_t^2 x(1 + x_t^2)^{-1/2}\right] = 0$$

Integration then gives

$$\frac{x}{(1 + x_t^2)^{1/2}} = a$$

Further integration yields the required function $x(t)$, where

$$x(t) = a \cosh\left(\frac{t}{a} + b\right) \tag{2.2-9}$$

The two constants of integration, a and b, can be determined from the specified end points of the curve. Equation (2.2-9) represents a catenary, a more complete and detailed solution of the problem being given in Forsyth [1960].

Example IX-3: Optimal sales policy

Suppose that x_0 items of a certain product are to be sold during a time interval θ. The selling price depends on the rate at which the item is marketed. If the gross profit per item (excluding storage costs) is given by

$$\frac{c_2}{1 + \alpha x_t} \qquad \text{dollars per item sold}$$

compute the optimal sales policy. Here x is the number of the original x_0 items which have been sold. Clearly, the net profit is the gross profit less the storage costs, or

$$y = \int_0^\theta \frac{c_2 x_t}{1 + \alpha x_t}\, dt - \int_0^\theta c_1(x_0 - x)\, dt \tag{2.2-10}$$

if the cost of storing these items is c_1 dollars per item per unit time. Since the optimal policy $x(t)$ must satisfy the Euler-Lagrange equation, $x(t)$ must satisfy

$$c_1 - c_2 \frac{d}{dt} (1 + \alpha x_t)^{-2} = 0$$

Integration gives

$$c_2(1 + \alpha x_t)^{-2} = c_1 t + A$$

where A is an integration constant. Thus

$$\frac{dx}{dt} = \frac{c_2^{1/2}}{\alpha(A + c_1 t)^{1/2}} - \frac{1}{\alpha}$$

the positive sign being selected since the number of items in storage must decrease with time. Since $x = 0$ at $t = 0$, the above equation can be integrated to yield

$$x = \frac{2c_2^{1/2}}{\alpha c_1} [(A + c_1 t)^{1/2} - A^{1/2}] - \frac{t}{\alpha} \tag{2.2-11}$$

Since all the items must be sold by $t = \theta$, then A is determined by

$$x_0 = \frac{2c_2^{1/2}}{\alpha c_1} [(A + c_1 \theta)^{1/2} - A^{1/2}] - \frac{\theta}{\alpha}$$

The optimal sales policy given by Eq. (2.2-11) is shown in Fig. IX-4. It is found to be best if sales are initially rapid, avoiding high storage

Fig. IX-4 Optimal sales policy.

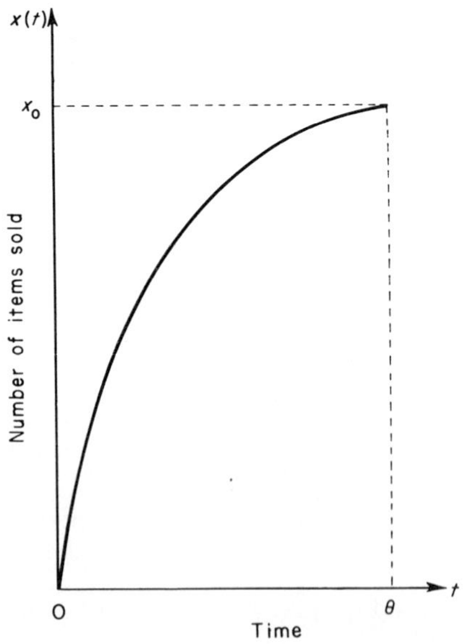

costs. However, one cannot sell too rapidly, since this would depress the selling price.

To show that an optimal policy has been found, one must prove that the stationary point defined by Eq. (2.2-11) yields a maximum profit (see Example IX-5).

It should be noted that the form of the Euler-Lagrange equation expressed by Eq. (2.2-6) would have to be modified if the integrand of the functional depended upon higher-order derivatives than the first. Equation (2.2-6) is a special case, but the one most commonly encountered. Problems IX-1 and IX-2 emphasize that the nature of the Euler-Lagrange equation will depend upon the structure of the integrand.

2.2.2 *Sufficient conditions: the second variation*

In the previous section, the first variation, δy, was defined as

$$\delta y = \epsilon \int_0^Y \left(\phi \frac{\partial Y}{\partial x} + \phi_t \frac{\partial Y}{\partial x_t} \right) dt$$

By analogy, from Eq. (2.2-2), the second variation, $\delta^2 y$, can be defined as

$$\delta^2 y = \frac{\epsilon^2}{2} \int_0^Y \left(\phi^2 \frac{\partial^2 Y}{\partial x^2} + 2\phi\phi_t \frac{\partial^2 Y}{\partial x\, \partial x_t} + \phi_t^2 \frac{\partial^2 Y}{\partial x_t^2} \right) dt \qquad (2.2\text{-}12)$$

Then if

$$\delta^2 y > 0$$

for all possible admissible variations, the stationary value of y will be a minimum, just as in the case of a function of a single independent variable. Similarly, the condition

$$\delta^2 y < 0$$

implies that the stationary value is a maximum. If $\delta^2 y$ is positive for some variations and negative for others, the stationary value is not an extremum. Of course, if $\delta^2 y$ vanishes identically for all variations, the character of the stationary point is still an open question. However, this problem need not be considered here.

The sufficiency condition represented by a positive-definite second variation is a very demanding condition, and it is usually not possible to establish the sufficiency of a particular solution. Because of this difficulty, it is of importance to note an additional necessary condition, the Legendre condition, which we shall not prove. This condition requires that

$$\frac{\partial^2 Y}{\partial x_t^2} \geq 0 \qquad \text{for all } 0 \leq t \leq \theta \qquad (2.2\text{-}13)$$

if the solution is to be a minimum. The inequality sign must be reversed if the solution is to be a maximum. This additional necessary condition arises because the derivative of a function (say, ϕ_t) can be large even when the function (say, ϕ) is small. Looking at Eq. (2.2-12), we see that, if ϕ_t becomes large with ϕ small, the last term dominates, and must therefore necessarily be positive if the second variation is to be positive for all possible variations. This additional necessary condition is easily investigated (Gel'fand and Fomin [1963]).

Example IX-4: The straight line is indeed a minimum

To show that the path $x_t =$ constant, defined in Example IX-1a, is a path of minimum length between two points, we must demonstrate that $\delta^2 y$ is positive everywhere within the interval. Let us check to see if this is the case. Substituting the integrand of Eq. (2-1) into Eq. (2.2-12) gives

$$\delta^2 y = \frac{\epsilon^2}{2}\int_0^\theta \phi_t^2\left(\frac{1}{(1 + x_t^2)^{3/2}}\right) dt$$

which is always positive. The derived path is therefore a minimum.

Example IX-5: Optimal sales policy—sufficient conditions

In Example IX-3 a sales policy was found which yielded a stationary value of the profit. Let us determine the nature of this stationary point.

Substitute the integrand of the objective function defined by Eq. (2.2-10) into the expression for the second variation. Equation (2.2-12) then gives the second variation as

$$\delta^2 y = -\epsilon^2 \int_0^\theta \phi_t^2 \frac{\alpha c_2}{(1 + \alpha x_t)^3}\, dt$$

Since $x_t > 0$, this quantity is always negative, so that the sales policy expressed by Eq. (2.2-11) does produce a maximum profit. This is a local maximum, and is not necessarily the global maximum. Equation (2.1-11) is the only sales policy which satisfies the Euler-Lagrange equation, but it should be recalled that the derivation of this equation depends on the continuity of $x(t)$ and its derivative $x_t(t)$. There may exist a discontinuous sales policy which yields more profit.

3 SEVERAL INDEPENDENT FUNCTIONS

Let us now suppose that Y depends on several functions, $\mathbf{x}(t) = \{x_1(t), \ldots, x_n(t)\}$. The functions must now be determined to render the objective function y stationary. Here y is given by

$$y = \int_0^\theta Y(t,\mathbf{x},\mathbf{x}_t)\,dt \tag{3-1}$$

Define variations so that

$$\delta x_i = x_i - x_i^* = \epsilon_i \phi_i(t) \qquad \text{for } i = 1, 2, \ldots, n \tag{3-2}$$

The difference in the objective function evaluated along some path slightly displaced from the optimal and the optimal path itself will therefore be given by

$$\delta y = y(\mathbf{x}) - y(\mathbf{x}^*) = \int_0^\theta Y(t,\mathbf{x},\mathbf{x}_t)\,dt - \int_0^\theta Y(t,\mathbf{x}^*,\mathbf{x}_t^*)\,dt$$

Expanding the first term on the right-hand side by the Taylor expansion, we get that

$$\int_0^\theta Y(t,\mathbf{x},\mathbf{x}_t)\,dt = \int_0^\theta Y(t,\mathbf{x}^*,\mathbf{x}_t)\,dt + \sum_{i=1}^n \epsilon_i \int_0^\theta \left(\frac{\partial Y}{\partial x_i}\phi_i + \frac{\partial Y}{\partial x_{i_t}}\phi_{i_t}\right) dt + \cdots$$

Thus δy, to the first order in ϵ_i, becomes

$$\delta y = \sum_{i=1}^n \epsilon_i \int_0^\theta \left(\frac{\partial Y}{\partial x_i}\phi_i + \frac{\partial Y}{\partial x_{i_t}}\phi_{i_t}\right) dt$$

Integrating by parts gives

$$\delta y = \sum_{i=1}^n \epsilon_i \int_0^\theta \left[\frac{\partial Y}{\partial x_i} - \frac{d}{dt}\left(\frac{\partial Y}{\partial x_{i_t}}\right)\right]\phi_i\,dt$$

assuming that the $x_i(t)$ are specified at the end points of the range of integration.

Since all the ϵ_i are independent and each can be positive or negative, then

$$\int_0^\theta \left[\frac{\partial Y}{\partial x_i} - \frac{d}{dt}\left(\frac{\partial Y}{\partial x_{i_t}}\right)\right]\phi_i\,dt = 0 \qquad \text{for } i = 1, 2, \ldots, n$$

if δy is to vanish for all ϵ_i. Moreover, ϕ_i is arbitrary, except for requirements of continuity, and therefore this equation has to be satisfied for all ϕ_i, so that

$$\frac{\partial Y}{\partial x_i} - \frac{d}{dt}\left(\frac{\partial Y}{\partial x_{i_t}}\right) = 0 \qquad \text{for } i = 1, 2, \ldots, n \tag{3-3}$$

Each function $x_i(t)$ must therefore satisfy a Euler-Lagrange equation as a necessary condition for the existence of an optimum value of the objective function.

The derivation here has been analogous to that exhibited in Sec. 2. Clearly, a set of sufficient conditions can also be developed following the approach presented in Sec. 2.2.2. However, this point will not be pursued further here.

4 BOUNDARY CONDITIONS AND VARIABLE LIMITS

Up to this point in this chapter the boundary conditions have been assumed to be specified; i.e., the function subject to variation has been fixed on the boundary. This assumption enters the mathematical arguments by forcing those admissible variations about the true solution to vanish on the boundary. This admissibility condition permitted, for example, the discarding of the boundary terms arising in the integration by parts which was a necessary step in the development of the Euler-Lagrange equation.

The second imposed boundary condition was the assumption of fixed limits of integration. This restriction must also on occasion be relaxed, since optimization problems are encountered in which the limits are not specified.

These two problems of free boundary conditions and variable limits are considered together here, since they are intimately related. First, however, let us develop some insight into the relationship of the boundary conditions to the variational problem.

4.1 Natural Boundary Conditions

Many problems will be encountered where the function to be varied is not specified on the boundary but other equally valid boundary conditions are imposed. Indeed, in some instances, no boundary conditions at all need be specified. It is important to understand that the boundary conditions are an integral part of the variational investigation, and any alteration of the boundary condition produces a corresponding change in the minimizing (or maximizing) function. Furthermore, if the function is not specified on the boundary, an appropriate number of boundary conditions will appear "naturally" during the course of the variational analysis. This is one of the particularly elegant features of the calculus of variations. To emphasize this point, let us consider a specific example. Consider the variational problem defined by the following equation:

$$y = \int_0^\theta [\tfrac{1}{2}x_t^2 - x\chi(t)]\,dt \tag{4.1-1}$$

where $\chi(t)$ is a specified function of t, and $x(t)$ is the function to be selected to produce a stationary value of y. For the present, leave the question of boundary conditions open. Thus, in the definition of a variation in $x(t)$,

$$\delta x = x(t) - x^*(t) = \epsilon\phi(t)$$

there is no reason to presume that $\phi(t)$ vanishes on the boundaries ($t = 0$ and $t = \theta$). It is only in those cases for which $x(t)$ is specified on the boundary that one can assert that the variation vanishes on the boundary. The first variation of Eq. (4.1-1) is

$$\delta y = \epsilon \int_0^\theta [\phi_t x_t - \phi \chi(t)]\, dt$$

On integrating by parts, the first term in the integrand gives

$$\delta y = -\epsilon \int_0^\theta \left[\frac{d^2x}{dt^2} + \chi(t)\right]\phi(t)\, dt + \epsilon \frac{dx}{dt}\phi\bigg|_{t=0}^{t=\theta}$$

If δy is to be zero for all admissible variations (restricted only by continuity requirements), then both the Euler-Lagrange equation

$$\frac{d^2x}{dt^2} + \chi(t) = 0 \tag{4.1-2}$$

and the boundary conditions

$$\frac{dx}{dt}\phi\bigg|_{t=\theta} = \frac{dx}{dt}\phi\bigg|_{t=0} = 0 \tag{4.1-3}$$

must be satisfied. The latter implies that the derivative must vanish on the boundary if the variation does not vanish. Note that this particular Euler-Lagrange equation (4.1-2) is a second-order differential equation and requires the specification of two boundary conditions to achieve a unique solution.

Since no boundary conditions have been specified and since Eq. (4.1-3) must be valid for all values of $\phi(0)$ and $\phi(\theta)$, it is clear that

$$\frac{dx}{dt} = 0 \qquad \text{for both } t = 0 \text{ and } t = \theta \tag{4.1-4}$$

These are the *natural boundary conditions* for the particular example under consideration, and are evolved directly from the variational problem.[1] A solution can be expected to exist only under very special circumstances. Integrating Eq. (4.1-2), we get

$$\frac{dx}{dt} = -\int_0^t \chi(\lambda)\, d\lambda$$

The condition

$$\frac{dx}{dt} = 0 \qquad \text{at } t = 0$$

is then satisfied. However, to satisfy the condition at $t = \theta$, the following equality must be satisfied:

$$0 = \int_0^\theta \chi(\lambda)\, d\lambda \tag{4.1-5}$$

[1] These natural boundary conditions represent perfectly insulated surfaces in a solid slab in which there is internal heat generation of amount $\chi(t)$.

This condition is necessary for the existence of a solution compatible with the boundary conditions which have been naturally imposed on the problem.

If the boundary conditions are thus left unspecified, a set of natural boundary conditions arises which is derived from the variational statement of the problem. Furthermore, the correct number of boundary conditions required to solve the Euler-Lagrange equation have appeared naturally. It is also clear that the specification of $x(t)$ on one boundary (say, at $t = 0$) would impose the restriction of a vanishing $\phi(t)$ on that boundary and would leave only one natural boundary condition. This boundary condition would be selected so as to satisfy Eq. (4.1-3), with

$$\phi(t) = \begin{cases} 0 & \text{at } t = 0 \\ \text{arbitrary} & \text{at } t = \theta \end{cases}$$

yielding the boundary conditions $x(t)$ at $t = 0$ and $dx/dt = 0$ at $t = \theta$. Note that again a number of boundary conditions have appeared appropriate to the order of the Euler-Lagrange equation. These are the two boundary conditions required for the present problem described by second-order differential equation (4.1-2).

What about the possibility of imposing other types of boundary conditions on the solution? For example, boundary conditions of the following type are frequently encountered in heat-conduction problems, where they represent the radiation boundary condition

$$-\frac{dx}{dt} + \alpha x = 0 \qquad \text{at } t = 0 \tag{4.1-6}$$

Here α is a constant. Thus the question arises as to how one can construct a variational principle which has the heat-conduction equation as its Euler-Lagrange equation and which is applicable with Eq. (4.1-6) as one of the boundary conditions. This type of problem can usually be handled by altering the variational problem. Consider the objective function

$$y = \int_0^\theta [\tfrac{1}{2}(x_t)^2 - x\chi(t)]\, dt + \frac{\alpha}{2} x^2 \Big|_{t=0} \tag{4.1-7}$$

where α is the same constant which appears in Eq. (4.1-6). Note that a term involving the function evaluated at the boundary has been added to Eq. (4.1-1) to obtain this new objective function. The first variation of Eq. (4.1-7) is

$$\delta y = -\epsilon \int_0^\theta \left[\frac{d^2x}{dt^2} + \chi(t)\right]\phi(t)\, dt + \epsilon\phi(t)\frac{dx}{dt}\Big|_{t=\theta} + \epsilon\left(\alpha x - \frac{dx}{dt}\right)\phi(t)\Big|_{t=0}$$

The last term of this expression is easily derived if it is recalled that the variation is defined as an infinitesimal change of the function $x(t)$ at a given value of t (Sec. 2). Since δy must vanish for all variations, one of the natural boundary conditions is

$$-\frac{dx}{dt} + \alpha x = 0 \qquad \text{at } t = 0$$

which is the desired result. It has been shown that the function $x(t)$ which results in a stationary state for the objective function y defined by Eq. (4.1-7) must satisfy the heat-conduction equation, the boundary condition expressed by Eq. (4.1-6), and have a zero derivative at $t = \theta$. Hence the natural boundary conditions for the variational problem have been modified by adding a term to the objective function. It is not difficult to visualize other possible modifications which give rise to different boundary conditions.

The essential result of our work in this section is the recognition that the answer derived from a variational formulation is dependent on the nature of the boundary conditions. In the absence of fixed boundary conditions, the variational problem will furnish itself with a sufficient number of restrictions to permit the solution of the Euler-Lagrange equation. Furthermore, suitable boundary conditions can often be built into the problem by artificially adding terms to the variational problem.

Example IX-6: The effect of boundary conditions

Consider the minimum of the objective function

$$y = \int_0^1 [x_t^2 - 2(\cos \pi t)x(t)]\, dt$$

If there are no boundary conditions imposed on the problem, then

$$\frac{dx}{dt} = 0 \qquad \text{at both } t = 0 \text{ and } t = 1$$

The Euler-Lagrange equation becomes

$$\frac{d^2x}{dt^2} + \cos \pi t = 0$$

which can be integrated to yield a solution satisfying these natural boundary conditions. This solution is

$$x = \frac{1}{\pi^2} \cos \pi t + B$$

where B is an undetermined constant. On substituting this result into the objective function, the smallest value of y which can be found is

$$y = -\frac{1}{2\pi^2}$$

The second variation can be used to check that this is a minimum value, assuming y is a continuous function. Any other function of t, for example,

$$x = at$$

yields a larger value of y, no matter what choice of a is made. Moreover, if boundary conditions are imposed, the minimum value of the objective function may increase, since $y = -1/2\pi^2$ is the smallest value of the objective function using any continuous function, $x(t)$, having a continuous first derivative.

If it is also necessary to include the boundary conditions

$$x = 0 \qquad \text{at } t = 0 \text{ and } t = 1$$

the solution to the Euler-Lagrange equation, then, is

$$x = \frac{1}{\pi^2}\cos \pi t + \frac{2}{\pi^2}t - \frac{1}{\pi^2}$$

The corresponding value of y is

$$(y)_{\text{subject to boundary conditions}} = \frac{4}{\pi^4} - \frac{1}{2\pi^2}$$

Therefore, in general, we may write that

$$(y_{\min})_{\text{subject to boundary conditions}} \geq (y_{\min})_{\text{natural boundary conditions}} \tag{4.1-8}$$

Thus, as in our previous work, the imposition of constraints yields a constrained minimum which cannot be less than an unconstrained one.

Example IX-7: Natural boundary conditions for maximum profit

Suppose the profit for a process is determined by

$$y = \int_0^\theta (x_t^2 + \beta x^2 - xx_t)\, dt$$

where $x(t)$ is the sales volume per unit of time, and β is a constant. Moreover, assume that the initial availability of manufactured product, $x(0)$, is determined by some outside influence. However, the end-of-the-year sales rate, $x(\theta)$, is not fixed by any limitation. Therefore the terminal sales policy (terminal in sense of being year-end) should be selected so that y is maximized. Therefore the terminal policy should be determined by the natural boundary conditions, because an unrestricted optimum is the best possible situation, as shown by Eq. (4.1-8). It can be shown that the Euler-Lagrange equation is then

$$-2\frac{d^2x}{dt^2} + 2\beta x = 0$$

and the natural boundary condition at $t = \theta$ is

$$2\frac{dx}{dt} = x \qquad \text{at } t = \theta$$

Since the specified condition can be stated as

$$x = x_0 \qquad \text{at } t = 0$$

being the x_0 required initial policy, the second-order Euler-Lagrange equation has two boundary conditions, one of which is natural. Imposing any other boundary condition at $t = \theta$ will decrease the profits.

4.2 General Boundary Conditions

Let us now generalize the preceding discussion. Consider the objective function

$$y = \int_0^\theta Y(t,x,x_t)\, dt \tag{4.2-1}$$

The $x(t)$, unrestricted by boundary conditions, is to be selected so as to optimize y. Moreover, the upper limit of integration is not specified.

For example, in the design of a chemical reactor, the length of the reactor may be subject to change between some practical limits. Here the length would be the upper limit of integration. If the upper limit is subject to variation, there exists some optimal length, θ^*, as well as some optimal function, $x^*(t)$. Therefore the difference between the optimal value of the objective function and a value generated by an infinitesimal change in $x(t)$ and θ is given as

$$y - y^* = \int_0^\theta Y(t,x,x_t)\, dt - \int_0^{\theta^*} Y(t,x^*,x_t^*)\, dt \tag{4.2-2}$$

If the first integral on the right is written[1]

$$\int_0^\theta Y(t,x,x_t)\, dt = \int_0^{\theta^*} Y(t,x,x_t)\, dt + \int_{\theta^*}^\theta Y(t,x,x_t)\, dt \tag{4.2-3}$$

then the resulting first term on the right can be expanded into a Taylor series, as was done in Sec. 2.2.1. Again, let

$$x(t) - x^*(t) = \epsilon\phi(t)$$

[1] Note that θ does not necessarily have to be larger than θ^*. If indeed θ^* is larger, then $x(t)$ must be extended so as to be defined in the range θ–θ^*. This can be done by drawing tangents to the curves, as shown in Fig. IX-5. For a rigorous treatment of this problem the reader is referred to Gel'fand and Fomin [1963, chapter 3].

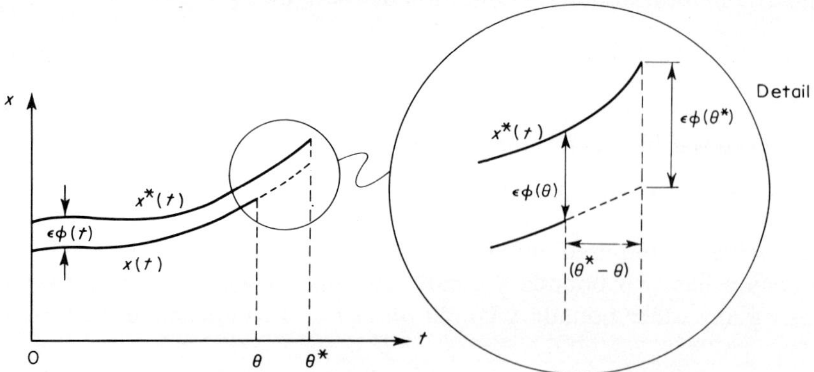

Fig. IX-5 Variation of the upper limit. If $\theta^* > \theta$, extend $x(t)$ linearly so that the integrals of Eq. (4.2-3) are all meaningful.

The second term also requires some attention. The integrand can certainly be expanded into a Taylor series, so that

$$\int_{\theta^*}^{\theta} Y(t,x,x_t)\,dt \approx \int_{\theta^*}^{\theta} \left[Y(t,x^*,x_t^*) + \frac{\partial Y}{\partial x}(\epsilon\phi) + \frac{\partial Y}{\partial x_t}(\epsilon\phi_t) \right] dt$$

neglecting powers of ϵ higher than the first. If the difference $\theta - \theta^*$ is small, i.e., of order ϵ, this expression can be further simplified using the first mean value theorem of iinegral calculus (see, for example, Love and Rainville [1962]), getting

$$\int_{\theta^*}^{\theta} Y(t,x,x_t)\,dt \approx (\theta - \theta^*)\,Y(\theta^*,x^*,x_t^*) \tag{4.2-4}$$

Using these two results, Eq. (4.2-2) therefore becomes

$$\delta y = \epsilon \int_0^{\theta^*} \left(\frac{\partial Y}{\partial x}\phi + \frac{\partial Y}{\partial x_t}\phi_t \right) dt + (\theta - \theta^*)\,Y(t,x^*,x_t^*)\,\big|_{t=\theta^*}$$

where $y - y^* = \delta y$ if all but the first-order terms are neglected. Finally, an integration by parts gives the result

$$\delta y = \epsilon \int_0^{\theta^*} \left[\frac{\partial Y}{\partial x} - \frac{d}{dt}\left(\frac{\partial Y}{\partial x_t}\right) \right] \phi\,dt - \epsilon \frac{\partial Y}{\partial x_t}\phi \bigg|_{t=0} + \epsilon \frac{\partial Y}{\partial x_t}\phi \bigg|_{t=\theta^*} + (\theta - \theta^*)\,Y(t,x^*,x_t^*)\,\big|_{t=\theta^*} \tag{4.2-5}$$

The terms evaluated at $t = 0$ and $t = \theta^*$ give rise to natural boundary conditions. Regardless of the form of these boundary conditions, however, the integrand of the integral must vanish for all ϕ if δy is to vanish, and so the Euler-Lagrange equation is still a necessary condition for the existence of an optimum, whatever the form of the boundary condition. The problem will be reduced no further, but the application of Eq. (4.2-5) is shown in Example IX-8. Of course, if $\theta - \theta^*$ and $\epsilon\phi(\theta^*)$ are independent, their coefficients must vanish individually. However, in some cases these variations are related. Example IX-8 considers such a case.

5 OPTIMIZATION WITH RESTRICTIONS

In the formulation of optimization problems, as discussed in Chap. II, Sec. 5, restrictions must often be considered in seeking the optimal policy. By and large, restrictions on continuous variables tend to be of precisely the same general form as those found in the discrete case. For example, there will be cases in which the $x(t)$ is to be found within certain limits, or the integral of $x(t)$ may be constrained to be a certain value. Thus the optimization may be restricted to values of $x(t)$ which satisfy

$$\int_0^\theta G[x(t)]\,dt = J \tag{5-1}$$

Such limitations arise, for example, in distributing a fixed total amount of solvent or heat in an optimal manner.

Of course, one may also require that the optimization be such that certain differential equations, material or energy balances, for example, be satisfied. The significance of these restrictions has been detailed in Chap. II. Restrictions of the form given by Eq. (5-1) are global-type restrictions, since the $x(t)$ must conform to an integrated restriction. Material-balance restrictions and limits are local restrictions, since they are imposed at every point in t.

This presentation considers the two classes of problem separately. Of course, the real physical problem may require both global and local restrictions to be imposed.

5.1 Global Restrictions

Let us examine first the effect of a single global restriction, and then the effect of more than one such restriction.

5.1.1 Single global restrictions

Our task is to find the function $x(t)$ which causes the objective function

$$y = \int_0^\theta Y(t,x,x_t)\,dt \tag{5.1-1}$$

to be stationary when subject to the global restriction

$$J = \int_0^\theta G(t,x,x_t)\,dt \tag{5.1-2}$$

the integrals having identical fixed limits and in which G is specified and the constant J is given. It is clear that one need not test all possible functions in Eq. (5.1-1) to determine the function which causes y to be the largest (or the smallest). For any admissible function, $x(t)$ must first satisfy Eq. (5.1-2), before the function need be considered as a possible solution to the problem. Thus Eq. (5.1-2) must be satisfied by all admissible functions. Because of this restriction, we can no longer use our previous definition of a variation as

$$\delta x = x - x^* = \epsilon\phi$$

This can be seen by noting that $x = x^* + \epsilon\phi$ gives a value $\int_0^\theta G(x,x_t,t)\,dt$, which in general depends on ϵ, rather than being the required constant. The one-parameter (ϵ) family of variations, δx, is in general not admissible.

Our problem is to construct a trial function x which can be made infinitesimally different from x^* and yet satisfy the imposed constraint. To satisfy this need we introduce a two-parameter family of variations,

$$\delta x = x - x^* = \epsilon_1\phi_1 + \epsilon_2\phi_2 \tag{5.1-3}$$

where both ϕ_1 and ϕ_2 are arbitrary differentiable functions which vanish at both limits of integration. If we substitute $x = x^* + \epsilon_1\phi_1 + \epsilon_2\phi_2$ into $\int_0^\theta G\,dt$, the value of the integral depends on both ϵ_1 and ϵ_2, and accordingly, we can write that

$$J(\epsilon_1,\epsilon_2) = \int_0^\theta G(x,x_t,t)\,dt$$

Clearly, ϵ_1 and ϵ_2 are related by this constraint and are not, therefore, independent. One of the parameters can then be expressed in terms of the other, but such an elimination is in general too complex to be practical. We shall make use of the method of Lagrangian multipliers.

As before, a necessary condition for the existence of an optimum is that

$$\delta y = \int_0^\theta [Y(x,x_t,t) - Y(x^*,x_t^*,t)]\,dt = 0$$

for all sufficiently small admissible variations. The admissible variations are those which satisfy the constraint

$$\delta J = 0 = \int_0^\theta [G(x,x_t,t) - G(x^*,x_t^*,t)]\,dt$$

As noted,. one of the ϵ's can be eliminated; instead, let us introduce the Lagrangian multiplier, as in the development of Chap. VII, Sec. 3.3. We therefore seek the condition which gives

$$\delta F = \delta y + \lambda\delta J = 0$$

for all sufficiently small ϵ_1 and ϵ_2. Expanding in a Taylor series as in Sec. 2.1, we obtain the result

$$\delta F = \int_0^\theta \left[\epsilon_1\left(\frac{\partial Y}{\partial x} - \frac{d}{dt}\frac{\partial Y}{\partial x_t}\right)\phi_1 + \epsilon_2\left(\frac{\partial Y}{\partial x} - \frac{d}{dt}\frac{\partial Y}{\partial x_t}\right)\phi_2\right] dt$$
$$+ \lambda \int_0^\theta \left[\epsilon_1\left(\frac{\partial G}{\partial x} - \frac{d}{dt}\frac{\partial G}{\partial x_t}\right)\phi_1 + \epsilon_2\left(\frac{\partial G}{\partial x} - \frac{d}{dt}\frac{\partial G}{\partial x_t}\right)\phi_2\right] dt$$

Since ϵ_1 and ϵ_2 are independent, then, by following the arguments set forth in Sec. 3, the necessary condition for δF to vanish will lead to the new Euler-Lagrange equation

$$\left(\frac{\partial Y}{\partial x} + \lambda\frac{\partial G}{\partial x}\right) - \frac{d}{dt}\left(\frac{\partial Y}{\partial x_t} + \lambda\frac{\partial G}{\partial x_t}\right) = 0 \qquad (5.1\text{-}4)$$

Clearly, this same result is obtained if the modified objective function

$$F = y + \lambda J = \int_0^\theta (Y + \lambda G)\, dt = \int_0^\theta F_s\, dt \qquad (5.1\text{-}5)$$

is made stationary. Thus a necessary condition for the function $x(t)$ to optimize the objective function y is that both Eqs. (5.1-4) and (5.1-2) be satisfied.

Example IX-8: A global restriction

Let us show that the sphere is the solid of revolution having the maximum volume for a given surface area.

The volume of the solid of revolution is found by summing the volume elements, such as shown in Fig. IX-6. The volume of the particular element shown in $2\pi x\, dx\, dt$, and the total volume is

$$y = \int_0^\theta \int_0^{x(t)} 2\pi x\, dx\, dt = \pi \int_0^\theta x^2\, dt$$

The surface area of the ring shown in Fig. IX-6 is $2\pi x\, ds$ ($ds = \sqrt{1 + x_t^2}\, dt$), and the total area is

$$J = 2\pi \int_0^\theta x\, ds = 2\pi \int_0^\theta x\sqrt{1 + x_t^2}\, dt$$

Define

$$F = y + \lambda J = \pi \int_0^\theta (x^2 + 2\lambda x\sqrt{1 + x_t^2})\, dt$$

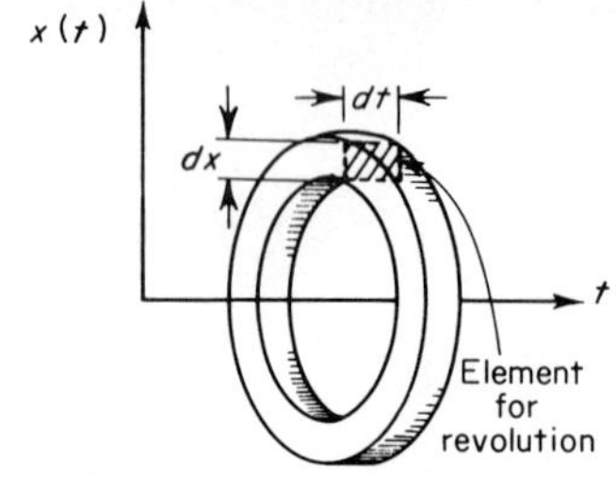

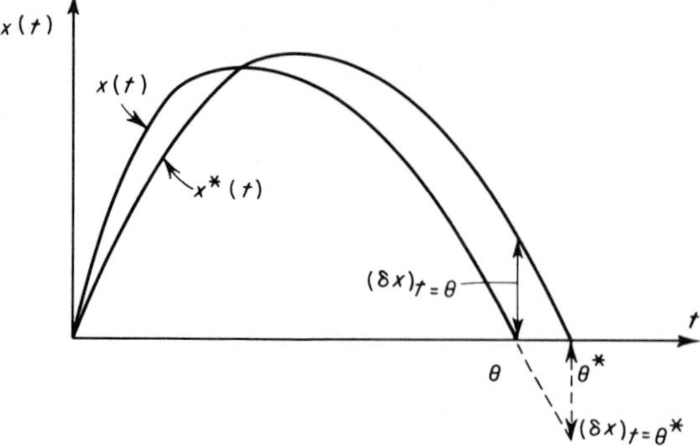

Fig. IX-6 Solid of revolution.

where λ is a Lagrangian multiplier. Since F must be stationary with respect to all admissible variations about x^*, Eq. (4.2-5) yields the necessary conditions

$$\lambda \frac{2x^* x_t^* \, \delta x}{\sqrt{1 + x_t^{*2}}} \bigg|_{t=\theta^*} + (\theta - \theta^*)(x^{*2} + 2\lambda x^* \sqrt{1 + x_t^{*2}}) \big|_{t=\theta^*} = 0$$

and

$$\frac{d}{dt}\left[x^{*2} + 2\lambda x^* \sqrt{(1 + x_t^{*2})} - \frac{2\lambda x^* x_t^{*2}}{\sqrt{1 + x_t^{*2}}}\right] = 0$$

where the latter expression is the modified form of the Euler-Lagrange equation (2.2-7), valid for those integrands which do not depend on t explicitly. The boundary condition arises at $t = \theta^*$ because the point at which the surface of revolution intersects the t axis is subject to variation; that is, θ is not specified in advance. Let us interpret this boundary condition in terms of the particular problem under consideration. Note that the detail of Fig. IX-6 indicates that, to the first order,

$$\frac{dx^*}{dt}\bigg|_{t=\theta^*} = \frac{\delta x|_{t=\theta^*}}{\theta^* - \theta}$$

Therefore $$\theta^* - \theta = \left.\frac{\delta x}{x_t^*}\right|_{t=\theta^*} \qquad \text{for } x_t^* \neq 0 \text{ at } t = \theta^*$$

In this problem the value of $\theta^* - \theta$ is then related to the variation in x, δx. This situation arises because both θ and x are geometric quantities, although, in general, the two quantities need not be related. Suppose, for example, $x(t)$ is a temperature profile or pressure profile in a tubular reactor, and θ represents the length of the reactor. Clearly, the $x(t)$ can be varied at the end of reactor in a way which is independent of the length of the reactor. In this case, there would not be a relationship between δx and $\delta\theta$. However, in the present example, a variation in shape of the revolution surface can produce a corresponding change in θ. The precise relationship depends on the derivative of the solution at the crossing point, as shown above. Thus the boundary condition becomes

$$\delta x \left.\left(\frac{2\lambda x^* x_t^*}{\sqrt{1 + x_t^{*2}}} - \frac{x^{*2} + 2\lambda x^*\sqrt{1 + x_t^{*2}}}{x_t^*}\right)\right|_{t=\theta^*} = 0$$

for $x_t^* \neq 0$ and for all δx.

The coefficient of δx vanishes, as it must, because $x^* = 0$ at $t = \theta^*$, with this coefficient still vanishing in the limit as x_t^* becomes infinite.

The Euler-Lagrange equation can be integrated to yield

$$x^2 - \frac{2\lambda x}{\sqrt{1 + x_t^2}} = C$$

Since there should be no confusion with $x(t)$ as the optimal configuration, the asterisks have been omitted. Since x vanishes at $t = 0$, C must also vanish. Therefore

$$x_t = \frac{dx}{dt} = \frac{\sqrt{4\lambda^2 - x^2}}{x}$$

Integrating a second time gives

$$-\sqrt{4\lambda^2 - x^2} = t + A$$

where A is a constant of integration. Since the surface of revolution emanates from the origin ($t = 0$, $x = 0$), then

$$A = -2\lambda$$

and $$4\lambda^2 = x^2 + (t - 2\lambda)^2$$

Such results are typical when using Lagrangian multipliers. The values of x which make F stationary are given in terms of the Lagrangian multiplier. How should λ be chosen? So far we have not attempted to ensure that the surface area is maintained at the prescribed value J. The λ must therefore be selected so that

$$J = 2\pi \int_0^\theta x(1 + x_t^2)^{1/2}\, dt = 4\pi\lambda\theta$$

Since $x = 0$ at $t = \theta$, we see that

$$\theta = 4\lambda$$

and

$$J = 16\pi\lambda^2$$

The problem is then solved. Clearly, the surface of revolution having the greatest volume for a fixed area J is a sphere of a radius 2λ, where λ is given in terms of J.

5.1.2 Multiple global restrictions

If there are s restrictions of the type

$$J_i = \int_0^\theta G_i(t,x,x_t)\, dt \qquad \text{for } i = 1, 2, \ldots, s \tag{5.1-6}$$

all dependent on the same variable, $x(t)$, and having identical limits, the analysis given in the preceding paragraph can be extended. We are thus interested in finding the stationary state of the objective function F,

$$F = y + \sum_{i=1}^{s} \lambda_i J_i = \int_0^\theta F_s\, dt$$

where

$$F_s = Y + \sum_{i=1}^{s} \lambda_i G_i$$

The Euler-Lagrange equation, to be satisfied, must therefore be

$$\frac{\partial F_s}{\partial x} - \frac{d}{dt}\left(\frac{\partial F_s}{\partial x_t}\right) = 0$$

as before.

5.2 Local Restrictions with "Simple" Boundary Conditions

In many optimization problems in continuous systems certain restrictions are often imposed at every point along the path of integration. Such restrictions, called *local restrictions*, can often be treated using methods involving Lagrangian multipliers and Pontryagin's maximum principle (Secs. 5.2.1 and 5.2.2) or dynamic programming (Sec. 5.2.3). Indeed, it will be shown here that both methods yield the same necessary conditions and can be considered to be identical when used to treat this class of problem. Since computer calculations use discrete models, however, these methods are computationally different.

The development is split into two parts. In the first part, we are concerned with the methods of handling local restrictions. In the second part, the same problem is again considered, with attention focused upon the boundary condition, being primarily concerned with the type of coupled boundary conditions arising in more complex systems, for example, recycle systems. While this division is not clearly defined, it does focus attention on the essential problem.

Before beginning a general approach to this important problem, let us consider an example which can be solved by an elimination procedure. In this way we can develop an understanding of the need for more powerful methods of studying such optimization problems.

Example IX-9: A simple problem with a local restriction

Let us maximize the yield of A_2 in the reaction

$$A_1 \rightarrow A_2$$

by selecting an appropriate temperature profile along the length of an existing reactor, in which the reaction kinetics are given by the rate expression

$$\frac{dC}{dt} = -kCe^{-x(t)}$$

Here C is the concentration of A_1, and $x(t)(= E/RT)$ is inversely proportional to the temperature. Thus the objective function

$$y = C_\theta - C_0 = -\int_0^\theta kCe^{-x(t)}\,dt$$

must be minimized, since reducing A_1 to a minimum in the effluent is equivalent to maximizing the production of A_2. At the same time the rate equation must be satisfied at every point t. Since variations in $x(t)$ give rise to variations in $C(t)$, these are not two independent functions. To state the problem properly, one must account for this coupling of x and C through the kinetic equation. One acceptable method is to solve for C in terms of x and substitute into the objective function. However, in this case, the elimination can be made far simpler by noting that

$$\ln\frac{C_\theta}{C_0} = -k\int_0^\theta e^{-x}\,dt \tag{5.2-1}$$

where C_0 is the inlet concentration of substance A into the reactor. Since the logarithm is a monotonic function of C_θ, then minimizing C_θ gives the same result as minimizing the logarithm of C_θ. Clearly, the solution to the problem is to make the x as small as possible, or the temperature as large as possible, since x is inversely proportional to the temperature.

In this example, the elimination of C from the integrand is analogous to the direct substitution of variables from an objective function when equality restrictions are applied, as discussed in Chap. VII, Sec. 3.1. However, this direct elimination of variables is generally quite complex, and the problem cannot always be put in the form of Eq. (5.2-1), in which the integrand depends only on the independent variables. Methods must now be derived to circumvent the need for direct elimination.

5.2.1 *The method of Lagrangian multipliers*

Let us consider the general problem of optimizing

$$y = \int_0^\theta Y(x,T)\,dt \tag{5.2-2}$$

subject to the local condition

$$\frac{dx}{dt} = f(x,T) \tag{5.2-3}$$

Here we shall attempt to select $T(t)$ and $x(t)$ so as to optimize y, although the two functions are related through the restriction. To treat such problems, Courant and Hilbert [1953] introduce a Lagrangian multiplier, so that the problem is now equivalent to finding the stationary point of F, where

$$F = \int_0^\theta \left\{ Y(x,T) - \lambda(t)\left[\frac{dx}{dt} - f(x,T)\right]\right\} dt \tag{5.2-4}$$

Here $\lambda(t)$ is an undetermined multiplier, which is, in general, a function of position. The undetermined multiplier is a function of position because the local condition must be satisfied at every point $0 \leq t \leq \theta$, rather than being a *global* restriction. In using the undetermined multiplier, the variables x and T are "uncoupled." If x and T are to optimize y, it is necessary that the two Euler-Lagrange equations

$$\frac{\partial Y}{\partial T} + \lambda(t)\frac{\partial f}{\partial T} = 0 \tag{5.2-5}$$

and

$$\frac{\partial Y}{\partial x} + \frac{d\lambda}{dt} + \lambda(t)\frac{\partial f}{\partial x} = 0 \tag{5.2-6}$$

with the two boundary conditions

$$x = x_0 \qquad \text{at } t = 0 \tag{5.2-7}$$

and

$$\lambda = 0 \qquad \text{at } t = \theta \tag{5.2-8}$$

must be satisfied. There are several important features to be emphasized. Equation (5.2-6) determines the $\lambda(t)$ as a function of t. Since λ was taken as being independent of variations in T and x, the functions Y and f appearing in Eq. (5.2-6) are understood to be evaluated at the optimal conditions. The λ is sometimes called the *adjoint function*. The properties of this adjoint function will be examined further. Note that the boundary condition on x is defined at $t = 0$, whereas the boundary condition on λ is at $t = \theta$. This state of affairs makes the problem more difficult, since, if the integration is begun at $t = 0$, the value of λ is unknown, whereas, if one tries to integrate backward from $t = \theta$, the value of x is unknown.

Example IX-10: Optimum under local restriction

Find functions $x(t)$ and $T(t)$ which minimize

$$y = \int_0^\theta (x^2 + T^2)\, dt \tag{5.2-9}$$

given that x and T are related by the differential equation

$$\frac{dx}{dt} = T - x \tag{5.2-10}$$

and the boundary condition is

$$x = 1 \qquad \text{at } t = 0$$

It is clear that Eq. (5.2-10) can be integrated to give x in terms of T, and the resulting expression used to eliminate x from Eq. (5.2-9). The appropriate minimizing function $T(t)$ can then be determined using the classical methods of variational calculus. However, let us apply Eqs. (5.2-5) to (5.2-8) to this particular problem. Equations (5.2-5) and (5.2-6) give, respectively,

$$2T + \lambda = 0 \tag{5.2-11}$$

and

$$2x + \frac{d\lambda}{dt} - \lambda = 0 \tag{5.2-12}$$

Eliminating λ, Eqs. (5.2-10) and (5.2-12) give

$$\frac{dT}{dt} - T = x$$

and

$$\frac{dx}{dt} + x = T$$

which are subject to the boundary conditions

$$x = 1 \qquad \text{at } t = 0$$
$$T = 0 \qquad \text{at } t = \theta$$

since $\lambda = 0$ at $t = \theta$. The latter boundary condition follows from Eqs. (5.2-8) and (5.2-11). This system of equations can be solved to give

$$x = C \exp(-\sqrt{2}t) + D \exp(\sqrt{2}t)$$

and $$T = C(1 - \sqrt{2}) \exp(-\sqrt{2}t) + D(1 + \sqrt{2}) \exp \sqrt{2}t \qquad (5.2\text{-}13a)$$

in which C and D are constants to be determined so as to satisfy the boundary conditions. These constants are

$$C = \frac{(1 + \sqrt{2}) \exp \sqrt{2}\theta}{2(\sinh \sqrt{2}\theta + \sqrt{2} \cosh \sqrt{2}\theta)} \qquad (5.2\text{-}13b)$$

and $$D = \frac{(\sqrt{2} - 1) \exp(-\sqrt{2}\theta)}{2(\sinh \sqrt{2}\theta + \sqrt{2} \cosh \sqrt{2}\theta)} \qquad (5.2\text{-}13c)$$

5.2.2 *Pontryagin's maximum principle*

In the preceding section the derived functions $x(t)$ and $T(t)$ satisfy Eqs. (5.2-3) and (5.2-5) to (5.2-8), which are necessary but not sufficient conditions for the objective function, Eq. (5.2-2), to be an optimum. Nothing has yet been said about sufficiency conditions, and indeed the same equations must be satisfied regardless of whether y is to be a maximum or a minimum. A much "stronger" set of necessary conditions has been developed by Pontryagin and his colleagues [1962]. Pontryagin's maximum principle is a very important result in the calculation of optimal results under local restrictions, a clear and concise derivation being given by Rozonoer [1959a–c]. To promote an appreciation of this important result, a plausible derivation of the maximum principle is given here.

Suppose that there are several values of T which are roots of Eq. (5.2-5). Which value should be selected? On the other hand, what if there are limits imposed on T, and suppose that no acceptable solution of Eq. (5.2-5) exists. What choice would then be best? The Pontryagin maximum principle states that for either case the choice must be such that the Hamiltonian function H, where H is defined as

$$H = Y + \lambda f$$

must be a maximum if y is to be a maximum and a minimum if y is to be a minimum. This is a necessary condition. Note that this principle dictates the appropriate choice of T for each of the situations posed above.

The plausibility of the maximum principle may be seen by considering a variation in T such that

$$T = \begin{cases} T^* + \epsilon\phi & \text{for } \bar{t} - h \leq t \leq \bar{t} \qquad (5.2\text{-}14a) \\ T^* & \text{everywhere else in interval } 0 \leq t \leq \theta \qquad (5.2\text{-}14b) \end{cases}$$

Here $\bar{t}$ is an arbitrary axial position, and h is a positive constant. Let us suppose that the problem is the maximization of the y defined by Eq. (5.2-2). The change in y resulting from the variation defined by these equations is then

$$y - y^* = \int_{\bar{t}-h}^{\theta} [Y(T,x) - Y(T^*,x^*)]\, dt \qquad (5.2\text{-}15)$$

Two Taylor series expansions will prove useful in the analysis. Both are assumed to converge properly, being, first,

$$Y(T,x) = Y(T,x^*) + \frac{\partial Y}{\partial x}(x - x^*) + \frac{1}{2}\frac{\partial^2 Y}{\partial x^2}(x - x^*)^2 + \cdots \qquad (5.2\text{-}16)$$

and second,

$$f(T,x^*) = f(T^*,x^*) + \frac{\partial f}{\partial T}(T - T^*) + \frac{1}{2}\frac{\partial^2 f}{\partial T^2}(T - T^*)^2 + \cdots \qquad (5.2\text{-}17)$$

Let us also note that the perturbation in T as defined by Eq. (5.2-14) gives rise to a change in x as follows:

$$x = x^* + \epsilon\mu \qquad \text{for } \bar{t} - h \leq t \leq \theta \qquad (5.2\text{-}18)$$

This function $\mu(t)$ is not, however, independent of the variation $\phi(t)$. Indeed, to the first order in ϵ, the two variations are related through the differential equation (5.2-3) as follows:

$$\epsilon\frac{d\mu}{dt} = \epsilon\frac{\partial f}{\partial x}\mu + \epsilon\frac{\partial f}{\partial T}\phi \qquad (5.2\text{-}19)$$

The ingredients necessary for the discussion of Pontryagin's maximum principle have now been assembled. The change in y resulting from the prescribed variation can now be written

$$\begin{aligned} y - y^* &= \int_{\bar{t}-h}^{\theta} [Y(T,x) - Y(T^*,x^*)]\, dt \\ &= \int_{\bar{t}-h}^{\theta} \left[Y(T,x^*) + \frac{\partial Y}{\partial x}(T,x^*)\epsilon\mu - Y(T^*,x^*) \right] dt \end{aligned}$$

to the first order in ϵ. In writing this expression, terms in Eq. (5.2-16) of second and higher orders have been neglected. Using Eq. (5.2-6), which defines the Lagrangian multiplier (the adjoint function), one can write that

$$\epsilon\mu\frac{\partial Y}{\partial x} = -\epsilon\mu\frac{d\lambda}{dt} - \epsilon\mu\lambda\frac{\partial f}{\partial x} \qquad (5.2\text{-}20)$$

Thus, after substituting Eqs. (5.2-19), we get

$$y - y^* = \int_{\bar{t}-h}^{\theta} \left[-\frac{d}{dt}(\epsilon\mu\lambda) + \epsilon\phi \frac{\partial f}{\partial T}\lambda + Y(T,x^*) - Y(T^*,x^*) \right] dt$$

The first term in the integrand vanishes because μ vanishes at $\bar{t} - h$ (no variation over region $0 \le t \le \bar{t} - h$) and λ vanishes at $t = 0$. The second term of the integrand can be reduced using the Taylor expansion given by Eq. (5.2-17). Thus to the first order in ϵ

$$y - y^* = \int_{\bar{t}-h}^{\theta} [\lambda f(T,x^*) - \lambda f(T^*,x^*) + Y(T,x^*) - Y(T^*,x^*)]\, dt$$

Since T differs from T^* only in the region $\bar{t} - h \le t \le \bar{t}$ and since h can be made very small, then, as a first approximation valid for very small h, we can write

$$\begin{aligned} y - y^* &= h[\lambda f(T,x^*) + Y(T,x^*) - \lambda f(T^*,x^*) - Y(T^*,x^*)] \\ &= h[H(T,x^*) - H(T^*,x^*)] \end{aligned} \tag{5.2-21}$$

Let us show that this is the desired result. Suppose y is to be a maximum. Then, clearly, the quantity $\lambda f(T^*,x^*) + Y(T^*,x^*)$ must be larger than $\lambda f(T,x^*) + Y(T,x^*)$, since otherwise y would be greater than y^*. Thus, in maximizing y, it is necessary to make the Hamiltonian function H as large as possible by selecting either T at one of the limits or an intermediate value of T. If the problem is to minimize y, then H must take the minimum possible value.

It can be also be shown that H is constant along the path of integration. This fact is an important computational check. To show that H is indeed constant, let us prove for continuous $T(t)$ that

$$\frac{dH}{dt} = 0$$

Following Rozonoer [1959b], let us first define ΔM to be

$$\Delta M = H[x(t + \Delta t),\, \lambda(t + \Delta t),\, T(t + \Delta t)] - H[x(t),\lambda(t),T(t)]$$

Let us consider that the H is to be a maximum. (The calculation for a minimum will be similar.) Since T at any point is selected so that H is a maximum, by hypothesis, then

$$H[x(t + \Delta t),\, \lambda(t + \Delta t),\, T(t + \Delta t)] \ge H[x(t + \Delta t),\, \lambda(t + \Delta t),\, T(t)]$$

and

$$H[x(t),\lambda(t),T(t)] \ge H[x(t),\lambda(t),T(t + \Delta t)]$$

Thus the following bounds are imposed on ΔM:

$$\begin{aligned} H[x(t + \Delta t),\, \lambda(t + \Delta t),\, T(t)] - H[x(t),\lambda(t),T(t)] &\le \Delta M \\ \le H[x(t + \Delta t),\, \lambda(t + \Delta t),\, T(t + \Delta t)] &- H[x(t),\, \lambda(t),\, T(t + \Delta t)] \end{aligned}$$

Since both $x(t)$ and $\lambda(t)$ are continuous, and if in the limit $T(t + \Delta t) \to T(t)$ as $\Delta t \to 0$, then, dividing the above expression by Δt and letting Δt tend to zero yields

$$\frac{dM}{dt} = \frac{\partial H}{\partial x}\frac{dx}{dt} + \frac{\partial H}{\partial \lambda}\frac{d\lambda}{dt} \equiv \frac{dH}{dt}$$

provided that t does not appear explicitly. However, we have assumed this.

Since the adjoint function is

$$\frac{d\lambda}{dt} = -\lambda\frac{\partial f}{\partial x} - \frac{\partial Y}{\partial x}$$

and

$$\frac{dx}{dt} = f$$

we can therefore write that

$$\frac{dH}{dt} = \frac{\partial H}{\partial x}f - \lambda\frac{\partial H}{\partial \lambda}\frac{\partial f}{\partial x} - \frac{\partial H}{\partial \lambda}\frac{\partial Y}{\partial x}$$

However, since H was defined as $H = \lambda f + Y$, then

$$\frac{\partial H}{\partial x} = \lambda\frac{\partial f}{\partial x} + \frac{\partial Y}{\partial x}$$

and

$$\frac{\partial H}{\partial \lambda} = f$$

Clearly, then, $dH/dt = 0$ or $H =$ constant. Since $\lambda = 0$ at $t = \theta$, then

$$H = Y\,|_{t=\theta} \tag{5.2-22}$$

It has been shown by Pontryagin et al. [1962, chap. 2] that the situation involving discontinuous $T(t)$ also gives a constant value of the Hamiltonian.

Example IX-11: Limits and the maximum principle

Previous problems showed no clear need for the maximum principle. Let us consider, however, a control problem in which it is desired to maintain y, at a minimum value, where

$$y = \int_0^1 (x + T)\,dt$$

subject to the restrictions

$$\frac{dx}{dt} = x - T$$

$$x = 5 \qquad \text{at} \qquad t = 0$$

$$\tfrac{1}{2} \le T \le 1$$

Here we have a limited choice of T. Since $Y = x + T$ and $f = x - T$, the function H as defined above can be formed. Thus;

$$H = Y + \lambda f = x + T + \lambda(x - T) = x(1 + \lambda) + T(1 - \lambda)$$

and the adjoint function (Lagrangian multiplier) becomes

$$\frac{d\lambda}{dt} = -1 - \lambda$$

This can be solved to yield, remembering $\lambda = 0$ at $t = 1$,

$$\lambda = e^{1-t} - 1$$

as shown in Fig. IX-7*a*. Now, if we insist that the optimal value of H lies at a stationary value, i.e., at

$$\frac{\partial H}{\partial T} = 0 = 1 - \lambda$$

it is clear that a solution in the sense of the classical variational calculus does not exist, since $1 - \lambda$ cannot be zero if $\lambda = e^{1-t} - 1$ is also to be satisfied. However, Pontryagin's maximum principle requires that T be selected so that H be as small as possible. For $\lambda < 1$, clearly, T should be as small as possible. Therefore T is $\frac{1}{2}$ for the region $\lambda < 1$. From $\lambda = e^{1-t} - 1$, λ becomes unity at $t = 1 - \ln 2 \approx 0.307$. From this point, T should be as large as possible. The resulting axial profile of T is shown in Fig. IX-7*b*.

The associated $x(t)$ function can then be obtained from the first restriction as

$$x = \begin{cases} 4e^t + 1 & \text{for } 0 \leq t \leq 0.307 \\ 4e^t + e^{t-1} + \frac{1}{2} & \text{for } 0.307 \leq t \leq 1 \end{cases}$$

shown in Fig. IX-7*c*. The optimal value of the objective function is then

$$y_{\min} = \int_0^{0.307} (x + 1)\, dt + \int_{0.307}^{1} (x + \tfrac{1}{2})\, dt$$

or

$$y_{\min} = -\tfrac{3}{2} - \ln 2 + 4e = 8.679$$

The value of H is easily shown to be a constant equal to $2 + 4e$ for all values of t.

Note that the conditions imposed here are necessary but not sufficient conditions, so that a few checks should be applied to ensure that the answer is reasonable. For example, let us assume the T is maintained at the upper level ($T = 1$) throughout the entire range of t. Then

$$x = 4e^t + 1 \qquad \text{for } 0 \leq t \leq 1$$

and

$$y = \int_0^1 (4e^t + 2)\, dt = 4e - 2 = 8.872$$

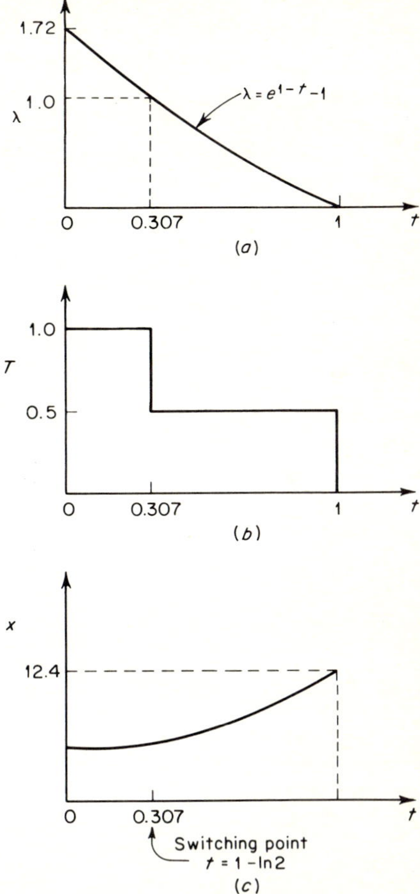

Fig. IX-7 The functions of Example IX-11.

which is a greater value than that obtained by using the maximum principle. Similarly, suppose $T = \frac{1}{2}$ for all t in the interval (0,1). Then

$$x = 4.5e^t + 0.5$$

and

$$y = 4.5e - 2.5 = 9.731$$

which is again greater. The original solution thus satisfies all the necessary conditions and seems reasonable. We shall accept it.

It will be seen that while the calculus of variations requires an optimum to lie at $\delta H/\delta T = 0$, a stationary point, the maximum principle provides a further necessary condition when multiple roots or limits are present by requiring that T be chosen so that H is optimal.

5.2.3 *Dynamic programming and continuous variables*

The problem treated in the preceding paragraph by the use of variable Lagrangian multipliers has been approached also by way of the principle of optimality (Chap. XIII, Sec. 3), the mathematical representation of which is called *dynamic programming*. The application of dynamic programming to systems involving continuous variables may be demonstrated for the present case to yield precisely the same necessary conditions as presented in Sec. 5.2.2. Once a $T(t)$ is selected, the $x(t)$ is determined. We must find the particular $T(t)$ which maximizes y. Let us call this maximum value y^*. Hence

$$y^*(x_0,\theta) = \max_{T(t)} \int_0^\theta Y(x,T)\,dt \tag{5.2-23}$$

The value of y^* depends on θ (the residence time of a reactor, the sales period in an inventory problem, etc.) and on the x_0 (the "feed" to the system).

Let us divide the range of integration into two parts so that

$$y^*(x_0,\theta) = \max_{T(t)} \left[\int_0^{\bar{t}} Y(x,T)\,dt + \int_{\bar{t}}^{\theta} Y(x,T)\,dt \right] \tag{5.2-24}$$

It is conceptually advantageous to interpret this expression in terms of a physical system which is divided into two parts, as shown in Fig. IX-8. The first term on the right of Eq. (5.2-24) might thus give the profit of part 1 associated with a processing time $\bar{t}$, and the second term gives the profit for part 2. The principle of optimality (Chap. XIII, Sec. 3) requires the $T(t)$ in part 2 to be selected so as to obtain the maximum profit with respect to the feed to that section regardless of the action [choice of $T(t)$] which has been taken in the first part. Thus, in part 2, we must select $T(t)$ so as to maximize

Fig. IX-8 Dynamic programming and the continuous system (1).

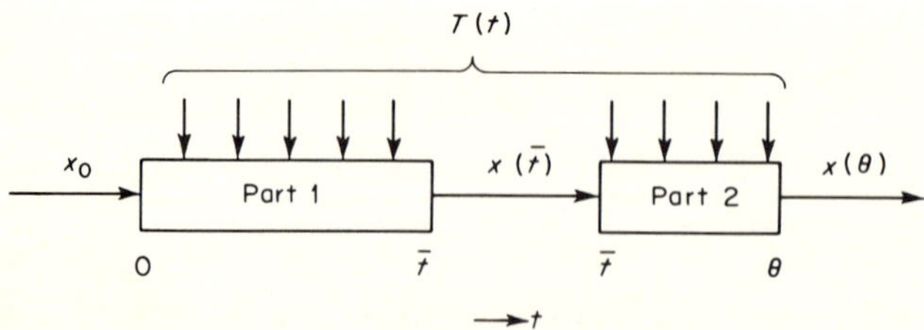

$\int_{\bar{t}}^{\theta} Y(x,T)\,dt$. This maximum will depend both on the processing time and the feed. If we therefore define

$$y^*(x(\bar{t}), \theta - \bar{t}) = \max_{T(t)} \left[\int_{\bar{t}}^{\theta} Y(x,T)\,dt \right]$$

then the principle of optimality requires that $T(t)$ be chosen in the range $0 \leq t \leq \bar{t}$ so that

$$y^*(x_0,\theta) = \max_{T(t)} \left[\int_0^{\bar{t}} Y(x,T)\,dt + y^*(x(\bar{t}), \theta - \bar{t}) \right] \tag{5.2-25}$$

Let us further subdivide the system as shown in Fig. IX-9. By arguments similar to those above, we can write that

$$y^*(x, \theta - t) = \max_{T(t)} \left[\int_t^{\bar{t}} Y(x,T)\,dt + y^*(x(\bar{t}), \theta - \bar{t}) \right] \tag{5.2-26}$$

The maximum profit $y^*(x(\bar{t}), \theta - \bar{t})$ can be expanded into a Taylor series, which, for small Δt, yields the following expression:[1]

$$y^*(x(\bar{t}), \theta - \bar{t}) = y^*(x, \theta - t) + \frac{\partial y^*}{\partial x}(x(\bar{t}) - x(t)) + \frac{\partial y^*}{\partial t}\Delta t \tag{5.2-27}$$

Since Δt is small, we have

$$\frac{dx}{dt} \approx \frac{x(\bar{t}) - x(t)}{\Delta t}$$

and from Eq. (5.2-3), it is evident that

$$x(\bar{t}) - x(t) = \Delta t\, f(x;T) \tag{5.2-28}$$

[1] The last term of Eq. (5.2-27) must be developed with care. Since y^* depends on $\theta - t$, we should form $\partial y^*/[\partial(\theta - t)]$ in writing the Taylor expression. However,

$$\frac{\partial y^*}{\partial(\theta - t)} = \frac{\partial y^*}{\partial t}\frac{\partial t}{\partial(\theta - t)} = -\frac{\partial y^*}{\partial t}$$

Fig. IX-9 Dynamic programming and the continuous system (2).

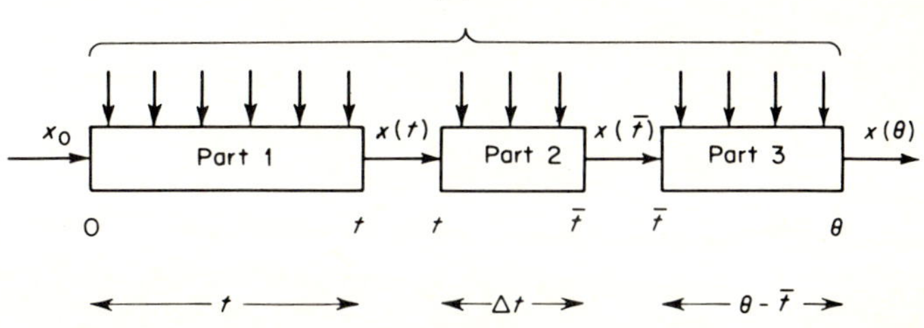

Using Eqs. (5.2-27) and (5.2-28), $y^*(x(\bar{t}), \theta - \bar{t})$ can be eliminated from Eq. (5.2-26) to give

$$y^*(x, \theta - t) = \max_{T(t)} \left[\int_t^{\bar{t}} Y(x,T)\, dt + y^*(x, \theta - t) + \frac{\partial y^*}{\partial x} f \,\Delta t + \frac{\partial y^*}{\partial t} \Delta t \right] \tag{5.2-29}$$

Again, since Δt is small,

$$\int_t^{\bar{t}} Y(x,T)\, dt = Y(x,T)\, \Delta t$$

so that Eq. (5.2-29) can be written

$$\frac{\partial y^*}{\partial t} = - \max_{T(t)} \left[Y(x,T) + f(x,T) \frac{\partial y^*}{\partial x} \right] \tag{5.2-30}$$

This is an essential result arising from the principle of optimality. To obtain a computational method of determining $T(t)$, we need to establish the precise way in which $\partial y^*/\partial x$ varies with t. Since y^* depends on both x and t, the total derivative along the processing path is as follows:

$$\frac{d}{dt}\left(\frac{\partial y^*}{\partial x}\right) = \frac{\partial}{\partial x}\left(\frac{\partial y^*}{\partial x}\right)\frac{dx}{dt} + \frac{\partial^2 y^*}{\partial x\, \partial t} \tag{5.2-31}$$

If we find the optimizing function $T^*(t)$, then Eq. (5.2-30) becomes

$$\frac{\partial y^*}{\partial t} = - \left[Y(x,T^*) + f(x,T^*) \frac{\partial y^*}{\partial x} \right] \tag{5.2-32}$$

Differentiating both sides of this expression partially with respect to x then gives

$$\frac{\partial^2 y^*}{\partial x\, \partial t} = - \left(\frac{\partial Y}{\partial x} + \frac{\partial f}{\partial x}\frac{\partial y^*}{\partial x} + f \frac{\partial^2 y^*}{\partial x^2} \right) \tag{5.2-33}$$

Substituting Eq. (5.2-33) into (5.2-31), we find the desired result

$$\frac{d}{dt}\left(\frac{\partial y^*}{\partial x}\right) = - \frac{\partial Y}{\partial x} - \frac{\partial f}{\partial x}\frac{\partial y^*}{\partial x} \tag{5.2-34}$$

This is a differential equation, giving the rate at which $\partial y^*/\partial x$ changes along the process. A boundary condition at $t = \theta$ can be provided for this differential equation. If no system remains to produce profit, then there can be no profit. Accordingly,

$$\left.\begin{aligned} y^*(x, \theta - t) &= 0 \\ \frac{\partial y^*}{\partial x} &= 0 \end{aligned}\right\} \text{at } t = \theta \tag{5.2-35}$$

Let us now collect the essential results for continuous dynamic programming in a slightly modified form. The maximizing temperature is given by that function $T(t)$ which maximizes a modified function at every point t; that is,

$$\max_{T(t)} [Y(x,T) + \lambda f(x,T)] \tag{5.2-30a}$$

where we have let $\lambda = \partial y^*/\partial x$ and rephrased Eq. (5.2-30). The λ satisfies the differential equation

$$\frac{d\lambda}{dt} = -\frac{\partial Y}{\partial x} - \lambda \frac{\partial f}{\partial x} \tag{5.2-34a}$$

and the boundary condition

$$\lambda = 0 \qquad \text{at } t = \theta \tag{5.2-35a}$$

In writing Eq. (5.2-34a) we must remember that both Y and f are to be evaluated at T^*, and the x are determined by the differential equation

$$\frac{dx}{dt} = f(x,T^*) \tag{5.2-3}$$

subject to the initial condition that

$$x = x_0 \qquad \text{at } t = 0 \tag{5.2-7}$$

Let us contrast these results with those already found. Clearly, Eq. (5.2-30a) is the Pontryagin maximum principle (Sec. 5.2.2) derived here by the application of the principle of optimality. The other results, Eqs. (5.2-34a) and (5.2-35a), are those found by the application of the classical calculus of variations to this same problem (Sec. 5.2.1), where λ was defined to be a variable Lagrangian multiplier or adjoint function. In addition, x is specified at the inlet to the system and λ at the end, as before. Therefore the two methods, dynamic programming and Pontryagin's maximum principle, yield the same results when applied to these continuous systems. Moreover, the adjoint function is seen to represent the gain in profit at any point of the system which could be accrued by increasing the feed x.

Bellman [1957] has suggested a finite difference scheme for computing the solution to Eq. (5.2-30). Aris [1961d] has also developed a method of solution based on characteristics. This, however, is restricted to systems in which the right-hand side of Eq. (5.2-30) can be represented by unimodal functions. Roberts [1964b] also discusses finite difference solutions, but his conclusions in a comparison of the calculus of variations and dynamic programming did not take into account the connecting link through Pontryagin's maximum principle. As Rozonoer [1959c] has pointed out, the correspondence between dynamic programming and the maximum principle holds when any limitation on $T(t)$ does not depend upon t. If such a condition is imposed, the maximum principle is inapplicable, while the dynamic programming method is still valid.

5.2.4 *General case for "simple" boundary conditions*

In the earlier sections covering "simple" boundary conditions, the integrand and restrictions involved only a single function $x(t)$ and a single independent variable $T(t)$. Let us now consider the general case. Suppose

$$y = \int_0^\theta Y(x_1, x_2, \ldots, x_n, T_1, T_2, \ldots, T_k)\, dt$$

or using an obvious notation,

$$y = \int_0^\theta Y(\mathbf{x},\mathbf{T})\, dt \tag{5.2-36}$$

The problem is to optimize y by an appropriate choice of $\mathbf{T}$ subject to the set of differential equations

$$\frac{dx_i}{dt} = f_i(\mathbf{x},\mathbf{T}) \qquad \text{for } i = 1, 2, \ldots, n \tag{5.2-37}$$

and the initial conditions

$$x_i = x_{i,0} \qquad \text{at } t = 0$$

Proceeding as in Sec. 5.2.1, define

$$F = \int_0^\theta \left\{ Y - \sum_{i=1}^{n} \lambda_i(t) \left[\frac{dx_i}{dt} - f_i(\mathbf{x},\mathbf{T}) \right] \right\} dt$$

The set of necessary conditions for the existence of an optimum can then be written from the Euler-Lagrange equations for this modified objective function as

$$\frac{d\lambda_i}{dt} = -\sum_{j=1}^{n} \frac{\partial f_j}{\partial x_i} \lambda_j - \frac{\partial Y}{\partial x_i} \qquad \text{for } i = 1, 2, \ldots, n \tag{5.2-38}$$

where $$\lambda_i = 0 \qquad \text{at } t = \theta,\ i = 1, 2, \ldots, n \tag{5.2-39}$$

Moreover, the $\mathbf{T}$ must be selected so that H is a maximum if the optimum is to be a maximum (or conversely, a minimum H if a minimum is required). Here H is defined as

$$H = Y + \sum_{i=1}^{n} \lambda_i f_i \tag{5.2-40}$$

These conditions, together with Eqs. (5.2-37) and the associated boundary conditions, form the basis of a computational scheme.

Such a method as applied to optimization of chemical reactors has been reported independently by Horn [1961a] and Katz [1960]. Moreover, Adler and Vortmeyer [1964] have shown that the method is applicable to the design of tubular flow reactors with axial diffusion. The equations describing such a system are second-order in t, and consequently the adjoint equations are also second-order equations.

Example IX-12: Multiple local restrictions

It is desired to maximize the production of substance A_2 when the reaction

$$A_1 \rightleftharpoons A_2 \rightarrow A_3$$

takes place in a plug-flow tubular reactor in which the local reaction rates are defined by the set of kinetic equations

$$\frac{dx_1}{dt} = k_2 x_2 - k_1 x_1 \equiv f_1 \tag{5.2-41}$$

and

$$\frac{dx_2}{dt} = k_1 x_1 - (k_2 + k_3) x_2 \equiv f_2 \tag{5.2-42}$$

where t = distance measured along reactor
x_1, x_2 = concentrations of A_1 and A_2, respectively
k_j = reaction velocity constants ($j = 1, 2, 3$)

Here the parameters can be identified with the usual kinetic equations by noting that k represents a function of the temperature, for example,

$$k_1 = e^{-E_1/RT} = k, \text{ say} \tag{5.2-43a}$$

and if $E_2/E_1 = 2$ and $k_3 = 2k_2$, then

$$k_2 = e^{-E_2/RT} = k^2 \tag{5.2-43b}$$

$$k_3 = 2k^2 \tag{5.2-43c}$$

For catalytic reactions, k could represent some other function of the temperature. However, for our purpose, it is sufficient to recognize that Eqs. (5.2-41) and (5.2-42) have the same structure as the kinetic equations usually applied to describe the reaction scheme considered in this example.

Since we wish to maximize the yield of A_2, that is, $x_2(\theta)$, we require to maximize y, where

$$y = x_2(\theta) - x_2(0) = \int_0^\theta \frac{dx_2}{dt}\,dt = \int_0^\theta f_2\,dt \tag{5.2-44}$$

Here $Y(\mathbf{x},\mathbf{T}) \equiv f_2$.

The adjoint functions, or Lagrangian multipliers, as defined by Eqs. (5.2-38), will then be

$$\frac{d\lambda_1}{dt} = -\sum_{j=1}^{n} \frac{\partial f_j}{\partial x_1}\lambda_j - \frac{\partial Y}{\partial x_1}$$

$$= k(\lambda_1 - \lambda_2) - k \tag{5.2-45}$$

and

$$\frac{d\lambda_2}{dt} = -\sum_{j=1}^{n} \frac{\partial f_j}{\partial x_2}\lambda_j - \frac{\partial Y}{\partial x_2}$$

$$= -k^2(\lambda_1 - 3\lambda_2) + 3k^2 \tag{5.2-46}$$

with

$$\left.\begin{aligned} \lambda_1 &= 0 \\ \lambda_2 &= 0 \end{aligned}\right\} \text{at } t = \theta \tag{5.2-47}$$

The temperature T must now be chosen to maximize H, as defined by Eq. (5.2-40). If this occurs at a stationary point, then

$$\frac{\partial H}{\partial T} = 0 = \frac{\partial}{\partial T}(Y + \lambda_1 f_1 + \lambda_2 f_2)$$

$$= \frac{\partial k}{\partial T}[(x_1 - 6kx_2)(1 + \lambda_2) + (2kx_2 - x_1)\lambda_1] \tag{5.2-48}$$

Let us make the following change of variable:

$$\bar{\lambda}_2 = \lambda_2 + 1 \tag{5.2-49}$$

so that

$$\frac{d\lambda_1}{dt} = k(\lambda_1 - \bar{\lambda}_2) \tag{5.2-50}$$

$$\frac{d\bar{\lambda}_2}{dt} = k^2(3\bar{\lambda}_2 - \lambda_1) \tag{5.2-51}$$

with boundary conditions

$$\lambda_1 = 0 \quad \text{at } t = \theta \tag{5.2-52}$$

$$\bar{\lambda}_2 = 1 \quad \text{at } t = \theta \tag{5.2-53}$$

Equation (5.2-48) then becomes

$$(\bar{\lambda}_2 - \lambda_1)x_1 + 2x_2k(\lambda_1 - 3\bar{\lambda}_2) = 0 \tag{5.2-54}$$

The problem is now completely defined if the inlet feed is specified. Thus, if the feed is pure A_1, at $t = 0$,

$$x_{1,0} = 1.0 \qquad \text{and} \qquad x_{2,0} = 0.0 \tag{5.2-55}$$

We must now solve simultaneously Eqs. (5.2-41), (5.2-42), (5.2-50), (5.2-51), and (5.2-54) to determine the optimal temperature function T

and the resulting compositions x_1 and x_2. The equations are nonlinear, and an analytic solution does not seem possible. It is instructive to follow, partially at least, a numerical computation (Example IX-13).

The numerical analysis of this type of problem has been discussed by Horn and Troltenier [1963], Storey and Rosenbrock [1964], and Lee [1964a]. A reasonable computational scheme (essentially, the gradient method in function space) is outlined in the following five-step sequence:

1. Assume values for the functions $\mathbf{T}(t)$ along the continuous system.
2. Once the values of $\mathbf{T}$ have been established as a function of t, the $\mathbf{x}(t)$ can be determined by solving Eqs. (5.2-37) numerically. Here the Runge-Kutta method (Chap. III, Sec. 4.2.2) might be used.
3. After establishing the $\mathbf{x}(t)$, the adjoint functions can be computed using an approximate numerical integration of Eqs. (5.2-38).
4. It is now necessary to determine a better approximation for the optimal $\mathbf{T}$. This improved approximation can be found by optimizing the Hamiltonian, Eq. (5.2-40). The adjoint functions found in step 3 and $\mathbf{x}(t)$ determined in step 2 can be substituted into the Hamiltonian, and a better value of the $\mathbf{T}(t)$ function computed.
5. Steps 2 to 4 should be repeated until the $\mathbf{T}(t)$ function remains unchanged.

The theoretical justification for this approach is based upon the concept of Green's functions as discussed by Kelley [1962] and Denn and Aris [1965c].

Example IX-13: Numerical solution

Let us follow this numerical procedure through for a single cycle in Example IX-12.

Step 1 Assume values for the temperature function. Such a set is shown in Table IX-1.

TABLE IX-1 TUBULAR REACTOR—FIRST-CYCLE CONCENTRATIONS

t	x_1	x_2	k
0.00	1.000	0.000	1.000
0.14	0.918	0.076	0.752
0.26	0.850	0.133	0.634
0.38	0.798	0.172	0.547
0.52	0.747	0.207	0.539
0.64	0.708	0.231	0.542
0.76	0.671	0.250	0.546
0.88	0.638	0.265	0.549
1.00	0.602	0.278	0.553

The values of the temperature function which are shown have been determined from a previous iteration. This, however, is not yet the optimal distribution, as will be seen when the improved temperature functions are computed. Also, it should be noted that the length of the reactor has been fixed so that θ is unity.

At the reactor inlet, since the feed contains only A_1, $x_2 = 0$ and $x_1 = 1$. Therefore

$$f_2 = -f_1 = k \qquad \text{at } t = 0$$

The Hamiltonian at $t = 0$ becomes

$$H = k(\bar{\lambda}_2 - \lambda_1)$$

where $\bar{\lambda}_2$ is tabulated in Table IX-2. Clearly, H does not possess a stationary value here, but either increases or decreases monotonically with increasing k, depending on the sign of $\bar{\lambda}_2 - \lambda_1$. We must therefore apply the Pontryagin maximum principle in selecting the proper k. In this case, seen from Table IX-2, $\bar{\lambda}_2 - \lambda_1$ is a positive quantity at $t = 0$, and therefore k should be made as large as possible to maximize H. Since $0 \leq k \leq 1$, the appropriate choice at $t = 0$ is $k = 1$. At all other positions the maximum of H is at a stationary point.

Step 2 The concentrations x_1 and x_2 can be computed as a function of position from Eqs. (5.2-41) and (5.2-42). The results are shown in Table IX-1.

Step 3 The adjoint functions are found from Eqs. (5.2-50) and (5.2-51), the results being shown in Table IX-2. Note that the integration begins at the outlet of the reactor, since the boundary conditions for the adjoint functions are specified at the outlet. The boundary conditions for the concentrations are, of course, given at the inlet. Thus we have a two-point boundary problem. It is this dual specification of

TABLE IX-2 TUBULAR REACTOR—FIRST-CYCLE ADJOINT FUNCTIONS

t	λ_1	$\bar{\lambda}_2$	k
1.00	0.000	1.000	0.553
0.88	0.070	0.882	0.549
0.76	0.118	0.795	0.546
0.64	0.159	0.719	0.542
0.52	0.192	0.653	0.539
0.38	0.223	0.586	0.547
0.26	0.245	0.526	0.634
0.14	0.265	0.453	0.752
0.00	0.278	0.390	1.000

the boundary conditions which makes the problem difficult. If both the concentration and the adjoint function were specified at the inlet (or the outlet of the reactor), the problem would not be of the trial-and-error type, but would be a straightforward numerical calculation.

Step 4 The new temperature function is computed using Eq. (5.2-54), rearranged in the form

$$k = \frac{(\lambda_1 - \bar{\lambda}_2)x_1}{2(\lambda_1 - 3\bar{\lambda}_2)x_2}$$

Thus, for example, at $t = 0.52$, the corrected temperature function is

$$k = \frac{(0.192 - 0.653)(0.747)}{2[0.192 - 3(0.653)](0.207)} = 0.4698$$

which would then replace the assumed value of 0.539.[1] A corrected temperature function must be computed at each axial position, and the entire process repeated.

The final solution is given in Table IX-3. In this table the optimal temperature function is very large near the inlet, since only substance A_1 is present. However, as A_2 is produced, the temperature function decreases so as to reduce the rate at which A_2 is degraded to A_3. Substance A_3 is a waste material, since the reaction scheme does not provide a path for producing A_2 from A_3.

TABLE IX-3 TUBULAR REACTOR—OPTIMAL REACTION PATH

t	x_1	x_2	λ_1	$\bar{\lambda}_2$	k
0.00	1.000	0.000	0.287	0.399	1.000
0.12	0.901	0.089	0.268	0.514	0.982
0.24	0.827	0.145	0.238	0.626	0.678
0.36	0.776	0.181	0.205	0.707	0.560
0.50	0.728	0.214	0.164	0.784	0.483
0.62	0.695	0.236	0.127	0.842	0.439
0.74	0.666	0.254	0.089	0.895	0.406
0.86	0.640	0.271	0.049	0.945	0.380
1.00	0.613	0.287	0.000	1.000	0.356

[1] Actually, it was found that a better convergence rate was achieved if the new temperature function used in the computation was an average of the previous value and the corrected temperature computed as shown above. Thus the new temperature would be $\frac{1}{2}(0.539 + 0.470)$. Of course, the better convergence using an average of the two may be true only in the special case given here. No other problems have been attempted. However, Storey and Rosenbrock [1964] report a similar experience with a reaction system. They suggested that this weighting factor should be larger than 1/2 at first and then reduced. Owing to the lack of computing experience which has been reported in the literature, this suggestion, while appearing to be quite reasonable, should be taken as being tentative.

In studying the results shown in Table IX-3, note that the maximum possible yield of $A_2(x_2)$ for the reaction system is quite small. It would seem that such a result, while not actually useful in practice because of the practical difficulty of achieving the optimal temperature profile, has use in giving the designer an ultimate yield.

Finally, one can check in part the computations by determining if H is a constant. Here it is not difficult to show that H is numerically about 0.11 for all times in Table IX-3.

A second numerical approach, more generally applicable than gradients in function space,[1] is the method of quasi-linearization, in which the differential equations are linearized and an iterative scheme similar to the Newton-Raphson type is used to approach the solution of the original nonlinear set of equations. This technique is described by Bellman and Kalaba [1965] and has been applied to chemical reactors by Lee [1966, 1967].

Briefly, the idea of quasi-linearization is to solve a sequence of linear differential equations, with the hope of converging to the true solution. To prove convergence is, in general, an impossible task, and although sufficient conditions have been established in a limited number of cases (see, for example, Kalaba [1959]), the proofs are complex and will be omitted. The computational algorithm is straightforward and relatively easy to understand. We start with the stationary conditions for the general problem, namely, Eqs. (5.2-37) to (5.2-39). These can be written as follows, noting that $\mathbf{T}$ can be eliminated by solving the conditions giving a stationary value of H†:

$$\frac{dx_i}{dt} = f_i(\mathbf{x},\boldsymbol{\lambda}) \tag{5.2-56}$$

and

$$-\frac{d\lambda_i}{dt} = \sum_{j=1}^{n} \frac{\partial f_j}{\partial x_i} \lambda_j + \frac{\partial Y(\mathbf{x},\boldsymbol{\lambda})}{\partial x_i} \tag{5.2-57}$$

subject to

$$x_i(0) = x_{i,0} \tag{5.2-58}$$

and

$$\lambda_i(\theta) = 0 \tag{5.2-59}$$

These are coupled ordinary differential equations which are difficult to solve numerically because of the two-point boundary conditions. As noted above, one can assume various initial values of $\lambda_i(0)$, integrate the differential equations forward, and see if Eq. (5.2-59) is satisfied. If not, new guesses as

[1] A situation in which gradients in function space cannot be applied is described by Coward [1967]. In this case, that of time-optimal batch distillation, the initial and final states are specified, and the Lagrangian multipliers have no associated boundary conditions.

† If this elimination cannot be accomplished analytically, then the $\mathbf{T}$ must be computed at each t in terms of $\mathbf{x}^{(p)}$ and $\boldsymbol{\lambda}^{(p)}$ and recalculated in each iteration.

to the appropriate initial values of λ_i are made, and the integration repeated.

As an alternative, let us imagine that $x_i^{(p)}$ and $\lambda_i^{(p)}$ are good approximations to the true solutions x_i and λ_i. The differences $x_i - x_i^{(p)}$ and $\lambda_i - \lambda_i^{(p)}$ are thus visualized as being small, and by a Taylor expansion,

$$\frac{dx_i}{dt} \approx f_i(\mathbf{x}^{(p)},\boldsymbol{\lambda}^{(p)}) + \sum_{j=1}^{n} \frac{\partial f_i}{\partial x_j}(x_j - x_j^{(p)}) + \sum_{j=1}^{n} \frac{\partial f_i}{\partial \lambda_j}(\lambda_j - \lambda_j^{(p)})$$

the approximation sign being invoked because the series has been truncated. We can use this approximation to generate a better guess at the true solution. If $x_i^{(p+1)}$ is the next approximation, then, expanding both Eqs. (5.2-56) and (5.2-57), we have

$$\frac{dx_i^{(p+1)}}{dt} = Q_i + \sum_{k=1}^{n} A_{i,k}x_k^{(p+1)} + \sum_{k=1}^{n} B_{i,k}x_k^{(p+1)} \tag{5.2-60}$$

and

$$-\frac{d\lambda_i^{(p+1)}}{dt} = P_i + \sum_{k=1}^{n} G_{i,k}x_k^{(p+1)} + \sum_{k=1}^{n} H_{i,k}\lambda_k^{(p+1)} \tag{5.2-61}$$

in which

$$Q_i = f_i(\mathbf{x}^{(p)},\boldsymbol{\lambda}^{(p)}) - \sum_{j=1}^{n} \frac{\partial f_i}{\partial x_j} x_j^{(p)} - \sum_{j=1}^{n} \frac{\partial f_i}{\partial \lambda_j} \lambda_j^{(p)}$$

$$A_{i,k} = \frac{\partial f_i}{\partial x_k}$$

$$B_{i,k} = \frac{\partial f_i}{\partial \lambda_k}$$

$$P_i = \frac{\partial Y}{\partial x_i} + \sum_{j=1}^{n} \frac{\partial f_j}{\partial x_i} \lambda_j^{(p)} - \sum_{j=1}^{n} \frac{\partial^2 Y}{\partial x_i\, \partial x_j} x_j^{(p)} - \sum_{j=1}^{n} \frac{\partial^2 Y}{\partial x_i\, \partial \lambda_j} \lambda_j^{(p)}$$
$$- \sum_{j=1}^{n}\sum_{k=1}^{n} \frac{\partial^2 f_j}{\partial x_i\, \partial x_k} \lambda_j^{(p)} x_k^{(p)} - \sum_{j=1}^{n}\sum_{k=1}^{n} \frac{\partial^2 f_j}{\partial x_i\, \partial \lambda_k} \lambda_j^{(p)} \lambda_k^{(p)} - \sum_{j=1}^{n} \frac{\partial f_j}{\partial x_i} \lambda_j^{(p)}$$

$$G_{i,k} = \sum_{j=1}^{n} \frac{\partial^2 f_j}{\partial x_i\, \partial x_k} \lambda_j^{(p)} + \frac{\partial^2 Y}{\partial x_i\, \partial x_k}$$

$$H_{i,k} = \sum_{j=1}^{n} \frac{\partial^2 f_j}{\partial x_i\, \partial \lambda_k} \lambda_j^{(p)} + \frac{\partial^2 Y}{\partial x_i\, \partial \lambda_k} + \frac{\partial f_k}{\partial x_i}$$

There are several essential points to be noted. First, the coefficients $A_{i,k}$, $B_{i,k}$, $G_{i,k}$, and $H_{i,k}$ are known functions of t. They are known because we have evaluated them using the known functions $\mathbf{x}^{(p)}$ and $\boldsymbol{\lambda}^{(p)}$. Equations (5.2-60) and (5.2-61) cannot be readily integrated because of the time-dependent nature of the coefficients. In general, such equations must still be integrated numerically, even though they are linear differential equations. Furthermore, to converge to the desired solution, it is necessary to force both $\mathbf{x}^{(p+1)}$ and $\boldsymbol{\lambda}^{(p+1)}$ to satisfy the imposed boundary conditions, namely, $x_i^{(p+1)} = x_{i,0}$ at $t = 0$ and $\lambda_i^{(p+1)} = 0$ at $t = \theta$. Therefore our linear problem also retains the two-point character of the original nonlinear problem. What, then, has been gained? The linear problem cannot, in general, be solved by other than numerical techniques, and it is still a two-point boundary-value problem. Our position has actually been improved in only one respect. Since the problem is linear, by adding solutions together properly (permitted for linear equations), we can construct a solution from the solution of a series of initial-value problems. This being the case, the approximations for $\mathbf{x}^{(p+1)}$ and $\boldsymbol{\lambda}^{(p+1)}$ proceed from $\mathbf{x}^{(p)}$ and $\boldsymbol{\lambda}^{(p)}$ without the trial-and-error calculation usually associated with the two-point boundary-value problem.

To understand the method of adding solutions that we have in mind, let us first consider the following two-point boundary-value problem:

$$\frac{dx}{dt} = Ax + B\lambda + P \tag{5.2-62}$$

and

$$\frac{d\lambda}{dt} = Gx + H\lambda + Q \tag{5.2-63}$$

with $x = x_0$ at $t = 0$ and $\lambda = 0$ at $t = \theta$.

The system of equations defined above can be solved by solving two sets of linear initial-value problems; first, we have a nonhomogeneous set of equations

$$\frac{dz}{dt} = Az + Bw + P \tag{5.2-64}$$

and

$$\frac{dw}{dt} = Gz + Hw + Q \tag{5.2-65}$$

with $z = Z_0$ and $w = W_0$ at $t = 0$. Second, we have the homogeneous set

$$\frac{dv}{dt} = Av + Bu \tag{5.2-66}$$

and

$$\frac{du}{dt} = Gv + Hu \tag{5.2-67}$$

with $v = V_0$ and $u = U_0$ at $t = 0$.

The solution of the two initial-value problems is one of straightforward integration, starting with $t = 0$. Call the final values of w and u, W_θ and U_θ, respectively. It can easily be tested that if the functions w, v, z, and u satisfy the differential equations stated above, a linear combination of them will also satisfy Eqs. (5.2-62) and (5.2-63). Thus, if the boundary conditions for the two-point problem can be satisfied, the linear combinations

$$x = \alpha_1 z + \beta_1 v$$

and

$$\lambda = \alpha_1 w + \beta_1 u$$

are solutions to the original two-point boundary problem. If we choose $Z_0 = x_0$ and $V_0 = 0$, the final solutions are

$$x = z - \frac{W_\theta}{U_\theta} v$$

and

$$\lambda = w - \frac{W_\theta}{U_\theta} u$$

provided $U_\theta \neq 0$. If U_θ happens to be zero, the set of equations could again be integrated, using a new value for U_0 other than the one used, and the first result discarded.

Thus we see that the price which must be paid in order to obtain the solution of a two-point boundary-value problem is that two initial-value problems must be solved and their solutions combined in such a way that the final-point boundary conditions (here the conditions at $t = \theta$) are satisfied. The cost is therefore not negligible, and the main question to be answered of course concerns the rate of convergence of the $\mathbf{x}^{(p)}$ to the final true solution of the original nonlinear differential equations.

We can now see the way to solve the set of differential equations defined by Eqs. (5.2-60) and (5.2-61). For each variable whose boundary condition is specified at $t = \theta$, a set of homogeneous differential equations must be solved. First, solve the initial-value problems

$$\frac{dz_i}{dt} = \sum_{k=1}^{n} A_{i,k} z_k + \sum_{k=1}^{n} B_{i,k} w_k + P_i \tag{5.2-68}$$

and

$$\frac{dw_i}{dt} = \sum_{k=1}^{n} G_{i,k} z_k + \sum_{k=1}^{n} H_{i,k} w_k + Q_i \tag{5.2-69}$$

with $z_i = x_{i,0}$ and $w_i = w_{i,0}$ at $t = 0$. Here $w_{i,0}$ is a nonzero number.

Also, the following $2n$ coupled equations must be solved n times, each time using a different set of initial conditions for the $u_{i,j}$:

$$\frac{dv_{i,j}}{dt} = \sum_{k=1}^{n} A_{i,k} v_{k,j} + \sum_{k=1}^{n} B_{i,k} u_{k,j} \tag{5.2-70}$$

and

$$\frac{du_{i,j}}{dt} = \sum_{k=1}^{n} G_{i,k} v_{k,j} + \sum_{k=1}^{n} H_{i,k} u_{k,j} \tag{5.2-71}$$

with $v_{i,j} = 0$ and $u_{i,j} = U_{i,j}$ at $t = 0$.

The initial conditions $U_{i,j}$ are selected so that if $U_{i,j}^{(\theta)}$ are the final values of $u_{i,j}$, that is, the values of $u_{i,j}$ at $t = \theta$, then

$$\begin{vmatrix} U_{1,1}^{(\theta)} & U_{1,2}^{(\theta)} & U_{1,3}^{(\theta)} & \cdots & U_{1,n}^{(\theta)} \\ U_{2,1}^{(\theta)} & U_{2,2}^{(\theta)} & \cdots & \cdots & U_{2,n}^{(\theta)} \\ \cdot & \cdot & \cdot & \cdot & \cdot \\ U_{n,1}^{(\theta)} & \cdots & \cdots & \cdots & U_{n,n}^{(\theta)} \end{vmatrix} \neq 0 \tag{5.2-72}$$

If this condition is satisfied, then there exists a solution of the form

$$x_i^{(p+1)} = z_i + \sum_{k=1}^{n} \beta_k v_{i,k} \tag{5.2-73}$$

and

$$\lambda_i^{(p+1)} = w_i + \sum_{k=1}^{n} \beta_k u_{i,k} \tag{5.2-74}$$

the β_k being selected so that

$$0 = w_i(\theta) + \sum_{k=1}^{n} \beta_k U_{i,k}^{(\theta)} \qquad \text{for } i = 1, 2, \ldots, n$$

A solution for the β_k exists when the condition (5.2-72) is satisfied.

The method of quasi-linearization can now be summarized. One first guesses a solution $\mathbf{x}^{(0)}$ and $\boldsymbol{\lambda}^{(0)}$, which satisfies the boundary conditions. This guessed solution is substituted so as to form the coefficients appearing in Eqs. (5.2-60) and (5.2-61), which are to be solved so as to generate new approximations to the exact solution. The method of obtaining new approximations is to solve the set of differential equations (5.2-68) and (5.2-69), first using an arbitrarily selected set of initial conditions, $w_{i,0}$. This solution is straightforward, since the problem is an initial-value problem and integration can start from $t = 0$. This solution is stored. Equations (5.2-70) and (5.2-71) must be solved n times, each time beginning with different initial conditions $U_{i,j}$. This is essential because using the same initial conditions will generate the same $u_{i,j}$, and therefore condition (5.2-72) will not be satisfied. Once the $n + 1$ sets of equations have been integrated, the next approximation is constructed from Eqs. (5.2-73) and (5.2-74). The β_k are of course selected so that the $\boldsymbol{\lambda}$ vanish at $t = \theta$. The entire process is repeated until successive approximations show no change.

Lee [1966] has considered a particular example and found that, for this case, the method of quasi-linearization converged much more rapidly than did the technique of assuming initial conditions for the $\boldsymbol{\lambda}$ and integrating the nonlinear equations until the correct final values for the $\boldsymbol{\lambda}$ are found. This problem also indicated an advantage in convergence rate for quasi-linearization when compared with the method of gradients in function space. Despite this latter observation, it is recommended here that gradients in function space be used where applicable because an improved control is obtained in each iteration when improvement is possible. This is not the case in using quasi-linearization, nor is convergence generally guaranteed. A proof of convergence in those cases where it exists is a complex task. Therefore the use of gradients in function space is recommended.

If the method of gradients in function space is not applicable, quasi-linearization seems to be the best technique to try.

It should be remembered that Pontryagin's maximum principle gives necessary but not sufficient conditions so that we could find a solution satisfying Eqs. (5.2-37) to (5.2-39), in addition maximizing or minimizing H as required, and yet find that this solution is still not the optimal one. Of course, the optimal solution will satisfy these equations. A discussion of this facet of the problem has been presented by Coward and Jackson [1965].

5.3 "Complex" Boundary Conditions

An example of complex boundary conditions is found in systems which involve recycle. The problem of recycle has been studied, for example, using dynamic programming, by Mitten and Nemhauser [1963a] and Aris [1964]. However, the work to be presented here will be confined to the method of Lagrangian multipliers, and Pontryagin's maximum principle applied to a rather general case. Moreover, the method is easily extendable to more complex situations. Indeed, it will be shown that one need modify the boundary conditions on the adjoint functions only, in order to treat the recycle case. Calculations of the type studied here have been reported by Beveridge and Schechter [1965] and Denn and Aris [1965b].

The system to be considered consists of a mixer, a tubular reactor, and a separator, arranged as shown in Fig. IX-10. Fresh feed is introduced into the reaction system at a constant rate q, with a fixed composition $\mathbf{x}_I$. [$\mathbf{x}_j$

Fig. IX-10 Reaction system with recycle.

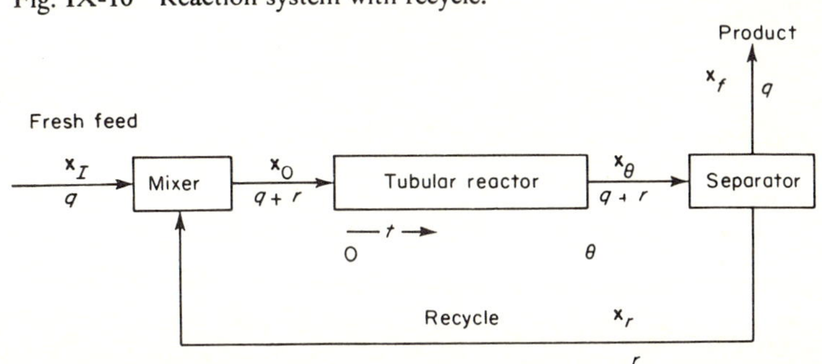

denotes the set of n weight fractions $(x_1, x_2, \ldots, x_n)$, which completely specify the chemical state of an $n + 1$ component stream j.]

The separator following the reactor produces a product stream and a recycle stream at the respective flow rates, q and r, with compositions $\mathbf{x}_f$ and $\mathbf{x}_r$. It is assumed that the system already exists, so that the reactor operation can be varied only by manipulating the external operating variables $\mathbf{T}(t)$. There are k of these variables which can be altered independently along the length of the reactor. The split in the separator can be characterized by the nonnegative separation factors $\mathbf{s}$, where

$$s_j = \frac{x_{j,r}}{x_{j,f}} \qquad \text{for } j = 1, 2, \ldots, n \tag{5.3-1}$$

so that, from a mass balance over the separator, we can write the composition of the jth component in the recycle stream, $x_{j,r}$, as

$$x_{j,r}\left(r + \frac{q}{s_j}\right) = (r + q)x_{j,\theta} \tag{5.3-2}$$

Since the feed to the reactor, $\mathbf{x}_0$, is the sum of the contributions from the fresh feed and the recycle streams, it follows from Eq. (5.3-2) that

$$x_{j,0} = \frac{q}{q + r} x_{j,I} + \frac{r}{r + q/s_j} x_{j,\theta} \qquad \text{for } j = 1, 2, \ldots, n \tag{5.3-3}$$

As in the previous section, the composition within the reactor, $\mathbf{x}(t)$, will vary along its length, the variation being determined by the n kinetic equations

$$\frac{dx_i}{dt} = f_i(\mathbf{x},\mathbf{T},r) \qquad \text{for } i = 1, 2, \ldots, n \tag{5.3-4}$$

The reaction rate depends upon the n local compositions, $\mathbf{x}(t)$, the level of the k control variables, $\mathbf{T}(t)$, and the mass flow rate, $r + q$. Given a feed rate of a fixed composition, then a given set of the operating variables, recycle rate, and separation factors will uniquely determine the product composition.

Having completed a design basis for the problem it is now necessary to obtain the best design. The objective function generally has the form

$$y = \int_0^\theta Y(\mathbf{x}(t), \mathbf{T}(t), r, \mathbf{s})\, dt + C(\mathbf{x}_\theta, \mathbf{s}, r) \tag{5.3-5}$$

C being a cost function associated with separation. Thus the problem is to maximize, say, y, subject to certain restrictions. As in our previous work, a set of local conditions is imposed which must be satisfied at every point t. For this case the restrictions are given by Eq. (5.3-4). A second set of restrictions is to be imposed also; i.e., the boundary conditions (5.3-2) must also be satisfied. Of course, in the usual case, there will be constraints to be imposed on the control variables **T**. However, let us assume that such constraints are absent.

In handling constraints it is frequently advantageous to use the method of Lagrangian multipliers. Thus a necessary condition for y to be a maximum is that the F, defined as

$$F = \int_0^\theta \left[Y - \sum_{i=1}^n \lambda_i \left(\frac{dx_i}{dt} - f_i \right) \right] dt + C(\mathbf{x}_\theta,\mathbf{s},r) + \sum_{i=1}^n \alpha_i \left(x_{i,0} - \frac{q}{q+r} x_{i,I} - \frac{r}{r + q/s_i} x_{i,\theta} \right) \tag{5.3-6}$$

be stationary for arbitrary variations in x_j, T_s, r, and s_j.

Of course, in this expression, the λ_i are Lagrangian multipliers (adjoint functions) which depend on position t. Moreover, the α_i are also Lagrangian multipliers, which have been added to ensure that the appropriate boundary conditions are satisfied.

Taking a variation of F with respect to x_j yields the following Euler-Lagrange equations and boundary conditions:

$$\frac{\partial Y}{\partial x_j} + \frac{d\lambda_j}{dt} + \sum_{i=1}^n \frac{\partial f_i}{\partial x_j} \lambda_i = 0 \qquad \text{for } 0 \le t \le \theta \tag{5.3-7}$$

$$\left(-\lambda_j + \frac{\partial C}{\partial x_j} - \alpha_j \frac{r}{r + q/s_j} \right) \delta x_j \big|_{t=\theta} = 0 \tag{5.3-8}$$

$$(\lambda_j + \alpha_j)\, \delta x_j \big|_{t=0} = 0 \tag{5.3-9a}$$

all for $j = 1, 2, \ldots, n$. Since $\delta x_j|_{t=0}$ does not necessarily vanish, owing to the recycle, then

$$\lambda_{j,0} = -\alpha_j$$

Equation (5.3-8) can be rewritten

$$\lambda_{j,\theta} = \frac{\partial C}{\partial x_j} + \lambda_{j,0} \frac{r}{r + q/s_j} \qquad \text{for } j = 1, 2, \ldots, n \tag{5.3-9b}$$

This gives the boundary conditions for the adjoint functions. These are defined by the same differential equations as in situations with simple boundary conditions. Computationally, these recycle problems are, however, more difficult, owing to the form of the boundary conditions for the adjoint functions [Eqs. (5.3-9*b*)] and the boundary conditions for the concentrations [Eqs. (5.3-3)]. The complexity arises because of coupling of variables in each case.

A variation with respect to T_j, one of the control variables, gives the equations

$$\frac{\partial Y}{\partial T_j} + \sum_{i=1}^{n} \lambda_i \frac{\partial f_i}{\partial T_j} = 0 \qquad 0 \le t \le \theta, \text{for } j = 1, 2, \ldots, k \qquad (5.3\text{-}10)$$

Equations (5.3-10) represent the requirement that the Hamiltonian be stationary as a necessary condition, where the Hamiltonian is defined, as before, as

$$H = Y + \sum_{i=1}^{n} \lambda_i f_i \qquad (5.3\text{-}11)$$

This is the weak principle, but what if there are bounds imposed on the choice of T_j? What if there are several stationary points satisfying Eq. (5.3-10)? Again, one can prove (Beveridge and Schechter [1965]) that the Pontryagin maximum principle is applicable, and if one is seeking a maximum, then the choice of $\mathbf{T}$ at some point t should be those which maximize H. The proof will not be given here.

Finally, F must be stationary with respect to variations in the recycle r and a separation factor s_j. These two conditions lead to the equations

$$\int_0^{\theta} \left(\frac{\partial Y}{\partial r} + \sum_{j=1}^{n} \lambda_i \frac{\partial f_i}{\partial r} \right) dt - \sum_{j-1}^{n} \lambda_{i,0} \left[\frac{q}{(q + r)^2} \right] x_{i,I}$$
$$+ \sum_{j=1}^{n} \frac{q s_i}{(q + s_i r)^2} \lambda_{i,0} x_{i,\theta} + \frac{\partial C}{\partial r} = 0 \qquad (5.3\text{-}12)$$

and

$$\int_0^{\theta} \frac{\partial Y}{\partial s_j} dt + \frac{qr}{(q + s_j r)^2} \lambda_{j,0} x_{j,\theta} + \frac{\partial C}{\partial s_j} = 0 \qquad (5.3\text{-}13)$$

Equations (5.3-3), a boundary condition; (5.3-4), the kinetic equations; (5.3-7), the definition of the adjoint function; (5.3-9*a*), a boundary condition; (5.3-10), maximum or minimum of H; (5.3-12), defining change with amount of recycle; and (5.3-13), defining variation with separation factor; form the basis of a computational scheme—a very complex computation in the general case.

Example IX-14: A recycle reactor

Consider the special case of a single irreversible reaction $A_1 \rightarrow A_2$ being carried out in an existing tubular reactor, with a constant fraction of the product being recycled, the separation factor s_1 being unity. Here the vector $\mathbf{x}$ has only one component, x, the weight fraction of A_1. The kinetic expression, Eq. (5.3-4), is assumed to be of the form

$$\frac{dx}{dt} = - \frac{K}{r + q} (Z x^{\alpha}) = f(x, T) \qquad (5.3\text{-}14)$$

where K is a constant, and Z a known function of temperature (for example, $Z = T^m$ or $Z = \exp^{(-E/RT)}$).

It is desired to optimize the profit from the process, this being defined as the income from the sale of product A_2 less the cost of the raw materials and the cost of operation. It is assumed that the cost of maintaining the level of temperature along the reactor length will be the only important operating cost (being proportional to the temperature level) and that all other costs are negligible or constant, thus being omitted. The objective function for the process is therefore

$$\begin{aligned} y &= \text{income} - \text{operating cost} - \text{raw material cost} \\ &= c_1 q(1 - x_\theta) - c_2 \int_0^\theta T\,dt - c_3 q x_I \end{aligned} \tag{5.3-15}$$

where c_1, c_2, c_3 are, respectively, the value of a unit quantity of B, the cost of providing the temperature per unit length of reactor, and the cost of a unit amount of raw material. Integrating the kinetic equation over the reactor length θ, we get

$$x_\theta - x_0 = \int_0^\theta f\,dt \tag{5.3-16}$$

We can also write the material balance, Eq. (5.3-3), at the inlet to the reactor, as

$$r x_\theta + q x_I = (r + q) x_0 \tag{5.3-17}$$

Our objective function may therefore be written, after substituting Eqs. (5.3-16) and (5.3-17) into Eq. (5.3-15), as

$$y = y_0 + \int_0^\theta Y\,dt$$

where

$$y_0 = c_1 q - (c_1 + c_3) q x_I$$

$$Y = -[c_1(r + q) f + c_2 T]$$

Since y_0 is a constant for a given feed, our problem is reduced to maximizing the functional $y - y_0$ by the determination of a suitable form of the independent variable T, which is a function of position along the reactor. In other words, we wish to find the function $T(t)$ so as to maximize

$$y - y_0 = \int_0^\theta Y\,dt$$

The adjoint functions defined by Eq. (5.3-7) are

$$\frac{d\lambda}{dt} = -\frac{\partial Y}{\partial x} - \lambda \frac{\partial f}{\partial x} \tag{5.3-18}$$

From Eq. (5.3-18) we can show that

$$\frac{d\lambda}{dt} = \frac{K\alpha}{r+q} Zx^{\alpha-1}[\lambda - c_1(r+q)] \tag{5.3-19}$$

since f was defined in Eq. (5.3-14). The adjoint function $\lambda(t)$ may be related to the local composition of the reacting mixture by taking the ratio of Eqs. (5.3-14) and (5.3-19) and then integrating the resultant expression. These steps lead to

$$[\lambda(t) - c_1(r+q)]x^{\alpha}(t) = [\lambda_0 - c_1(r+q)]x^{\alpha}(0) \tag{5.3-20}$$

If the value of T which makes the Hamiltonian as large as possible is used, then Eq. (5.3-10) will be satisfied, so that

$$[\lambda(t) - c_1(r+q)]\left[-\frac{K}{r+q} Z'x^{\alpha}(t)\right] = c_2 \tag{5.3-21}$$

where $Z' = \partial Z/\partial T$ is the only function of temperature for the example under consideration. Rearranging and substituting for $\lambda(t)$ from Eq. (5.3-20), we have that

$$Z' = \frac{-c_2(r+q)}{x_0{}^{\alpha}K[\lambda_0 - c_1(r+q)]} \tag{5.3-22}$$

It will be seen that the right-hand side of the equation is independent of the position t along the reactor, being only a function of the inlet and outlet conditions of the reactor. This equation cannot be solved directly for the temperature, but it is obvious that if the temperature is independent of the reactor position, it must be a constant throughout the reactor. The optimal operating policy in this case is therefore to operate the tubular reactor isothermally. This will produce the greatest profit from the system. The cost of achieving the desired temperature level, the rate of recycle, the flow and composition of the fresh feed, and the value of the final product, all serve to change the level of the optimal temperature, but in any problem the temperature is independent of the position along the axis of the reaction vessel.

Example IX-15: A numerical illustration

Let us consider the consecutive reactions

$$A_1 \rightleftharpoons A_2 \rightarrow A_3$$

being carried out in an existing tubular reactor, a constant fraction of the product being recycled. Thus the separation factors are all unity. The vector $\mathbf{x}$ has two independent components, x_1 and x_2. As in Example

IX-12, our aim is to maximize the production of A_2. The kinetic equations are assumed to be

$$\frac{dx_1}{dt} = \frac{1}{q+r}(k^2x_2 - kx_1) \equiv f_1$$

and
$$\frac{dx_2}{dt} = \frac{1}{q+r}(kx_1 - 3k^2x_2) \equiv f_2$$

These equations differ from those studied in Example IX-12 only in that the flow rate $q + r$ is included. It is necessary to identify the flow, since changing the recycle rate will of course change the flow. The boundary conditions for the adjoint functions, Eq. (5.3-9*b*), can be written

$$\lambda_{1,\theta} = \frac{r}{r+q}\lambda_{1,0} \tag{5.3-23}$$

and
$$\bar{\lambda}_{2,\theta} = \frac{r}{r+q}\bar{\lambda}_{2,0} + 1 \tag{5.3-24}$$

The adjoint equations are then, with $\bar{\lambda}_2 = \lambda_2 + 1$,

$$\frac{d\lambda_1}{dt} = \frac{k}{r+q}(\lambda_1 - \bar{\lambda}_2) \tag{5.3-25}$$

and
$$\frac{d\bar{\lambda}_2}{dt} = \frac{k^2}{q+r}(3\bar{\lambda}_2 - \lambda_1) \tag{5.3-26}$$

The temperature function is determined by maximizing H, which, except at the inlet, is at a stationary point, defined by

$$0 = (\bar{\lambda}_2 - \lambda_1)x_1 + 2kx_2(\lambda_1 - 3\bar{\lambda}_2) \tag{5.3-27}$$

To complete the specification of the problem, we may define the boundary conditions on the compositions as

$$x_{1,0} = \frac{q}{q+r}(1.0) + \frac{r}{r+q}x_{1,\theta} \tag{5.3-28}$$

and
$$x_{2,0} = \frac{r}{r+q}x_{2,\theta} \tag{5.3-29}$$

since the feed to the reaction system is entirely composed of substance $A_1(x_1 = 1.0,\ x_2 = 0$ in the feed to the reactor system). The equations are nonlinear and coupled, so that it does not seem possible to achieve an analytical solution. As in Example IX-13, we shall resort to the use of numerical techniques.

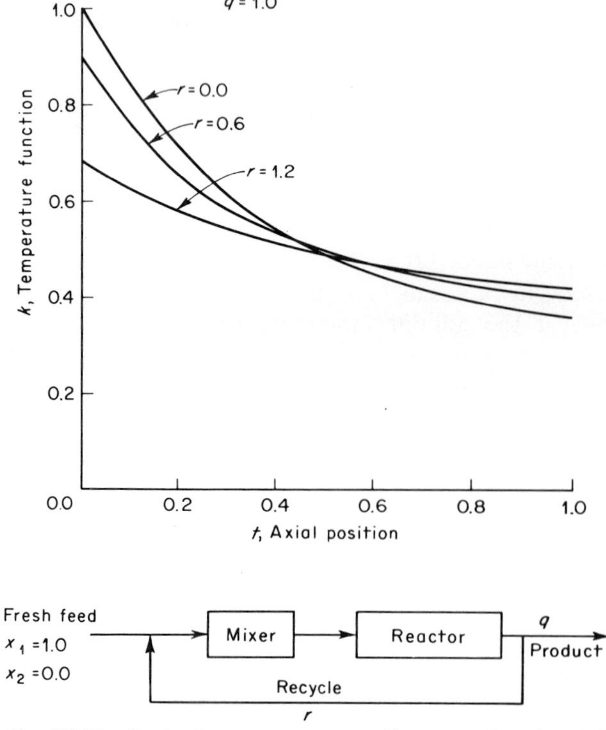

Fig. IX-11 Optimal temperature profile as a function of recycle rate.

The computational procedure presented in Sec. 5.2.4 is still applicable to the reaction system with recycle, and we shall not detail the procedure again. However, it is worth noting that steps 2 and 3 are more difficult if there is recycle. Even if the temperature function is specified at every point along the reaction path, the determination of the compositions is not a straightforward calculation, owing to the nature of the boundary conditions expressed by Eqs. (5.3-28) and (5.3-29). In general, a trial-and-error procedure is required to find the concentrations; however, judging from this experience, developed in performing a few calculations, the determination of the concentrations is not difficult. The same difficulties arise in the computation of the adjoint functions. Typical results are shown in Figs. IX-11 to IX-13.

It should be noted with recycle that the inlet and outlet conditions for both the adjoint functions and the concentrations "float," so as to satisfy Eqs. (5.3-23), (5.3-24), (5.3-28), and (5.3-29). The variation of the terminal conditions is clearly shown in Figs. IX-12 and IX-13.

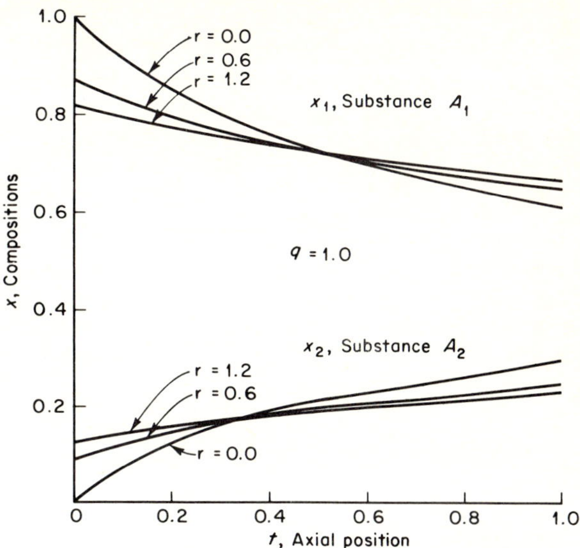

Fig. IX-12 Compositions under optimal conditions.

The problems which have a more complex structure (such as found in recycle systems) are not difficult to treat formally by introducing any new boundary conditions via the method of Lagrangian multipliers, thus allowing the appropriate necessary conditions to be developed. However, the actual numerical solution of these complex problems is itself more difficult and

Fig. IX-13 Adjoint functions.

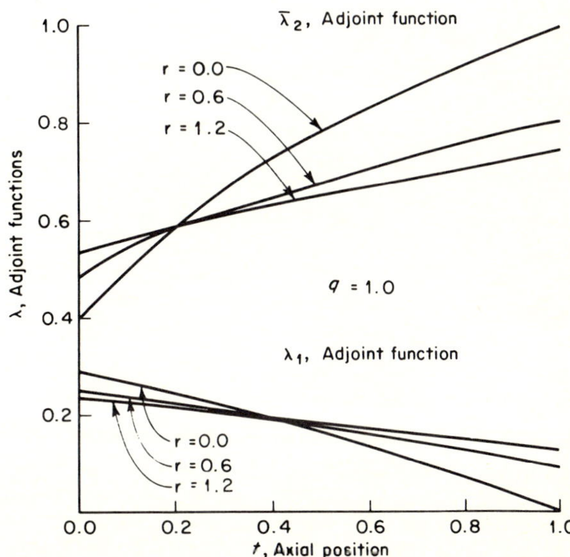

requires more ingenious programming and more computer time. Moreover, the convergence of the methods, which seems to be satisfactory in cases with simple boundary conditions, might not be realized in the general case. All these factors lead us to conclude that the more complex systems, while not more difficult to set up, are often most difficult to solve, due to difficulties in the numerical computation.

6 CONCLUSIONS

The analytical solution of problems, depending on a choice of continuous variables, proceeds in much the same fashion as in the case of a finite number of variables, the derivatives being replaced by a functional variation in the continuous case. For an optimum to exist in the continuous case we have seen that the first variations must vanish, just as for systems of finite dimensionality. This stationary requirement was seen to lead to a set of differential equations, the Euler-Lagrange equations, one for each of the independent variables appearing in the integrand of the objective function. In the general case these differential equations must be solved simultaneously, just like the set of algebraic equations which define the stationary points in an n-dimensional space.

The Euler-Lagrange equations are usually nonlinear, and exact solutions can be found only in the very simple examples. The solution of the differential equation is generally attempted using numerical methods, this task being complicated by the two-point nature of the boundary conditions. The numerical integrations therefore tend to be trial-and-error calculations. Most often the numerical search, which is discussed in Chap. X, is preferred.

If restrictions are present, we have used Lagrangian multipliers to aid in the calculation. Global restrictions are those in which the continuous variable is subject to constraints which have an integral form, and for this classic problem in the calculus of variations the Lagrangian multipliers have much the same character as found in Chap. VII. If, however, restrictions are imposed from point to point within the system, the associated Lagrangian multipliers will also vary from point to point within the system. Aside from the usual stationary conditions, an additional necessary condition was found when the restriction was a local one. This important restriction, the Pontryagin maximum principle, helps to determine an appropriate choice of the continuous variable in cases for which several values are admissible within the framework of the usual necessary conditions defining a stationary point, or if no acceptable choice is found, the maximum principle then helps to guide the choice usually found at a boundary. This principle is a very important aid in computation.

Dynamic programming also leads to the same necessary conditions as those found using the calculus of variations, together with Pontryagin's maximum principle. The two methods are, however, normally applied in quite different computational schemes. Using dynamic programming, one usually begins at the end of the system and stores solutions to a range of problems while progressing toward the front of the system. The final result is embedded within the stored tables, and there are no trial-and error calculations. The approach using the calculus of variations is to attempt to solve the Euler-Lagrange equations by a trial-and-error procedure. This method seems to be quite effective, and is especially recommended for those cases in which computer storage is a problem.

The trial-and-error approach recommended here is the gradients in function space technique in which the optimizing profile is corrected after each iteration, using the maximum principle. Unfortunately, this scheme is not always applicable. It cannot be applied to cases in which the continuous variables are specified at both the start and the completion of the process. For example, in examining a batch distillation column, where the contents of the still at both the start and the finish of the process are given, or in planning a trip to the moon, where the initial and final coordinates are specified, gradients in function space do not apply. In such problems, it is usual to assume the initial values of the adjoint functions and to perform the integrations to see if these assumed values produce the desired end result. If not, new initial values must be assumed and the integrations again performed. This is continued until the specified terminal conditions are achieved.

The required computations associated with the solution of problems associated with continuous variables are usually complex, and most often some direct numerical procedure is adopted, such as those discussed in the following chapter.

Problems

IX-1 Find the necessary condition for the functional

$$y = \int_0^\theta Y(x, x_t, x_{tt})\, dt$$

to be an optimum relative to all choices of $x(t)$ which are continuous and also having continuous first four derivatives. Here

$$x_t = \frac{dx}{dt} \qquad \text{and} \qquad x_{tt} = \frac{d^2x}{dt^2}$$

The function x, as well as its first derivatives, is specified at both end points.

IX-2 Find the necessary condition for the functional

$$y = A\iint_0^\theta K(z,t)x(z)x(t)\,dz\,dt - \int_0^\theta [x(t)]^2\,dt + 2\int_0^\theta x(t)G(t)\,dt$$

to be stationary. In these equations A is a constant, $G(t)$ is a specified function of position, and $K(z,x)$ is a specified symmetric kernel $[K(t,z) = K(z,t)]$.

IX-3 Show that if x is the function which optimizes

$$y = \int_0^\theta (p^2x_t^2 + q^2x^2)\,dt$$

where $p(t)$ and $q(t)$ are known functions, then y has the value

$$p^2x_tx\big|_{t=\theta} - p^2x_tx\big|_{t=0}$$

IX-4 Develop a set of sufficient conditions for the objective function of Prob. IX-1 to be a maximum.

IX-5 What conditions on $p(t)$ and $q(t)$ will ensure that the objective function of Prob. IX-3 is a minimum?

IX-6 What continuous function with continuous first and second derivatives minimizes the functional

$$y = \int_0^\theta \left[x^2 + 2tx + \left(\frac{dx}{dt}\right)^2\right] dt$$

given $x(0) = 2$.

IX-7 Repeat Prob. IX-6 using the boundary conditions

$$x(0) = 2 \qquad x(\theta) = 0$$

Which function (solution to Prob. IX-6 or solution to this problem) yields the smallest value y? Why? Demonstrate the validity of your answer by computing y, using both solutions and comparing the numerical result.

IX-8 What continuous function with continuous first and second derivatives minimizes the functional

$$y = \int_0 (x^2 + 2tx^2 + x_t^2)\,dt$$

IX-9 Show that the following inequalities are inconsistent:

$$\int_{-1}^1 [f(x) - x^3]^2\,dx < 1/9 \qquad \text{and} \qquad \int_{-1}^1 [f(x) - x^4]^2\,dx < 1/9$$

(Synge [1957]).

{*Hint:* Let $y = \int_{-1}^1 [f(x) - x^3]^2\,dx$ and minimize y subject to restriction

$$G = \int_{-1}^1 [f(x) - x^4]^2\,dx - d_1 = 0$$

show that for any acceptable choice of $d_1 (0 \le d_1 < 1/9)$, y exceeds $1/9$.}

IX-10 If $\int_{-1}^{1} [f(x) - x]^2 \, dx = 4$ what are the bounds of $\int_{-1}^{1} x^2 f(x) \, dx$? (Synge [1957]).

IX-11 Minimize

$$y = \int_0^\theta x_t^2 \, dt$$

subject to the constraint

$$\int_0^\theta x^2 \, dt = 1$$

together with $x(0) = x(\theta) = 0$.

The solution to this problem is typical of the Sturm-Liouville class (Weinstock [1952, chap. 8]), and we see that here the Lagrangian multiplier is the eigenvalue associated with the eigenfunction $x(t)$.

IX-12 For a space vehicle moving in a rarefied atmosphere the force of the molecules is normal to the surface if reflections are specular and given by

$$p = 2\rho v^2 \sin^2 \theta$$

where p is the normal pressure acting on an element of surface inclined at an angle θ relative to direction of motion. Develop the necessary condition for a surface of revolution to give minimum drag. The surface has zero radius at the leading edge and is constrained to have a radius of W at trailing edge. Do not attempt to solve the problem. An approximate solution has been given by Miele [1962].

IX-13 Repeat Prob. IX-12, except in this case find the surface of revolution giving minimum drag, having a fixed volume V and zero radius at its leading edge. The length θ of the vehicle is specified.

IX-14 Minimize the function

$$y = \int_0^\theta (x^2 + T^2) \, dt$$

subject to the restrictions that

$$\frac{dx}{dt} = T - 4x$$

and

$$x(0) = 1.0$$

IX-15 Minimize the function

$$y = \int_0^\theta (x^2 + T^2) \, dt$$

$$\frac{dx}{dt} = T - x$$

subject to the restrictions

$$x(0) = 1 \qquad x(\theta) = 0$$

IX-16 Find the minimum value of $x(\theta)$, given that

$$\frac{dx}{dt} = ax + T$$

$$x(0) = c_0$$

$$\int_0^\theta T^2(t) \, dt \leq V$$

(Roberts [1964*b*].)

IX-17 Minimize the objective function

$$y = \int_0^1 (3x + 2T)\, dt$$

subject to the restrictions $\dfrac{dx}{dt} = 3x - T$

$$x = 5 \qquad \text{at } t = 0$$

$$\tfrac{1}{2} \le T \le 2$$

Show from your solution that the Hamiltonian is a constant.

IX-18 Suppose that the following reaction is carried out in an existing reactor of length θ:

$$A \rightleftarrows B \rightarrow C$$

The kinetics of this reaction system

$$\frac{dc_1}{dt} = k_2 c_2 - k_1 c_1 \qquad \frac{dc_2}{dt} = k_1 c_1 - (k_2 + k_3) c_2$$

$$\frac{dc_3}{dt} = k_3 c_2$$

in which t is the axial position, c_1 is the concentration of A, c_2 is the concentration B, c_3 is the concentration of C, and

$$k_i = k_i^0 e^{-E_i/RT} \qquad \text{for } i = 1, 2, 3$$

What differential equations determine the temperature distribution giving the maximum profit of the objective function

$$y = \sum_{i=1}^{3} \alpha_i c_i(\theta)$$

Assume that the concentrations are fixed at the inlet. Describe a step-by-step iterative procedure you would use to find a solution.

IX-19 Repeat Prob. IX-18, except that the optimization must be carried out in such a way that the concentration of c_i at the reactor outlet is fixed. Therefore maximize

$$y = \sum_{i=1}^{2} \alpha_i c_i(\theta)$$

subject to

$$c_3(\theta) = \text{constant}$$

IX-20 It is desired to find the temperature to give the maximum concentration of product C in the reaction scheme

$$\begin{array}{l} A \xrightarrow{k_1} B \xrightarrow{k_3} C \xrightarrow{k_4} D \\ {\scriptstyle k_2}\downarrow \\ D \end{array}$$

If c_1, c_2, c_3 are the concentrations of A, B, and C, the kinetic equations are

$$\frac{dc_1}{dt} = -c_1(k_1 + k_2)$$

$$\frac{dc_2}{dt} = k_1c_1 - k_3c_2$$

$$\frac{dc_3}{dt} = k_3c_2 - k_4c_3$$

where T = absolute temperature

t = axial position

k_i = usual Arrhenius form $k_i = k_i^0 e^{-E_i/RT}$

The initial concentrations, the reactor feed, may be taken as $c_1 = 1$, $c_2 = 0$, $c_3 = 0$ for $t = 0$. The total length of the reactor is $\frac{1}{2}$. Describe a procedure whereby you would find an optimizing temperature profile along the reaction path. This problem has been solved numerically by Storey [1964] and Storey and Rosenbrock [1964].

IX-21 The reaction

$$A \to B \to C$$

is represented by the kinetic equations

$$\frac{dc_1}{dt} = -Hc_1$$

$$\frac{dc_2}{dt} = Hc_1 - H^2c_2$$

Prove that H must be selected so that

$$\frac{dH}{dt} = -\frac{H^2c_1}{c_2}$$

if $c_2(\theta)$ is to be maximized, where θ is the length of the reactor (Coward and Jackson [1965]).

IX-22 The reaction

$$A \to B \to C \to D$$

takes place in a reactor, the concentrations of A, B, C, and D being represented by c_i ($i = 1, 2, 3, 4$), respectively. The system kinetics may be represented by

$$\frac{dc_1}{dt} = -\alpha_1c_1 + \beta_1T$$

$$\frac{dc_2}{dt} = \alpha_1c_1 - \alpha_2c_2 + \beta_2T^2$$

$$\frac{dc_3}{dt} = \alpha_2c_2 - \alpha_3c_3 + \beta_3T^3$$

where $\sum_{i=1}^{4} c_i = 1$; the reaction rate constant α_i are constants; the temperature effects are given by the β_i terms, the β_i being constants; and t is the distance measured along the reactor.

Show that the analytical solution for the temperature profile $T(t)$ is

$$T(t) = \frac{\beta_1}{2\beta_2}\left(e^{(\alpha_1-\alpha_2)(t-\theta)} - 1\right)$$

if it is desired to maximize the concentration c_2 of B at the exit from the reactor of length θ, using a feed of pure A; that is, $c_1(0) = 1.0$.

IX-23 Minimize the functional

$$y = \int_0^l (x^2 + T^2)\, dt$$

subject to the conditions

$$\frac{dx}{dt} = T - x$$

and

$$x(0) = 1 + \tfrac{1}{2}x(l)$$

This latter condition, connecting the value of x at the start with that at the end, is the equivalent of a recycle system.

IX-24 Find the minimum value of $x(\theta)$, given that

$$\frac{dx}{dt} = ax + T$$

$$x(0) = 1 + \tfrac{1}{2}x(\theta)$$

$$\int_0^\theta T^2(t)\, dt \le V$$

CHAPTER X

FUNCTIONS OF A CONTINUOUS VARIABLE—NUMERICAL METHODS

1 THE NEED FOR NUMERICAL SEARCH

In Chap. IX we examined the fundamental problem of the calculus of variation: that of selecting the function $x(t)$ which optimizes the function

$$y = \int_a^b Y(x,x_t,t)\,dt \tag{1-1}$$

There we examined the conditions under which y was stationary, a process exactly equivalent to the n-dimensional analytical approach of Chap. VII, except that now a differential equation arises instead of n simultaneous algebraic equations. In Chap. IX we attempted to integrate the Euler-Lagrange equation, representing the necessary conditions, so that our approach can be properly called an analytical one. The material given in Chap. IX parallels that of Chap. VII, and we should therefore inquire as to the existence of possible numerical methods as an alternative approach to infinitely dimensional problems, thus corresponding to the methods of Chap. VIII.

Perhaps a more basic question is to ask whether there is a need for this second approach. We recall that one need for numerical search in n dimensions arises when the simultaneous algebraic equations defining the stationary points are so complex as to be unsolvable for all practical purposes. Similar situations are found in the continuous case, when the Euler-Lagrange equation becomes so complex that it cannot be integrated by analytical means. When faced with such a problem, we could resort to numerical techniques for solving differential equations, such as the Euler or the Runge-Kutta methods (Chap. III, Sec. 4.2). There is the difficulty, however, that the differential equation must be integrated subject to boundary conditions specified at both ends of the interval of interest—a two-point boundary-value problem—and trial-and-error calculations are usually required. We must therefore resort to numerical search techniques for circumventing this difficulty. Unrestricted problems are considered first (Sec. 2). The two types of restrictions, global and local, are then treated separately in Secs. 3 and 4, respectively, since the numerical approach to these two cases can be quite different.

2 UNRESTRICTED PROBLEMS

Our task is to determine the function $x(t)$ which optimizes the objective function defined by Eq. (1-1). The previous approach, in Chap. IX, was through the development of a differential equation which x had to satisfy. Here we intend to take a different approach and make direct calculations based on the integral form of the objective function, the resultant solution being an approximation to the true solution. Two essentially different methods will be discussed, namely, the Rayleigh-Ritz technique (Sec. 2.1) and the method of finite differences (Sec. 2.2), although both techniques reduce the original problem to one of n-dimensional search. Since these methods, especially the former, are often very effective, they are sometimes used for approximating differential equations (Sec. 2.3).

2.1 The Rayleigh-Ritz Method

The essential basis of this method is the substitution of an approximation in place of the true minimizing function in the objective function, the approximate function being selected in advance, apart from a number of undetermined constants. Thus an approximating function might have the form

$$\tilde{x}(t) = \sum_{i=1}^{n} C_i u_i(t) \tag{2.1-1}$$

Here the approximation $\tilde{x}$ consists of a sum of known functions, $u_i(t)$, each multiplied by an undetermined constant C_i. If the solution is specified on one or both of the boundaries, the approximating function should be constructed in such a way that the given conditions will be satisfied for all values of the constants. If the boundary conditions are natural boundary conditions, no such special precaution is required.

The approximation is then substituted into the objective function, Eq. (1-1), thus determining y as a function of the n constants $C_1, C_2, \ldots, C_n$, instead of being a functional of the continuous function $x(t)$. Our task is thus simplified to one of n-dimensional search for the values of the n constants which optimize y, using the most suitable of the techniques discussed in Chap. VII or VIII.

Once the n values of C_i have been found, we have, then, an approximation to the true solution. How good is this approximation? How can it be improved? The approximation can certainly be improved by increasing the number of undetermined constants appearing in the approximation, as represented by an increase in the number of terms in the series expressed by Eq. (2.1-1). If y_j is the objective function found using j terms in the series, we know that

$$y_1 \leq y_2 \leq y_3 \leq \cdots \leq y_n \leq y^*$$

when a maximum is sought. This series of inequalities is valid, since the value of y cannot increase if fewer terms are used. The use of a smaller number of terms is a form of restriction, and a restricted optimum can never be better than one subject to lesser restrictions. In other words, our solution can never become worse as more undetermined constants are introduced into the approximating function.

Since our results will generally improve as the number of constants is increased, we are certain to find a better and better approximation, although we are not assured that the true answer will be reached even with an infinite number of terms. This is a difficult question and one which cannot be answered in general, although proofs have been developed in special cases (Kantorovich and Krylov [1958]). The greatest confidence can be gained by using an approximating sequence drawn from a complete set of functions in the interval of interest. Thus, for example, in a certain interval one might select terms from an infinite Fourier sine series, knowing that the infinite series is complete in the sense that all well-behaved functions can be expressed as a sum of such terms. Of course, we should never use an infinite number of terms, since the task of determining the constants would then be as difficult as tackling the original problem, but it is comforting to know that the approximation is being constructed in such a way that the true solution can be adequately represented by the approximating function. Other complete sets include the power series, the Bessel series, and Legendre polynomials. Again we note that our reason for selecting members of a complete set in constructing a suitable trial function is that we are virtually assured of being able to describe the true solution exactly, provided a sufficient number of terms is used in the approximation.

The approximation generally becomes better and better as the number of terms is increased, but of course the labor in determining the value of the constants likewise increases, so that we should like to obtain a satisfactory approximation using as few terms as is possible. The number of constants required depends on the complexity of the problem and on the shrewdness with which the form of the approximation is selected. Obviously, if an approximation having very nearly the same structure as the true solution is assumed, very few constants may be required, but a poor choice might require a much larger number of constants before suitable accuracy is obtained. This situation is not different from any of those previously encountered, where we have seen that experience is very helpful in guiding the selection of an effective means of solution.

Our succession of approximations is normally terminated when the value of a particular y_j is sufficiently close to that of y_{j-1} for all practical purposes. Of course, this stopping criterion can give erroneous results, and the remarks of Chap. XI, Sec. 5.3.2, regarding the termination of a search, are applicable here.

Example X-1: The Rayleigh-Ritz method

Suppose that we are required to find the function $x(t)$ which minimizes the integral

$$y = \int_1^2 \left[\frac{1}{2}\left(\frac{dx}{dt}\right)^2 - 9/2x^2\right] t\,dt \tag{2.1-2}$$

and satisfies the boundary conditions

$$x(t) = \begin{cases} 1 & \text{at } t = 1 \quad (2.1\text{-}3a) \\ 0 & \text{at } t = 2 \quad (2.1\text{-}3b) \end{cases}$$

This problem could be solved by determining the Euler-Lagrange equation and integrating that equation subject to the stated boundary conditions. Instead, we shall use the Rayleigh-Ritz method, and assume a trial function of the form

$$\tilde{x}(t) = C_1(t - 2)(t - 1) + 2 - t \tag{2.1-4}$$

where C_1 is an undetermined constant. The boundary conditions given by Eqs. (2.1-3) are satisfied, irrespective of the value of C_1. The choice of this particular approximating function was guided only by the fact that the integrations are easy to perform using this series, and the power series is a complete set. Another choice might have been

$$\tilde{x}(t) = C_1 \sin \pi t + 2 - t$$

This approximation may actually be better, but we have no real way of knowing in advance, and the integrations involving sines and cosines are more difficult to perform, so that we shall use Eq. (2.1-4) as our first approximation.

Substituting Eq. (2.1-4) for the $x(t)$ in Eq. (2.1-2), we find

$$y_1 = \int_1^2 \{1/2[C_1(2t - 3) - 1]^2 - 9/2[C_1(t - 2)(t - 1) + 2 - t]^2\}t\,dt$$

Integrating this equation, we obtain

$$y_1 = \frac{C_1^2}{40} + \frac{53}{60}C_1 - \frac{9}{8} \tag{2.1-5}$$

Our task is to select C_1 so as to minimize the value of y. To find the appropriate values of C_1, we can use either analytical or numerical search techniques. Here it seems more convenient to use an analytical approach and find the stationary point. Thus C_1 is determined by the equation

$$\frac{\partial y_1}{\partial C_1} = \frac{C_1}{20} + \frac{53}{60} = 0$$

or $C_1 = -17.667$. Using this value of C_1, our estimate y_1 of the minimum value of the function is -8.9278.

Is this an acceptable result or is the true minimum significantly less than this value? Without some physical insight into the particular problem, we must make further calculations in order to be able to make a decision regarding the quality of the result. Indeed, our situation can be compared with that in numerical search, in which one function evaluation has been made. Clearly, a second approximation is required to measure the quality of the result. Let us take as our second approximation

$$\tilde{x}(t) = C_1(t-2)(t-1) + C_2t(t-2)(t-1) + 2 - t \qquad (2.1\text{-}6)$$

Substituting this trial function into the objective function for $x(t)$ and integrating, we find

$$y_2 = \frac{C_1^2}{40} + \frac{41}{420}C_1C_2 + \frac{67}{560}C_2^2 + \frac{53}{60}C_1 + \frac{5C_2}{4} - \frac{9}{8}$$

Solving for the minimizing values C_1 and C_2, we find that $C_1 = -36.69498$, $C_2 = 9.74621$, and $y_2 = -11.2405$. This value of the objective function is less than the previous value of $y_1 = -8.9278$, as we should expect, since an additional degree of freedom is included in the second of the approximating functions by adding the second undetermined constant. Should we terminate our search? The second of the objective functions is considerably less than the first, so that an additional approximation might prove of further advantage. A suitable form for the next approximating function would be

$$\tilde{x} = C_1(t-2)(t-1) + C_2(t-2)(t-1)t + C_3(t-2)(t-1)t^2 + 2 - t$$

This approximation would provide a better estimate of the true solution. We shall not, however, make this third approximation, since the computational labor increases considerably as the number of undetermined constants is increased. The true minimum obtained by substituting the true solution

$$x = -15.860J_0(3t) - 8.291\,Y_0(3t)$$

where J_0 and Y_0 are Bessel functions of zero order and the first and second kind, respectively, into the objective function, Eq. (2.1-2) is found to be $y = -12.108$, which is, of course, less than both of the previous approximate values.

A further comparison of the approximate solutions with the true solution is made in Table X-1, the values of the minimizing functions being compared at various values of t. Here we see that even the second

TABLE X-1 A COMPARISON OF THE APPROXIMATE AND TRUE SOLUTIONS TO EXAMPLE XI-1

t	*One-constant approximation* $\tilde{x}(t)$	*Two-constant approximation* $\tilde{x}(t)$	*True solution* $x(t)$
1.0	1.00	1.00	1.00
1.2	3.63	4.80	4.99
1.4	4.84	6.13	6.75
1.6	4.64	5.46	6.07
1.8	3.03	3.26	3.47
2.0	0.00	0.00	0.00
Objective function y	−8.93	−11.24	−12.11

approximation is considerably different from the true solution, although the values of the objective functions differ by only about 7 percent. This is typical of many variational problems, although not generally true. Thus one can often achieve an objective very nearly equal to that produced by the true minimizing function. This fact can be of great practical importance since the true minimizing function is often not attainable in practice, and thus it is comforting to know that the optimal objective can be very nearly attained with a suboptimal policy, which is practically acceptable.

2.2 The Method of Finite Differences

An integral is, by definition, the sum of a series of terms, the sum being computed in a limiting sense (Kaplan [1952]). We intend to use this definition and to approximate the integral objective function with a finite series of terms, the quality of the approximation improving as the number of terms in the series is increased. In approximating the integral by a finite sum, we gain the advantage of having replaced the continuous function $x(t)$ by a finite set of discrete values, these values being taken at certain nodes, and thus leaving us with a search problem of finite dimensionality. Again we can resort to a suitable n-dimensional method to solve the problem.

Let us divide the interval of interest into $n + 1$ equal subintervals, thus introducing the points $t_0, t_1, t_2, \ldots, t_n, t_{n+1}$, where $t_0 = a$, $t_{n+1} = b$, and $\Delta t = t_{i+1} - t_i$. At the same time we shall replace the continuous function $x(t)$ with the set of discrete values $x_0, x_1, \ldots, x_{n+1}$, where x_i is an estimate of $x(t_i)$. Using this notation, the integral

$$y = \int_a^b Y(x,x_t,t)\,dt$$

can be approximated as

$$y_n = \sum_{i=0}^{n} Y\left(x_i, \frac{x_{i+1} - x_i}{\Delta t}, t_i\right)\Delta t \tag{2.2-1}$$

Our task is now to select the $n + 2$ values of x_i which optimize y_n. If the values of x_i at the boundaries, x_0 and x_{n+1}, are specified however, there will be only n variables to be determined.

Example X-2: The method of finite differences

Let us apply this method to the problem of optimizing the objective function

$$y = \int_0^1 (x_t^2 - 9x^2)\, dt$$

Replace this integral by the sum

$$y_9 = \sum_{i=0}^{9}\left[\left(\frac{x_{i+1} - x_i}{\Delta t}\right)^2 - 9x_i^2\right]\Delta t \tag{2.2-2}$$

If the original problem is stated so that the boundary conditions are $x = 0$ at $t = 0$ and $x = 1$ at $t = 1$, we should fix $x_0 = 0$ and $x_{10} = 1$ in the above summation.

In writing Eq. (2.2-2) we have assumed that the interval of interest, $0 \leq t \leq 1$, has been divided into 10 equally spaced intervals, so that $n = 9$ and $\Delta t = 0.1$. All the quantities appearing in Eq. (2.2-2) are known, except $x_1, x_2, \ldots, x_9$. In this particular simple problem, we find it easiest to seek a stationary solution from the nine simultaneous algebraic equations

$$\frac{\partial y_9}{\partial x_i} = 0 = 2\Delta t\left(\frac{x_i - x_{i-1}}{\Delta t^2} - \frac{x_{i+1} - x_i}{\Delta t^2} - 9x_i\right) \qquad \text{for } i = 1, 2, \ldots, 9 \tag{2.2-3}$$

Solving these equations, we find that $x_1 = 2.284$, $x_2 = 4.363$, $x_3 = 6.048$, $x_4 = 7.190$, $x_5 = 7.684$, $x_6 = 7.487$, $x_7 = 6.616$, $x_8 = 5.150$, $x_9 = 3.220$. These values of x_i yield an approximate value of the objective function of $y_9 = -22.197$. These values of x_i should be compared with the exact solution,

$$x = \frac{\sin 3t}{\sin 3}$$

giving a value of $y = -21.046$.†

† The approximation to y apparently gives a smaller value of y than is found with the true minimizing function. Actually, this paradox arises because the approximating piecewise function is continuous but has discontinuities in the first derivatives at each of the nodes.

A more precise approximation can be found by increasing the number of terms in the approximation (2.2-2), but of course the computational labor also increases.

2.3 The Solution of Differential Equations

The direct approach to the solution of variational problems is often a very convenient and rapidly convergent process and as a consequence has found considerable application in the solution of certain differential equations (Kantorovich and Krylov [1958] and Schechter [1967]). We simply recognize that each time an approximate solution to a given variational problem is found, we are at the same time approximating the solution to an associated differential equation, namely, the Euler-Lagrange equation. Thus Eq. (2.1-4), together with the optimizing value of C_1, represents an approximation to the Euler-Lagrange equation associated with the particular variational problem of minimizing the y defined by Eq. (2.1-2). Indeed, the quality of this approximation is shown in Table X-1. Thus the direct approach, using the Rayleigh-Ritz method, provides a means of approximating the solution of differential equations.

Finally, we note that the method of finite differences yields a difference equation by differentiating with respect to one of the discrete values, say, x_i. This difference equation is an analog to the Euler-Lagrange differential equation, Varga [1962] recommending that such difference analogs be developed, starting, wherever possible, with the variational formulation.

3 GLOBAL RESTRICTIONS

As noted in Chap. IX, Sec. 5, the optimization may be restricted to values of $x(t)$ which satisfy the global restriction

$$J = \int_a^b G(x,t)\,dt \tag{3-1}$$

Such restrictions were successfully treated in Chap. IX, Sec. 5.1, using the method of Lagrangian multipliers. Both the method of Rayleigh-Ritz (Sec. 3.1) and finite differences (Sec. 3.2) are easily extended to account for global restrictions when used in the Lagrangian formulation of the problem.

3.1 The Rayleigh-Ritz Method and Global Restrictions

In the face of global restrictions, we previously found it correct to use a Lagrangian formulation of the problem. The same approach can be used, together with direct methods, as long as the undetermined constants are sought analytically. Numerical search techniques cannot be safely used, since we have already established that the character of the augmented Lagrangian function $F = y - \lambda g$ is not necessarily in correspondence with that of the original objective function y. Thus F may be a maximum, minimum, or saddle point when y is a minimum. Numerical search techniques cannot therefore be used in seeking the values of the undetermined constants. Apart from this reservation, there is no difference in the application of the Rayleigh-Ritz method to restricted problems and those which are unrestricted. This is shown in the following example.

Example X-3: A Rayleigh-Ritz solution to a globally restricted problem

It is desired to minimize the objective function

$$y = \int_0^1 x_t^2\, dt \tag{3.1-1}$$

subject to the global restriction

$$\int_0^1 x^2\, dt = 2 \tag{3.1-2}$$

and the boundary conditions

$$x = 1 \qquad \text{at } t = 0 \tag{3.1-3a}$$

$$x = 0 \qquad \text{at } t = 1 \tag{3.1-3b}$$

A suitable approximation might have the form

$$\tilde{x} = C_1 \sin \pi t + C_2 \sin 2\pi t + (1 - t) \tag{3.1-4}$$

This expression is drawn from a complete set in the interval of interest, being the first two terms in a Fourier sine expansion, the boundary conditions being satisfied for all values of the two undetermined constants, but no other considerations guiding its selection. Many other forms would probably serve just as well. The calculations for any other form are the same as those shown here, except in the actual integrations.

The Lagrangian form of the problem is

$$F = \int_0^1 (x_t^2 - \lambda x^2)\, dt \tag{3.1-5}$$

Substituting Eq. (3.1-4) into this expression gives

$$F = \frac{\pi^2 C_1^2}{2} + 2\pi^2 C_2^2 + 1 - \lambda\left(\frac{C_1^2}{2} + \frac{2C_1}{\pi} + \frac{C_2^2}{2} + \frac{C_2}{\pi} + \frac{1}{3}\right)$$

Solving for the values of C_1 and C_2 which make the $\partial F/\partial C_1$ and $\partial F/\partial C_2$ both vanish simultaneously, we obtain

$$C_1 = \frac{2\lambda}{(-\lambda + \pi^2)\pi} \qquad \text{and} \qquad C_2 = \frac{\lambda}{(-\lambda + 4\pi^2)\pi}$$

To complete the solution we need only compute the value of λ. This quantity must be determined so that the imposed global restriction is satisfied. Thus, substituting these values of C_1 and C_2, together with the approximating function, Eq. (3.1-4), into Eq. (3.1-2), we obtain the following equation, which determines the value of λ:

$$\frac{2\lambda^2}{(\lambda - \pi^2)^2\pi^2} + \frac{4\lambda}{(-\lambda + \pi^2)\pi^2} + \frac{\lambda^2}{2(-\lambda + 4\pi^2)^2\pi^2} + \frac{\lambda}{(-\lambda + 4\pi^2)\pi^2} + \frac{1}{3} = 2$$

The value of λ satisfying this equation is 6.600, the corresponding value of the objective function, y_2, being 9.233.

To further improve this value, a possible approximation such as

$$\tilde{x} = C_1 \sin \pi t + C_2 \sin 2\pi t + C_3 \sin 3\pi t - (1 - t)$$

might be introduced.

The true solution to this problem is

$$x = \frac{-\cos\sqrt{\lambda}\,\sin\sqrt{\lambda}t}{\sin\sqrt{\lambda}} + \cos\sqrt{\lambda}t$$

TABLE X-2 Comparison of the Approximate and Exact Solutions to Example X-3

t	*Approximate* $\tilde{x}$	*Exact* x
0.0	1.000	1.000
0.2	1.616	1.632
0.4	1.860	1.842
0.6	1.585	1.577
0.8	0.895	0.906
1.0	0.000	0.000
Objective function		
y	9.233	9.213
λ†	6.600	6.595

† It is no accident that the two values of the Lagrangian multipliers are about the same, since this approach is often used to compute the multipliers when they have physical significance (Temple and Bickley [1956]).

with a value of λ of 6.595. A comparison between the approximate solution and the exact solution is shown in Table X-2. The approximation is satisfactory.

3.2 The Method of Finite Differences and Global Restrictions

The method of finite differences, described in Sec. 2.2, can be applied to the augmented objective function, observing, however, the same precaution as in using the Rayleigh-Ritz method (Sec. 3.1); namely, the x_i must be determined by seeking stationary values of F analytically as opposed to the unrestricted case, where either analytical or numerical search methods may be applied.

4 LOCAL RESTRICTIONS

Restrictions imposed on variational problems that have to be satisfied at all points within the interval of interest are called *local restrictions*. This is to be contrasted with global restrictions, which are in the form of integrals. Local restrictions were accounted for previously (Chap. IX, Sec. 5.2), using Lagrangian multipliers, these multipliers having been found not to be constants, but dependent on position. Thus, in generating approximate solutions, we would be faced with the task of approximating the variable Lagrangian multipliers as well. There has been little development of reliable methods for making both of these approximations simultaneously. The only procedure that will be discussed here entails finding an approximating function satisfying the local restrictions for all values of the undetermined constants. Our approach would thus be much the same as that used in connection with boundary conditions, where we required our approximating functions to satisfy the imposed conditions for all values of the arbitrary constants. This is discussed in Sec. 4.1.

4.1 Differential Restrictions

Let us consider the problem (Chap. IX, Sec. 5.2.1) of optimizing

$$y = \int_0^\theta Y(x,T)\,dt \tag{4.1-1}$$

subject to the local restriction

$$\frac{dx}{dt} = f(x,T) \tag{4.1-2}$$

Although y is a functional of two variables, x and T, they are related through the local restriction, Eq. (4.1-2), and are not, therefore, independent. We are not at liberty to assume two different approximating functions—one for x and the other for T—because we should have no assurance that the local restriction would indeed be satisfied. Therefore we shall assume a form for one of the two interrelated functions and determine the second by integrating Eq. (4.1-2). In this case it is easiest to assume a functional form for T, with the assumed function having some degrees of freedom available. The values of x as a function of t must then be determined by integrating Eq. (4.1-2). The assumed T and the x satisfying Eq. (4.1-2) can then be substituted into the objective function, and the undetermined constants or degrees of freedom are then computed so as to optimize y. A procedure such as this has been applied by Storey and Rosenbrock [1964] in computing optimal temperatures in a chemical reactor. They found that this method was not as effective as those discussed in Chap. IX, Sec. 5.2.4, based on the Pontryagin maximum principle. However, to make these general remarks more understandable, we shall consider such an example.

Example X-4: The Rayleigh-Ritz method applied to local restrictions

Consider the problem of minimizing

$$y = \int_0^1 (x^2 + T^2)\, dt \qquad (4.1\text{-}3)$$

subject to the restriction

$$\frac{dx}{dt} = T - x \qquad (4.1\text{-}4)$$

and the boundary condition

$$x = 1.0 \qquad \text{at } t = 0 \qquad (4.1\text{-}5)$$

Let us assume the approximating function is

$$\tilde{T} = C_1 + C_2 t + C_3 t^2 \qquad (4.1\text{-}6)$$

Once this approximating function is selected and there are no restrictions on the form of its choice, the function of x is fixed by the local restriction. Here we can determine the x by integrating Eq. (4.1-4) analytically, although in the usual practical situation this step must be done numerically. In this case

$$\frac{dx}{dt} = C_1 + C_2 t + C_3 t^2 - x$$

Integrating and applying the imposed boundary condition on x, we find that

$$x = (1 - C_1 + C_2 - 2C_3)e^{-t} + C_3 t^2 + (C_2 - 2C_3)t + C_1 - C_2 + 2C_3 \qquad (4.1\text{-}7)$$

The local restriction is thus clearly shown to provide a relationship between x and T. Substituting both Eqs. (4.1-6) and (4.1-7) into Eq. (4.1-3), the objective function is seen to be the following function of the undetermined constants:

$$y_3 = 0.43233 + 0.39958C_1 + 0.12891C_2 + 0.06340C_3 + 1.16809C_1^2 + 1.13534C_1C_2 + 0.74146C_1C_3 + 0.36324C_2^2 + 0.53491C_2C_3 + 0.21055C_3^2$$

The result of applying the Rayleigh-Ritz method is, as before, to yield an expression for the objective function in terms of a finite number of parameters. The problem of selecting C_1, C_2, and C_3 so as to minimize y_3 is a far easier one than that of the selection of a minimizing function $T(t)$.

Solving for the C_1, C_2, and C_3, which minimize y_3, we obtain $C_1 = -0.3793$, $C_2 = 0.5323$, and $C_3 = -0.1588$, giving as an approximation

$$T = -0.3793 + 0.5323t - 0.1588t^2$$

which may be compared with the exact solution found in Eq. (5.2-13) of Example IX-10. The exact solution yields $y = 0.3855$, whereas the approximation gives $y_3 = 0.3858$. Here the approximate value of the objective function is very nearly the best that can possibly be achieved. From a practical viewpoint, such a finding may be very important, since the control required to produce and maintain the true minimizing $T(t)$ may be much more complex than that required for the approximation, and if there is little lost in adopting the approximate solution, it may then be the "best" solution.

5 CONCLUSIONS

The analytical solution to continuous-variable problems entails the integration of one or more differential equations, which in practical cases may be hopelessly complex. Even numerical methods of integrating such equations are complicated by the two-point nature of the boundary conditions, so that a direct numerical calculation is sometimes an easier and more effective approach.

There are, generally, two different approaches to these direct calculations, one being the Rayleigh-Ritz technique and the other the method of finite differences. Both of these methods, however, reduce the original problem to one of n-dimensional search, and at this point the optimizer must choose a suitable method of solving this subproblem.

To solve problems subject to global constraints, the method of Lagrangian multipliers can be used, together with either the Rayleigh-Ritz or the finite difference method, with, however, the restriction that the resulting n-dimensional problem be solved by seeking a stationary value, since the augmented objective function need not have the same character as the original objective function. Numerical search methods would therefore have to be used very cautiously.

The direct numerical computation in the face of local restrictions is much less clearly defined than the other two cases. The use of Lagrangian multipliers is not recommended, because the multipliers vary with position and would therefore have to be approximated, as would the other functions appearing in the objective function. One approach is to approximate the independent functions and then to use the local restrictions to generate the dependent functions. The objective function is at this point dependent on the parameters appearing in the approximating functions of the independent variables, and these parameters can then be selected to optimize the objective function. This approach is probably not as effective as computations based on the maximum principle, and the latter is recommended where applicable.

An interesting aspect of the direct calculations resides in the fact that often the objective obtained with an approximate solution, a crude approximation, is very nearly the same as that found with the true minimizing function. This is an important point when putting the results of an optimization study into practice.

Problems

X-1 Estimate the minimum value of the objective function

$$y = \int_0^1 \left[\left(\frac{dx}{dt} \right)^2 - 2xt^2 \right] dt$$

subject to the conditions $dx/dt = 0$ at $t = 0$ and $x = 0$ at $t = 1$, by taking a trial function of the form $x(t) = \beta(1 - t^2)$ and selecting β appropriately. Note that the boundary conditions are satisfied for all values of the constant β. Compare the value of y found using the approximate trial function with the exact solution. Suggest a method of improving the agreement between the two solutions.

X-2 Using the method of finite differences, estimate the objective function defined in Prob. X-1. Approximate $x(t)$ by

$$x(t) = (x_i - x_{i-1}) \frac{t - i\,\Delta t + \Delta t}{\Delta t} + x_{i-1} \qquad \text{for } (i - 1)\,\Delta t \leq t \leq i\,\Delta t$$

Compare the approximate value of the objective function with the true value for the case $\Delta t = 0.2$. How would you improve the agreement?

X-3 What changes in the trial function used in Prob. X-1 would you recommend if

(*a*) Boundary conditions at $t = 0$ were left unspecified.

(*b*) Boundary conditions at $t = 1$ were left unspecified.

X-4 Select a trial function and approximate the following objective function:

$$y = \int_0^1 \left[\left(\frac{dx}{dt} \right)^2 + x(t - 1) \right] dt$$

subject to the conditions that $x = 3$ at $t = 1$. Find the approximate value of the objective function and compare your result with the exact solution of the problem.

X-5 Consider the task of optimizing

$$y = \int_0^1 \left[\left(\frac{dx_1}{dt}\right)^2 + \left(\frac{dx_2}{dt}\right)^2 + 2x_1x_2 \right] dt$$

subject to the conditions that $x_1 = x_2 = 0$ at $t = 0$ and $x_1 = 1, x_2 = 2$ at $t = 1$. Approximate the independent variables as

$$x_1 = t + a_1 t(t - 1)$$

and

$$x_2 = 2t + a_2 t(t - 1)$$

Compare the approximate value with the true solution. [*Hint:* Try $x_1 = \alpha e^t - \beta e^{-t} - \gamma \sin t - \delta \cos t$ and $x_2 = \alpha e^t + \beta e^{-t} + \gamma \sin t + \delta \cos t$ as solutions to the Euler-Lagrange equations.]

X-6 Let $x(t) = b_1 + b_2 t$ be a trial solution minimizing the objective function

$$y = \int_0^1 x^2 \, dt$$

subject to the condition

$$16 = \int_0^1 (x^2 - 6x + 9) \, dt$$

The constants b_1 and b_2 are to be determined. What is the true solution?

X-7 Let $x(t) = b_1 + b_2 t$ be a trial function to minimize

$$y = \int_0^1 x^2 \, dt$$

subject to the condition

$$16 = \int_0^1 \left(\frac{dx}{dt} - 3\right)^2 dt$$

What is the approximate value of the objective function?

X-8 The objective function

$$y = \int_0^1 (12x^2 + T^2) \, dt$$

is to be minimized subject to the condition

$$\frac{dx}{dt} = 2x + T \qquad \text{for } 0 \leq t \leq 1$$

Obtain the best approximation if T is a constant. How much does the objective function determined in this way differ from that found when T is permitted to be an arbitrary function of t?

X-9 Repeat Prob. X-8, assuming T to be a linear function of t.

PART THREE

OPTIMIZATION IN PRACTICE

Many difficulties arise in the application of optimization techniques to practical systems. Some consideration of these difficulties is now presented. The particularly complexing problems of uncontrollable variables and of large systems are also discussed.

CHAPTER XI

A CRITIQUE OF TECHNIQUES AND THEIR APPLICATION

1 INTRODUCTION

In Parts One and Two we considered in some detail the organization of optimization problems and their subsequent solution. This breakdown, summarized in Fig. XI-1, was adopted to allow the essential ideas of each to be examined without the confusion of additional detail. In practice, both aspects are closely interlinked. For example, optimization is valueless without an adequate system model, and the actual optimization technique is largely determined by the complexity and form of the system. On the other hand, the extent of specification required in the definition of the system will be affected by the level of optimization, whether, for instance, in a final analysis resulting in closely defined optimum conditions, or in an initial general feasibility study.

The various practical problems which might arise are so widespread in interest and application that no attempt is made to refer to all that has been done, although some typical applications are listed in Table XI-1. Our intention in this chapter is rather to bring together the ideas and techniques which have been developed in earlier sections of the book. Emphasis will be laid upon the interaction between the various facets of a practical optimization.

Let us briefly summarize the important stages in an optimization study, first introduced in Chap. II. First, the problem must be defined, and a need for investigation identified. Second, a criterion or objective for making any decisions or selection must be chosen. Third, information must be acquired about the problem, a system or systems identified, the parameters, variables, and constraints in a given system specified, and a model postulated before the objective can be expressed in terms of the system variables to give an objective function. Fourth, a given system may be improved and optimum conditions attained by variation of the system variables subject to possible restrictions, and the optimization completed by a comparison of the best results from each system initially specified.

The validity of any optimum values must then be evaluated against

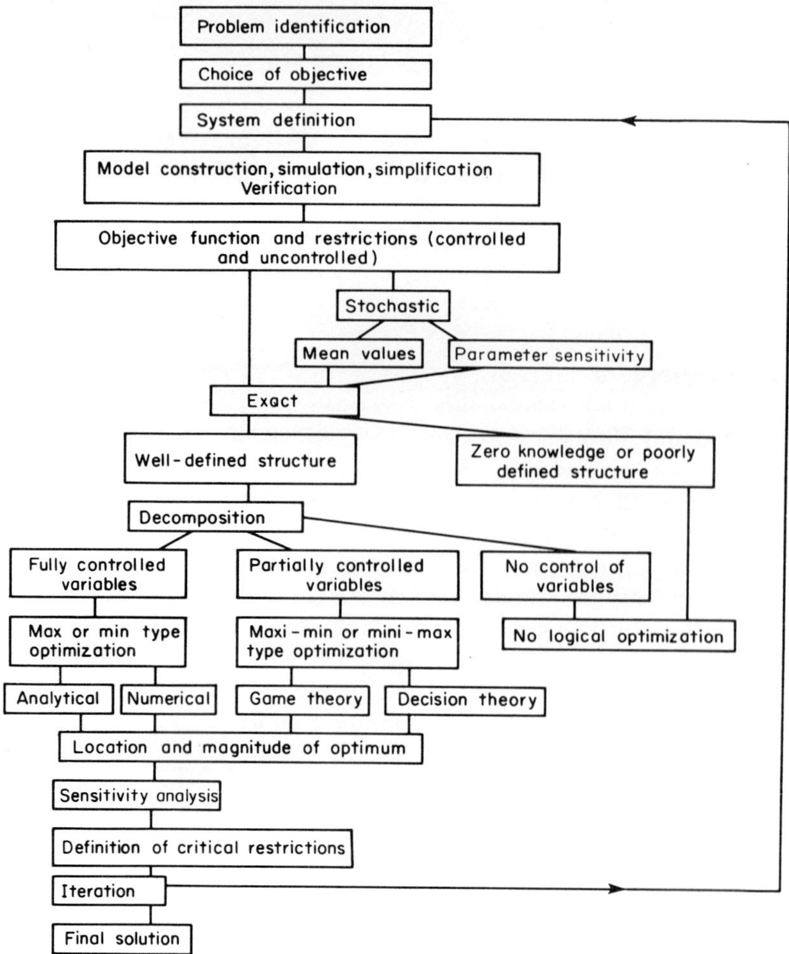

Fig. XI-1 Approach to optimization.

experimental information when this is available, and the initial statement of the original problem suitably modified as necessary.

The validity of any solution is dependent in the first instance upon a suitable choice of objective (Sec. 2). If this is not correctly defined, there is little point in determining the optimal values, since these may not be optimal with respect to other criteria. The validity of optimization results is also affected by the choice of systems for examination (Sec. 3), since the usual procedure is to identify a particular configuration. The optimal system might not be recognized, and thus not included in the study. By far the most important feature in any optimization is the need for a representative model of the system (Sec. 4). If this is inaccurate due to poor data or to an ill-defined structure, any subsequent simulation and optimization studies are unlikely to yield results which are optimal in actual application. Even if an adequate objective criterion and model are available, optimization difficulties can also arise (Sec. 5).

The system structure can affect optimization in that it can produce a deterministic or a stochastic form of the objective function. The stochastic form, however, does not imply a vastly different approach, since average, or mean, values might be used or the probability terms might be included in the objective function expression, both modifications being then treated by the optimization techniques for exact functions.

Previously, we have only considered well-defined systems, in which optimization had to be carried out with respect to a particular objective function and constraints expressed uniquely in terms of certain independent but controlled variables. Such systems are not, however, the only ones encountered. Let us consider the situations arising from various degrees of knowledge about the system structure, data, and variables. The different cases are indicated in Table XI-2.

When a system structure is poorly defined due to lack of information, this would have to be improved (Sec. 4), but when there is zero knowledge of the system or no control over the system variables, no logical optimization techniques can be applied. In fact, since a zero knowledge of design data, system structure, or the uncontrollable variables means that no optimization is possible, we need not consider these cases further. This situation would mean that any decisions could equally well be made by a random choice of conditions, or, as Johnston [1963] has said, by "pin-sticking."

When the system structure and design data are known and the objective can be expressed solely in terms of the independent variables which can be chosen at the discretion of the optimizer, one or more of the methods proposed in Part Two might be used to determine the optimum conditions. These methods are summarized in Sec. 5, and some practical difficulties considered.

Most realistic systems will not, however, fit into this classification, since at least a few variables will be external variables whose values are outside the jurisdiction of the optimizer. These variables are uncontrollable, the resultant uncertainty giving rise to further difficulties, which, however, can often be resolved by a special approach. In particular, the straightforward maximization or minimization approach cannot be used, but a different policy of reducing the worst which might happen must be adopted. This leads to a different philosophy of optimization, using maxi-min or mini-max criteria, as discussed in Chap. XII.

When the structure is well defined, optimization techniques can be found, even though most practical systems will be large and involve a large number

TABLE XI-1 SOME APPLICATIONS IN OPTIMIZATION[1]

(a) Declining performance and replacement
Radner and Jorgenson [1963], Roberts [1960, 1964b], Westbrook [1961], McCall [1963], Wen [1963].

(b) Network analysis
Pollack [1961], Boldyreff [1955], Ford and Fulkerson [1962], Welsh [1966].

(c) Theory of the firm
Dorfman [1951], Boulding and Spivey [1960].

(d) Distribution
Land [1957], Baumol [1961].

(e) Scheduling
Hall and Eilon [1963], Ackoff [1955].

(f) Inventory
Bellman [1957], Resh and Naor [1963], Roberts [1964b], Bellman and Dreyfus [1962], Zehna [1962].

(g) Control
Kalman, Lapidus, and Shapiro [1959, 1960], Roberts [1964b], Siebenthal and Aris [1964a, b], Aoki [1961], Bellman and Kalaba [1961], Lee and Aris [1963b], Bellman [1961], Douglas [1966], Douglas and Denn [1965], Blakemore and Aris [1962].

(h) Allocation
Rudd [1962a, b], Boot [1964], Young [1963].

(i) Strategy
Dresher [1961]

(j) Reliability
Rudd [1962c].

(k) Contract bidding
Peterson [1965], Churchman, Ackoff, and Arnoff [1957].

(l) Management decisions
Park [1966], Post and Viavant [1962], Gaver [1965], Thorne and Miller [1966], Bates [1966], Kantorovich [1960].

(m) Start-up problems
Jackson [1966b].

(n) Some recycle systems
Van Cauwenberghe [1964], Beveridge and Schechter [1965], Rudd and Blum [1962], Jackson [1963, 1964a], Chien [1966], Jackson and Horn [1965a, b], Horn and Jackson [1965a, b], Fan and Wang [1964a, c, e,], Brosilow and Lasdon [1965], Lovland [1964], Katz [1962a], Fan, Erickson, Sucher, and Mathad [1965].

(o) Some chemical-process and equipment systems
Aris, Rudd, and Amundson [1960], Jenson and Jeffreys [1961], Boreskov and Slin'ko [1965], Srygley and Holland [1965], King [1965], Jackson [1965], Hwa [1965], Himmelblau [1963], Barnes and Himmelblau [1966], Hills [1961, 1963], Di Bella and Stevens [1965], Chu and Hougen [1961], Denton [1960], Lee [1963], Hartland [1967a, b], Kern and Seaton [1959], Schrage [1958], Mitten and Nemhauser [1963a], Happel [1958], Enyedy [1962], Boas [1963a, b], Taborek [1960a, b], Ahn, Chen, Fan, and Wang [1964], Almasy, Hay, and Pallai [1966], Moorehead and Himmelblau [1962], Berg [1960], Ahlgren and Stevens [1966], Luss [1965], Kuo and Rubin [1962], Lermoyez [1963], Lester [1961], Skelland [1962], Youle [1960], Zahradnik and Archer [1963], Brosilow and Nunez [1968], Malengé and Villermaux

[1] Various applications have also been given in Tables VII-8 and XIII-12 for the respective linear and dynamic programming techniques.

TABLE XI-1 CONTINUED

[1967], Chen and Ceaglske [1966], Itahara and Stiel [1966, 1968], Pan and Roth [1968], Rafal and Dranoff [1966], Kuo, Rubin, and Wright [1966], Avriel and Wilde [1967].

(p) Systems for single reactions
  Single continuous stirred tank reactor (CSTR)
    $A \to B$: Denbigh [1944].
  Series of CSTR
    $A \to B$: Denbigh [1944, 1961], Szepe and Levenspiel [1964], Leclerc [1953], Pis'men and Ioffe [1962], Kubota et al. [1961].
    $A \rightleftharpoons B$: Bakemeier and Krabetz [1962], Aris, Bellman, and Kalaba [1960], Aris [1961a, b, c].
  Tubular or batch
    $A \to B$: Denbigh [1944], Blakemore and Aris [1962].
    $A \rightleftharpoons B$: Lee and Aris [1963a, b], Denbigh [1944, 1958], Adler and Vortmeyer [1963, 1964], Pings [1963, 1965], Aris [1961a, b], Horn [1961a].
  Multibed adiabatic series
    $A \rightleftharpoons B$: Horn and Kuchler [1959], Aris [1960b], Horn [1961c].
  Mixed
    $A \rightleftharpoons B$: Aris [1962a].

(q) Systems for multiple reactions
  Single CSTR

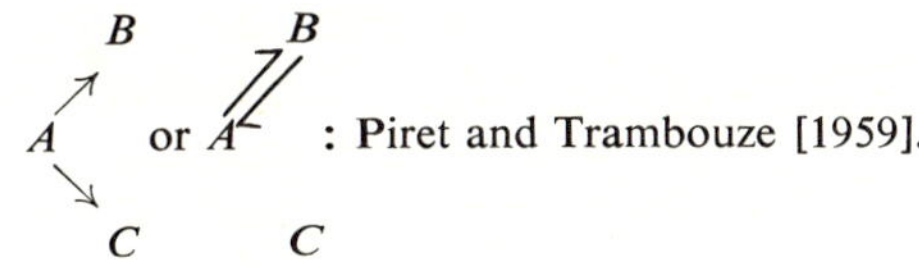

: Piret and Trambouze [1959].

Others : Mah and Aris [1964].

  Series of CSTR

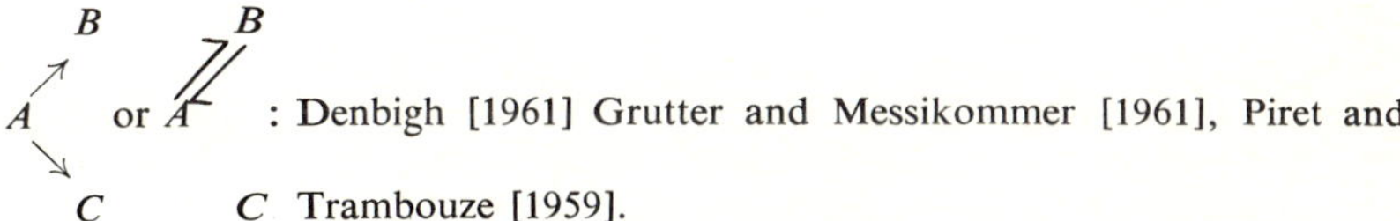

: Denbigh [1961] Grutter and Messikommer [1961], Piret and Trambouze [1959].

$A \to B \to C$ or $A \rightleftharpoons B \rightleftharpoons C$: Denn and Aris [1965a].

Others : Denbigh [1958], Horn [1961c], Aris [1960d].

  Tubular or batch

$A \nearrow B$, $A \searrow C$ or $A \rightleftharpoons B$, $A \rightleftharpoons C$: Denbigh [1944], Horn [1961a], Grutter and Messikommer [1961], Chermin and van Krevelen [1961].

$A \to B \to C$ or $A \rightleftharpoons B \rightleftharpoons C$: Dahlin and Nelson [1964], Bilous and Amundson [1956a, b], Aris [1960c], Katz [1960].

Others : Denbigh [1958], Lee [1963], Horn and Troltenier [1960].

TABLE XI-2 SITUATIONS ARISING IN SYSTEM AND MODEL DEFINITION

		Prediction ability for uncontrollable variables		
		Complete	*Partial*	*Zero*
Prediction ability for system structure and data	*Complete*	Part Two, Chap. XI, Sec. 5	Chap. XII	
	Partial	Chap. XI, Sec. 4		
	Zero	No logical approach		

After Johnston [1963].

of variables. By simplifying and breaking down the system into a set of subsystems, each of which is of a lower dimensionality and can be optimized separately (Sec. 5.1), the computational aspects of optimization might be greatly eased. The proper choice of these subsystems is very important, a full discussion of the appropriate decomposition methods being presented in Chap. XIII.

The final stage in any study is that of iteration, to check that the solution obtained is relevant to the original problem and whether or not further improvement can be achieved by a restatement of the problem (Sec. 3). Once optimization is complete, the problem of control of optimum conditions must be investigated, by examining the sensitivity of the solution to changes in location and to changes in parameter value (Sec. 6).

2 OBJECTIVES IN DECISION MAKING

One of the major stumbling blocks in any analysis is the initial definition of the problem. A first proposal can be as vague as "find an alternative use for product *X*" or "improve the profitability of plant *Y*." Yet, until a need for investigation has been identified and the problem clearly stated, little or no progress can be made. Most problem specifications are obtained by an iterative procedure in which information derived from an earlier study of the problem is fed back to improve the statement of the problem. This aspect is considered further in discussing system specification (Sec. 3).

Once a problem has been posed, it is necessary to decide upon a selection criterion by which all decisions can be made. This will allow the results of specific actions to be evaluated, preferably in a quantitative manner. If the correct choice of objective is not made, the subsequent optimization analysis may not result in worthwhile solutions, since different solutions may be found with different criteria, even though the same technical model is used (see, for example, Shahbendarian [1964]). A realistic objective is, normally, an economic one (Chap II. Sec. 2). Technical forms may, on occasion, be convenient, as in the fitting of the best curve to data, but most, such as the improvement of reaction yields, are utilized only within an economic framework, so that they actually represent reduced forms of economic criteria, this reduction often being accomplished in the simplification of a system model or in a subsequent decomposition into smaller subsystems. The decision as to an appropriate economic basis is dependent, first, on the need for knowledge of the capital required. Thus, in existing systems, capital is not involved as a variable, so that a profit or cost objective can be used. Systems in which both physical specifications such as plant layout, sizes, and location and operating conditions have to be defined are more complex, requiring a knowledge of both profit and capital charges. One of the rate-of-return formulations has then, generally, to be chosen as a suitable profit-investment criterion. A minimum cost is sometimes used in this case, but is not wholly satisfactory, since capital only in its depreciated form as an operating cost is considered, and the relationship between profit and investment cannot be examined. It is necessary to consider, secondly, the effect of time. There are several forms of rate-of-return criteria, which include the effect of the time value of money. While it is simpler to use time-averaged values, this reduction might produce misleading results. When, however, time is being taken into account, the main difficulty lies in the choice of a suitable interest rate for the capital employed.

More than one objective criterion may on occasion need to be satisfied. In this not uncommon situation some compromise must be made between the different aims presented by these criteria, and a solution obtained which is in some sense jointly best. Simple illustrations of this problem are found when a project has to produce a maximum profit but be operated in a safe and legal manner at a minimum investment level; when a chemical product yield has to be maximized at a minimum cost; and when production has to be improved, but with an impurity held below a specified level. The first situation was shown in Fig. II-1, with higher profits obtained, but of gradually decreasing increments, up to a maximum value, after which the profits began to decline with increasing capital investment. The concept of marginal rate of return was then introduced as one satisfactory manner of including both profit and capital in a criterion. In the case of multiple objectives it is usually impossible to optimize all at the same time. One has to be chosen as the primary objective, and optimization carried out, taking care that satisfactory levels of the others are attained. An example of this is presented by McEwan and Beveridge [1965], who found the optimal profit-investment policy, using a marginal-rate-of-return criterion for a system of one, two, and three stirred tank reactors, as shown in Fig. XI-2. The final optimal solution had then to be chosen subject to the secondary objectives of satisfactory exit concentration level, profit level, and absolute rate of return, as

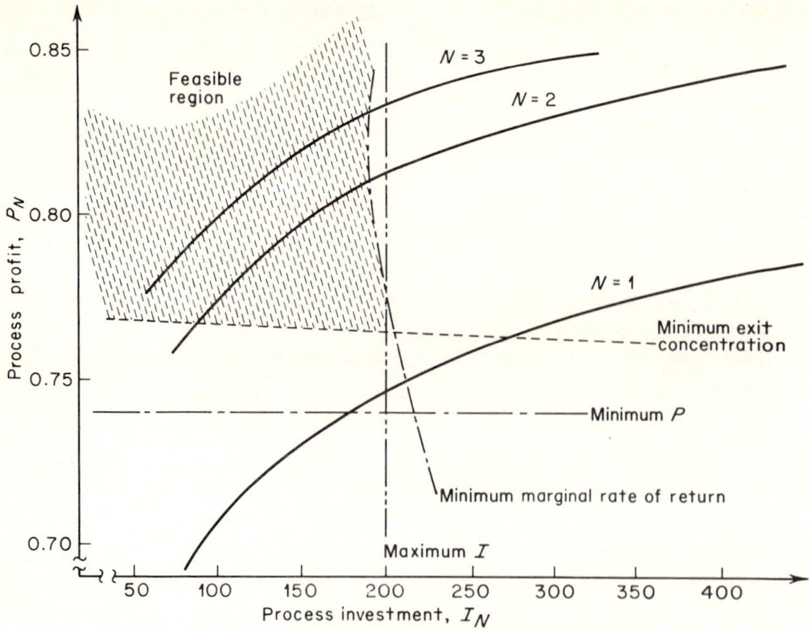

Fig. XI-2 Optimal profit-investment relations for continuous stirred tank reactor systems (McEwan & Beveridge [1965]).

well as an upper limit on investment, the valid region being shown in the figure to exclude the one-tank arrangement.

The influence of uncertainty in the problem specification is important in the choice of a suitable criterion. If the project can be defined exactly, in the sense that all links with other systems are rigidly fixed and all the variables are within the control of the optimizer, it is likely that a straightforward minimization or maximization would be required. If, however, competition exists, some of the variables in our problem will lie in the control of an outside agency, and the usual type of objective cannot be used. As will be seen in Chap. XII, we cannot then maximize our profits, but must rather consider the problem of minimizing our losses. We are thus adopting a very conservative approach and ensuring that we do attain a certain level of performance. Even here the difficulty of defining competition and its effect remains. Certainly, competition influences such decisions and results as pricing policy, inventory level, the reliability and quality of the required product, and also the trading policies of the companies concerned (for instance, the traditional "old boy" influence of personal contacts and a preference for trade with known concerns). Not all aspects of competition can be expressed in mathematical terms. Nevertheless, the essential influences of competitive action must be included in a meaningful optimization study on any systems in which competition can affect the results. Game theory (Chap. XII, Sec. 2) is an example of the way in which competition is introduced into economic studies. Even here, however, it has to be assumed that competition acts in a directly hostile form, an assumption which is not always realistic.

Not all forms of the objective, even if correctly chosen and defined, are suitable for use in optimization. Thus, if subproblems have to be examined in isolation, as in some decomposition methods (Chap. XIII), a desirable form of objective function might be $y = \sum_{i=1}^{N} y_i$. This is feasible if y is a profit or cost-type function, but is not directly so when a rate of return on investment criterion is being considered, where

$$y = \frac{\sum_{i=1}^{N} P_i}{\sum_{i=1}^{N} I_i}$$

P_i and I_i being the profit and investment associated with the ith subproblem unit. McEwan and Beveridge [1965] have considered the modification of y for this situation.

It is thus not always possible to determine clearly which objective is appropriate for a given purpose or which should be selected if more than one is necessary, and in some cases it is not obvious how an objective can, in fact, be expressed in quantitative terms. It is nevertheless necessary to define unambiguously an objective before optimization can commence.

3 CHOICE OF SYSTEM

If a problem has been posed, identified, and deemed worthy of further investigation, an objective will have been specified. It is now necessary to define a system which will provide the basis for quantitative work, any study requiring a basic system structure which has been or can be modeled.

The definition of a representative system structure may not immediately follow from the initial statement of a problem, and usually depends upon the depth of analysis desired. While optimization starts with problem identification, it terminates with a system iteration in which information gained in the analysis will be fed back to modify as necessary the initial statement of the problem, the procedure not ending until satisfactory agreement is achieved. As long as more than one form of system is feasible—the usual situation—some type of iterative approach should always be adopted. In the early stages, the problem statement is usually vague, with little subsequent attempt at optimization. Instead, there is a relatively crude comparison of the essential features to decide upon the best systems for further examination. With continuing flow of information back to the early system specifications with regard to their accuracy, feasibility, and potential improvement, a more definitive system structure can be formulated, allowing optimization to play a more important part (Sec. 3.1).

In typical engineering problems, the need for system definition might arise at any one of the levels of investigation described in Table I-1, and the technology might accordingly be couched in business, engineering, or scientific language. Thus different levels of investigation might in one case require the definition of a proposed system involving both design and operating variables, and in yet another, an existing system might be examined, an analysis involving only the operating variables. This type of distinction affects the choice of system, there being a range of possible systems for study in the former situation, while in the latter the system is seemingly already specified. However, even with an existing system, it can be found that, while in one situation it is best to operate under steady-state conditions, it might prove better in other cases to use some time-dependent condition (for example, see Horn [1967], Horn and Lin [1967], May and Horn [1968], and Horn and May [1968]), which requires a different system specification and model (Sec. 3.2).

3.1 Iteration in System Specification

Iteration in design or system specification is not to be confused with any iterative approach to optimization. The latter is usually an essential part of any design and optimization study, but is concerned only with a cyclic search of a given system or between systems. Here we are interested in the iteration which increases the accuracy and relevance of the problem and associated systems. Thus it might be found that the original problem and system is capable of improvement by relaxing a constraint. The depth of the analysis is also important, with the problem initially considered in general terms, detail being added stage by stage, the starting stage depending upon the ultimate use to which the results may be put, the availability of accurate data, and the time available for the inquiry. Such a wide range of approaches can be illustrated by the iteration in process design, in which initial appraisals check whether the proposal appears feasible, and subsequent predesign and preliminary and final design studies (see, for example, Vilbrandt and Dryden [1959]) add further detail and accuracy, but only at the expense of time and money. Obviously, optimization in the early stages, with a crude specification, will be confined to rough comparisons in the elimination of unrealistic and unfavorable projects, and only later, with more complex and more accurately defined systems, will elegance in optimization become important. The later stages would, of course, be carried out only if the early analyses showed that one or more of the proposals had merit.

The connection between the depth of the study and the levels of investigation, as shown in Table I-1, has already been noted. The more detailed the system specification, the greater will be the attention paid to the lower levels of the problem. Thus an initial appraisal might be concerned only with alternative chemical routes, but a final design would require a system which allows not only a comparison of these routes, but also an analysis of design and operating conditions for process equipment. Let us consider some examples of the requirements in system specification, concentrating attention mainly on the iterative aspect, rather than attempting to carry each stage through to its conclusion.

Example XI-1a: A new company product

Management has expressed an interest in diversifying the range of company products by the introduction of a new chemical. Phenol has been suggested as a possible product which fits into the company trading policy and its know-how. We are asked to investigate this possibility, market research having shown that phenol could indeed be sold up to about 20,000 tons annually.

The choice of starting raw material, of chemical route, of process, of flowsheet, of items of equipment, and of the optimal design and operating specifications for an appropriate plant capacity must now be made. First, however, we must define some systems for examination.

Commercial production of phenol has been found from an examination of the literature to be made from four main raw materials, one main chemical route being used from each. These and all other potential raw materials and routes should be considered in the early investigations. Let us, however, illustrate a first stage in the choice and definition of a system for optimization by considering the cumene route to phenol.

Example XI-1b: An initial appraisal of the cumene route to phenol

At this stage we are interested in choosing suitable feasible systems for detailed study. Should we proceed to further design studies on the cumene process? Here cumene (isopropyl benzene) is oxidized by air to cumene hydroperoxide, which reacts with sulfuric acid to give phenol and a by-product acetone, according to

$$C_6H_5CH(CH_3)_2 \xrightarrow{O_2} C_6H_5C(CH_3)_2{-}O{-}OH \xrightarrow{H_2SO_4} C_6H_5OH + (CH_3)_2C{=}O$$

Cumene → Cumene hydroperoxide → Phenol + Acetone

First, we must find out whether the main raw material, cumene, is available (although in later stages we should also want to know the maximum quantity available and, if necessary, make a reallocation of this raw material among the processes under our control which use it). The other materials, air and acid, will normally present no limitation upon the design. Second, we must confirm that this route is technically feasible. Both of these objects might be achieved from an examination of the literature or company records. It remains only to check the economic feasibility of the project. It might suffice in an initial appraisal to check whether a loss is inevitable or not. Thus we might determine the cost of the raw materials and compare these with the value of the product. Assuming the best yield possible, that of 100 percent, even though this might not be practically attainable, the stoichiometric equation above shows that 1.276 tons of cumene is required to produce 1 ton of phenol. Since the cost of cumene is \$224 per ton,[1] the corresponding cost associated with each ton of phenol produced will be \$286. This can be compared with a selling price of phenol of \$240 per ton. One hundred percent conversion is of course the most favorable situation, and in practice the economic situation would be worse. On the present evidence we should seemingly have to reject the cumene route from further consideration. In this case, however, a by-product is produced, namely, 0.611 ton acetone per ton phenol. If this can be sold with a value greater than \$75/ton acetone, the cost of raw materials will be less than the total income, and the project might become economically feasible.

Since the selling price of acetone is \$166/ton, it might be possible to obtain this required income from acetone sales. This simple initial appraisal has, however, demonstrated that attention must be devoted not only to phenol market research, but also to the potential acetone sales. If this by-product cannot be disposed of in sufficient quantities at an adequate price level, the cumene route will not be feasible.

Optimization, if the expression can be used here, is at this stage merely a decision as to the retention or rejection of the route in subsequent deliberations.

If additional information is available on conversions and costs, we might be able to make a more realistic analysis. Thus since, on the average, operating costs are about half the sales value (Dybdal [1950]), this provides a rough guide, which might allow additional processes to be eliminated if the product values do not provide a sufficient margin over the raw material costs. Faced with an initial selection of a few feasible projects, this and

[1] Costs are taken as the market prices (*European Chemical News*, June 30, 1967), values which may not be strictly correct for internal sales, etc.

similar crude evaluations would be carried out for all potential chemical routes to phenol. Our intention at this stage is to weed out the less desirable systems from further consideration by little expenditure of effort. The more attractive systems will then be reexamined in successive stages, incorporating more detail into each as information becomes available as a result of these investigations. Such a study might produce the situation next described.

Example XI-1c: Predesign study of the cumene process

More information having been derived from a knowledge of existing processes, published literature, or company files or from small-scale experimentation, we may find ourselves in a position to produce a more detailed economic analysis of the cumene route to phenol. Typical costs for a production capacity equal to the estimated sales of 20,000 tons/year are shown in column 1 of Table XI-3, and it is known that all company projects must earn at least 25 percent rate of return on all capital employed. It is assumed that the full market price does not have to be paid for the cumene raw material, so that even if no acetone is sold, the project will still make some profit. Under what circumstances should we retain the cumene route for more detailed design and optimization studies?

Assuming that all the by-products can be sold, the cumene rate, as seen from Table XI-3, column 1, will yield a total profit of $\$1.72 \times 10^6$

TABLE XI-3 ESTIMATED ECONOMIC DATA FOR SYSTEMS TO PRODUCE 20,000 ANNUAL TONS OF PHENOL

System	(1) *Cumene*	(2) *Toluene*	(3) *Benzene*	(4) *Cyclohexane*
Production costs, \$/ton phenol (including depreciation and interest)	197	205	160	199
Income (\$/ton phenol):				
Phenol sales	240	240	240	240
By-product sales	43			11
Profit (\$/ton phenol):				
All by-products sold	86	35	80	52
No by-products sold	43	35	80	41
Capital investment, $\$ \times 10^6$	3.6	5.2	4.0	3.1
Rate of return %:				
All by-products sold	48	13	40	34
No by-products sold	26	13	40	26

for a capital investment of $\$3.6 \times 10^6$. This represents an absolute rate of return of 48 percent, which is quite acceptable compared with the average rate of return within the company. If the acetone by-product cannot be sold, however, the profits decrease, and the rate of return becomes 26 percent. Since this is essentially the acceptable rate of return, further study of the cumene route to phenol is justified only if the by-product income is significant.

We might continue in this way, each iteration increasing the degree of accuracy of the study, along with the amount of information required and the complexity of the analysis. Further investigations might include the projected capacity of the plant and the choice among alternative routes, using estimated data, while later iterations will rely on more definite information. At this point, however, optimization is still a question of retain or discard, but does allow the significant features of a system to be distinguished.

Example XI-1d: The choice of alternative routes to phenol

In an initial design appraisal, a literature search shows that four major chemical routes are used in the production of phenol. These start from the raw materials toluene, benzene, cyclohexane, and cumene, respectively. The estimated costs for each route are given in Table XI-3 for a design capacity of 20,000 tons of phenol per year. If the selling price of phenol is \$240/ton, determine which route would be most suitable for further design investigation, assuming that by-products can be sold at the value shown and that all projects must earn at least 25 percent rate of return on capital employed. The toluene and benzene routes have no major by-products, but the cyclohexane and cumene routes produce hydrogen and acetone, respectively. If these by-products cannot be sold at the values given, how does this affect the previous decision? What will happen if only 75 percent of the acetone by-product income is available in the cumene route?

All by-products sold Let us consider the system with the lowest investment—route 4. The rate of return is then 34 percent, which is acceptable. Route 1, however, produces a higher profit level and should be examined as a possible choice, even though it requires a higher investment. Its absolute rate of return on investment is 86 percent, but while this is satisfactory, we must also consider the rate of return on the increment of capital required over and above that for route 4 to obtain the route 1 system. This incremental rate of return (Chap. II, Sec. 2.2) is

$$\frac{\Delta P}{\Delta I} = \frac{(86 - 52)(20{,}000)}{(3.6 - 3.1)(10^6)} \times 100 = 136\%$$

which indicates that this incremental investment is preferable to that offered by the alternative minimum of 25 percent. The other two routes show lower profits for greater investments, and are thus not acceptable. When all by-products can be sold, the cumene route is thus the best on the evidence available.

No by-products sold When no income can be derived from the by-products in routes 1 and 4, the picture changes. Starting again with route 4, offering a feasible rate of return of 26 percent, we find that further investment to give route 1, with an accompanying marginal rate of return of 8 percent is unacceptable. Route 3, however, offers an increased profit, achieved at a marginal rate of return of 87 percent, and would therefore be chosen. The remaining route 2 shows a loss on further investment, and would not be considered. When no income is available from by-products, the benzene route will then be preferred over the other possibilities.

Some acetone sales When 75 percent of the possible income from acetone sales is available, the cumene route will show a profit of \$75/ton phenol sold. This route 1 then shows an incremental rate of return of 92 percent over route 4, and even though route 4's absolute rate of return is acceptable, we should choose route 1. When route 1 is compared with route 3, however, the latter requiring a further increase in investment, a return of only 25 percent is achieved on this investment increment. Even though route 3, the benzene rate, offers a higher profit level, the same result could be obtained by installing the cumene system at a lower investment and placing the additional capital at 25 percent in the company's alternative minimum-level source of income. Route 1 or 3 could then be installed, and both would have to be considered further, the final choice being controlled by the amount of capital available and the future of the by-product market.

The cyclohexane route, system 4, might, however, be the only possible choice if the capital available were severely restricted.

Having decided on the basis of such appraisals that a few particular systems are worthy of further investigation, detailed preliminary technical design studies would be carried through for each. The early stages considered only general system structures, each in little detail, but subsequent proposals will contain increasingly more detailed specifications. Thus, once it has been decided to produce a specific product by a certain chemical route from given raw materials, the next stage is the determination of a suitable process and a process flowsheet with full mass and energy balances for attaining the desired ends. Each flowsheet arrangement is then a system which can be used as a basis for process design and a full cost analysis, and it is at this level that optimization techniques will find their main application.

This iteration in the problem will now, of course, involve the company in extensive expenditure of time and money.

Our object, continuing the chemical process illustration, is the definition of a range of flowsheets, each being a potential system for study. Each will represent, in general, a sequence (simple or complex, depending upon the need for recycling or bypassing mass or energy) of operations such as material storage, purification and preparation, conversion of raw materials to products, and the separation, purification, and storage of products. Each set of operations will require the selection of process equipment or sets of equipment for each flowsheet duty or operation. Thus, for a given flowsheet layout, we shall have to choose alternatives such as tray or packed columns for distillation, adsorption or extraction operations, centrifugal versus reciprocating compressors, and so on, while the choice of reactor may be made from batch or continuous, and if the latter, from stirred tank, tubular, or various parallel or series combinations of these. For each of these choices we should then have to examine different modes of operation—time-dependent or -independent—and should then have to consider different methods of control, all this having to be carried out to define a system which can be used to provide a model as a basis for quantitative work.

In this way the choice of systems might, even for a single flowsheet arrangement, include many variations. If the choice of a system cannot be made so that the system itself is also a variable in the investigation, the only alternative is the specification of a set of different systems, each of which has to be investigated separately, and their results brought together in the final comparison. From this final comparison the successful chemical route, process, and flowsheet will pass to the final design stage. In this ultimate stage we shall consider only a single project, and we shall prepare the necessary detailed design and operation specifications ready for construction and start-up. Factors such as site location may now be included, since these will affect transportation costs, inventories, labor, power availabilities and costs, and also the raw-material sources and product markets (Vilbrandt and Dryden [1959]). The choice of plant capacity[1] will have an upper limit, due to the availability of raw material and to the market demand for the product

[1] Even if the plant capacity is specified, the problem of under- or overdesign must be examined. Overdesign is often introduced to allow for uncertainties and unknown circumstance, by increasing equipment sizes or improving specifications. In any case, this may increase capital and operating costs. This allowance is sometimes regarded as an advantage, since it can allow the plant at some later date to be operated at greater than design throughputs. However, if this is the future intention, it should be brought out at the system specification stage of the investigation. Underdesign may also result in additional expenditure, due to penalties incurred by not meeting production requirements. It is also possible that projects have to be designed so as to operate over a range of conditions, such as 35 to 110 percent of design capacity. Kittrell and Watson [1966], among others, have considered this problem of overdesign.

and a lower limit imposed by the production break-even point when the project operating costs are balanced by the product sales income. However, consumer demand can be improved by advertising and user education; raw materials might be in absolute short supply if the plant is their sole user or in relative shortage when several projects are competing for the same upper-limited quantity; and the relative profitability for the company as a whole must be taken into account (Prob. XI-3). Limitation on capital, again relative or absolute, must be taken into consideration. All the uncertainties due to competition or natural forces may combine in this general illustration to make it very difficult to define a system for study. Optimization is unlikely to find, at the present, as much scope in this general problem as in some of the intermediate stages of a design, due to the complexity of the total problem. This situation may be compared with the earliest stages, in which optimization was necessarily frugal due to the lack of information.

In such different depths of problem analysis as we have indicated, a somewhat similar approach is being applied at all times, with all available information utilized to form a judgment, its quality determining the reliability of the decision. In successive iterations information is being fed forward, and there is also a feedback, since decisions concerning the higher levels (Fig. XI-1) cannot be made until data from the later, more detailed stages are available. Many system iterations and stages might thus have to be examined before adequate dependency is obtained. The choice of a system for optimization is thus not an entirely straightforward matter.

3.2 Different Systems

As discussed in Sec. 3.1, a given problem can be tackled in almost an infinite number of ways, each requiring a different system specification. Each simulated model can then be optimized, and the individual optima compared in order to decide upon the best system. In most cases the greatest difficulty is in deciding on the representative systems to be included in the evaluation. Certainly, once a selection has been proposed and analyzed, the best one can be found. But how can it be ensured that the best system has been, in fact, included in the first place?

If the system itself can be included in the optimization as a variable, this will be an advantage. It is, however, rare for this to be done. Thus, for example, the flowsheeting of chemical processes is rarely defined in full; instead, optimization is usually carried out for specific equipment or arrangements of equipment within a fixed flowsheet layout. We should thus consider a basic flowsheet, and alternatives would be posed which would contain different layouts or arrangements of the original scheme. For example, in Fig. XI-3*a*, the basic scheme may represent a stream preheated, reacted, and then cooled. Different arrangements may lead to more

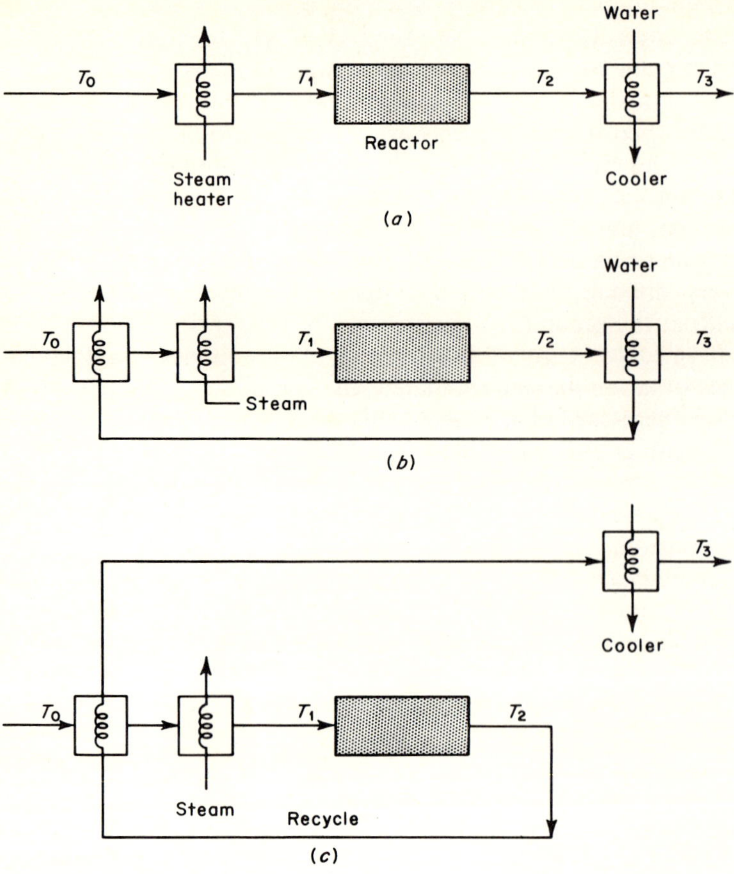

Fig. XI-3 Flowsheet arrangements. (*a*) Basic flowsheet; (*b*) heat recycle using second fluid; (*c*) heat recycle using process stream.

economic results. Thus, in Fig. XI-3*b*, some of the product-stream heat is recovered and used to preheat the feed stream, the heat exchange being effected through an intermediate fluid. Another possibility is the direct use of the product stream to heat the feed stream, as shown in Fig. XI-3*c*. The possibility of system variation allowing for the rearrangement of the units or links in a system, or for their replacement by alternatives, is in general beyond the present limits of optimization. Present techniques appear to require the separate proposal of distinct systems, so that the optimization of a chosen flowsheet, for instance, and comparison with other flowsheets and flowsheet arrangements is the normal procedure. As discussed in Sec. 6, however, the relaxation of system constraints is a step toward this ultimate, as is also the iteration in successive stages of a design, in an attempt to define an appropriate system and model for study.

Even for a given system, it may be found necessary to optimize different but equivalent mathematical representations or systems. Thus dual systems (Chap. VII, Sec. 5) might prove easier to handle, while the question of ideality versus practicality might cause the potentially better ideal system to be replaced by the more realistic but sub-optimal system. Thus, for the case of the tubular reactor and its optimum temperature profile, we have noted that temperature control cannot be maintained at all points in a bed. In practice, the reactor has to be divided into a finite number of reactor beds, which may be essentially adiabatic, and the temperature of the interconnecting streams is controlled outside the reactors by heat exchangers or by mixing the stream with fresh but cold feed. Since the introduction of each additional reactor will involve additional costs, there will be some optimal number of practical subdivisions which can be introduced in this situation. A comparison with the ideal profile shows, then, how close the results will be to the ultimate possible. An analogous situation is the approach to thermodynamic equilibrium, which must be considered as a limit in the evaluation of chemical reaction yields.

4 MODELS AND SIMULATION

Once a system has been specified, a suitable model must be derived. The construction of a model forms the technical design basis for the expression or simulation of the specified objective criterion in terms of the system variables. Data must be collected, and the important parameters, variables, and constraints defined. Each unit, or collection of units, making up the system can be represented in two possible ways. First, it can be represented by conservation balances to yield an analytical model, as with mass balances in a chemical reactor, but a solution must be possible from the resulting relations before this approach can be used. Second, the system can be represented by a black box, as found in using existing systems, the model providing sets of output data for given inputs. A complete model representing the whole system can then be assembled on the basis of such analytical or black-box approaches or on a combination of both. However, irrespective of the initial model used, the final model can be either analytical or black-box; the interchangeability of these is discussed in Chap. III.

Since we are dealing with a given system, model building is relatively straightforward, because the structure is known, and it is necessary only to have a detailed model representation to enable design and operating conditions to be determined. The formulation of a model can, as we have said, be one of two types. First, if a system exists, it can be used as a black box, and the relevant output information for specified inputs can be used to obtain a corresponding value of the objective function. Thus technical data, such as raw-material requirements, product yields, and heating and cooling demands, can be used with suitable cost data to obtain the system profit. Second, if the black-box approach is not convenient, or is invalid for existing systems, or is not possible in the case of proposed designs, the analytical model must be one in which mathematical relations are used to derive the necessary outputs. Typical illustrations of the analytical model are found in the chemical-process systems described by Williams and Otto [1960] and by Gunness [1951]. The former has been discussed in detail and used in optimization by Di Bella and Stevens [1965] with a rate-of-return objective. Stout [1959, 1960] has presented further details on another system, Himmelblau [1963] optimizing this with respect to a profit function. Computer programs are now readily available for handling the design of most individual items of equipment, and more complex routines, such as Opcon, Optimat, and Cheops, have been formulated for flowsheet calculation.[1] The organization of the design calculations is important, and the work of Rudd and Stewart [1965], Rosen [1962], Nagiev [1957], Naphthali [1964], Ravicz and Norman [1964], Norman [1965], Rubin [1962], Rudd and Watson [1968], Himmelblau [1966, 1967], and Sargent and Westerberg [1964] is of interest.

We have already mentioned that, ideally, the system should be incorporated within the analysis as a variable. This is difficult in many cases; certainly so if no common variables exist. It may be possible, however, to define a model which will incorporate such features. Thus, in the storage-tank specification of Example I-1, instead of separate models for each shape, a single model could be postulated, using an arbitrary shape (see also Example IX-8) as a variable. Yet another illustration is found in the field of chemical-reactor design. Instead of specifying various separate sequential combinations of reactors, such as continuous tubular and continuous stirred tanks, a continuous reactor unit with a variable diffusion or mixing coefficient defined as a function of reactor length can be used, the coefficient being locally zero for a tubular plug-flow unit and infinity for a perfectly mixed stirred tank.

Having obtained a model, the objective has now to be expressed in terms of the system variables free for manipulation, as must also the various restrictions, internal or external. This simulation usually involves cost estimation, and even with computer routines, this has to be performed

[1] Opcon, developed by Westinghouse Corporation, has been used in the yield maximization of a flowsheet for the catalytic dehydrogenation of ethylbenzene to styrene by variations in the stream flow, reactor temperature, or feed rate (Anon [1959] and Archer [1960]) and also in the optimization of distillation-column operation (Weiss, Archer, and Burt [1961]). Optimat is a similar device, due to Elliot Automation Ltd. (Bell [1962]). Cheops, of the Shell Development Co. (Cochran [1960] and Anon [1961a, b]) can design a complete ethylene oxide production unit, but requires additional subroutines for particular items such as distillation columns. Similar programs exist in other industrial concerns.

manually, the appropriate parameters being fed into any design calculations. The errors in application of this economic material vary, depending upon the stage of the investigation and the amount of information available (Nichols [1951]), from about 100 to 5 percent for chemical plant. The accuracy is related to the design effort as indicated in Fig. II-7.

Having chosen a system, defined a model, whether black-box or analytical, and used it to get a technical design basis which allows the objective to be expressed in terms of the independent variables, attention must be turned to the uncertainty in its reliability. We must be able to predict, design, and construct our model with a known degree of confidence, and we must ask how far the model will represent the actual system and how much faith can be placed in the models, simulation, and the data. Difficulties arise due to inaccuracies, past or future, in cost estimates and in data introduced as parameters into an analytical model; to the inclusion or absence of extraneous or pertinent data; and to the fact that most models are oversimplified and idealistic. Simplification is an essential feature in model construction, otherwise a bewildering variety of major and minor variables would hopelessly complicate or make impossible any analysis, although they might be necessary for a complete design specification at the construction stage. We must erase the less important variables from consideration, although they are retained in the sense that they are needed to maintain the general level of costs and performance in a system. A balance between loss of realism and gain in reduced computational difficulty is required. Caution and judgment must be exercised, and what can be omitted may be as important as what is put in. Any mathematical rigor can produce no better result than the confidence given by the model. An illustration can be found in the fitting of a surface to a set of data, an approximation being adequately fitted locally, using only the first-order terms in a Taylor's expansion. This form would, however, be rather inaccurate in the vicinity of the optimum, since there the first-order terms approach zero, and it would be necessary to include second-order, and perhaps higher-order, terms to obtain an adequate representation. Once data have been produced using a particular model, these can be checked against actual values, if such exist, and the model adjusted as necessary. Bellman [1961], for instance, discusses the improvement of a model by the use of adaptive control methods.

The expression for the objective function resulting from model building may be exact or stochastic in form. An exact relationship here implies that a unique value of the objective function is determined for each set of the system variables. This is the form most commonly found in the literature on optimization, but it does not represent the commonest problem. Most practical problems must account for some degree of uncertainty. While the exact form can provide a reasonable starting solution, a further level of sophistication in the analysis would be the inclusion of data which are no longer single-valued with exact or expected values, but which have a probabilistic distribution of values associated with each variable. These probabilities can sometimes (in, for example, linear systems) be introduced into the objective function, the resulting stochastic model being more realistic and finding widespread application. In most problems of practical interest a knowledge of the distribution of inputs cannot be translated into a knowledge of the distribution of outputs, and hence knowledge of the objective function. For such cases Monte Carlo techniques (Chap. III, Sec. 4.4) are attractive in defining the probabilistic nature of the objective function. In most situations, however, as indicated in Chap. II, Sec. 7.1, such stochastic problems are generally handled in a manner similar to the treatment given the deterministic problems, by using mean or expected values of the variables. Caution must, however, be exercised in the use of expected values. When problems involve repeated actions at frequent intervals, such a criterion can be expected to give plausible results, but when a decision can be made only once, difficulties can arise. Such problems as the latter are common, as in the single large investment of capital in new industrial projects, these usually not being repeated in similar circumstances, but the consequences having an important influence upon the well-being of the optimizer.

The state of knowledge of the system structure, data, and model may be complete, partial, or zero, as discussed in Sec. 1. The degree of knowledge in this classification will obviously affect the subsequent optimization techniques which might be employed. From the point of view of optimization, it is perhaps unfortunate that most practical systems do not fall into the first classification, because they will probably contain incompletely known structures and data, and knowledge of the uncontrollable factors may be incomplete. If mathematical models are not available or knowledge of the data is scarce, it is sometimes possible to make statistical estimates of these using, for example, the Monte Carlo method (Chap. III, Sec. 4.4). This approach might be satisfactory if past experience is available or if sufficient experimentation can be carried out to yield good statistical estimates. In other cases informed quantitative guesses may have to be utilized.

At the moment, the procedure for the solution of these problems depends very much upon the reduction of the problem by assessing the relative importance of the various contributing factors and discarding those which seem of less value. There is no comprehensive theory for tackling these problems, and we must therefore reduce them to one of the situations discussed earlier. Thus, if there is an imperfect knowledge of the structure, optimization techniques such as discussed in Part Two might be used, but it would be desirable to supplement these by, for example, working out the problem several times, trying out different sets of coefficient values in the structure, and investigating how the proposed actions and likely objectives depend upon the various factors in the structure. On the other hand, if we think that a reasonably good estimate of the structure and data has been found and that the main difficulties center on the possible actions of competitors or the likely values of other uncontrollable variables, then the methods discussed in Chap. XII might be adopted. In fact, while statistical decision theory was noted there as being of use when the structure was known exactly and the only uncertainty existed in the uncontrolled variables, there is no reason why it should not be applied to the general problem, in which the uncertainty would include the parameters of the system model, as well as the uncontrollable variables. Sampling, as in the use of Monte Carlo methods, within a defined framework might then be used.

5 OPTIMIZATION

Once a problem including a relevant objective has been defined, a system and model specified, and simulation carried out, an appropriate optimization method must be chosen. We have seen that we can consider maximization and minimization of an objective function in well-defined systems and that the presence of uncontrollable variables requires other strategies. Since the latter are discussed in Chap. XII, we shall confine our attention in this section to the former situation.

If we have complete knowledge of the system data and structure and the resultant model, and if the variables are within our control, subject only to the system constraints, we may find the value of the objective function for any set of independent variables, at least within the accuracy of the model and data. An exact or expected solution might then be found, depending upon the form of the system and data used in the optimization. The discussion of suitable methods in Part Two was necessarily separated into distinct, although somewhat arbitrary, categories for the purpose of explanation and illustration. This distinction is normally blurred since, as discussed in Chap. III, numerical data can be handled by analytical methods through suitable surface fitting, while analytical expressions can be treated numerically. In addition, most multidimensional methods require local one-dimensional searches, and analytically derived results may need to be compared by search among possible optima at discontinuities and boundaries. No general technique can be specified in advance, but it will often be found that, while particular techniques are favored in specific applications, most practical problems can be solved in several ways.

Let us now consider some of the aspects of optimization, as summarized in Fig. XI-4. Having defined the objective and restrictions in terms of the system variables, large or complex systems might require some form of simplification of structure or a modified approach (Sec. 5.1), but such decomposition techniques will merely provide a convenient framework within which we must apply a suitable optimization method. Various

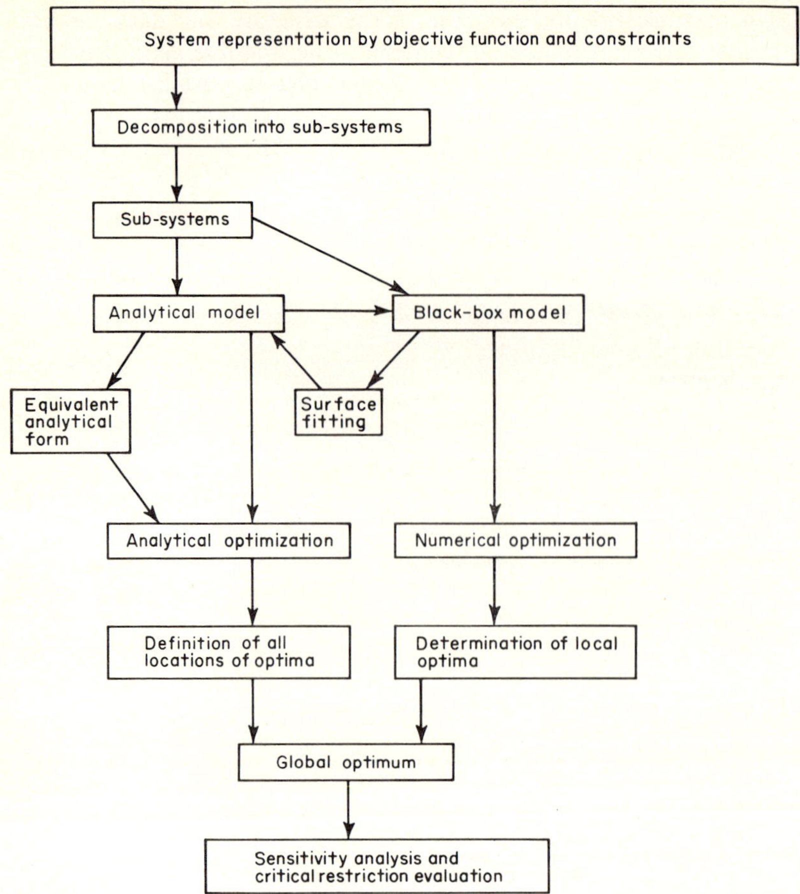

Fig. XI-4 Optimization organization in well-defined systems.

optimization methods have been discussed in Part Two, the actual technique to be used being controlled by the form and complexity of the objective function and attendant restrictions. While it may seem preferable to use analytical techniques (Sec. 5.2), it should be remembered that the complexity of practical systems makes it highly likely that even here numerical methods (Sec. 5.3) will be required, the actual approach being determined by circumstances (Sec. 5.4).

5.1 Dimensionality—The Bane of Optimization

Most realistic systems require that a large number of conditions be specified in a completed design, and if this design is to be optimal, the associated optimization studies must be carried out with respect to a large number of variables. Many difficulties arise as the number of variables increases, since the effectiveness of many of the methods, both analytical and numerical, deteriorates rapidly, and there is a greatly increased demand for computer storage. It would be generally convenient, therefore, if optimization could be carried out with respect to a relatively small number of variables. Normally, there are but two courses of action available to us for combating this perplexing problem. We must either simplify the original problem in some way (Sec. 5.1.1) or decompose the system into smaller systems (Sec. 5.1.2).

It is obvious that any solution to a modified problem will be invalid unless the modifications retain the essential features of the original system and problem, so that all simplifications and decomposition approaches must be used with this in mind.

5.1.1 Simplification

In many cases it will be found that some practical variables, while required for a detailed specification, are irrelevant to the optimization in hand. We might therefore be able to simplify our problem as discussed in Chap. II, Sec. 8, by keeping the general level of these variables fixed and varying only those which significantly affect the particular objective in use. By simplification we mean a reduction in the number of independent variables, an elimination of constraints, or an alteration in the original problem. The actual approach is governed primarily by the particular problem under study and the optimization techniques which can be used in its solution. If, for example, a solution is required to a problem involving a large number of variables, all having more or less the same influence on the objective function, it may not be possible to eliminate any of the variables or to discard any of the imposed constraints. However, it may still prove fruitful to modify the problem. Thus the objective function and constraints may be approximated locally by linear functions, and linear programming techniques then applied, the latter being capable of handling many variables efficiently. In making the linear approximation (Chap. III, Sec. 5.1) one must, however, ensure that the behavior of the modified problem is very nearly the same as the original problem; otherwise the work of optimization is wasted.

5.1.2 Decomposition

A preferable approach to problems of high dimensionality is by the decomposition of the original complex system. This is defined as the reduction of the original problem into a collection of subproblems, each of which is of a lower dimensionality and can be legitimately optimized separately. This is a very important aspect of the overall problem of optimization, and we have devoted a full chapter to the development of such methods. Their discussion is therefore deferred until Chap. XIII.

5.2 Analytical Optimization

Let us first generally consider optimization by analytical methods (Sec. 5.2.1) and then look at some computational problems (Sec. 5.2.2).

5.2.1 Analytical methods—a survey

For pedagogic reasons, the discussion of analytical methods of optimization was divided into three chapters (V, VII, and IX), although there is clearly a unity of approach applicable to all methods in these chapters.

Fig. XI-5 Analytical optimization.

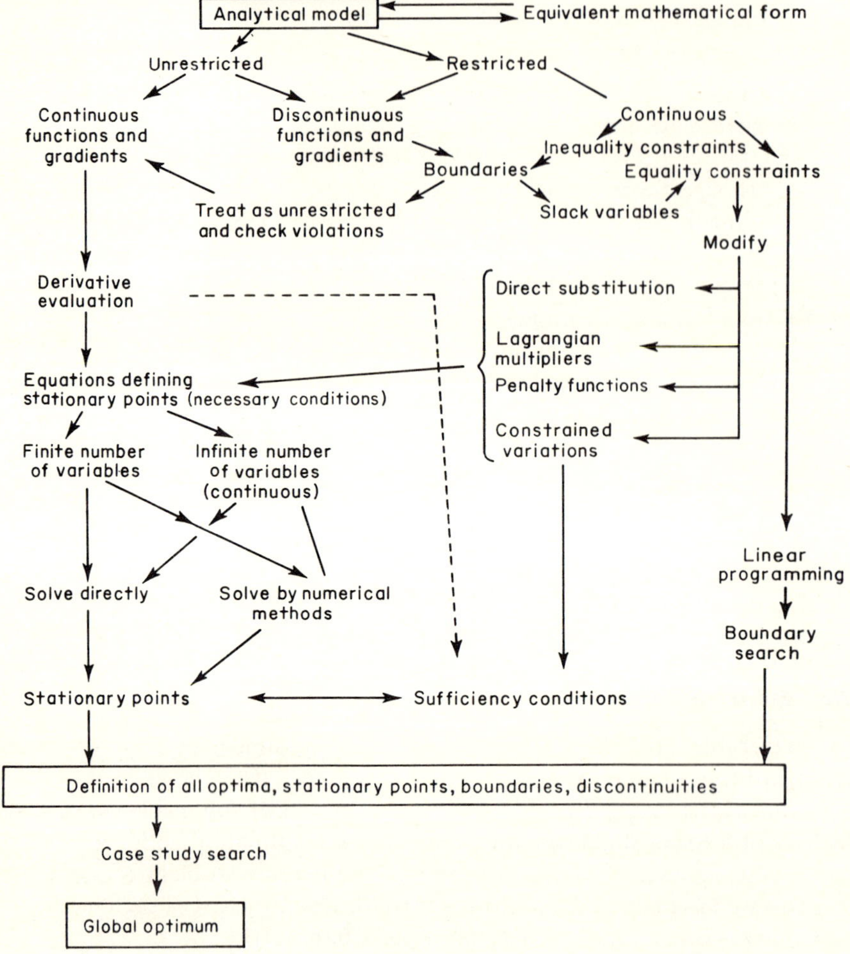

The flow diagram shown in Fig. XI-5 depicts the hierarchy of logical operations required to complete the optimization steps. Let us examine this diagram more fully as a basis for a unified discussion of the analytical optimization techniques.

When an analytical expression exists, whether derived from analytical or black-box models, we may assume that derivatives of the function exist. When the function is continuous, with no discontinuities in gradient or function, and is unrestricted, optima can exist only at stationary points. These can be found by setting the derivatives equal to zero and solving the resultant n simultaneous equations (Chap. V, Sec. 2.1, and Chap. VII, Sec. 2.2) or, if n becomes very large, the Euler-Lagrange equation (Chap. IX, Sec. 2.2.1). This latter equation, a necessary condition for the existence of an optimum in the continuous case, is a differential equation. These solutions may sometimes be found directly, but may be so complex as to require numerical methods (Chap. III, Sec. 4). The roots may also be found numerically by some suitable search method (Chap. VIII, Sec. 3). Sufficient conditions must then be found from the derivatives to establish the nature of each of the stationary points so determined (Chap. V, Sec. 2.2, Chap. VII, Sec. 2.2.3, and Chap. IX, Sec. 2.2.2).

When the objective function or its derivatives are discontinuous, artificial boundaries may be imposed, and the problem treated as a restricted optimization subject to inequality constraints.

Restricted functions are more difficult to treat, since the solutions can lie on the boundary without being at a stationary point. In general, they will be subject to inequality or equality constraints or both. Inequalities can be handled either by treating the function as unrestricted and, once the stationary points have been located, checking the feasibility of their location (Chap. VII, Sec. 4.1.1) or by the use of slack variables (Chap. VII, Sec. 4.1.2) to convert the problem into one constrained by equalities. In the Kuhn-Tucker approach (Chap. VII, Sec. 4.3), the slack variables need not be incorporated, the method having the advantage of yielding an additional necessary condition which can aid in identifying the character of a particular stationary point of the augmented Lagrangian function. These conditions call for the signs of the Lagrangian multipliers to be nonnegative, for example, if the particular stationary point is to be a minimum.

With equalities, four methods are described. The method of direct substitution (Chap. VII, Sec. 3.1), if possible, eliminates one variable for each restriction, and hence solves a lower-dimensionality problem. Difficulties arise, however, since elimination is not often easy, and restrictions may still be required in the reduced problem. The method of constrained variations, on the other hand (Chap. VII, Sec. 3.2), eliminates variations after differentiation, the location of the resulting stationary point being defined by the equations derived from a set of determinants set equal to zero. The main difficulty here lies in the evaluation of these determinants if there are more than about four variables. The introduction of Lagrangian multipliers produces an unconstrained optimization problem of higher dimensions and finds extensive application throughout a wide range of optimization problems (Chap. VII, Sec. 3.3, and Chap. IX, Sec. 5.1). An alternative approach to transforming the problem into an unrestricted system, that of penalty functions (Chap. VII, Sec. 3.4), also proves useful in numerical search (Chap. VIII, Secs. 4.4.2 and 5.4.1). All these techniques define the possible locations of optima. Only the method of constrained variations provides a full set of sufficiency conditions (Chap. VII, Sec. 3.2.3) to test the nature of these locations. In the continuous case the calculus of variations finds only stationary points after solving a two-point boundary-value problem, but cannot cope with bounds on the variables, although equality restrictions can be taken into account. In contrast, the maximum principle handles both stationary points and bounds, the latter helping the solution. The method of dynamic programming is in the limit equivalent to the maximum principle, the difference between the two methods lying in the approach to computation. In the former, the calculations start at the rear of the system and pass through it once, storing all necessary information (Chap. XIII, Sec. 4.1), while the maximum principle, on the other hand, iterates and does not store more information than is then needed. These analytical techniques result in the definition of possible locations for optima, only stationary points, discontinuities, or boundaries needing to be investigated. The nature of the stationary point is then checked by sufficiency conditions where possible. A case study must now be instigated, each local optimum value being checked against the others until the global result is attained. The special case of linear equality or inequality constraints, together with linear functions, results in a search conducted along the boundaries from vertex to vertex (Chap. VII, Sec. 4.4).

Instead of considering the original analytical expressions, equivalent mathematical formulations might have been considered whose solutions yield the same optimal results, arrived at through an easier calculation. This may be done through the use of dual systems (Chap. VII, Sec. 5), which at the moment find most application in the field of linear systems, for instance, as used in the mixed strategies of game theory (Chap. XII, Sec. 2). Instead of altering the formulation, it is often appropriate to reorganize the approach to optimization. Thus, in the optimization of a tubular reactor, it is easier to calculate the optimal reactor length for a given conversion than tackle the equivalent problem of optimizing conversion for a given length.

5.2.2 *Computational problems in analytical optimization*

While analytical methods can define the location of potential optima, it is not always possible to obtain the actual location by solving the resulting relations without resorting to numerical methods. Let us consider several examples.

In Example VII-26 the position of potential optima lying along a particular boundary was defined by the relation

$$3{,}000 - 100p^2 - 4p^5 - 6p^6 = 0$$

The roots of this equation can be found only by one of the numerical methods of Chap. III, an actual solution being illustrated in Example III-8. While there is no doubt, once a solution is found, that we have attained a stationary point, the accuracy of solution will be wholly dependent upon the numerical root-finding technique. In this illustration it was noted that an estimate of the roots was also desirable. This situation may be compared with numerical search methods, which again determine the accuracy of the location, but may not now be able to ensure that the solution will truly be an optimum (Sec. 5.3.2).

A more common problem is the solution of n simultaneous equations, defining the location of optima. If these necessary conditions produce linear equations, as in the method of constrained variation, the solution can be written down by Cramer's rule (Appendix 1). This, however, requires the evaluation of $n + 1$ determinants, where each determinant is of order n. When n is greater than 3 or 4, the main difficulty lies in the calculation of each determinant. More suitable numerical methods have then to be sought, such as Gauss' method of solution (Scarborough [1955] and Faddeeva and Faddeeva [1963]). With nonlinear equations, we may illustrate some of the problems by an example.

Example XI-2: Optimal design of a heat-exchanger train—numerical methods in analytical optimization

Let us consider the problem of selecting the areas of the four heat exchangers shown in Fig. XI-6 in such a way that the outlet temperature is maximized, the total heat-exchange surface being restricted to 200. To simplify the calculations we shall assume that the mass flow rates of

Fig. XI-6 Heat-exchanger train used in Example XI-2.

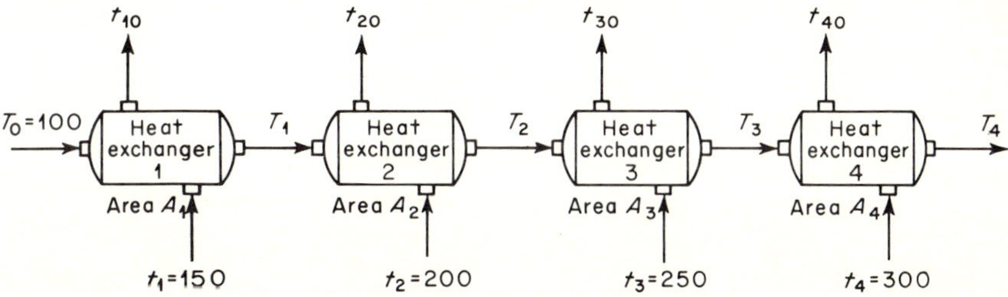

all streams are equal, as are their heat capacities. The inlet temperatures of the heating fluids are fixed by other operations and cannot be changed from the values shown in Fig. XI-6.

To approach any problem in optimization, we require a model for the system, so that our approach here is to relate the temperature produced by the ith exchanger in the train, T_i, to the area of that heat exchanger, A_i. An energy balance yields

$$WC_p(T_i - T_{i-1}) = WC_p(t_i - t_{i0})$$

or since the mass flow rates W and the heat capacities of all streams, C_p, are respectively equal throughout, we have

$$T_i - T_{i-1} = t_i - t_{i0} \tag{5.2-1}$$

If U is the overall heat-transfer coefficient, the amount of heat transferred per unit of time is also $UA_i(t_i - T_i)$, where the latter temperature difference is assumed to be the mean driving force. Thus we can also write

$$t_i - t_{i0} = \frac{UA_i}{WC_p}(t_i - T_i) = \alpha(t_i - T_i)A_i \tag{5.2-2}$$

Eliminating t_{i0} between Eqs. (5.2-1) and (5.2-2) gives, then, the transformation equation

$$T_i = \frac{T_{i-1} + \alpha A_i t_i}{1 + \alpha A_i} \tag{5.2-3}$$

We also have the further restriction that

$$200 - \sum_{i=1}^{4} A_i = 0 \tag{5.2-4}$$

Following the procedure used in Chap. VII, Sec. 3.3.3, we can formulate the problem in terms of Lagrangian multipliers as follows:

$$F = T_4 + \lambda_0\left(200 - \sum_{i=1}^{4} A_i\right) + \sum_{i=1}^{4} \lambda_i\left(\frac{T_{i-1} + \alpha A_i t_i}{1 + \alpha A_i} - T_i\right)$$

The necessary conditions for an optimum are derived by taking first derivatives of F with respect to both T_i and A_i. These eight equations are

$$\lambda_i = (1 + \alpha A_i)\lambda_{i-1} \qquad \text{for } i = 1, 2, \text{ and } 3$$

$$\lambda_4 = 1$$

and finally,

$$\alpha\lambda_i(t_i - T_{i-1}) = \lambda_0(1 + \alpha A_i)^2 \qquad \text{for } i = 1, 2, 3, \text{ and } 4$$

These eight equations, together with the transformation equations and the restriction on the areas, must be solved simultaneously. In general, the system of algebraic equations is too complex to find a solution by simple elimination, so that a numerical calculation is warranted. The calculation is simplified considerably if we note that the ratios of the Lagrangian multipliers to λ_0 are all that is required. As will be seen λ_4 does not enter the calculation. Let us define

$$\lambda_i^* = \frac{\lambda_i}{\lambda_0}\alpha$$

The relevant equations can be put in the form

$$A_1 = \frac{\sqrt{\lambda_1^*(t_1 - T_0)} - 1}{\alpha} \tag{5.2-5}$$

and $$A_i = \lambda_{i-1}^*(t_i - T_{i-1}) - \frac{1}{\alpha} \qquad \text{for } i = 2, 3, \text{ and } 4 \tag{5.2-6}$$

$$\lambda_i^* = (1 + \alpha A_i)\lambda_{i-1}^* \tag{5.2-7}$$

Equations (5.2-3) to (5.2-7) are now to be solved simultaneously, using a trial-and-error procedure. We shall assume a value of λ_1^* and, using Eq. (5.2-5), compute A_1. This first step is shown in Table XI-4, where, for the special case of $\alpha = 0.1$ and the assumed value of $\lambda_1^* = 0.2$, we find A_1 as 0. The transformation equation (5.2-3) then determines the outlet temperature of the first exchanger as 100. This is the entry

TABLE XI-4 THE OPTIMIZATION OF A HEAT-EXCHANGE SYSTEM

	Iteration				
	1	2	3	4	5
λ_1^*	0.2	0.3	0.219	0.2238	0.2249
A_1	0.0	2.25	0.4643	0.5781	0.6042
T_1	100.0	109.2	102.3	102.7	102.8
A_2	10.0	17.24	11.40	11.70	11.86
T_2	150.0	166.7	154.3	155.3	155.5
λ_2^*	0.4	0.8172	0.4687	0.4872	0.4916
A_3	30.0	58.0	34.85	36.14	36.46
T_3	225.0	237.8	228.7	229.5	229.7
λ_3^*	1.6	5.557	2.102	2.248	2.284
A_4	110.0	335.6	139.9	148.5	150.6
T_4	293.7	298.0	295.2	295.6	295.6
$\sum_i A_i$	150.0	413.1	186.6	197.0	199.5
Comments	Total area too small; try larger Lagrangian multiplier.	Area too large; correct multiplier is straddled; interpolate.	Inter-polate again.	Inter-polate again.	Good enough agreement.

for T_1 in the first iteration shown in Table XI-4. Proceeding, we then determine A_2 from Eq. (5.2-6), T_2 from Eq. (5.2-3), and λ_2^* from Eq. (5.2-7). The method of computing in succession A_3, T_3, and λ_3^* is exactly the same as that for the second exchanger. Once the conditions of the third heat exchanger are established, the fourth can be computed. Finally, we add up the areas in the last row of Table XI-4 and find that our assumed value of λ_1^* was erroneous, since the sum of the areas does not total to the required 200. We therefore make a new guess for λ_1^*, which is the 0.3 shown starting the second iteration in Table XI-4. The same sequence of calculations detailed in the first iteration is again used, and the resulting total area is 413.1. We have now straddled the correct value of the Lagrangian multiplier, since we have a guess giving an area that is too small and one giving an area that is too large. To determine our next guess, we use the method of interpolation detailed in Chap. III, Sec. 4.1.1. The new guess for the Lagrangian multiplier is given by

$$\lambda_1^{(N)*} = \lambda_1^{(1)*} - f_1 \frac{\lambda_1^{(2)*} - \lambda_1^{(1)*}}{f_2 - f_1} \tag{5.2-8}$$

where $\lambda_1^{(N)*}$ = new estimate of correct value of Lagrangian multiplier

$\lambda_1^{(1)*}, \lambda_2^{(2)*}$ = two values straddling correct value, as shown in Fig. XI-7

f_1, f_2 = functions $200 - \Sigma A_i$ evaluated at the two Lagrangian multipliers $\lambda_1^{(1)*}$ and $\lambda_1^{(2)*}$, respectively

Fig. XI-7 Interpolation to obtain new estimate of first Lagrangian multiplier.

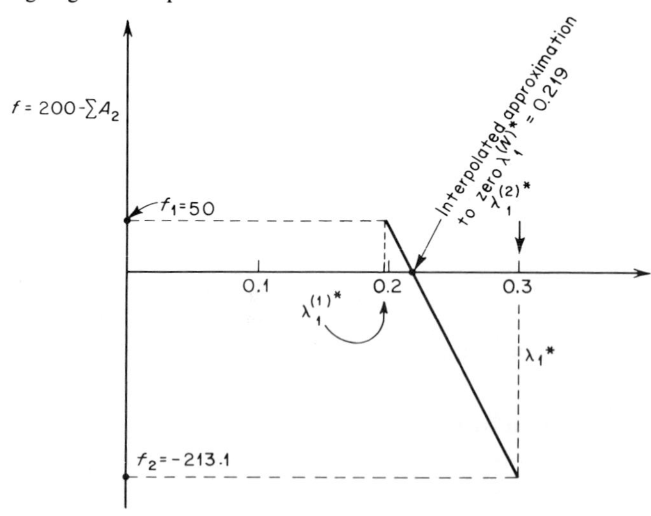

This is clearly a problem in one-dimensional root finding, since the sum of the areas depends on the value of λ_1^*. Equation (5.2-8) is then just the expression for interpolation developed in Chap. III, Sec. 4.1.1. This new guess is found as follows:

$$\lambda_1^{(N)*} = 0.2 + 50 \frac{0.3 - 0.2}{213.1 + 50} = 0.219$$

The calculation is then completed by successive approximations until the correct area total is found. The results of the fifth iteration are sufficiently accurate for most purposes.

The results also indicate that the last area added does not really enhance the outlet temperature very much; for example, increasing the total area from 150 to 200 changed the outlet temperature by only two units. The result is therefore not sensitive to the areas.

Another important step is to attempt to check to see if a maximum has truly been found, for after all, we have converged to a stationary point, but we have no assurance that this is indeed a maximum. To check the sufficiency conditions is all but impossible in this case, and therefore we can make only a few calculations which give us some confidence in the result. For example, if all the area were used in the last exchanger, the outlet temperature would be $T_4 = 290.5$, and if the areas were equally distributed among all the exchangers, an outlet temperature $T_4 = 290$ would result. From these calculations we see that our stationary point is certainly not the global minimum, and since there is only one stationary point, we hope that it is the desired maximum. The more checks we perform, the more certain we shall become that our answer is a maximum; but only by looking at second derivatives will we ever be absolutely positive.

With continuous variables, the greatest difficulty may lie in the solution of the differential equations which define the necessary conditions. Suitable methods for numerical solution have been indicated in Chap. III, Secs. 4.2 and 4.3.

5.3 Numerical Optimization

Let us survey the numerical search methods (Sec. 5.3.1) before considering some of the associated problems of convergence (Sec. 5.3.2).

5.3.1 Numerical methods—a survey

The general interaction of the various numerical search methods discussed in Chap. VIII, with special cases considered in Chaps. VI and X, is shown in

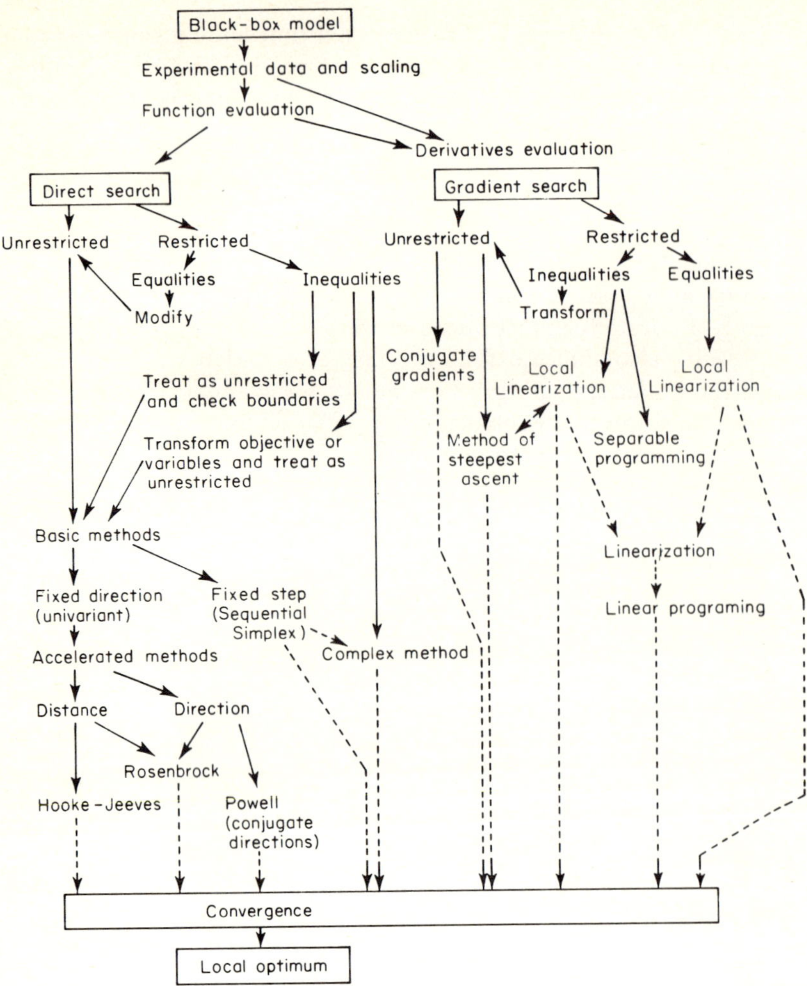

Fig. XI-8 Numerical optimization.

Fig. XI-8. Irrespective of the source of the information, the function value $y(\mathbf{x})$ can be found for a set of particular values of the variables, $\mathbf{x}$. Two classes of search techniques can be found, the distinction being dependent upon the availability of the function derivatives. If the latter can be found directly from a mathematical relation, gradient methods would be preferred over those, the direct search methods, which do not require derivative evaluation.

In unrestricted function search the basic direct search methods are those of the Sequential Simplex method (Chap. VIII, Sec. 3.1.2) and the univariant technique (Chap. VIII, Sec. 3.1.1), which use fixed step size and fixed search directions, respectively. The Simplex method is a convenient technique, which may be recommended, using as it does only a few simple rules to move rapidly. The univariant method, on the other hand, cannot effectively cope with ridges or sharp valleys. Both of these basic methods may suffer from the disadvantage imposed by fixed step size or directions. Accelerated forms based upon the univariant method are often effective. The Hooke-Jeeves, the Powell conjugate directions, and the Rosenbrock methods (Chap. VIII, Sec. 3.1.3) may be mentioned, representing accelerations in distance, direction, and both distance and direction, respectively. The Powell method is particularly effective, based as it is on conjugate directions.

When equality restrictions are present, direct search methods are ineffective unless the problem can be transformed into an unrestricted search, by, for example, the use of penalty functions (Chap. VIII, Sec. 4.4.2).

With inequality constraints three direct search approaches might be considered. The problem could be treated as unrestricted, and the boundaries later checked, which is not particularly effective, especially if the optimum does lie at or near a boundary; the objective function or the constrained variables could be transformed (Chap. VIII, Secs. 5.4.1 and 5.4.2) so that the problem is unrestricted; and the problem could also be handled by the Complex method of Box, which is an extension (Chap. VIII, Sec. 5.2.1) of the ideas behind the Sequential Simplex technique. This Complex method is, again, conceptually simple, and can be recommended for general use.

Corresponding to these direct search methods, the Ritz method (Chap. X) is the equivalent approach in the continuous variable, using various functional forms to approximate to the desired solution, the approximation being fitted by some multidimensional search technique. Direct search methods in one dimension were considered in Chap. VI.

The most effective methods of search will, however, become available if the function derivatives are easily obtained. The resulting gradient methods may attempt to follow the function gradient closely, as in the method of steepest ascent or descent (Chap. VIII, Sec. 3.2.1), even though the gradient is a local property, or use it in guiding the search as in the conjugate gradient methods (Chap. VIII, Sec. 3.2.2). The latter form a very efficient approach, the Fletcher and Reeves technique being particularly recommended.

With restricted functions, the problem can be transformed into an equivalent search either by modifying the variables (Chap. VIII, Sec. 5.4.2) or by introducing a modified objective function, using penalty functions (Chap. VIII, Secs. 4.4.2 and 5.4.1). A difficulty here lies in the choice of magnitude of the penalty factor since its size can affect the form of the modified function, care having to be taken that narrow ridges are not formed in the early stages which might increase difficulties in the search. In particular, the Carroll method has found favor. With both inequalities and equalities, the gradient projection methods (Chap. VIII, Secs. 4.3.1 and 5.3.3) can be used on the unmodified problem, these techniques being essentially based upon the method of steepest ascent in the feasible region, constraints being linearized when the search approaches a boundary and movement along the line of the constraints being then handled by linear programming. A major problem at this stage is the difficulty of ensuring that the search remains within the feasible region, since a linearized version of a constraint may allow the search to move outside the original boundary.

In all constrained problems in which the constraints are not removed by some device, an initial obstacle is the choice of a first feasible search location. This may be handled through an objective function formed from any violated constraints (Chap. VIII, Secs. 4.1 and 5.1).

There is no generally effective and acceptable numerical search method, there being counterexamples to confuse most of the available techniques. All are personal methods devised to suit particular users and problems. Efficient search requires a good n-dimensional search routine, together with an excellent one-dimensional search method. We have seen in Chap. VI that while, in one-dimensional restricted search, a criterion can be defined to allow a choice of the most efficient optimization method in this category—the Fibonacci method (Chap. VI, Sec. 4.2.3)—in the unrestricted situation an accelerated form of direct search of gradient search should be utilized. The methods such as Powell's direct search technique (Chap. VI, Sec. 3.2) and that of Fletcher and Reeves (Chap. VI, Sec. 3.3), based upon Davidon's gradient approach, find wide acceptance.

The question of convergence of any search to the local optimum (Sec. 5.3.2) should always be examined closely.

5.3.2 *Optima and convergence*

With analytical methods, the possible locations of the desired optimum are defined by analytical expressions, and a direct comparison of the objective function at these points will allow the best value to be found. With numerical techniques, on the other hand, we start from some point, $\mathbf{x}_j$, and move along some direction, $\boldsymbol{\xi}_j$, to find a new point, $\mathbf{x}_{j+1}$, such that

$$\mathbf{x}_{j+1} = \mathbf{x}_j + \boldsymbol{\xi}_j s$$

in order that $y(\mathbf{x}_{j+1})$ may be better than $y(\mathbf{x}_j)$. This procedure does not guarantee to reach an optimum or, when one is found, that it is a global value. As we have seen in Chap. VIII, numerical searches can become stranded before reaching a local optimum, although this is rare unless constraints are present. The problem of ensuring a global optimum is more difficult. The starting of new numerical searches from different base points in an unrestricted search gives us some confidence in having found the global optimum if all searches eventually lead to the same result. However, we cannot exclude the possibility of the global optimum lying elsewhere. The presence of constraints can make this starting procedure more difficult, with an initial task the search for a feasible starting point.

In a special class of functions, the unimodal functions, only one optimum will be found in the unrestricted problem, and this must be the global result. The commonest of these are the concave and convex functions (Chap. IV, Sec. 2.3), containing, respectively, a global maximum and minimum. With constraints, however, a global result will be found only if the search lies in a convex region. In the general problem, when the form of the function and constraints is unknown, it can only be initially assumed that the system is well behaved, until otherwise disproved.

The rate of approach to the optimum can be greatly influenced by the choice of scales for the independent variables. In many problems of optimization, the independent variables are expressed in completely unrelated units. For example, a reaction yield might well depend on temperature, pressure, and composition, these dimensions being unrelated to each other. In contrast, in geometric problems, all the variables are measured in terms of the same basic unit of length. In the former situation, when the units are measured on different yardsticks, it is possible to change the measure for one variable while leaving the others unchanged. For example, a temperature can be expressed in either degrees Centigrade or degrees Fahrenheit. The choice of a particular scale of units is dictated primarily by convenience, convention, and experience. None of these factors have any connection with optimization. However, it is important to note that, given precisely the same problem, the effectiveness of many search methods will partly depend upon the choice of dimensions. To clarify this point, suppose that the response function in one system of units has the form

$$y = x_1^2 + x_2^2 \tag{5.3-1}$$

Another investigator might, however, use a different set of units and, although measuring the same function, y, would record one of the variables on a different basis, obtaining, for example,

$$y = a\bar{x}_1^2 + b\bar{x}_1 + x_2^2 + c \tag{5.3-2}$$

where $a = 1.8^2$

$b = 3.6(492)$

$c = 492^2$

Here y and x_2 are identical in both investigations, so that y has the same dependence on x_2 in both cases. However, the dependence on the other variable, $\bar{x}_1$, seems quite different. This difference could be due solely to a difference in the scale of measurement. If x_1 were the absolute temperature in degrees Rankine and $\bar{x}_1$ were the temperature in degrees Centigrade, the objective function in Eq. (5.3-1) would transform into a form identical with Eq. (5.3-2). This statement can be verified by noting that the conversion between the two temperature scales is

$$\bar{x}_1 = \frac{x_1 - 492}{1.8}$$

Equations (5.3-1) and (5.3-2) thus represent the same objective function, the temperature variable being in two different scales. What has this well-known fact to do with optimization? The method of steepest ascent, for example, will converge in a single trial if the contours of the objective function are concentric circles, as in the case using Eq. (5.3-1). This is true since any perpendicular to the contour tangent passes through the center if the contours are, in general, spherical or hyperspherical. Thus, in this case, the direction of steepest ascent from any base point must pass through the optimum. If, however, the contours are elliptical, as in Eq. (5.3-2), the convergence of the search might be slow and require many experiments with most direct search or gradient methods. Since we have just demonstrated that an objective function can be distorted from spherical to elliptical contours by a scale change, we must conclude that the rate of convergence of at least some methods may depend to a large extent on the choice of scales.

This important fact then raises the question of how to make a good choice of scales. The best choice would make the contours as nearly spherical as possible, since in the perfectly spherical situation all gradient lines would pass through the optimum. We should therefore normally select scales of measurement such that unit changes in any independent variable would give the same change in the objective function. It is also preferable to use a representation in which the variables do not interact; in other words, there are no terms containing more than one variable. There is also no reason why there should not be a change of scales during a search. Thus, for example, if, after initial moves, a ridge type of region were entered, this would mean that a search method might oscillate back and forth across the ridge, most moves being essentially perpendicular to the crest instead of parallel to it. The scales could now be changed so as to speed the convergence of the search in this region. The inability of some gradient and direct search methods to deal efficiently with ridges formed by dissimilar scales is an important disadvantage in their use.

The problem of ensuring convergence on any local optimum before stopping a numerical search is a very real one, and several different convergence criteria have been proposed. One simple criterion is to stop the

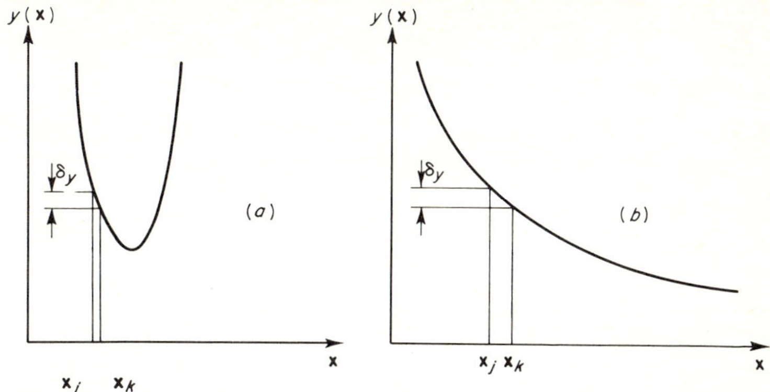

Fig. XI-9 Criterion on y for stopping search. (*a*) Steep valley; (*b*) gently sloping function.

search when the difference between values of the objective function remains small enough after several successive moves, so that

$$|y(\mathbf{x}_j) - y(\mathbf{x}_k)| < \delta_y \tag{5.3-3}$$

where δ_y is the accuracy to which numerical values of the function are to be found, and $k = j + 1, j + 2, \ldots, j + l$, with l being some predefined integer. This is sufficient for most steep-sided valley-type functions, as in Fig. XI-9*a*. It is not satisfactory, however, for gently sloping functions, as in Fig. XI-9*b*, in which the search may become halted prematurely, since for this function, considerable variations in $\mathbf{x}$ cause only small changes in y, even though the optimal $\mathbf{x}^*$ can be far from $\mathbf{x}_k$, and $|y(\mathbf{x}_k) - y(\mathbf{x}^*)|$ might still be much larger than δ_y. Another possible criterion is to stop the search when the distance moved in the last few iterations is less than the location accuracy required, which might be different for each variable, depending upon the scales adopted. Thus we might use

$$|x_{i,j} - x_{i,k}| < \delta_i \qquad \text{for } i = 1, 2, \ldots, n;\ k = j + 1, j + 2, \ldots, j + l \tag{5.3-4}$$

Difficulties can also arise when small changes in x_i cause large changes in y, as indicated in Fig. XI-10, where $y(\mathbf{x}^*)$ can be far from $y(\mathbf{x}_k)$ and $|x_{i,k} - x_i^*|$ much larger than δ_i. Perhaps a better approach would be the use of such criteria on both y and the x_i, so that convergence to an optimum would be assumed when both Eqs. (5.3-3) and (5.3-4) hold. Even in this situation, a search might be halted before the optimum is attained, as, for example, in Fig. XI-11, when $\mathbf{x}_k$ is still far from $\mathbf{x}^*$. In such cases a random search around the vicinity of $\mathbf{x}_k$, with varying step size, might indicate the possibility of a false optimum.

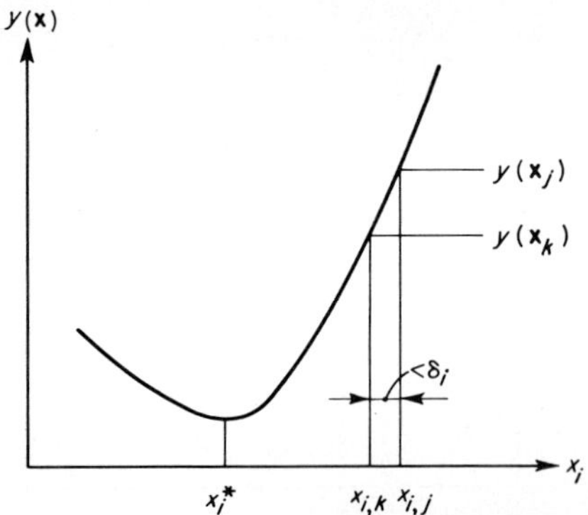

Fig. XI-10 Criterion on x for stopping search.

Numerical methods, if they do attain an optimum value, will generally reach only local optima, failure being possible in some cases on ridges or against constraints, a halt in the search also being possible in a saddle-point or plateau-type region, as discussed in the previous paragraph. While the repetition of a search from a series of initial starting positions may give us confidence in retaining a local value as a global solution, in general, there is no guarantee that a numerical search will have found more than a local result. There are special cases, of course, which do guarantee globally valid results (Chap. IV, Sec. 4.4). Generally, once a local result has been found, it is wise to consider checks against the possibility of other optima.

5.4 Analytical and Numerical Optimization—A Comparison

Optimization might utilize many different techniques. We have already seen that the two basic types of system model are interchangeable, so that

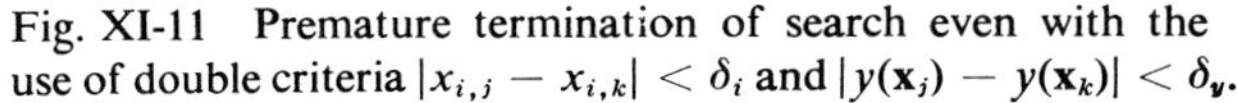

Fig. XI-11 Premature termination of search even with the use of double criteria $|x_{i,j} - x_{i,k}| < \delta_i$ and $|y(\mathbf{x}_j) - y(\mathbf{x}_k)| < \delta_y$.

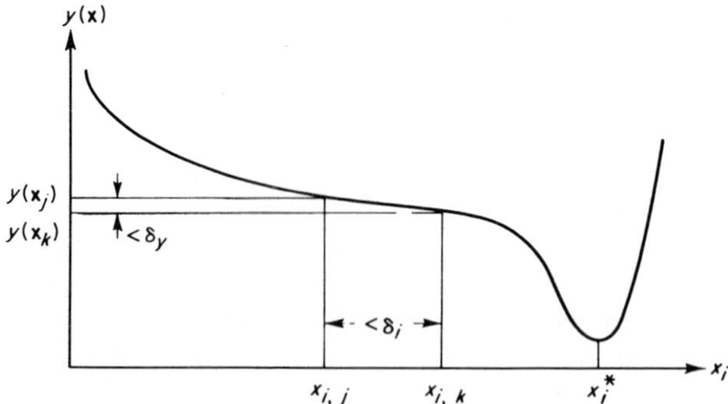

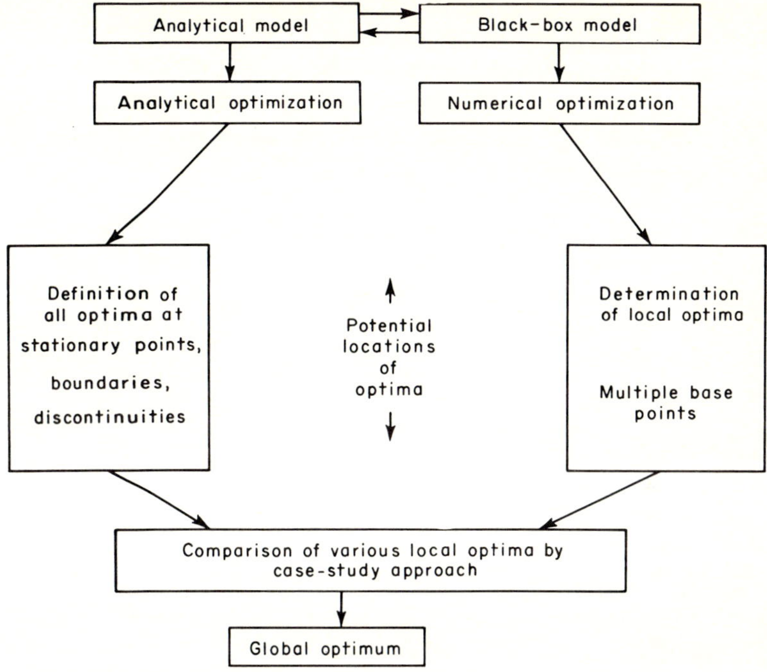

Fig. XI-12 General interaction between optimization approaches.

both analytical and numerical methods could be brought to bear upon the same problem. Some of the interlinking features of the two approaches are indicated in Fig. XI-12.

Analytical methods do have an advantage if an analytical solution is obtainable and similar problems with different sets of data are being investigated. However, in the analytical approach, we must find solutions to the relations defining the potential optima. Such solutions may have to be determined by numerical root-finding methods, a possible optimization problem in itself, and the question must always be asked, should numerical methods not be used directly upon the problem, especially if the function and its derivatives are not easily obtainable? The main disadvantage is that numerical searches are local in nature while analytical methods can define all the points at which optima might lie. On the other hand, while all analytical methods allow stationary points to be located, not all these techniques produce sufficiency conditions to test the nature of the resultant point. Such sufficiency conditions are not available in numerical methods, but experimental information can be equivalently used.

While many of the analytical techniques become less convenient as the number of variables rises, especially when inequality restrictions are imposed, the advent of large-scale computers has also diverted attention to numerical methods of optimization. Numerical methods also possess an important advantage over analytical, in the general use of a computer program written for a particular numerical search routine which can be applied independently of the actual function being optimized. Trouble with convergence and scaling may arise in numerical methods, but when an optimum is achieved, this is only a local result, and it may not be certain that the true global result is included in the final comparative search among local optima.

When a function originates directly from a black-box model, numerical direct search methods will be normally used. With an analytical model, it would be preferable to locate the positions of the various potential optima by analytical methods. If, however, it proves difficult to solve the relations defining the necessary conditions, it is likely that numerical methods using gradients would be advantageous, since the function derivatives would probably be conveniently available.

Troubles which might arise in analytical or numerical optimization when constraints are present can be partly avoided by replacing the original restricted objective function by one in which the constraints can be included and which can then be handled by unrestricted methods. Penalty function and Lagrangian multiplier methods both work well, the former finding appropriate use in numerical application and the latter in analytical optimization. The use of penalty functions results in a sequence of iterations, which, however, does not necessarily imply a longer search, while the introduction of Lagrangian multipliers produces a problem of a higher demensionality.

Although various methods have been tested against each other on various functions and the results published (for example, see Appendix 2), there is no point in reproducing these findings in detail. Most problems are different, and in general there is no one efficient optimization method. The most we have attempted is to draw attention to those methods which seem to have general promise. It is then up to the individual optimizer to determine the most suitable technique for his own problem and for his own preference.

6 THE GLOBAL OPTIMUM AND SENSITIVITY

Having carried out an optimization analysis, we must ensure that we have found globally valid results (Sec. 6.1); that the optimal solution can be maintained (Sec. 6.2); that an optimum can in fact be achieved in practice (Sec. 6.3); that the solution cannot be further improved (Sec. 6.4); that the effect of poorly defined parameter values is taken into account (Sec. 6.5); and that time-dependent optima can be followed (Sec. 6.6).

6.1 Global Optimum

Analytical methods can be used to define the location of all stationary points, the sites of potential local optima, whether or not constraints are present. The nature of these stationary points can be found, although not every method can provide sufficiency criteria for this purpose. When these maxima or minima have been located, they can be compared, as necessary, with boundary and discontinuity values, to find the desired global result.

With numerical methods the problem is not so easy. These methods use local searches and, in trying to reach a local optimum, might become stranded on ridges or boundaries, while convergence criteria may also cause the search to stop short of the optimum (Sec. 5.3.2). The presence of such false optima might, however, be detected by random searching around these locations or by restarting the search from a new base point. Once a local optimum is located, there is no certain way of determining, in general, whether it is a global value. As we have seen (Chap. IV, Sec. 4.4), a guarantee of a global result is obtainable only when the search region is convex and when the function is of an appropriate unimodal form, for instance, a concave function in maximization. In general, we can start the search only at various base points, and if the same global result is achieved, we must accept the solution with a degree of confidence, depending upon the number of iterations attempted.

6.2 Function Sensitivity at the Optimum

In practical cases it is important not only to locate the optimum, but also to examine the nature of the function in this neighborhood, since it is unlikely that we can exactly maintain these optimal conditions. Certainly, each adjacent point should have a worse value of the objective function than that at the optimum, but this is not all. It is also necessary to know the sensitivity of the system in this region. In other words, how will the value of the function change for given variations in the independent variables?

For example, consider the functions shown in Fig. XI-13. In (*a*) the optimum lies at a fairly flat plateau on the function curve, so that it is not so important that the location of the optimum be strictly observed. Any small variation of the independent variable would produce little change in magnitude of the objective function. In (*b*), on the other hand, since the functions have very steep peaks, any deviation from the optimal value of the variable will cause a marked change in the optimal result. In (*c*) a function discontinuity would cause a large decrease in value of the objective function on decreasing x from its best value, so that it is essential that x be maintained above a certain critical value. A similar argument holds in (*d*), where the

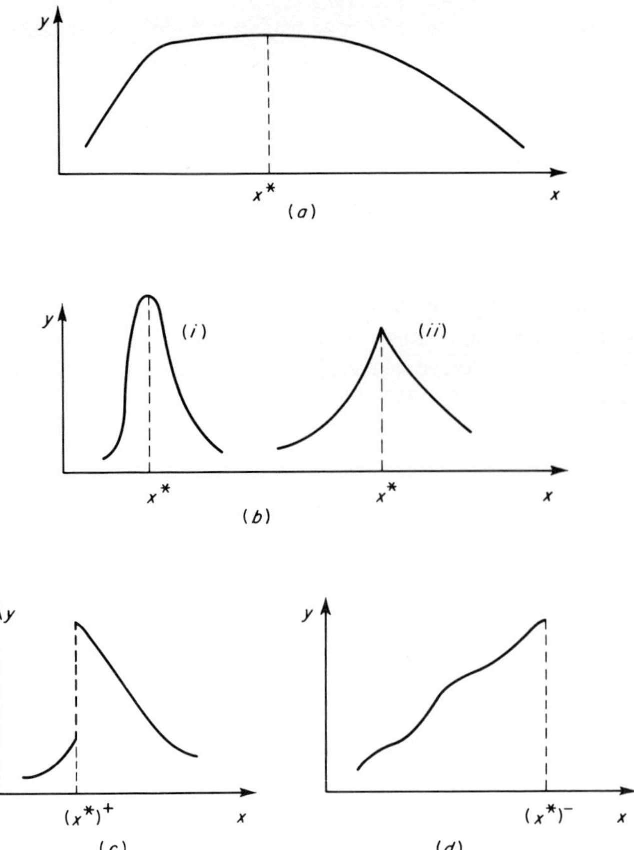

Fig. XI-13 Function sensitivity at the optimum.

monotonic function gives an optimum value at its upper boundary. If the latter is violated, failure of some form may occur in the system.

It will be seen, therefore, that the system might have to be maintained accurately at certain design conditions for optimal or near-optimal results to be found. Any deviation may prove uneconomic or, in some situations, disastrous. It is therefore essential in any optimization study not to stop investigations at the location of the optimal solution, but to continue the analysis until some information and understanding of the surrounding neighborhood are obtained. In analytically defined systems such sensitivity analyses can often be performed using the quadratic form of the Taylor's expansion in the region of the optimum. For numerical investigations the general approach is to consider the sensitivity of the optimal value to changes in the basic data and variables. This information might be gleaned directly by experimentation, with each variable altered in turn by some fraction of its value. This approach is not entirely satisfactory, however, since a combination of changes may yield quite different results from a series of single changes, but it does allow us to examine the effect of inaccurate data. Such changes of variable to produce set differences in the objective function value can also be used to define those parameters whose variation is significant in a simplification study. Alternatively, surfaces may be fitted to experimental results, and the subsequent analysis carried out analytically.

6.3 Ideality

Even more important than a sensitivity analysis is the problem of trying to achieve the "ideal" optimal conditions. An excellent illustration of the attendant difficulties is found in the field of tubular chemical reactor design. Here a continuous optimal temperature profile might be defined (Chap. IX) to hold along the length of the tubular system. This can be attained only by the continuous addition or removal of heat at each point along the system. In practice, this is an impossible situation, and another potential suboptimal system would have to be postulated, which, however, might prove easier to achieve. Thus a series of isothermal or adiabatic reactors might be used as a practical approximation to the ideal situation requiring continuous control of the temperature profile. The optimal solution, however, stands as the ultimate possible, and allows an evaluation of any approximations imposed by mechanical and similar complications.

6.4 Improvement by Restriction Relaxation

Most systems are subject to some form of constraint, and it is often found that no interior optimum exists, so that the optimum lies on a boundary. The resulting optimum would then not be likely to be the best achievable, if this boundary or restriction were relaxed. In general, we can conclude that a restricted optimum is either less favorable than or equal in value to an unrestricted optimum. Thus

$$(y_{\min})_{\text{restricted}} \geq (y_{\min})_{\text{unrestricted}}$$

and

$$(y_{\max})_{\text{restricted}} \leq (y_{\max})_{\text{unrestricted}}$$

In certain cases, therefore, the removal or relaxation of a critical restriction could result in the further improvement of the determined optimum. While this may not be possible in a rigorously defined system, we have seen (Sec. 4) that most systems and their models are approximations, and once a critical restriction has been recognized, it might be possible to effect additional improvements by altering this constraint. Since most systems are iteratively developed (Sec. 3), the system might be suitably redefined in subsequent calculations. The presence of critical constraints can be found immediately in most numerical methods, while in analytical techniques the vanishing of a slack variable or a nonzero Kuhn-Tucker Lagrangian multiplier would serve as indicators.

6.5 Parameters Defined Only within a Range

The problem of certain parameters being ill defined is often encountered in practice. It may thus frequently happen that a certain quantity will be known to be restricted to given limits, but it will not even have a well-defined average value. In such cases we frequently will vary this parameter through its range of values to discover the effect of its variation on the optimum, or if the statistics of the variation about some mean value were known, we could use Monte Carlo techniques to evaluate some of the statistical properties of the solution. In either case we must use the important result, when seeking, for example, the minimum of a function, y that

$$(y_{\min})_{\text{less restricted}} \leq (y)_{\text{more restricted}}$$

a principle which can save computational effort, since in many cases the choice of one particular valuc of the parameter will result in the most restrictive of conditions governing the choice of admissible sets of the independent variables. This value will then give the maximum of all possible minima; i.e., it will give the poorest result.

Let us consider the linear problem of minimizing the function y (Vajda [1961]), where

$$y = \sum_{i=1}^{n} c_i x_i$$

subject to the restrictions that $x_i \geq 0$ and

$$\sum_{i=1}^{n} b_{j,i} x_i - d_j \leq 0$$

in which the parameters are known only between limits, such as

$$c_i^- \leq c_i \leq c_i^+$$

$$b_{j,i}^- \leq b_{j,i} \leq b_{j,i}^+$$

$$d_j^- \leq d_j \leq d_j^+$$

What can be said about the optimum with respect to variations of these parameters? Which set of parameters will yield the worst possible case? Which the best possible?

Let us consider three possible sets of solutions, namely, those satisfying

$$\sum_{i=1}^{n} b_{j,i}^{+} x_i - d_j^{-} \leq 0 \tag{6.5-1}$$

or

$$\sum_{i=1}^{n} b_{j,i} x_i - d_j \leq 0 \tag{6.5-2}$$

or

$$\sum_{i=1}^{n} b_{j,i}^{-} x_i - d_j^{+} \leq 0 \tag{6.5-3}$$

The most difficult set of restrictions to satisfy is inequality (6.5-1). Any set of $\mathbf{x}$ satisfying this inequality will automatically satisfy both inequalities (6.5-2) and (6.5-3). This is clearly true, for if one of the $b_{j,i}$ is replaced by a smaller value, the left-hand side becomes even smaller, thus reinforcing the inequality. Since $b_{j,i}^{+}$ is the largest possible value, we see that inequality (6.5-1) is indeed the most restrictive. By the same reasoning we see that inequality (6.5-3) is the least restricitve, there being sets of $\mathbf{x}$ which satisfy this inequality which will not satisfy either inequality (6.5-1) or (6.5-2). We can now make some general statements about the possible range of solutions to the optimization problem. Let us consider the five possible solutions

y_1 is the minimum of $\sum_{i=1}^{n} c_i^{-} x_i$ subject to inequality (6.5-3)

y_2 is the minimum of $\sum_{i=1}^{n} c_i^{-} x_i$ subject to inequality (6.5-2)

y_3 is the minimum of $\sum_{i=1}^{n} c_i x_i$ subject to inequality (6.5-2)

y_4 is the minimum of $\sum_{i=1}^{n} c_i^{+} x_i$ subject to inequality (6.5-2)

y_5 is the minimum of $\sum_{i=1}^{n} c_i^{+} x_i$ subject to inequality (6.5-1)

Since inequality (6.5-2) is more restrictive than inequality (6.5-3), then $y_1 \leq y_2$. We must use a different line of reasoning to order y_2 and y_3. The smaller the coefficients in the objective function, the smaller will be its value, since all the $\mathbf{x}$ are positive. Of course, if the $\mathbf{x}$ are not restricted to be positive, the conclusions will be incorrect. Therefore we note that, since the same $\mathbf{x}$ are feasible in solution 2 as in solution 3, then $y_2 \leq y_3$. By the same logic $y_3 \leq y_4$ and $y_4 \leq y_5$, since inequality (6.5-1) is more restrictive than inequality (6.5-2). We can summarize our results as

$$\begin{bmatrix} \min \sum_{i=1}^{n} c_i^- x_i \\ \sum_{i=1}^{n} b_{j,i}^- x_i - d_j^+ \leq 0 \\ x_i \geq 0 \end{bmatrix} \leq \begin{bmatrix} \min \sum_{i=1}^{n} c_i x_i \\ \sum_{i=1}^{n} b_{j,i} x_i - d_j \leq 0 \\ x_i \geq 0 \end{bmatrix} \leq \begin{bmatrix} \min \sum_{i=1}^{n} c_i^+ x_i \\ \sum_{i=1}^{n} b_{j,i}^+ x_i - d_j^- \leq 0 \\ x_i \geq 0 \end{bmatrix}$$

It should be noted that the possible range of solutions is fixed between the two limits given by solutions 1 and 5. Solution 1 gives the best that can be achieved, and solution 5 the worst that can happen. We can determine the probability of what might happen only when we know the probability distribution of the values of the parameters.

The reasoning used here on linear systems is also applicable to more complex problems, but with the primary difference that the more complex the problem, the more difficult it is to decide which is the more restrictive of any of the possible choices of parameter value.

Example XI-3: The range of optimal conditions for a linear problem with uncertain coefficients

Consider the minimization of the function

$$y = c_1x_1 + c_2x_2 + c_3x_3$$

subject to $x_i \geq 0$

$$b_{1,1}x_1 + b_{1,2}x_2 + b_{1,3}x_3 - d_1 \leq 0$$
$$b_{2,1}x_1 + b_{2,2}x_2 + b_{2,3}x_3 - d_2 \leq 0$$

where

$$\begin{array}{ll} -1 \leq c_1 \leq 3 & 2 \leq b_{1,1} \leq 4 \\ 2 \leq c_2 \leq 6 & 4 \leq b_{1,2} \leq 5 \\ -2 \leq c_3 \leq -2 & 1 \leq b_{1,3} \leq 2 \\ 1 \leq d_1 \leq 2 & 1 \leq b_{2,1} \leq 1 \\ 2 \leq d_2 \leq 3 & 1 \leq b_{2,2} \leq 5 \\ & 1 \leq b_{2,3} \leq 2 \end{array}$$

Solution 1, giving the best possible result, is found from the problem

$$y = -x_1 + 2x_2 - 2x_3$$

subject to

$$2x_1 + 4x_2 + x_3 - 2 \leq 0$$
$$x_1 + x_2 + x_3 - 3 \leq 0$$

with all the $x_i \geq 0$. The solution of this problem can then be found by the Simplex method, or otherwise, to be $x_1 = 0$, $x_2 = 0$, $x_3 = 2$, with $y_1 = -4.0$. This is the least value which can possibly be expected, since the parameters range between their limits.

Let us now find the largest possible value of the minimum. From solution 5, as above, we see that the problem to be solved is

$$y = 3x_1 + 6x_2 - 2x_3$$

subject to the restrictions

$$4x_1 + 5x_2 + 2x_3 - 1 \leq 0$$
$$x_1 + 5x_2 + 2x_3 - 2 \leq 0$$

with the $x_i \geq 0$. The solution to this problem is $x_1 = 0$, $x_2 = 0$, $x_3 = 0.5$, with $y_5 = -1.0$. This is the maximum of the possible minima and therefore represents the worst which can befall us as the parameters vary.

The solution to the original problem is thus that

$$-4.0 \leq y_{\min} \leq -1.0$$

6.6 Time-dependent Optima

Situations also occur in practice in which the systems can be completely defined and optimized but in which some temporal variation can occur affecting performance, so that the question of continuing the optimization at later periods must be considered. Even if an optimal design were achieved, updating in construction, start-up, and later in operation would still be essential, not only to correct technical conditions, but also to follow changing trends. The latter might include catalyst activity decline, with a falling off in reactor yields; fouling and scaling in heat exchangers; old age, affecting maintenance and replacement policies; raw material and product specifications and economics; transportation costs; modes of shipping; labor charges; government policy; customer location and market demands; and the general level of sales relative to the original design capacity of the system. While these factors need not always be taken into account in the initial optimization, it is advantageous to check the optimum conditions periodically and to make suitable modifications to keep the system in its optimal state. Some form of adaptive optimization technique might be used (Bellman [1961], Rudd, Aris, and Amundson [1961], Kipiniak [1961], Aris [1962b], Box and Chanmugan [1962]), or by a systematic variation of the system variables an "evolutionary operation" search can be made for any improvements during plant operation (Box [1957, 1960], Box and Hunter [1959], Box and Chanmugan [1962], Carpenter and Sweeny [1965], Lowe [1964, 1967] including a select bibliography).

7 GENERAL CONCLUSIONS

We have seen that there are three important facets to any optimization study. First, a suitable objective must be defined. This will normally present little or no difficulty, except in the case of systems about which there is only imperfect knowledge. Second, the statement of the problem must be converted into terms of specified systems capable of definition by mathematical models from which the objective function can be simulated in terms of the system variables. Only when this stage is complete can optimization begin. The model building and simulation stage is at the moment the weakest link, and, together with the inaccuracy of data, limits the usefulness of optimization studies. Much work still requires to be carried out in the construction of realistic models, the fields of business operations and chemical reactor design having so far received much attention. The former are usually very general in their representation, with simplifying assumptions. More realistic are the detailed reactor studies, but here too the system is greatly idealized in behavior. The third stage is that of optimization. This can be carried out with respect to the system variables or to the system itself. Most well-defined problems can now be handled, although optimization may not be particularly rapid in attaining solutions and may be limited by the practicability of the technical designs available. Difficulties do arise if a problem has variables or restrictions partly or wholly under the influence of uncontrollable external agencies.

Arguments have sometimes been put forth attempting to minimize the importance of optimization in present-day industrial activities. Thus one hears that the lack of realistic system models makes the study of optimization techniques academic, that the simplifications assumed in the theoretical analysis can not be realized in practice, or that data are too inaccurate for meaningful calculation.

It should be noted, however, that precisely the same system models are required in optimization as in design, so that it is certainly true that unrealistic results would be obtained with inadequate design techniques. In many cases, however, good design procedures are available, and optimization under these circumstances would be quite advantageous. Simplifications unattainable in practice can be justified by noting that an optimization solution can provide an ultimate limit against which practical achievement can be judged. A similar situation is experienced when equilibrium thermodynamics imposes a bound on the ultimate performance of irreversible processes. The validity of any "optimal solution" depends, of course, on the accuracy of the numbers used to generate the answer. Optimization can again play the role of providing limits, associated with a particular set of data, the data being varied as desired to produce a range of solutions to test the sensitivity of the optimum.

How far can we go in achieving a complete optimization solution? Practical optimization problems fall into groups of varying complexity. The simplest may be defined as those representing single components or items of equipment. Then follows the combination of these into a process; the combination of these to form a complex with integrated feeds and products; and finally, a combination within a larger network, such as a company or business group. It may seem that the latter type of problem would be so detailed and complicated that it would not be capable of analysis. Certainly, this would be the case if the problem included all possible variables, right down to the placing of the last nut and bolt or the accounting for the last penny, although the system could be specified and a design proposed, with an appropriate objective determined for a given set of values for these variables. The complete system might include many thousands of variables, and it could be optimized in its entirety only by some case study approach, an approach which is unlikely to achieve the desired result.

Normally, such problems can be reduced in complexity by simplification and by decomposition into subsystems. This allows some degree of optimization, involving fewer variables to be attempted. In general, however, a complete optimization is rarely attempted. There are several reasons, including the difficulty of defining an adequate model, of controlling variables and obtaining sufficiently accurate data, and of optimization by altering the system. At the moment, two areas in which optimization is very successful are found in the optimization of mathematically defined systems, whether analytic or black-box, and, taking examples from the chemical industry, in the design of individual items of process equipment. Even in the latter, the largest systems considered have essentially been straight-chain optimization problems, relatively little work having been attempted on more general systems involving loops and branches. The subsequent step of considering an arrangement of such items linked in a mathematical structure, forming the flowsheet for a chemical process, for instance, and economically optimizing the arrangement with respect to the number and dimensions of the various items of equipment, is quite feasible, provided that the layout and types of the components have been specified in advance. Such optimization of a flowsheet may be carried out, of course, based upon models derived from sets of computer programs, one for each of the units within the system.

The higher level of investigation of specifiying the flowsheet itself or rearranging the units within a given flowsheet during optimization would not appear to be reasonable for the present, the selection or generation of different arrangements and different types of process components for comparative study being usually the work of a skilled engineer. These levels cannot be handled in full detail, but can be treated by the use of approximating models, especially common in the branch of optimization known as *operations research*.

Problems

XI-1 In an analysis of air-cooled heat exchangers, Anderson [1961] considered three possible units. The appropriate cost data were as follows:

	A	*B*	*C*
Finned surface area, ft^2	12,360	14,735	16,425
Plot area, ft^2	2,520	2,940	4,320
Fan, hp	221	118	72
Price (including motors), $	91,200	102,000	124,200

Hence he derived:

	A	*B*	*C*
Kilowatts (hp × 0.746)	164.5	88	53.7
Fixed offsite power investment (kw × $45)	7,410	3,960	2,418
Operating cost per year (8,000 hr)	16,440	8,820	5,370

Assuming that the purchaser specifies the payout period to be 2.5 or 5 years, Anderson then calculates:

	2.5 years			5 years		
	A	*B*	*C*	*A*	*B*	*C*
Quoted price	$91,200	$102,000	$124,200	$91,200	$102,000	$124,200
Power investment	7,410	3,960	2,418	7,410	3,960	2,418
Operating power cost	41,100	22,050	13,425	82,200	44,100	26,850
	$139,710	$128,010	$140,043	$180,810	$150,060	$153,468

Unit *B* is thus seen to be the most feasible for installation. Is the basis for Anderson's calculations [i.e., taking the sum of capital investment (unit and power) and operating power cost for the period of the payout time] a justifiable one? What other economic tests might be applied, and what is the result?

XI-2 A chemical plant requires a supply of 1,000 gal of purified process water hourly. Ion-exchange and distillation purification units have been proposed as alternatives to provide a satisfactory product. Accounting for the time value of money at 6 percent rate of interest, determine which of these units is the more economical, based upon a 10-year life, a 350-day year, and a 24-hr day for both. State clearly which method is being used to make the necessary decisions, justifying any assumptions.

Data: Distillation unit, installation costs $6,000, operating costs 30 cents/1,000 gal water produced; ion-exchange unit, installation costs $6 per gal/hr of feed rate, operating costs 15 cents/1,000 gal water treated, plus resin replacement at 30 cents/lb (1 lb resin used per

10^4 gal of treated water). After an hour's operation, the resin must be regenerated for half an hour before pure enough water can be obtained again.

XI-3 Two chemical manufacturing companies, X and Y, make common products from the same raw material. For the problems specified below, determine the most profitable production plans for X company

(*a*) If both companies are competitors and company X controls the market.

(*b*) If both companies are noncompeting members of an industrial group.

Both companies make the same main product α and the same by-product β, company X using raw materials A and C, while company Y uses raw materials A and B. The specific technological and economic information for each company may be taken as follows:

	Company			
	X		*Y*	
Raw material	*A*	*C*	*A*	*B*
Yield of main product α, ton/ton raw material processed	0.6	0.2	0.8	0.4
Yield of by-product β, ton/ton raw material processed	0.4	0.8	0.2	0.6
Profit, \$/ton processed	18	8	22	15

The production in these companies is limited by the potential product sales. Both companies compete in selling their main product α in the same sales area, but due to high transportation charges, the secondary product, β, must be sold in the vicinity of the production unit, and there is therefore no competition in the latter market. The total monthly demand for product α is 180 tons in all areas, while 200 tons of β may be sold by company X and 100 tons of β by company Y.

The suppliers of the raw materials can provide up to 160 tons of A, 80 tons of B (available to company Y only), and 180 tons of C (available to company X only) each month.

XI-4 A homogeneous liquid-phase reaction $A \xrightarrow{k_1} B \xrightarrow{k_2} C$, which is first-order and irreversible, is to be carried out in a single "ideal" stirred tank reactor operated under steady-state flow conditions. The reaction constants k_i are given by $k_i = k_i^0{}^{-E_i/RT}$. Here the system temperature T is restricted to the range $T_L \leq T \leq T^L$, the limits being physically imposed by the available temperature of cooling water and the decrease of material physical strength and corrosion resistance at higher temperatures. The reactor comes in standard sizes, each of which is fitted with standard-size cooling coils.

The feed to the system is pure A, and it has been suggested that the yield of intermediate product B should be maximized, the final product C being waste and undesirable. Making suitable assumptions, construct a model for this system, and discuss whether the maximization of yield is a valid objective in the specification of the operating temperature, the reactor volume, and the heat duty. How sensitive will the optimal solution be to fluctuations in the main-reactant flow rate and the supply of heating or cooling agent?

XI-5 A continuous stirred tank reactor system consists of N tanks in series. A continuous liquid flow of L ft³/hr passes through all the tanks in sequence. The feed to the first tank in the series contains no component B, but has a concentration of component A of C_0 moles/ft³. In each tank an irreversible first-order reaction $A \xrightarrow{k} B$ occurs, the rate of reaction of A being given by

$$-\frac{dC_n}{dt} = KC_n \qquad \text{moles/hr} \cdot \text{ft}^3$$

Here K is the reaction rate constant, the same for all tanks, and v_n is the volume of the nth tank. Each tank may be assumed to be well stirred. Show that the highest concentration of component B will be produced by the tank system in which all the tanks are of equal size, that is, $v_1 = v_2 = \cdots = v_N$, provided that the total volume of the system remains constant, so that

$$\sum_{i=1}^{N} v_n = V$$

XI-6 A cylindrical vessel closed at both ends is made of steel plate and is to be lagged to cut down the heat loss. Given the cost of the steel plate, the cost of the insulation, and the cost of a heat loss from the system, determine an expression which will specify (*a*) the dimensions for a fixed volume of vessel and (*b*) the lagging thickness which will minimize the total cost (Amundson et al. [1960]).

XI-7 A company owns three plants and sells to four main customers. Use the following production and purchasing data and costs of freight between various plants and customers to minimize company transportation charges.

Plant	*Production, tons/day*
A	40
B	100
C	120

Customers	*Customers' requirement, tons/day*
a	50
b	60
c	45
d	105

TRANSPORTATION COSTS, $/TON

Plant	*A*	*A*	*A*	*A*	*B*	*B*	*B*	*B*	*C*	*C*	*C*	*C*
Customer	*a*	*b*	*c*	*d*	*a*	*b*	*c*	*d*	*a*	*b*	*c*	*d*
Cost	4	9	10	6	10	7	18	18	9	21	14	20

Does this model represent a realistic system? Suggest improvements and show the effect of these on the optimal solution.

XI-8 A company produces three products, A, B, and C, to satisfy a known fixed demand D_i (units of product i per month), each product being produced in batches of size N_i (units of i per batch). It is desired to minimize the cost of manufacture and storage, which depends upon the costs of storage, processing, labor, and raw materials.

If the processing costs are P_i per batch of i, the raw material and labor charges are R_i per unit of i, and the cost of storage is S times the value of the average inventory (where the value of unit inventory can be taken as the manufacturing cost of that unit, and the inventory varies linearly from a maximum to zero between the times of successive batches entering the store), show that the monthly manufacturing and storage costs M are given by

$$M = \sum_{i=A,B,C} \left(\frac{P_i}{N_i} + R_i\right)\left(D_i + \frac{SN_i}{2}\right)$$

the fixed charges being omitted.

Determine the optimal operating conditions if the total available storage space is restricted to 20,000 ft³, when the demand, cost, and storage-volume (V_i the volume of a unit of i) parameters are given as follows:

		Costs			
Product i	*Product demand* D_i, *units*	*Processing* P_i, *\$/batch*	*Raw material* R_i, *\$/unit*	*Storage* S, *\$/\$ unit value*	*Storage volume* V_i, *ft³/unit*
A	100	40	1	5	250
B	300	10	3	5	100
C	150	30	2	5	400

Indicate the underlying approximations present in this model's representation of the practical system, and show how these could lead to an erroneous solution. Show how this might be avoided.

XI-9 In a given oil refinery it is desired to find the most suitable allocation of products from a specified crude so as to achieve a minimum production cost y. The minimum specifications of each product, the production rate of each, and the necessary technical data on the capacity and performance of the various refinery units have been written into the problem defined below (Deam [1959]).

Minimize the production cost y, where

$$y = 16.50x_1 + 5.25x_3 + 5.25x_4 + 3.00x_{13}$$

subject to the restrictions

(*a*) Constraint upon the upper level of the crude distillation plant,

$$x_1 \leq 750{,}000$$

(*b*) Disposal of components (mass balances from the technical design basis),

$$0.178x_1 - x_2 - x_3 = 0$$
$$0.048x_1 - x_4 - x_5 - x_6 = 0$$
$$0.069x_1 - x_7 - x_8 - x_9 = 0$$
$$0.184x_1 - x_{10} - x_{11} - x_{12} = 0$$
$$0.241x_1 - x_{13} - x_{14} = 0$$
$$0.266x_1 - x_{15} - x_{16} = 0$$

(*c*) Production requirements,

Motor spirit: $x_2 + 0.865x_3 + 0.85x_4 + 0.373x_{13} = 124{,}400$

Gas oil: $x_5 + x_7 + x_{10} = 18{,}800$

Diesel oil: $x_8 + x_{11} + x_{16} = 90{,}700$

Fuel oil: $x_6 + x_9 + x_{12} + 0.331x_{13} + x_{14} + x_{15} = 291{,}600$

(*d*) Quality restrictions,
Motor spirit (minimum octane number of 85):

$$15x_2 - 6.75x_3 - 8.5x_4 - 5.97x_{13} \leq 0$$

Diesel-oil quality (residiuum 90:10):

$$10x_8 + 10x_{11} - 90x_{16} \leq 0$$

Fuel oil (minimum viscosity 55.1):

$$11.9x_{15} - 25.7x_6 - 25.7x_9 - 16.1x_{12} - 4.4x_{13} - 5.1x_{14} \leq 0$$

(The minimum cost is \$9,593,755.2, with $x_1 = 552{,}120.0$ tons of crude being processed.)

XI-10 A fertilizer company purchases nitrates, phosphates, and potash as raw materials and mixes them in different proportions to form three compound fertilizers, which it then packs and distributes.

The percentages by weight of the nitrates, phosphates, and potash, respectively, in the three fertilizers are 5-10-5, 5-10-10, and 10-10-10, the remainder in each case being made up from an inert chalky base.

At the beginning of a certain month, it is found that no more than 1,000 tons of nitrate, 1,800 tons of phosphate, and 1,200 tons of potash will be available during the month. The firm is committed to supplying 6,000 tons of fertilizer 5-10-5 to various customers before the end of the month, but it is otherwise free to manufacture the three compounds in any quantities it pleases. Given the cost information below, calculate what quantity of each fertilizer should be produced during the month in order to maximize the company's profit.

Cost of mixing, packaging, and distribution is the same for all three products, at \$15 per ton.

COSTS OF RAW MATERIALS PER TON

Nitrates	\$160
Phosphates	40
Potash	100
Inert chalk base	5

SELLING PRICES PER TON OF FERTILIZERS SOLD

Compound	
5–10–5	\$40
5–10–10	50
10–10–10	60

(Use the Simplex method of solution.)

What is the effect on the solution if it is found that the raw-material costs are liable to a ±10 percent fluctuation and the selling prices to a ±5 percent variation?

XI-11 In a chemical plant four raw materials are available, *A*, *B*, *C*, *D*, and five processes are available to produce four products *E*, *F*, *G*, *H*. The raw materials are in limited supply, the processing costs are fixed but vary from process to process, and the selling prices of the products are fixed.

Raw materials	*Max available supply, lb/day*	*Cost/lb*
A	400	\$1.50
B	300	2.00
C	100	4.50
D	250	2.50

The data on the processes are as follows:

Process	*Pound of raw material used*				*Pound of product*	*Selling price of product per pound*
	A	*B*	*C*	*D*		
1	2	1	0	0	3 (*E*)	\$3.00
2	2	1	0	0	3 (*F*)	2.33
3	3	1	0	2	6 (*G*)	3.75
4	2	7	3	3	15 (*H*)	5.00
5	2	7	3	3	15 (*H*)	5.00

Processes 4 and 5 differ only in the processing cost. Process 4 has a maximum capacity of 75 lb of *H* per day. If more than 75 lb of *H* is produced, the additional amount of *H* must be produced in process 5. The processes are as follows:

		Reaction cost per pound
Process 1	$A + B \rightarrow E$	\$1.50 (*A*)
Process 2	$A + B \rightarrow F$	0.50 (*A*)
Process 3	$A + B \rightarrow E$	1.50 (*A*)
	$E + A + D \rightarrow G$	1.00 (*G*)
Process 4		
Max capacity 75 lb *H*/day	$A + B \rightarrow F$	0.50 (*A*)
	$F + B + C + D \rightarrow H$	2.00 (*H*)
Process 5		
Excess capacity *H*	$A + B \rightarrow F$	0.50 (*A*)
	$F + B + C + D \rightarrow H$	2.20 (*H*)

Derive a mathematical model for the problem. *Hint:* Define x_j for $j = 1, 2, 3, 4, 5$ as the net profit from the jth process per day. Then the profit function is

$$P = \sum_j x_j$$

Derive the restrictions on the problem. How should the processes be operated for maximum profit? Solve by the Simplex method (Fenech and Acrivos [1956]). (Solution gives a maximum possible profit of $P = \$307.59$.)

XI-12 An oil company has two refineries, R_1 and R_2, from which a given product is distributed to customers C_1, C_2, C_3, and C_4. The relative transportation costs are given as follows:

Refinery	R_1	R_1	R_1	R_1	R_2	R_2	R_2	R_2
Customer	C_1	C_2	C_3	C_4	C_1	C_2	C_3	C_4
Relative costs, \$($\times$ 10^6 tons)	20	55	70	45	15	45	80	55

The refinery production rates and customer demand rates are fixed and known to be as follows:

Refinery/customer	R_1	R_2	C_1	C_2	C_3	C_4
Rate (10^6 tons)	1.3	0.5	0.7	0.5	0.2	0.4

Determine the optimum distribution policy to minimize the company's freight bill.

Using this linear first approximation, it is convenient to let $x_{j,i}$ be the number of millions of tons shipped from R_j to C_i.

Discuss a more realistic model and specify what additional information would be required for its use in optimization.

CHAPTER XII

SYSTEMS WITH PARTIAL KNOWLEDGE OF SOME VARIABLES

1 INTRODUCTION

Practical systems are rarely encountered in which the objective function can be expressed in terms of independent variables, all of which are completely specifiable in exact or stochastic terms. It is commonly found that one or more variables can take values within a range, but the actual values cannot be predicted in advance. We have referred to these variables in Chap. II, Sec. 6.1.1, calling them external, or uncontrollable, since in any optimization study they are out of our jurisdiction. Examples include the effect of weather; of government or competitors' policies; of trade-union wage demands; of restrictive practices; of trends in market demand; and of variations in prices and specifications of future feedstocks for projected processes. An illustration of one approach to the problem is given in Example XI-3.

The uncontrollable nature of such variables will be obvious, although limits can probably be placed upon the range of variation. Such problems as the level of government taxation in the next fiscal year might affect future capital investment in new projects, while the size of the project will be largely determined by sales forecasts produced by market research, and the type of equipment to be installed may be affected by the minor components or impurities in any material fed to the process. The resultant optimally designed and operated plant can scarcely be as profitable if the sales demand, for example, fails to meet, or perhaps even exceeds, the sales estimates by any significant factor. These variables can be conveniently treated, depending upon whether they are hostile or neutral in effect.

Since the system structure and data will be known, we can express the objective function y in the form

$$y = y(\mathbf{x},\mathbf{x}')$$

where $\mathbf{x}$ represents the internally controlled variables, and $\mathbf{x}'$ the external, or uncontrollable, variables. In treating this type of problem it is difficult to do other than consider representative sets of possible actions resulting from sets of values for each uncontrollable variable, x_i'. The number of sets required will vary. Thus, for example, a single variable might represent the unknown factor of whether or not the government will enter a trade agreement with another nation, removing all tariff barriers next year, so that we need consider only the two cases of "yes" and "no." On the other hand, if an infinite number of choices are available, as in the influence on profits of changes in the average annual rate of a nation's economic growth in the next decade, we could take representative values corresponding to the probable range 1, 2, 3, . . . percent. For any given set of uncontrollable variable values we cannot, in the general case, optimize the objective function with respect to the internal variables x_i. As we shall see in discussing the decomposition of systems (Chap. XIII), we can optimize with respect to specific variables only when the others are fixed or temporarily fixed, the latter being still within our control. In the present situation we do not have control of these external variables, and we cannot thus expect to obtain an optimum result by optimizing the system with respect to the internal variables x_i, for a given set of uncontrollable x_i'. This point will be illustrated below. For this reason it is usual not to consider a continuous range of x_i, but instead to choose a discrete set of representative values. The objective function can then be evaluated for each combination of values, its value

$$y_{i,j} = y(x_i, x_j')$$

being conveniently represented in a matrix, such as shown in Table XII-1 for a single controlled and a single uncontrolled variable. This matrix, called a *payoff* matrix in operations research literature, can of course be extended to higher dimensions.

Let us consider the specific results presented in Table XII-2 to show that we cannot optimize freely for a given value of the uncontrollable variable. Such an optimization, maximization for definiteness, would produce, by suitable choices of x_i, the column maxima shown in the last row of the table. It will be seen that if we specify a value x_2 for our system, this will certainly be the optimal value if the unknown, uncontrolled variable has a value of x_2' or x_3'. The best possible decision would result from a choice of x_3, but would

TABLE XII-1 AN OBJECTIVE FUNCTION MATRIX IN TWO DIMENSIONS

$y_{i,j} = y(x_i, x_j')$		Uncontrollable variable x_j'				
		x_1'	x_2'	x_3'	. . .	x_J'
	x_1	$y_{1,1}$	$y_{1,2}$	$y_{1,3}$	. . .	$y_{1,J}$
Controllable	x_2	$y_{2,1}$	$y_{2,2}$	$y_{2,3}$	. . .	$y_{2,J}$
variable x_i	x_3	$y_{3,1}$	$y_{3,2}$	$y_{3,3}$	. . .	$y_{3,J}$
	. . .	. . .	. . .	. . .	. . .	. . .
	x_I	$y_{I,1}$	$y_{I,2}$	$y_{I,3}$	. . .	$y_{I,J}$

TABLE XII-2 A PARTICULAR ILLUSTRATION

$\mathbf{x}_i \equiv x_i \qquad \mathbf{x}_j \equiv x_j'$

$y(x_i, x_j')$		Uncontrollable variable x_j' x_1'	x_2'	x_3'	x_4'	Row minimum
	x_1	11	4	6	1	1
Controlled	x_2	3	5	10	8	3†
variable x_i	x_3	−2	1	5	12	−2
Column maximum		11	5‡	10	12	

† Row maxi-min.
‡ Column mini-max.

be achieved only if x_4' were available, and in fact this design choice could end up with the worst solution if the uncontrolled value turned out to be x_1'. The conservative approach is not to attempt a maximization, since the results are so greatly affected by the unknown variable value, but rather to minimize our losses. This has been done in the last column of Table XII-2, where the row minima are listed, and where the maximum of these values is chosen as the optimal policy. This ensures that, irrespective of the uncontrolled actions, we shall achieve a value of 3 for the objective function, and if the uncontrolled value is any other than x_1', the result will be more favorable. It will thus be seen that the conventional optimization techniques cannot be directly applied; and we certainly cannot optimize the problem for specific levels of the uncontrollable variables.

Two main approaches are used in this type of problem—game theory and decision theory—the choice between them depending mainly upon the nature of the uncontrolled variables. If these are determined by the action of a competitor—a cunning but malevolent intelligence determined to thwart our realization of a particularly favorable objective—we should use game theory (Sec. 2). If, on the other hand, these uncontrollable variables are determined by a neutral agency which is not hostile but neutral—customers, weather, government, etc.—this situation can be handled by decision theory (Sec. 3).

2 UNCERTAINTY AND HOSTILE COMPETITION—GAME THEORY

When the objective function depends on a set of controlled as well as uncontrolled variables, then, if the latter are controlled by the strategy of some competitor, we can solve the resulting optimization problem by game theory. We can no longer, however, ensure that the best is obtained from our system; the problem is to reduce the worst happening to its minimal level. A variety of strategies to handle this type of problem have been presented or reviewed by, for example, von Neumann and Morgenstern [1953], McKinsey [1952], Luce and Raiffa [1957], Williams [1965a], Vajda [1960], and Baumol [1961, chap. 18]. Let us now illustrate the underlying principles by two examples.

Example XII-1: Game theory—a two-person zero-sum saddle-point game

Let us consider the situation when a competitor B can choose a strategy x'_j, which is unknown when company A makes its decisions x_i, the objective function to be maximized being $y_{i,j} = y(x_i,x'_j)$. It is assumed that representative values of B's actions might be x'_1, x'_2, x'_3, and x'_4, while a representative set of A's decisions might be x_1, x_2, and x_3. The resulting values of $y_{i,j}$ are given in Table XII-3, the profit for A being equal to the loss for B, as might be the case when both are competing for the same market. This is the two-person zero-sum game in game theory nomenclature.

If A picks x_3, which gives him a chance of obtaining his greatest profit of 12, B might discover this action and pick any other x'_j than x'_4, all of which produce worse results for A. In general, we can argue that if x_1 is chosen, B would prefer x'_4; if x_2 is chosen, B would take x'_2; and if x_3 is taken, B would take x'_1, B always maximizing his own return, which here is equivalent to minimizing A's results. From this it is seen that the worst A could expect from each of his decisions is 1, 5, -2, respectively. He would thus prefer the decision x_2, since this gives him his largest profit regardless of B's actions. This result is achieved by maximizing the possible row minima, as shown in the final column. On the other hand, B loses 11, 5, 10, and 12, respectively, if A were to make the best moves opposing B's actions. B's best policy is therefore to choose the decision x'_2, which minimizes A's best result; in other words, to choose the minimum of the column maxima as shown in the last row.

TABLE XII-3 A TWO-PERSON ZERO-SUM GAME

Objective function values $y_{i,j}$		*Representative strategy for B*				*A's decision, row minimum*
		x'_1	x'_2	x_3	x'_4	
Representative	x_1	11	4	6	1	1
decisions for A	x_2	6	5	10	8	5†
	x_3	−2	1	5	12	−2
B's decisions, column maximum		11	5‡	10	12	

† Maxi-min.
‡ Mini-max.

Our general approach is for A to maximize the row minima and for B to minimize the column maxima. In this problem both A and B will be satisfied by the same solution, $y(x_2,x_2') = 5$. This means that A will always get a profit of at least 5, but will make more if B makes a wrong decision. Similarly, B's losses will not exceed 5 and may be smaller if A errs. This solution is called a *mini-max solution* at the saddle point of the matrix.

This type of solution is not always found. The more general situation requiring the use of mixed strategies is now examined.

Example XII-2: Another type of game—mixed strategies

Let us now consider the data of Table XII-2, a two-person zero-sum game in which the type of solution found in Example XII-1 is not satisfactory to both sides. Here A's decisions in maximizing his results will be x_i, and B's actions x_j'. As before, we cannot optimize directly, but can only ensure that A's profits will lie above a certain minimum level.

Let us find the minimum of each row as before. From the final column we see that the greatest of these values, 3, is found with the decision x_2. This represents the lowest profit which A would receive, but it will be attainable only if A makes the decision x_2. Similarly, B will maximize his gains (equal to minimizing A's profits) by choosing the minimum, 5, of the column maxima, as given in the last row for x_2'. The two solutions no longer coincide, and if x_2' is chosen, A will obtain a better result than he had expected. There is now no saddle-point-type solution.

We can show that it is better not to use such an invariant strategy. We see that if A uses x_2, he is guaranteed a profit of at least 3, and B's best policy is then to use x_1', and A will get 3. However, if B uses x_1', it is better for A to use x_1, since he will then get a profit of 11. But then B's policy will be to use x_4', with a profit of only 1. A might counter with x_3 to achieve 12. Then B plays x_1', which limits A to -2, unless A changes to x_1, getting 11. We are now back where we started. This situation could not occur with a saddle point, since that is an equilibrium position from which neither competitor can profit by a change of decision if the opponent does not move. This problem can best be handled by the use of a mixed strategy. This, however, would be feasible only for repeated actions, the "one-shot" action[1] requiring a pure strategy.

Let us define a_i as the probabilities with which A should use the

[1] See note in Chap. II, Sec. 7.1, on "one-shot" actions in practice.

decisions x_i. Here $\sum_i a_i = 1$ and $a_i \geq 0$, with the a_i chosen so that the minimum expected profit is as large as possible. The expected profit for A can then be written for each action of B as follows:

B's action	*Expected profit for A*
x_1'	$11a_1 + 3a_2 - 2a_3 \geq P$
x_2'	$4a_1 + 5a_2 + a_3 \geq P$
x_3'	$6a_1 + 10a_2 + 5a_3 \geq P$
x_4'	$a_1 + 8a_2 + 12a_3 \geq P$

We have expressed this expected profit as being greater than or equal to P, the minimum value of these profits. While $a_2 = 1$, $a_1 = 0 = a_3$ is a feasible solution, yielding $P = 3$, we want to find the a_i which maximize P.

Similarly, we could examine B's situation, getting $b_j \geq 0$, $\sum_j b_j = 1$, and

$$11b_1 + 4b_2 + 6b_3 + b_4 \leq L$$

$$3b_1 + 5b_2 + 10b_3 + 8b_4 \leq L$$

$$-2b_1 + b_2 + 5b_3 + 12b_4 \leq L$$

where L is the highest possible expected loss which has to be minimized by a suitable choice of probabilities, b_j. It will be seen that the system for B is the dual to the problem for A (Chap. VII, Sec. 5).

These two problems may be separately solved, for example, by linear programming (see Probs. VII-44 and VII-45), getting

A: $a_1 = 0.2222$ $\quad a_2 = 0.7778$ $\quad a_3 = 0$ $\quad P = 4.778$

B: $b_1 = 0.1111$ $\quad b_2 = 0.8889$ $\quad b_3 = 0$ $\quad b_4 = 0$ $\quad L = 4.778$

With A and B in opposition, an equilibrium situation exists in the form of an equilibrium of mixed strategies, equivalent to the saddle-point equilibrium described in Example XII-1.

This result represents the mini-max theorem of von Neumann and Morgenstein [1953], which states that A's expected gain is at least P, and B's expected loss is at the worst L, where $P = L$. If A plays his optimal strategy but B is not using the best method, then A may get more than P, while similarly, B can lose less than L. The value of P (and L) lies between the row maxi-min and the column mini-max.

When there are more than two participants or when the game is not zero-sum, the game theory technique is more complicated. While "bribes" and "threats" (e.g., strikes) can be taken into account, it might also be found profitable for a group of participants to form a cartel, which results in a new game with fewer competitors and a smaller matrix. Even with two competitors, some form of working rules is usually established between them to mitigate competition, although this is feasible only in non-zero-sum games—the normal situation, however. The game theory literature is divided into those non-constant-sum games in which there is cooperation and those with no cooperation. Baumol [1961, chap. 18] presents a good introduction to this topic.

3 UNCERTAINTY AND NONHOSTILE FACTORS—DECISION THEORY

In the game theory approach, at least in the two-person zero-sum situation, there is a major element of predictability in the behavior of a hostile opponent, since he will do everything to minimize one's own profit. On the other hand, when the second participant is not such an opponent but acts in a neutral manner, like nature affecting all operations to the same degree, then nothing will be known about his actions. The subject of decision theory has been built up mainly covering this so-called "complete-ignorance" situation, when no clues are available as to the possible levels of the uncontrolled variables. A clear introductory exposition of the essential ideas is given by Baumol [1961, chap. 19]. Other references include Luce and Raiffa [1957], Ackoff [1962], Schlaifer [1961], and Raiffa and Schlaifer [1961], the last being the most detailed in mathematical presentation.

There is no universally valid rule for choosing a suitable strategy, and the appropriate objective function might vary with the user and the situation, just as attitudes to gambling and financial circumstances will also vary from person to person or company to company.

Most of the literature discusses the so-called complete-ignorance situation in which absolutely no clues are available to the actions of the uncontrollable element. The decision theory analyses do require the problem to be described in terms of objective function, or payoff, matrices, which list the objective values resulting from sets of specified strategies for both oneself and the uncontrollable agencies. This is not always possible. The actual objectives take the usual forms, such as costs or profits, the difficulty in this section being how to handle them. Should we maximize a profit, reduce our possible losses to their lowest level, or take some other decision? Some approaches which are commonly available in decision theory are now discussed.

The most conservative decision rule is that of the maxi-min criterion. As in game theory, the worst that can happen is found, and the "least worst" strategy is chosen. In this case, however, the maxi-min strategy is not so attractive, since the uncontrollable agency is not directly opposed to our decisions, but is rather neutral in effect.

Example XII-3: A maxi-min criterion

Consider the information given in Table XII-4. If we use strategy x_1, the worst that can happen is 2, while for decision x_2, the worst will be 1. Thus the most conservative choice, which might on occasion be appropriate from the point of view of self-preservation, is to take the maximum value of the worst, the maximum value of 2.

This criterion considers only the worst that can happen, and does not consider the other possibilities. Since the two worst cases are similar in this illustration, very little would be lost if the wrong one were chosen, but a great deal of difference would result if the uncontrollable agent happened to choose strategy x_2' when it would be seen that the second strategy, x_2, would have been our better choice.

Baumol [1961, chap. 19] has suggested the maxi-max criterion, lying at the other end of the scale from the maxi-min, as another possibility. This considers the most attractive rewards available and ignores the dangers inherent in neglecting the other objective values. It is suited to the gambling instinct. Thus, in Table XII-4, strategy x_1 would be chosen, since the maximum in this row is greater than the maximum of any other row. As indicated in this example, the gambling-orientated maxi-max rule can lead to the same result as the conservative maxi-min criterion, and like the latter, it also ignores all intermediate choices, even if these show marked advantages with one strategy, as in this situation.

As a compromise between these two criteria, Hurwicz [1951] has proposed that a weighted average of the minimum and maximum objective function values be used as a decision criterion. Thus, if in Table XII-4, the conservative minimum is weighted at $\alpha = 3/4$ and the gambling maximum at $1/4$,

TABLE XII-4 OBJECTIVE FUNCTION MATRIX

	Uncontrollable agency—B's strategies			
	x_1'	x_2'	x_3'	*Minimum for A*
A's strategy x_1	67	10	2	2 *maxi-min*
x_2	65	58	1	1

then the Hurwicz α criterion would evaluate the x_1 strategy as

$$2(3/4) + 67(1/4) = 20.5$$

and for the x_2 strategy as

$$1(3/4) + 65(1/4) = 17.0$$

so that we should again select the first strategy. This is to be expected for this example, since we found the same solution with both previous criteria. In general, however, this Hurwicz criterion reflects the player's psychology by a suitable choice of weightings, while neglecting the intermediate values.

Another approach in handling neutral-type uncontrollable variables is to attach probabilities of occurrence to the various possible strategies of the uncontrollable agency. When we have absolutely no information about the relative probabilities of these strategies taking place, we can assign equal probabilities to them and adopt as our strategy that which produces the greatest expected objective value. This is the basis of the Bayes criterion. Thus, in Table XII-4, we could assign probabilities of 1/3 to each strategy x_i', getting for x_1

$$1/3(67) + 1/3(10) + 1/3(2) = 26.33$$

and for x_2

$$1/3(65) + 1/3(58) + 1/3(1) = 41.33$$

so that the second strategy would be chosen on this basis. The Bayes criterion is the only one so far considered which takes all the information into account. The main difficulty in its use is the problem involved in deciding in advance the groups of possibilities having equal probability. Schlaifer [1961] has considered these points in detail. Another variant is the assignation of estimated probabilities to each of the uncontrollable strategies, based upon past experience. Thus, in a given problem, it might be thought that an industrial process will function with 95 percent probability and 5 percent that it fails. Using such probability weightings, the weighted average value of the objective function can be found for all the results possible for a given decision. In both methods it will be seen that the averaging is over the uncontrolled values, and not over the self-determined variables, as in the mixed strategy of game theory.

Yet another criterion, the mini-max regret, has been suggested by Savage [1951]. This concentrates upon the cost of a lost opportunity as the result of an incorrect decision. This approach protects one from the excessive cost of mistakes. The first step is the formation of a new payoff matrix of regrets to replace the original matrix of objective function values. Taking each uncontrollable strategy in turn, each entry in the matrix is replaced by the regret which equals the loss entailed by that decision, as opposed to the

TABLE XII-5 REGRET MATRIX CORRESPONDING TO TABLE XII-4

		B's strategies x_1'	x_2'	x_3'	*Maximum for A*
A's	x_1	0	48	0	48
strategy	x_2	2	0	1	2 mini-max

best which could be obtained under that uncontrollable strategy value. Thus, in the x_1' column of Table XII-5, the best result is that found with our decision x_1, so that the regret is 0. On the other hand, the choice of x_2 would result in a loss of $67 - 65 = 2$, a regret of 2. The full regret matrix corresponding to Table XII-4 is given in Table XII-5. To protect ourselves from excessive losses, we can use the conservative mini-max criterion, getting a minimum value of the maximum regret of 2 resulting from the x_2 strategy. This method does not suffer all the disadvantages of the first three methods, since the intermediate values are taken into account, but only to the extent of considering the extreme value of the row regrets. Since it is not clear that the difference between two objective values, the regret, is a good measure of our loss in receiving the smaller of the two, this criterion is somewhat arbitrary, but gives an indication of losses.

Just as in game theory (Sec. 2) we need not use a pure strategy, but a combination of strategies, attaching a probability of occurrence to each. Thus we might use each of our strategies in Table XII-4 with 50 percent probability—a mixed strategy representing a 50 percent chance of either x_1 or x_2 being used, so that we get

$$1/2(67) + 1/2(65),\ 1/2(10) + 1/2(58),\ 1/2(2) + 1/2(1)$$

as the expected objective values resulting from each of the uncontrollable actions. Using a maxi-min criterion, we could then change the probabilities so as to get an optimal mixed strategy. A mixed strategy results in a higher security, since it considers all the objective values.

Having chosen a suitable criterion, a second step in the application of statistical decision theory is based upon the recognition that, by sampling or other means, the estimate of the likely uncontrollable-variable values might be improved, and in this way the reliability of any decision can be improved. A sampling procedure is, however, expensive and involves some degree of uncertainty. A complete statistical decision theory technique thus requires a much more detailed specification than given by the payoff matrix, since we require not only the objective values for each possible uncontrollable action, but also additional information from sampling experiments. See, for example, Raiffa and Schlaifer [1961].

Problems

XII-1 Solve the zero-sum game represented by the matrix

		B's actions			
		x'_1	x'_2	x'_3	x'_4
	x_1	10	24	6	14
A's	x_2	15	21	27	18
actions	x_3	12	4	22	4
	x_4	9	11	26	19

if A wishes to maximize his gains represented by the matrix elements and B is playing in opposition.

XII-2 Solve the zero-sum game represented by

		B's actions				
		x'_1	x'_2	x'_3	x'_4	x'_5
A's	x_1	35	14	26	1	7
actions	x_2	34	17	21	30	19

A wishing to maximize and B minimize the payoff from the game.

XII-3 A zero-sum game is represented by the matrix

		B's actions				
		x'_1	x'_2	x'_3	x'_4	x'_5
A's	x_1	5	25	26	1	8
actions	x_2	38	14	9	30	17

What pure strategy should be adopted by A in playing the game once and wishing to maximize his profit represented by the matrix elements? B will try to reduce A's gains wherever possible.

XII-4 What mixed strategy could be adopted in Prob. XII-3 if the game is played on many occasions?

XII-5 The profits accruing to A from a particular zero-sum game are given by

		B's actions		
		x'_1	x'_2	x'_3
A's	x_1	45	35	72
	x_2	50	70	16
actions	x_3	27	62	39

Write down the equivalent linear programming formulation and determine the mixed strategy which should be adopted by A.

XII-6 Write down the payoff matrix for the two-person game in which each person selects an integer from the set $\{1,2,3\}$. The player with the lower number wins two points, unless his number is only one less than his opponent, when he will instead lose four points. When the numbers are identical, there is zero payoff.

XII-7 Write down the equivalent linear programming formulation for Prob. XII-6, and hence find the best strategy (Gass [1964, p. 255]).

XII-8 Management must decide what size of production capacity must be chosen for future development, but they are faced with large uncertainties concerning the rate of market penetration. Feasible forecasts can be made for high, medium, and low values of the nonhostile uncontrollable penetration rates, as given in the table below, in the form of cumulative earnings over a 10-year period. On this basis find the most suitable plant size if a maxi-min criterion is used (Andersen [1961]).

Alternative decisions as to plant size, lb/year	*Feasible forecasts of market-penetration rates* x'_1 (high)	x'_2 (medium)	x'_3 (low)
5×10^6 x_1	50	32	17
10×10^6 x_2	70	48	12
15×10^6 x_3	86	41	9

XII-9 Which capacity should be chosen in Prob. XII-8 if a gambling attitude is adopted, a maxi-max criterion being used in the selection?

XII-10 Find the solution to Prob. XII-8 if a Bayesian criterion is used.

XII-11 Use the method of regrets to obtain the solution to Prob. XII-8.

XII-12 Is there any advantage in using a mixed strategy in Prob. XII-8?

XII-13 Andersen and Haight [1961] have presented a problem involving two decision variables, x_1 and x_2, and two neutral-type uncontrollable variables, x'_1 and x'_2. The former might be the pricing policy and the plant size, respectively, and the latter, the levels of optimism about the size of the market and the rate of sales penetration, on one hand, and the duration of the demand, on the other. Using three representative values for each of the four variables, the resulting values of an appropriate profit objective are given below. Use the maxi-min criterion to determine the best solution.

Uncontrolled variable x'_1	*Uncontrolled variable* x'_2: *Low*			*Medium*			*High*		
Smallest	2	3	5	−2	−2	1	−3	0	−3
	3	2	5	−2	−2	0	−3	−2	−5
	3	2	4	−4	−2	−1	−4	−3	−6
Small	10	11	12	5	6	7	6	7	6
	10	11	13	5	6	8	6	7	5
	10	10	13	5	5	7	6	6	4
Large	17	17	17	10	11	12	11	11	13
	17	18	19	11	12	13	12	12	11
	17	18	19	11	12	13	12	11	9
Largest	22	22	25	16	17	19	17	17	19
	22	23	24	17	17	19	18	18	19
	23	25	24	17	19	18	18	18	16

Note: Each subarray represents the results of varying the controllable variables:

Controlled variable x_2 ↑ ⟶ Controlled variable x_1

XII-14 Find the solution to Prob. XII-13 using a Bayesian criterion.

XII-15 Use the method of regrets to obtain a solution to Prob. XII-13.

CHAPTER XIII

THE APPROACH TO OPTIMIZATION OF COMPLEX SYSTEMS

1 INTRODUCTION

Almost any optimization problem can be solved, at least in principle, by one or more of the methods described in Part Two. As already indicated in Chap. XI, however, computational difficulties can arise when the dimensionality of the system becomes large. This chapter discusses the methods by which the overall optimization problem can be reduced to a set of subproblems, each of which is of smaller dimensionality than the original problem. In this manner complex systems can be broken down into separate smallcr systems which can be more easily handled.

Methods which reduce the dimensionality by setting up subproblems, each of lower dimensionality, are called *decomposition techniques*. Such methods do not carry out the optimization itself, but provide a framework within which any suitable optimization method, such as described in Part Two, can be applied. Several decomposition methods have been proposed, but these are far from being generally applicable. Certainly, no method has yet been developed for handling the general complex system, but due to the importance of this facet of optimization application in practical systems, it is still the subject of active investigation.

At first thought, the problem of optimizing large, possibly complex systems may seem relatively simple. The total system can certainly be quite easily partitioned into a set of subsystems, each containing only a smaller number of variables. However, it would be misleading to optimize an arbitrarily defined subproblem in isolation from the main system. We must take into account the interaction between the various subsystems and ensure that any subproblems optimized separately are so chosen that decisions taken within them do not affect other portions of the total system. Let us look at two examples of an erroneous approach.

Example XIII-1: An erroneous approach (1)

Let us consider the situation in which several processes exist within a given company, each process being separately managed but contributing to the gross profits of the company. If each process can operate independently of the other units, with no actions affecting the other units, then each process can be treated as a subsystem whose profit can be separately optimized. Let us, however, examine the case in which the process products are all derived from the same basic raw material, which is in short supply. Problem XI-3 has, in particular, shown for two processes that optimizing each process subsystem separately results in process X making a profit of $4,240 if it has control over the distribution of raw material usage, while process Y, using the remaining material, can make a profit of $1,200, a total profit to the company of $5,440. This result is not, however, optimal for the company, even though it may be so for one of the processes. We have made a mistake in considering one process separately, since the profits from the two processes are in fact linked through the limited availability of raw material.

If an integrated production plan for both processes is used, as discussed in Prob. XI-3, it can be shown that the optimal company profit is, in fact, $5,840, with $2,880 coming from process X and $2,960 from Y. It will be seen that process X no longer makes its optimal profit, but by suffering an apparent loss, it allows process Y to increase its profits, yielding the best results on the company level.

Example XIII-2: An erroneous approach (2)

Let us consider the sequential system depicted in Fig. VII-9 and discussed in Example VII-15. Here the objective function is given by

$$y = \sum_{i=1}^{N} y_i$$

where

$$y_i = p_{i-1} - p_i - Bq_i \tag{1-1}$$

and the transformation equation for each stage in the sequence is

$$p_{i-1} = p_i(1 + m'q_i) \tag{1-2}$$

It is desired to maximize y when the feed to the N-stage system is fixed at p_0.

If each stage is taken as a subproblem which is optimized separately for a given feed p_{i-1} to that stage, then, when p_i is eliminated using Eq. (1-2), we find that Eq. (1-1) gives

$$y_i = p_{i-1} - \frac{p_{i-1}}{1 + m'q_i} - Bq_i$$

and we should get the stationary point defined by

$$\frac{dy_i}{dq_i} = 0 = \frac{m'p_{i-1}}{(1 + m'q_i)^2} - B$$

The apparent optimal conditions are then

$$q_i^* = \frac{1}{m'}\left[\left(\frac{m'p_{i-1}}{B}\right)^{1/2} - 1\right]$$

$$p_i^* = \frac{p_{i-1}}{[(m'p_{i-1})/B]^{1/2}} = \left(\frac{Bp_{i-1}}{m'}\right)^{1/2}$$

Using these recurrence relations and working through the N stages, we should find the following apparent optimal system profit for a feed p_0:

$$y^* = \sum_{i=1}^{N} y_i^* = \frac{B}{m'}\left[\left(\frac{m'p_0}{B}\right) - \left(\frac{m'p_0}{B}\right)^{1/2} - \left(\frac{m'p_0}{B}\right)^{1/4} - \cdots - \left(\frac{m'p_0}{B}\right)^{(1/2)^{N-1}} - 2\left(\frac{m'p_0}{B}\right)^{(1/2)^N} + N\right]$$

This is not, however, the optimal result, since, as shown by Example VII-15, we found the better value of

$$y^* = \frac{B}{m'}\left[\left(\frac{m'p_0}{B}\right) - (1 + N)\left(\frac{m'p_0}{B}\right)^{1/1+N} + N\right]$$

when optimizing the whole system directly. The incorrect result is found here by erroneously considering each stage as a subproblem to be separately optimized. It is obvious that any decision q_i taken at the ith stage will affect all the succeeding stages because it determines p_i, the feed to these remaining stages. The correct decomposition approach to this problem is the subject of Example XIII-6.

In both of these illustrations, arbitrarily chosen subsystems (here based on a process unit or a stage) were optimized separately. Since these subsystems are in fact linked, decisions made in one subsystem affected the possible profit in the linked sections. This type of interaction is very common, and must be taken into account if the global optimum for a complete system is to be attained. Let us now consider the proper approach to choosing a set of suboptimization problems.

Essentially, we may use two basic principles in choosing subproblems from a complex system. These two principles are so simple as to be obviously true:

1. A subsystem operating between specified interconnecting links (fixed links) with other subsystems must be operated optimally. This we may call the *principle of splitting*, its use being discussed in Sec. 2.

2. Whatever the course in the initial stages of a sequential system, the remaining stages must be conducted in an optimal manner with respect to the feed to those stages. This is the *principle of optimality for sequential systems*, discussed in Sec. 3. Its mathematical formulation, called *dynamic programming*, is described in Sec. 4.

Further applications of both principles to various complex systems are presented in Sec. 5.

2 SPLITTING

One approach to the handling of complex systems is the splitting of the main system into subsystems. The operation of each such subsystem may then be carried out freely as long as the split is made at a point where the interlinking streams are fixed in value. Certainly, any subsystem operating between such fixed links must be operated optimally. This principle will help us to reduce the labor of computation in many cases by dividing a complex system into a set of easier subproblems which can be optimized separately.

Let us now examine the significance of this splitting procedure, with the primary intention of recognizing its application potential. Its fuller application will be deferred to Sec. 5. Our present discussion will first consider those links which are truly fixed (Sec. 2.1), and second, those which are only temporarily fixed (Sec. 2.2).

2.1 Fixed Links

In general, the optimum result from a system will not equal the sum of the optima determined from each of the subsystems comprising the total system. The fallacy is clearly exposed in Examples XIII-1 and XIII-2. A subsystem can, however, be optimized separately if it is linked to other parts of the main system by invariant, or fixed, links. The system optimum can then be found as the sum of such local optima as long as the subsystems are correctly chosen. Thus, for example, in the design of an integrated ammonia–nitric acid–fertilizer manufacturing complex, we may consider the optimization of the ammonia–nitric acid production section independently of the fertilizer manufacturing facility only if the interconnecting link—the quality and quantity of the nitric acid feed to the fertilizer section—is held constant during the study. Any changes then made during optimization will not affect this link or units, other than the unit under examination. On the other hand, if a fixed link were not specified, we could not freely carry out this separate optimization, since changes in one section would then affect the other unit. It cannot be emphasized too strongly that the connections of the subsystem with the whole must be fixed before splitting can be valid.

The use of such fixed links is not so common as the temporarily specified links discussed in the next section. Fixed links can be found, however, in most systems, wherever units have to perform specified duties. Examples might include the input and output streams for heat exchangers on a fixed cooling load, and for compressors designed for a fixed compression ratio. Care must always be taken, however, that these links are not subject to variation, since, in contrast with the next section, these links are not taken into account in any optimization technique used.

2.2 Temporarily Fixed Links

Systems can often be decomposed into smaller subsystems, even if fixed links are not present. Thus, if certain connections between subsystems are temporarily regarded as being specified by constants, each of the subsystems so isolated may be separately optimized as in Sec. 2.1. Finally, of course, a search must be conducted to determine the best feasible values of these links.

Example XIII-3: A comparison of the direct and splitting approaches

Let us consider the system depicted in Fig. XIII-1. Here the product stream $\mathbf{p}_1$ from some process is split into a system product $\mathbf{r}$ and a recycle stream $\mathbf{p}_2$. This recycle stream is then mixed with the system feed $\mathbf{q}_f$, the combined streams $\mathbf{p}_0$ then entering the process. There is, therefore, a link between the inlet and outlet process streams. Let us assume that the feed and recycle stream flow rates are fixed at F_f and F_r and that all state vectors $\mathbf{r}$, $\mathbf{p}_j$, and $\mathbf{q}_f$, each represent only a single component, x_j. If the process can be adjusted by varying the three-component decision

Fig. XIII-1 A typical recycle system.

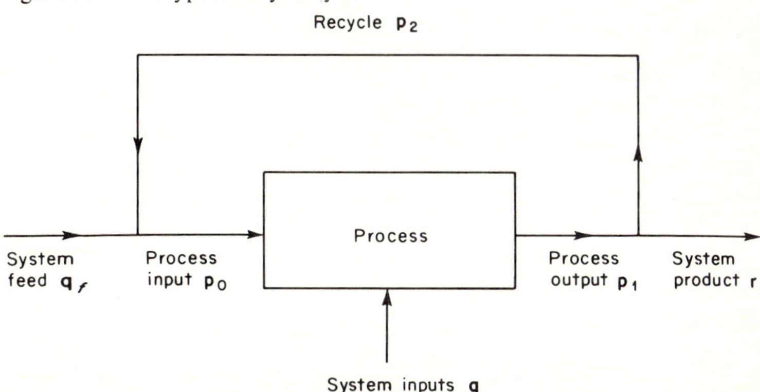

vector $\mathbf{q}$, let us consider which choice of $\mathbf{q}$ should be made to optimize y, where

$$y = y(x_f, x_1, \mathbf{q}) \tag{2.2-1}$$

The process transformation equation may be written

$$g_1 = x_1 - f(x_0, \mathbf{q}) = 0 \tag{2.2-2}$$

Further mass-balance restrictions will be imposed by the mixing of recycle and feed streams:

$$g_2 = F_f x_f + F_r x_2 - (F_f + F_r)x_0 = 0 \tag{2.2-3a}$$

and by the continuity of compositions at the recycle takeoff point:

$$x_1 = x_2 = x_r \tag{2.2-3b}$$

Equation (2.2-3a) could be written

$$g_2 = F_f x_f + F_r x_1 - (F_f + F_r)x_0 = 0$$

so that y must be optimized subject to $g_k = 0$ ($k = 1, 2$).

Let us first examine the *direct approach* to the recycle problem, in which all the independent variables are allowed to change at one time. These variables are x_0, x_1, q_1, q_2, and q_3, so that we can write

$$y = y(x_0, x_1, q_1, q_2, q_3)$$

This will be subject to the two restrictions

$$g_1 = x_1 - f(x_0, q_1, q_2, q_3) = 0$$
$$g_2 = F_f x_f + F_r x_1 - (F_r + F_f)x_0 = 0$$

on the process and mixer, respectively. Using the method of constrained variations (Chap. VII, Sec. 3.2), the necessary conditions for an optimum in y will be

$$J_k\left(\frac{y, g_1, g_2}{q_k, x_1, x_0}\right) = 0$$

for $k = 1, 2$, and 3, if $J_4\left(\frac{g_1, g_2}{x_1, x_0}\right) \neq 0$. Here

$$J_4 = \begin{vmatrix} \dfrac{\partial g_1}{\partial x_1} & \dfrac{\partial g_1}{\partial x_0} \\ \dfrac{\partial g_2}{\partial x_1} & \dfrac{\partial g_2}{\partial x_0} \end{vmatrix} = \begin{vmatrix} 1 & -\dfrac{\partial f}{\partial x_0} \\ F_r & -(F_r + F_f) \end{vmatrix} \neq 0$$

which will be assumed to be satisfied. We may now evaluate the Jacobian, J_k for $k = 1, 2, 3$, getting

$$J_k = \begin{vmatrix} \dfrac{\partial y}{\partial q_k} & \dfrac{\partial y}{\partial x_1} & \dfrac{\partial y}{\partial x_0} \\ \dfrac{\partial g_1}{\partial q_k} & \dfrac{\partial g_1}{\partial x_1} & \dfrac{\partial g_1}{\partial x_0} \\ \dfrac{\partial g_2}{\partial q_k} & \dfrac{\partial g_2}{\partial x_1} & \dfrac{\partial g_2}{\partial x_0} \end{vmatrix} = \begin{vmatrix} \dfrac{\partial y}{\partial q_k} & \dfrac{\partial y}{\partial x_1} & \dfrac{\partial y}{\partial x_0} \\ -\dfrac{\partial f}{\partial q_k} & 1 & -\dfrac{\partial f}{\partial x_0} \\ 0 & F_r & -(F_r + F_f) \end{vmatrix}$$

$$= -\frac{\partial y}{\partial q_k}\left[(F_r + F_f) - F_r \frac{\partial f}{\partial x_0}\right] - \frac{\partial f}{\partial q_k}\left[\frac{\partial y}{\partial x_1}(F_r + F_f) + \frac{\partial y}{\partial x_0} F_r\right] = 0$$

These equations can now be rearranged in the form

$$\frac{\partial y/\partial q_1}{\partial f/\partial q_1} = \frac{\partial y/\partial q_2}{\partial f/\partial q_2} = \frac{\partial y/\partial q_3}{\partial f/\partial q_3} = \frac{(F_r + F_f)(\partial y/\partial x_1) + F_r(\partial y/\partial x_0)}{F_r(\partial f/\partial x_0) - (F_r + F_f)} \tag{2.2-4}$$

which, together with the restrictions $g_1 = 0$ and $g_2 = 0$, will define the necessary conditions for the optimal values of q.

Let us now examine a *decomposition approach* to this problem through the principle of splitting. First, we may define a subproblem for separate study by fixing $x_1 = x_2 = x_r$ at some value $x_1^{(0)}$. From Eqs. (2.2-3*a*), it will be seen that this corresponds to a specified process outlet state $x_1^{(0)}$ and a specified process inlet state $x_0^{(0)}$. The process is now bounded by fixed limits, the subproblem being shown in Fig. XIII-2.

We now want to optimize y, where

$$y = y(x_0^{(0)}, x_1^{(0)}, q_1, q_2, q_3)$$

subject to the restriction that

$$g_1 = x_1^{(0)} - f(x_0^{(0)}, q_1, q_2, q_3) = 0$$

y and g now being functions of $\mathbf{q}$ only, since these functions otherwise contain only fixed parameter values, $x_1^{(0)}$ and $x_0^{(0)}$, and these parameters are the fixed links.

Fig. XIII-2 The first subproblem.

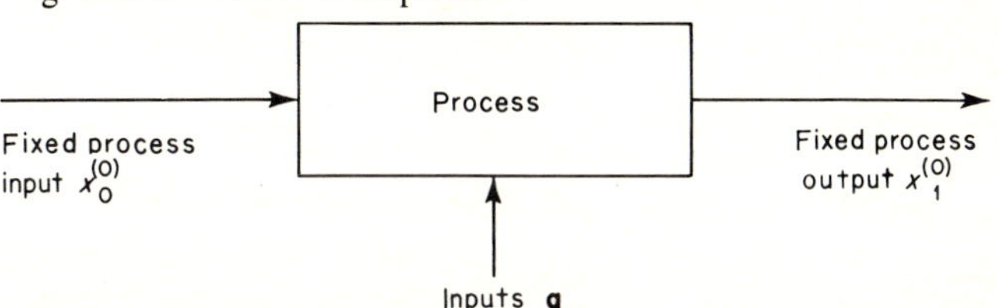

This problem can now be handled by any of the methods of Chap. VII, Sec. 3. Thus, for example, using the method of constrained variation, we should find the necessary conditions for an optimum value of y to exist from the Jacobians as

$$J_k\left(\frac{y,g_1}{q_k,q_1}\right) = 0 \qquad \text{for } k = 2, 3$$

provided $J(g_1/q_1) \neq 0$, which may be assumed to be satisfied. Thus we get

$$J_k = -\frac{\partial y}{\partial q_k}\frac{\partial f}{\partial q_1} + \frac{\partial y}{\partial q_1}\frac{\partial f}{\partial q_k} \qquad \text{for } k = 2, 3 \tag{2.2-5}$$

both of which can be solved with $g_1 = 0$ to obtain the $\mathbf{q}$ which yield the optimum value of y when the recycle stream is fixed at $x_1^{(0)}$. Sufficiency conditions could also be checked (Chap. VII, Sec. 3.2.3).

Since the $x_1^{(0)}$ value was temporarily fixed, the second stage in the calculation is to repeat the analysis for various values of x_1, thus obtaining the optimum value of y. We now wish to optimize

$$y = y[x_0, x_1, q_1(x_0,x_1), q_2(x_0,x_1), q_3(x_0,x_1)] \tag{2.2-6}$$

with respect to x_0 and x_1, subject only to the restriction that

$$g_2(x_0,x_1) = 0$$

Thus, by constrained variation, the necessary conditions are that

$$J_4\left(\frac{y, g_2}{x_0, x_1}\right) = 0$$

which has to be solved with $g_2 = 0$. This completes the decomposition solution.

In this simple illustration, we have carried out a direct calculation on the whole problem, this being a three-dimensional optimization. When, however, we introduced temporarily fixed links, we obtained one subproblem in two dimensions, and another in one dimension. This approach may therefore simplify a problem, especially for an original system of high dimension.

Let us now show that the two approaches are in fact equivalent.

It will be seen that Eqs. (2.2-5) are the same as the two equalities on the left of Eq. (2.2-4). From Eq. (2.2-6) we can write that $dy = 0$, so that

$$\frac{dy}{dx_0} = 0 = \frac{\partial y}{\partial x_0} + \frac{\partial y}{\partial x_1}\left(\frac{dx_1}{dx_0}\right) + \frac{\partial y}{\partial q_1}\left(\frac{dq_1}{dx_0}\right) + \frac{\partial y}{\partial q_2}\left(\frac{dq_2}{dx_0}\right) + \frac{\partial y}{\partial q_3}\left(\frac{dq_3}{dx_0}\right)$$

But from $g_2 = 0$, $dx_1/dx_0 = (F_r + F_f)/F_r$, and substituting this, with $\partial y/\partial q_2$ and $\partial y/\partial q_3$ from Eqs. (2.2-5), we get

$$0 = \frac{\partial y}{\partial x_0} + \frac{F_r + F_f}{F_r}\frac{\partial y}{\partial x_1} + \frac{\partial y}{\partial q_1}\left(\frac{dq_1}{dx_0} + \frac{dq_2}{dx_0}\frac{\partial f/\partial q_2}{\partial f/\partial q_1} + \frac{dq_3}{dx_0}\frac{\partial f/\partial q_3}{\partial f/\partial q_1}\right)$$

Multiplying through by $\partial f/\partial q_1$ then gives

$$\frac{\partial f}{\partial q_1}\left(\frac{\partial y}{\partial x_0}+\frac{F_r+F_f}{F_r}\frac{\partial y}{\partial x_1}\right)+\frac{\partial y}{\partial q_1}\left(\sum_{k=1}^{3}\frac{\partial f}{\partial q_k}\frac{dq_k}{dx_0}\right)=0 \tag{2.2-7}$$

Now, from $g_1 = 0$ and $g_2 = 0$, we can eliminate x_1, getting

$$g=\frac{F_r+F_f}{F_r}x_0-\frac{F_f}{F_r}x_f-f(x_0,q_1(x_0),q_2(x_0),q_3(x_0))=0$$

so that $dg = 0$ gives

$$\frac{dg}{dx_0}=0=\frac{F_r+F_f}{F_r}-\frac{\partial f}{\partial q_1}\frac{dq_1}{dx_0}-\frac{\partial f}{\partial q_2}\frac{dq_2}{dx_0}-\frac{\partial f}{\partial q_3}\frac{dq_3}{dx_0}-\frac{\partial f}{\partial x_0}$$

This can be used to eliminate the term $\sum_k (\partial f/\partial q_k)(dq_k/dx_0)$ in Eq. (2.2-7) so that

$$\frac{\partial y/\partial q_1}{\partial f/\partial q_1}=\frac{\partial y/\partial x_0+(\partial y/\partial x_1)(F_r+F_f)/F_r}{\partial f/\partial x_0-(F_r+F_f)/F_r} \tag{2.2-8}$$

Equations (2.2-5) and (2.2-8), derived from the decomposition method, are then equivalent to Eqs. (2.2-4), found directly.

One point should be stressed in the application of the principle of splitting. We must leave at least one variable to be determined,[1] so that a fixed link can be truly constant. Failure to observe this essential requirement may lead to a computational scheme which might be incorrect. Take, for example, the following scheme (referring to the previous recycle illustration): Assume x_0 to be fixed, but allow q_1, q_2, and q_3 to vary so as to optimize y. Of course, once the optimal values of q_1, q_2, q_3 are selected in this way, the resulting value of x_1 is unlikely to yield a value of x_0 [using Eq. (2.2-3*a*)] which would agree with that value of x_0 initially assumed. A second value of x_0 might then be guessed, and the procedure repeated until a match between the assumed and computed value is found. That this is the wrong approach is easily seen, for the above scheme leads to, at each iteration, the necessary conditions

$$\frac{\partial y}{\partial q_i}=0 \qquad \text{for } i=1,2,3 \tag{2.2-9}$$

assuming that the optimum is always at a stationary point. The iterations cease when the recycle equation (2.2-3*a*) is satisfied simultaneously with Eq. (2.2-9). However, from Example XIII-3, the correct necessary conditions were found to include

$$\frac{\partial y}{\partial q_i}\frac{\partial g_1}{\partial x_0}=\frac{\partial g_1}{\partial q_i}\frac{\partial y}{\partial x_0} \qquad \text{for } i=1,2,3 \tag{2.2-10}$$

[1] See Chap. II, Sec. 4.4.1, on degrees of freedom.

Since in general neither $\partial y/\partial x_0$ nor $\partial g_1/\partial q_i$ vanish when Eqs. (2.2-10) are satisfied, we see that Eqs. (2.2-9), arising from the iteration scheme, do not lead to the same solution as that found from Eqs.(2.2-10), which were derived directly. This method of iterating through nonfeasible states is incorrect. Such an iteration scheme, proposed, for example, by Rudd and Blum [1962], as pointed out by Jackson [1963], is incorrect in the general case because the links are not truly fixed. The outlet values have been allowed to change while presuming to hold x_0 constant. This is not physically possible, and the calculation produces nonfeasible states.

A true splitting calculation always produces a feasible state, i.e., one which can be operationally realized. Thus the calculation of a subproblem must be performed in such a way that an operationally possible state with fixed end conditions is achieved. If feasible states are not used, the computational scheme must be closely inspected to ensure the convergence of the method.

Splitting, for the present purpose, will always be used in a *feasible context at fixed links.* Its application will be considered further in Sec. 5.

3 THE PRINCIPLE OF OPTIMALITY FOR SEQUENTIAL SYSTEMS

The optimization of sequential systems can be considerably simplified by the use of the principle of optimality first formally stated by Bellman [1957]. While the principle itself may be regarded as self-evident, its application has had a far-reaching importance. Before discussing the principle (Sec. 3.1), it may be instructive to examine a simple illustration.

Example XIII-4: The need for the principle of optimality

To illustrate the application of the principle, let us consider the sequential process depicted in Fig. XIII-3. Here a chemical reactor R, a separator S, and a distillation column D are arranged in series. Let us assume that, for a given feed to the system, the reactor may be operated in an R_ith manner, producing a product specified by i. For each of these product specifications, the separator may be operated in the S_{ij}th manner to give a product specified by ij. Similarly, the distillation column can be operated in the D_{ijk}th manner, giving a system product specified by ijk. While each unit might be operated in an infinite number of ways, let us take representative values so that the reactor and separator each have three modes of operation and the distillation column has two. The 18 possible modes of system operation are shown in the "tree" diagram of Fig. XIII-4.

If operational costs are associated with each sequence of operations, let us consider which sequence provides the lowest cost of operation for

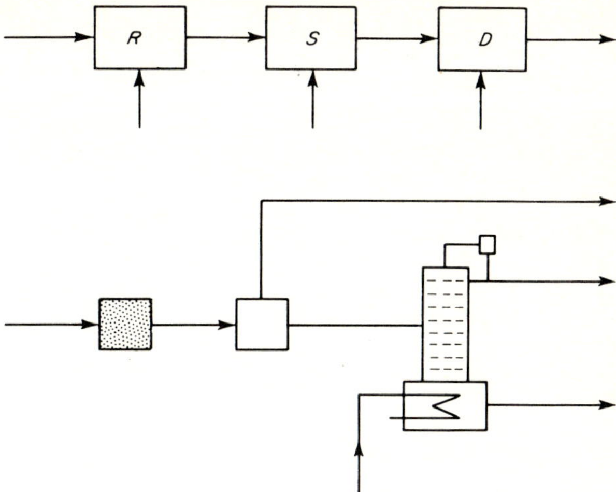

Fig. XIII-3 A sequential processing system. (*After Mitten and Nemhauser* [1963a]).

this system. Such costs are shown as parameters in Fig. XIII-4. Thus, for operation R_1, the cost is 5 units, while for a feed of $i = 1$ to the separator, the cost of operation by S_{12} will be 2, giving a total cost of operation of 7 for the first two stages by this route, and so on.

How can we find the desired lowest cost? We could, of course, evaluate every route shown in the diagram, a total of $3 \times 3 \times 2 = 18$ ways, and choose the best. Can we not, however, use some decomposition approach to reduce the number of calculations?

Suggestion 1 If we optimize each stage (R, S, and D) separately in turn, we should find a total cost of $4 + 1 + 1$, a minimum of 6 units. Unfortunately, this sequence cannot be obtained, and as is also shown in Examples XIII-1 and XIII-2, this approach is quite wrong. It will be seen here that if we choose route R_2, it is impossible to achieve costs of 1 in the second and third stages.

Suggestion 2 Let us choose the best first-stage policy. This is R_2, yielding a cost of 4. From the products ($i = 2$), we can now choose the best result from the second stage S_{22}, yielding a cost of 4, while in the last stage, for a feed $i = 2$, $j = 2$, we get a minimum cost of 4, with D_{221}. This is still *not* the optimum result—we have made a mistake at the first stage because we made a decision, R_i, which affects all our later choices. In this case we have excluded routes beginning with R_1 and R_3, which might yield better results. We must make choices which are independent of any succeeding decisions.

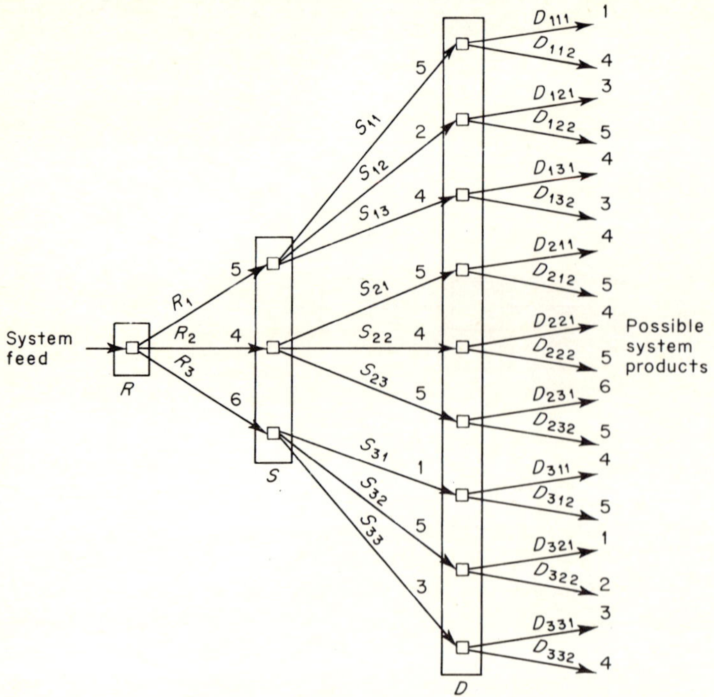

Fig. XIII-4 Arrow diagram, with arrows indicating modes of operation R_i, S_{ij}, or D_{ijk} and numerical parameter indicating the associated operating costs. (*Based on Mitten and Nemhauser* [1963a].)

Suggestion 3 Let us start with the last stage and consider a given feed to this stage. We may then choose the best operating conditions for that feed. Indeed, we *must* choose the best operating condition for that feed, since all other operations necessary to achieve that feed are independent of this choice. Thus, if the path S_{11} has been used to produce the feed to the distillation column, then obviously, D_{111} is the best choice, since this gives the minimum cost 1 to transform this feed into final process stream. The other choice, D_{112}, results in a higher cost, 4, and need never be considered. By beginning at the last stage, we can make decisions which, for any given feed, do not affect any other stages, and we can here eliminate half of the distillation column modes of operation. The results are tabulated in Table XIII-1. If in a subsequent-stage calculation we wish to know the effect of using a feed ij, the resultant operating policy and cost can be immediately found.

Consider now the last two units together. If the feed to the separator is R_1, we can consider the alternative methods of processing and obtain

TABLE XIII-1 SINGLE-STAGE POLICY

Feed ij	*Optimal decision policy*	*Cost resulting from this optimal decision*
11	D_{111}	1
12	D_{121}	3
13	D_{132}	3
21	D_{211}	4
22	D_{221}	4
23	D_{232}	5
31	D_{311}	4
32	D_{321}	1
33	D_{331}	3

the resultant costs. Thus, for a feed R_1, we shall have:

Mode of separator operation S_{ij}	*Cost in separator*	*Product*	*Distillation cost* (Table XIII-1)	*Total cost*
S_{11}	5	11	1	6
S_{12}	2	12	3	5
S_{13}	4	13	3	7

The cheapest way of operating with a feed R_1 is thus to use S_{12} at a two-stage cost of 5. (This implies, from Table XIII-1, that we should also use a policy D_{121} in the third stage.) Proceeding in this way, we may consider each feed to the second stage in turn, and draw up the results in the form of Table XIII-2. At this point in the calculation we have eliminated two-thirds of the possible operation conditions in the separator from further consideration, as well as half the distillation modes of operation.

We are now in a position to complete the solution by considering our third subproblem, the three stages together. If we select a route R_1, we know that the total system cost of operation will be $5 + 5 = 10$ (the second cost coming from Table XIII-2, which gives the costs resulting from a feed i to the last two stages). Similarly, the total costs for

TABLE XIII-2 TWO-STAGE POLICY

Feed i	*Policy*	*Cost*
1	$S_{12} \rightarrow D_{121}$	5
2	$S_{22} \rightarrow D_{221}$	8
3	$S_{31} \rightarrow D_{311}$	5

routes R_2 and R_3 will be 12 and 11, respectively. The optimum policy for the system is thus

$$R_1 \qquad S_{12} \qquad D_{121}$$

resulting in a minimum cost of 10.

Let us now summarize this computational scheme. We start at the rear of the process. The final stage (the first subproblem) is optimized with respect to a fixed feed to that stage. This is repeated for a range of possible feeds. Next the last two units in the system are considered (the second subproblem), and the two-stage process is optimized for a range of feeds into the penultimate stage. It should be noted that changing the feed to the second last stage results, in general, in a change in the feed to the last stage, but the optimal policy for the last stage has already been found as a function of the feed to that last stage. Thus we need only vary the feed to the second last stage to find a two-stage optimal policy as a function of that feed. This procedure is represented diagrammatically in Fig. XIII-5, wherein the optimization of a sequential system is shown to proceed in steps formed by the decomposition of the problem into subproblems, in each of which a smaller number of decisions had to be made.

This procedure outlined above employs, in fact, a basic principle, first formally stated by Bellman [1957] in the principle of optimality.

Fig. XIII-5 The formation of optimization subproblems in a sequential system.

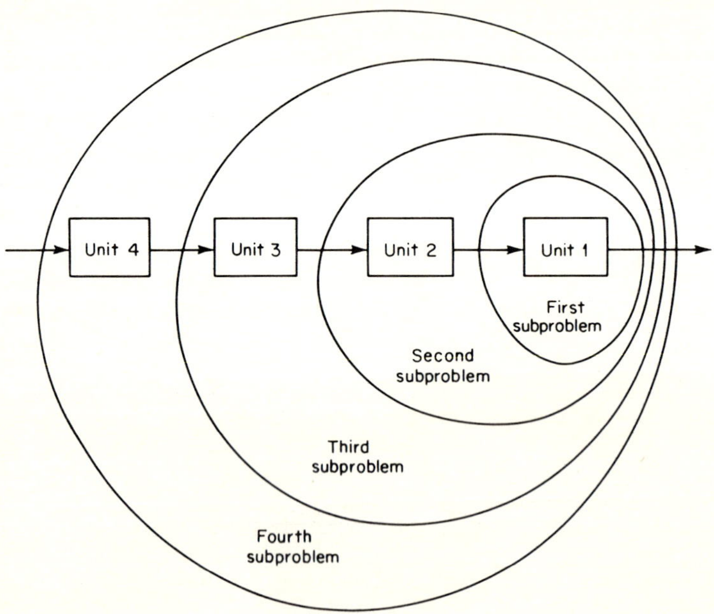

3.1 Statement of the Principle

Bellman [1957, p. 83] states the principle of optimality as follows "An optimal policy has the property that, whatever the initial state and the initial decision are, the remaining decisions must constitute an optimal policy with regard to the state resulting from the first decision."

Let us examine this statement in light of the illustration considered in Sec. 3. Once we understand its meaning, the validity of this principle is immediately obvious. It is applicable only to sequential processes for which a choice (or decision) made at one stage can influence only what follows, but has no effect on the preceding events in the sequence. Consider again Fig. XIII-4. We see that decision D_{112} was rejected from consideration. Why? Because no matter what the previous choices have been, once the feed to the last unit is specified as resulting from section S_{11}, we must do the best we can with that feed. Therefore we should never select D_{112}, since this decision would not change any of the prior costs and is clearly more expensive than D_{111}. In accordance with the principle of optimality, the choice D_{112} must be rejected. Of course, if this were not a sequential system, but a recycle system, the principle would not apply, because it is possible that the more expensive distillation decision would give rise to a saving in the earlier parts of the processing. This is why we have termed this principle the *principle of optimality for sequential systems.*

Example XIII-5: The longest path—critical-path scheduling

Consider the network shown in Fig. XIII-6. If, for example, each section of a path represents a task to be performed, then the network represents all the work which has to be completed, from its initiation at A to completion at B. Each job, or task, will require a certain period of time for completion, as given by the parameters on each path, and the sequence of tasks will be shown by the arrows. No job can be started until all preceding jobs on that path are completed, and so on. This type of arrow diagram forms the basis for the programming of design, construction, and maintenance work. While many methods[1] have been

[1] The first method of scheduling control was PERT (program evaluation and review technique), developed by the U.S. Navy, being adapted by DuPont to the needs of operating plants, these methods being called CPS (critical path scheduling) and LESS (least cost estimating and scheduling). Other critical-path methods include FAST (Fluor analytical scheduling technique), PEP (program evaluation procedure), and RAMPS (reserve allocation and multiproject scheduling). Some typical applications are discussed, for example, by Davidmann [1959a, b], Salisbury [1958], Kast [1962], and Frischberg [1962], while Imbrie [1962], Berman [1962], Foord [1962], Bigelow [1962], and Lerda-Olberg [1966], among others, discuss the methods.

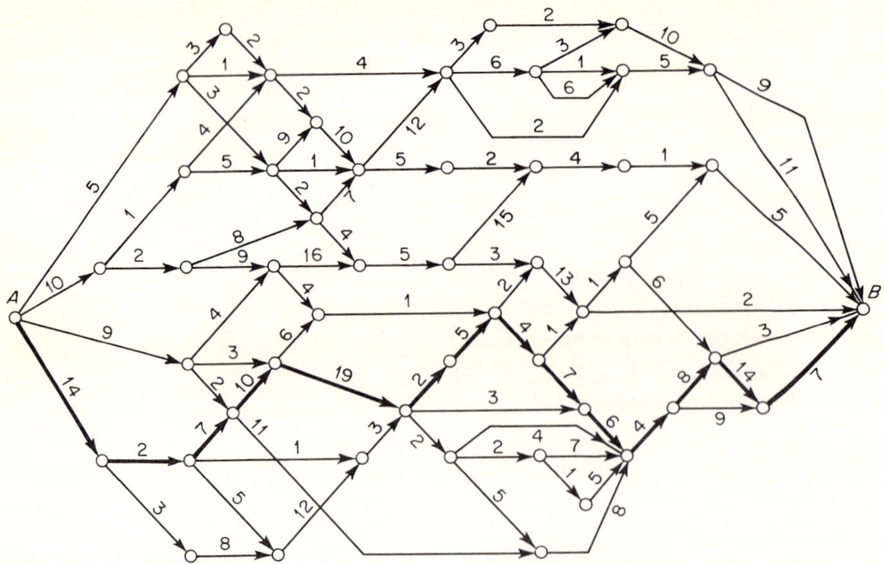

Fig. XIII-6 Path network.

proposed, the essential feature is the need to determine the longest path. This will, of course, determine the period in which the overall task can be achieved, and in application attention can be devoted to reducing this period or to keeping the work along this critical path on schedule. The choice of the longest path or route is of course an optimization problem.

Let us consider the network shown in Fig. XIII-6. Take as an illustration the section in the upper right-hand corner, this detail being depicted in Fig. XIII-7. For convenience the nodes are labeled α, $\beta, \ldots, \delta$, B, and we are interested in finding the *longest* path from node α to node B.

Let us consider each node in turn, starting with the rear node δ. From this node we may travel to node B by paths of length 9 or 11. The latter will be the largest and should be adopted. We need no longer consider the other path. Let us now look at the next nodes, γ and ϕ,

Fig. XIII-7 Detail of the main network shown in Fig. XIII-6.

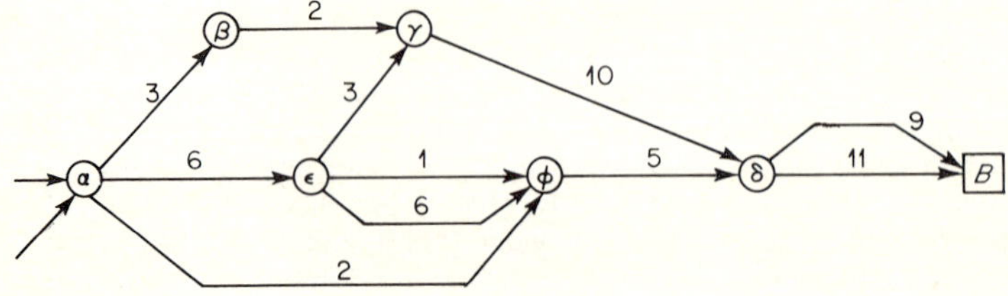

TABLE XIII-3 SUMMARY OF LONGEST PATH LENGTHS FROM ANY NODE

Starting node	*Possible paths and lengths*	*Total length to B*	*Optimal route from starting node*
δ	9 to B	$9 + 0 = 9$	
	11 to B	$11 + 0 = 11$	*
γ	10 to δ	$10 + 11 = 21$	*
ϕ	5 to δ	$5 + 11 = 16$	*
	3 to γ	$3 + 21 = 24$	*
ϵ	1 to ϕ	$1 + 16 = 17$	
	6 to ϕ	$6 + 16 = 22$	
β	2 to γ	$2 + 21 = 23$	*
	3 to β	$3 + 23 = 26$	
α	6 to ϵ	$6 + 24 = 30$	*
	2 to ϕ	$2 + 16 = 18$	

in the sequence moving from the rear—first of all, node γ. Only a single path of length 10 leads from node γ to δ. The total path length to B will then be 10 plus the maximum from the last stage (δ to B of length 11), which has already been found. Continuing in this manner, the results for each node can be found, using the principle of optimality to obtain the maximum total path lengths. It is obvious that no results can be found unless all the succeeding nodes on all possible paths have already been examined. Table XIII-3 summarizes the information obtained at each stage (node), each entry making use of earlier entries in the table.

It will be seen, for this detail of the main problem, that if we start at node α, the longest path will be found of length 30 units by passing through nodes α, ϵ, γ, δ, B. The longest path (the critical path) for the complete problem is shown in Fig. XIII-6 by a heavy line.

The mathematical formulation of the principle of optimality is called dynamic programming. This is described in Sec. 4.

4 DYNAMIC PROGRAMMING

Dynamic programming is a method of decomposition based upon Bellman's principle of optimality (Sec. 3.1). It can be applied to situations in which many decisions are required in defining the optimum of a system composed of distinct stages, as long as decisions at the later stages do not affect the performance of earlier stages. We shall therefore confine our attention here to sequential systems. In Sec. 5, however, it will be seen that more complex systems can be handled if the principle of splitting is also used.

4.1 Mathematical Formulation

Let us consider a discrete sequential system of units such as that depicted in Fig. XIII-8. As will be seen later, it is convenient to label these units, starting from the rear. In each stage, or unit, the transformation of the state vector $\mathbf{p}_i$ is represented by an input-output relationship of the form

$$\mathbf{p}_i = \mathbf{T}_i(\mathbf{p}_{i+1},\mathbf{q}_i) \qquad \text{for } i = 1, 2, \ldots, N \tag{4.1-1}$$

in which $\mathbf{p}_{i+1}$ is the interconnecting link and feed to any unit, and $\mathbf{q}_i$ the input vector. In addition, the system feed and product streams are

$$\mathbf{p}_{N+1} \equiv \mathbf{q}_{N+1}$$

$$\mathbf{p}_1 \equiv \mathbf{r}_1$$

Suppose, further, that the objective function can be written

$$y = \sum_{i=1}^{N} y_i(\mathbf{p}_i,\mathbf{p}_{i+1},\mathbf{q}_i) \tag{4.1-2}$$

it being assumed that there is a certain profit or cost, $y_i(\mathbf{p}_i,\mathbf{p}_{i+1},\mathbf{q}_i)$, associated with each stage.[1]

Let us consider the first subproblem by starting at the final stage ($i = 1$) in the sequence. If the feed $\mathbf{p}_2$ to this stage is specified, then the principle of optimality dictates that the input, or decision vector, $\mathbf{q}_1$, must be selected to optimize y_1. Irrespective of what happens in the other stages, $\mathbf{q}_1$ is to be chosen so that $y_1(\mathbf{p}_1,\mathbf{p}_2,\mathbf{q}_1)$ is an optimum for that feed, $\mathbf{p}_2$. Call this optimum value M_1. Then

$$M_1(\mathbf{p}_2) = \underset{\mathbf{q}_1}{\text{opt}} \{y_1(\mathbf{p}_2,\mathbf{p}_1,\mathbf{q}_1)\} \tag{4.1-3}$$

[1] Although most objective functions can be written in this form, it is not completely general. Thus, for example, a rate-of-return criterion could not be so written as a summation over N stages (Chap. XI, Sec. 2).

Fig. XIII-8 A sequential system.

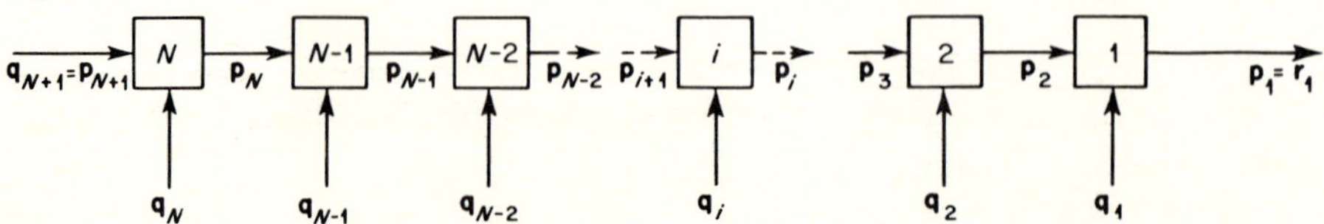

This equation expresses the information that, for a given feed, the optimum value of y_1 is M_1, with $\mathbf{q}_1$ so adjusted as to yield this value. This is called a *one-stage policy*, since once the feed $\mathbf{p}_2$ is specified, M_1, $\mathbf{q}_1$, and $\mathbf{p}_1$ will be completely defined, the values depending parametrically upon the feed.

Next, let us consider the second subproblem, taking the last two stages, and define M_2 to be the optimal profit

$$M_2(\mathbf{p}_3) = \underset{\mathbf{q}_1,\mathbf{q}_2}{\text{opt}} \{y_1(\mathbf{p}_2,\mathbf{p}_1,\mathbf{q}_1) + y_2(\mathbf{p}_3,\mathbf{p}_2,\mathbf{q}_2)\} \tag{4.1-4}$$

Here M_2 is the optimum value of the objective function for the last two stages determined for a specified feed to the second-to-last stage. Now the principle of optimality requires that the $\mathbf{q}_1$ be selected so as to optimize y_1 for a given $\mathbf{p}_2$, since this is a sequential process. Therefore we can write the two-stage policy as

$$M_2(\mathbf{p}_3) = \underset{\mathbf{q}_2}{\text{opt}} \{M_1(\mathbf{p}_2) + y_2(\mathbf{p}_2,\mathbf{p}_3,\mathbf{q}_2)\} \tag{4.1-5}$$

While Eq. (4.1-4) is a function of both $\mathbf{q}_1$ and $\mathbf{q}_2$ [the transformation equations (4.1-1) being used to eliminate $\mathbf{p}_1$ and $\mathbf{p}_2$], the principle of optimality has allowed us to reduce this to a function of $\mathbf{q}_2$ only, as shown in Eq. (4.1-5). We have now decreased the dimensionality of the optimization problem by a half. This may be seen more clearly by using Eq. (4.1-1) to rewrite Eq. (4.1-5) as

$$M_2(\mathbf{p}_3) = \underset{\mathbf{q}_2}{\text{opt}} \{M_1(\mathbf{T}_2(\mathbf{p}_3,\mathbf{q}_2)) + y_2(\mathbf{T}_2(\mathbf{p}_3,\mathbf{q}_2), \mathbf{p}_3, \mathbf{q}_2)\} \tag{4.1-6}$$

In this form we see that, for a given feed $\mathbf{p}_3$, the optimum is determined solely by an appropriate choice of $\mathbf{q}_2$.

Some of the advantages of dynamic programming will now be obvious. $M_1(\mathbf{p}_2)$ was studied in isolation from the rest of the system, so that $\mathbf{q}_N, \ldots, \mathbf{q}_2$ did not need to be considered. This calculation is possible because we must do the best we can with what we have. Once M_1 is determined, M_2 then requires only a study of possible $\mathbf{q}_2$ values, since $\mathbf{p}_2$ is determined by Eq. (4.1-1) for given values of $\mathbf{p}_3$ and $\mathbf{q}_2$. The optimization problem posed by Eq. (4.1-4), in which both $\mathbf{q}_1$ and $\mathbf{q}_2$ are to be simultaneously varied to produce an optimum, is therefore reduced to two subproblems defined by Eqs. (4.1-3) and (4.1-5). In each of these subproblems we need only consider a single vector $\mathbf{q}_i$, so that the optimization is, in principle, simpler.

In general, we can rewrite the problem expressed by

$$M_i(\mathbf{p}_{i+1}) = \underset{\mathbf{q}_i,\mathbf{q}_{i-1},\ldots,\mathbf{q}_1}{\text{opt}} \{y_1 + y_2 + \cdots + y_i\} \tag{4.1-7}$$

in the form

$$M_i(\mathbf{p}_{i+1}) = \underset{\mathbf{q}_i}{\text{opt}} \{M_{i-1}(\mathbf{p}_i) + y_i(\mathbf{p}_{i+1},\mathbf{q}_i)\} \tag{4.1-8}$$

Here y_i is the profit from the ith unit, and M_{i-1} is the optimal profit from the succeeding $(i - 1)$ stages. It will be seen that, in general, we can consider the i-stage sequence as being made up of two effective stages—the current stage i and the remaining $i - 1$ stages.

The original problem depended upon the choice of i sets of variables, $\mathbf{q}_1, \mathbf{q}_2, \ldots, \mathbf{q}_i$. We have now decomposed this problem into the consideration of i separate problems, each of lower dimensionality. Each subproblem requires only the specification of a single decision vector. Since at any unit it is necessary to know the best policy resulting from the unit product which enters the set of succeeding units, the approach starts at the rear end of the choice and moves stage by stage toward the beginning; this is why there is some notational convenience in numbering the units from the rear. The unit number then corresponds to the subproblem being treated.

Example XIII-6: Cross-current extraction

Let us rework the problem given in Examples VII-7 and VII-15. The system is shown in Fig. XIII-8 (but with each vector having only a single component, so that $\mathbf{p}_i \equiv p_i$), being identical with that shown in Fig. VII-9, except for the direction of the stage numbering. The transformation relationship is

$$p_i = \frac{p_{i+1}}{1 + m'q_i} \tag{4.1-9}$$

and the objective function may be written as the summation

$$y = \sum_{i=1}^{N} (p_{i+1} - p_i) - B \sum_{i=1}^{N} q_i \tag{4.1-10}$$

from Example VII-15. Here we can identify the y_i of Eq. (4.1-2) as

$$y_i(\mathbf{p}_i,\mathbf{p}_{i+1},\mathbf{q}_i) \equiv y_i(p_i,p_{i+1},q_i) = p_{i+1} - p_i - Bq_i \tag{4.1-11}$$

since the vectors have only single components.

We wish to select a set of values of $q_1, q_2, \ldots, q_N (q_i \geq 0)$ which maximize the objective function. Two direct approaches have been examined in Chap. VII in which all the q_i play an equal role. If we use dynamic programming on this sequential problem, we can decompose it into a series of N subproblems, in each of which only one q_i is required.

In the first subproblem we consider the final stage. Thus, as required by Eq. (4.1-3),

$$M_1(p_2) = \max_{q_1} \{y_1\} = \max_{q_1} \{p_2 - p_1 - Bq_1\} \tag{4.1-12}$$

We must now choose the $q_1(q_1 \geq 0)$ which maximizes y_1 for a fixed value of p_2. If we assume that this maximum lies at a stationary point of y_1,† the appropriate value of q_1 is defined by

$$\left(\frac{\partial y_1}{\partial q_1}\right)_{p_2} = 0 = -\frac{\partial p_1}{\partial q_1} - B \tag{4.1-13}$$

The material balance, Eq. (4.1-9), then gives

$$\left(\frac{\partial p_i}{\partial q_i}\right)_{p_{i+1}} = -\frac{m' p_{i+1}}{(1 + m' q_i)^2} = -\frac{m' p_i^2}{p_{i+1}} \tag{4.1-14}$$

From Eqs. (4.1-13) and (4.1-14) we then get the location of the stationary point

$$p_1 = \left(\frac{B p_2}{m'}\right)^{1/2} \tag{4.1-15}$$

Checking the nature of this stationary point, Eq. (4.1-13) can be differentiated with respect to q_1, getting

$$\frac{\partial^2 y_1}{\partial q_1^2} = -\frac{\partial^2 p_1}{\partial q_1^2}$$

From Eq. (4.1-14) we find that

$$\frac{\partial^2 p_i}{\partial q_i^2} = \frac{2(m')^2 p_i^3}{p_{i+1}^2} \tag{4.1-16}$$

so that

$$\frac{\partial^2 y_1}{\partial q_1^2} < 0$$

provides the sufficient condition for a maximum. From Eqs. (4.1-9) and (4.1-15), the optimal one-stage policy is then

$$q_1 = \frac{1}{m'}\left(\frac{p_2}{p_1} - 1\right) = \frac{1}{m'}\left[\left(\frac{p_2 m'}{B}\right)^{1/2} - 1\right] \tag{4.1-17a}$$

so that the maximum profit from the last stage for a feed p_2 will be

$$M_1(p_2) = p_2 - 2\left(\frac{B p_2}{m'}\right)^{1/2} + \frac{B}{m'} \tag{4.1-17b}$$

This maximum was found at a stationary point. The only other possible location is at a boundary, here $q_1 = 0$. This would result in $p_1 = p_2$, and the profit would vanish. Since $M_1 = [(p_2)^{1/2} - (B/m')^{1/2}]^2 > 0$, it is therefore the desired global maximum.

† It is essential that we later check this assumption, since we must find the global maximum.

The first subproblem is now completely defined by Eqs. (4.1-17) for any given feed. Let us now look at the second subproblem, formed by the last two stages in the system. The optimal profit from these stages is then found from Eq. (4.1-6) as

$$M_2(p_3) = \max_{q_1,q_2} \{y_1 + y_2\} = \max_{q_2} \{M_1(p_2) + y_2\}$$

$$= \max_{q_2} \left\{p_3 - 2\left(\frac{Bp_2}{m'}\right)^{1/2} + \frac{B}{m'} - Bq_2\right\}$$

This is a function only of q_2, since p_2 can be eliminated using Eq. (4.1-9). Searching for the two-stage maximum at a stationary point, we get

$$\frac{\partial(M_1 + y_2)}{\partial q_2} = 0 = -\left(\frac{B}{m'p_2}\right)^{1/2} \frac{\partial p_2}{\partial q_2} - B$$

Using Eq. (4.1-14), we may then solve for p_2, getting

$$p_2 = \left(\frac{Bp_3^{\,2}}{m'}\right)^{1/3} \tag{4.1-18}$$

This value of p_2 gives the global maximum of the objective function (after checking the nature of the stationary point and the boundary on q_2) as

$$M_2(p_3) = p_3 - 3\left(\frac{B}{m'}\right)^{2/3} p_3^{1/3} + \frac{2B}{m'} \tag{4.1-19}$$

This completes the solution of the two-stage subproblem for any given feed to this subproblem. From the principle of optimality, we must use this result if we wish to include more stages in the optimization.

Continuing in this manner, we can show for the ith subproblem that the i-stage policy is

$$M_i(p_{i+1}) = p_{i+1} - (i + 1)\left(\frac{B}{m'}\right)^{i/1+i} p_{i+1}^{1/(1+i)} + \frac{iB}{m'} \tag{4.1-20}$$

found at

$$p_i = \left(\frac{B}{m'}\right)^{1/1+i} p_{i+1}^{i/(1+i)} \tag{4.1-21}$$

The profit from the entire system can be found here by setting $i = N$. Since p_{N+1}, the feed to the system, is fixed by some external agency, this maximum profit M_N is a well-defined number. The state of each unit's product stream can be found from Eq. (4.1-21) to be

$$p_{N-i} = \left(\frac{B}{m'}\right)^{(1+i)/(1+N)} p_{N+1}^{(N-i)/(N+1)} \tag{4.1-22}$$

The amount of solvent, q_i, to be used in each stage can then be found from Eqs. (4.1-16) and (4.1-22) as

$$q_i = \frac{1}{m'}\left[\left(\frac{m'p_{N+1}}{B}\right)^{1/(1+N)} - 1\right] \tag{4.1-23}$$

Since the right-hand side is independent of i, we have q_i constant and equal in each stage.

These results are, of course, exactly the same as found by other methods in Examples VII-7 and VII-15.

This simple illustration, involving only a single component link, was mainly concerned with the introduction of the dynamic programming concept based upon subsystems. Let us now consider some of the computational aspects.

4.2 Computational Aspects of Dynamic Programming

We have seen that dynamic programming is a method of decomposition. The essential feature is the simplification of the organization of optimization. Dynamic programming does not indicate how the optimization can or will be carried out. Since the actual optimization procedure used to resolve each of the subproblems is not, therefore, dictated by the decomposition method, we should adopt the most efficient search technique available. This can then be used within the framework set up by dynamic programming, whether or not constraints have been imposed. Such constraints would have to be included in each subproblem (Sec. 4.2.3). The computations associated with dynamic programming are often facilitated by the preparation of certain tables of data (Sec. 4.2.1). Since the use of such tables corresponds closely to the logic required for digital-computer solution of the dynamic programming equations, these tables are an important computational feature. It will also be noted (Sec. 4.2.2) that most sequential problems become unmanageable by dynamic programming when more than a few variables are required to specify the state of each interconnecting link between stages.

4.2.1 Tabular representation and single-component links

When an analytical expression is available, the optimal i-stage profit, $M_i(\mathbf{p}_{i+1})$, can be represented in a functional form as in Example XIII-6. In general, however, such a representation cannot be found, and an alternative format is required to store information, it being convenient to use tables for this purpose. While the calculations may appear tedious when performed manually, they are easily programmed for automatic calculation, the tabular basis still being required. We shall here confine our attention to situations in which the link between subproblems is a single component vector, the more general case being deferred to Sec. 4.2.2.

In a sequential system, the principle of optimality has allowed us to write the optimal value of the objective function for the i-stage subproblem, from Eq. (4.1-8), as

$$M_i(p_{i+1}) = \underset{q_i}{\text{opt}} \{y_i(p_{i+1},q_i) + M_{i-1}(p_i)\}$$

In order to find any value of the i-stage objective function $\{y_i + M_{i-1}\}$, we must be able to find the optimal value M_{i-1} for the remaining $i - 1$ stages. This optimal value is a function only of p_i. Since the required values of p_i would not be known in advance when M_{i-1} was being evaluated, it is necessary to determine M_{i-1} for a range of values of p_i, the resultant numerical information being stored in a table.

Let us illustrate the procedure by referring to an example.

Example XIII-7: Cross-current extraction with a nonlinear equilibrium relationship

Consider again the much overworked but instructive cross-current liquid-liquid extraction problem of Example XIII-6. For a three-stage system, the objective function to be maximized will be, again,

$$y = \sum_{i=1}^{3} y_i = \sum_{i=1}^{3} (p_{i+1} - p_i) - B \sum_{i=1}^{3} q_i \tag{4.2-1}$$

but the transformation equations (4.1-9) will no longer hold, since these were derived for linear equilibrium. The more general form of material balance,

$$p_{i+1} - p_i = q_i\phi(p_i) \tag{4.2-2}$$

must be used, where $\phi(p_i)$ is the equilibrium relation given by $\phi(p_i) = \sqrt{p_i}$. Let us assume that the system feed p_4 has a value of 1.5 and that $B = 0.25$.

Using dynamic programming, we shall start at the last stage and consider the optimal one-stage profit

$$M_1(p_2) = \max_{q_1} \{y_1\} = \max_{q_1} \{p_2 - p_1 - 0.25q_1\} \tag{4.2-3}$$

The best value of q_1 will, obviously, depend upon the value of p_2. In Example XIII-6, under linear equilibrium, we obtained this optimal value of q_1 using analytical methods. In general, however, we must seek this subsystem optimum by numerical methods. Let us eliminate q_1 from Eq. (4.2-3), using Eq. (4.2-2). Then

$$M_1(p_2) = \max_{p_1} \left\{(p_2 - p_1)\left(1 - \frac{0.25}{\phi(p_1)}\right)\right\} \tag{4.2-4}$$

TABLE XIII-4 A ONE-STAGE POLICY FOR EXAMPLE XIII-7

Given feed	Calculated values		
p_2	$M_1(p_2)$	p_1^*	q_1^*
1.5	0.66791	0.381	1.811
1.4	0.60672	0.365	1.713
1.3	0.54856	0.350	1.608
1.2	0.49139	0.331	1.510
1.1	0.43534	0.315	1.399
1.0	0.38052	0.297	1.292
0.9	0.32708	0.279	1.176
0.8	0.27524	0.259	1.063
0.7	0.22526	0.239	0.943
0.6	0.17747	0.219	0.814
0.5	0.13234	0.197	0.685
0.4	0.09056	0.173	0.546
0.3	0.05324	0.146	0.404
0.2	0.02234	0.116	0.247
0.1	0.00232	0.080	0.071
0.0	0.00000	0.000	0.000

Thus, for a given value of p_2, we can find the value of p_1 to yield $M_1(p_2)$. This procedure may now be carried out for a range of values of p_2. Since at this point in the calculations it will not be known what value of p_2 will be needed in the final analysis, we must store these data until the calculation is almost complete. Since the greatest possible value of p_i will be given by p_4, we need to consider the range of feeds $0 \leq p_2 \leq 1.5$, the optimal values of p_1 being conveniently summarized in a table, such as Table XIII-4. When this table is complete, we shall know everything about the one-stage subproblem, once a feed to that stage is specified. The analogy with the analytical approach will be obvious, with a table or data store replacing an analytical expression for M_1.

Let us now examine the two-stage subproblem. We wish to maximize the total profit from the last two stages so that, by the principle of optimality, we have

$$\begin{aligned} M_2(p_3) &= \max_{q_1,q_2} \{y_2 + y_1\} = \max_{q_2} \{y_2 + M_1(p_2)\} \\ &= \max_{q_2} \{p_3 - p_2 - 0.25q_2 + M_1(p_2)\} \\ &= \max_{p_2} \left\{(p_3 - p_2)\left(1 - \frac{0.25}{\phi(p_2)}\right) + M_1(p_2)\right\} \end{aligned} \quad (4.2\text{-}5)$$

TABLE XIII-5 A TWO-STAGE POLICY FOR EXAMPLE XIII-7

Given feed	*Calculated values*		
p_3	$M_2(p_3)$	p_2^*	q_2^*
1.5	0.78784	0.640	1.075
. . .	. . . .	. . . .	. . .
1.0	0.45658	0.463	0.789
0.9	0.39393	0.434	0.709
0.8	0.33248	0.406	0.621
0.7	0.27401	0.354	0.582
0.6	0.21695	0.323	0.487
0.5	0.16332	0.268	0.448
0.4	0.11307	0.237	0.336
0.3	0.06765	0.177	0.292
. . .	. . . .	. . .	. . .

upon substituting for q_2 from Eq. (4.2-2). It is now necessary to optimize the function { } only with respect to p_2 for each of a set of representative feeds, p_3. The results are drawn up in Table XIII-5. For each value of p_3 in the range $0 \leq p_3 \leq p_4$, the optimal value of p_2 is determined, the value of M_1 for any p_2 being found from Table XIII-4, using linear interpolation where necessary. We now have built up the two-stage policy for the second subsystem, containing all the necessary information needed for the last two units to be operated optimally for a given feed.

Proceeding in this manner, we can construct successive tables of information. For the three-stage and final subproblem, however, we need only consider a single value of the feed, that specified as $p_4 = 1.5$ (Table XIII-6).[1]

[1] Care has to be taken in the necessary interpolation of data. If we had investigated a four-stage system using the same approach, part of the new Table XIII-6 would read

p_4	$M_3(p_4)$	p_3^*	q_3^*
1.6	0.90986	0.860	0.798
1.5	0.83706	0.829	0.738
1.4	0.76561	0.750	0.751

the peculiar result in the last column being obtained due to linear interpolation from Table XIII-5. Accurate results could be found only by using more entries in each table or by some better interpolation scheme. The use of quadratic interpolation, for example, would lead to the modified entry of

p_3	$M_2(p_3)$	p_2^*	p_2^*
0.8	0.33246	0.402	0.628

in Table XIII-5.

TABLE XIII-6 A THREE-STAGE POLICY FOR EXAMPLE XIII-7

Given feed	*Calculated values*		
p_4	$M_3(p_4)$	p_3^*	q_3^*
1.5	0.83706	0.829	0.738

Complete information on the system is now available in the three derived tables, although at any stage in the calculation only the data in the preceding table are required. In the whole investigation, changes need not be made in more than one variable at a time. This tabular approach is the only feasible approach to the general solution of such multistage sequential processes. It will be noted that the particular solution required is embedded in the optimal results derived. Little extra work would be entailed if the feed state were altered, since only the final subproblem would be reinvestigated.

4.2.2 Multivariable links

So far, our illustrations of dynamic programming decomposition have been limited to systems in which the links between the sequential stages can be specified by single variables. Let us now consider the effect of having such links defined by a general vector $\mathbf{p}_{i+1}$. It will be obvious from the discussion of Sec. 4.2-1 that when $M_i(\mathbf{p}_{i+1})$ is calculated, it must be evaluated for a range of values of $\mathbf{p}_{i+1}$. Thus, if $\mathbf{p}_{i+1}$ has two components, sufficient information must be stored so that a two-dimensional interpolation is possible within the desired accuracy. Clearly, the number of storage tables required will increase rapidly as the number of variables increases. For example, if 10 entries are needed to specify the possible feed states for a single variable, then 10^n entries are required when an n-component vector defines the feed to any stage. A feed containing four or more variables cannot be feasibily handled by purely dynamic programming techniques. Bellman has referred to this problem of multivariable feed states as involving a "curse of dimensionality." Such problems might be solved by the method of Lagrangian multipliers (Chap. VII, Sec. 3.3).

The main problem in the application of the ideas of dynamic programming thus lies in the handling of a high number of feed-state dimensions and in the reduction of these for the purposes of computation. Bellman [1957], Bellman and Dreyfus [1962], Aris [1964], and Roberts [1964b], among others, have discussed various computational aspects of reducing the dimensionality of the problem. The principal method is that of the Lagrangian multiplier (Bellman and Dreyfus [1962, chap. 2]). Advantage can also be taken of any linearity of equations, it then being possible to work with ratios of variables rather than the variables themselves, an example of this being found in cross-current extraction with partly immiscible solvents (Aris, Bellman, and Kalaba [1960]). Bellman and Dreyfus [1962, p. 322] have also developed the idea of a polynomial approximation to replace the tabular representation.

4.2.3 *Constraints*

In most practical situations, limitations will be imposed upon the independent variables. Decomposition by dynamic programming can still be carried out in the presence of these constraints, since they are taken into account in each subproblem. In numerical studies these limits may prove advantageous in defining a region of search.

Example XIII-8: Constraints in a sequence of stirred reaction vessels

The kinetics of production of a certain chemical from its raw materials may be represented by the first-order irreversible reaction rate equation

$$\frac{dp}{dt} = -kpe^{-E/RT} \tag{4.2-6}$$

where $p =$ concentration of raw material
$T =$ absolute temperature
$E, k =$ kinetic parameters

We want to reduce the concentration of the raw material in the feed from p_4 to p_1, using the three-stage system depicted in Fig. XIII-9. The system is in the design stage, so that the external variables $\mathbf{q}_1$, $\mathbf{q}_2$, and $\mathbf{q}_3$ each have two components—the residence time θ_i and the reaction temperature T_i. The problem is to select these external variables $\mathbf{q}_i$ so as to minimize the sum of the residence times

$$y = \theta_1 + \theta_2 + \theta_3 \tag{4.2-7}$$

subject to the restrictions that the outlet concentration be p_1 (specified), that the temperatures be fixed between an upper limit T^L and a lower limit T_L, and that the residence times be positive. A material balance about the jth stage gives

$$p_{j+1} - p_j = k\theta_j p_j e^{-E/RT_j} \tag{4.2-8}$$

Decomposing this three-stage system by dynamic programming, we shall begin with the final reactor. The minimum residence time is then

$$M_1(p_2) = \min_{q_1} \{\theta_1\} \tag{4.2-9}$$

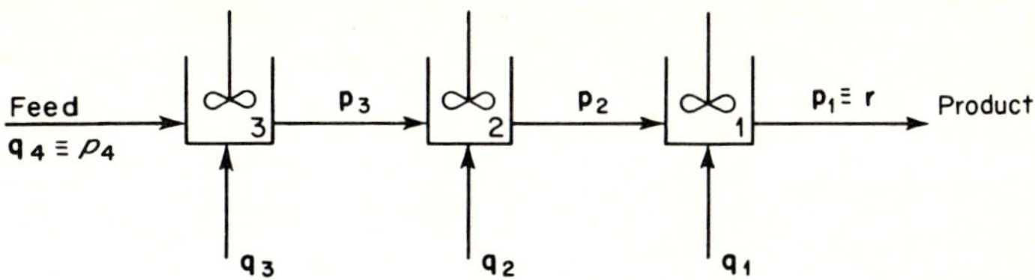

Fig. XIII-9 Sequence of stirred tank reactors. Feed $\mathbf{q}_4 = \{$concentration of raw material$\} = \{p_4\}$; reactor product streams $\mathbf{p}_i = \{$concentration of raw material in stream leaving reactor$\} = \{p_i\}$; external inputs to each reactor $\mathbf{q}_i = \{$residence time, reaction temperature$\} = \{\theta_i, T_i\}$.

subject to the imposed restrictions. Since this minimization has to be carried out for a range of fixed feeds to the last reactor, and since the system product is specified, then, from Eq. (4.2-8) we may write that

$$k\theta_1 e^{-E/RT_1} = \frac{p_2 - p_1}{p_1} = \text{constant} \tag{4.2-10}$$

Only one of the two variables θ_1 and T_1 is therefore independent. Thus, for a given temperature, the reactor residence time is fixed. The smallest value of the residence-time objective function M_1 is found when the temperature takes its largest possible value. The solution to the first subproblem is then

$$T = T^L$$

and

$$M_1(p_2) = \min \{\theta_1\} = \frac{(p_2 - p_1)e^{E/RT^L}}{p_1 k} \tag{4.2-11}$$

For any feed we should therefore use the highest possible temperature and find the residence time from Eq. (4.2-11). This Aris [1961d] calls a "disjoint" policy.

Let us now consider the two-stage policy. Thus

$$\begin{aligned} M_2(p_3) &= \min_{\mathbf{q}_1, \mathbf{q}_2} \{\theta_1 + \theta_2\} \\ &= \min_{\mathbf{q}_2} \{M_1(p_2) + \theta_2\} \end{aligned} \tag{4.2-12}$$

by the principle of optimality. Solving Eq. (4.2-8) for p_2,

$$p_2 = \frac{p_3}{1 + k\theta_2 e^{-E/RT_2}}$$

so that from Eqs. (4.2-11) and (4.2-12) we get

$$M_2(p_3) = \min_{\mathbf{q}_2} \left\{ \frac{p_3 e^{E/RT^L}}{kp_1(1 + k\theta_2 e^{-E/RT_2})} - \frac{e^{E/RT^L}}{k} + \theta_2 \right\} \tag{4.2-13}$$

This is a function only of $\mathbf{q}_2 = \{\theta_2, T_2\}$. We must minimize the terms in the braces by selecting $\mathbf{q}_2$ which satisfies the constraints $\theta_2 > 0$ and $T_L \leq T_2 \leq T^L$. T_2 appears in the first term, but in no others. This first term will take its smallest value if its denominator is as large as possible. This will occur when the highest temperature, T^L, is used, since the denominator is a monotonic increasing function of temperature. We see that the constraints are treated in the solution of the subproblem, but do not affect the method of decomposition.

Having set $T_2 = T^L$, we may now consider the task of minimizing the right-hand side of Eq. (4.2-13) by an appropriate choice of θ_2. The required minimum will lie at a stationary point, so that, after differentiating, setting the resultant expression equal to zero, and solving for θ_2, we get

$$\theta_2 = \frac{1}{k} e^{E/RT^L}\left[\left(\frac{p_3}{p_1}\right)^{1/2} - 1\right] \tag{4.2-14}$$

The minimum residence time for the two-stage subproblem is thus related to the feed p_3 by

$$M_2(p_3) = \frac{2e^{E/RT^L}}{k}\left[\left(\frac{p_3}{p_1}\right)^{1/2} - 1\right] \tag{4.2-15}$$

Finally, we may determine the optimal total residence time in the three stages as

$$M_3(p_4) = \min_{\mathbf{q}_3} \{M_2(p_3) + \theta_3\}$$

Since

$$p_3 = \frac{p_4}{1 + k\theta_3 e^{-E/RT_3}} \tag{4.2-16}$$

we have

$$M_3(p_4) = \min_{\mathbf{q}_3}\left\{\frac{2e^{E/RT^L}}{k}\left[\left(\frac{p_4}{p_1}\right)^{1/2}\left(\frac{1}{1 + k\theta_3 e^{-E/RT_3}}\right)^{1/2} - 1\right] + \theta_3\right\}$$

By the same arguments as before, the optimal values of $\mathbf{q}_3$ are

$$T_3 = T^L$$

$$\theta_3 = \frac{e^{E/RT^L}}{k}\left[\left(\frac{p_4}{p_1}\right)^{1/3} - 1\right] \tag{4.2-17}$$

The problem is now solved. All the temperatures are to be set at their maximum permitted value. The θ_3 is to be computed from Eq. (4.2-17), since all the quantities appearing on the right are known. Once θ_3 has been established, p_3 can be computed from Eq. (4.2-16), and following that, θ_2 computed by Eq. (4.2-14). In fact, it is not difficult to prove that these two steps lead to

$$\theta_2 = \frac{1}{k} e^{E/RT^L}\left[\left(\frac{p_4}{p_1}\right)^{1/3} - 1\right] = \theta_3 \tag{4.2-18}$$

Repeating these same calculations, we could also show that $\theta_1 = \theta_2 = \theta_3$.

The residence times and the temperatures for the system are now established. The essential point of this example is to indicate clearly that the presence of restrictions is accounted for in the solution of the subproblem, but does not affect the decomposition of a system by dynamic programming.

If a restriction is placed upon the input variables so that

$$\sum_{i=1}^{N} q_i \leq Q^L$$

this type of constraint can be handled within the framework of dynamic programming by the use of Lagrangian multipliers. An illustration of this will be found in the cross-current extraction problem of Aris, Rudd, and Amundson [1960], in which only a limited quantity of wash solvent is to be allocated among N stages, so as to maximize the solute extraction. Aris [1964] discusses the formal application of the Lagrangian multiplier. However, care must be taken in its use, since, as we have seen in Chap. VII, Sec. 3.3, there is no guarantee in the general case that a maximum of the modified objective function corresponds to a maximum of the original function. Roberts [1964b] also discusses the use of Lagrangian multipliers and gives extensive references to the original works of Bellman in this field.

When only discrete values of the independent variables are permitted, this proves an advantage in limiting the range of search in each subproblem.

Example XIII-9: Discrete-valued variables and dynamic programming

Maximize

$$y = \sum_{i=1}^{3} (5x_i - ix_i^2)$$

subject to the restrictions that

$$\sum_{i=1}^{3} x_i = 4 \qquad x_i = 0, 1, 2, 3, \text{ or } 4$$

This problem could be handled by a three-dimensional numerical search, using only discrete values of x_i (Chap. VIII, Sec. 6), but for illustration it will be treated by dynamic programming.

Consider the three-stage sequential system depicted in Fig. XIII-10. Here the decision vector is $\mathbf{q}_i$ (the quantity x_i to be removed at each stage), and the stage vector $\mathbf{p}_i$ is given by

$$\mathbf{p}_i \equiv p_i = p_{i+1} - x_i$$

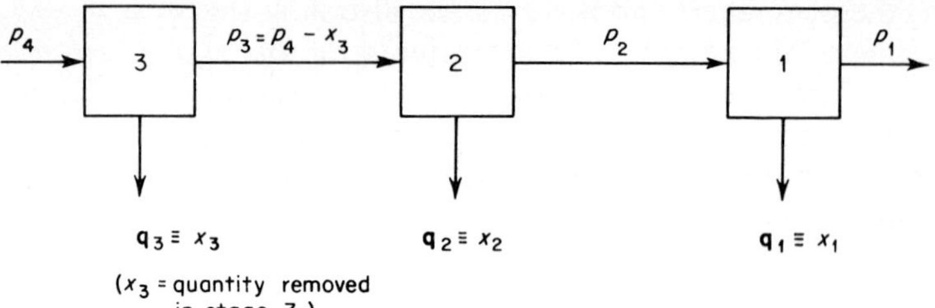

Fig. XIII-10 Three-stage sequential process.

a system feed being specified as

$$p_4 = 4$$

The objective function for each stage is

$$y_i = 5x_i - ix_i^2$$

Consider the final stage. Then

$$M_1(p_2) = \max_{x_1} \{y_1\} = \max_{x_1} \{5x_1 - x_1^2\}$$

The optimization of y_1 can best be carried out for each feed p_2 in a tabular form, since x_1 can take only discrete values 0, 1, 2, 3, or 4. Since $p_4 = 4$ and $\sum_{i=2}^{3} x_i$ must be an integer, then p_2 can take only integer values in the range $0 \le p_2 \le 4$. The values of y_1 for any valid p_2 and x_1 will then be:

	p_2	0	1	2	3	4
x_1	0	0	0	0	0	0
	1		4	4	4	4
	2			6	6	6
	3				6	6
	4					4

The optimal values M_1 of y_1 can be found and summarized in the one-stage-policy table, Table XIII-7.

Now examine the second subproblem, comprising the last two stages. Here

$$M_2(p_3) = \max_{x_1, x_2} \{y_2 + y_1\} = \max_{x_2} \{y_2 + M_1(p_2)\}$$

$$= \max_{x_2} \{5x_2 - 2x_2^2 + M_1(p_3 - x_2)\}$$

TABLE XIII-7 THE ONE-STAGE POLICY

Feed p_2	$M_1(p_2)$	x_1
0	0	0
1	4	1
2	6	2
3	6	2 or 3
4	6	2 or 3

We may again form a calculation table to find the objective function $(y_1 + y_2)$ as a function of x_2 and p_3. For any valid pair of values of p_3 and x_2 we can find y_2 and $M_1(p_3 - x_2)$, the latter being read from the one-stage-policy table, Table XIII-7. Hence:

	p_3	0	1	2	3	4
x_2	0	0	4	6	6	6
	1		3	7	9	9
	2			2	6	8
	3				−3	1
	4					−12

The optimal values M_2 can then be found and summarized in Table XIII-8 for the two-stage policy.

Finally, for the three-stage policy, we need consider only a single value of the feed, p_4, of value 4. Thus

$$M_3(p_4) = \max_{x_3} \{y_3 + M_2(p_3)\}$$

$$= \max_{x_3} \{5x_3 - 3x_3^2 + M_2(p_4 - x_3)\}$$

TABLE XIII-8 THE TWO-STAGE POLICY

Feed p_3	$M_2(p_3)$	x_2
0	0	0
1	4	0
2	7	1
3	9	1
4	9	1

TABLE XIII-9 THE THREE-STAGE POLICY

Feed p_4	$M_3(p_4)$	x_3
4	11	1

For $p_4 = 4$, the terms within the braces will be

x_3	0	1	2	3	4
$\{y_3 + M_2(p_3)\}$	9	11	5	−8	−28

so that the optimal value of $M_3(4) = 11$ at $x_3 = 1$ can be inserted in Table XIII-9.

The solution can now be found by entering Table XIII-9 at $p_4 = 4$, getting a maximum value of 11 at $x_3 = 1$. Thus $p_3 = 3$, and we can enter Table XIII-8 at this value to get $x_2 = 1$. Hence $p_2 = 2$, so that the third table, Table XIII-7, will give $x_1 = 2$. This is the desired solution. While elementary, the illustration serves to show the procedure for discrete variables and, in addition, indicates how a mathematical summation can be set up in the discrete-stage form.

In most cases the decision variables are altered during optimization in order to choose the appropriate values to yield an optimal result. In some situations, however, the decision may be restricted, not to a range of values of a given variable, but to the choice between two systems. An illustration is found in the keep-replace decision required in the periodic maintenance review of equipment whose performance deteriorates with use. Here the decision made might change the form of the objective function. This is an example of a k^N policy, here with $k = 2$, k being the number of decisions to be made at each stage. (Another illustration was given in Sec. 3.)

Example XIII-10: Optimal equipment-replacement policies

A process has a particular item of equipment which is t years old and which has no salvage or resale value. The annual income from this item, after all running expenses have been met, is given by

$$S = \begin{cases} 25 - t^2 & \text{for } 0 \leq t \leq 4 \\ 0 & \text{for } t > 4 \end{cases}$$

The cost of replacement I_0 is 21, and the process still has a life of 5 operating years. What is the optimal future equipment replacement policy if the present item is 2 years old and if a decision must be made annually on the basis of a maximum total 5-year profit?

The annual profit $P(t)$ will be, then, the difference between income and costs

$$P(t) = \begin{cases} 25 - t^2 & \text{for } t \leq 4 \\ 0 & \text{for } t > 4 \end{cases}$$

if we keep the unit, but

$$P(t) = 25 - 0 - 21 = 4$$

if we replace it where the additional charge is the cost of replacement. Since the unit has no resale value, the net profit from a new unit is independent of the age of the unit being replaced, and we may write $P(0) = 4$ as the resultant first-year profit.

We must now find the best set of decisions, keep (K) or replace (R), to maximize the profit from the last 5 years of the process's life. Thus we wish to maximize

$$y = \sum_{i=1}^{N} P_i(t)$$

with respect to D_i, where D_i can take the value K_i or R_i for $i = 1, 2, \ldots, N$.

By a direct approach we could write all possible decision considerations and find the best result. Thus, for one arbitrary combination *KRKKR*, we have the result shown in Table XIII-10. This approach requires 2^N evaluations, here 32, all of which need to be compared with each other to determine the highest level of profits for the 5-year period.

The problem can, however, be carried out more simply by dynamic programming, since it is a multistage process, as depicted in Fig. XIII-11. By the principle of optimality we may write that

$$M_N(t_{N+1}) = \max_{\text{all } K,R} \{P_N(t_{N+1})\} = \max_{K_N,R_N} \{P_N(t_{N+1}) + M_{N-1}(t_N)\} \quad (4.2\text{-}19)$$

TABLE XIII-10 DIRECT CALCULATION OF EQUIPMENT PROFIT FOR POLICY *KRKKR*

Age at end of previous period	*Policy*	*Age at start of period, t*	*Profit*
2	*K*	2	21
3	*R*	0	4
1	*K*	1	24
2	*K*	2	21
3	*R*	0	4
		Total profit =	74

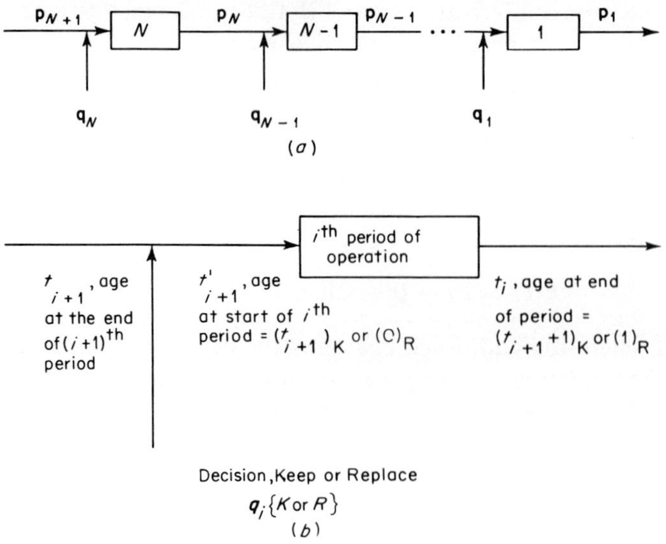

Fig. XIII-11 A periodic decision.

where $t_N = t_{N+1} + 1$ if we keep the unit, and $t_N = 1$ if we replace it. Starting with 1 year of operation left (the final stage of Fig. XIII-11), we have that

$$M_1(t_2) = \max_{K_1, R_1} \{P(t_2)\}$$

The results are shown in the first column of Table XIII-11 for a range of t values. For convenience the value of the optimal decision variable (K or R) is also included. Equation (4.2-19) can now be used, together with these results, to generate the 2-year policy, the results being shown

TABLE XIII-11 DYNAMIC PROGRAMMING SOLUTION

M_N, profit from N years operation	*Years of process operation left, N* 1	2	3	4	5
Age of unit, t_{N+1}, at end of previous period					
1	24K	45K	61K	73K	94K
2	21K	37K	49K (or R)	70K	86K
3	16K	28R	49R	65K (or R)	81K
4	9K	28R	49R	65R	77R
5	4R	28R	49R	65R	77R
6	4R	28R	49R	65R	77R

in column 2. For example, with a 3-year-old item and 2 years to go, we shall have the form

$$M_2(3) = \max \begin{cases} P(3) + M_1(4) & \text{if we keep it} \\ P(0) + M_1(1) & \text{if we replace it} \end{cases}$$

Since from column 1, $M_1(4) = 9$ and $M_1(1) = 24$ (both resulting from a keep decision), we have the two possible results

$$K: \sum_{i=1}^{2} P(t) = 16 + 9 = 25$$

$$R: \sum_{i=1}^{2} P(t) = 4 + 24 = 28$$

It is therefore more profitable to replace this unit in this situation. Results for other situations are shown in the complete table.

The desired solution may now be found by entering the table at $N = 5$, $t = 2$, from which it will be seen that an optimal policy of *KKRKK* (or *KRKKK*) would yield a profit of 86. Each of the two equivalent policies is indicated in the table by a sequence of arrows. It will be noted from, for example, the first policy, that while the 5-year operation profit is 86, 65 of this was obtained in the final 4 years, 49 in the last 3, 45 in the last 2, and 21 in the last period. A comparison of the two policies can thus show that more profit is returned in the first 2 years of the final 5 years left in the process with the first optimal policy than the second, and this might be preferred if the accuracy of the cost estimates might be questioned after several years of operation.

For this table, 38 calculations[1] were made compared with the $2^N (= 32)$ in the direct approach. Dynamic programming would therefore find favor only if more than one unit had to be analyzed over varying periods of operation, since the table can be used for any N and t as long as the system satisfies the original functions.

4.3 The Advantages of Dynamic Programming

In many cases, as Aris [1961d, 1964], for example, has noted, there is a distinct computational saving in using dynamic programming as compared with other methods. The general (and the worst possible) approach in an N-stage process may require r decisions at each stage, α values of each decision variable being investigated. The number of objective function evaluations will then be α^{Nr}. By dynamic programming, on the other hand,

[1] Actually 30×2 calculations were required for the table shown, but the problem posed needed only 19×2 of these.

α^r calculations would be required in each stage for each feed state. If the latter is defined by s variables, then each subproblem except the last needs α^{r+s} evaluations (the last requiring only α^r, since the system feed is specified), giving a total of $\alpha^r[1 + (N - 1)\alpha^s]$ for the whole system. This is quite a saving, even for small calculations. Thus, if $N = 4$, $r = 1$, $s = 1$, $\alpha = 10$, the savings are about 4×10^4 calculations, while if $N = 5$, $r = 3$, $s = 1$, $\alpha = 10$, the savings are 5×10^{15}—an astronomical difference! However, these savings depend also on the dimensionality s of the feed, and they may disappear with increasing values of s when storage problems arise. (In

TABLE XIII-12 SOME REPRESENTATIVE APPLICATIONS OF DYNAMIC PROGRAMMING

General texts:	Bellman [1957], Aris [1961d], Bellman [1961], Bellman and Dreyfus [1962], Aris [1964], Roberts [1964b].
General applications:	
(a) Equipment-replacement policies	White [1965], Bellman [1955], Dreyfus [1960b], Bellman and Dreyfus [1962, chap. 3], Roberts [1964b, chap. 2].
(b) Economic planning periods	McDowell [1960], White [1962].
(c) Optimal lot-size policy	Chambers, Bond, and Leake [1958], White [1965].
(d) Investment problems	White [1963], Cord [1964], White and Norman [1965].
(e) Production control	White [1962, 1964].
(f) Inventory	Bellman [1956], Dreyfus [1957], Bellman and Dreyfus [1962, chap. 3], Roberts [1964b, chap. 8], White [1966], Sasieni [1960].
(g) Allocation problems	Simpson [1959], Aris, Rudd, and Amundson [1960], White [1961], Roberts [1964b, chap. 3], Rudd [1962a, b].
(h) Transportation problems	Bellman and Dreyfus [1962, chap. 2]
(i) Reliability theory	Rudd [1962c].
(j) Control	Bellman and Dreyfus [1962, chap. 8], Roberts [1964b, chap. 6].
(k) Labor distribution	Bishop and Rockwell [1958].
Chemical reactor design (Multistage):	
(a) Stirred tank reactors	
Single reaction—	Aris [1961d, p. 64; 1961b], Aris, Bellman, and Kalaba [1960].
Consecutive reactions—	Aris [1961d, p. 78].
Multiple reactions—	Aris [1961d, p. 85; 1960a].
Bypass feed—	Aris [1961d, p. 96; 1960d, 1961c].
(b) Multibed adiabatic reactors	
Heat-exchange intercooling—	Aris [1961d, p. 106; 1960b], Lee and Aris [1963a, b].
Cold-shot cooling—	Aris [1961d, p. 121; 1961c].
(c) Catalyst replacement	Roberts [1960; 1964b, chap. 2].

TABLE XIII-12 (CONTD.)

Other multistage process problems:	
(a) Cross-current liquid-liquid extraction	Aris, Rudd, and Amundson [1960], Aris, Bellman, and Kalaba [1960].
(b) Cross-current extraction with recycle	Rudd and Blum [1962], Jackson [1963, 1964a], Van Cauwenberghe [1964], Lovland [1964].
(c) Multistage gas compression	Aris, Bellman, and Kalaba [1960].
(d) General chemical process	Nemhauser [1961], Aris [1961d], Mitten and Nemhauser [1963a, b, c].
(e) Heat-exchange systems	Westbrook [1961].
(f) Distillation	Converse and Gross [1963], Converse and Huber [1965].
(g) Heavy-water distillation	Bellman [1957], Cerri, Silestri, and Villon [1956].
Continuous processes in time or space:	
(a) Catalyst replacement	Roberts [1964b].
(b) Control	Roberts [1964a; 1964b, chap. 6] Bellman [1961].
(c) Batch or tubular reactor	Aris [1960c; 1961d, chap. 7; 1961a, 1961b] Aris, Bellman, and Kalaba [1960], Katz [1960], Pis'men and Ioffe [1962].
Stochastic processes:	
(a) Processes subject to random failure	Rudd [1962c], Aris [1961d, p. 160].
(b) Raw-material sources	Aris [1961d p. 163], Rudd [1962a].
(c) Catalyst replacement	Roberts [1960; 1964b, chap. 9], Aris [1961d].
(d) Equipment replacement	Bellman and Dreyfus [1962].
(e) Allocation	Roberts [1964b, chap. 9].
(f) Averaging control	Roberts [1964b, chap. 9].

addition, of course, it is unlikely that we should adopt the worst method as an alternative approach.)

Some representative applications of dynamic programming are shown in Table XIII-12. Dynamic programming can also be used in the optimization of continuous systems (Chap. IX, Sec. 5.2.3), which might be regarded as an infinite number of discrete units in sequence. It will be remembered that dynamic programming forms a framework for optimization, in which any suitable optimization technique can be used. Most convenient for single component links between subproblems, dynamic programming becomes increasingly more unmanageable with greater dimensionality of these links. It cannot, of course, directly handle more complex situations than a straight sequential chain. This aspect is taken up in Sec. 5.

Instead of solving any individual problem, dynamic programming considers a family of related problems. This invariant bedding technique also has its limitations, since we must consider any possible condition beginning at any stage, leading to a high redundancy in calculations because of the introduction of additional but finally unused members of the original family. However, when answers are required to a class of problems, as in the general study proceeding from an initial state to a final product, when both of these might cover a range of values, the method has advantages.

Dynamic programming is conceptionally simple, however, and can be applied to any number of units in series, even if these have different forms of variables and constraints from stage to stage. It can easily accommodate constraints, the inequality form being an advantage in limiting the region of search, especially when disjoint policies are found. Since dynamic programming provides a framework for optimization, appropriate optimization techniques can be used in any stage as needed.

5 COMPLEX SYSTEMS

In the previous section we confined our attention to sequential chains and showed how dynamic programmng was a possible aid in optimization. Let us now consider more general situations which might contain loops, branches, and recycle streams, in addition to sequences of units. Some examples of these more realistic arrangements are shown in Fig. XIII-12. Optimization of such general situations is difficult. The principle of optimality (Sec. 3) cannot always be used for decomposition, since, for example, recycle operations have the property that a decision in the last stage can influence previous results, in contrast to the purely sequential operations in which the final-stage decision has no effect on prior operations.

The decomposition of complex optimization problems into a series of simpler subproblems is still a fruitful and important area of optimization research. The available techniques may be classified either as methods of feasible states or of non-feasible states.

The *methods of feasible states* (Sec. 5.1) are all based upon the principle of splitting (Sec. 2) and differ only in their detailed approach to a solution. They are characterized by the fact that each computation aims at improving the objective function, using only admissible sets of independent variables—in other words, a set which could be used in the real system.

The *methods of nonfeasible states* (Sec. 5.2), on the other hand, approach the optimum, often using impossible or invalid states as a basis of computation.

In any splitting problem, it is important, as seen, for example, in Sec. 4.2.2, that a minimum of interactions exist between subproblems. Logical procedures for deciding how to decompose large systems by splitting at preferred locations have been considered by Aris, Nemhauser, and Wilde [1964], Graves and Wolfe [1963], Jackson [1964a], Vajda [1961], and Wilde [1964] in connection with optimization schemes; by Lasdon and

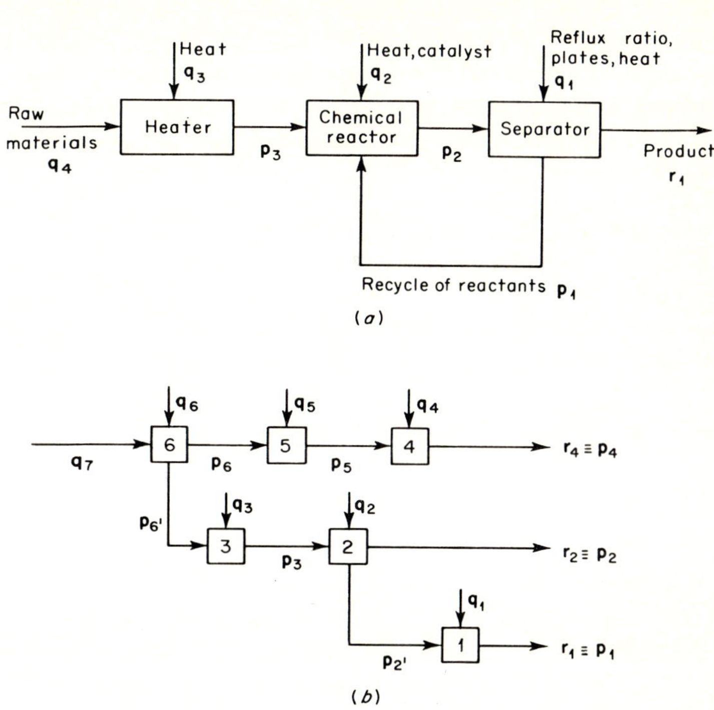

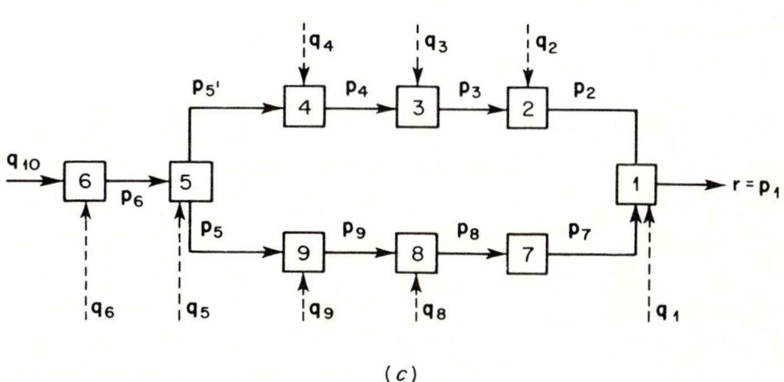

Fig. XIII-12 (*a*) A recycle system; (*b*) a branching system; (*c*) a loop system.

Schoeffler [1964] in process control; and by Rudd and Stewart [1965], Sargent and Westerberg [1964], and Himmelblau [1966] in design.

5.1 Methods of Feasible States

Based upon the principle of splitting, each iterative cycle of search, using one of these methods, seeks to enhance the objective function by locating a new set of admissible independent variables. Each such policy, or set of variables, will then constitute a valid solution to the problem, although only the final set will be optimal. Two main approaches might be used.

First, in the approach which is here called *diagrammatic feasible state decomposition* (Sec. 5.1.1), the principle of splitting is used to fix temporarily certain links, so that the main system is reduced to a set of subsystems connected by these links. Each subsystem is now independent and can be optimized separately. In the choice of these subsystems it is advantageous to obtain an arrangement of the structure so that sequential chains are formed.

Second, a mathematical approach can be used (Sec. 5.1.2), its essential idea being the fixing of certain linking variables so as to form a particular mathematical, as opposed to structural, arrangement.

It can be shown that the resulting mathematical and structural arrangements are related, since the decompositions are achieved by cutting links of lower dimensionality (in other words, splitting into sections which are "weakly coupled").

5.1.1 Diagrammatic feasible-state decomposition

The main theme of diagrammatic decomposition is the division of a complex system by splitting it into a number of subsystems, each of which can be readily optimized. Sequential subsystems seem preferable in most cases, since established methods of solution such as dynamic programming and Lagrangian multipliers are readily applicable to most sequential systems. The methods of Mitten and Nemhauser [1963a] and Aris, Nemhauser, and Wilde [1964] are of this type.

Let us consider the recycle system of Fig. XIII-12*a* as an illustration. Here certain decision variables represented by $\mathbf{q}_i$ have to be selected so as to optimize a given form of objective function, usually economic in nature. These variables could, for example, be the reactor volume, the heat input, the reflux ratios in the separation plant, and so on. One approach would be the direct one of applying a numerical search technique in which all the dimension variables $\mathbf{q}_i$ are varied independently. It is likely, however, that the dimensionality of the system would render this approach hopeless. A more realistic approach may be found through application of the principle of splitting.

Let us temporarily fix the recycle-stream specification at some preselected state, $\mathbf{p}_1^{(0)}$. If this stream is fixed, the resulting system is the sequential one depicted in Fig. XIII-13. Here the circle against a stream denotes a new invariant link, temporarily fixed at $\mathbf{p}_1^{(0)}$. Since two such circles are required to represent the severing of an internal link, such as a recycle stream, we see that the recycle stream leaving stage 1 is fixed at the same value as it enters stage 2. For a given feed $\mathbf{p}_2$ to the last stage, at least some of the variables in

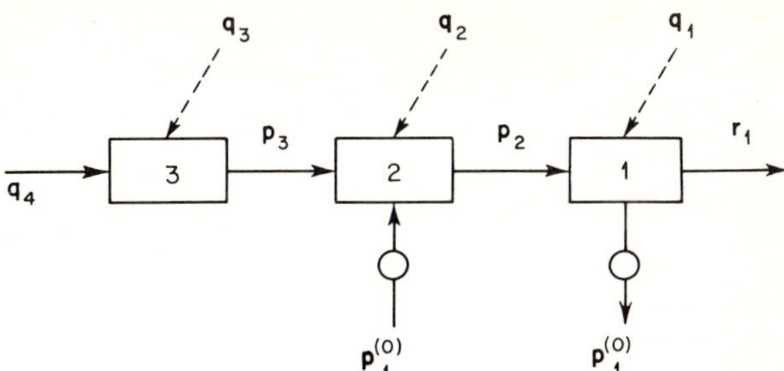

Fig. XIII-13 The subsystem resulting from a fixed recycle stream.

$\mathbf{q}_1$ must be manipulated so as to achieve the preselected value of the recycle stream, and so are not free for variation during optimization.

Let us therefore denote those variables of the vector $\mathbf{q}_1$ which are free to be manipulated as $\mathbf{q}_{1,f}$, remembering that the list of free variables changes as the feed to the last stage changes and as the external variables themselves are varied. To clarify this idea, we might, for a given feed to the separation section, stage 1, be able to keep the recycle stream at the preselected state by appropriately varying a reflux ratio while freely varying the number of distillation plates in a certain column so long as the number of plates remains between fixed limits. However, when the number of plates exceeds one of these limits, it might be necessary to adjust a second external variable, say, the boil-up rate, together with the reflux ratio, to obtain the preselected value of the recycle stream. The essential point is that the recycle stream must be maintained at the preselected value. If there is no combination of the $\mathbf{q}_1$ which yields $\mathbf{p}_1^{(0)}$ for a given $\mathbf{p}_2$, then that particular feed is not admissible as a possible feed to the first stage when the recycle is $\mathbf{p}_1^{(0)}$.

Remembering that we must maintain $\mathbf{p}_1^{(0)}$ at its specified value, we are now in a position to optimize the subsystem resulting from the decomposition. Since this is now a sequential problem, we could apply dynamic programming to provide a framework for optimization. However, the most convenient method should always be adopted. Although not the sole possibility, let us proceed using dynamic programming, assuming that the objective function can be written

$$y(\mathbf{q}_4,\mathbf{q}_3,\mathbf{q}_2,\mathbf{q}_1,\mathbf{r}_1) = y_3(\mathbf{q}_4,\mathbf{q}_3,\mathbf{p}_3) + y_2(\mathbf{p}_3,\mathbf{p}_2,\mathbf{p}_1,\mathbf{q}_2) + y_1(\mathbf{p}_2,\mathbf{p}_1,\mathbf{r}_1,\mathbf{q}_1) \quad (5.1\text{-}1)$$

Let us also assume that $\mathbf{q}_4$ is a fixed state feed and that the transformation equations (Chap. II, Sec. 4.3.1) are

$$\mathbf{p}_3 = \mathbf{T}_3(\mathbf{q}_4,\mathbf{q}_3) \quad (5.1\text{-}2a)$$

$$\mathbf{p}_2 = \mathbf{T}_2(\mathbf{p}_3,\mathbf{p}_1,\mathbf{q}_2) \quad (5.1\text{-}2b)$$

$$\mathbf{p}_1 = \mathbf{T}_1(\mathbf{p}_2,\mathbf{q}_1) \quad (5.1\text{-}2c)$$

$$\mathbf{r}_1 = \mathbf{T}_1'(\mathbf{p}_2,\mathbf{q}_1) \quad (5.1\text{-}2d)$$

Since $\mathbf{p}_1$ is specified at a value $\mathbf{p}_1^{(0)}$, then Eq. (5.1-2*c*) dictates that at least some of the components of $\mathbf{q}_1$ are not free, with $\mathbf{q}_{1,f}$ representing those which are free for a given feed $\mathbf{p}_2$. We can then define the optimal value of the objective function, y_1, for the last stage as

$$M_1(\mathbf{p}_2,\mathbf{p}_1^{(0)}) = \underset{\mathbf{q}_{1,f}}{\text{opt}} \{y_1(\mathbf{p}_2,\mathbf{p}_1^{(0)},\mathbf{r}_1,\mathbf{q}_1)\} \tag{5.1-3}$$

This defines the optimal value when both $\mathbf{p}_2$ and $\mathbf{p}_1$ are fixed. By the principle of optimality we can then define

$$\begin{aligned} M_2(\mathbf{p}_3,\mathbf{p}_1^{(0)}) &= \underset{\mathbf{q}_2,\mathbf{q}_{1,f}}{\text{opt}} \{y_1 + y_2\} = \underset{\mathbf{q}_2}{\text{opt}} \{M_1(\mathbf{p}_2,\mathbf{p}_1^{(0)}) + y_2(\mathbf{p}_3,\mathbf{p}_2,\mathbf{p}_1^{(0)},\mathbf{q}_2)\} \\ &= \underset{\mathbf{q}_2}{\text{opt}} \{M_1[\mathbf{T}_2(\mathbf{p}_3,\mathbf{p}_1^{(0)},\mathbf{q}_2),\mathbf{p}_1^{(0)}] + y_2[\mathbf{p}_3,\mathbf{T}_2(\mathbf{p}_3,\mathbf{p}_1^{(0)},\mathbf{q}_2),\mathbf{p}_1^{(0)},\mathbf{q}_3]\} \end{aligned} \tag{5.1-4}$$

and

$$M_3(\mathbf{q}_4,\mathbf{p}_1^{(0)}) = \underset{\mathbf{q}_3}{\text{opt}} \{M_2(\mathbf{p}_3,\mathbf{p}_1^{(0)}) + y_3(\mathbf{p}_3,\mathbf{q}_4,\mathbf{q}_3)\} \tag{5.1-5}$$

Since the feed to the system $\mathbf{q}_4$ is specified, $M_3(\mathbf{q}_4,\mathbf{p}_1^{(0)})$ is the three-stage profit for the sequential system for a given $\mathbf{p}_1^{(0)}$. This link must now be changed to some value $\mathbf{p}_1^{(1)}$, which, it is hoped, will improve M_3. Thus $M_3(\mathbf{q}_4,\mathbf{p}_1)$ is a function of a single vector, $\mathbf{p}_1$, and we must now search for the value of $\mathbf{p}_1$ which gives the optimum value of M_3. In other words, the system optimum is given by

$$M_{\text{system}} = \underset{\mathbf{p}_1}{\text{opt}} \{M_3\} \tag{5.1-6}$$

Using some search technique from Part Two, we should logically select various values of $\mathbf{p}_1$ until the most suitable value is found.

Example XIII-11: A recycle process

Consider the system shown in Fig. XIII-14*a*. Raw material entering the system is heated to a temperature T before entering the reaction section. Here one of two alternative reactors, A_2 or B_2, might be chosen. The product stream containing some unconverted raw material then enters a separation unit in which pure reaction product is withdrawn for sale. The separation unit can be operated in one of two ways (A_1 or B_1). The unconverted raw material and some unseparated product are sent to a recovery unit, which purifies the stream, sending pure raw material back to join the heater inlet stream. The recovery unit can be one of two units, A_4 or B_4.

It is desired to maximize the system profit by a suitable choice of the $\mathbf{q}_i$, where $\mathbf{q}_1 = A_1$ or B_1; $\mathbf{q}_2 = A_2$ or B_2; $\mathbf{q}_3 = T$; $\mathbf{q}_4 = A_4$ or B_4. The system feed x_5 is fixed at 10^4 lb raw material per day; x_i is the total weight flow rate per day of the ith stream; w_2 is the fraction by weight of the

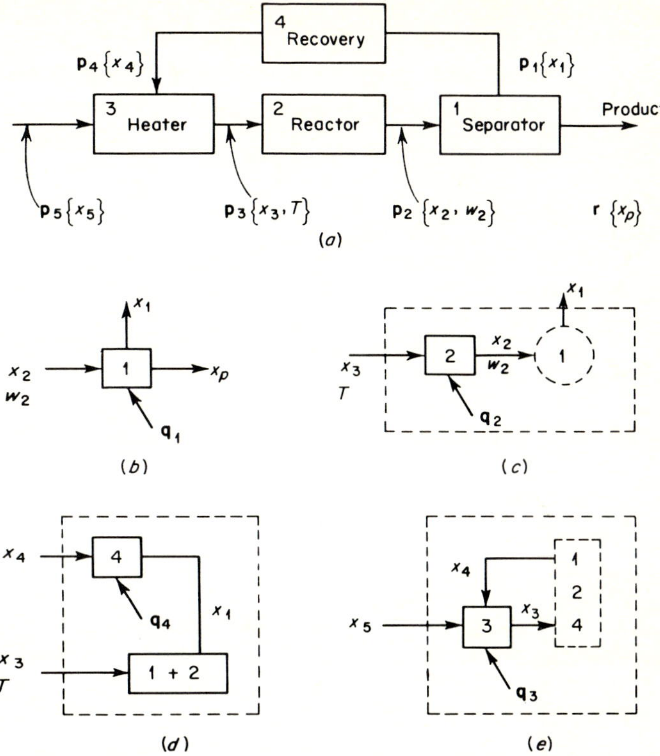

Fig. XIII-14 A recycle system. (*a*) The system; (*b*) stage 1; (*c*) stage 2; (*d*) stage 3; (*e*) stage 4.

product component in the reactor exit stream; T is the temperature to which the heater heats up the feed $(x_4 + x_5)$, within the range $200 \le T \le 450°\text{C}$; x_p, the system product, consists of pure reaction product. Maximize the system profit if the income from the sale of the product is $40x_p$ and the daily costs of operation are $10^{-7}x_3T^2$ for the heater; $3x_3$ for reactor A_2, giving a conversion of $w_2 = 0.002T$ lb product per pound raw material processed; $2x_3$ for reactor B_2, giving a conversion of $w_2 = 0.001T$; $3x_2 + 100/w_2$ for separator A_1; $x_2 + 200/w_2$ for separator B_1; $4x_4$ for recovery unit A_4, converting 75 percent input into pure recycled raw material; and $3x_4$ for recovery unit B_4, converting 50 percent input into pure recycled raw material.

Stage 1 Let us fix x_1. From Fig. XIII-14*b*, mass balances give

$$x_2 = x_1 + x_p$$

$$x_p \le w_2 x_2$$

and the profit from unit 1 is given by ($40x_p$ − cost of operating unit 1 as A_1 or B_1). Hence on eliminating x_p

$$M_1(x_2,w_2,x_1) = \max_{A_1 \text{ or } B_1} \begin{Bmatrix} \left(37x_2 - 40x_1 - \dfrac{100}{w_2}\right)_{A_1} \\ \left(39x_2 - 40x_1 - \dfrac{200}{w_2}\right)_{B_1} \end{Bmatrix}$$

Now profit A_1 is less than profit B_1 if $x_2w_2 > 50$, which is true. Hence, using unit B_1, we have

$$M_1(x_2,w_2,x_1) = 39x_2 - 40x_1 - \frac{200}{w_2}$$

Stage 2 This is shown in Fig. XIII-14*c*. We can write that

$$M_2(x_3,T,x_1) = \max_{A_1 \text{ or } B_1,\ A_2 \text{ or } B_2} \{\text{profit 1} + \text{profit 2}\}$$

$$= \max_{A_2 \text{ or } B_2} \{\text{profit from 2} + M_1(x_2,w_2,x_1)\}$$

Now the stage 2 profit, using A_2, is $36x_3 - 40x_1 - 10^5/T$, and, using B_2, is $37x_3 - 40x_1 - (2 \times 10^5)/T$, so that the profit with A_2 is less than the profit with B_2 when $x_3T > 10^5$, which is true. There is, however, an additional restriction on x_1, that if $x_1 \geq (1 - 0.002T)x_3$, we must use A_2, or if $x_1 \geq (1 - 0.001T)x_3$, we must use B_2. Hence, if $(1 - 0.002T)x_3 \leq x_1 \leq (1 - 0.001T)x_3$, we use A_2, with $M_3(x_3,T,x_1) = 36x_3 - 40x_1 - 10^5/T$; but if $(1 - 0.001T)x_3 \leq x_1$, we use B_2, and $M_3(x_3,T,x_1) = 37x_3 - 40x_1 - (2 \times 10^5)/T$.

Stage 3 Using the system shown in Fig. XIII-14*d*, we have

$$M_3(x_3,x_4,T) = \max_{A_4 \text{ or } B_4} \{\text{profit from 4} + M_2\}$$

Four possibilities arise, which can be summarized as follows:

	$(1 - 0.002T)x_3 \leq x_1 \leq (1 - 0.001T)x_3$	$(1 - 0.001T)x_3 \leq x_1$
Profit using A_4 ($x_4 = 0.75x_1$)	$-57.33x_4 + 36x_3 - \dfrac{10^5}{T}$	$-57.33x_4 + 37x_3 - \dfrac{2 \times 10^5}{T}$
Profit using B_4 ($x_4 = 0.50x_1$)	$-83x_4 + 36x_3 - \dfrac{10^5}{T}$	$-83x_4 + 37x_3 - \dfrac{2 \times 10^5}{T}$

We should use A_4 in both cases, so that $M_3(x_3,x_4,T)$ has one of two forms, depending on the restrictions imposed.

Stage 4 Looking at the system of Fig. XIII-14*e*, we can write

$$M_4(x_5) = \max_{T,x_3,x_4}\{-10^{-7}T^2x_3 + M_3(x_3,T,x_4)\}$$

subject to restrictions $200 \leq T \leq 450$; $x_3 = x_4 + x_5$; and one of the pair

$$(1 - 0.002T)x_3 \leq \tfrac{4}{3}x_4 \leq (1 - 0.001T)x_3$$
$$(1 - 0.001T)x_3 \leq \tfrac{4}{3}x_4$$

The first of this pair gives the profit relation

$$y = -10^{-7}T^2(x_4 + x_5) - 21.3x_4 + 36x_5 - \frac{10^5}{T}$$

which has the solution $T = 450$, $x_4 = 810$, $y = 345{,}000$; the second gives the function

$$y = -10^{-7}T^2(x_4 + x_5) - 20.3x_4 + 37x_5 - \frac{2 \times 10^5}{T}$$

which has the solution $T = 450$, $x_4 = 8{,}800$, $y = 230{,}000$. Hence $M_4 = 345{,}000$, using the first of these solutions.

Our desired solution and specification is, then,

$$T = 450; \qquad x_4 = 810; \qquad x_3 = 10{,}000 + 810 = 10{,}810 = x_2;$$

$$x_1 = \frac{x_4}{0.75} = 1{,}080; \qquad w_2 = 0.002T = 0.9; \qquad x_p = x_2 - x_1 = 9{,}730;$$

using B_1, A_2, and A_4; and yielding an income of $40x_p = \$389{,}200$, with a total profit of \$345,000. (Based on Mitten and Nemhauser [1963a].)

This technique can also be used to handle loops and branches, such as shown in Fig. XIII-12. For example, the simple branching system of Fig. XIII-12*b* can be decomposed by fixing a single link, $\mathbf{p}_6$, as shown in Fig. XIII-15 to get two parallel sequential systems which can be optimized separately. The global optimum can then be found as before by varying $\mathbf{p}_6$ in each iteration, using some suitable numerical search method.

To decompose the loop system of Fig. XIII-12*c*, two links must be fixed temporarily. The resulting pair of parallel sequential systems is depicted in Fig. XIII-16. In this case, the iteration to find a global optimum has to be made on two vectors, one at each fixed link, and the dimensionality of the numerical search is thus increased. If a choice is available, we might therefore consider splitting at the links defined by vectors with smaller numbers of components.

It is clear that in very complex systems a large number of links may have to be fixed temporarily to achieve the structure of parallel sequential systems.

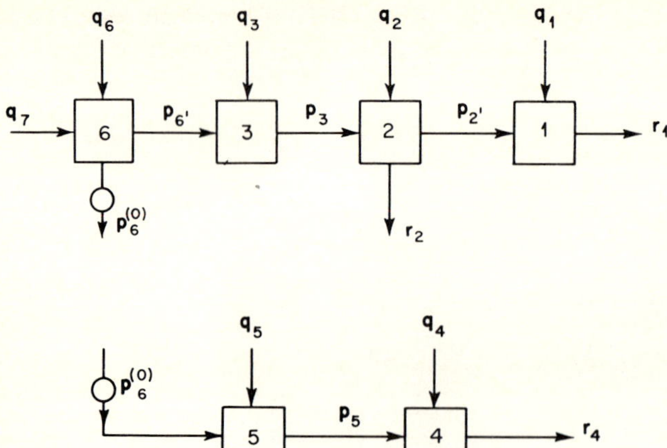

Fig. XIII-15 The decomposed branching system.

The dimensionality of the direct search is accordingly very large, and the convergence to a solution slow and costly. Such systems are presently beyond our capability for solution, and one usually makes simplifications (hopefully, valid) which reduce the problems to manageable proportions.

5.1.2 *Mathematical feasible state decomposition*

In Sec. 5.1.1 we saw how the principle of splitting was used to uncouple complex unit arrangements into a set of parallel sequential subsystems. In many cases these subsystems could then be further decomposed by dynamic programming. In certain problems, however, such a decomposition scheme might not produce a manageable subproblem or, indeed, due to the structure of the objective function, might not be susceptible to this form of

Fig. XIII-16 The decomposition of a loop system.

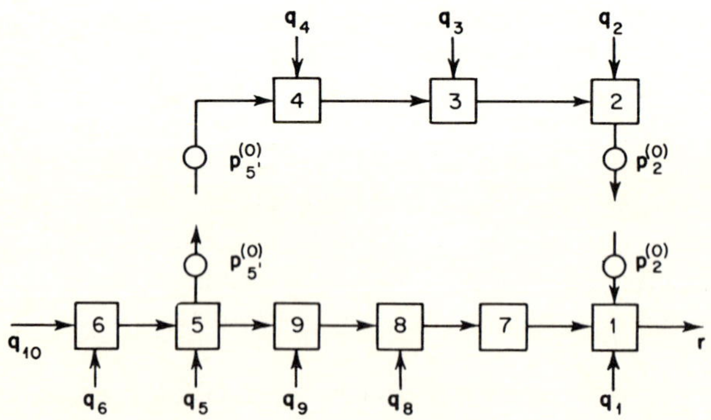

decomposition. An example of this latter situation would be found if the objective function could not be decomposed into a sum of objective functions, one for each subsystem. The methods of Sec. 5.1.1 cannot then be readily applied. A second approach to decomposition is found in the examination of the mathematical structure of the equations defining the objective function, the input-output relations, and the restrictions and the subsequent simplification of the problem—a mathematical decomposition.

Partition Programming One approach is that due to Rosen and Ornea [1963] (see also Rosen [1963] and Ornea and Eldredge [1965]), called *partition programming*. The method depends upon the existence of certain mathematical features (as opposed to the topological features required in Sec. 5.1.1). These features can be considered in two classes.

First, a distinction is made between the independent variables which enter the problem linearly and those which are nonlinear. Neither group can be empty if partition programming is to be effective. In other words, we must have both linear and nonlinear forms in the system. The problem is then decomposed so that, for any set of fixed values of the nonlinear variables $(z_1, z_2, \ldots, z_N)$, the resultant subproblem must have completely linear constraints and objective function, linear in the variables $(x_1, x_2, \ldots, x_N)$. The z_i may appear in the objective function as

$$y = \sum_{i=1}^{N} c_i(\mathbf{z})x_i + \phi(\mathbf{z}) \tag{5.1-7}$$

the function ϕ being then constant, and in the constraints as

$$\sum_{j=1}^{N} a_{i,j}(\mathbf{z})x_j \leq b_i(\mathbf{z}) \qquad \text{for } i = 1, 2, \ldots, m \tag{5.1-8}$$

Here $\mathbf{z}$ represents the entire collection of nonlinear variables.

Second, for very large systems, the constraints must be in a form such that the m constraint equations reduce to

$$\sum_{j=1}^{N_1} a_{1,j}(\mathbf{z})x_j \leq b_1(\mathbf{z}) \tag{5.1-9a}$$

$$\sum_{j=N_1+1}^{N_2} a_{2,j}(\mathbf{z})x_j \leq b_2(\mathbf{z}) \tag{5.1-9b}$$

$$\cdots\cdots\cdots\cdots\cdots$$

$$\sum_{j=N_{m-1}+1}^{N_m} a_{m,j}(\mathbf{z})x_j \leq b_m(\mathbf{z}) \tag{5.1-9c}$$

with $\sum_{i=1}^{m} N_i = N$. The first constraint thus involves only the first N_1 of the linear variables x_i; the second involves only $N_2 - N_1$ of the x_i; and so on. If this structure does not occur naturally, the problem can be put into this form by letting some of the linear variables be classed in the nonlinear group and transferring the terms involving those linear variables to the right-hand side of the inequalities (5.1-8).

Any problem exhibiting these features can then be solved by an iterative procedure. An initial, feasible choice for the $\mathbf{z}$ is first made, and the linear problems are separately solved by linear programming. There are m such linear subproblems, the first being

$$y_1 = \sum_{i=1}^{N_1} c_i(\mathbf{z})x_i \tag{5.1-10a}$$

subject to the constraints set forth by inequalities (5.1-9*a*). The second subproblem is

$$y_2 = \sum_{i=N_1+1}^{N_2} c_i(\mathbf{z})x_i \tag{5.1-10b}$$

subject to the constraints given by inequalities (5.1-9*b*). The other subproblems can be similarly defined.

These optimal solutions are then used to obtain improved values of $\mathbf{z}$. The procedure is repeated until an acceptable approximation to the optimum is achieved. The gradient projection method (Chap. VIII, Sec. 5.3) has been used to search for these new locations of $\mathbf{z}$, the optimal solutions to the subproblems being required, together with the partial derivatives of y.

This method appears to be of some value when a large number of the variables enter the problem in a linear fashion. In this case the method should be capable of optimizing large and complex systems. If, however, all or almost all the decision variables enter the problem nonlinearly, one could not expect much gain in using partition programming.

Decomposition in Linear Programming If all the variables enter the objective function and constraints linearly, the decomposition method of Dantzig and Wolfe [1960] for linear programming may be used to form subsystem problems. This approach is closely related to that of partition programming. The essential idea is the solution of local linear programming problems (Chap. VII, Sec. 4.4), in which additional penalty constraints have been added to account for their relationship with the main system (Dantzig [1963]). Beale, Hughes, and Small [1965] have discussed some of the computational aspects of this technique, while Wilde [1965] has, for example, applied it in production planning, using not the original matrix algebraic formulation, but only simple differentiation, to obtain the penalty constraints.

5.2 A Method of Nonfeasible States

As we have seen in Part Two (Chap. VII, Sec. 3.3), the method of Lagrangian multipliers can be effectively used to optimize certain types of system.

Jackson [1964b], Katz [1962a], Horn [1961d], Fan and Wang [1964b], and Brosilow and Lasdon [1965] have considered various aspects of this important method applied to complex systems. We shall now discuss one such method of decomposition, based upon the Lagrangian formulation, which entails an iteration over nonfeasible states. In contrast to the methods of Sec. 5.1, this approach will iterate toward the optimum, often involving invalid or impossible values of the independent variables as the basis for computation. As noted in Sec. 2, such iterative methods require extreme care in use if erroneous results are to be avoided.

Before discussing the method in Secs. 5.2.2 and 5.2.3, it is convenient to introduce some modified notation.

5.2.1 A modified notation

Our original notation (Chap. II, Sec. 4.2) defined the state of an interlinking stream, leaving one process unit (and feeding another unit) by a vector, $\mathbf{p}_j$. It is now desirable to modify this notation by the introduction of a dual notation. We may thus label all the inputs to the ith unit which originate from other units by the vectors $\mathbf{p}_1^{(i)}, \mathbf{p}_2^{(i)}, \ldots, \mathbf{p}_k^{(i)}$, assuming that there are k such inputs or interlinking streams to the ith unit. Under this procedure, an interlinking stream will have two names. It will be called $\mathbf{p}_j$ as it leaves the jth unit, and it will also be called $\mathbf{p}_l^{(i)}$ if it is the lth input feeding the ith unit. We can, obviously, then write for this situation that

$$\mathbf{p}_j \equiv \mathbf{p}_l^{(i)}$$

The reasons for introducing this purely artificial distinction will become apparent later.

For our present purposes, we shall assume that each unit produces only a single interlinking stream, so that $\mathbf{p}_j$ can then be used to designate the interlinking stream leaving the jth unit of the total N units in the system. This assumption is prompted by a desire for simplicity in notation, and is not essential for the argument. Indeed, the more general case can be treated by the same mathematical techniques used here, although a more complex notation would be required (see, for example, both Jackson [1964a] and Brosilow and Lasdon [1965]).

Because of the dual designation for each stream, we can write that

$$\mathbf{p}_l^{(i)} = \sum_{j=1}^{N} \epsilon_{l,j}^{(i)} \mathbf{p}_j \tag{5.2-1}$$

where the $\epsilon_{l,j}^{(i)}$ are a set of numbers having the property

$$\epsilon_{l,j}^{(i)} = \begin{cases} 0 & \text{if } l\text{th input to } i\text{th stage is not } \mathbf{p}_j \\ 1 & \text{if } l\text{th input to } i\text{th stage is } \mathbf{p}_j \end{cases}$$

This set of numbers, which might be called a *connection matrix*, gives us a table of correspondence from which the origins of the inputs to a particular stage can be identified. For any given l and i, only one of the N numbers $\epsilon_{l,j}^{(i)}$ will be nonzero. The nonzero element will identify the origin of the lth input to stage i. For example, if $\epsilon_{2,3}^{(5)} = 1$ and $\epsilon_{2,1}^{(5)} = \epsilon_{2,2}^{(5)} = \epsilon_{2,4}^{(5)} = \cdots = \epsilon_{2,N}^{(5)} = 0$, then we know that the second input to the fifth unit is the output of the third unit.

5.2.2 *Mathematical formulation*

Having introduced this dual notation, let us now formulate the Lagrangian function. The original objective function will be assumed to have the form

$$y = \sum_{i=1}^{N} y_i(\mathbf{p}_1^{(i)}, \mathbf{p}_2^{(i)}, \ldots, \mathbf{p}_k^{(i)}, \mathbf{q}_i) \tag{5.2-2}$$

This is not in a generally applicable form, as has already been noted, but it has the important feature that the profit y_i associated with the ith stage is supposed to be expressible in terms of the stage inputs, both interlinking and external. In addition, the system is assumed to be subject to certain restrictions of the form

$$g_{s,i}(\mathbf{p}_1^{(i)}, \mathbf{p}_2^{(i)}, \ldots, \mathbf{p}_k^{(i)}, \mathbf{q}_i) \leq 0 \tag{5.2-3}$$

where $s = 1, 2, \ldots, m_i$. We therefore have m_i inequalities associated with the ith stage, it being essential that only inputs for the ith stage are included as relevant variables. If a certain restriction should involve the inputs to more than one stage, our method of decomposition would be weakened, if not destroyed.

The transformation equations for each stage can now be written in terms of the dual notation as

$$\mathbf{p}_j = \mathbf{T}_j(\mathbf{p}_1^{(j)}, \mathbf{p}_2^{(j)}, \ldots, \mathbf{p}_k^{(j)}, \mathbf{q}_j) \tag{5.2-4}$$

since the output $\mathbf{p}_j$ of the jth stage depends only on the inputs to that stage.

We are now in a position to write the modified objective function as

$$F = \sum_{i=1}^{N} \left[y_i + \sum_{n} \lambda_{n,i}(T_{n,i} - p_{n,i}) + \sum_{s=1}^{m_i} \alpha_{s,i} g_{s,i} - \sum_{l=1}^{k} \sum_{n} \chi_{n,l}^{(i)} \left(p_{n,l}^{(i)} - \sum_{j=1}^{N} \epsilon_{l,j}^{(i)} p_{n,j}\right) \right] \tag{5.2-5}$$

or

$$F = \sum_{i=1}^{N} F_i(\mathbf{p}_1^{(i)}, \mathbf{p}_2^{(i)}, \ldots, \mathbf{p}_k^{(i)}; \mathbf{p}_1, \mathbf{p}_2, \ldots, \mathbf{p}_N; \mathbf{q}_i) \tag{5.2-6}$$

where $\lambda_{n,i}$, $\alpha_{s,i}$, and $\chi_{n,l}^{(i)}$ are Lagrangian multipliers. The summations on n extend over all the components of a particular output vector. Since these vectors can be of varying dimensionality, it is not possible to define the range of these summations without first knowing which vector is involved.

The important feature of our discussion thus far has been the expression of the Lagrangian formulation in terms of the inputs to a single stage, together with a possible connection, through the connection matrix, to the outputs of all the other N stages. It will be seen that this seemingly elegant situation has been achieved by the use of a dual notation, and in fact the original complex problem has not been simplified. What can we hope to gain by the use of this dual notation? Our intention is the decomposition of the system by "disconnecting" the interlinking streams by not satisfying the identities (5.2-1), then to consider each subsystem separately, and by a process of correction, to bring the identities (5.2-1) more nearly into agreement. By such a procedure we shall have decomposed the system, each subsystem having been treated separately. However, at any particular iteration, we shall not have ensured that the assumed feed streams to a specific unit are identically the corresponding outputs, as they ought to be. The result is that nonfeasible states are often used in the iteration, since the overall mass balance, for example, need not be satisfied until convergence is complete. Let us see how these goals can be accomplished.

The necessary conditions for the existence of an optimum are that F must be stationary with respect to the arguments. The necessary conditions are therefore

$$\frac{\partial F}{\partial p_{n,j}} = -\lambda_{n,j} + \sum_{l=1}^{k}\sum_{i=1}^{N} \chi_{n,l}^{(i)} \epsilon_{l,j}^{(i)} = 0 \tag{5.2-7}$$

for $j = 1, 2, \ldots, N$ and all possible components n;

$$\frac{\partial F}{\partial p_{m,l}^{(i)}} = \frac{\partial y_i}{\partial p_{m,l}^{(i)}} + \sum_n \lambda_{n,i} \frac{\partial T_{n,i}}{\partial p_{m,l}^{(i)}} + \sum_{l=1}^{m_i} \alpha_{s,i} \frac{\partial g_{s,i}}{\partial p_{m,l}^{(i)}} - \chi_{n,l}^{(i)} = 0 \tag{5.2-8}$$

for $l = 1, 2, \ldots, k$, $i = 1, 2, \ldots, N$, and all possible components m; and

$$\frac{\partial F}{\partial q_{m,i}} = \frac{\partial y_i}{\partial q_{m,i}} + \sum_n \lambda_{n,i} \frac{\partial T_{n,i}}{\partial q_{m,i}} + \sum_{s=1}^{m_i} \alpha_{s,i} \frac{\partial g_{s,i}}{\partial q_{m,i}} = 0 \tag{5.2-9}$$

for $i = 1, 2, \ldots, N$ and all possible components m of $\mathbf{q}_i$. In addition to these necessary conditions, we must satisfy the Kuhn-Tucker conditions (Chap. VII, Sec. 4.3), so that

$$\sum_{i=1}^{N}\sum_{s=1}^{m_i} \alpha_{s,i} g_{s,i} = 0 \tag{5.2-10}$$

and $\qquad \alpha_{s,i} \leq 0$ for a maximum (5.2-11a)

or $\qquad \alpha_{s,i} \geq 0$ for a minimum (5.2-11b)

Apart from introducing the dual notation, the approach has been that of the classical Lagrangian formulation. It remains to be shown that the present structure can be of computational value.

5.2.3 *A method of iteration*

We must now select the external (decision) variables $\mathbf{q}_1, \mathbf{q}_2, \ldots, \mathbf{q}_N$ so as to satisfy the necessary conditions given by Eqs. (5.2-7) to (5.2-11), the connection Eq. (5.2-1), the inequalities (5.2-3), and the transformation Eq. (5.2-4). Any method of iteration which yields a simultaneous solution to all those equations is of interest, but the preferred method should have the virtue of simplicity. An inviting, but not yet fully justified, approach[1] is to relax, initially, the requirement that Eq. (5.2-1) must be satisfied and to select arbitrary values for the Lagrangian multipliers $\chi^{(i)}_{n,l}$. From Eq. (5.2-7) we see that the $\lambda_{n,j}$ can be computed. Since Eqs. (5.2-8) to (5.2-11) involve only $\mathbf{q}_i$, $\mathbf{p}^{(i)}_l$, and $\boldsymbol{\alpha}_i$, these quantities can be determined stage by stage. This is the essential simplification which has been achieved. Each of the N units can be tackled individually to find the $\mathbf{q}_i$, $\mathbf{p}^{(i)}_l$, and $\boldsymbol{\alpha}_i$ for the ith stage without considering the interaction with other stages. Thus we see the basic decomposition which has been achieved by starting with assumed values of $\chi^{(i)}_{n,l}$. After studying each individual unit, the outputs $\mathbf{p}_j$ can be computed from the transformation Eq. (5.2-4). If the assumed values of the Lagrangian multipliers, $\chi^{(i)}_{n,l}$, were the correct ones, the computed $\mathbf{p}_j$ would agree with their dual counterparts, as found by the connection equations (5.2-1). If the assumed values are not the correct Lagangian multipliers, the two values of $\mathbf{p}_j$ will not agree, and new values of the Lagrangian multipliers will have to be selected.

Before detailing this next step in the iteration, let us try to see clearly the simplifications which we have achieved, and also let us inquire as to the reasons for classifying this approach as a method of nonfeasible states.

The major computational simplification stems from the fact that if Eq. (5.2-1) is not satisfied (relaxed), the inputs to a given stage, say, the ith, are no longer equivalent to the corresponding interlinking feed. In this way the ith stage is isolated from the rest of the system and is optimized separately. The decomposition has been achieved by a process resulting in the stages being disconnected. The method therefore produces nonfeasible states, since the identities (5.2-1) are not satisfied and there is no mode of operation which will achieve this behavior. The streams leaving one process unit and entering another need not have the same state as it leaves as when it enters. Hence the computed states are all nonfeasible.

We must now improve our guesses of the $\chi^{(i)}_{n,l}$ so that Eq. (5.2-1) is more

[1] See, for example, the discussion following the paper by Brosilow and Lasdon [1965].

nearly satisfied. Under certain conditions, namely, when each of the subproblems is an optimum for all values of $\chi_{n,l}^{(i)}$ and $\lambda_{n,i}$, Brosilow and Lasdon [1965] suggest that a better guess at the Lagrangian multipliers can be found from the differential equations

$$\frac{d\chi_{n,l}^{(i)}}{dt} = -\beta E_{n,l}^{(i)} \tag{5.2-12}$$

where $$E_{n,l}^{(i)} = \sum_{j=1}^{N} \epsilon_{l,j}^{(i)} p_{n,j} - p_{n,l}^{(i)}$$

Here t is a parameter which essentially marks the number of iterations, and β is a positive constant selected so as to make the $E_{n,l}^{(i)}$ vanish as rapidly as possible. The $E_{n,l}^{(i)}$ represent the excess of the nth component of the stream $\mathbf{p}_j$, because $\sum_{j=1}^{N} \epsilon_{l,j}^{(i)} p_{n,j}$ is the output feeding the ith stage and $p_{n,l}^{(i)}$ is the amount that arrives. If $E_{n,l}^{(i)}$ is positive, more is being supplied than is consumed, so that an excess is available.

This tentative means of correcting the $\chi_{n,l}^{(i)}$ so as more nearly to approach the optimum has not as yet been tested. Brosilow and Lasdon do give an example in their paper, but it does not follow this method of iteration, and in view of the very restrictive requirements necessary to ensure convergence, the method should be used with caution. Indeed, the requirement necessary to ensure convergence is that the Lagrangian function be maximized when a maximum is sought. Of course, the hope is that the method will in fact converge under far less stringent conditions. This method of decomposition is therefore a promising one, but many problems remain to be resolved, the most obvious being the method of improving the $\chi_{n,l}^{(i)}$.

6 CONCLUSIONS

It will have been seen that the analysis of complex systems is not straightforward. Indeed, the problem may become completely intractable if the dimensionality of the connecting links becomes too great.

If fixed links are available, or if temporarily fixed links are introduced, the system can be decomposed into a set of subproblems which can be examined separately. The optimum for the system would then be the sum of the individual subsystem optima if the connecting links were truly fixed. However, if these links were only temporarily specified, the global optimum would have to be found by optimizing the sum of the subsystem optima with respect to the values of the temporarily fixed links. For sequential systems, the method of dynamic programming is one approach which provides a framework of subproblems that can be separately optimized. Dynamic programming has trouble, however, in coping with systems in which the subproblems are linked with vectors of high dimension. No powerful technique for general decomposition exists at present capable of handling all problems.

Problems

XIII-1 Determine the shortest path through the network of Fig. XIII-6 in Example XIII-5.

XIII-2 Rework Prob. XI-5 on the maximization of yield in a sequence of stirred tank reactors using dynamic programming.

XIII-3 A simple network consists of nine nodes, the common start being at the node labeled 0 and the common finish at 8. The network represents a set of activities. The duration of each activity represented by the path between two nodes i and j is listed in the following table. Determine the longest path through the network by a decomposition method.

Path nodes		*Path length*
Activity starting point i	*Activity finishing point j*	*Duration of activity*
0	1	1
0	2	3
0	3	7
1	2	2
1	4	16
2	4	10
3	5	21
3	5	10
3	7	25
4	6	12
4	5	16
5	7	5
6	7	9
6	8	7
7	8	2

XIII-4 Determine the shortest path through the network of Prob. XIII-3 using a decomposition approach.

XIII-5 Maximize the function

$$y = \sum_{i=1}^{3} (4x_i - ix_i^2)$$

subject to the restrictions that

$$\sum_{i=1}^{3} x_i = 4$$

and that x_i takes only the discrete value 0, 1, 2, 3, or 4.

XIII-6 A chemical process unit has one item of equipment which is t years old and has no salvage value. The allocated profit from a year's operation of a t-year-old piece of this equipment is given by the expression $13 - t$ for an item less than 14 years old and by zero for an older item. The cost of purchase and installation of a new replacement is 11, and it is estimated that the process still has N years of operational life left. If $N = 5$ and the equipment under examination is 2 years old, what is the optimal replacement policy and profit for this item, the annual decisions being made at the beginning of each year?

XIII-7 Minimize y, where

$$y = \epsilon \sum_{i=1}^{N} (p_i^2 + q_i^2)$$

with respect to $\mathbf{q}$, subject to

$$p_i = \frac{p_{i+1} - \epsilon q_i}{1 - \epsilon}$$

(This is the finite difference form of $dp/dl = p - q$; Roberts [1964b].)

XIII-8 Figure XIII-17 represents a flow system in which material is withdrawn from source A and passes through various plants, represented by nodes in the figure, to sink B. If the parameters against each path arrow represent distances, determine the shortest route through the network.

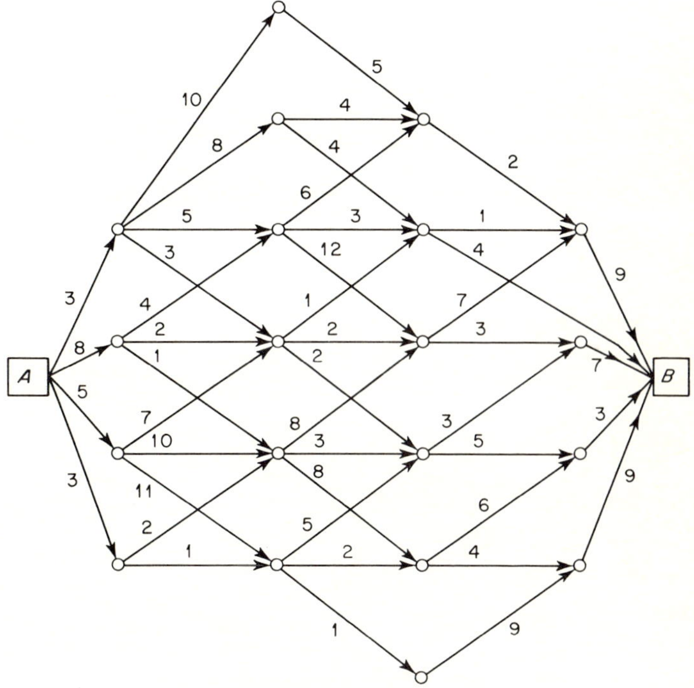

Fig. XIII-17 A flow network (Probs. XIII-8 and XIII-9).

XIII-9 Using the network of Prob. XIII-8 and assuming that the parameters represent some form of profit resulting from the use of the stream, determine that single route through the system which yields the greatest profit.

XIII-10 Find the maximum in Prob. XIII-5 if the summation is over four stages, instead of three.

XIII-11 A company has a total capital of \$1 million to use for plant maintenance and replacement over the next 10-year period. At the beginning of each year, the plant manager must decide how much capital should be allocated in that year. The annual profit from plant operation in a particular year, the jth, has been found to be

$$x_j[j(10^6)] - x_j^2$$

where x_j is the capital allocated to plant maintenance in the jth year. Assuming that all the available capital, together with all the accumulated interest in the unused capital, must be used on maintenance in the 10-year period, and that the unallocated funds earn compound interest at the rate of 5 percent per year (i.e., the money allocated for the jth year will earn no interest for that year), find the allocation policy which will maximize the total 10-year-period profit.

XIII-12 A simplified model of the economy has been proposed in which a dollar invested in industry A at the beginning of the year yields \$2 of capital and \$1 product b at the end of that year, and similarly, an investment of \$1 in industry B yields \$1.8 of capital and \$4 of product b. If it is desired to maximize the total value of the production of product b in the next 5 years, how much of the \$$C$ capital now available should be spent in industries A and B each year? (See, for example, Bellman and Dreyfus [1962, chap. VII] for discussion of the general problem of this type.)

XIII-13 Consider the reaction system as depicted in Fig. XIII-18, in which each of the stream states p_j and decision variable vectors q_i has only a single component. Here the product stream from a two-stage process is linked to the system feed by a recycle stream.

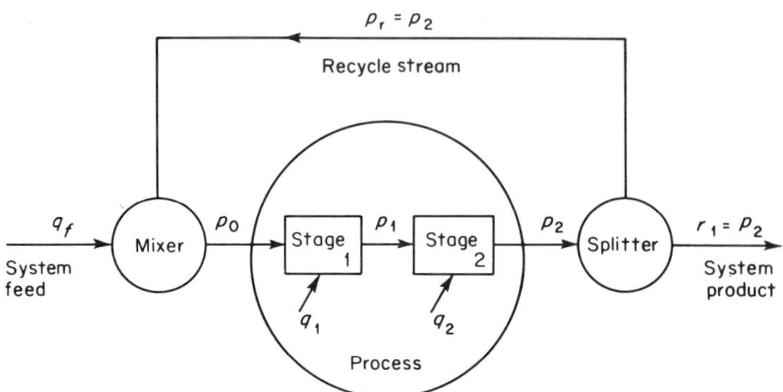

Fig. XIII-18 Process flowsheet for Prob. XIII-13.

The volume flow V through the process stages, the feed flow V_f, and the recycle flow V_r are all fixed so that

$$V = V_f + V_r$$

The stream states p_j may be taken as the mole fraction of the important stream component, so that the usual mass balances can be written on the mixer and splitter in the system. The process transformation in the ith stage can be written

$$p_i = f_i(p_{i-1}, q_i) \qquad \text{for } i = 1 \text{ and } 2$$

Obtain the necessary conditions, using some decomposition approach, such as the principle of splitting, to define the values of q_1 and q_2 which will maximize the objective function

$$y = (q_f - r_1)V - T_1(q_1) - T_2(q_2)$$

where $T_i(q_i)$ are functions, each solely dependent on q_i. Compare this result with that obtained directly.

XIII-14 Using the reaction system described in Prob. XIII-13, show that a decomposition approach based on nonfeasible states in which the iterative procedure assumes that p_0 is held fixed at suitable values, and both the q_i are varied so as to maximize y, will not necessarily yield the correct solution.

XIII-15 Determine the optimal solution to Prob. XIII-13 if the parameters have the following numerical values:

$$V = 1.0 \qquad V_r = 0.5 \qquad V_f = 0.5 \qquad q_f = 5 \qquad T_1 = q_1^4 \qquad T_2 = 0.5q_2^4$$

and

$$f_1 = p_0 e^{-(q_1^2)/10} \qquad f_2 = p_1 e^{-(q_2^2)/20}$$

Does the direct or decomposition approach yield the easier solution here?

XIII-16 Based upon the system of Fig. XIII-1 and Example XIII-3, find the maximum value of

$$y = x_f - 10x_1^2 + x_0 q_1^2 + 3q_2^2 - 5q_3^3$$

for a fixed value of the feed, $x_f = 0.6$, a flow rate ratio of $(F_r + F_f)/F_r = 1.5$, and the two restrictions

$$g_1 = 1.5x_0 - 0.5x_f - x_1 = 0$$
$$g_2 = x_1 - (2x_0 + q_1 + 3q_2 - q_3^2) = 0$$

using the *principle of splitting* approach.

XIII-17 Using a chain system similar to that of Example XIII-7, maximize y, where

$$y = \sum_{i=1}^{3} (p_{i+1} - p_i) - 0.25 \sum_{i=1}^{3} q_i$$

subject to the restriction on each unit that

$$\frac{p_{i+1}}{p_i} = (1 + 0.5q_i)^2$$

for a system feed p_4 of 2.

XIII-18 A components manufacturer produces a specific component in lots, the setup cost being \$300 per lot. The variable production cost per unit is \$0.90, and the cost of storage is \$0.03 per component unit per month. The company sales forecast is for the following quantities to be required each month for the next year:

16,000, 25,000, 21,000, 16,000, 12,000, 7,000,

6,000, 12,000, 14,000, 25,000, 31,000, 9,000

Ignoring the time required to produce the units, and assuming that decisions are made at the beginning of each month, find the optimal times at which lots should be produced and the size of these lots, if 1,000 units are at present in storage and it is desired to have 6,000 in store at the end of the year.

XIII-19 A caterer knows that in connection with the meals he has arranged to serve during the next n days he will need r_j fresh napkins on the jth day, for $j = 1, 2, \ldots, n$.

There are two types of laundry service available; one type requires p days and costs b cents per napkin; a faster service requires q days, $q < p$, but costs c cents per napkin, $c > b$. Beginning with no usable napkins on hand or in the laundry, the caterer meets the demands by purchasing napkins at a cents per napkin. How does the caterer purchase and launder napkins so as to minimize the total cost for n days? Set up the necessary equations for a solution by dynamic programming for $q = 1, p = 2$ (Bellman and Dreyfus [1962, chap. 3], Bellman [1957]).

XIII-20 Find a solution to Prob. XIII-19 for

$$a = 9 \qquad b = 3 \qquad c = 6 \qquad n = 6 \qquad q = 1 \qquad p = 2$$

(Solution is $r_j = \{100, 120, 150, 80, 90, 130\}$.)

XIII-21 Given the array

7	6	4	8	2
3	5	10	5	0
9	2	8	9	7
1	3	4	6	11
7	5	3	3	8

find the route from the top left-hand corner to the bottom right, so that the sum of the numbers along the route is the least, the path step being horizontally to the right, or vertically down, one move at a time.

XIII-22 Find the longest path in Prob. XIII-21.

XIII-23 Show by dynamic programming that the maximum value of

$$y = 2x_1 + 4x_2 + x_3 + x_4$$

subject to the constraints

$$\begin{aligned} x_1 + 3x_2 + x_4 &\leq 4 \\ 2x_1 + x_2 &\leq 3 \\ x_2 + 4x_3 + x_4 &\leq 3 \end{aligned}$$

is $y = 6.5$, found at $x_1 = 1$, $x_2 = 1$, $x_3 = 0.5$, $x_4 = 0$ (Amundson, Aris, Kalman, and Lapidus [1960]).

XIII-24 Find the solution to Prob. XIII-23 by another method.

XIII-25 A positive quantity Q has to be divided into N portions so as to maximize the product of the N parts. Set up the dynamic programming formulation of this problem.

XIII-26 Find the solution to Prob. XIII-25 when $N = 2$ and 3, and hence deduce the general solution.

XIII-27 A manufacturing plant may be considered as being essentially a sequence of four processing units. The flow of materials through the system is such that the system raw material enters at one end of the plant, and the product from any one unit is used as the raw material for the next unit in the sequence. Some of the product from any particular unit may be withdrawn and sold.

If the units are numbered from the rear and the flow of product to be sold from a given unit is x_i, the profit from the system can be written

$$y = \sum_{i=1}^{4} (6x_i - ix_i^2)$$

Determine the quantity of intermediate product which should be withdrawn from each unit in the sequence if it is desired to maximize the system profit, when the system feed is equivalent to a total flow of 10.

XIII-28 The replacement and operating costs and associated income for a particular item of equipment are given in Table XIII-13 for each of 10 years, the estimated life of the process within which it is employed. At the beginning of each year it must be decided whether to keep the equipment item for another period of operation or to replace it by a new item. What is the best 10-year policy if we start with a 3-year-old item which has the characteristics given in Table XIII-14? (Bellman and Dreyfus [1962, p. 119], Roberts [1964b, p. 33].) Solution is 310, with a policy of *RKKKRKKKKK*.

TABLE XIII-13 EQUIPMENT CHARACTERISTICS FOR VARIOUS YEARS OF MANUFACTURE

(*a*) Equipment made in year 1

Age	1	1	2	3	4	5	6	7	8	9
Income	90	85	80	75	70	70	70	60	60	60
Operating costs	20	20	25	25	30	30	35	40	45	50
Replacement costs	200	220	240	250	255	260	265	270	270	270

(*b*) Equipment made in year 2

Age	0	1	2	3	4	5	6	7	8
Income	100	90	80	75	70	65	65	65	65
Operating costs	15	20	20	25	25	30	30	35	35
Replacement costs	200	220	240	250	255	260	265	270	270

(*c*) Equipment made in year 3

Age	0	1	2	3	4	5	6	7
Income	110	105	100	95	90	80	70	60
Operating costs	15	15	20	20	25	25	30	30
Replacement costs	200	220	240	250	255	260	265	270

(*d*) Equipment made in year 4

Age	0	1	2	3	4	5	6
Income	115	110	100	90	80	70	60
Operating costs	15	15	20	20	25	25	30
Replacement costs	210	215	220	225	230	235	240

(*e*) Equipment made in year 5

Age	0	1	2	3	4	5
Income	120	115	115	110	105	100
Operating costs	10	10	15	15	20	20
Replacement costs	210	215	220	225	230	235

(*f*) Equipment made in year 6

Age	0	1	2	3	4
Income	125	120	110	105	100
Operating costs	10	10	10	15	15
Replacement costs	210	220	230	240	250

TABLE XIII-13 (CONTD.)

(*g*) Equipment made in year 7

Age	0	1	2	3
Income	135	125	110	105
Operating costs	10	10	10	10
Replacement costs	210	220	230	240

(*h*) Equipment made in year 8

Age	0	1	2
Income	140	135	125
Operating costs	5	10	10
Replacement costs	220	230	240

(*i*) Equipment made in year 9

Age	0	1
Income	150	140
Operating costs	5	10
Replacement costs	220	225

(*j*) Equipment made in year 10

Age	0
Income	155
Operating costs	5
Replacement costs	220

TABLE XIII-14 INCUMBENT EQUIPMENT CHARACTERISTICS

Age	3	4	5	6	7	8	9	10	11	12
Income	60	60	50	50	50	40	40	40	30	30
Operating costs	55	55	55	60	60	60	60	65	65	70
Replacement costs	250	260	270	280	280	290	290	300	300	310

XIII-29 What is the effect on the solution to Prob. XIII-28 if it proves impossible to replace the unit at the start of the 10-year period? Find the new optimal replacement-keep policy. (Solution is 290 with policy of *KKKKRKKKKK*.)

EPILOGUE

In this presentation an integrated account of the general organization required before optimization can take place and of the full range of optimization methods has been given. While optimization theory is essentially complete for well-defined systems with specified objectives and models, there is much room for improvement in the areas of model building, objective specification, and the handling of uncertainty and competition, the accuracy and reliability of any study being wholly dependent upon their correct formulation. The successful conclusion of any optimization analysis will however, always depend upon the critical familiarity of the user.

"This is not the end.
It is not even the
beginning of the end.
But it is, perhaps, the
end of the beginning."
(*Sir Winston Churchill*, *Nov.* 10, 1942)

APPENDIX 1: SOME NOTES ON DETERMINANTS

1 DEFINITION OF THE DETERMINANT

The determinant is a number obtained by applying certain rules to a set of numbers arranged in rows and columns, there being precisely the same number of rows as there are columns. The set of numbers so arranged is called a *square matrix*, and the determinant of the matrix is represented as follows:

$$A = \det |a_{i,j}| = \begin{vmatrix} a_{1,1} & a_{1,2} & \cdots & a_{1,n} \\ a_{2,1} & a_{2,2} & \cdots & a_{2,n} \\ \cdot & \cdot & \cdot & \cdot \\ a_{n,1} & a_{n,2} & \cdots & a_{n,n} \end{vmatrix}$$

The determinant is a sum of products:

$$A = \sum \pm a_{1,\alpha}, a_{2,\beta}, \ldots, a_{n,\delta} \tag{1-1}$$

This symbolism implies that the single number A is found by forming certain products and then summing them. These products are all the possible different products formed so that exactly one, and only one, element from each row and each column appears in a particular product. All these products are to be assigned a sign and then summed, the sign being fixed as follows: A product is given a positive sign if the indices denoting the column from which the element is selected, $\alpha, \beta, \ldots, \delta$, form an even permutation of the first n integers, $1, 2, \ldots, n$. A permutation $\alpha, \beta, \ldots, \delta$ of the first n integers is said to be even if an even number of position interchanges is required to restore the natural order of the integers. Thus 4, 2, 1, 3 is a permutation of the first four integers, and the natural order can be restored by the successive interchanges 2, 4, 1, 3; 2, 1, 4, 3; 2, 1, 3, 4; and 1, 2, 3, 4. Here four interchanges are required to restore the natural order, and thus we say that 4, 2, 1, 3 is an even permutation of 1, 2, 3,4. Since each term in a particular product of Eq. (1-1) contains only one element from each column, the $\alpha, \beta, \ldots, \delta$ are a permutation of the first n integers, and as noted above, the product is assigned a positive sign in the summation if the permutation is even and a negative sign if the permutation is odd. These rules establish a single numerical value for the determinant of each square matrix. Thus, as an illustration, we can write

$$A = \begin{vmatrix} a_{1,1} & a_{1,2} & a_{1,3} \\ a_{2,1} & a_{2,2} & a_{2,3} \\ a_{3,1} & a_{3,2} & a_{3,3} \end{vmatrix} = a_{1,1}a_{2,2}a_{3,3} + a_{1,2}a_{2,3}a_{3,1} + a_{1,3}a_{2,1}a_{3,2} - a_{1,1}a_{2,3}a_{3,2} - a_{1,2}a_{2,1}a_{3,3} - a_{1,3}a_{2,2}a_{3,1}$$

Here each product is formed as required. Thus the fifth term contains one element from the first row ($a_{1,2}$), one from the second ($a_{2,1}$), and one from the third ($a_{3,3}$), and each of these three terms can be seen to be drawn from a different column, all three of the columns being represented in the product. The sign of this term is negative, since one interchange is required to restore the natural order, and therefore the column numbers 2, 1, 3 are an odd permutation of the first three integers.

It is easily seen that the multiplication of a single row or column by a scalar number is equivalent to multiplying the original determinant by the scalar. Thus, if A is the determinant of Eq. (1-1), then

$$\alpha A = \begin{vmatrix} \alpha a_{1,1} & \cdots & a_{1,n} \\ \cdots & \cdots & \cdots \\ \alpha a_{n,1} & \cdots & a_{n,n} \end{vmatrix} \tag{1-2}$$

where α is a scalar.

2 PROPERTY I

Property I states that the sign of the determinant is changed if any row of the matrix is exchanged with any other row or if any column is interchanged with any other column. We thus can write, for example, that

$$\begin{vmatrix} a_{1,1} & a_{1,2} & a_{1,3} \\ a_{2,1} & a_{2,2} & a_{2,3} \\ a_{3,1} & a_{3,2} & a_{3,3} \end{vmatrix} = - \begin{vmatrix} a_{3,1} & a_{3,2} & a_{3,3} \\ a_{2,1} & a_{2,2} & a_{2,3} \\ a_{1,1} & a_{1,2} & a_{1,3} \end{vmatrix}$$

Let us consider the case of row interchange, the proof for a column interchange being developed in a similar fashion. To see that an interchange of rows leads to a sign change, let us first establish that if only two numbers are out of their natural order in a permutation, the permutation is odd. If i and j are permuted, we have the sequence $1, 2, \ldots, j-1, i, j+1, \ldots, i-1, j, i+1, \ldots, n$, with i and j having interchanged positions and $j < i$. The number of interchanges required to obtain $1, 2, \ldots, j-1, i, j, j+1, \ldots, i-1, i+1, \ldots, n$ is $i-j-1$. With one additional interchange, we have $1, 2, \ldots, j, i, j+1, \ldots, n$. Then an additional $i-j-1$ interchanges are required to put the i back into its appropriate place, giving a total of $2(i-j)-1$ interchanges necessary to go from the original permuted form to the natural order. If only two numbers are out of order, the permutation is odd, since $2(i-j)-1$ is always odd for all i and j, with $i>j$.

The sign of the product is determined by whether or not the permutation is even or odd. Suppose that $\alpha, \beta, \ldots, \delta$ is an even permutation of the first n integers; then if any two numbers are interchanged, the resulting permutation is odd, by the reasoning shown above, and similarly, if the original permutation is odd, an interchange of two elements gives rise to an even permutation.

In interchanging any two rows in a matrix, the determinant is affected only in that two elements are interchanged, since precisely the same products will appear. Since two elements are interchanged, the signs, however, will be just the opposite of the original ones, i.e., the sign of the corresponding term prior to the interchange. This reversal in sign clearly stems from the fact that two elements have traded positions and an odd number of interchanges are required to obtain the original order. The sign of each term in the sum, and hence the sign of the determinant, is then reversed by interchanging two rows of the matrix. This is property I.

This important property can be used to show that if any two columns or rows of the matrix are the same, its determinant vanishes. Assume that two rows are exactly the same. Clearly, they can be interchanged without changing the numerical value of the determinant, but using property I, we see that the negative of the original must result from such an interchange. We have, therefore, a determinant which is equal to the negative of itself. The only possible value having this property is zero. Therefore, if any two rows or two columns are the same, the determinant vanishes, and is said to be singular. Of course, if any row is a scalar multiple of another row, the determinant again vanishes, since, by a two-step process, we can first factor the scalar from the determinant, as indicated in Sec. 1, leaving two identical rows. Then, applying property I, we can deduce that the determinant must vanish.

3 PROPERTY II

The $n!$ terms constituting the sum in Eq. (1-1) can be rearranged as follows:

$$\begin{aligned}\sum \pm a_{1,\alpha} a_{2,\beta}, \ldots, a_{n,\delta} &= a_{1,1} \sum \pm a_{2,\beta}, \ldots, a_{n,\delta} \\ &- a_{1,2} \sum \pm a_{2,\beta}, \ldots, a_{n,\delta} + \cdots + (-1)^{n+1} a_{1,n} \sum \pm a_{2,\beta}, \ldots, a_{n,\delta} \qquad (3\text{-}1)\end{aligned}$$

the sign of each term being selected in the same way as before; i.e., if $\beta, \ldots, \delta$ is an even permutation of the $n-1$ integers, the positive sign is selected, but if it is not, the negative sign is chosen. Let us show that this is indeed the case.

The sum of terms multiplied by $a_{1,i}$ are the $(n-1)!$ terms in Eq. (1-1), in which this particular element appears. Suppose we have the permutation $i, j, \ldots, n$, and suppose, further, that the permutation is even. Then we can inquire about the permutation of the $n-1$ terms which remain if i is removed from consideration, as is the case when $a_{1,i}$ is factored from the product. If i happens to be odd, an even number of interchanges are required to put it in its proper place, and if i is even, an odd number of interchanges are required. For example, $5, 1, k, j, \ldots, n$ requires four interchanges to move the 5 to the proper spot—an even number of interchanges for an odd number. We can therefore conclude that if the permutation $i, j, \ldots, n$ is even, the permutation $j, \ldots, n$ is also even, so long as i is odd, since an even number of interchanges were required to put the i in the proper position. On the other hand, if $i, j, \ldots, n$ is an odd permutation of the first n integers, then $j, \ldots, n$ is an odd permutation of the $n-1$ integers formed by deleting i, so long as i is odd. If a sign is to be governed by the permutation $i, j, \ldots, n$, we can summarize by saying that the proper sign is given by $(-1)^{i+1}$ times the sign found by examining whether the permutation $j, \ldots, n$ is even or odd.

These arguments are intended to clarify the reason for the alternating sign seen in Eq. (3-1). The form of Eq. (3-1) is very interesting and represents the basis for many expansions of determinant into simpler products. In the sum multiplied by $a_{1,i}$, we shall find all possible combinations of terms, with one and only one element selected from each row and each column, excluding the column and row in which we find $a_{1,i}$. The sum therefore represents in this case the determinant of the matrix

$$A_{1,i} = \begin{vmatrix} a_{2,1} & a_{2,2} & \cdots & a_{2,i-1} & a_{2,i+1} & \cdots & a_{2,n} \\ \cdot & \cdot & \cdot & \cdot & \cdot & \cdot & \cdot \\ a_{n,1} & a_{n,2} & \cdots & a_{n,i-1} & a_{n,i+1} & \cdots & a_{n,n} \end{vmatrix} \tag{3-2}$$

which is simply the matrix found by omitting the column and the row in which $a_{1,i}$ appears. This matrix is called the *cofactor* of $a_{1,i}$. Equation (3-1) can then be written in the particularly useful form

$$A = \sum_{i=1}^{n} (-1)^{i+1} a_{1,i} A_{1,i} \tag{3-3}$$

A determinant can then be represented by the sum of the products of a row times the determinant of the cofactor of the element, with the sign alternating, depending on whether the element occupies an even or an odd column.

By similar reasoning it can be shown that the same form is valid for any column; i.e.,

$$A = \sum_{i=1}^{n} (-1)^{i+1} a_{i,1} A_{i,1} \tag{3-4}$$

This expansion of the determinant in a series of the determinants of the cofactors is called property II, which is used in Chap. VII.

4 CRAMER'S RULE

The linear equations

$$\begin{array}{cccc} a_{1,1}x_1 & a_{1,2}x_2 & \cdots & a_{1,n}x_n = b_1 \\ \cdots & \cdots & \cdots & \cdots \\ a_{n,1}x_1 & a_{n,2}x_2 & \cdots & a_{n,n}x_n = b_n \end{array}$$

have the solution

$$x_i = \frac{\begin{vmatrix} a_{1,1} & \cdots & a_{1,i-1} & b_1 & a_{1,i+1} & a_{1,n} \\ \cdots & \cdots & \cdots & \cdots & \cdots & \cdots \\ a_{n,1} & \cdots & a_{n,i-1} & b_n & a_{n,i+1} & a_{n,n} \end{vmatrix}}{\begin{vmatrix} a_{1,1} & \cdots & a_{1,n} \\ \cdots & \cdots & \cdots \\ a_{n,1} & \cdots & a_{n,n} \end{vmatrix}}$$

provided the determinant in the denominator is not singular. Note that if this determinant is singular, it implies that at least two of the n algebraic equations are not linearly independent, and therefore that no solution exists.

APPENDIX 2: OPTIMIZATION OF NONLINEAR OBJECTIVE FUNCTIONS SUBJECT TO EQUALITY AND INEQUALITY CONSTRAINTS—A COMPARISON

Stocker [1969] has compared five search methods in solving each of the 15 problems described briefly in Table A-I. A complete description of these problems is to be found in Stocker [1969].

The five search codes tested were

1. Paviani (*P*), which is described in Chap. VIII, Sec. 4.4.2.
2. A modification due to Barnes and Himmelblau [1966] (*B*) of the basic method proposed by Di Bella and Stevens [1965].
3. A modification of the Rosenbrock (*R*) method (Chap. VIII, Sec. 3.1.3) to allow for constraints.
4. The method of Fiacco and McCormick (*F*) using penalty functions (Chap. VIII, Secs. 4.4.2 and 5.4.1).
5. The program developed by H. V. Smith (*S*) and distributed through the IBM Share General Program Library. This program uses a local linearization and iteration procedure similar to that of Glass and Cooper [1965] and Zoutendijk [1960] (Chap. VII, Sec. 5.3.3).

The results of each search are summarized in Table A-II. The numbers entered in the table are the times in seconds required for the termination of each search. Although these have no absolute significance, they give some measure of the relative merits of each technique.

In conclusion, we note that program *S* does not appear to be useful, and Rosenbrock's method is not recommended except for unconstrained problems. Of the three remaining techniques, it should be noted that program *F* requires second derivatives which can be extremely laborious to obtain. Indeed we have taken the attitude that in practical situations these derivatives are impossible to reliably compute. Program *P* is a direct search technique and is therefore quickly and easily prepared. It suffers from requiring very long convergence times for problems having more than five dimensions. Program *B* converged rapidly as long as there were three or fewer nonlinear equality constraints. This program is a gradient technique and is, therefore, more difficult to prepare than direct search techniques.

Of the five methods tested by Stocker, program *P* is the most reliable direct-search method, and Program *B* was the gradient method giving the best overall performance of the methods tested.

TABLE A-I THE TEST PROBLEMS

Category[1]	*Test problem number*	*Dimensions*	*Brief description (for more detail, see Stocker [1969])*
Un-C	1	4	Quartic objective function
l-i	2	2	Cubic objective function; eight linear inequality constraints
l-i	3	5	Cubic objective function; ten linear inequality constraints
l-i, n-i	4	2	Quartic objective function; four linear inequality and three nonlinear inequality constraints
n-i	5	5	Quartic objective function; six quadratic inequalities and upper and lower bounds
n-i	6	3	Complex objective function with fourteen nonlinear inequality constraints
n-i	7	15	Cubic objective function with five cubic constraints
l-e, n-i	8	5	Complex objective function; five linear equality constraints and thirty nonlinear inequalities
l-e, n-i	9	2	Quartic objective function; one linear equality constraint and one nonlinear inequality
l-e	10	16	Fourth-order polynomial objective function; eight linear equalities together with upper and lower bounds on all variables
l-e	11	10	Logarithmic objective function; three linear equality constraints and lower bounds on all variables
l-e	12	45	Complex, logarithmic objective function; sixteen linear equality constraints together with lower bounds

TABLE A-I CONTINUED

Category[1]	*Test problem number*	*Dimensions*	*Brief description (for more detail, see Stocker [1969])*
n-e, l-e	13	3	Quadratic objective function with one linear and one nonlinear equality constraint
n-e, n-i	14	3	Discontinuous first derivative in objective function and sines and cosines in equality constraints; one nonlinear equality and six nonlinear inequality constraints
n-e, n-i l-e	15	24	Linear objective function with twelve nonlinear and two linear equality constraints; six nonlinear equality constraints and bounds in all variables

[1] *Un-C* = unconstrained, *i* = inequality, *e* = equality, *l* = linear, and *n* = nonlinear

TABLE A-II RESULTS SUMMARY[1]

Test problem	*Code*				
	P	*B*	*F*	*R*	*S*
1	0.27	0.10	0.45	1.26	1.06
2	0.50	14.04	1.80	0.70	*t*
3	7.58	1.62	2.79	*t*	*t*
4	0.34	0.27	0.82	1.31	0.63
5	2.67	2.31	1.05	1.71	*t*
6	4.23	1.60	*t*	0.52	1.72
7	*t*	91.42	5.58	*t*	*t*
8	103.36	4.88	*t*	*t*	*t*
9	0.43	0.07	0.51	4.52	*t*
10	142.31	55.35	15.59	*t*	*t*
11	27.94	5.37	2.44	*t*	*t*
12	*t*	187.40	29.40	*t*	*t*
13	0.84	0.33	*t*	*t*	*t*
14	4.71	7.76	*u*	*t*	*t*
15	1083.0	*t*	*t*	*t*	*t*

[1] The entry *t* denotes termination prior to convergence, and *u* denotes a problem which was not attempted because of the impossibility of obtaining second derivatives.

APPENDIX 3: REFERENCES

Ackoff, R. L. 1955. Production and Inventory Control in a Chemical Process, *J. Opn. Res. Soc. Am.*, **3**:319–338.

———. 1962. "Scientific Method," John Wiley & Sons, Inc., New York.

Adler, J., and D. Vortmeyer. 1963. Effect of Axial Diffusion Processes in the Optimum Yield of Tubular Reactors, Part I, Isothermal $A \rightleftharpoons C$, *Chem. Eng. Sci.*, **18**:99–108.

——— and ———. 1964. Effect of Axial Diffusion Processes in the Optimum Yield of Tubular Reactors, Part II, Variable Temperatures, Simple Reversible Reaction $A \rightleftharpoons C$, *Chem. Eng. Sci.*, **19**:413–422.

Ahlgren, T. D., and W. F. Stevens. 1966. Process Optimization in the Presence of Error, *Ind. Eng. Chem.* (*Process Design Develop.*) **5**:290–297.

Ahn, Y. K., H. C. Chen, L. T. Fan, and C. S. Wang. 1964. A Modified Moving Bed Grain Dryer, *Ind. Eng. Chem.* (*Process Design Develop.*), **3**:96–100.

Almasy, G. A., J. J. Hay, and I. M. Pallai. 1966. Ammonia Synthesis Loop Optimization, *Brit. Chem. Eng.*, **11**:188–192.

Amey, L. R. 1964. The Allocation and Utilization of Resources, *Opn. Res. Quart.*, **15**:87–100.

Amundson, N. R. 1966. "Mathematical Methods in Chemical Engineering: Matrices and Their Application," Prentice-Hall, Inc., Englewood Cliffs, N.J.

———, R. Aris, R. E. Kalman, and L. Lapidus. 1960. Applied Mathematics in Chemical Engineering, *Am. Inst. Chem. Engrs. Spec. Lecture Ser.*, Washington, D.C., December 3rd.

Andersen, S. L. 1961. Venture Analysis, *Chem. Eng. Progr.*, **57**(3):80–83.

———, and H. G. Haight. 1961. A 2 × 2 Risk Decision Problem, *Chem. Eng. Progr.*, **57**(5):70–73.

Anderson, R. L. 1953. Recent Advances in Finding Best Operating Conditions, *J. Am. Statist. Assoc.*, **48**:789–798.

Anderson, R. J. 1961. An Introduction to Air Cooled Heat Exchangers, *Brit. Chem. Eng.*, **6**:468–473.

Anon. 1959. Progress Report on Opcon, *Control Engr.*, **6** (Nov.):124.

———. 1961a. *Chem. Eng. News*, **39**(10):82.

———. 1961b. *Chem. Weekly*, **88**(8):73.

Aoki, M. 1961. Stochastic Time Optimal Control Systems, *I.E.E.E. Trans.* (*Appl. Ind.*), **80**:41–44.

Archer, D. H. 1960. An Optimizing Control for the Chemical Process Industries, *Brit. Chem. Eng.*, **5**:88–94.

Aries, R. S., and R. D. Newton. 1955. "Chemical Engineering Cost Estimation," McGraw-Hill Book Company, New York.

Aris, R. 1960a. On Denbigh's Optimum Temperature Sequence, *Chem. Eng. Sci.*, **12**:56–64.

———. 1960b. Part I, Optimum Design of Multi-bed Adiabatic Reactors, *Chem. Eng. Sci.*, **12**:243–252.

———. 1960c. Part II, Optimum Temperature Gradients in Tubular Reactors, *Chem. Eng. Sci.*, **13**:18–29.

———. 1960d. Part III, Optimum Conditions in Sequences of Stirred Tank Reactors, *Chem. Eng. Sci.*, **13**:75–81.

———. 1961a. Part IV, Optimum Conditions for a Single Reaction, *Chem. Eng. Sci.*, **13**:197–206.

———. 1961b. The Determination of Optimum Operating Conditions by the Methods of Dynamic Programming, *Z. Elektrochem.*, **65**:229–244.

———. 1961c. Optimal Bypass Rates for Sequences of CSTR, *Can. J. Chem. Eng.*, **39**:121–126.

———. 1961d. "The Optimal Design of Chemical Reactors: A Study in Dynamic Programming," Academic Press Inc., New York.

———. 1962a. On Optimal Adiabatic Reactors of Combined Types, *Can. J. Chem. Eng.*, **40**:87–92.

———. 1962b. Adaptive Control, *Brit. Chem. Eng.*, **7**:896–900.

———. 1964. "Discrete Dynamic Programming," Blaisdell Publishing Co., Waltham, Mass.

———, R. Bellman, and R. E. Kalaba. 1960. Some Optimum Problems in Chemical Engineering, *Chem. Eng. Progr. Symp. Ser.* 31, **56**:95–102.

———, G. L. Nemhauser, and D. J. Wilde. 1964. Optimization of Multistage Cyclic and Branching Systems by Serial Procedures, *Am. Inst. Chem. Engrs. J.*, **10**:913–919.

———, D. F. Rudd, and N. R. Amundson. 1960. On Optimum Cross-current Extraction, *Chem. Eng. Sci.*, **12**:88–97.

Atkinson, A. C., and W. G. Hunter. 1968. The Design of Experiments for Parameter Estimation, *Technometrics*, **10**:271–289.

Avriel, M., and D. J. Wilde. 1967. Optimal Condenser Design by Geometric Programming, *Ind. Eng. Chem. (Process Design Develop.)*, **6**:256.

Bakemeier, H., and R. Krabetz. 1962. Ein Beitrag zur Optimalisierung von Abschnitts Reaktoren, *Chem. Ingr. Tech.*, **34**:1.

Balakrishnan, A. V., and L. W. Neustadt (eds.). 1964. "Computing Methods in Optimization Problems," Academic Press, Inc., New York.

Balas E. 1965. Solution of Large Scale Transportation Problems through Aggregation, *J. Opn. Res. Soc. Am.*, **13**:82–93.

Balinski, M. L., and R. E. Quandt. 1964. On an Integer Program for a Delivery Problem, *J. Opn. Res. Soc. Am.*, **12**:300–304.

Ball, W. E., and L. C. D. Groenweghe. 1966. Determination of Best-fit Rate Constants in Chemical Kinetics, *Ind. Eng. Chem. Fundamentals*, **5**:181–184.

Barankin, E. W., and R. Dorfman. 1955. "Towards Quadratic Programming," Office of Naval Research Logistics Projects at Columbia University, New York, and University of California, Berkeley, Calif.

Barnes, G. K., and D. M. Himmelblau. 1966. Optimization of a Kraft Pulping Mill, paper at Canadian Institute of Chemistry Conference, October, 17.

Barnett, S. 1962. Stability of the Solution to a Linear Programming Problem, *Opn. Res. Quart*, **13**:219–228.

Bates, A. G. 1966. Managing Capital Expenditure, Part 3, *Hydrocarbon Process Petrol. Refiner*, **45**(3):181–186.

Battersby, A. 1958. Piping and Plant Design: Can Linear Programming help? *Manuf. Chemist*, **29**(9):368.

Bauman, H. C. 1964. "Fundamentals of Cost Engineering in the Chemical Industry," Reinhold Publishing Corporation, New York.

Baumol, W. 1961. "Economic Theory and Operations Analysis," Prentice-Hall, Inc., Englewood Cliffs, N.J.

Beale, E. M. L. 1954. An Alternative Method of Linear Programming, *Proc. Cambridge Phil. Soc.*, **50**:513–523.

———. 1955a. On Minimizing a Convex Function Subject to Linear Inequalities, *J. Royal Statist. Soc.*, **17B**:173-184.

———. 1955b. Cycling in the Dual Simplex Algorithm, *Naval Res. Logistics Quart.*, **2**:269–276.

———. 1959. On Quadratic Programming, *Naval Res. Logistics Quart.*, **6**:227–244.

———, P. A. B. Hughes, and R. E. Small. 1965. Experiences in Using a Decomposition Program, *Computer J.*, **8**:13–19.

Beckman, F. S. 1960. The Solution of Linear Equations by the Conjugate Gradient Method, in A. Ralston and H. S. Wilf (eds.), "Mathematical Methods for Digital Computers," John Wiley & Sons, Inc., New York.

Beightler, C. S., and D. J. Wilde. 1965. Sensitivity in Linear Programming, *Hydrocarbon Process Petrol. Refiner.*, **44**(2):111–126.

Bell, D. A. 1962. "Intelligent Machines," pp. 62–63, Blaisdell Publishing Co., Waltham, Mass.

Bellman, R. 1955. Equipment Replacement Policy, *J. Soc. Ind. Appl. Math.*, **3**:133–136.

———. 1956. On the Theory of Dynamic Programming: A Warehousing Problem, *Management Sci.*, **2**:272–276.

———. 1957. "Dynamic Programming," Princeton University Press, Princeton, N.J.

———. 1961. "Adaptive Control Processes: A Guided Tour," Princeton University Press, Princeton, N.J.

———, and S. Dreyfus. 1962. "Applied Dynamic Programming," Princeton University Press, Princeton, N.J.

———. and R. Kalaba. 1961. Reduction of Dimensionality, Dynamic Programming and Control Processes, *Trans. ASME J. Basic Eng.*, **83D**:82–84.

———, and ———. 1965. "Quasi-linearization and Non-linear Boundary Value Problems," American Elsevier Publishing Company of New York.

Berezin, I. S., and N. P. Zhidkev. 1965. "Computing Methods" (transl. by O. M. Blum), Oxford University Press, London.

Berg, C. 1960. Optimization in Process Development, *Chem. Eng. Progr.*, **56**(8):42–48.

Berman, H. 1962. The Critical Path Method, *Chem. Process Eng.*, **43**:57–58.

Beveridge, G. S. G., and R. S. Schechter. 1965. Optimal Reaction Systems with Recycle, *Ind. Eng. Chem. Fundamentals*, **4**:257–260.

Bigelow, C. G. 1962. Bibliography on Project Planning and Control by Network Analysis: 1959–1961, *J. Opn. Res. Soc. Am.*, **10**:728–731.

Bilous, O., and N. R. Amundson. 1956a. Optimum Temperature Gradients in Tubular Reactors, Part I, General Theory and Methods, *Chem. Eng. Sci.*, **5**:81–92.

———, and ———. 1956b. Optimum Temperature Gradients in Tubular Reactors, Part II, Numerical Study, *Chem. Eng. Sci.*, **5**:115–126.

Bird, R. B., W. E. Stewart, and E. N. Lightfoot. 1960. "Transport Phenomena," John Wiley & Sons, Inc., New York.

Birkhoff, G., and S. MacLane. 1953. "A Survey of Modern Algebra," pp. 192–193, The Macmillan Company, New York.

Bishop, A. B. and T. H. Rockwell. 1958. A Dynamic Programming Computational Procedure for Optimal Manpower Loading in a Large Aircraft Company, *J. Opn. Res. Soc. Am.*, **6**:835–848.

Blakemore, N., and R. Aris. 1962. Bang-bang Control of a Batch Reactor, *Chem. Eng. Sci.*, **17**:591–598.

Blum, J. R. 1954. Multi-dimensional Stochastic Approximation Methods, *Ann. Math. Statistics*, **25**:737–744.

Boas, A. H. 1963a. How Search Methods Locate Optima in Univariable Problems, *Chem. Eng.*, **70**(Feb. 4):105–108.

———. 1963b. Optimization via Linear and Dynamic Programming, *Chem. Eng.*, **70**(Apr. 1):85–88.

Boldyreff, A. W. 1955. Determination of the Maximal Steady State Flow of Traffic through a Railroad Network, *J. Opn. Res. Soc. Am.*, **3**:443–465.

Boot, J. C. G. 1964. "Quadratic Programming," N. Holland Publishing Co., Amsterdam.

Booth, A. D. 1955. "Numerical Methods," Academic Press Inc., New York; also Butterworths & Co. (Publishers), London, 1957.

Boreskov, G. K., and M. G. Slin'ko. 1965. Simulation and Optimisation of Chemical Reactors, *Brit. Chem. Eng.*, **10**(3):170–174.

Boulding, K. E., and W. A. Spivey. 1960. "Linear Programming and the Theory of the Firm," The Macmillan Company, New York.

Bowman, E. H. 1956. Production Scheduling by the Transportation Method of Linear Programming, *J. Opn. Res. Soc. Am.*, **4**:100–103.

Box, G. E. P. 1954. Exploration and Exploitation of Response Surfaces: Some General Considerations and Examples, *Biometrics*, **10**:16–60.

———. 1956. In Davies, [1956, chap. 11].

———. 1957. Evolutionary Operation: A Method of Increasing Industrial Productivity, *Appl. Statistics*, **6**:3–22.

———. 1960. Some General Considerations in Process Optimization, *Trans. Am. Soc. Mech. Eng.*, **82D**:113–119.

Box, G. E. P., and D. W. Behnken. 1960. Simplex-sum Designs: A Class of Second-order Designs Derivable from Those of First Order, *Ann. Math. Statistics*, **31**:838–864.

———, and J. Chanmugan. 1962. Adaptive Optimization of Continuous Processes, *Ind. Eng. Chem. Fundamentals*, **1**:2–16.

Box, G. E. P., and W. J. Hill. 1967. Discrimination among Mechanistic Models, *Technometrics*, **9**:57–71.

———, and J. S. Hunter. 1957. Multifactor Experimental Designs for Exploring Response Surfaces, *Ann. Math. Statistics*, **28**:195–241.

———, and ———. 1959. Condensed Calculations for Evolutionary Operation Programs, *Technometrics*, **1**:77–95.

Box, G. E. P., and H. L. Lucas. 1959. Design of Experiments in Nonlinear Situations, *Biometrika*, **46**:77–90.

———, and K. B. Wilson. 1951. On the Experimental Attainment of Optimum Conditions, *J. Roy. Statist. Soc.*, **13B**:1–45.

Box, M. J. 1965. A New Method of Constrained Optimization and a Comparison with Other Methods, *Computer J.*, **8**:42–52.

———. 1966. A Comparison of Several Current Optimization Methods, and the Use of Transformations in Constrained Problems, *Computer J.*, **9**:67–77.

Box, M. J., D. Davies and W. H. Swan. 1969, "Nonlinear Optimization Techniques," I.C.I. Monograph on Mathematics and Statistics, Oliver and Boyd, Edinburgh.

Brand, L. 1958. "Advanced Calculus," John Wiley & Sons, Inc., New York.

Brander, J. F. R., E. J. Cannon, and B. D. Dagnall. 1964. Optimization Procedure for a Complex Plant, *Brit. Chem. Eng.*, **9**:745–748.

Brooks, S. H. 1958. A Discussion of Random Methods for Seeking Maxima, *J. Opn. Res. Soc. Am.*, **6**:244–251.

———, and M. R. Mickley. 1961. Optimal Estimation of Gradient Direction in Steepest Ascent Experiments, *Biometrics*, **17**:48–56.

Brosilow, C., and L. S. Lasdon. 1965. A Two-level Optimization Technique for Recycle Processes, *Am. Inst. Chem. Eng./Inst. Chem. Eng. (London) Joint Symp.*, **4**:75–83.

———, and E. Nunez. 1968. Multilevel Optimization Applied to a Catalytic Cracking Plant, *Can. J. Chem. Eng.*, **46**:205–212.

Buehler, R. J., B. V. Shah, and O. Kempthorne. 1961a. Some Problems of Steepest Ascent and Related Procedures for Finding Optimum Conditions, *Statist. Lab. Tech. Rept.* 1, Office of Naval Res. Contract Non-r. -530 (05) NR-042-207, Iowa State, Ames, Iowa, April, 1961.

———, ———, and ———. 1961b. Some Further Properties of the Methods of Parallel Tangents and Conjugate Gradients, *ibid.*, *Rept.* 3, September, 1961.

———, ———, and ———. 1962. The Method of Parallel Tangents (PARTAN) for Finding an Optimum, *ibid.*, *Rept.* 2 (April, 1961), rev. August, 1962.

Calderbank, P. H. 1953. Contact Process Converter Design, *Chem. Eng. Progr.*, **49**:585–590.

Campey, I. G., and D. G. Nickols. 1961. Simplex Minimization, *I.C.I. Ltd. (Central Instr. Lab.) Rept.*, London, August, 1961.

Carpenter, B. H., and H. C. Sweeny. 1965. Process Improvement with Simplex Self Directing Evolutionary Operation, *Chem. Eng.*, **72**(July):117–126.

Carr, N. L. 1960. Kinetics of Catalytic Isomerization of an *n*-pentane, *Ind. Eng. Chem.*, **52**:391–396.

Carroll, G. W. 1961. The Created Response Surface Technique for Optimizing Nonlinear Restrained Systems, *J. Opn. Res. Soc. Am.*, **9**:169–184.

Cauchy, A. 1847. Méthode générale pour la résolution des systèmes d'équations simultanées, *Compt. Rend. Sci. Paris*, **25**:536–538.

Cerri, E., M. Silestri, and S. Villon. 1956. Dynamic Programming: Heavy Water Distillation, *Z. Naturforsch.*, **11a**:694.

Chambers, J. C., A. F. Bond, and J. H. Leake. 1958. Optimum Lot Size for Parts Used in Aircraft Production, *J. Opn. Res. Soc. Am.*, **6**:385–398.

Chang, S. S. L. 1961. "Synthesis of Optimum Control Systems," p. 102, McGraw-Hill Book Company, New York.

Chanmugan, J., and G. M. Jenkins. 1963. Optimum Experimentation in the Process Industries, *Chem. Eng. Progr. Symp., Ser.* 42, **59**:108.

Charnes, A. 1952. Optimality and Degeneracy in Linear Programming, *Econometrica*, **20**:160–170.

Charnes, A. and W. W. Cooper. 1953. Scheduling of Manufactured Product, *Econometrica*, **21**:480–481.

———, and ———. 1955. Generalizations of the Warehouse Model, *J. Opn. Res. Soc. Am.*, **6**:131–172.

———, and ———. 1957. Management Models and Industrial Applications of Linear Programming, *Management Sci.*, **4**:38–92.

———, and ———. 1961. "Management Models and Industrial Applications of Linear Programming," 2 vols., John Wiley & Sons, Inc., New York.

———, ———, and D. Farr. 1953. Linear Programming and Profit Preference Scheduling for a Manufacturing Firm, *J. Opn. Res. Soc. Am.*, **1**:114–129.

———, ———, and R. C. Ferguson. 1954–1955. Optimal Estimation of Executive Compensation by Linear Programming, *Management Sci.*, **1**:138–151.

———, ———, and B. Mellon. 1952. Blending Aviation Gasolines: a Study in Programming Inter-dependent Activities in an Integrated Oil Company, *Econometrica*, **20**:135–159.

———, ———, and ———. 1954. A Model for Programming and Sensitivity Analysis in an Integrated Oil Company, *Econometrica*, **22**:193–217.

———, and C. E. Lemke. 1954. Computational Theory of Linear Programming, Part 1, The Bounded Variables Problem, *Carnegie Inst. Tech. ONR Res. Mem.* 10, Pittsburgh (Jan. 7, 1954).

Chen, T. C., and N. H. Ceaglske. 1966. Optimum Cross-current Extraction with Miscible Solvents, *Ind. Eng. Chem. Fundamentals*, **5**:422–426.

Chermin, H. A. G., and D. W. van Krevelen. 1961. Selectivity in Consecutive Reactions, *Chem. Eng. Sci.*, **14**:58–71.

Chien, H. H-Y. 1966. Optimization of Recycle Problem, *Ind. Eng. Chem. Fundamentals*, **5**:66–68.

Christensen, J. H., and D. F. Rudd. 1969. Structuring Design Computations, *Am. Inst. Chem. Eng. J.*, **15**:94–100.

Chu, C., and O. A. Hougen. 1961. Optimum Design of a Catalytic Nitric Oxide Reactor, *Chem. Eng. Progr.*, **57**(6):51–58.

Chung, A. 1963. "Linear Programming," Charles E. Merrill, Inc., Englewood Cliffs, N.J.

Chung, K. L. 1954. On a Stochastic Approximation Method, *Ann. Math. Statistics*, **25**:463–483.

Churchman, C. W., R. L. Ackoff, and E. L. Arnoff. 1957. "Introduction to Operations Research," Chapman & Hall, Ltd., London and John Wiley & Sons, Inc., New York.

Cochran, W. O. 1960. Procedures for Selection of Optimum Conditions, *Chem. Eng. Progr. Symp. Ser.* 31, **56**:88. Also in Hsieh (*ed.*) "Optimization Techniques in Chemical Engineering," New York University Symposium, May 18, 1960.

Converse, A. O., and G. D. Gross. 1963. Optimal Distillate Rate Policy in Batch Distillation, *Ind. Eng. Chem. Fundamentals*, **2**:217–221.

———, and C. I. Huber. 1965. Effect of Hold-up on Batch Distillation Optimization, *Ind. Eng. Chem. Fundamentals*, **4**:475–477.

Cord, J. 1964. A Method for Allocating Funds to Investment Projects When Returns Are Subject to Uncertainty, *Management Sci.*, **10**:335–341.

Courant, R. 1936. "Differential and Integral Calculus," Interscience Publishers, Inc., New York.

———. 1943. Variational Methods for the Solution of Problems of Equilibrium and Vibrations, *Bull. Am. Math. Soc.*, **49**:1–23.

———, and D. Hilbert. 1953. "Methods of Mathematical Physics," vol. 1, p. 222, Interscience Publishers, Inc., New York.

Coward, I. 1967. The Time-optimal Problem in Batch Distillation, *Chem. Eng. Sci.*, **22**:503–516.

———, and R. Jackson. 1965. Optimum Temperature Profiles in Tubular Reactors: An Exploration of Some Difficulties in the Use of Pontryagin's Maximum Principle, *Chem. Eng. Sci.*, **20**:911–920.

Dahlin, E. B., and J. M. Nelson. 1964. Simulation and Optimal Control of Chemical Processes, *Chem. Eng. Progr.*, **60**(3):49–56.

Dakin, R. J. 1965. Tree Search Algorithm for Mixed Integer Programming Problems, *Computer J.*, **8**:250–255.

Dantzig, G. B. 1949. Programming of Independent Activities, II, Mathematical Model, *Econometrica*, **17**:200–211.

———. 1951a. Programming of Interdependent Activities, II, Mathematical Model, in Koopmans [1951].

———. 1951b. Application of the Simplex Method to a Transportation Problem, in Koopmans [1951, chap. 23].

———. 1954. Notes on Linear Programming, VIII, Upper Bounds, *RAND Rept.* RM1367 (October, 1954), *Econometrica*, **23**:174–183, 1955.

———. 1959. On a Linear Programming Combinatorial Approach to the Travelling Salesman Problem, *J. Opn. Res. Soc. Am.*, **7**:58–66.

———. 1960. On the Significance of Solving Linear Programming Problems with Some Integer Variables, *Econometrica*, **27**:30–45.

———. 1963. "Linear Programming and Extensions," Princeton University Press, Princeton, N.J.

———, L. R. Ford, and D. R. Fulkerson. 1956. A Primal-dual Algorithm for Linear Programs, in Kuhn and Tucker [1956, pp. 171–181].

———, D. R. Fulkerson, and S. M. Johnson. 1954. Solution of a Large Scale Travelling Salesman Problem, *J. Opn. Res. Soc. Am.*, **2**:393–410.

———, S. M. Johnson, and W. White. 1958. A Linear Programming Approach to the Chemical Equilibrium Problem, *Management Sci.*, **5**:38–43.

———, and P. Wolfe. 1960. Decomposition Principle for Linear Programs, *J. Opn. Res. Soc. Am.*, **8**:101–111.

Davidmann, M. 1959a. Problems in Organizing Chemical Plant Projects, *Brit. Chem. Eng.*, **4**:403–408.

———. 1959b. Planning the Work of a Design Group, *Brit. Chem. Eng.*, **4**:589–593.

Davidon, W. C. 1959. Variable Metric Method for Minimization, *AEC Res. Develop. Rept.* ANL-5990, December, 1959 (rev.).

Davies, O. L. (ed.). 1956. "The Design and Analysis of Industrial Experiments," Oliver & Boyd, Ltd. Edinburgh and London; also Hafner, New York.

Deam, R. J. 1959. Use of Linear Programming in Oil Refinery Operations, *Process Control Automation* (Feb.): 59–64.

Debreu, G. 1952. Definite and Semidefinite Quadratic Forms, *Econometrica*, **20**:295–300.

Denbigh, K. G. 1944. Velocity and Yield in Continuous Reaction Systems, *Trans. Faraday Soc.*, **40**:352–373.

———. 1958. Optimum Temperature Sequences in Reactors, *Chem. Eng. Sci.*, **8**:125–132.

———. 1961. Instantaneous and Overall Reaction Yields, *Chem. Eng. Sci.*, **14**:25–28. Also in Second European Symposium on Chemical Reaction Engineering, April, 1960, Amsterdam.

Denn, M. M. 1965. Discrete Maximum Principle, *Ind. Eng. Chem. Fundamentals* **4**:240.

———, and R. Aris. 1965a. Green's Functions and Optimal Systems: Necessary Conditions and Iterative Technique, *Ind. Eng. Chem. Fundamentals*, **4**:7–16.

———, and ———. 1965b. Green's Functions and Optimal Systems: Complex Interconnected Structures, *Ind. Eng. Chem. Fundamentals*, **4**:248–257.

———, and ———. 1965c. Green's Functions and Optimal Systems: Gradient Direction in Decision Space, *Ind. Eng. Chem. Fundamentals*, **4**:213–221.

Denton, W. H. 1960. Overall Economics of Plate-fin Heat Exchangers, *Brit. Chem. Eng.*, **5**:40–48.

Di Bella, C. W., and W. F. Stevens. 1965. Process Optimization by Non-linear Programming, *Ind. Eng. Chem.* (*Process Design Develop.*), **4**:16–20.

Dorfman, R. 1951. "Applications of Linear Programming to the Theory of The Firm," University of California Press, Berkeley, Calif.

Dorn, W. S. 1958. Duality in Quadratic Programming, *N.Y. Univ. Rept.* NY08676 (*Physics*). See also Dorn [1960].

———. 1960. Duality in Quadratic Programming, *Quart. Appl. Math.*, **18**:155–162.

———. 1963. Non-linear Programming: A Survey, *Management Sci.*, **9**(2):171–208.

Douglas, J. M. 1966. The Use of Optimization Theory to Design Simple Multi-variable Control Systems, *Chem. Eng. Sci.*, **21**:519–532.

———, and M. M. Denn. 1965. Optimal Design and Control by Variational Methods, *Ind. Eng. Chem.*, **57**(11):18–31.

Draper, N. R., and W. G. Hunter. 1966. Design of Experiments for Parameter Estimation in Multiresponse Situations, *Biometrika*, **53**:525–533.

Draper, N. R., and H. Smith. 1966. "Applied Regression Analysis," John Wiley & Sons, Inc., New York.

Dresher, M. 1961. "Games of Strategy: Theory and Application," Prentice-Hall, Inc., Englewood Cliffs, N.J.

Dreyfus, S. E. 1957. An Analytic Solution of the Warehousing Problem, *Management Sci.*, **4**:99–104.

———. 1960. A Generalized Equipment Study, *J. Soc. Ind. Appl. Math.*, **8**:425–435.

Duffin, R. J. 1962a. Cost Minimization Problems Treated by Geometric Means, *J. Opn. Res. Soc. Am.*, **10**:668–675.

———. 1962b. Dual Programs and Minimum Cost, *J. Soc. Ind. Appl. Math.*, **10**:119–123.

———, and E. L. Peterson. 1964. Constrained Minima Treated by Geometric Means, *Westinghouse Sci. paper* 64-158-129-P3, March.

———, ———, and C. M. Zener. 1967. "Geometric Programming," John Wiley & Sons, Inc., New York.

Dvoretzky, A. 1956. On Stochastic Approximation, in Neyman [1956, pp. 39–55].

Dybdal, E. C. 1950. Engineering and Economic Evaluation of Projects, *Chem. Eng. Progr.*, **46**(2):57–66.

Eben, C. D., and J. R. Ferron. 1968. A Conjugate Inequality for General Means with Applications to Extremum Problems, *Am. Inst. Chem. Eng. J.*, **14**:32–37.

Edelbaum, T. N. 1962. Theory of Maxima and Minima, in Leitmann [1962, chap. 1].

Eisemann, K. 1957. The Trim Problem, *Management Sci.*, **3**:279–284.

Enyedy, G. 1962. Maximum Yield of Crystals in a Continuous Vacuum Crystallization Process, *Chem. Eng. Progr. Symp. Ser.* 37, **58**:10.

Faddeeva, V. N., and D. K. Faddeeva. 1963. "Computational Methods of Linear Algebra," Freeman Publications, Inc., San Francisco (trans. by R. C. Williams).

Fan, L. T., L. E. Erickson, R. W. Sucher, and G. S. Mathad. 1965. Optimal Design for a Sequence of Continuous Flow Stirred Tank Reactors with Product Recycle, *Ind. Eng. Chem. (Process Design Develop.)*, **4**:431–434.

———, and C. S. Wang. 1964a. Multi-stage Optimization by the Generalized Discrete Maximum Principle, *J. Electron. Control*, **16**:441–450.

———, and ———. 1964b. "The Discrete Maximum Principle: A Study of Multistage Systems Optimization," John Wiley & Sons, Inc., New York.

———, and ———. 1964c. Optimization of Multistage Processes with Product Recycle, *Chem. Eng. Sci.*, **19**:86–87.

———, and ———. 1964d. Optimization of One-dimensional Multi-stage Processes, *Z. Angew. Math. Phys.*, **15**:46–57.

———, and ———. 1964e. On the Optimization of Multi-stage Feedback Processes, *J. Soc. Ind. Appl. Math.*, **12**:226–232.

———, and ———. 1965. The Discrete Maximum Principle, *Ind. Eng. Chem. Fundamentals*, **4**:239.

Fenech, R. R., and A. Acrivos. 1956. The Application of Linear Programming to Design Problems, *Chem. Eng. Sci.*, **5**:93–98.

Ferguson, R. C., and G. B. Dantzig. 1956. The Allocation of Aircraft to Routes: An Example of Linear Programming under Uncertain Demand, *Management Sci.*, **3**:45–73.

Fiacco, A. V. 1961. Comments on Carroll's [1961] paper, *J. Opn. Res. Soc. Am.*, **9**:184–185.

Fiacco, A. V., and G. P. McCormick. 1964a. The Sequential Unconstrained Minimization Technique for Non-linear Programming: A Primal-dual Method, *Management Sci.*, **10**:360–366.

———, and ———. 1964b. Computational Algorithm for Sequential Unconstrained Minimization Technique for Non-linear Programming, *Management Sci.*, **10**:601–617.

———, and ———. 1966. Extension of Sequential Unconstrained Minimization Techniques for Non-linear Programming: Equality Constraints and Extrapolation, *Management Sci.*, **12**:816–828.

———, N. M. Smith, and D. Blackwell. 1960. A More General Method for Non-linear Programming, *J. Opn. Res. Soc. Am.*, **8** (suppl. 1), B36.

Fletcher, R. 1965. Function Evaluation without Evaluating Derivatives: A Review, *Computer J.*, **8**:33–41.

———, and M. J. D. Powell. 1963. A Rapidly Convergent Descent Method for Minimization, *Computer J.*, **6**:163–168. Also *AERE*, March, 1963, HL63/1280 TP100.

———, and C. M. Reeves. 1964. Function Minimization by Conjugate Gradients, *Computer J.*, **7**:149–154.

Flood, M. M. 1954. Application of Transportation Theory to Scheduling a Military Tanker Fleet, *J. Opn. Res. Soc. Am.*, **2**:150–162.

———. 1956. The Travelling Salesman Problem, *J. Opn. Res. Soc. Am.*, **4**:61–75. See also McCloskey and Coppinger [1956, p. 340].

Foord, J. W. 1962. Project Planning Techniques, *Chem. Proc. Eng.* (*Cost Eng.*), **43**:613–614.

Ford, L. R., and D. R. Fulkerson. 1956. Solving the Transportation Problem, *Management Sci.*, **3**:24–32.

———, and ———. 1962. "Flows in Networks," Princeton University Press, Princeton, N.J.

Forsyth, A. R. 1960. "Calculus of Variations," Dover Publications, Inc., New York.

Forsythe, G. E., and T. S. Motzkin. 1951. Acceleration of the Optimum Gradient Method, *Bull. Am. Math. Soc.*, **57**:304–305.

Frank, M., and P. Wolfe. 1956. An Algorithm for Quadratic Programming, *Naval Res. Logistics Quart.*, **3**:95–110.

Friedman, M., and L. S. Savage. 1947. "Selected Techniques of Statistical Analysis," McGraw-Hill Book Company, New York.

Frischberg, M. C. 1962. LESS Tells You How Project Is Doing, *Hydrocarbon Process Petrol Refiner*, **41**(2):130–138.

Frost, A. A., and R. G. Pearson. 1961. "Kinetics and Mechanisms," 2d ed., John Wiley & Sons, Inc., New York.

Gaddum, J. W., A. J. Hoffman, and D. Sokolowsky. 1954. On the Solution of the Caterer Problem, *Naval Res. Logistics Quart.*, **1**:223–229.

Gardner, M. 1969. Mathematical Games, *Sci. Amer.*, **220**(3):116–120.

Gass, S. I. 1964. "Linear Programming," McGraw-Hill Book Company, New York.

Gaver, D. P. 1965. Models for Appraising Investments Yielding Stochastic Returns, *Management Sci.*, **11**:815–830.

Gel'fand, I. M., and S. V. Fomin. 1963. "Calculus of Variations," Prentice-Hall, Inc., Englewood Cliffs, N.J.

Gilmore, P. C., and R. E. Gomory. 1961. A Linear Programming Approach to Cutting Stock Problem, *J. Opn. Res. Soc. Am.*, **9**:849–859.

———, and ———. 1963. A Linear Programming Approach to Cutting Stock Problem, Part II, *J. Opn. Res. Soc. Am.*, **11**:863–888.

———, and ———. 1964. A Solvable Case of the Travelling Salesman Problem, *Proc. Natl. Acad. Sci.*, **51**:178–181.

———, and ———. 1965. Multistage Cutting Stock Problem of Two and More Dimensions, *J. Opn. Res. Soc. Am.*, **13**:94–120.

Glass, H., and L. Cooper. 1965. Sequential Search: A Method for Solving Constrained Optimization Problems, *J. Assoc. Computing Machinery*, **12**(1):71–82.

Gomory, R. E. 1958. Outline of an Algorithm for Integer Solutions to Linear Programs, *Bull. Am. Math. Soc.*, **64**:275–278.

———. 1965. On the Relation between Integer and Non-integer Solutions to Linear Programs, *Proc. Natl. Acad. Sci.*, **53**:260–265.

Graves, R. L., and P. Wolfe. 1963. "Recent Advances in Mathematical Programming," McGraw-Hill Book Company, New York.

Grutter, W. F., and B. H. Messikommer. 1961. Systematische Ausbeuteberechnung für isotherme Reaktoren in Falle zusammengesetzter Reaktionen, *Chem. Eng. Sci.*, **14**:231–239.

Guin, J. A. 1968. Modification of the Complex Method of Constrained Optima, *Computer J.*, **10**:416–417.

Gunness, R. C. 1951. In Progress in Petroleum Technology, *Advan. Chem. Ser.* 5, pp. 109–119, Am. Chem. Soc., Washington, D.C.

Gunther, P. 1955. Use of Linear Programming in Capital Budgeting, *J. Opn. Res. Soc. Am.*, **3**:219–224.

Hadley, G. H. 1962. "Linear Programming," Addison-Wesley Publishing Company, Inc., Reading, Mass.

———. 1964. "Non-linear Programming and Dynamic Programming," Addison-Wesley Publishing Company, Inc., Reading, Mass.

Hall, R., and S. Eilon. 1963. Controlling Production Processes, *Opn. Res. Quart.*, **14**:279–289.

Hancock, H. 1960. "Theory of Maxima and Minima," Dover Publications, Inc., New York. (Original publication 1917).

Hanson, M. A. 1963. An Algorithm for Convex Programming, *Australian J. Statistics*, **5**:14–19.

Happel, J. 1958. "Chemical Process Economics," John Wiley & Sons, Inc., New York.

Harris, P. M. J. 1964. An Algorithm for Solving Mixed Integer Linear Programmes, *Opn. Res. Quart.*, **15**(2):117–132.

Hartland, S. 1967a. The Optimization of Forward and Back Extraction, Part I, Countercurrent Flow, *Trans. Inst. Chem. Eng.* (*London*), **45**:T82–89.

———. 1967b. The Optimization of Forward and Back Extraction, Part II, Crosscurrent Flow with Recirculation, *Trans. Inst. Chem. Eng.* (*London*), **45**:T90–93.

Hartley, H. O. 1960. Non-linear Programming for Separable Objective Functions and Constraints, RAND Symposium on Mathematical Programming, *RAND Corp. Rept.* R-351, Santa Monica, Calif.

Henley, E. J., H. K. Staffin, and R. Staffin. 1962. Frequency Response Techniques and Dynamic Analysis for the CPI, *Ind. Eng. Chem.*, **54**(5):30-34.

Hensgen, C. 1966. Process Optimization by Non-Linear Programming, Rev. Autom. **8**(3): 99–104.

Hess, S. W., and H. A. Quigley. 1963. Analysis of Risk in Investments Using Monte Carlo Techniques, *Chem. Eng. Progr. Symp. Ser.* 42, **59**:55.

Hestenes, M. R., and E. Stiefel. 1952. Methods of Conjugate Gradients for Solving Linear Systems, *J. Res. Natl. Bur. Std.*, **49**:409–436.

Hildebrand, F. B. 1956. "Introduction to Numerical Analysis," McGraw-Hill Book Company, New York.

Hildreth, C. 1957. A Quadratic Programming Procedure, *Naval Res. Logistics Quart.*, **4**:79–85.

Hill, W. J., W. G. Hunter, and D. W. Wichern. 1968. A Joint Design Criterion for the Dual Problem of Model Discrimination and Parameter Estimation, *Technometrics*, **10**:145–160.

Hills, B. A. 1961. Volume Determination of Heterogeneous Catalytic Gas Reactors, *Brit. Chem. Eng.*, **6**:856–859.

———, 1963. Determination of Optimum Pressure for Operating Catalytic Gas Reactors, *Brit. Chem. Eng.*, **8**:683–687.

Himmelblau, D. M. 1963. Process Optimization by Search Techniques, *Ind. Eng. Chem. (Process Design Develop.)*, **2**:296–300.

———. 1966. Decomposition of Large Scale Systems, I, Systems Composed of Lumped Parameter Elements, *Chem. Eng. Sci.*, **21**:425–438.

———. 1967. Decomposition of Large Scale Systems, II, Systems Containing Non-linear Elements, *Chem. Eng. Sci.*, **22**:883–895.

———, and K. E. Bischoff. 1968. "Process Analysis and Simulation," John Wiley & Sons, Inc., New York.

Himsworth, F. R. 1962. Empirical Methods of Optimization, *Trans. Inst. Chem. Eng. (London)*, **40**:345–349.

Hitchcock, F. L. 1941. The Distribution of a Product from Several Sources to Numerous Localities, *J. Math. Phys.*, **20**:224–230.

Hobbs, J. A. 1966. Linear Programming with Costs and Incomes in Two or more Separated Currencies, *Opn. Res. Quart.*, **17**:133.

Hodges, J. L., and E. L. Lehman. 1956. In Neyman [1956, pp. 95–104].

Hoffman, A. J. 1953. Cycling in the Simplex Algorithm, *Natl. Bur. Std. Rept.* 2974, December, 1953.

Hooke, R., and T. A. Jeeves. 1961. Direct Search Solution of Numerical and Statistical Problems, *J. Assoc. Computer Machines*, **8**:212–229.

Horn, F. J. M. 1961a. Optimale Temperatur-und konzentrations Verlaufe, *Chem. Eng. Sci.*, **14**:77–89.

———. 1961b. Uber die optimale Temperatur führung bei kontinuierlichen chemischen Prozessen, *Z. Elektrochem.*, **65**:209.

———. 1961c. Zur Berechnung von adiabatischen Abschnitts Reaktoren, I, *Z. Elektrochem.*, **65**:295.

———. 1961d. Uber das Problem der optimalen Ruhrkesselkaskade fur chemische Reaktionen, *Chem. Eng. Sci.*, **15**:176–187.

———. 1967. Periodic Countercurrent Processes, *Ind. Eng. Chem. (Process Design Devel.)*, **6**:30–35.

———, and R. Jackson. 1965a. Discrete Maximum Principle, *Ind. Eng. Chem. Fundamentals*, **4**:110–112.

———, and ———. 1965b. Discrete Maximum Principle, *Ind. Eng. Chem. Fundamentals*, **4**:239.

———, and L. Kuchler. 1959. Probleme bei reacktions technischen Berechnungen, *Chem. Ing. Tech.*, **31**:1–11.

———, and R. C. Lin. 1967. Periodic Processes: a Variational Approach, *Ind. Eng. Chem. (Process Design Devel.)*, **6**:21–30.

———, and R. A. May. 1968. Effect of Mixing on Periodic Countercurrent Processes, *Ind. Eng. Chem. Fundamentals*, **7**:349–354.

———, and U. Troltenier. 1960. Uber der optimale Temperaturverlauf im Reaktionsrohr, *Chem. Ing. Tech.*, **32**:382–392.

———, and ———. 1963. Zur Berechnung von Ruhrkesselkaskaden mit Hilfe eines programmgesteurten Rechenautomaten, *Chem. Ing. Tech.*, **35**:11–18.

Houthakker, H. S. 1955. On the Numerical Solution of the Transportation Problem, *J. Opn. Res. Soc. Am.*, **3**:210–214.

———. 1960. The Capacity Method of Quadratic Programming, *Econometrica*, **28**: 62–87.

Hunter, J. S. 1960. Some Applications of Statistics to Experimentation, *Chem. Eng. Progr. Symp. Ser.* 31, **56**:10.

Hunter, W. G. 1967. Estimation of Unknown Constants from Multiresponse Data, *Ind. Eng. Chem. Fundamentals*, **6**:461–463.

———, and A. C. Atkinson. 1966. Statistical Designs for Pilot Plant and Laboratory Experiments—II, *Chem. Eng.*, **73**(June 6):159–164.

———, W. J. Hill, and T. L. Henson. 1969. Designing Experiments for Precise Estimation of all or some of the Constants in a Mechanistic Model, *Can. J. Chem. Eng.*, **47**:76–80.

———, and M. E. Hoff. 1967. Planning Experiments to Increase Research Efficiency, *Ind. Eng. Chem.*, **59**(3):43–48.

———, and R. Mezaki. 1964. A Model Building Technique for Chemical Engineering Kinetics, *Am. Inst. Chem. Eng. J.*, **10**:315–322.

Hurwicz, L. 1951. Optimality Criteria for Decision Making under Ignorance, *Cowles Commission Discussion Paper* 370.

Hwa, C. S. 1965. Mathematical Formulation and Optimization of Heat Exchanger Networks Using Separable Programming, *Proc. Am. Inst. Chem. Eng./Inst. Chem. Eng. (London) Joint Meeting, London Symp.*, June, **4**:101–106.

Imbrie, W. 1962. Critical Path Method for Scheduling and Control, *Chem. Process Eng. (Cost Eng.)*, **43**:377–379.

Itahara, S., and L. I. Stiel. 1966. Optimal Design of Multiple effect Evaporators by dynamic programming, *Ind. Eng. Chem. (Process Design Develop.)*, **5**:309–315.

———, and ———. 1968. Optimal Design of Multiple Effect Evaporators with Vapor Bleed Streams, *Ind. Eng. Chem. (Process Design Develop.)*, **7**:6–11.

Jackson, R. 1963. Comments on "Optimum Cross-current Extraction with Product Recycle" by Rudd and Blum [1962], *Chem. Eng. Sci.*, **18**:215–217.

———. 1964a. Some Algebraic Properties of Optimization Problems in Complex Chemical Plants, *Chem. Eng. Sci.*, **19**:19–31.

———. 1964b. A Generalized Variational Treatment of Optimization Problems in Complex Chemical Plants, *Chem. Eng. Sci.*, **19**:253–260.

———. 1965. A Variational Solution of Unsteady State Optimization Problems in Complex Chemical Plants, *Chem. Eng. Sci.*, **20**:405–413.

———. 1966a. Class Notes, Rice University, Dept. of Chemical Engineering, Houston, Texas.

———. 1966b. Optimum Start-up Procedures for an Autothermic Reaction System, *Chem. Eng. Sci.*, **21**:241–260.

———, and F. J. M. Horn. 1965a. Discrete Maximum Principle, *Ind. Eng. Chem. Fundamentals*, **4**:487–488.

———, and ———. 1965b. On Discrete Analogues of Pontryagin's Maximum Principle, *Intern. J. Control*, **1**:389–395.

Jacobs, W. W. 1954. The Caterer Problem, *Naval Res. Logistics Quart.*, **1**:154–165.

Jenson, V. G., and G. V. Jeffreys. 1961. Economic Analysis of Liquid-Liquid Extraction Processes, *Brit. Chem. Eng.*, **6**:676–679.

Johnson, S. M. 1956. Optimal Search for a Maximum Is Fibonaccian, RAND *Corp. Rept.* P-856, Santa Monica, Calif.

Johnston, J. 1963. Decision Theory, *Unilever Progr.*, Spring, pp. 164–171. Also *Centre Business Res. Reprint Ser.* 1, *Univ. Manchester.*

Kalaba, R. E. 1959. On Non-linear Differential Equations, the Maximum Operation and Monotone Convergence, *J. Math. Mech.*, **8**:519.

Kalman, R. E., L. Lapidus, and E. Shapiro. 1959. Optimal Control of Chemical and Petroleum Processes, *Proc. Joint Symp. on Instrumentation Computation Process Develop. Plant Design, Inst. Chem. Eng., Soc. Instrument. Tech., British Computer Soc.*, London, May, 1959, p. 6.

———, ———, and ———. 1960. Computer Control of Chemical Processes, *Chem. Eng. Progr.*, **56**(2):55–61.

Kantorovich, L. V. 1960. Mathematical Methods of Organizing and Planning Production, *Management Sci.*, **6**:366–422.

———, and V. I. Krylov. 1958. "Approximate Methods of Higher Analysis," (*transl. by* Benster), Erven P. Noordhoff, NV, Groningen, Netherlands.

Kaplan, W. 1952. "Advanced Calculus," Addison-Wesley Publishing Company, Inc., Reading, Mass.

Kast, W. G. 1962. Critical Path Method for Plant Construction, *Hydrocarbon Process Petrol. Refiner*, **41**(2):123–130.

Katz, S. 1960. Best Temperature Profiles in Plug Flow Reactors: Methods of the Calculus of Variations, *Ann. N.Y. Acad. Sci.*, **84**:441–478.

———, 1962a. Best Operating Points for Staged Systems, *Ind. Eng. Chem. Fundamentals*, **1**:226-240.

———. 1962b. A Discrete Version of Pontryagin's Maximum Principle, *J. Electron. Control*, **13**:179.

———. 1964. A General Maximum Principle for End Point Control Problems, *J. Electron. Control*, **16**:189.

Kelley, H. J. 1962. Method of Gradients, In Leitmann [1962, chap 6].

Kelley, J. E. 1960. The Cutting Plane Method for Solving Convex Programs, *J. Soc. Ind. Appl. Math.*, **8**:703–712.

Kern, D. Q., and R. E. Seaton. 1959. Optimum Trim Cooler Temperature, *Chem. Eng. Progr.*, **55**(7):69–70.

Kesten, H. 1958. Accelerated Stochastic Approximation, *Ann. Math. Statistics*, **29**:41–59.

Kiefer, J. 1953. Sequential Minimax Search for a Maximum, *Proc. Am. Math. Soc.*, **4**:502–506.

———. 1957. Optimum Sequential Search and Approximation Methods under Minimum Regularity Assumptions, *J. Soc. Ind. Appl. Math.*, **5**:125.

———, and J. Wolfowitz. 1952. Stochastic Estimation of the Maximum of a Regression Function, *Ann. Math. Statistics*, **23**:462–466.

King, R. P. 1965. Calculation of the Optimal Conditions for Chemical Reactors of the Combined Type, *Chem. Eng. Sci.*, **20**:537–544.

Kipiniak, W. 1961. "Dynamic Optimization and Control: A Variational Approach," The M.I.T. Press, Cambridge, Mass., and John Wiley & Sons, Inc., New York.

Kittrell, J. R., and J. Erjavec. 1968. Response Surface Methods in Heterogeneous Kinetic Modeling, *Ind. Eng. Chem.* (*Proc. Design Develop.*), **7**:321–327.

———, and C. C. Watson. 1966. Don't Overdesign Process Equipment, *Chem. Eng. Progr.*, **62**(4):79–83.

Klingman, W. R., and D. M. Himmelblau. 1964. Non-linear Programming with the Aid of a Multiple Gradient Summation Technique, *J. Assoc. Computer Machines*, **11**:400–415.

Koenigsberg, E. 1961. Some Industrial Applications of Linear Programming, *Opn. Res. Quart.*, **12**:105–114.

Koopmans, T. C. (ed.). 1951. "Activity Analysis of Production and Allocation," *Cowles Commission Monogr.* 13, John Wiley & Sons, Inc., New York.

Korn, G. A., and T.M. Korn. 1964. "Electronic Analog and Hybrid Computers," 2d ed., McGraw-Hill Book Company, New York.

Kubota, H., S. Namkoong, T. Akehata, and M. Shindo. 1961. Optimum Process Conditions for Completely Mixed Multi-stage Reactor, *Can. J. Chem. Eng.*, **39**:64–66.

Kuhn, H. W., and A. W. Tucker. 1951. Non-linear Programming, in Neyman [1951, pp. 481–492].

———, and ———. (eds.). 1956. "Linear Inequalities and Related Systems," *Ann. Math. Study* 38, Princeton University Press, Princeton, N.J.

Kunz, K. S. 1957. "Numerical Analysis," McGraw-Hill Book Company, New York.

Kuo, M. T., and D. I. Rubin. 1962. Optimization Study of Chemical Processes, *Can. J. Chem. Eng.*, **40**:152–156.

———, D. I. Rubin, and B. S. Wright. 1966. Optimization of Complex Chemical Processes, Comparison of Variational and Steepest Descent Methods, *Ind. Eng. Chem. (Process Design Develop.)*, **5**:404–408.

Kwauk, M. 1956. A System for Counting Variables in Separation Processes, *Am. Inst. Chem. Eng. J.*, **2**:240–248.

Laidler, K. J. 1965. "Chemical Kinetics," 2d ed., McGraw-Hill Book Company, New York.

Lanczos, C. 1966. "The Variational Principles of Mechanics," 3d ed., p. 162, University of Toronto Press, Toronto.

Land, A. H. 1957. An Application of Linear Programming to the Transport of Coking Coals, *J. Roy. Statist. Soc.*, 120**A**:308.

———, and A. G. Doig, 1960. An Automatic Method of Solving Discrete Programming Problems, *Econometrica*, **28**:497–521.

Lang, H. J. 1948. Simplified Approach to Preliminary Cost Estimates, *Chem. Eng.*, **55**(6):112–113.

Lapidus, L. 1962. "Digital Computation for Chemical Engineers," McGraw-Hill Book Company, New York.

———, and T. I. Peterson. 1965. Analysis of Heterogeneous Catalytic Reactions by Nonlinear Estimation, *Am. Inst. Chem. Eng. J.*, **11**:891–897.

———, E. Shapiro, S. Shapiro, and R. E. Stillman. 1961. Optimization of Process Performance, *Am. Inst. Chem. Eng. J.*, **1**:288–294.

Lasdon, L. S., and J. D. Schoeffler. 1964. *Instr. Soc. Am. 19th Ann. Conf.*, Oct. 12, 1964, *Preprint* 3.3-3-64,

Law, V. J., and R. H. Fariss. 1969. Sufficiency Conditions for Constrained Extrema, *Am. Inst. Chem. Eng. J.*, In press.

Leclerc, V. R. 1953. Le reacteur étage: outil fondamental de génie chimique, *Chem. Eng. Sci.*, **2**:213–219.

Lee, E. S. 1963. Optimum Design and Operation of Chemical Processes, *Ind. Eng. Chem.*, **65**(8):30–39.

———. 1964a. Optimization by Pontryagin's Maximum Principle on the Analogue Computer, *Am. Inst. Chem. Eng. J.*, **10**:309–315.

———. 1966. Quasi-linearization, Non-linear Boundary Value Problems and Optimization, *Chem. Eng. Sci.*, **21**:183–194.

———. 1967. Quasi-linearization in Optimization: A Numerical Study, *Am. Inst. Chem. Eng. J.*, **13**:1043–1045.

Lee, K-Y, and R. Aris. 1963a. Optimal Adiabatic Bed Reactors for SO_2 with Cold Shot Cooling, *Ind. Eng. Chem. (Process Design Develop.)*, **2**:300–306.

———, and ———. 1963b. Control of Cold Shot Adiabatic Reactor with Decaying Catalyst, *Ind. Eng. Chem. (Process Design Develop.)*, **2**:306–309.

Lee, W., J. H. Christensen, and D. F. Rudd. 1966. Design Variable Selection to Simplify Process Calculations, *Am. Inst. Chem. Eng. J.*, **12**:1104–1110.

Leitmann, G. (ed.). 1962. "Optimization Techniques," Academic Press Inc., New York.

Lemke, C. E. 1954. The Dual Method of Solving the Linear Programming Problem. *Naval Res. Logistics Quart.*, **1**:36–47.

Lerda-Olberg, S. 1966. Bibliography on Network-based Project Planning and Control Techniques: 1962–1965, *J. Opn. Res. Soc. Am.*, **14**:925–931.

Lermoyez, M. 1963. Generalized Optimization Programme—for Ethylene Plant, *Genie Chim.*, **89**:101–110.

Lester, J. T. 1961. Optimization in Equipment Design, *Brit. Chem. Eng.*, **6**:851–855.

Leung, P. K., and D. Quon. 1966. The Use of Dual Linear Programming in Formulating Approximating Functions by Using the Chebyshev Criterion, *Am. Inst. Chem. Eng. J.*, **12**:596–598.

Levenberg, K. 1944. A Method for the Solution of Certain Non-linear Problems in Least Squares, *Q. Appl. Math.*, **2**:164.

Love, C. E., and E. D. Rainville. 1962. "Differential and Integral Calculus," The Macmillan Company, New York.

Lovland, J. 1964. Dynamic Programming Applied to Multi-stage Processes with Recycle, *Chem. Eng. Sci.*, **19**:843–844.

Lowe, C. W. 1964. Some Techniques of Evolutionary Operation, *Trans. Inst. Chem. Eng. (London)*, **42**:T334–T344.

———. 1967. A Report on a Simplex Evolutionary Operation for Multiple Responses, *Trans. Inst. Chem. Eng. (London)*, **45**:T3–T8.

Luce, R. D. and H. Raiffa. 1957. "Games and Decisions; Introduction and Critical Survey," John Wiley & Sons, Inc., New York.

Luss, D. 1965. Optimum Volume Ratios for Residence Time in Stirred Tank Reactor Sequences, *Chem. Eng. Sci.*, **20**:171.

McCall, J. J. 1963. Operating Characteristics of Opportunistic Replacement and Inspection Policies, *Management Sci.*, **10**:85–97.

McCloskey, J. F., and J. M. Coppinger (eds.). "Operations Research for Management," vol. 2, Johns Hopkins Press, Baltimore, Md.

McCormick, G. P., W. C. Mylander, and A. V. Fiacco. 1965. Computer Program Implementing the Sequential Unconstrained Minimization Technique for Non-linear Programming, *Res. Analysis Corp. Advanced Res. Div. Tech. Paper* RAC-TP-151.

McDowell, I. 1960. The Economical Planning Period for Engineering Works, *J. Opn. Res. Soc. Am.*, **8**:533–542.

McEwan, M. W., and G. S. G. Beveridge. 1965. Process Optimization at the Design Stage with an Objective Function Linear in Profit and Investment, *Chem. Eng. Sci.*, **20**:987–995.

McKinsey, J. C. C. 1952. "Introduction to the Theory of Games," McGraw-Hill Book Company, New York.

Mah, R. S. H., and R. Aris. 1964. Optimal Policies for First Order Consecutive Reversible Reactions, *Chem. Eng. Sci.*, **19**:541–553.

Malengé, J. P., and J. Villermaux. 1967. Optimal Adiabatic Bed Reactor with Cold Shot Cooling, *Ind. Eng. Chem. (Process Design Develop.)*, **6**:535–536.

Manne, A. S. 1956. "Scheduling of Petroleum Refinery Operations," chap. 6, Harvard University Press, Cambridge, Mass.

Margenau, H., and G. M. Murphy. 1961. "The Mathematics of Physics and Chemistry," 2d ed., vol. 1, D. Van Nostrand Company, Inc., Princeton, N.J.

———, and ———. 1964. "The Mathematics of Physics and Chemistry," vol. 2, p. 209, D. Van Nostrand Company, Inc., Princeton, N.J.

Markowitz, H. 1956. The Optimization of a Quadratic Function Subject to Linear Constraints, *Naval Res. Logistics Quart.*, **3**:111–133.

Marquardt, D. W. 1959. Solution of Non-linear Chemical Engineering Models, *Chem. Eng. Progr.*, **55**(6):65–70.

Masso, A. H., and D. F. Rudd. 1969. The Synthesis of System Designs, II, Heuristic Structuring, *Am. Inst. Chem. Eng. J.*, **15**:10–17.

May, R. A., and F. J. M. Horn. 1968. Stage Efficiency of a Periodically Operated Distillation Column, *Ind. Eng. Chem. (Process Design Develop.)*, **7**:61–64.

Merrett, A. J., and A. Sykes. 1966. "Capital Budgeting and Company Finance," Longmans, Green & Co., Inc., New York.

Metzger, R. W. 1958. "Elementary Mathematical Programming," John Wiley & Sons, Inc., New York.

Meyer, H. A. (ed.). 1956. Symposium on Monte Carlo Methods, John Wiley & Sons, Inc., New York.

Mezaki, R., and J. B. Butt. 1968. Estimation of Rate Constants from Multiresponse Kinetic Data, *Ind. Eng. Chem. Fundamentals*, **7**:120–125.

Mickley, H. S., T. K. Sherwood, and C. E. Reed. 1957. "Applied Mathematics in Chemical Engineering," 2d ed., McGraw-Hill Book Company, New York.

Miele, A. 1962. The Calculus of Variations in Applied Aerodynamics and Flight Mechanics, *in* Leitmann [1962, Chap. 4].

Mitten, L. G., and G. L. Nemhauser. 1963a. Multistage Optimization, *Chem. Eng. Progr.*, **59**(1):52–60.

———, and ———. 1963b. Optimise Multistage Processes with Dynamic Programming, *Chem. Eng.*, **70** (Oct. 14): 195--200.

———, and ———. 1963c. Optimization of Multistage Separation Processes by Dynamic Programming, *Can. J. Chem. Eng.*, **41**:187–194.

Moorehead, D. H. and D. M. Himmelblau. 1962. Optimization of Operating Conditions in a Packed Liquid-Liquid Extraction Column, *Ind. Eng. Chem. Fundamentals*, **1**:68–72.

Mugele, R. A. 1962. A Non-linear Digital Optimizing Program for Process Control Systems, *IBM Systems J.*, **1**:2.

Nagiev, M. F. 1957. Material Balance in Complex and Multistage Recycle Chemical Process, *Chem. Eng. Progr.*, **53**(6):297–303.

Naphthali, L. M. 1964. Process Heat and Material Balances, *Chem. Eng. Progr.*, **60**(9): 70–74.

Nelder, J. A., and R. Mead. 1964. A Simplex Method for Function Minimization, *Computer J.*, **7**:308–313.

———, and ———. 1965. A Simplex Method for Function Minimization—Errata, *Computer J.*, **8**:27.

Nemhauser, G. L. 1961. A Dynamic Programming Approach for Optimal Design and Operation of Multistage Systems in the Process Industries, Ph.D. dissertation, North Western University, Evanston, Ill.

Neyman, J. (ed.). 1951. *Proc. 2d Berkeley Symp. on Math. Statistics Probl.*, University of California Press, Berkeley, Calif.

———. (ed.). 1956. *Proc. 3d Berkeley Symp. on Math. Statistics Probl.*, University of California Press, Berkeley, Calif.

Nichols, W. T. 1951. Next Time Give the Boss a Precise Capital Cost Estimate, *Chem. Eng.*, **58**(6):248.

Norman, R. L. 1965. A Matrix Method for Location of Cycles of a Directed Graph, *Am. Inst. Chem. Eng. J.*, **11**:450–452.

Orden, A. 1956. The Transhipment Problem, *Management Sci.*, **2**:276–285.

Ornea, J. C., and G. G. Eldredge. 1965. Non-linear Partitioned Models for Plant Scheduling and Economic Evaluation, *Proc. Am. Inst. Chem. Eng./Inst. Chem. Eng. (London), Joint Meeting, London, Symp.*, June, **4**:119–130.

Osgood, W. F. 1925. "Advanced Calculus," The Macmillan Company, New York.

Pan, Y. K., and R. G. Roth. 1968. Optimization of Yield through Feed Composition: HCN Process, *Ind. Eng. Chem. (Process Design Develop.)*, **7**:53–61.

Park, W. R. 1966. Managing Capital Expenditures, Part 2, The Divergent Path to Money Spending, *Hydrocarbon Process Petrol. Refiner*, **45**(2):167–170.

Paull, A. E. 1957. Linear Programming: A Key to Optimum Newsprint Production, *Pulp Paper Mag. Can.*, **57**:85–90.

Paviani, D. A. 1969. Nonlinear programming, Ph.D. dissertation, University of Texas, Austin, Texas.

Peterson, R. W. 1965. Use Probability in Your Management, *Hydrocarbon Process Petrol. Refiner*, **44**(5):261–266.

Pings, C. J. 1963. Maximization of Yield in Isothermal Gas Phase Reactions in Presence of Inerts, *Ind. Eng. Chem. Fundamentals*, **2**:321–323.

———. 1965. Optimization of Initial Composition: Adiabatic Equilibrium Gas-phase Reactions in Presence of Inerts, *Ind. Eng. Chem. Fundamentals*, **4**:260–264.

Piret, E. L., and P. J. Trambouze. 1959. CSTR: Designs for Maximum Conversions of Raw Material to Desired Product: Homogeneous Reactions, *Am. Inst. Chem. Eng. J.*, **5**:384–390.

Pis'men, L. M., and I. I. Ioffe. 1962. Calculation of Optimal Processes in Chemical Reactors by the Method of Dynamic Programming, *Intern. Chem. Eng.*, **2**:567–574.

Pollack, M. 1961. *k*th Best Route through a Network, *J. Math. Anal. Applic.*, **3**:547, and *J. Opn. Res. Soc. Am.*, **9**:578–580.

Pontryagin, L. S., V. G. Boltyanskii, R. V. Gamkrelidze, and E. F. Mishchenko. 1962. "The Mathematical Theory of Optimal Processes" (transl. by K. N. Trircgoff), Interscience Publishers, Inc., New York.

Post, M. H., and W. Viavant. 1962. A Management Game for Petroleum Industry, *Chem. Eng. Progr. Symp. Ser.* 37, **58**:83.

Powell, M. J. D. 1962. An Iterative Method for Finding Stationary Values of a Function of Several Variables, *Computer J.*, **5**:147–151.

———. 1964. An Efficient Method for Finding the Minimum of a Function of Several Variables without Calculating Derivatives, *Computer J.*, **7**:155–162.

Prager, W. 1955. On the Role of Congestion in Transportation Problems, *Z. Angew. Math. Mech.*, **35**:264–268.

———. 1956. On the Caterer Problem, *Management Sci.*, **3**:15–27.

Radner, R., and D. W. Jorgenson. 1963. Opportunitistic Replacement of a Single Part in the Presence of Several Monitored Parts, *Management Sci.*, **10**:70–84.

Rafal, M. D., and J. S. Dranoff. 1966. Efficient Algorithm for Optimization of a Multibed Adiabatic Reactor Sequence, *Ind. Eng. Chem.* (*Process Design Develop.*), **5**:129–135.

Raiffa, H., and R. Schlaifer. 1961. "Applied Statistical Decision Theory," Harvard University Press, Cambridge, Mass.

Ravicz, A. E., and R. L.Norman. 1964. Heat and Mass Balancing on a Digital Computer, *Chem. Eng. Progr.*, **60(5)**:71–76.

Resh, M., and P. Naor. 1963. An Inventory Problem with Discrete Time Review and Replenishment by Batches of Fixed Size, *Management Sci.*, **10**:109–118.

Riley, V., and S. I. Gass. 1958. "Linear Programming and Associated Techniques: A Comprehensive Bibliography on Linear, Non-linear and Dynamic Programming," Johns Hopkins Press, Baltimore, Md.

Robbins, H., and S. Munro. 1951. A Stochastic Approximation Method, *Ann. Math. Statistics*, **22**:400–407.

Roberts, S. M. 1960. Dynamic Programming Formulation of Catalyst Replacement Problem, *Chem. Eng. Progr. Symp. Ser.* 31, **56**:103–110.

———. 1964a. Terminal and Averaging Control of Pyrolysis in Tubular Reactors, *Ind. Eng. Chem.* (*Process Design Develop.*), **3**:14–18.

———. 1964b. "Dynamic Programming in Chemical Engineering and Process Control," Academic Press Inc., New York.

———, and H. I. Lyvers. 1961. The Gradient Method in Process Control, *Ind. Eng. Chem.*, **53**:877–882.

Rohde, F. V. 1957. A Bibliography of Linear Programming, *J. Opn. Res. Soc. Am.*, **5**:45–62.

Rosander, A. C. 1956. The Use of Linear Programming to Improve the Quality of Decisions, *Ind. Quality Control*, **12**:11–16.

Rosen, E. M. 1962. A Machine Computation Method for Performing Material Balances, *Chem. Eng. Progr.*, **58(10)**:69–73.

Rosen, J. B. 1960. The Gradient Projection Method for Non-linear Programming, Part I, Linear Constraints, *J. Soc. Ind. Appl. Math.*, **8**:181–217.

———. 1961. The Gradient Projection Method for Non-linear Programming, Part II, Non-linear Constraints, *J. Soc. Ind. Appl. Math.*, **9**:514–532.
———. 1963. Convex Partition Programming, in Graves and Wolfe [1963, p. 159–176].
———, and J. C. Ornea. 1963. Solution of Non-linear Programming Problems by Partitioning, *Management Sci.*, **10**:160–173.
Rosenbrock, H. H. 1960. An Automatic Method for finding the Greatest or Least Value of a Function, *Computer J.*, **3**:175–184.
Rozonoer, L. I. 1959a. Pontryagin Maximum Principle in the Theory of Optimal Systems, Part I, *Automation Remote Control*, **20**:1288–1302.
———. 1959b. Pontryagin's Maximum Principle in Optimal System Theory, Part II, *Automation Remote Control*, **20**:1405–1421.
———. 1959c. The Maximum Principle of Pontryagin in Optimal System Theory, Part III, *Automation Remote Control*, **20**:1517–1532.
Rubin, D. I. 1962. Generalized Mass Balance, *Chem. Eng. Progr. Symp. Ser.* 37, **58**:54.
Rubin, H., and P. Ungar. 1957. Motion under a Strong Constraining Force, *Commun. Pure Appl. Math.*, **10**:65–87.
Rudd, D. F. 1962a. On Design Policies for the Optimal Use of Limited Resources, *Chem. Eng. Sci.*, **17**:609–617.
———. 1962b. On Optimal Resource Allocation in Multistage Feedback Processes, *J. Soc. Ind. Appl. Math.*, **10**:448–453.
———. 1962c. Reliability Theory in Chemical System Design, *Ind. Eng. Chem. Fundamentals*, **1**:138–143.
———. 1968. The Synthesis of System Designs, I, Problem Decomposition, *Am. Inst. Chem. Eng. J.*, **14**:343–349.
———, R. Aris, and N. R. Amundson. 1961. A Study of Iterative Optimization, *Am. Inst. Chem. Eng. J.*, **7**:376–384.
———, and E. D. Blum. 1962. Optimum Cross-current Extraction with Product Recycle, *Chem. Eng. Sci.*, **17**:277–281.
———, and D. V. Stewart. 1965. On Information Flow and Process Design Calculations, *Paper Am. Inst. Chem. Eng. Natl. Meeting*, Sept., 1965, Las Vegas, Nev.
———, and C. C. Watson. 1968. "Strategy of Process Engineering," John Wiley & Sons, Inc., New York. (Also preliminary edition, 1966.)
Sacks, J. 1958. Asymptotic Distribution of Stochastic Approximation Procedures, *Ann. Math. Statistics*, **29**:397–398.
Salisbury, E. H. 1958. Work Study in the Design of Chemical Plants, *Chem. Process Eng.*, **39**(3):96–98.
Sargent, R. W. H., and A. W. Westerberg. 1964. Speed-up in Chemical Engineering Design, *Trans. Inst. Chem. Eng. (London)*, **42**:T190–T197.
Sasieni, M. 1960. Dynamic Programming and Inventory Problems, *Opn. Res. Quart.*, **11**:41–49.
Savage, L. J. 1951. The Theory of Statistical Decisions, *J. Am. Statist. Assoc.*, **46**:55–56.
Scarborough, J. B. 1955. "Numerical Mathematical Analysis," 3d ed., Johns Hopkins Press, Baltimore, Md.
Schechter, R. S. 1967. "The Variational Method in Engineering," McGraw-Hill Book Company, New York.
———, and G. S. G. Beveridge. 1966. Sufficiency Conditions in Constrained Variations, *Ind. Eng. Chem. Fundamentals*, **5**:571–573.
Schlaifer, R. 1961. "Introduction to Statistics for Business Decisions," McGraw-Hill Book Company, New York.
Schrage, R. W. 1958. Optimizing a Catalytic Cracking Operation by the Method of Steepest Ascents, *J. Opn. Res. Soc. Am.*, **6**:498–515.

Schwan, H. T. and J. J. Wilkinson. 1956. Linear Programming: What Can it Do? *Chem. Eng.*, **63** (Aug.): 211–214.

Schweyer, H. E. 1955. "Process Engineering Economics," McGraw-Hill Book Company, New York.

Sears, G. W. 1963. Linear Programming: Allocation Model, *Trans. Inst. Chem. Eng. (London)*, **41**:CE285–289.

Shah, B. V., R. J. Buehler, and O. Kempthorne. 1964. Some Algorithms for Minimizing a Function of Several Variables, *J. Soc. Ind. Appl. Math.*, **12**:74–92.

Shahbendarian, A. P. 1964. Influence of Profit Criterion on Chemical Reactor System, *Chem. Proc. Eng.*, **45**:647–649.

Shetty, C. M. 1961. On Analyses of the Solution to a Linear Programming Problem, *Opn. Res. Quart.*, **12**:89–104.

Siebenthal, C. D., and R. Aris. 1964a. Studies in Optn. VI: Application of Pontryagin's Methods to the Control of a Stirred Reactor, *Chem. Eng. Sci.*, **19**:729–746.

———, and ———. 1964b. Studies in Optn. VII: Application of Pontryagin's Method to the Control of Batch and Tubular Reactors, *Chem. Eng. Sci.*, **19**:747–761.

Simpson, K. F. 1959. Theory of Allocation of Stocks to Warehouses, *J. Opn. Res. Soc. Am.*, **7**:797–806.

Singer, E. 1962. Simulation and Optimization of Oil Refinery Design, *Chem. Eng. Progr. Symp. Ser.* 37, **58**:62.

Skelland, A. H. P. 1962. Optimum Filtration Cycles, *Chem. Proc. Eng.*, **43**:78.

Smith, C. S. 1962. The Automatic Computation of Maximum Likelihood Estimates, *Natl. Coal Board Sci. Dept. Rept.* S.C. 846/MR/40, London.

Smith, G. W., and R. C. Wood. 1959. "Principles of Analogue Computation," McGraw-Hill Book Company, New York.

Smith, S. B. 1965. Planning Transistor Production by Linear Programming, *J. Opn. Res. Soc. Am.*, **13**:132–139.

Sokolnikoff, I. S., and R. M. Redheffer. 1966. "Mathematics of Physics and Modern Engineering," 2d ed., McGraw-Hill Book Company, New York.

Southwell, R. V. 1956. "Relaxation Methods in Theoretical Physics," Oxford University Press, London.

Spendley, W., G. R. Hext, and F. R. Himsworth. 1962. Sequential Applications of Simplex Designs in Optimization and Evolutionary Operation, *Technometrics*, **4**:441–461.

Srygley, J. M., and C. D. Holland. 1965. Optimum Design of Conventional and Complex Distillation Columns, *Am. Inst. Chem. Eng. J.*, **11**:695–701.

Stanley, E. D., D. P. Honig, and L. Gainen. 1954. Linear Programming in Bid Evaluation, *Naval Res. Logistics Quart.*, **1**:48–54.

Stocker, D. C. 1969. "A Comparative Study of Nonlinear Programming Codes," M. S. Thesis, University of Texas, Austin, Texas.

Storey, C. 1962. Application of Hill Climbing Method of Optimization, *Chem. Eng. Sci.*, **17**:45–52.

———. 1964. A Review of What Can Be Achieved by Optimization Techniques, Assuming They Exist, *Trans. Inst. Chem. Eng. (London)*, **42**:T345–351.

———, and H. H. Rosenbrock. 1964. On the Computation of Optimal Temperature Profiles in a Tubular Reactor, in Balakrishnan and Neustadt [1964, pp. 23–64]. Also in *Proc. Conf. Computing Methods in Optimizing Processes*, University of California, Los Angeles, January, 1964.

Stout, T. M. 1959. System Considerations in Computer Control of a Semicontinuous Chemical Process, *Trans. Am. Inst. Elec. Eng. (Appl. Ind.)*, **77**:640–651.

———. 1960. Evaluating Control System Payout from Process Data, *Control Eng.*, **7(2)**:93.

Strong, R. E. 1965. A Note on the Sequential Unconstrained Minimization Technique for Non-linear Programming, *Management Sci.*, **12**:142–144.

Swann, W. H. 1964. Report on the Development of a New Direct Search Method of Optimization, *I.C.I. Ltd. Central Instr. Res. Lab.*, *Res. Note* 64/3, London.

Symonds, G. H. 1955. "Linear Programming: Solution of Refinery Problems," Esso Standard Oil Co., New York.

———. 1956. Linear Programming Solves Gasoline Refinery and Blending Problems, *Ind. Eng. Chem.*, **48**:394–401.

Synge, J. L. 1944. A Geometric Interpretation of the Relaxation Method, *Quart. J. Appl. Math.*, **2**:87.

———. 1957. "The Hypercircle in Mathematical Physics," Cambridge University Press, London.

Szepe, S., and O. Levenspiel. 1964. Optimization of Backmix Reactors in Series for a Single Reaction, *Ind. Eng. Chem.* (*Process Design Develop.*), **3**:214–217.

Taborek, J. J. 1960a. Optimization in Process Equipment Design, *Chem. Eng. Progr.*, **56** (Aug.):37–41.

———. 1960b. Design of Process Equipment to Economic Optimum, *Brit. Chem. Eng.*, **5**:784–788.

Temple, G., and W. G. Bickley. 1956. "Rayleigh's Principle and Its Applications to Engineering," Dover Publications, Inc., New York.

Thorne, H. O., and W. E. Miller 1966. Managing Capital Expenditures, Part I, Know How to Request Money, *Hydrocarbon Process Petrol. Refiner*, **45**(1):172–177.

Truxal, J. G. 1955. "Automatic Feedback Control System Synthesis," McGraw-Hill Book Company, New York.

Turnbull, H. W. 1960. "The Theory of Determinants, Matrices and Invariants," Dover Publications, Inc., New York.

Vajda, S. 1956. "Theory of Games and Linear Programming," Methuen & Co., Ltd., London.

———. 1960. "An Introduction to Linear Programming and the Theory of Games," Methuen & Co., Ltd., London, and John Wiley & Sons, Inc., New York.

———. 1961. "Mathematical Programming," Addison-Wesley Publishing Company, Inc., Mass.

Van Cauwenberghe, A. R. 1964. Optimization of Multistage Feedback Processes by the Method of Dynamic Programming, *Chem. Eng. Sci.*, **19**:514–517.

Van Heerden, C. 1961. *Chem. Eng. Sci.*, **14**:101–102. (Discussion, Second European Symposium on Chemical Reaction Engineering, Amsterdam 1960.)

Van Krevelen, D. W., and P. J. Hoftyzer. 1962. Process Optimization: Principles and Methods, *Trans. Inst. Chem. Eng.* (*London*), **40**:337–344.

Varga, R. S. 1962. "Matrix Iterative Analysis," Prentice-Hall, Inc., Englewood Cliffs, N.J.

Vilbrandt, F. C., and C. E. Dryden. 1959. "Chemical Engineering Plant Design," 4th ed., McGraw-Hill Book Company, New York.

Von Neumann, J., and O. Morgenstern. 1953. "Theory of Games and Economic Behavior," 3d ed., Princeton University Press, Princeton, N.J.

Vorob'ev, N. N. 1966. "Fibonacci Numbers" (transl. by H. Moss), Pergamon Press, New York.

Weinstock, R. 1952. "Calculus of Variations," McGraw-Hill Book Company, New York.

Weiss, P. A., D. H. Archer, and D. A. Burt. 1961. Computer Sets Tower for Best Run, *Hydrocarbon Process Petrol. Refiner*, **40**(10):169–174.

Welsh, D. J. A. 1966. Supercritical Arcs of a PERT Network, *J. Opn. Res. Soc. Am.*, **14**:172–174.

Wen, C. Y. 1963. Transient Activity and Replacement of Catalyst in Multi-stage Reactors, *Can. J. Chem. Eng.*, **41**:28–29.

Westbrook, G. 1961. Use This Method to Size Each Stage for Best Operation, *Hydrocarbon Process Petrol. Refiner*, **40**(9):201.

White, D. J. 1961. Comments on paper by McDowell, 1960, *J. Opn. Res. Soc. Am.*, **9**:580–584.

———. 1962. Optimal Revision Periods, *J. Math. Anal. Applic.*, **4**:353–365.

———. 1963. "Dynamic Programming," *Univ. Manchester Reprint Ser. 2. Centre Business Res.*

———. 1964. Dynamic Programming and Equipment Resetting Decision Rules, *Opn. Res. Quart.*, **15**:133–137.

———. 1965. Dynamic Programming and Systems of Uncertain Duration, *Management Sci.*, **12**:37–67. Also *Univ. Manchester Reprint Ser. 7. Centre Business Res.*

———. 1966. Forecasts and Decision Making, *J. Math. Anal. Applic.*, **14**:163–173.

White, D. J. and J. M. Norman, 1965. Control of Cash Reserves, *Opn. Res. Quart.*, **16**:309–328.

Wilde, D. J. 1963. Optimization by the Method of Contour Tangents, *Am. Inst. Chem. Eng. J.*, **9**:186–190.

———. 1964. "Optimum Seeking Methods," Prentice-Hall, Inc., Englewood Cliffs, N.J.

———. 1965. Jacobians in Constrained Non-linear Optimization, *J. Opn. Res. Soc. Am.*, **13**:848–856.

———, and C. S. Beightler. 1967. "Foundations in Optimization," Prentice-Hall, Inc., Englewood Cliffs, N.J.

Williams, A. C. 1962. A Treatment of Transportation Problems by Decomposition, *J. Soc. Ind. Appl. Math.*, **10**:35–48.

———. 1963. A Stochastic Transportation Problem, *J. Opn. Res. Soc. Am.*, **11**:759–770.

———. 1965. On Stochastic Linear Programming, *J. Soc. Ind. Appl. Math.*, **13**:927–940.

Williams, B. R. 1964. Information and Criteria in Capital Expenditure Decisions, *Centre Business Res., Univ. Manchester Rept.*

Williams, J. D. 1965a. "The Compleat Strategyst," rev. ed., McGraw-Hill Book Company, New York.

Williams, T. J. 1959–1965. Computers, Automation and Process Control: Annual Reviews, *Ind. Eng. Chem.*, **51**:432, 1959; **52**:183, 1960; **53**:166, 1961; **55**(11):43, 1963; **56**(11):47, 1964; **57**(11):33, 1965.

———, and R. E. Otto. 1960. A Generalized Chemical Processing Model for the Investigation of Computer Control, *Trans. Am. Inst. Elec. Eng.* (*Commun. Electronics*), **79**:458–473.

Wolfe, P. 1959. The Simplex Method for Quadratic Programming, *Econometrica*, **27**:382–398.

———. 1961. Accelerating the Cutting-plane Method for Non-linear Programming, *J. Soc. Appl. Math.*, **9**:481–488.

———. 1963. Method of Non-linear Programming, in Graves and Wolfe [1963, p. 67–86].

Wolfowitz, J. 1952. On the Stochastic Approximation Method of Robbins and Munro, *Ann. Math. Stat.*, **23**:457–461.

Yefimov, N. V. 1964. "Quadratic Forms and Matrices: An Introductory Approach," trans. by A. Shenitzer, Academic Press Inc., New York.

Youle, P. V. 1960. Simulation of Full-scale Multistage Batch-wise Chemical Plant, *Computer J.*, **3**:150–157.

Young, H. A. 1963. On the Optimum Location of Checking Stations, *J. Opn. Res. Soc. Am.*, **11**:721–731.

Zahradnik, R. L., and D. H. Archer. 1963. Applications of Discrete Maximum Principle to Crosscurrent Extractions, *Ind. Eng. Chem. Fundamentals*, **2**:238–240.

Zehna, P. W. 1962. Inventory Depletion Problems, in Arrow, K. J., *et al.* (eds), "Studies in Applied Probability and Management Science," pp. 263–287, Stanford University Press, Stanford, Calif.

Zellnik, H. W., N. E. Sondal, and R. S. Davis, 1962. Gradient Search Optimization, *Chem. Eng. Progr.*, **58**(8):35–41.

Zener, C. M. 1961. A Mathematical Aid in Optimizing Engineering Design, *Proc. Natl. Acad. Sci.*, **47**:537–539.

Zoutendijk, G. 1960. "Methods of Feasible Directions," Elsevier Publishing Company, Amsterdam, and D. Van Nostrand Company, Inc., Princeton, N.J.

NAME INDEX

SUBJECT INDEX